THE DENT DICTIONARY OF MEASUREMENT

THE DENT DICTIONARY OF MEASUREMENT

MIKE DARTON AND
JOHN O. E. CLARK

J. M. Dent
London

Produced by Sheila Dallas Publishing
Edited, designed, and typeset by Curtis Garratt Limited
Printed and bound in the United States of America
for J. M. Dent
The Orion Publishing Group Ltd
Orion House
5 Upper St Martin's Lane
London WC2H 9EA

A catalogue record for this book is available from the British Library

ISBN 0460 861379

FOREWORD

We are surrounded in everyday life by measurement and quantification. At any moment, if we so desire, we can tell the time, gauge the atmospheric temperature and pressure, see how fast we are moving, check the weight or volume of the contents of a bottle or bag, monitor our heartbeats or blood-sugar level, or count up our loose change. In business and professional life, measurement is even more diverse and often of critical importance.

As the scope of activities in all walks of life has widened dramatically over the past fifty years – due in part to the opening of further fields of technology, to the great increase in leisure time and opportunity, and to the development of close cultural links with distant parts of the world – so the units of quantification and measurement have themselves proliferated. At the same time, certain systems of measurement have been officially adopted over large areas of the world, in the process replacing older, local systems of measurement that were perfectly adequate and may also have had linguistic and cultural meaning for the people who used them.

This book has thus to take account of an enormous number of disciplines, of worldwide quantification systems that may not be familiar to its readers, and of units that may now be discovered only in historical contexts. Science and mathematics feature largely, of course. But the book also contains (for example) the units of analytical linguistics, of music, of textiles and cloth, of paper and publishing, of minerals and metals, and of military ranks and formations. The measurements of sports and games constitute a sizeable element of the text, as do the listing and history of coins and currencies of the present and of the past.

The historical entries prove two things in particular: that not all measurements remain the same over time, and that even when a unit is discarded as antiquated, it may remain embedded in the language in odd proverbs or even individual words (just as, for example, the old linear measure the *ell* is represented in the first syllable of *elbows* – between which it was originally measured).

To confirm the first point – that measurements may change – we need only instance all the various *pounds* and *feet* that existed in Europe, North Africa, and the Near East until quite recently. Although some were very different from the pounds and feet well known in Britain and North America, they all derived from the ancient Roman system of measurement. Even when the spelling remains constant, the unit may change: the British (UK) gallon is larger than the American (US) gallon – and the US liquid gallon is not even the same as the US dry gallon. This means that all those measurements and units that relate to gallons (the *pint*, the *quart*, the *peck*, the *bushel*, the *hogshead*, the *barrel*, and so on, including *miles per gallon* and *gallons per minute*) are not the same in Britain as in the United States. Part of the basic function of this book, therefore, is to provide the means of readily converting units in one system to units in another.

The Dictionary in this way also provides the means to relate historical units to the standard units in use today. For many readers this may bring new meaning to well-known texts (for instance, those passages in the Bible that make mention of ancient measures such as the *cubit* or *talent*).

As often as possible we have cross-referenced entries (cross-references appear in SMALL CAPITALS) so that little information is repeated, yet complete information can be traced through the entries that are cross-referred. A second means of thematic access to entries is provided by the **Index** at the back of the book (pages 526–538).

Finally, it is our hope that – unlike most dictionaries – this one may prove worthy not only of use as a comprehensive reference work, but also of sitting down to read through. Certainly, the intention has been to provide information that is not restricted solely to technicalities, but that also includes further interesting background, in particular how a unit or measure came to be given its name.

Mike Darton and John Clark
January 1994

Acknowledgement
The lengthy process of compiling this book was considerably expedited by the assistance of our respective wives, Ann Darton and Gill Clark, whose help we duly and gratefully acknowledge.

A

ab- [quantitatives: prefix] Prefix used in the CENTIMETRE-GRAM-SECOND (CGS) system of measurement and denoting a practical quantification of a theoretical electromagnetic unit. Effectively, the prefix links the CGS system with the SI unitary system, rendering conversions from one to the other unnecessary.

abampere [physics] Unit of electric current in the CENTIMETRE-GRAM-SECOND (CGS) system of measurement.

$$1 \text{ abampere } = 10 \text{ AMPERES}$$

It is the current, flowing through two infinitely long parallel conductors 1 centimetre apart, that produces a force of 2 DYNES per centimetre between them. *See also* AB-.

abcoulomb [physics] Unit of electric charge in the CENTIMETRE-GRAM-SECOND (CGS) system of measurement.

$$1 \text{ abcoulomb } = 10 \text{ COULOMBS}$$

It is the charge that each second passes any cross-section of a conductor through which a steady current of 1 AMPERE is flowing. *See also* AB-.

aberration [astronomy] An apparent shift in the position of a heavenly object because the relative speed of the object and the observer is a significant fraction of the velocity of light.

abfarad [physics] Unit of electric capacitance in the CENTIMETRE-GRAM-SECOND (CGS) system of measurement.

$$1 \text{ abfarad } = 1{,}000 \text{ million FARADS } (10^9 \text{ farads})$$

It is the capacitance of a capacitor carrying a charge of 1 ABCOULOMB and with a potential difference of 1 ABVOLT between its plates. *See also* AB-.

abhenry [physics] Unit of electric inductance in the CENTIMETRE-GRAM-SECOND (CGS) system of measurement.

$$1 \text{ abhenry } = \text{ one thousand-millionth of a HENRY } (10^{-9} \text{ henry})$$

It is the inductance produced when a rate of change of current of 1 ABAMPERE per second generates an induced electromotive force (EMF) of 1 ABVOLT. *See also* AB-.

abohm [physics] Unit of electric resistance in the CENTIMETRE-GRAM-SECOND (CGS) system of measurement.

$$1 \text{ abohm } = \text{ one thousand-millionth of an OHM } (10^{-9} \text{ ohm})$$

See also AB-.

abscissa [maths] In coordinate geometry, the perpendicular distance of a point from the y-axis – that is, its x coordinate. The term derives directly from the Latin *linea abscissa* 'line (that has been) cut off'. *See also* CARTESIAN COORDINATES.

absolute magnitude *see* MAGNITUDE

absolute scale [physics] An absolute scale is any system of measuring temperature in units that begin at ABSOLUTE ZERO (-273.16°C, -459.67°F). The most commonly used absolute scale is the KELVIN SCALE (boiling point of water 373 K); another absolute scale is the RANKINE SCALE (boiling point of water 672°Rankine).

absolute zero [physics] Temperature at which a substance has no heat at all, and its constituent molecules are stationary. In theory it is the lowest possible temperature. It has the values:

-273.16°Celsius	-459.67°Fahrenheit
0 kelvin	0 °Rankine

See also ABSOLUTE SCALE.

absorbance [chemistry; physics] The logarithm of the intensity of light falling on a sample, such as a solution, divided by the amount that passes through it. Absorbance is proportional to the concentration of the substance dissolved in solution.

absorbed dose *see* X-RAYS

absorptance [physics] The absorptance (formerly 'absorptivity') of an object is calculated as the amount of radiation it absorbs divided by the total radiation incident on it, and is thus expressed as a numerical ratio without units.

absorptiometer [physics; chemistry] An instrument for measuring (a) the solubility of a gas, or (b) the absorption of light by a solution (which in certain circumstances is a measure of the concentration of the dissolved substance).

absorption hygrometer [physics: meteorology] An instrument for measuring the amount of water vapour in air (by determining the increase in weight of a drying agent over which a sample of air is drawn).

absorptivity *see* ABSORPTANCE

abundance [physics; chemistry] The percentage or proportion of a given isotope to the total in a naturally occurring sample of an element. It is known also as the *abundance ratio*.

abundant year [time] In the Jewish calendar, a year of 355 days, representing the longest form of the three common types of year. It is sometimes alternatively called a perfect year. The other two types are the REGULAR YEAR and the DEFECTIVE YEAR. *See also* YEAR.

abvolt [physics] Unit of potential difference or electromotive force in the CENTIMETRE-GRAM-SECOND (CGS) system of measurement.

$$1 \text{ abvolt} = \text{one hundred-millionth of a VOLT } (10^{-8} \text{ volt})$$

It is the potential difference that exists between two points when 1 ERG of work must be done to transfer 1 ABCOULOMB of charge from one of the points to the other. *See also* AB-.

abwatt [physics] Unit of power in the CENTIMETRE-GRAM-SECOND (CGS) system of measurement.

$$1 \text{ abwatt} = \text{one ten-millionth of a watt } (10^{-7} \text{ watt})$$

It is the power dissipated when a current of 1 ABAMPERE flows across a potential difference of 1 ABVOLT. *See also* AB-.

accelerando [music] Musical instruction: gradually increase tempo.

acceleration [physics] Acceleration – the rate of change of VELOCITY or speed – is measured in units of distance per second per second (distance/second^2), such as metres per second per second (m/sec^2).

Negative acceleration (in which velocity decreases over time) is called deceleration, and is measured in exactly the same fashion.

The Latin word *celer*, from which both the English words *acceleration* and *deceleration* mostly derive, means 'fast' but, in turn, derives from an older root of which the basic meaning was, appropriately here, 'to push on', 'to drive'.

acceleration of free fall [physics; astronomy] The acceleration given to an object by the force of gravity (the gravitational attraction of the Earth), equal to 9.80665 metres per second per second (32.1740 feet per second per second). Strictly, it applies to an object falling freely in a vacuum, and it varies slightly with the object's distance from the centre of the Earth.

accidental [music] A note to be played or sung that is not in the chromatic scale indicated by the key signature at the beginning, and that therefore has to be preceded on the musical stave by a symbol declaring it to be either flat of (a semitone/ half-step below) or sharp of (a semitone/ half-step above) the note as written; alternatively the symbol might instead require the note to be sounded as a 'natural' as opposed to the sharp or flat required by the key signature. *See also* KEY SIGNATURE; FLAT; SHARP; NATURAL.

accordian, accordion *see* HARMONICAS' AND ACCORDIANS' RANGE

accumulator [sporting term; comparative values] In Britain and elsewhere, one or a series of gambles or bets in which the stake is all the money won on an initial and successful gamble or bet (plus the initial stake) immediately beforehand; in the United States it is known as a *parlay*.

ace [sport: cards/tennis] In cards, the ace is the number one card of each suit, its value – depending on the game being played – either the lowest or (more commonly, and with the exception of a very few games) the highest of its suit, above that of the king. In certain games (such as pontoon and poker), the ace can represent the lowest *or* the highest value, at the discretion of the player.

As an extension of its meaning of value unbeatable in a hand or trick at cards, an ace in tennis is a serve that lands correctly within the service court on the other side of the net at such a speed and in such a direction that the receiver cannot get racket to it at all.

By derivation, the ace – initially, *as* – is properly the first unit of a duodecimal system of weight that predated ancient Rome but is most famous for being used there as part of the coinage: *see* as *under* LIBRA.

acetabulum [volumetric measure] In ancient Rome, a measure of liquid capacity – especially in the preparation or preservation of wine and oil – and of dry capacity, approximating roughly to one-eighth of a modern pint.

1 acetabulum	=	6 ligulae
	=	66.406 millilitres, 0.066406 litre
	=	0.1169 UK pint, 0.1403 US pint
	=	2.338 UK fluid ounces, 2.243 US fluid ounces
	=	0.1169 UK dry pint, 0.1205 US dry pint
	=	66.406 cubic centimetres
	=	4.0526 cubic inches
4 acetabula	=	1 HEMINA
8 acetabula, 2 heminae	=	1 SEXTARIUS
48 acetabula, 12 heminae, 6 sextarii	=	1 CONGIUS

The name of the unit derives from its being the (liquid and dry) capacity of the standard Roman vinegar-pot (Latin *acetum* 'vinegar'), shaped like a rather deep egg-cup (which is why in modern medical terminology the acetabulum is the socket of the hip joint). *See also* COTYLA, KOTYLE; OXYBATHON.

acidity [chemistry] In chemistry and many of its applications, acidity (and ALKALIN-ITY) are often expressed on the pH scale (acidity is caused by the presence of hydrogen ions, and the pH of a solution is the negative logarithm of its hydrogen ion concentration). A neutral solution, such as pure water, has a pH of 7. A pH of less than 7 indicates an acidic solution; a pH of more than 7 indicates an alkaline solution. The lower the pH, the more acid it is.

For any specific acid – such as hydrochloric acid – acidity can also be expressed in terms of the concentration of that acid: for example in molality (moles per kilogram), normality (gram-equivalents per litre), grams per litre, and so on.

The term acid was coined as an English word by Francis Bacon (1561-1626), but derives fairly directly from Latin *acidus* 'sharp' (*see* the etymology of ACRE below).

acoustic absorption coefficient [physics] The acoustic energy absorbed by a surface divided by the total energy incident on it. It depends on the frequency (pitch) of the sound.

acoustic ohm [physics] Unit of acoustic resistance (or reactance or impedance).

1 acoustic ohm	=	100,000 pascal seconds per cubic metre

$$(10^5 \text{ Pa sec/m}^3)$$

acre [square measure] Unit of area in the imperial scale, first defined around AD 1300 and said to be the area of land that one yoke of oxen could plough in the course of a day's work.

1 acre	=	4 roods, 10 square chains, 160 square rods
	=	4,840 square yards
	=	43,560 square feet
	=	0.0156 square mile, 0.1 square furlong
	=	4,047 square metres, 0.4047 hectare
		(1 hectare, 100 ares, 10,000 square metres
		= 2.471 acres; 2 hectares = 4.942 acres)
5 acres	=	2.035 hectares
10 acres	=	1 square furlong (220 x 220 yards)
247.1 acres	=	1 SQUARE KILOMETRE, 100 hectares
640 acres	=	1 SQUARE MILE

The etymological background is Old English *æcer* 'ploughed field'; Latin *ager* 'field'; ancient Greek *agora* 'flattened open space'; probably ultimately akin to the Indo-European root **akh-* 'sharp point' as in English *acute* and, as a diminutive, *angle*, but in this case referring to the sharpened stick used either to till the ground or to goad on the animals pulling the plough [Latin *agere* 'to drive', 'to make go', and thus 'to go', 'to operate', 'to lead a life', 'to do': English *agent*, *act*-].

acre-foot [volumetric measure] Volume of water that would be 1 foot deep if it occupied an area of 1 ACRE, used for expressing the volumes of lakes and reservoirs.

$$\begin{aligned} 1 \text{ acre-foot} \quad &= \quad 43{,}560 \text{ cubic feet} \\ &= \quad 1{,}233.774 \text{ cubic metres} \\ &= \quad 271{,}400 \text{ UK gallons} \\ &= \quad 325{,}803 \text{ US gallons} \\ &= \quad 1{,}233{,}774 \text{ litres} \end{aligned}$$

acre-inch [volumetric measure] Volume of water that would be 1 inch deep if it occupied an area of 1 ACRE; one-twelfth of an ACRE-FOOT.

$$\begin{aligned} 1 \text{ acre-inch} \quad &= \quad 3{,}630 \text{ cubic feet} \\ &= \quad 102.815 \text{ cubic metres} \\ &= \quad 22{,}616 \text{ UK gallons} \\ &= \quad 27{,}150 \text{ US gallons} \\ &= \quad 102{,}815 \text{ litres} \end{aligned}$$

activation energy [chemistry] The amount of energy necessary to initiate a chemical reaction (by starting to break and re-form chemical bonds).

active mass [chemistry] The concentration of a substance involved in a chemical reaction (which may be less than the total concentration of that substance – not all of which may react). For a substance in solution it is given in moles per cubic decimetre; for a gas, it is expressed as its partial pressure. *See also* ACTIVITY.

activity [chemistry] The theoretically ideal (thermodynamic) concentration of a substance involved in a chemical reaction. When it is replaced by the actual concentration, the law of mass action is valid. The activity may alternatively be called the ACTIVE MASS.

actus [linear measure; square measure] Slightly confusing measure of distance-with-breadth in ancient Rome, used primarily to define the lengths of bridleways and paths down which cattle could be driven in single file (which had therefore to be of a specific width).

$$\begin{aligned} 1 \text{ actus simplex} \quad &= \quad 120 \text{ 'feet' x 4 'feet'} \\ &= \quad 116 \text{ feet } 9.6 \text{ inches x 3 feet } 10.7 \text{ inches} \\ &= \quad 35.604 \text{ metres x } 1.187 \text{ metre} \\ &\qquad (= 480 \text{ square 'feet'}) \end{aligned}$$

The name of the unit derives from the theme of cattle driving and horse riding: the word is Latin for 'driven', 'ridden', thus 'drove', 'ride'.

But the greater measure of the actus simplex was also the basis for a system of square measurement, for the most part again involving constant width with variable length.

$$\begin{aligned} 1 \text{ actus quadratus} \quad &= \quad 120 \text{ 'feet' x 120 'feet', 14,400 square 'feet'} \\ &= \quad 116 \text{ feet } 9.6 \text{ inches x 116 feet } 9.6 \text{ inches} \\ &\qquad (13{,}642.24 \text{ square feet, } 1{,}515.8 \text{ square yards}) \\ &= \quad 35.604 \text{ metres x } 35.604 \text{ metres} \\ &\qquad (1{,}267.645 \text{ square metres, } 0.1268 \text{ hectare}) \\ 2 \text{ acti quadrati} \quad &= \quad 1 \text{ jugerum ('yoke-area')} \\ &= \quad 120 \text{ 'feet' x 240 'feet', 28,800 square 'feet'} \\ 4 \text{ acti quadrati, } 2 \text{ jugera} \quad &= \quad 1 \text{ heredium ('estate')} \\ &= \quad 120 \text{ 'feet' x 480 'feet'} \end{aligned}$$

The term *actus quadratus* means 'square actus'. *See also* JUGERUM; HEREDIUM.

acute angle [maths] An angle of between 0° and 90°.

adagio [music] Musical instruction on tempo: slow, in a leisurely fashion. Italian: 'at ease' (*ad agio*). A movement marked *adagio* is generally played more slowly than an ANDANTE, but in a freer and less dignified way than a LARGO.

addition [maths] Mathematical operation that produces the *sum* of two or more quantities, usually denoted by the symbol + .

Addition is commutative – the order in which the two or more quantities are totalled does not matter (so $a + b$ is the same as $b + a$).

adjacent (angle, side) [maths] Adjacent angles lie, as a neighbouring pair of angles, on the same side of a line intersected by another line; together they add up to 180°. (Angles on different sides of the line are opposite angles.)

In a right-angled triangle, the longest side (opposite the right-angle) is known as the hypotenuse. If one of the two other angles of the triangle is taken as the subject, the side of the triangle that forms the angle with the hypotenuse is known as the adjacent side (and the third side is then the opposite side).

adjective [literary] Word descriptive of a NOUN or PRONOUN or another adjective. Examples:

(qualifiers:)	red ugly poor clean little saintly
(quantifiers:)	numerous little quite three more
(demonstratives:)	that yonder those this these
(possessives:)	your my our his her their its
(interrogatives:)	which whose what
(participles:)	bored forgiving unwitting

The standard form of adjectival comparison is:

(positive:)	small	free	feeble	dirty
(comparative:)	smaller	freer	feebler	dirtier
(superlative:)	smallest	freest	feeblest	dirtiest

although there is a substantial group in which comparison is instead like this:

beautiful	more beautiful	most beautiful
convenient	more convenient	most convenient
usual	more usual	most usual
wrong	more wrong	most wrong

and there are quite a few outright exceptions – examples:

bad	worse	worst
far	farther/further	farthest/furthest
good	better	best
little/few	less	least
much/many	more	most

admiral [military rank] Very senior commissioned rank in the navy. In the Royal (British) Navy, an admiral ranks between a vice-admiral and an admiral of the fleet (the most senior rank in the navy), and is the equivalent of a general in the British army and an air chief marshal in the Royal Air Force.

In the US Navy, an admiral ranks between a vice-admiral and a fleet admiral (the most senior rank in the navy), and is the equivalent of a general in the US Army and Airforce.

The title derives through a slight confusion from medieval Arabic *amir al-* 'commander of the – ', the first two elements of an Arab title that would have been followed by the Arabic equivalent of 'army', 'navy', 'camel-team', or similar.

admiral of the fleet [military rank] The most senior rank in the Royal Navy, ranking higher than an admiral, and equivalent to field marshal of the British army and marshal of the Royal Air Force.

The title of the equivalent rank in the US navy is fleet admiral.

advantage [sporting term] In tennis, once deuce (40-40) has been reached in a game, the next point scored is the advantage (because one player or side holds the advantage over the other). Another point scored by the same player or side wins the game; a point lost brings the score back to deuce.

The word derives from the medieval French for 'being in front'.

adverb [literary] Word descriptive of a VERB or of another adverb. Most describe how, why, or when the action of the sentence takes place, and end in the suffix -ly. Examples:

quickly easily stupidly thinly peculiarly especially
funnily surprisingly ludicrously decidedly apprehensively
very today how well soon alone

Some authorities define a further type of adverb as the *conjunctive adverb*, conjoining two statements. Such adverbs include:

besides furthermore however indeed likewise
moreover nonetheless therefore when while

The standard form of adverbial comparison is:

(positive:)	quickly	handsomely	surprisingly
(comparative:)	more quickly	more handsomely	more surprisingly

(superlative:)   most quickly   most handsomely   most surprisingly
but those that do not end in -ly may be adjectival in form:

fast	faster	fastest
soon	sooner	soonest

and there are some outright exceptions – examples:

badly	worse	worst
early	earlier	earliest
well	better	best

Aeolian mode [music] Medieval form of key, the SCALE of which is characterized on a modern piano by the white notes between one A and the next. In technical terms this corresponds to the 'natural' minor key of A (as opposed to the harmonic or melodic minor keys of A: *see* explanation *under* KEY, KEYNOTE). This mode was one of the most commonly used during medieval times, and is particularly associated with the monastic incantation known as Gregorian chant. It is perhaps no wonder, then, that it retains an importance even after modal music in general has virtually vanished.

Its name derives from one of the musical modes used in ancient Greece, and held to be particularly stately and imposing – but the ancient Greeks had completely different musical referents.

aeolian tone *see* STROUHAL NUMBER

aeon, eon [time] Technically, an aeon/eon is not a specific length of time, simply an extreme one, involving thousands of years. In astronomy and geology, however, it has come to mean a period of precisely 1 US billion (i.e. 1,000 million) years.

The term derives from ancient Greek through Latin *aion* 'life-span', 'age'; cognate with Latin *aevum* 'age', Dutch *eeuw* 'age', 'century', Sanskrit *ayus* 'life', English *age*, akin to *act-* and *agent*.

afghani [comparative values] Unit of currency in Afghanistan.

          100 puls = 1 afghani

See also COINS AND CURRENCIES OF THE WORLD.

age [astrology; astronomy; biology; geology] The astrological age is determined by the precise location in the night sky of the intersection of the ecliptic and the equator – 'the First Point of Aries', the 'beginning' of the zodiac. It was defined as such in the time of Hipparchus in about 150 BC, when the constellation Aries was the nearest. Since that time, the intersection has moved westwards because of the PRECESSION of the Earth's axis, and has all but run through the sector allocated to the constellation Pisces and entered the sector of the constellation Aquarius. This is why 'the dawning of the age of Aquarius' became famous during the 1960s.

In astronomy the most important measurements relating to age are:
    the Big Bang: 15,000-20,000 million years ago
    our (Milky Way) Galaxy formed around 14,000 million years ago
    the Solar System came into existence around 4,600 million years ago

In biology, all organic life is subject to ageing, mostly through the non-replacement of cells following the peak of maturity, or the inelasticity of those that remain. There is no infallible method by which to judge the age of an adult human, either at sight or forensically, to any exactitude closer than the nearest eighteen months. Some adult animals do exhibit more finely tuned indications of their precise age, and so do most woody plants older than seven years.

In geology, the word *age* is not a technical term. Geological time is properly divided into Eras, Periods, Epochs, and Stages: *see* GEOLOGICAL TIME.

age [physics; geology; archaeology] Age is measured in units of time. For a sub-atomic particle that has only a fleeting existence, age may be measured in nano-seconds; for a geological era or period, it is expressed in millions of years.

agonic line [geology] A line drawn on a map joining places at which magnetic declination is zero (that is, at which magnetic north and true north are the same).

agora [comparative values] Unit of currency in Israel: singular agora, plural agorot.

          100 agorot = 1 Israeli shekel

See also COINS AND CURRENCIES OF THE WORLD.

A horizon [chemistry; physics] *see* SOIL HORIZONS, SOIL LAYERS

air chief marshal [military rank] Very senior rank in the Royal Air Force, ranking between an air marshal and a marshal of the royal air force (the most senior rank in the airforce), and an equivalent of general in the British army and admiral in the Royal Navy.

In the US Airforce the equivalent rank is that of an airforce GENERAL.

air commodore [military rank] Senior rank in the Royal Air Force, ranking between a group captain and an air vice-marshal. The rank has no exact equivalent in the British army or Royal Navy, but is regarded as senior to a colonel or brigadier in the army, and to a captain or commodore in the navy.

In the US Airforce the equivalent rank is that of a BRIGADIER-GENERAL.

air marshal [military rank] Senior rank in the Royal Air Force, ranking between an air vice-marshal and an air chief marshal, and the equivalent of a lieutenant-general in the British army and a vice-admiral in the Royal Navy.

In the US Airforce the equivalent rank is that of a LIEUTENANT-GENERAL.

air speed [aeronautics] The speed of an aircraft or rocket relative to the air in which it is flying (and different from the ground speed, which is its speed relative to the ground). For most aircraft, it is measured in kilometres per hour, mph, or knots; for high-speed aircraft and rockets, it may be stated in metres per second, feet per second, or as a Mach number.

air vice-marshal [military rank] Senior rank in the Royal Air Force, ranking between an air commodore and an air marshal, and the equivalent of a major-general in the British army and a rear admiral in the Royal Navy.

In the US Airforce, the equivalent rank is that of a MAJOR-GENERAL.

alcohol content [chemistry; physics; medicine] The alcohol content of a liquid is generally expressed in terms of percentage by volume or of percentage by weight. The alcohol content of blood (in someone who has been drinking) is frequently stated in milligrams of alcohol per 100 millilitres of blood.

The liquor trade has, by tradition, used the proof system to express alcohol content: 100 proof is approximately 52% of alcohol by volume. The name of the system came about through the fact that measurement of the alcohol content of wines and spirits in Britain was formerly the responsibility of the same proof houses that tested the safety of firearms. A 100-proof spirit is a mixture of alcohol and water in precisely the right proportions that when it is poured over gunpowder the gunpowder can still just be ignited with a flame.

The alcohol content of an alcohol-water mixture can also be indicated by its SPECIFIC GRAVITY.

alexandrine [literary: poetic measure] A line in verse made up of six usually iambic feet with a CAESURA (a pause) after the third foot. The stress pattern is thus:

u —/ u —/ u —// u —/ u —/ u —

The term derives from Old French, in which language a considerable amount of poetry using such lines was written on the subject of Alexander the Great. *See also* METRE IN VERSE.

Alfven number [physics] In magnetohydrodynamics (the study of the movement of a conducting fluid, such as a plasma, in a magnetic field), a number that characterizes the flow of such a fluid past an obstruction in a uniform magnetic field aligned parallel to the direction of flow.

It was named after the Swedish astrophycisist Hannes Alfvèn (1908-).

algebra [maths] A major branch of mathematics in which symbols (letters) are used to represent unknown quantities, so that mathematical operations can be carried out on them (often with a view to determining the value of the unknown quantity). Algebraic equations can represent lines, curves, and shapes when they are plotted as a graph in the related branch of mathematics called coordinate (or analytic) geometry. *See also* BOOLEAN ALGEBRA.

alignment chart *see* NOMOGRAM, NOMOGRAPH

alkalinity [chemistry] In chemistry and many of its applications, alkalinity and acidity are often expressed on the PH SCALE (acidity is caused by the presence of hydrogen ions, and the pH of a solution is the negative logarithm of its hydrogen ion concentration). A neutral solution, such as pure water, has a pH of 7. A pH of less than 7 indicates an acidic solution; a pH of more than 7 indicates an alkaline solution. The higher the pH, the more alkaline it is.

For any specific alkali – such as sodium hydroxide (caustic soda) – alkalinity can also be expressed in terms of the concentration of that alkali, for example in molality (moles per kilogram), normality (gram-equivalents per litre), grams per litre, and so on.

The word 'alkali' acknowledges chemistry's debt to medieval Arabic science, coming from Arabic *al-qili* 'the ashes', originally the ashes of the saltwort, used for medicinal or household purposes.

allegro, allegretto [music] Musical instructions on tempo: allegro – brisk, lively; allegretto – fairly briskly, quite lively (but not so much as to be allegro). Italian: 'brisk' (from Latin *alacer*, English *alacrity*).

almude [volumetric measure] A unit used in Spain and Portugal, in both liquid and dry measure. The unit differs by some margin, however, between the two countries.

1 Spanish almude	=	16 OCTAVILLOS, 4 CUARTILLOS (dry measure only)
	=	4.625 litres
		(1 litre = 0.216 Spanish almude)
	=	1.02 UK gallons, 8.16 UK pints
		(1 UK gallon = 0.98 Spanish almude)
	=	1.22 US gallons, 9.76 US pints
		(1 US gallon = 0.82 Spanish almude)
12 Spanish almudes	=	1 Spanish FANEGA (55.50 litres, 12.21 UK gallons, 14.66 US gallons)
1 Portuguese almude	=	16.7 litres
		(1 litre = 0.06 Portuguese almude)
	=	3.67 UK gallons, 29.4 UK pints
		(1 UK gallon = 0.27 Portuguese almude)
	=	4.41 US gallons, 35.3 US pints
		(1 US gallon = 0.23 Portuguese almude)
3.315 Portuguese almudes	=	1 Portuguese fanega (55.364 litres, 12.18 UK gallons, 14.63 US gallons)

From its name the unit would seem to derive from an obsolete Arabic measure.

alpha particle [physics] A particle consisting of two protons and two neutrons (and therefore equivalent to a nucleus of a helium atom), with two positive charges and a mass number of 4. Alpha particles are produced during certain types of radioactive decay. A stream of them is called alpha rays, or alpha radiation. They have only little penetrating power.

alpha radiation *see* ALPHA PARTICLE

alpha waves [medicine] Slow electrical waves produced in the human brain in a person who is awake but inactive. They have a frequency of about 10 hertz and can be recorded by an electroencephalograph. *See also* BETA WAVES; DELTA WAVES.

altazimuth [astronomy] Astronomical instrument consisting of a telescope that can be rotated against angular scales both vertically and horizontally, for measuring the ALTITUDE and AZIMUTH of a celestial object.

alternate angles [maths] When two straight lines are cut by another line that crosses both (a transversal), each of the two angles between the transversal and the two lines, on opposite sides of the transversal, is an alternate angle. If the original two straight lines are parallel, the alternate angles are equal.

altimetry [meteorology] The measurement of altitude, usually by means of an ANEROID BAROMETER, an instrument that measures atmospheric pressure (which decreases with increasing altitude).

altitude [physics; geography; meteorology] Altitude, or height, is a distance and is therefore measured in standard linear units as appropriate, such as the foot, the metre, the kilometre, and the mile. The altitude of a mountain can be measured in several mathematical ways (such as trigonometry), and through a number of mechanical methods (such as orometry: the use of an aneroid barometer to gauge elevation).

altitude [astronomy] For a celestial object, its angular distance (measured north-wards) from the horizon along the great circle that passes perpendicularly through the object and the zenith. It is analogous to geographical latitude: *see* LATITUDE.

The term derives ultimately from Latin *altus* 'high', 'deep'.

alto range [music] The range of the human voice that is between a soprano and a tenor or baritone is generally considered to span from the note F below middle C to the note B above middle C: a range of about an OCTAVE and a half. A woman who sings in the alto range professionally may be called a contralto; a man who sings in the same range is known as a counter-tenor.

Various families of instruments – recorders and saxophones, for example – include an alto member with a range between the soprano and tenor of the same family. Ironically, the term derives ultimately from Latin *altus* 'high', 'deep'.

amagat [physics; volumetric measure] Primarily a unit of gas density at 0°C (273.16 K) and normal pressure (760 millimetres of mercury).

$$1 \text{ amagat} \quad = \quad 1 \text{ mole per } 22.4 \text{ dm}^3$$
$$= \quad 0.04464 \text{ mole per litre}$$

Alternatively a unit of volume: the volume occupied by 1 mole of gas at 0°C (273.16 K) and normal pressure (760 mm Hg). For an ideal gas, this volume – also known as the *standard volume* – corresponds to 22.4 litres.

The units were named after the Dutch physicist E. H. Amagat (1841-1915), a contemporary of J. D. van der Waals (1837-1923), who did much theoretical work on ideal and non-ideal gases.

American football measurements, units, and positions [sport] In the United States, American football is known simply as 'football' (elsewhere in the world, 'football' is more usually soccer, although rugby, Australian Rules and Gaelic football may also be known by the simple term).

The principal endeavour of the game is to get the ball into the opponents' end zone. To do this, a team with the ball at the line of scrimmage has four chances to advance 10 yards up the field, or lose possession. If 10 yards is gained, the next (down at the) scrimmage counts as the 'first-down' (of the next four).

The dimensions of the field:
>   length: 120 yards (109.8 metres), comprising 100 yards of playing surface (marked off every 5 yards) and two end zones each 10 yards deep (corresponding to the try area in rugby, in which touchdowns are scored)
>   width: 160 feet (48.8 metres)

Positions of players:
>   in scrimmage, from center outward:

offense:	center
	guard
	(nose-)tackle
	tight end
	end
defense:	defense tackle
	defense end
	line backer

>   behind scrimmage:

offense:	quarter-back
	backs
defense	line backers
	defense backs

>   Eleven players of each team are on the field at any time

Timing: 60 minutes playing time (the timing stops when the ball is not in play), in four quarters
>   intervals: 2 minutes between quarters; 15 minutes at half-time
>   overtime (when tied at normal game end): 15-minute sudden-death play till the first score (if then no score, tie stands)

Points scoring:
>   touchdown (legal possession of the ball in the opponents' end zone): 6 points
>   extra point (kick through the goalposts after touchdown): 1 point
>   field goal (place- or drop-kick through the goalposts): 3 points
>   safety (possession by opponent in own end zone when play halts): 2 points

Method of winning:
>   scoring more points than opponents

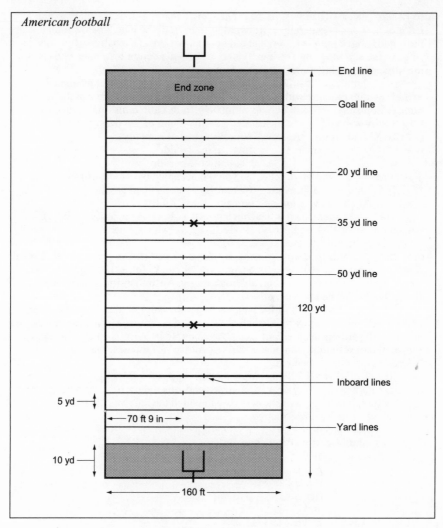

American football

Dimensions of equipment:

goal: height of posts – 10 feet (3.05 metres) above ground to 30 feet (9.14 metres) high

                           distance between posts – 18 feet 6 inches (5.64 metres)

ball: overall length – 11-11.25 inches (27.9-28.6 centimetres)

                           circumference around middle – about 21.375 inches (54.29 centimetres)

During the 1980s, American football became popular all over the world, although the initial flush of enthusiasm has since then rather faded (especially in Europe).

American run [textiles] Unit for the thickness of yarn equal to the weight in ounces of 100 yards of yarn. *See also* DENIER; DREX; TEX.

ampere, amp [physics] Unit of electric current in the SI system of measurement. It is the current, flowing in a pair of infinitely long parallel conductors of negligible cross-section located 1 metre apart in vacuum, that produces a force of 2×10^{-7} NEWTONS per metre between them.

It was named after the French physicist André Marie Ampère (1775-1836).

ampere hour [physics] Unit of electric charge equivalent to a current of 1 AMPERE flowing for 1 hour.

                1 ampere hour   =   3,600 COULOMBS

It is often used to indicate the quantity of electricity that can be delivered by a fully charged battery or accumulator (secondary cell).

ampere-turn [physics] Unit of magnetomotive force in the SI system of measurement, equal to the force produced by a current of 1 AMPERE flowing round one turn of a conductor.

amphibrach *see* FOOT [literary]

amphora [volumetric measure] Large unit of liquid and dry capacity in ancient Greece and of liquid capacity (only) in ancient Rome, originating as the name for a two-handled jar or urn.

In ancient Greece,

1 amphora	=	144 COTYLAI, 72 SEXTE or xestes, 12 KHOES
	=	38.8356 litres
	=	8.5429 UK gallons, 10.2596 US gallons
	=	8.5429 UK dry gallons, 8.8127 US dry gallons
	=	1.0679 UK bushels, 1.1016 US bushels
	=	38,835.6 cubic centimetres, 0.03884 cubic metre
	=	1.3711 cubic feet
1¼ amphora	=	1 MEDIMNOS (dry measure only)

In ancient Rome,

1 amphora	=	48 sextarii, 8 CONGII, 3 MODII, 2 URNAE
	=	25.50 litres
	=	5.6094 UK gallons, 6.7366 US gallons

The term derives from the shape of the original Greek jar or pitcher, which had a handle (*-phor-*) on both (*amph-*) sides.

amplitude (of a wave) [physics] For a quantity that varies in periodic cycles (such as electromagnetic radiation, alternating current, or sound waves), the maximum value attained by that quantity, corresponding to its maximum displacement from its mean position – usually half the total displacement.

For a sound wave, amplitude corresponds to loudness or, loosely, to volume.

AMU, amu [physics; chemistry] *see* ATOMIC MASS UNIT

analytic geometry [maths] *see* COORDINATE GEOMETRY

anap(a)est/anap(a)estic foot *see* METRE IN VERSE

andante [music] Musical instruction on tempo: moderately slow. Italian: '(at) walking (speed)' (*andare* 'to walk'). A movement marked andante is generally played slightly faster than an ADAGIO, but in a freer and less dignified way than a LARGO.

anemometer [meteorology] An instrument for measuring WIND SPEED, usually consisting of a vertical, rotatable shaft with three or four horizontal arms carrying hemispherical cups. The wind blows round the arms and hence the shaft, and the speed of rotation is a measure of wind speed.

The first element of the term derives from ancient Greek *anemos* 'wind', 'puff', cognate with Latin *animus* 'breath', 'wind', and *anima* 'spirit', 'soul'. It is akin to Greek *atmos* 'breath', 'vapour', and probably also to Latin *elemens* 'living', 'being' (cf. Finnish *ilma* 'air').

aneroid barometer [meteorology] A type of BAROMETER with an evacuated bellows made of thin corrugated metal. The change in shape of the bellows following changes in atmospheric pressure is made, by means of levers, to move a pointer along a scale calibrated in pressure units.

The term aneroid derives from ancient Greek words that together mean 'without being wet', referring to the fact that the device is totally dry, using neither liquid nor fluid metal (mercury).

aneuploid [medicine; biology] Description of a cell from which one or more chromosomes are missing or in which there are one or more extra chromosomes.

The term derives from ancient Greek elements meaning 'not well formed'.

angle [maths] An angle – the inclination of one line in relation to another – may be measured in degrees (°) or in radians. In a complete revolution there are 360° or 2π radians. Degrees may be subdivided into minutes and seconds:

$$1 \text{ degree} = 60 \text{ minutes, } 3{,}600 \text{ seconds}$$

To prevent confusion with time units of the same names, these minutes and seconds are often called minutes of arc and seconds of arc.

angle of advance [engineering] In a spark-ignited internal combustion engine, the angle between dead centre and the ignition position (as selected for the most efficient use of the fuel).

angle of attack [aeronautics] Angle that an aerofoil (such as an aircraft wing) makes to the airflow, measured relative to the aerofoil's chord line.

angle of contact [physics] At the place where a liquid is in contact with a solid (a liquid-solid interface), the angle between the surface of the liquid and the solid (measured against the vertical side of the container, within the liquid). It is an acute angle (less than 90 degrees) for liquids that wet the solid; for liquids that do not wet the solid, the angle of contact is obtuse (more than 90 degrees).

angle of deviation [physics] When a ray of light passes through a prism, the angle between the incident ray and the emergent ray.

angle of dip *see* MAGNETIC ELEMENTS

angle of incidence [physics] When a ray of light or other radiation is reflected from a surface (for example, a mirror) or refracted as it passes from one optical medium to another, the angle between the incident ray and the normal (a line at right-angles) at the surface of the mirror or interface between the optical media.
The term is also incorrectly used for the ANGLE OF ATTACK.

angle of reflection [physics] When a ray of light or other radiation is reflected from a surface (for example, a mirror), the angle between the reflected ray and the normal (a line at right-angles to the surface). It is equal to the ANGLE OF INCIDENCE.

angle of refraction [physics] When a ray of light or other radiation is refracted as it passes from one medium to another, the angle between the refracted ray and the normal (a line at right-angles) at the interface between the media.
The ratio of the sines of the angles of incidence and refraction is the *refractive index*.

angle of repose [geology] When loose material (such as scree or sand) is piled up into a heap, the largest angle between the horizontal and the sloping side of the heap.

angle of stall [aeronautics] For an aerofoil (such as an aircraft wing), the ANGLE OF ATTACK that provides maximum lift.

Angstrom/angstrom unit [physics; linear measure] In spectroscopy, an angstrom unit is a unit of measurement of the wavelength of light, corresponding to a linear measurement of

1 ten-thousand millionth of a metre (10^{-10} m)
1 hundred-millionth of a centimetre (10^{-8} cm)
1 ten-millionth of a millimetre (10^{-7} mm)
1 ten-thousandth of a micron (10^{-4} µ)
0.000000003937008 inch

Used also in atomic measurements, the unit is named after the Swedish scientist Anders Jonas Ångström [pronounced 'Ong-strerm'] (1814-74) whose 1868 'map' of the solar spectrum introduced the unit. The unit was formally accepted in international terms only in 1927. It has been largely superseded by the NANOMETRE.
The unit's symbol is , Å, A., or A.U.

angular acceleration [physics] For an object moving in a curved path, the rate of change of ANGULAR VELOCITY, generally expressed in radians per second per second.

angular distance [astronomy] The angle between two objects as perceived by an observer at a distance (for example, the angle between two stars as viewed from Earth).

angular frequency [physics] For an oscillating object or system, the number of vibrations in unit time multiplied by 2π, generally expressed in radians per second. The angular frequency is known also as the *pulsatance*.

angular momentum [physics] For a rotating object, the moment of its linear momentum about an axis (which generally passes through its centre of mass). For an object moving in a curved path, its angular momentum about an external axis (such as the centre of the curve) is called its orbital angular momentum.

angular velocity [physics] For an object moving in a curved path, the rate of change in its angular displacement, generally expressed in radians per second.

anker [volumetric measure] An ancient measure, the anker is still used in some north

European countries as a volumetric unit relating to wines and spirits. It was in use in England until at least the 1750s, when the term could also be spelled *anchor*.

> 1 anker = 8.5 imperial (UK) gallons, 10 medieval wine-gallons
>
> = 10.2 US gallons
>
> = 38.64 litres

The term derives from the Dutch and is the same word as English *anchor*, which suggests that the measure might originally have been one of weight rather than of volume.

anna [comparative values] Former unit of currency in India and Pakistan.

> 1 anna = 4 PICE
>
> 16 annas = 1 RUPEE

Despite the fact that there were 4 pice to the anna, the name of the unit derives from Hindi *ana* 'insignificant'.

anniversaries [time] Many people celebrate wedding anniversaries by presenting gifts of a specific nature according to the number of years that have passed since the initial event. The list of types of gift is not definitive: it varies from country to country. In general, however, it runs:

after 1 year – cotton; 2 years – paper; 3 years – leather; 4 years – iron or fruit/flowers; 5 years – wood; 6 years – sugar; 7 years – wool; 8 years – bronze; 9 years – copper; 10 years – tin; 11 years – steel; 12 years – silk and/or fine linen; 13 years – lace; 14 years – ivory; 15 years – crystal; 20 years – china; 25 years – silver; 30 years – pearl; 35 years – coral; 40 years – ruby; 45 years – sapphire; 50 years – gold; 55 years – emerald; 60 years – diamond; 70 years – platinum.

The list is evidently intended to correspond to a notion of graded values – the older the anniversary, the more valuable the gift to mark its passing. Since its compilation, however, many of the values of the items listed have changed, some notably increased (leather, wood, copper, tin) and some vastly reduced (sugar, wool, china, coral). One is actually now virtually unobtainable new (ivory).

annual percentage rate (of interest) [comparative values] A rate of interest payable weekly, monthly, bimonthly, quarterly, or every six months, expressed as a rate per year overall. Also sometimes known as the annualized percentage rate or APR, an annual percentage rate may be quoted so that it may be compared with other annually payable rates of interest or with yearly indexes (such as the rate of change in the cost of living).

antilogarithims [maths] An antilogarithm is a number of which the LOGARITHM is that number. Example: the logarithm of 2 (in the ordinary base 10 sense) is 0.3010; the number 0.3010 is thus the antilogarithm of 2.

antinode [physics] The position of maximum displacement or vibration in a standing wave. *See also* NODE.

antiparticle [physics] A subatomic particle of equal mass as another particle but of opposite electric charge (for example the positron is the antiparticle of the electron). Antimatter (an hypothetical concept) is made up entirely of antiparticles.

Apgar score [medicine] The Apgar score is a measure of a baby's wellbeing one minute after birth, at which time the obstetrician, midwife, or attendant physician tests five specific vital signs. The signs are:

> respiratory effort   heart rate
>
> skin colour   muscle tone
>
> reflex reaction to olfactory stimulation of the nose

Each sign is rated between 0 and 2; the maximum possible total is thus 10. The resultant score is an immediate indicator of whether the baby is likely to be healthy or to need medical assistance. If the initial score is low, the test may be repeated at intervals of five or ten minutes to check progress, especially if medical treatment has been given.

The system is named after the American anaesthetist Virginia Apgar, who devised it during the 1950s.

aphelion [astronomy] The point on the orbit of a planet at which it is at its maximum distance from the Sun (Greek *ap-helio-* 'from the Sun'). *See also* PERIHELION.

apostilb [physics] Unit of luminance corresponding to the luminance of a uniformly diffusing surface that gives off 1 LUMEN per square metre.

$$1 \text{ apostilb} = 10^{-4} \text{ lambert}$$
$$= 1/\pi \text{ candela}$$

See also BLONDEL.

apothecaries' weight system [weight] The apothecaries' weight system was just that – a system used by apothecaries in weighing out medicines and pharmaceutical potions during the seventeenth to nineteenth centuries – fundamentally based on the weight of the ounce in the TROY WEIGHT SYSTEM.

1 grain	=	0.0020833 troy ounce (0.06479891 gram)
20 grains	=	1 scruple, 0.04166 troy ounce
3 scruples	=	1 drachm (US dram), 0.125 troy ounce
24 scruples	=	1 troy ounce [1 ounce ap(oth.)], 480 grains
1 ounce apoth.	=	1.09709 avoirdupois ounce
		(1 oz avdp. = 0.9115 oz apoth., 437.5 grains)
	=	31.103475 grams
12 ounces apoth.	=	1 pound apoth.
	=	373.2417 grams
	=	13.1657 ounces avdp., 0.82286 pound avdp.

The apothecaries' weight system is now obsolete, although the troy ounce still remains a unit in the measurement of the weight of precious metals. The AVOIRDUPOIS WEIGHT SYSTEM, devised slightly earlier even than the troy system, has been the standard for almost all everyday weight measurements in the English-speaking world from around AD 1340 until the present, although the METRIC WEIGHT SYSTEM has now eclipsed it in world terms.

Appleton layer [physics] Another name for the F-LAYER.

Arabic numbers [maths] Ordinary counting numbers (1, 2, 3, 4, and so on) as opposed to ROMAN NUMERALS (I, II, III, IV, and so on). They originated in India but came to Western mathematics via Arabia (and one or two numbers bear a resemblance to the numbers used in Arabic script). *See also* CARDINAL NUMBERS; NUMBER; ORDINAL NUMBERS.

arc [maths] An arc is part of a curve, and therefore a length – and is therefore expressed in linear units. It may also be defined in terms of the angle (in degrees or radians) that subtends the arc. For example: in a circle, an angle of 90° at the centre subtends an arc equal to one-quarter of the circumference.

Two arcs are generated on a circle by a line that intersects it. The larger of these is the major arc, the smaller the minor arc. (In the special case in which the intersector is a diameter, the two arcs are semicircles and equal in length.)

arc cos, arcos [maths] The trigonometrical function that is the inverse of COSINE, also written $\cos^{-1}$.

Archaeozoic era *see* PRECAMBRIAN ERA

archery measurements and units [sport] In target archery, competitors shoot one or two rounds of arrows (in tournaments often 144 per round) in consecutive sessions at a target, each session comprising a set proportion of the total number of arrows (in tournaments there are often four rounds of 36 arrows) from a different distance. Men and women do not compete against each other.

The dimensions of the field:

men's lanes: maximum 90 metres (99 yards); lines across the lanes also at 50 metres (55 yards) and 30 metres (33 yards)

women's lanes: maximum 70 metres (77 yards); lines across the lanes also at 50 metres (55 yards) and 30 metres (33 yards)

Sequence:

men's lanes: men shoot one to three (but normally two) per lane at distances of 90, 70, 50, and then 30 metres

the score is called after every three arrows ('end'), after which the archer is replaced at the shooting line by a team-member or an opponent

women's lanes: women shoot one to three (but normally two) per lane at distances of 70, 60, 50, and then 30 metres

the score is called after every three arrows ('end'), after which the archer is replaced at the shooting line by a team-member or an opponent

each archer has 2½ minutes to shoot the three arrows of the end
Points scoring:
    the target consists of four concentric rings around a circle, each element of
which is divided into an inner and an outer portion:
    scoring from the outside inwards –
        white outer   – 1 point   white inner   – 2 points
        black outer   – 3 points   black inner   – 4 points
        blue outer   – 5 points   blue inner   – 6 points
        red outer   – 7 points   red inner   – 8 points
        gold outer   – 9 points   gold inner   – 10 points
arrows that rebound off the target or pass straight through are scored only if there
is genuine identificatory evidence of precisely where they struck the target
Method of winning: achieving the highest score
    in the event of a tie, the winner is the archer with the greatest number of
    scoring hits; if the result is still a tie, the winner is the archer who has hit
    most golds
Dimensions of the target:
    diameter: 122 centimetres (4 feet), used for distances greater than 50 metres, or
            80 centimetres (2 feet 7½ inches), used for distances of 50 metres and less

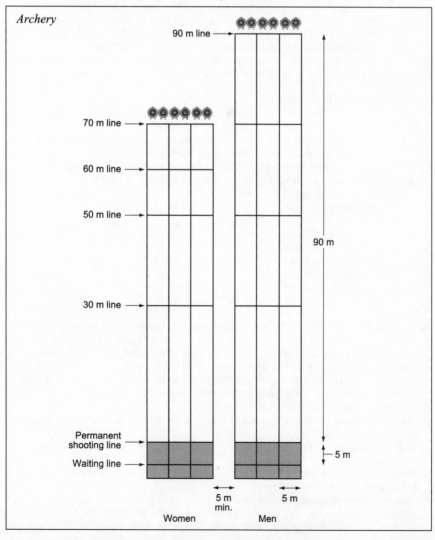

Archery

90 m line →

70 m line →

60 m line →

50 m line →

90 m

30 m line →

Permanent
shooting line →

Waiting line →

5 m

5 m
min.

5 m

Women

Men

target angle: the target is set at 15° off the vertical

height: the centre of the target is set at 130 centimetres (4 feet 3 inches) above the ground

Field archery is a different form of archery, involving progression around a course, shooting 224 arrows at four different targets in a complex sequence and from a huge variety of distances. Contests are comparatively rare.

Archimedes' principle [physics] When an object is immersed in a fluid (such as water), it has an apparent loss in weight, equal to the weight of fluid it displaces. It was named after the Greek mathematician, Archimedes (*c*. 287-212 BC).

arcos *see* ARC COS, ARCOS

arc sin, arcsin [maths] The trigonometrical function that is the inverse of SINE, also written $\sin^{-1}$.

arc tan, arctan [maths] The trigonometrical function that is the inverse of TANGENT, also written $\tan^{-1}$.

are [square measure] Unit of area in the metric system, more commonly found in units of 100, or HECTARES.

1 are	=	10 x 10 metres, 100 square metres
	=	119.6 square yards, 1,076.4 square feet
	=	0.02471 ACRE
		(1 acre = 40.47 ares, 4,047 square metres)
100 ares	=	1 hectare, 10,000 square metres, 2.471 acres
10,000 ares, 100 hectares	=	1 square kilometre, 1,000,000 square metres

The term derives through French from the same root as English *area*.

area [square measure; maths] Areas are calculated by multiplying two lengths, and are generally reckoned in units of length squared (such as square yards, square metres).

Specific units of area include

the acre	–	4,840 square yards
the are	–	100 square metres
the barn	–	10^{-28} square metre (for nuclear cross-sections)
the UK perch	–	30¼ square yards

The following formulae may be used to calculate the areas of some common geometrical figures [l = length/base; h = height/altitude; r = radius]:

circle	πr^2
cone, curved surface	πrs (s = slant height)
total surface	$\pi rs + \pi r^2$
cube (total surface)	$6l^2$
cylinder, curved surface	$2\pi rh$
total surface	$2\pi rh + 2\pi r^2$
parallelogram	lh
rectangle/oblong	lh
sphere	$4\pi r^3$
triangle	½lh

Argand diagram [maths] Method of representing a complex number graphically (*see* NUMBER). On a set of CARTESIAN COORDINATES, the imaginary part of the number is plotted on the vertical *y*-axis, and the real part of the number on the horizontal *x*-axis. For example, the complex number $3 + 5i$ is represented by the point (3, 5) plotted on the grid. The angle between a line from the origin to this point and the *x*-axis is the ARGUMENT of the complex number. It was named after the Swiss mathematician Jean Argand (1768-1822).

argument [maths] In any mathematical function, an independent variable term (for example *x* in the function $y = 3x^3$).

In the complex number $x + iy$, the angle whose tangent is y/x. It can be constructed on an ARGAND DIAGRAM.

arithmetic mean [maths] An AVERAGE calculated by adding a group of numbers and dividing the total by the quantity of numbers in the group.

arithmetic progression/series [maths] A sequence of numbers such that the difference between any two successive terms is always the same (i.e. there is a constant common difference). For example, the numbers 2, 8, 14, 20, 26, 32 are in arithmetic progression (with a common difference of 6).

armillary sphere [astronomy] An ancient astronomical instrument consisting of a framework of intersecting circles representing the sky (celestial globe), with the Earth (incorrectly) located at the centre. It was used for measuring the positions of stars.

The term armillary derives from the Latin *armilla* 'armlet', 'bracelet', and refers to the circular framework of the instrument.

aroura, arura [square measure] Unit of square measure in ancient Egypt, based upon the standard ancient Egyptian unit of length, the (ordinary, not royal) cubit.

1 aroura	=	100 cubits x 100 cubits, 10,000 square cubits
	=	45.11 metres x 45.11 metres, 2,034.91 square metres, (0.2035 hectare)
	=	148 feet x 148 feet, 2,433.777 square yards (just over half an acre)

The unit is known only by its being mentioned by ancient Greek writers.

arpent [square measure] The arpent was a French unit of area that was apparently borrowed from the imperial system and was later imported into Francophone Canada and Louisiana as a legal technical term relating to a different size.

1 arpent	=	100 square perches/rods, 3,025 square yards
	=	0.625 acre (five-eighths of an acre)

redefined as: 51.07 ares, 0.5107 hectare, 5,107 square metres

	=	1.262 acres
		(1 acre = 0.79 arpent)

The confusion may have come about through the fact that the term in French originally applied to a linear measure, not a square measure, corresponding to 11.5 perches/rods (189 ft 9 in, 57.83 metres) or 180 French (Canadian) *pieds* ('feet').

The term itself derives from late Latin, possibly from a Celtic source, and certainly referred to a square measure, but its original size is now unknown.

arshin [linear measure] In western Russia and Estonia, the arshin represents a linear measure formerly used almost as an equivalent of a yard or metre.

1 arshin	=	16 verchoki		
	=	28 inches, 2 ft 4 in (exactly)		
	=	71.12 centimetres, 0.7112 metre		
		(1 foot	=	0.43 arshin
		1 yard	=	1.29 arshin
		1 metre	=	1.41 arshin)
3 arshin	=	1 SADZHEN'	=	7 feet (exactly)
			=	2.1336 metres
1,500 arshin	=	500 sadzhen'	=	1 VERST or versta

Turkey also uses the arshin (alternatively spelled archin) but only as a name for the standard metre (39.37 inches). In Turkey,

1 arshin	=	1 metre (= 100 khats = 100 centimetres)
1,000 arshins	=	1 mill = 1 kilometre
10,000 arshins	=	10 mills = 1 pharoagh

artaba [volumetric measure] In Iran, a large unit of liquid capacity.

1 artaba	=	8 COLLOTHUN, 48 CHENICA
	=	66 litres
	=	14.520 UK gallons, 17.436 US gallons

The unit is ancient, deriving ultimately from Babylonian times; it is mentioned by some ancient Greek writers, who refer to it as being equal to 1 MEDIMNOS and 3 KHOINIKES (that is, 1.0625 medimnos).

article [literary: grammar] In English grammar there are two articles: the *definite article* corresponds to the word *the*; the *indefinite article* corresponds to the word *a*, or before a vowel, *an*. In languages in which there are genders, the articles are the most evident method of identifying the gender of a noun; some languages (like Latin and Finnish) have no articles at all.

By derivation, the th- of *the* is the definitive indicating particle evident also in *this, those, there*, and so forth; *a/an* is essentially identical with the word *one*.

arura *see* AROURA, ARURA

as *see* DENARIUS; LIBRA

ASA system of emulsion speed indicators [chemistry] An arithmetic scale of film emulsion speeds in which, for example, a 400 ASA film is twice as fast as a 200 ASA film, and four times as fast as a 100 ASA film. It is named after the American Standards Association.

Technically, it is defined as 0.8 divided by the exposure of a point that is 0.1 density units greater than the fog level on the characteristic curve of the emulsion. *See also* BSI SYSTEM OF EMULSION SPEED INDICATORS; DIN SYSTEM OF EMULSION SPEED INDICATORS.

A series [paper] A series of paper sizes in the ISO system.

code	millimetres			inches		
A0	841	x	1,189	33⅛	x	46¾
A1	594	x	841	23⅜	x	33⅛
A2	420	x	594	16½	x	23⅜
A3	297	x	420	11¾	x	16½
A4	210	x	297	8¼	x	11¾
A5	148	x	210	5⅞	x	8¼
A6	105	x	148	4⅛	x	5⅞
A7	74	x	105	2⅞	x	4⅛
A8	52	x	74	2	x	2⅞
A9	37	x	52	1½	x	2
A10	26	x	37	1	x	1½

See also B SERIES; C SERIES.

aspect ratio [aeronautics] The span of an aerofoil (usually a wing) divided by its mean chord (width), equal to the square of the span divided by the area of the wing. Low aspect ratios are less than 6; high aspect ratios are greater than 9.

assai [music] Adverb of degree used in musical instructions – Italian: 'very' (Latin *ad satis* 'to sufficiency'). Thus,

allegro assai = very briskly

adagio assai ma non troppo = very slowly, but not too much so

(The French equivalent *assez* more reasonably means 'quite'.)

assay ton [minerals] Mass used in expressing the richness of a precious metal ore, calculated in milligrams per assay ton. It is equal to 29.19 grams for the short ton (2,000 pounds) or 32.67 grams for the (long) ton (2,240 pounds). It is chosen so that the value in milligrams per assay ton is the same as when it is expressed in troy ounces per ton.

associative [maths] Describing a two-stage mathematical operation of which the result is independent of the order in which the operation is carried out. For example, the addition $6 + (3 + 2)$ gives the same result as $(6 + 3) + 2$, and so addition is associative (as is multiplication). *See also* COMMUTATIVE.

astrolabe [astronomy] An early astronomical instrument, the forerunner of the sextant, which was used to observe the positions of heavenly objects and to measure their altitudes.

The term derives from ancient Greek elements meaning 'star-taker'.

astronomical twilight [physics; time] Part of evening during sunset, when the Sun's centre is optically 18 degrees below the horizon (as opposed to CIVIL TWILIGHT, when the Sun's centre is optically 6 degrees below the horizon).

astronomical unit [linear measure] The average distance between the planet Earth and the Sun, used as a unit in relation to linear distances within the Solar System.

1 astronomical unit (a.u.)	=	1.496×10^{11} metres
	=	149,600,000 kilometres
	=	92,957,130 miles
	=	0.000016 light year
	=	0.000005 parsec
63,271.47 a.u.	=	1 light year
206,265 a.u.	=	1 parsec, 3.26 light years

asymptote [maths] In coordinate geometry and geometry, a line approached closer and closer (but never reached) by a curve, equivalent to a tangent to the curve at infinity.

The term derives from ancient Greek elements meaning 'not falling together' – that is, never meeting.

athletics field events area, height, and distance measurements [sport] The usual field events included in an athletics meeting are: javelin, shot-put, discus, hammer-throw, pole-vault, high jump, long (or broad) jump, and triple jump (hop, skip, and jump).

Javelin

    minimum weight of the javelin is 800 grams (1 pound 12.218 ounces) for men, and 600 grams (1 pound 5.163 ounces) for women

    the overall shape and balance of the javelin were officially changed twice during the late 1980s and early 1990s in an endeavour to reduce the length of competitive throws (notably by Hungarian, Finnish, and British competitors)

    the javelin is thrown from behind an arc subtended by an angle of 29° between radii of 8 metres (26 feet 3 inches) length

    the javelin must land point first within a sector of the field marked out with flags, usually of a maximum distance of 110 metres

    measurement of a throw is from the mark on the ground made by the point in landing to the inner edge of the throwing arc on a line that extends through the arc's centre of radius

    the longest throw wins

Shot-put

    minimum weight for the shot is 16 pounds (7.257 kilograms) for men, and 4 kilograms (8 pounds 13 ounces) for women

    the diameter of the men's shot must be between 110 and 130 millimetres (4.33 and 5.12 inches); the diameter of the women's shot must be between 95 and 110 millimetres (3.74 and 4.33 inches)

    the shot is put (thrown) from within a circle (diameter 7 feet/2.13 metres) which at the front has a raised stopboard

    the shot must land within a sector of the field marked out for the event

    measurement of a put is from the innermost edge of the mark on the ground made by the shot in landing to the nearest inner edge of the throwing circle

    the longest throw wins

Discus

    minimum weight for the discus is 2 kilograms (4 pounds 6.547 ounces) for men, and 1 kilogram (2 pounds 3.247 ounces) for women

    the diameter of the men's discus must be between 219 and 221 millimetres (8.62 and 8.70 inches); its maximum width must be between 44 and 46 millimetres (1.73 and 1.81 inches)

    the diameter of the women's discus must be between 180 and 182 millimetres (7.09 and 7.17 inches); its maximum width must be between 37 and 39 millimetres (1.46 and 1.54 inches)

    the discus is thrown from within a circle (diameter 2.50 metres/8 feet 2½ inches), itself within a safety cage of wire netting

    the discus must land within a sector of the field marked out for the event

    measurement of a throw is from the innermost edge of the mark on the ground made by the discus in landing to the nearest inner edge of the throwing circle

    the longest throw wins

Hammer-throw

    minimum weight for the hammer (including the steel wire attachment that is the handle) is 16 pounds (7.257 kilograms)

    total overall length of the hammer must be between 117.5 and 121.5 centimetres (46.26 and 47.83 inches); the diameter of the spherical hammer must be between 102 and 120 millimetres (4.02 and 4.72 inches)

    the hammer is thrown from within a circle (diameter 7 feet/2.135 metres), itself within a safety cage of wire netting

    the hammer must land within a sector of the field marked out for the event

    measurement of a throw is from the innermost edge of the mark on the ground made by the hammer in landing to the nearest inner edge of the throwing circle

    the longest throw wins

Pole-vault
> pole may be of any length, diameter, or weight approved by judges
> usual maximum run-up length: 45 metres (147 feet 6 inches)
> run-up track width: 4 feet (1.22 metre)
> maximum depth of box (in which pole is planted): 9 inches (22.4 centimetres);
>> the back wall of the box angled at 15° beyond the vertical; overall length of
>> box: 1 metre (3 feet 3 inches)
> crossbar dimensions:
>> maximum weight – 5 pounds (2.27 kilograms)
>> thickness – 3 centimetres (1.125 inch)
>> length – 12 feet to 14 feet 2 inches (3.66 to 4.32 metres)
> the highest vault wins – or the highest vault together with the fewest failures

High jump
> run-up length unlimited, although approach must be at an angle of more than
>> 16° to the crossbar
> crossbar dimensions:
>> maximum weight – 2 kilograms (4 pounds 6½ ounces)
>> thickness – 3 centimetres (1.125 inch)
>> length – 4 metres (13 feet 2 inches), on a further 10 millimetres of crossbar
>>> support each side
> the highest jump wins – or the highest jump together with the fewest failures

Long jump, broad jump
> run-up length unlimited
> run-up track width: 4 feet (1.22 metres)
> take-off board width (length): 8 inches (20 centimetres)
> Plasticine tray beyond take-off board: 4 inches (10 centimetres)
> measurement of a jump is from the innermost edge of the marks in the sand
>> made by the competitor in landing to the nearest edge of the take-off board
> the longest jump wins

Triple jump (hop, skip, and jump)
> run-up length unlimited (but usually 45 metres/147 feet 6 inches)
> run-up track width: 4 feet (1.22 metres)
> take-off board width (length): 8 inches (20 centimetres)
> Plasticine tray beyond take-off board: 4 inches (10 centimetres)
> minimum track length between tray and sandpit: 13 metres (42 feet)
> measurement of a jump is from the innermost edge of the marks in the sand
>> made by the competitor in landing to the nearest edge of the take-off board
> the longest jump wins

athletics track events race distances [sport] The usual distances for track
(running) events are:

60 yards [54.86 metres]	60 metres [65.64 yards]
110 yards [100.58 metres]	100 metres [109.4 yards]
220 yards [201.17 metres]	200 metres [218.8 yards]
	300 metres [328.2 yards]
440 yards [402.34 metres]	400 metres [437.6 yards]
880 yards [804.67 metres]	800 metres [875.2 yards]
1 mile [1,609.34 metres]	1,500 metres [1,641 yards, = 0.932 mile]
	3,000 metres [1.864 miles]
3 miles [4,828.03 metres]	5,000 metres [3.107 miles]
	10,000 metres [6.214 miles]

Off the track there are three standard longer running races:
> 10 miles [16.0934 kilometres]
> half-marathon (13 miles 192 yards; 21.0975 kilometres)
> marathon (26 miles 385 yards; 42.195 kilometres)

See also HURDLES EVENTS.

atmosphere [meteorology] Unit of pressure.
> 1 atmosphere = 101,325 pascals
> = 1.01325 x 10^5 newtons per square metre

$$= \quad 1.01325 \text{ bar}$$
$$(1 \text{ bar} = \ 0.9869 \text{ atmosphere})$$

It is equivalent also to 760 millimetres of mercury or 14.72 pounds per square inch.

The term derives from ancient Greek elements meaning 'the sphere of breathing' (*see* ANEMOMETER).

atmosphere, composition of [meteorology] The chemical composition by weight of the atmosphere at the Earth's surface is:

> 78.1% nitrogen
> 20.8% oxygen
> 0.9% argon
> 0.2% carbon dioxide, hydrogen, water vapour, and inert gases

The layers of the atmosphere (from those nearest the planetary surface outwards/upwards) are:

	km	miles	
troposphere	0-12	0-8	(dust, water vapour: the weather)
[tropopause]			
stratosphere	12-55	8-34	(ozone concentration at top)
[stratopause]			
mesosphere	55-77	34-48	(chemical reactions with sunlight)
[mesopause]			
thermosphere	77-400	48-250	(temperature rises steadily)
[thermopause]			
exosphere	400-	250-	(He, H_2 escape into space)

In a different sense, the mesosphere and thermosphere together may alternatively be called the ionosphere, for it is within those layers that the air is ionized by solar ultraviolet radiation (which makes longwave radio transmission possible by reflecting the waves).

atmosphere of the Sun, of a star [astronomy] The Sun – presumably like other stars of its astronomical type – has three major atmospheric layers surrounding it. They are:

> the photosphere
> the chromosphere
> the corona

In relation to the Sun, the photosphere is the visible 'surface' which is, in fact, a gaseous envelope, cooler patches in which constitute sunspots. The chromosphere represents a stratum 8,000-16,000 kilometres (5,000 to 10,000 miles) in depth, the lower part of which – the reversing layer – is where the dark lines in the solar spectrum are produced. Above that is the extremely rarefied corona, from which solar prominences and the solar wind may escape into space.

atmospheric pressure [meteorology] The pressure exerted (on everything on Earth) by the weight of the atmosphere, generally measured using a BAROMETER. It undergoes local variations (because of the weather), and decreases steadily with altitude. Its standard, mean value is 1 atmosphere.

atomicity [chemistry] The number of atoms in a single molecule of an element in its natural state (for example, helium has an atomicity of 1, oxygen 2, ozone 3, and sulphur up to 8).

atomic mass unit (AMU), dalton [chemistry; physics] Unit for expressing the mass of individual atoms, based on one-twelfth of the mass of the carbon-12 isotope as standard. Masses expressed in AMU are known as relative atomic masses (equivalent to what were formerly called atomic weights).

The alternative name for the unit, the dalton, commemorates the English chemist and physicist John Dalton (1766-1844) who was responsible also for DALTON'S LAW and for first describing colour blindness.

atomic number [physics; chemistry] For a particular chemical element, the number of protons in its atomic nucleus (equal to the number of electrons in the un-ionized atom), responsible for the element's chemical identity. *See also* PERIODIC TABLE OF ELEMENTS.

atomic volume [physics; chemistry] For a chemical element, its relative atomic mass (formerly atomic weight) divided by its density. This quantity varies periodically with increasing atomic number, and was one of the factors that led to the idea of the PERIODIC TABLE OF THE ELEMENTS.

atomic weight *see* atomic mass unit (AMU), dalton; relative atomic mass

attenuation [telecommunications] A decrease in the strength of a physical quantity resulting from absorption in or scattering by a medium. A device designed deliberately to weaken a signal (without distortion) is called an *attenuator*.

atto- [quantitatives] Prefix which, when it precedes a unit, reduces the unit to 1 UK trillionth/1 US quintillionth (10^{-18}) of its standard size or quantity.
> Example: 1 attofarad   =   0.000000000000000001 farad
> The prefix derives from *atten*, the Danish for 'eighteen'. (The only other internationally accepted quantitative prefix deriving from Danish is FEMTO-.)

audio frequency [physics] The frequency of an acoustic wave (sound) that is audible, for young adults in the general range 20 hertz to 20 kilohertz (20 to 20,000 cycles per second).

audiometer [medicine] An instrument for measuring the acuity of a person's hearing, generally in terms of the minimum intensity (quietest sound) that can be heard at various frequencies (pitches). The results can be plotted as a chart called an *audiogram*, commonly used as an aid to diagnosis and treatment of hearing disorders.

aune *see* ELL

aur, aurar [comparative values] Unit of currency in Iceland, the equivalent of the *öre* of other Scandinavian countries.
> 100 aurar   =   1 króna
> *See also* COINS AND CURRENCIES OF THE WORLD.

Australian rules football measurements, units, and positions [sport]
Australian Rules is played in very few other countries of the world, and virtually only by expatriates even then. It is a fast and furious game.
> The dimensions of the pitch:
>> the oval field: width – 110 to 155 metres (120.3 to 169.6 yards)
>>> length – 135 to 185 metres (147.6 to 202.3 yards)
>> the central square: 50 x 50 yards (45.72 x 45.72 metres)
>> the centre circle: diameter – 3 metres (9 feet 10 inches)
>> the posts: distance between – 6.4 metres (20 feet 10 inches)
>> goal area: depth – 9 metres (29 feet 6 inches)
> Positions of players:
>> from backs to forwards:
>>> full back (goal)
>>> right full back/left full back
>>> right half back/centre half back/left half back
>>> follower (2)
>>> rover
>>> right centre (wing)/centre/left centre (wing)
>>> right half forward/centre half forward/left half forward
>>> right full forward/left full forward
>>> full forward
>> 18 players of each team are on the field at any time
> Timing: 100 minutes, in four quarters
>> intervals: 3 minutes between first and second quarters; 15 minutes at half-time; 5 minutes between third and fourth quarters
> Points scoring:
>> goal (through central posts): 6 points
>> behind (between outer posts, or when touched or carried over between the central posts or outer posts): 1 point
> Method of winning: scoring more points than opponents
> Dimensions of equipment:
>> ball: maximum circumference – 29 inches (73.6 centimetres)
>>> circumference around middle – 22½ inches (57.2 centimetres)

average [maths] Number that is representative of a collection of numbers or quantities: *see* ARITHMETIC MEAN; GEOMETRIC MEAN; MODE; MEDIAN; ROOT MEAN SQUARE (RMS).

The word 'average' derives from medieval Arabic *awariyah* through Italian: it originally meant 'the proportion of goods inevitably damaged or having perished during transportation by sea', thus 'a set proportion' in any sense.

avo [comparative values] Unit of currency in Macao.

$$100 \text{ avos} = 1 \text{ pataca}$$

See also COINS AND CURRENCIES OF THE WORLD.

Avogadro's law [physics; chemistry] At the same temperature and pressure, equal volumes of all gases contain the same number of molecules (or atoms or ions). The number of molecules in 22.4 litres of any gas is AVOGADRO'S NUMBER. The law is also known as *Avogadro's hypothesis*.

Avogadro's number [physics; chemistry] The number of molecules (or atoms or ions) in a mole (gram molecular weight) of a substance, equal (for all substances) to 6.02253×10^{23}. It is known alternatively as Avogadro's constant, and was named after the Italian nobleman and physicist Amedeo Avogadro (1776-1856).

avoirdupois weight system [weight] The avoirdupois ('having weight' in medieval French) weight system was introduced in England in about AD 1340 and has been the system in everyday use in the English-speaking world virtually ever since, although eclipsed in modern times around the world by the metric weight system.

1 dram	=	27.344 grains (1.772 grams)
		(1 gram = 0.03527396 oz)
16 drams	=	1 ounce [1 oz, 1 oz av(dp.)]
		(1 oz = 28.349523125 grams
	=	437.5 grains, 0.9115 troy ounce)
		[1 troy ounce = 1.09709 oz av(dp.)]
16 ounces	=	1 pound (1 lb)
		(1 lb = 453.59237 grams, 0.45359 kilogram)
		(1 kilogram = 2.204623 lb)
14 pounds	=	1 stone (6.35 kilograms)
28 pounds, 2 stone	=	1 quarter (i.e. of a hundredweight)
100 pounds	=	1 short hundredweight (45.359 kilograms), in some countries known as a cental (50 kg = 1 centner, 110.23 lb)
112 pounds, 4 quarters	=	1 hundredweight (1 cwt) or quintal, in some countries also known as a centner or sentner (1 cwt = 50.80208 kilograms)
2,000 pounds, 20 short hundredweight/centals	=	1 short ton (1 short ton = 907.18474 kilograms = 0.90718474 tonne) (1 tonne = 1.1023113 short tons)
2,240 pounds, 20 hundredweight	=	1 ton (US long ton) (1 ton = 1,016.0416 kilograms = 1.0160416 tonnes, 10.160416 metric quintals) (1 tonne = 0.9842116 ton) (1,000 tons = 1,016.0416 tonnes) (1,000 tonnes = 984.2116 tons)

The TROY WEIGHT SYSTEM was introduced a little later than the avoirdupois system and, concentrating on small weights, is now in evidence only in the form of the troy ounce used in measuring weight in precious metal. Based on the troy ounce, the APOTHECARIES' WEIGHT SYSTEM also had a vogue for a brief couple of centuries.

axis [maths; physics] Reference line for a graph of geometric figure; plural: axes. In two-dimensional coordinate geometry, for example, points are defined in terms of their distances from the x-axis and the y-axis, one of which is horizontal and the other vertical, and which meet at a point called the origin. A plane geometric figure may have an axis of symmetry about which it is symmetrical. A solid figure

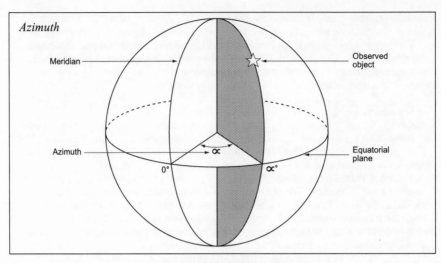

Azimuth

Meridian

Observed object

Azimuth

Equatorial plane

0°   ∝°

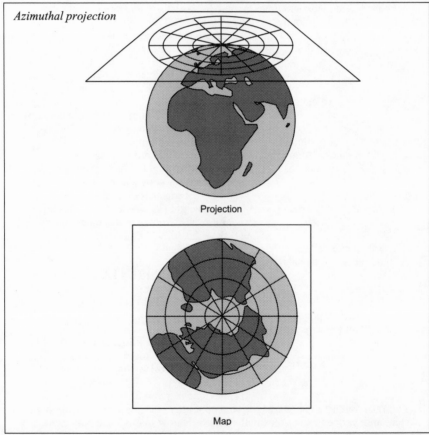

Azimuthal projection

Projection

Map

generated by rotating a two-dimensional figure has an axis of rotation. The shapes
of crystals are defined with reference to three axes.

azimuth [physics: astronomy; surveying; maths] The angle between a defined
  vertical plane and a plane containing an observed object or position. In astronomy it
  is conventionally measured eastwards from the MERIDIAN. In surveying, the azimuth
  is measured westwards from the reference.
    The term derives from the medieval Arabic *as-sumūt* 'the path'.

azimuthal projection [physics: cartography] Type of map projection in which the locations of points on the Earth are projected on to a plane placed tangentially to it. Great circles passing through the centre of the projection are projected as straight lines, and angles between points on the resulting map are their true compass bearings. Azimuthal projection is alternatively known as *zenithal projection*.

B

Babcock test [chemistry] The Babcock test measures the amount of butterfat in milk. The method – introduced in 1890 – is to mix a measured amount of milk with sulphuric acid and so liberate fat globules. Skimmed into a graduated container, the volume of globules immediately indicates the percentage of butterfat in the milk. It was named after the American chemist Stephen M. Babcock (1843-1931, the 'father of scientific dairying'), who devised it.

Babo's law [physics] Dissolving a substance (solute) in a solvent lowers the vapour pressure of the liquid. Babo's law states that the lowering of the vapour pressure is proportional to the amount of solute. It was named after the German chemist Clemens von Babo (1818-99). *See also* BLAGDEN'S LAW.

back, beck [volumetric measure] A large vat or tub, of no specific size, used in soap-making, brewing, dyeing, and other industries.

The term derives through Dutch or German but may ultimately represent the Old English word of which *bucket* is a diminutive.

back focus [photography] In a camera focused at infinity, the distance between the rear surface of the lens and the film (image) plane.

back titration [chemistry] A technique in VOLUMETRIC ANALYSIS in which excess reagent is added (passing the end-point), and the excess is determined by TITRATION.

badminton measurements and units [sport] The object of the game is to hit the shuttle(cock) over the net and on to the floor within the opponent's court, or so that an opponent touches it but is unable to return it.

The dimensions of the court:

overall length (doubles court): 44 feet (13.4 metres)
                (singles court): 39 feet (11.89 metres)
overall width   (doubles court): 20 feet (6.1 metres)
                (singles court): 17 feet (5.18 metres)
service court depth: 13 feet (3.96 metres)
net area depth: 6 feet 6 inches (1.98 metre)
depth between doubles and singles long service lines: 2 feet 6 inches (76 centimetres)
width between doubles and singles sidelines: 1 foot 6 inches (46 centimetres)
height of top of net: 5 feet 1 inch (1.55 metre)

Points scoring:

only the server/serving side scores; service changes when the server loses a play
games are to 21 or 15 points (although in women's singles, games may be to only 11 points); there is a sort of tie-break arrangement ('setting') available if scores are tied as the end of the game approaches

Dimensions of the equipment:

racket weight: usually 4 to 5 ounces (113.4 to 141.75 grams)
shuttle(cock) weight: 1/6 to 1/5 ounce (4.73 to 5.50 grams)

The name of the game is borrowed from the large estate in Gloucestershire, England, now more famous as the venue for an important annual international (equestrian) three-day event. *See over* for illustration.

baht [comparative values] Unit of currency in Thailand, derived primarily as a unit of weight corresponding to 15 grams, 0.53 ounce (231.5 grains) and formerly alternatively called a *tical*.

The '-h-' is not pronounced and is there merely as a phonetic device to lengthen the vowel '-a-'.

See also COINS AND CURRENCIES OF THE WORLD.

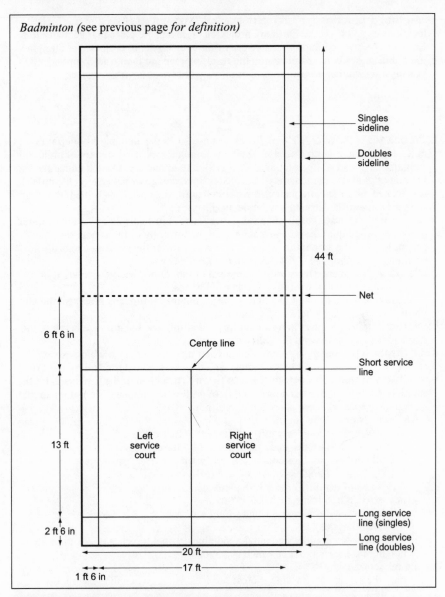

*Badminton (*see previous page *for definition)*

Singles sideline

Doubles sideline

44 ft

Net

6 ft 6 in

Centre line

Short service line

13 ft

Left service court

Right service court

Long service line (singles)

2 ft 6 in

Long service line (doubles)

20 ft

17 ft

1 ft 6 in

Bailling degree [physics] One of many obsolete units relating to the specific gravity of a liquid, for use with a hydrometer. Technically a 'degree Bailling' on the Bailling scale, it was named by and after its inventor, who devised it in 1835 for measuring the specific gravity of aqueous sugar solutions.

baker's dozen [quantitatives] A baker's dozen is by tradition 13. It would seem to be the result of the bakers' wanting to avoid the heavy fine liable in England from the fifteenth century for giving short measure. Surplus loaves in general were known as *inbread*; the thirteenth loaf was the *vantage loaf*.

baktun [time] According to the religion-oriented calendar of the Maya people who flourished in Mexico and Central America during the AD 200s-800s, a cycle of 400 years, each year of 360 days made up of eighteen twenty-day 'weeks'. It may be totally coincidental, but 400 years of this kind amount to precisely 144,000 days – a number that in souls, years, or other units is significant in several different religions.

balance of payments, balance of trade [comparative values] The net profit or loss of a country when annual financial totals of exports, imports, investments, grants, and tourist expenditures are set against one another. This measurement may alternatively be narrowed down to a comparison between one country and another, and the overall annual profit or loss of one in relation to the other in terms of exports, imports, investments, grants, and tourist expenditures.

balboa [comparative values] Unit of currency in Panama, so named after the Spanish explorer Vasco N. Balboa (*c.*1475-1519).

$$1 \text{ balboa} = 100 \text{ centesimos}$$

See also COINS AND CURRENCIES OF THE WORLD.

ballad, ballade *see* VERSE FORMS

ballistic galvanometry [physics] The use of a ballistic galvanometer to measure surges in electric current, which are expressed in AMPERES or their subdivisions.

balthazar [volumetric measure] Extremely large bottle (containing alcoholic drink) with a capacity of about 2.5 UK gallons, 3 US gallons (11.356 litres). Reckoning on a standard wine bottle containing 75 centilitres (0.75 litre),

1 magnum	=	the contents of	2 bottles
1 jeroboam	=		4 bottles
1 rehoboam	=		6 bottles
1 methuselah	=		8 bottles
1 salmanazar	=		12 bottles
1 balthazar	=		16 bottles
1 nebuchadnezzar	=		20 bottles

Some of these terms relate more specifically to a net weight of the drink in fluid ounces, although the approximation to the overall volume by contents of bottles (as listed above) is remarkably accurate.

The balthazar is named after the King of Babylon described in the Biblical book of *Daniel* as the son of Nebuchadnezzar, at whose feast the first manifestation of the writing on the wall occurred, duly interpreted by Daniel (who was himself known at the time as Belteshazzar). In historical fact, his name was Bel-shar-uzur (Akkadian for 'Lord, protect the king', adapted in Greek to *Baltasar* and now most often spelled Belshazzar), he was the son of the almost-as-famous King Nabonidus, and he was never crowned king. By tradition, Balthazar was also the name of one of the three Wise Men or Magi, who came to worship Jesus Christ shortly after his birth, and specifically so named because he came 'from the east' – that is, from Babylon, where the Zoroastrian magi were renowned for astrological wisdom.

ban [comparative values] Unit of currency in Romania: singular ban, plural bani.

$$100 \text{ bani} = 1 \text{ leu}$$

See also COINS AND CURRENCIES OF THE WORLD.

bandwidth [physics] In telecommunications, the difference between the maximum and minimum frequencies that can be handled by an apparatus or component. It may be a limitation of a particular design, or deliberately defined for a specific purpose (such as the individual channels allocated to radio or television stations).

Bandwidth is expressed in HERTZ.

bandy measurements and units [sport] Bandy is similar to ice hockey, but played with a plastic ball instead of a puck, with eleven players from each side on the ice, and on a larger rink that does not extend behind the goals. There is a movable barrier down the sides of the rink.

The dimensions of the rink:
maximum area: 110 x 65 metres (120.3 x 71.08 yards)
minimum area: 100 x 55 metres (119.6 x 60.15 yards)
penalty area (semicircle) radius: 17 metres (18.6 yards)
goal: width – 3.5 metres (11 feet 6 inches)
    height – 2.25 metres (7 feet 4½ inches)
    depth – 1-2 metres (3 feet 3 inches to 6 feet 6 inches) top to bottom
Timing: 90 minutes, in two halves, separated by an interval of 5 minutes
Points scoring:
the team that scores the greater number of goals wins
Dimensions of equipment:
stick: overall length (including hook) 1.2 metres (3 feet 11¼ inches)

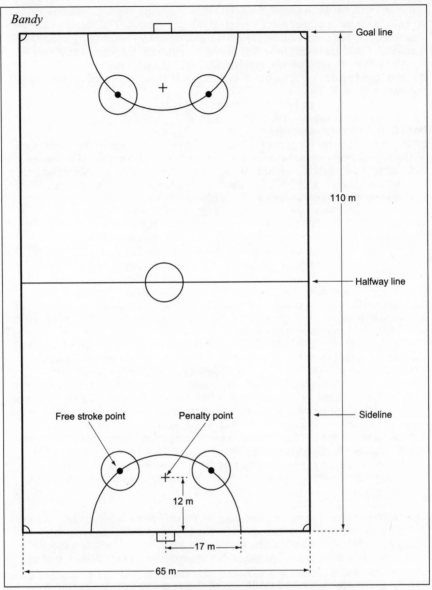

Bandy

Goal line

110 m

Halfway line

Free stroke point   Penalty point   Sideline

12 m

17 m

65 m

ball: weight – 58-62 grams (2.05-2.19 ounces)
         minimum-maximum bounce when dropped from a height of 1.5 metres
         (4 feet) – 15-30 centimetres (6-12 inches)

The game is especially popular in Sweden and Finland (where it is often considered a particularly English game, although most English people have never heard of it).

The name of the game derives from Norman French *bandé* 'bent' (with which English word it is cognate), and originally referred to a game that was more like tennis in which the ball was 'bandied' from side to side across a net (hence the notion of bandying insults or wisecracks to and fro). Ultimately, the word is akin to *band* and *bind* and *bond* and other such forms that refer to a joining of two sides.

bar [physics: meteorology] Unit of pressure used primarily in measuring atmospheric pressure and in engineering contexts.

$$1 \text{ bar} = 750.07 \text{ millimetres of mercury at } 0°C \text{ (750.07 TORR)}$$
$$= 10^5 \text{ newtons per square metre } (10^5 \text{ pascals})$$

The bar is also approximately equivalent to 1 ATMOSPHERE:

$$1 \text{ bar} = 0.9869 \text{ atmosphere}$$

$$(1 \text{ atmosphere} = 1.01325 \text{ bar, } 101,325 \text{ pascals})$$

In defining atmospheric pressure, the bar is ordinarily used in the form of the MILLIBAR (one-thousandth of a bar: 100 newtons per square metre). In engineering, the form is generally that of the HECTOBAR (100 bar: 10^7 newtons per square metre).

The term derives as an abbreviation from 'barye' which, in turn, derives ultimately from ancient Greek *baros* 'pressure', 'weight'.

bar [music] The bar is represented on a sheet of music by a length of the stave (the five horizontal lines, either as a single set of five or as two or more sets of five) bordered at both ends by a vertical line. It is in effect a unit of tempo, in that the first note or chord in it (almost always) corresponds to the main beat, and other notes or chords in it thereafter correspond to all the remaining notes or chords before the next main beat. Where music is written in a three-based rhythm (as for example a waltz or other rhythm with a TIME SIGNATURE in which the upper figure is 3, 6, 9, or 12), the bar's overall length comprises the equivalent of 3, 6, 9, or 12 notes (the duration of each note also determined by the time signature, as the lower figure). In a four-based rhythm, the bar contains the equivalent of 4, 8, 12, or 16 notes; and in a two-based rhythm, the bar contains the equivalent of 2, 4, 6, 8, 10, or 12 notes.

barad [physics] Obsolete unit of pressure.

$$1 \text{ barad} = 1 \text{ DYNE per square centimetre}$$

$$= 1 \text{ microbar (one-millionth of a BAR)}$$

The term derives as an analogue of the *farad*, a unit of capacitance in electricity (named after the English chemist and physicist Michael Faraday), its first syllable meant to represent instead the ancient Greek *baros* 'pressure', 'weight'. *See also* BARIE.

bar code [maths] Coded sequence of thick and thin lines and numbers corresponding to a particular item or product in a system of computerized inventory or sales, and recognizable at speed by an electric scanner linked to a microcomputer. It includes code terms for the country of manufacture, the manufacturer, and the type of product – a combination specific enough ordinarily to identify any product.

barie [physics] Obsolete French unit of gaseous pressure.

$$1 \text{ barie} = 1 \text{ DYNE per square centimetre}$$

$$= 1 \text{ microbar (one-millionth of a BAR)}$$

The name derives from a French variant on the ancient Greek *baros* 'pressure', 'weight'. *See also* BARAD.

baritone [music] The male voice of which the register is between tenor and bass; the range is therefore something like from A an octave and a third below middle C to the E immediately above middle C.

The word is often said to derive from Greek *barys* 'heavy', as if 'heavy' meant 'bass'. But there is a possibly equally relevant Latin term *baritus* 'the war-cry of the German tribes', derived from an ancient Germanic verb *baren* 'to raise the voice'.

Barker index [physics] In crystallography, a way of classifying and thereby identifying crystals by measuring their interfacial angles.

barn [chemistry; physics] Extremely small unit of area, used for expressing the effective cross-sectional area of the nucleus of an atom.

$$1 \text{ barn} = 10^{-28} \text{ square metre}$$

$$= 1.55 \times 10^{-25} \text{ square inch}$$

$$= 10^{-22} \text{ square millimetre}$$

This measure expresses the probability that a particular subatomic particle can be captured by a nucleus.

The unit was devised in 1942 by the American nuclear physicists C. P. Baker and H. G. Holloway, and so named – apocryphally – because, compared with a subatomic particle, an atomic nucleus is 'as big as a barn door': an unmissable target. *See also* SHED.

barograph [meteorology] An ANEROID BAROMETER that makes a continuous record of atmospheric pressure changes on a chart or graph.

barometer [meteorology] An instrument for measuring ATMOSPHERIC PRESSURE. The most accurate readings are given by a mercury barometer, but an aneroid barometer is more compact and convenient.

barometric pressure [meteorology] The pressure of the atmosphere as recorded by a BAROMETER, of great significance in weather forecasting. *See also* ATMOSPHERE; ATMOSPHERIC PRESSURE.

barrel [volumetric measure] Thanks to the vagaries of history, the volume of barrels tends to differ greatly according to the commodities they hold.

For dry goods and materials, notably grain and fruit,
1 barrel is usually 0.1156 m^3 = 4.083 cu. ft, 7,056 cu. in.
= 3.180 UK bushels, 3.283 US bushels
or 115.62 litres, 25.43 UK gallons,
30.55 US gallons

For most ordinary liquids,
1 barrel is usually 119.24 litres, 26.23 UK gallons, 31.50 US gallons
For oil and petroleum products,
1 barrel is usually 158.98 litres, 35.00 UK gallons, 42.00 US gallons
For beer, as transported internationally,
1 barrel is usually 163.66 litres, 36.00 UK gallons, 43.24 US gallons

barrel bulk [volumetric measure] Obsolete cubic measure, formerly used in shipping stowage.

1 barrel bulk = roughly 20½ x 20½ x 20½ inches
= 5 cubic feet, 8,640 cubic inches
= 0.1850 cubic yard
(1 cubic yard = 5.4054 barrels bulk)
= 141,500 cubic centimetres, 0.1415 cubic metre
(1 cubic metre = 7.0671 barrels bulk)
= 3.894 UK bushels, 4.0195 US bushels
7 barrels bulk = 1 DISPLACEMENT TON
8 barrels bulk = 1 FREIGHT TON
20 barrels bulk = 1 REGISTER TON

barye [physics] Obsolete unit of pressure, now known as the BAR.

baryon number [physics] In nuclear physics, a baryon is a member of a group of subatomic particles that are involved in strong interactions with other particles; they include hyperons (short-lived particles heavier than a neutron), neutrons themselves, protons, and their respective antiparticles. The baryon number of all baryons is +1 ; that of antibaryons is -1 . Baryon numbers are also ascribed to other particles that do not interact strongly, such as leptons (for example, electrons, negative muons, tau-minus particles and their neutrinos), MESONS (pi-, K-, and eta-) and gauge bosons (photons, gluons, gravitons, and intermediate vector bosons) – all of which have a baryon number of 0. Quarks have a baryon number of +⅓, and for antiquarks it is -⅓ . In all reactions between particles, baryon number is conserved.

The term derives from the ancient Greek *barys* 'heavy'.

base [maths: logarithms] A number is expressed as a LOGARITHM in terms of a base (the logarithm of a number is the power to which the base must be raised to equal the number). For example: $100 = 10^2$ and so the logarithm of 100 to the base 10 is 2 (that is, $\log_{10} 100 = 2$). Napierian or natural logarithms (written log$_e$) use the irrational number e as their base.

base [maths: arithmetic] Number on which an exponent or index operates. For example: in the expression 6^3, 6 is the base (and 3 the exponent).

base [maths: numeral systems] The base, or *radix*, of a numeral system is the number of one-digit numbers in it, including 0 . In normal everyday arithmetic, the sequence 0 to 9 involves ten numbers and the base is 10. But the numeral 10 in any number system indicates its base: for example, 10 in the BINARY SYSTEM represents the number 2 – which is the base of the binary system.

In history there have been other bases used at various times, echoes of which remain in vestigial form. Elements of a numeral system with base 12 are evident in such 'units' as a DOZEN and a GROSS, and at one time were also prominent in British coinage. It is possible that the system derived from a time when most calculation was by religious astronomers who used a notation founded upon divisions of the solar year into twelve months. And it is quite commonly suggested that base 12 is in some respects more convenient than base 10 in that 12 is exactly divisible by 2, 3, 4, and 6, whereas 10 is divisible only by 2 and 5. But units such as the SCORE suggest a base 20, and the Babylonians had an even higher base in 60, the origin of such measurements as degrees in a circle, and seconds and minutes as fractions of an hour. *See also* DECIMAL; DUODECIMAL SYSTEM.

baseball measurements, units, and positions [sport] In baseball, the object of the game is to make complete runs around three bases and back to home base, either stage by stage or preferably all at once after hitting the ball to an area within the field from which fielders find difficulty in retrieving the ball, or having hit the ball so hard that it goes high across the field and into the crowd.

The dimensions of the pitch:

infield: 90 x 90 feet (27.45 x 27.45 metres), bases at the corners

outfield: grass area surrounding the infield, then as big as the surrounding stadium

pitcher's circle: diameter – 18 feet (5.49 metres)

batter's box (each side of the home base): 6 x 4 feet (1.83 x 1.22 metres)

catcher and umpire's box (behind home base): extends to 8 feet (2.44 metres) behind point of home base, 4 feet 3 inches (1.29 metres) wide

distance from pitcher's plate to farthest point on home base (over which batter hits): 60 feet 6 inches (18.45 metres)

home base plate: maximum width – 17 inches (43 centimetres)

maximum length (tapers to a point from halfway) – 17 inches (43 centimetres)

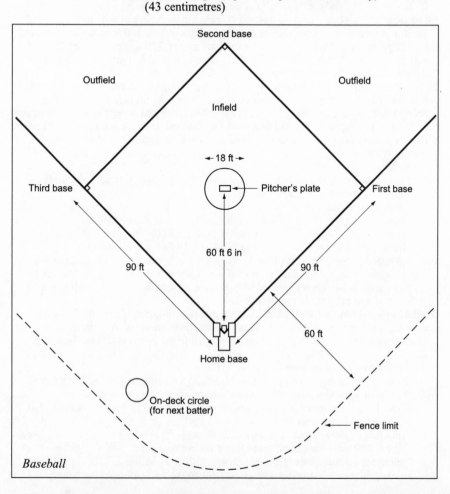

Baseball

bag at each base: 1 foot 3 inches (38 centimetres) square
Positions of fielding players:
pitcher
catcher
first baseman
second baseman
shortstop (between second and third basemen)
third baseman
left (out)fielder
center (out)fielder
right (out)fielder
Timing: duration of game depends on whether batters strike out frequently, and
whether all innings (maximum: 9) are played out by both teams
Points scoring:
completed runs score
(three STRIKES end an inning, whether there are men on bases or not)
the team with the higher number of runs wins
Dimensions of equipment:
bat:　　maximum overall length – 3 feet 6 inches (1.06 metre)
　　　　maximum diameter – 2¾ inches (6.99 centimetres)
ball:　　weight – 5 to 5½ ounces (141.75 to 155.92 grams)
Forms of baseball are popular all over the world, some variants peculiar to their
own country (such as the baseball played in Finland; compare also rounders in
Britain). The US style of baseball is particularly popular in Japan.

base box [chemistry] Obsolete unit expressing the thickness of a metallic coating
(as, for example, in galvanizing or electroplating). A base box was the area formed
by 122 plates each 10 x 14 inches – a total area of 31,360 square inches (217.777
square feet, 20.232 square metres) – and the thickness of a coating was expressed as
the mass in pounds per base box.

base unit [quantitatives] Any of the units on which a particular unit system is based.
For example, the international SI system has seven base units (metre, kilogram,
second, kelvin, ampere, mole, and candela). Multiples and submultiples of these are
constructed using prefixes, and DERIVED UNITS (for measuring quantities other than
the seven base ones) are formed by combining two or more of the base units. *See
also* CENTIMETRE-GRAM-SECOND (CGS) SYSTEM; METRE-KILOGRAM-SECOND (MKS) SYSTEM;
SI UNITS.

basic size [paper] A series of paper sizes in the United States, used for specifying
thicknesses in terms of the BASIS WEIGHT.
The common types and sizes are:
bond　　　　　　　　　17 x 22 inches (431.8 x 558.8 millimetres)
cover　　　　　　　　　20 x 26 inches (508.0 x 660.4 millimetres)
book, offset and text　25 x 38 inches (635.0 x 965.2 millimetres)

basis weight [paper] American method of effectively specifying the thickness of
paper, equal to the weight (in pounds) of a 500-sheet ream in the BASIC SIZE for the
type of paper. It has been superseded in Britain and elsewhere by specifications
stated in grams per square metre (gsm).

basketball measurements, units, and positions [sport] Basketball is a fast-
moving game for two teams of five players on the court at any one time. The object
is to throw or propel the ball through the ring (into the cord net) on the opponents'
side.
The dimensions of the court:
overall length and width proportionate to 26 x 14 metres (84 feet x 45 feet
9 inches), and within 7% of those dimensions
free-throw line distance from backboard: 5.8 metres (18 feet 11 inches), also
partly defining goal area limit
ring height above floor: 10 feet (3.05 metres)
ring diameter: 1 foot 6 inches (45 centimetres)
backboard dimensions: 1.8 x 1.2 metres (5 feet 11 inches x 3 feet 11 inches);
ring attached 20 centimetres (8 inches) above the bottom of the backboard,

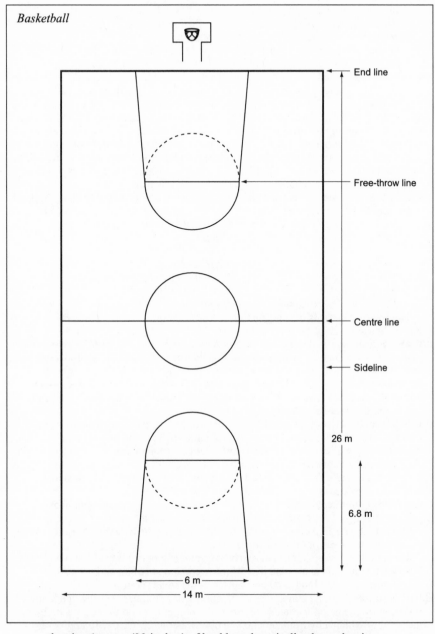

Basketball

End line

Free-throw line

Centre line

Sideline

26 m

6.8 m

6 m

14 m

leaving 1 metre (39 inches) of backboard vertically above the ring
Timing: 40 minutes, in two halves, with an interval of 10-15 minutes
    the game watch stops for time-outs and for dead ball situations
    in a tied game, a sudden-death playoff may continue for as many 5-minute
        extra periods as necessary for a team to score
Points scoring:
    a basket from out of the goal area: 2 points
    a basket from within the goal area during play: 2 points
    a basket from a free throw: 1 point
    there are several variations on this scoring scheme relating to international and
        college basketball games, for some of which the goal area is enlarged
Method of winning: the team that scores the higher number of points wins

Dimensions of ball:
    circumference: 75-78 centimetres (29.5-30.7 inches)
    weight: 600-650 grams (21.16- 22.93 ounces)
    minimum-maximum bounce when dropped from a height of 1.8 metre (5 feet
    10.9 inches) – 1.2-1.4 metre (3 feet 11¼ inches to 4 feet 7 inches)
Outside the United States, Russia, Israel, and a very few other countries, basketball has yet to achieve the sporting super-status it has attained in those nations.

bass range [music] The lowest range of the human voice is generally considered to
span from middle C down to the note F about an OCTAVE and a half below. A man
who specializes in the lower part of this range and obtains true resonance at notes
even below low F may be called a *basso profundo*. Various families of instruments
– clarinets, bassoons, trombones, saxophones, and drums, for example – include a
bass member with a range lower than that of the corresponding TENOR member.
(The term derives ultimately from Latin *bassus* 'deep'.)

bath, batos [volumetric measure] In ancient Israel, a unit of liquid capacity applied
especially to large measures of oil.

    1 bath = 3 SEAH, 6 HIN, 18 CAB or kab, 72 LOGS
            = 40.23 litres
               (40 litres = 0.9943 bath)
            = 8.8506 UK gallons, 10.6279 US gallons
               (9 UK gallons =1.0169 bath
               11 US gallons =1.035 bath)
    10 bath = 1 KOR or homer

The spelling *batos* is the ancient Greek transliteration found in some Biblical
commentaries.

battalion [military] Today's army battalion corresponds to a large force, comprising
two or more companies, with headquarters. Part of a REGIMENT, it generally has a
specific function – infantry, engineers, signals, armoured vehicles, or equivalent –
and is intended to be virtually self-sufficient in combat. The term derives as a
borrowing from a French word (ultimately from Latin) describing enough troops to
go into battle (French *bataille*).

batting average [sporting term] The number of runs a batter or batsman makes
averaged over the number of innings or of games in a season or series. In baseball,
the average is expressed as a three-figure decimal: a batter who makes three runs in
12 games has an average of (3 ÷ 12 =) 0.250. In cricket, the average is expressed as
a figure with single decimal thereafter: a batsman who makes 1,200 runs in 25
innings has an average of (1,200 ÷ 25 =) 48.0.

baud [telecommunications] Unit of speed of transmission in telecommunications.
    1 baud = 1 pulse or bit per second
The actual data signalling rate may not be as fast as the baud rate because of the
inclusion also of various control signals.

The unit was named after the French telegraph engineer J. M. E. Baudot (1845-
1903), who gave his name also to a 5-bit telegraph code of on and off pulses
(contrasting with the long and short pulses of the MORSE CODE).

Baumé scale [physics] Units of specific gravity of a liquid, for use with a hydrometer, in frequent application in continental Europe. Zero on the scale is defined as
the point to which a hydrometer sinks in pure water at 25°C, and 10 degrees Baumé
corresponds to its sinking point in 10 per cent sodium chloride (common salt)
solution.

The scale was originally proposed by a physicist called Lunge, but it is now
known only as the Baumé scale after the French chemist Antoine Baumé (1728-
1804), who adopted it in 1784, originally for measuring the specific gravity of
aqueous sugar solutions.

BB shot [sport] Standard size of shot – about 0.18 inch (4.57 mm) in diameter –
used in a shotgun cartridge or singly in an air weapon.

bce *see* ERA, CALENDRICAL.

beamwidth [physics] The angular width of a radar or radio beam, measured in
degrees.

Beaufort scale of windspeed [physics: meteorology] Scale of wind strength devised originally by Admiral Sir Francis Beaufort (1774-1857) to provide some graded system for the range of sailing conditions from a complete calm to a hurricane.

His system, adopted by the International Meteorological Committee in 1874, named twelve sailing conditions numbered 1 to 12, but did not specify wind speeds. In 1939, the same Committee allocated a range of wind speeds to each numbered condition, in relation to an ANEMOMETER at a height of 6 metres (20 feet). In the UK and the United States, an earlier table of wind speeds is ordinarily in use, and relates to an elevation of 36 feet (11 metres). Furthermore, the US Weather Bureau in 1955 added Beaufort scale numbers 13 to 17 to be able to announce projected wind speed in greater detail in hurricane forecasts.

	International scale 1939		normal UK/US scale		
Beaufort		wind speed		wind speed	
no.	description	km/h	description	mph	km/h
0	calm	0 - 1	light	0 - 1	0 - 2
1	light air	1 - 5	light	1 - 3	2 - 5
2	light breeze	6 - 11	light	4 - 7	6 - 11
3	gentle breeze	12 - 18	gentle	8 - 12	12 - 19
4	moderate breeze	19 - 26	moderate	13 - 18	20 - 29
5	fresh breeze	27 - 34	fresh	19 - 24	30 - 38
6	strong breeze	35 - 43	strong	25 - 31	39 - 50
7	moderate gale	44 - 53	strong	32 - 38	51 - 61
8	fresh gale	54 - 64	gale	39 - 46	62 - 74
9	strong gale	65 - 77	gale	47 - 54	75 - 86
10	whole gale	78 - 90	whole gale	55 - 63	87 - 101
11	storm	91 - 104	whole gale	64 - 72	102 - 115
12	hurricane	105+	hurricane	73 - 82	116 - 131
13			hurricane	83 - 92	132 - 147
14			hurricane	93 - 103	148 - 165
15			hurricane	104 - 114	166 - 182
16			hurricane	115 - 125	183 - 200
17			hurricane	126 - 136	201 - 218

In some Scandinavian countries the Beaufort numbers are called *Bofors* or *Boforia*, an assimilation to a more common (but quite different) surname in that region of the world.

The *mean windspeed* (V) in miles per hour can be derived from the Beaufort wind force (B) by applying the formula

$$V = (1.52B)^{3/2}$$

Beckmann thermometer [physics] A large-bulbed mercury thermometer used for making accurate measurements of small temperature changes (6 to 7 degrees). It was named after the German chemist Ernst Beckmann (1853-1923).

becquerel [physics] Unit of radioactivity in the SI system, equal to the number of nuclei in a radioactive element that disintegrate each second.

It has replaced the former unit, the curie.

$$1 \text{ becquerel} = 2.7 \times 10^{-11} \text{ curies}$$

The unit was named after the French physicist Henri Becquerel (1852-1908), who discovered radioactivity in 1896, and who shared the 1903 Nobel Prize for physics with Pierre and Marie Curie for doing so.

Beer's law [physics] For a given wavelength of light, the fraction of incident light absorbed by a solution is related to the thickness of the absorbing layer and the molality (concentration) of the solution. Beer's law is alternatively known as the Beer-Lambert law.

behind *see* AUSTRALIAN RULES FOOTBALL MEASUREMENTS, UNITS, AND POSITIONS

bekah [weight] In ancient Israel, a very small unit of weight.

1 bekah	=	2 rebahs, 10 gerahs
	=	just under 0.15 ounce avdp. (2.4 drams avdp.)
	=	about 4.2 grams
2 bekahs	=	1 (ordinary or 'holy') SHEKEL
120 bekahs, 60 shekels	=	1 (ordinary or 'holy') MINAH

bel [physics] Unit of sound intensity, used in measuring differences in intensity levels.

$$1 \text{ bel} = 10 \text{ decibels}$$

The unit is named after the Scottish-born US physicist and inventor Alexander Graham Bell (1847-1922), inventor of the telephone and one-time president of the National Geographic Society.

bell curve, bell-shaped curve [maths] In statistics, a symmetrical bell-shaped graph that represents the normal (or Gaussian) distribution of the values of a random variable. Any point on the curve indicates the probability that a particular value will occur. The value corresponding to the highest point of the curve is the mean value (identical in this case with the MEDIAN and the MODE), and the most probable. It is alternatively known as the *normal curve*.

bell-ringing *see* CHANGE-RINGING, CHANGES

bells [time] On board ship, a system of announcing the time in half-hours for all to hear. Eight bells, sounded as four paired chimes, represents 4, 8, and 12 o'clock both day and night. The next half-hour – 4.30, 8.30, and 12.30, a.m. and p.m. – is one bell; the following half-hour – 5, 9, and 1 o'clock – is two bells; and so on.

1 bell	4.30	8.30	12.30
2 bells	5.00	9.00	1.00
3 bells	5.30	9.30	1.30
4 bells	6.00	0.00	2.00
5 bells	6.30	0.30	2.30
6 bells	7.00	11.00	3.00
7 bells	7.30	11.30	3.30
8 bells	8.00	12.00	4.00

In this way, the hours are immediately audible as complete pairs of chimes; half-hours always have one final unpaired chime. *See also* WATCH.

belts [sporting term] In many forms of martial art, the colour of the belt worn as part of the sporting uniform represents the grade of proficiency or skill in the art of the wearer. The numbers of grades and the sequences of belt colours differ between the martial arts, however, and even in the same martial art between competitive countries. There are, in any case, additional titles and ranks once a belt of the highest grade has been obtained.

In Judo, for example, of the six *kyu* ('student') grades, the first three in Japan and the United States are represented by a white belt, and the second three by a brown belt; in other countries, each of the six grades has its own colour in the sequence white, yellow, orange, green, blue, brown. The 12 *dan* ('leader') grades thereafter have a further system of black (1st to 5th dan), black or red-and-white (6th to 8th dan), and red (9th dan and above) belts.

bench mark [surveying] In surveying, a fixed reference point indicating a particular level (usually relating to mean sea-level). The mark resembles a vertical arrowhead below a horizontal line, and is generally carved into an immovable rock or piece of masonry. The same mark was formerly used to identify government property (and is the origin of the arrowheads once used on convicts' clothing).

On British Ordnance Survey maps, the location of a bench mark is indicated by the letters BM and a number corresponding to the level (originally in feet above sea-level).

Bender Gestalt test [medicine: psychology] Psychological test designed to monitor a child's ability to perceive and to conceptualize in a practical way. It comprises a series of drawings of varying complexity that the child is asked to copy.

Bernoulli's law [physics] For a perfect (that is, incompressible and non-viscous) fluid, at any point in a tube through which it flows, the sum of the kinetic, potential, and pressure energies per unit volume is constant. It was named after the Swiss mathematician and physicist Daniel Bernoulli (1700-82).

beta particle [physics] A particle consisting of an electron (and therefore carrying a single negative electrical charge). Beta particles are produced in certain types of radioactive decay. A stream of them is called beta rays, or beta radiation.

beta radiation *see* BETA PARTICLE

beta waves [medicine] Comparatively high-frequency electrical waves produced in the human brain of a person who is awake and active. They have a frequency of between 13 and 50 hertz and can be recorded by an electroencephalograph. *See also* ALPHA WAVES; DELTA WAVES.

betting odds *see* ODDS, CALCULATION OF; PROBABILITY; RACING ODDS

bezant *see* SOLIDUS

B horizon *see* SOIL HORIZONS, SOIL LAYERS

bhp, brake horsepower *see* HORSEPOWER

bi- [quantitatives] Prefix implying the number two in various meanings:

    biannual: two times every year
    biennial: once every two years, lasting for two years
    bisect: divide into two parts
    bipartite: having two parts, in two parties

    The prefix is often quoted as being derived from Latin *bis* 'twice', but this is to ignore the fact that even that derives from an earlier *dvis* in which the ordinary Indo-European stem *duo-* 'two' is clearly evident.

biathlon [sport] Biathlon combines cross-country skiing with rifle shooting. The course is circular, and the start, the shooting area, and the finish are generally all in the same place.

    The dimensions of the course:

        20 kilometres (12.427 miles), including four bouts of shooting (two standing and two lying prone); or
        10 kilometres (6.214 miles), including two bouts of shooting (one standing and one lying prone)
        target range: 150 metres (164 yards)
        five targets to aim at, all to be hit
        target dimensions:
        prone-position target – 25-centimetre (10-inch) diameter
        standing target – 45-centimetre (18-inch) diameter (for convenience, the two targets may be superimposed)

    Points scoring:

        competitors are judged on a total time, from start to finish (including shooting time, and any penalty minutes – or the time taken to complete penalty circuits on a special 150-metre track – for missing targets)

    Races for relay teams are as common and as popular as races for individual competitors. The biathlon was first included in the Winter Olympics in 1960.

biennial [time] Once every two years; lasting for two years.

bifilar micrometer [astronomy] A measuring instrument on the eyepiece of an astronomical telescope which can be used to measure the angular separation of heavenly objects (such as the components of a double star).

Big Bang, posited age of *see* AGE

billiards measurements and units [sport] The English game of billiards is played either by two players or by four players in two teams. The object is to score points by using one white ball to knock either or both of another white ball and a red ball into one of the pockets at the edge of the table, or to go into a pocket itself having struck one of them, or to strike them both consecutively.

    The dimensions of the table:

        full-size table (table-top measurements, inside cushion of green baize): 12 feet x 6 feet 1½ inches (3.66 x 1.86 metres); smaller tables in proportion
        string line: across the width 2 feet 5 inches (73.7 centimetres) from the top cushion
        semicircle ('D') on string line: diameter: 3 inches (7.6 centimetres)
        pocket widths:
        corner pockets – 3½ inches (8.89 centimetres)
        middle pockets – 4 inches (10.16 centimetres)

    Timing: only important if players decide to play to a specific time

    Points scoring:

        cue ball hits other white ball into pocket: 2 points
        cue ball hits red ball into a pocket: 3 points

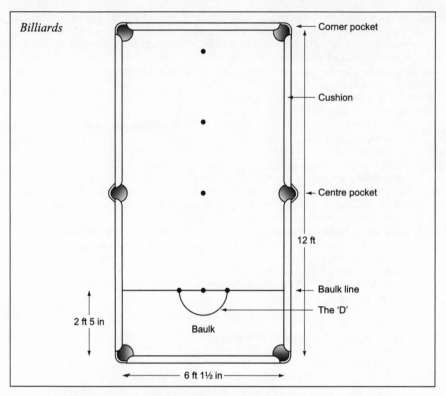

Billiards

Corner pocket

Cushion

Centre pocket

12 ft

Baulk line

The 'D'

2 ft 5 in

Baulk

6 ft 1½ in

(red ball may be pocketed consecutively only 5 times)
cue ball goes into pocket 'in off' white: 2 points
cue ball goes into pocket 'in off' red: 3 points
 (only 15 of all four methods of scoring above allowed consecutively)
cue ball strikes both other balls (cannon): 2 points
cue ball strikes both other balls and goes into pocket: 2 + 2 points if white ball
 hit first; 2 + 3 points if red ball hit first
 (only 7 such cannons may be scored consecutively)
once the opponent's white is off the table, it stays off until the player's break
 ends
scores against:
 touching the cue ball more than once – lose 1 point
 pushing the ball instead of striking it – lose 1 point
 when the cue ball misses other balls altogether – lose 1 point
 when the cue ball misses other balls and goes into a pocket – lose 3 points
 if a ball leaves the table-top surface – lose 3 points
 (in each case, that is also the end of the player's break)
the player or team with the more points at the end of the game wins
Dimensions of equipment:
 cue: minimum length – 3 feet (0.9144 metre)
 ball: diameter – 2¹⁄₁₆ inches (5.3 centimetres)

A form of this game exists using a table without pockets, and thus with cushions all the way down each side and around the corners. In that game, the object is to glance the cue ball off one ball on to another. Such a cannon (or, in the United States, carom) scores 1 point. The game is almost entirely restricted to North America, and is called carom billiards.

The name of the game 'billiards' is taken from a French diminutive of *bille* 'log', thus meaning 'stick', 'cue', although probably ultimately of Celtic derivation (taken also into English via another French diminutive as *billet*).

billion [quantitatives] Historically, in the UK a billion is a million million (10^{12}, or

1,000,000,000,000), whereas in North America and parts of continental Europe, notably France, a billion is a thousand million (10^9, or 1,000,000,000). But from the 1970s and 1980s, thanks to transcultural influences, the US billion was applied in the UK in certain specific contexts, in particular finance and currency markets, balance of trade statements, and measurements in ecological debates (trees felled, cubic feet of ozone, square measures of polluted coastline, etc.).

billionth [quantitatives] Just as the UK and US BILLION differ (*see above*), so does the reciprocal fraction, a billionth. In the UK, the fraction corresponds to 1 over 1,000,000,000,000 or 0.0000000000001 or 10^{-12}; in the USA (and elsewhere) a billionth is 1 over 1,000,000,000 or 0.000000001 or 10^{-9}.

bimestrial [time] Once every two months; lasting for two months.

binary [chemistry] Describing a chemical compound that consists of two elements.

binary system [maths] Numeral system in which only two different symbols are used, and corresponding therefore to a system of BASE 2. The symbols or numbers used are generally represented as 0 and 1:

binary/base 2 system	equivalent in base 10 system
1	1
10	2
11	3
100	4
101	5
110	6
111	7
1000	8
1001	9
1010	10
1011	11
1100	12

and so on. From a human point of view, the system appears clumsy and inevitably involves huge strings of symbols for any number above, say, 50. But, from a technical viewpoint, the system requires a distinction between only two symbols, represented in computing by on and off pulses of current.

binding energy [physics] In nuclear physics, the energy required to remove a particle from an atom or nucleus (for example, an electron from an atom). Alternatively, it is the difference between the energy of a nucleus and the combined energies of the separate particles (nucleons) of which it is composed (equivalent to the mass decrement or defect).

Binet-Simon test/scale [medicine] An intelligence test for determining the mental age of a subject by means of the subject's response to questions and the subject's speed of thought in fulfilling specific tasks. The test or scale was first devised in 1905 by the French psychologists Alfred Binet (1857-1911) and Théodore Simon (1873-1964) but was subsequently revised several times. Sometimes called merely the Binet test or scale, it has at various times been used in schools and in the armed forces, but has been criticized for being culture-oriented.

binomial [maths] In algebra, a mathematical expression consisting of two terms. For example: $5a - 2b$.

bioassay [medicine] In medicine, determining the strength and effect of a drug by using it in clinical tests on laboratory animals, and by comparing the results with those of other, standard drugs.

biological oxygen demand (BOD) [medicine; biology] The amount of oxygen needed fully to oxidize biological material in a sample of water in the environment, equal to the oxygen concentration (in parts per million, ppm) chemically equivalent to all the reducing agents present.

biomass [biology] The total mass (or weight) of all living organic material, animal and plant, within a specified area.

biomolecule [biology; medicine] A theoretical molecule of living matter, intended to represent the smallest possible unit of matter in which life is discernible.

biosphere [biology: ecology] All the parts of the Earth (above and below ground) and in its oceans and atmosphere that can support living organisms.

biot [physics] Some crystals, such as tourmaline, appear to have different colours depending on the direction of the light that falls on them – a property known as dichroism. If the phenomenon is due to differences in the absorption of right- and left-handed circularly polarized light, it is called circular dichroism. The biot is a unit for·measuring the rotational strength of a solid that exhibits circular dichroism, and is equal to the difference between the respective absorption coefficients (which vary with the wavelength of the light).

It was named after the French theoretical physicist Jean Baptiste Biot (1774-1862).

biot number *see* NUSSELT NUMBER, BIOT NUMBER

bipolar coordinates [maths] Two-dimensional coordinate system in which a point is specified in terms of its distances from two fixed reference points. *See also* POLAR COORDINATES.

bipyramid [maths; crystallography] A solid figure (POLYHEDRON) that consists of two pyramids joined base to base (for example, a regular octahedron). Among bipyramidal crystals, all the triangular faces are alike (and may be 6, 8, 12, 16, or 24 in number).

biquadratic [maths] Containing a mathematical expression of the fourth power, an expression to the power ^{4} .

biquinary [maths] Making use of a numeral system that is alternately of base 2 and base 5, as occurs in some digital computers. *See also* BINARY SYSTEM.

birdie [sporting term] In golf, a birdie is one under par (one shot less than the standard number for the hole); an eagle is two under par; and an albatross is no fewer than three under par. Presumably the terms 'eagle' and 'albatross' represent gradations in size (and status) based on the initial term 'birdie' – but why that first term was chosen, no one knows. (Is it the little bird that gets up/in early, and catches the worm in the hole?)

birr [comparative values] Unit of currency in Ethiopia since 1976.

$$1 \text{ birr } = 100 \text{ cents}$$

See also COINS AND CURRENCIES OF THE WORLD.

birth rate [medicine] The number of live human births in one year within a specified area or territory, as compared with the total population, with previous years' figures, with the area's DEATH RATE, or with any or all of these figures of another area.

bis [quantitatives] Twice; repeated once over. The Latin *bis* 'twice' itself stems from an earlier *dvis* in which the ordinary Indo-European element *duo* 'two' is clearly evident, and which is also very close to English 'twice'.

bisect [maths] In geometry, to divide an angle or a line into two equal parts, using a line called the bisector.

bisque [sporting term] A free point or an extra turn awarded to a player by another player or players, by agreement, whenever necessary during a match.

bit [electronics] Basic unit of information in a digital computing system, corresponding to one or other symbol of the binary system of numeral notation (that is, 1 or 0).

$$8 \text{ bits } = 1 \text{ byte}$$
$$1,000 \text{ bits } = 1 \text{ kilobit, } 0.125 \text{ kilobyte}$$

The term is a contraction of 'binary digit'.

bivalent [chemistry] Having a valence, or valency, of two – that is, able to combine with two hydrogen atoms or their equivalent. Bivalent elements, ions, or radicals also have an oxidation number of two.

Alternative expression: *divalent*.

black belt *see* BELTS

black box *see* FLIGHT RECORDER

blackjack *see* PONTOON SCORES

Blagden's law [physics; chemistry] Dissolving a substance (solute) in a solvent lowers the freezing point of the liquid. Blagden's law states that the depression of freezing point is proportional to the concentration (molality) of the solute.

blank verse *see* VERSE FORMS

block coefficient [volumetric measure] In shipping, the underwater volume of a ship divided by the rectangular volume (block) that would contain it.

blondel [physics] Unit of luminance (the brightness of a source of light) equal to π times the luminance per square metre per STERADIAN.

1 blondel = 1 APOSTILB

The unit was named after the French physicist A. Blondel (1863-1938).

blood clotting factors [medicine] Various factors that contribute to the coagulation (clotting) of blood after blood vessels are ruptured or damaged. Their principal function is to convert factor II (prothrombin, in blood) to thrombin, which factor I (fibrinogen) then converts to the insoluble protein fibrin. The fibrin forms a network of fibres that entangle blood cells and form a clot.

factor I	fibrinogen
factor II	prothrombin
factor III	tissue factor
factor IV	calcium ions
factor VII	serum prothrombin conversion accelerator (proconvertin)
factor VIII	antihaemolytic factor
factor IX	plasma thromboplastin
factor X	Prower (or Stuart) factor
factor XI	plasma thromboplastin antecedent
factor XII	glass (or Hageman) factor
factor XIII	fibrin stabilizing (or Laki-Lorand) factor

The blood clotting factors are alternatively known simply as coagulation factors.

blood content, measurement of [medicine] There are three principal methods of analysing a sample of blood, each undertaken for specific reasons.

A haematological/hematological test examines the overall components of the blood, with regard to the blood cells' number, shape, size, and appearance; the function of the clotting factors is also involved: *see* BLOOD COUNT; BLOOD GROUPING; TISSUE-TYPING.

A biochemical test investigates the presence and quantity of various chemical substances within the blood; elements of particular importance include sodium and potassium, the products of food digestion (including glucose and uric acid), and, of course, the gases oxygen and carbon dioxide. A few of these tests do not even require a blood sample but can provide results externally: an oximeter, for example, is a photoelectric device that records the oxygen saturation of the blood's haemoglobin/hemoglobin (the 'oxygen tension' in the blood).

A microbiological test relies on such techniques as culturing in a medium and IMMUNOASSAY, or even on simple examination under a microscope, to detect such organisms as bacteria, viruses, worms, or fungal parasites, and to discern the quantity and quality of the antibodies formed by the body against them.

blood count [medicine] The number of red blood cells (erythrocytes) or white blood cells (leucocytes) in a cubic millimetre of blood. A differential count gives the proportions (as a percentage) of each of the various leucocytes in every 100 or 200 cells. Normal counts are:

erythrocytes (male)	5 million per cu. millimetre
erythrocytes (female)	4 million per cu. millimetre
leucocytes	5,000 to 10,000 per cu. millimetre
basophils	0 to 1 per cent
eosinophils	1 to 3 per cent
lymphocytes	20 to 40 per cent
monocytes	4 to 8 per cent
neutrophils	40 to 60 per cent
platelets	200,000 to 300,000 per cu. millimetre

blood grouping [medicine] Blood types vary, and there are several characteristics of blood that can be identified.

The most common method of blood typing is the ABO system, in which there are four groups – A, B, AB, and O – depending on the presence or absence of A and B antigens in the blood. Antigens occur on the surface of red blood cells, and blood serum contains the antibodies (specifically agglutinins) corresponding to those antigens. Blood group A has A antigens (and therefore B antibodies); group B has

B antigens (and A antibodies); group AB has both A and B antigens (and neither type of antibody); and group O has neither antigen (and both antibodies). For a successful blood transfusion there must be a perfect match between the blood of the donor and the blood of the recipient – that is, the recipient's blood must not contain antibodies to the donor's blood, or clotting (agglutination) will occur. A person with group A blood must thus be transfused with either group A or group O blood; group B blood is compatible with group B or group O; group AB is compatible with all groups; and group O is compatible only with group O. For these reasons group O is known as 'the universal donor'– it is compatible with all blood types – and group AB a 'the universal recipient'.

Another common method of blood typing, especially significant to mothers and babies, is according to the Rhesus (Rh) factor – whether the blood is Rh-positive or Rh-negative. Again it is a matter of whether certain factors are or are not present in the blood, particularly an antigen known as D antigen. Rh-positive blood contains the D antigen; Rh-negative does not.

Both methods of typing were discovered as a result of the work of the German pathologist Karl Landsteiner between 1900 and 1940: the ABO system was the earlier, the Rhesus system was a result of his work on the blood of monkeys of that species. Blood banks mostly combine the two methods so that, for example, a blood may be classified as 'AB negative' or 'O positive', and so on.

Much more rarely, the MNS method of blood grouping is used, in which the blood types are M, N, and S. People of European and Asiatic origin are predominantly of groups M and N.

blood pressure [medicine] The pressure of the blood in a person's circulatory system, measured in a large artery (usually the brachial artery in the upper arm). It is measured at systole (when the heart muscle contracts) and at diastole (when the heart muscle relaxes), and expressed as the systolic pressure over the diastolic pressure. 'Normal' values are up to 140 for systolic, up to 100 for diastolic, 130/80 being a typical value for a healthy young adult male. Blood pressures are generally slightly higher for men than for women, lower for children, and higher for older people, and may be increased at any time by physical activity or emotion.

blouse and shirt sizes *see* SHIRT AND BLOUSE SIZES

board foot [cubic measure] A unit of timber in measuring logs and planks.

1 board foot	=	1 foot x 1 foot x 1 inch (a square foot, one inch thick)
	=	30.48 cm x 30.48 cm x 2.54 cm
	=	144 cu. in, 0.083 cu. ft
	=	2,360 cm^3, 0.00236 m^3
12 board feet	=	1,728 cu. in, 1 cu. ft
1 m^3	=	423.776 board feet

board lot [comparative values] The unit of trading at a trading exchange, expressed as a number of shares.

board of trade (BOT) unit *see* KILOWATT-HOUR

Bode's law [astronomy] Bode's law, sometimes alternatively known as the Titius-Bode law, describes an apparent (and coincidental) relationship between the mean distances from the Sun of the major planets up to and including Uranus. If the sequence 0, 3, 6, 12, 24 and so on is applied to the planets from the Sun outwards, adding 4 to each and then dividing by 10 gives the distance in ASTRONOMICAL UNITS (Sun–Earth =1). This formula spurred the search for planets beyond Saturn after it was suggested in 1766 by the German astronomer Johann Titius and published by his compatriot Johann Bode in 1772. But it does not work for Neptune or Pluto.

body weight *see* WEIGHT, BODY (STANDARDS AND NORMS)

bogey [sporting term] In golf, either the same as par – the standard number of shots for a player to get the ball in the hole – or, especially, one over (more than) par for the hole or for the entire round (of 9 or 18 holes).

The latter definition may account for the use of the term in the first place: the word is the same as *bogy* 'something to be feared'.

boiling point [physics] The temperature at which a liquid boils, when its vapour pressure equals the external (atmospheric) pressure and the liquid freely turns into a vapour. It varies with external pressure and, for this reason, boiling points are generally stated at standard atmospheric pressure (101.325 kilonewtons per square metre). *See also* STANDARD PRESSURE.

bolivar [comparative values] Unit of currency in Venezuela, named after the Venezuelan soldier and statesman Simon Bolívar (1783-1830) who liberated much of northern South America from Spanish rule.

$$1 \text{ bolivar} = 100 \text{ centimos}$$

See also COINS AND CURRENCIES OF THE WORLD.

bolometer [physics] Instrument for measuring the intensity of radiant energy, especially from a distant or feeble source.

The first element of the word relates to Greek *bolē* 'ray'.

bolt [textiles] A roll of woven fabric on a spool (the original 'bolt' or shaft), the fabric's width governed by that of the spool, but its length of a specified measure. In former centuries that length might have been 28 ells (35 yards, 105 feet, 32 metres), but today the length in the UK is most often 30 yards (90 feet, 27.43 metres) and in the USA is most often 40 yards (120 feet, 36.58 metres) in cloth or canvas and 16 yards (48 feet, 14.63 metres) in wallpaper.

Boltzmann's constant [physics] Fundamental constant in physics, given by the ideal gas constant divided by AVOGADRO'S NUMBER (and equal to 1.3805×10^{-23} joules per kelvin).

bomb calorimeter *see* CALORIMETER

bond angle [chemistry] In a molecule with three or more atoms, the angle between the bonds joining one of them to two others.

bond length [chemistry] The distance between two conjoined atoms in a (covalently bonded) molecule, generally of the order of a tenth of a nanometre. It is alternatively known as the *bond distance*.

book sizes [printing] There are certain preferred sizes for books, related to the size of paper (sheet) on which they are printed – they are slightly smaller than the sheet sizes because of trimming.

size	inches	millimetres
crown octavo	7¼ x 4⅞	86 x 123
crown quarto	9¾ x 7⅜	246 x 186
demy octavo	8½ x 5½	216 x 140
demy quarto	11 x 8⅝	279 x 219
foolscap octavo	6½ x 4⅛	165 x 105

Increasingly, with the advent of web printing presses, book sizes are chosen to give the best 'fill' on the machine – that is, to minimize waste of paper. *See also* PAPER SIZES.

Boolean algebra [maths] A type of ALGEBRA with only TRUE or FALSE as its elements, and AND (= multiplication) and OR (= addition) as its operations. Its chief modern application is to the internal logic of computers, particularly in the use of gates (electronic circuits with two or more input terminals but one output terminal).

The system was named after the British mathematician and logician George Boole (1815-64), who first devised it.

bore *see* CALIBRE, CALIBER

Bose-Einstein statistics [maths; physics] The statistical mechanics that are applicable to systems of identical particles whose WAVE FUNCTIONS remain the same if any two particles are interchanged.

The statistics are named after the Indian physicist Satyendra Bose (1894-1974) and the German-born US physicist Albert Einstein (1879-1955).

boson [physics] A subatomic particle, so called because it obeys BOSE-EINSTEIN STATISTICS (but not the pauli exclusion principle). Bosons include alpha particles, photons, and atomic nuclei with even mass numbers (that is, the total number of protons and neutrons is even).

BOT unit *see* KILOWATT-HOUR

boundary *see* CRICKET MEASUREMENTS, UNITS, AND POSITIONS

bound vector [maths] A vector that has a specific point of application. It is alternatively known as a *localized vector*.

Bourdon gauge [physics; engineering] Type of pressure gauge consisting of a
flattened metal tube, closed at one end and bent into a curve. Fluid pressure (gas or
liquid) applied to the open end tends to straighten the tube, movement of which
makes a pointer move round a scale calibrated in pressures. It was named after the
French engineer Eugène Bourdon (1808-84).

boutylka [volumetric measure] Unit of liquid volume in Russia, closely approximating to the capacity of a standard European wine-bottle (75 centilitres).

$$
\begin{aligned}
1 \text{ boutylka} \quad &= \quad 6.25 \text{ CHARKI} \\
&= \quad 76.9 \text{ centilitres, } 0.76895 \text{ litre} \\
&\qquad (1 \text{ litre} = 1.3005 \text{ boutylka}) \\
&= \quad 1.353 \text{ UK pints, } 1.625 \text{ US pints} \\
&\qquad (1 \text{ UK pint} = 0.7391 \text{ boutylka} \\
&\qquad 1 \text{ US pint} = 0.6154 \text{ boutylka}) \\
16 \text{ boutylki} \quad &= \quad 1 \text{ VEDRO}
\end{aligned}
$$

The name of the unit is a diminutive of the Russian word meaning 'bottle'.

bowling average [sporting term] In cricket, the number of wickets to fall while the
specified person is bowling (whether the batsmen involved are bowled out or get out
in some other way, except if run out) in comparison with the number of runs made off
his bowling, over one season or during a series. This can in turn be made proportional
to the number of matches in which the bowler played during the same season or series.

bowls measurements and units [sport] There are two forms of the game of
bowls: flat green bowling, and crown green bowling. Flat green bowling is by far
the more common style of game played around the world, and is the type of bowls
that has, since the early 1980s, also been taken under a roof as 'indoor bowling'. It
is played by two competitors or by two teams of up to four a side.

The dimensions of the green:
a square, level grass area: minimum – 33 x 33 yards (30.175 x 30.175 metres)
                              maximum – 44 x 44 yards (40.233 x 40.233 metres)
the green is divided into rinks 14-19 feet (4.27-5.79 metres) wide, marked by
    pegs at the corners and green thread between
each game of bowls takes place on a single rink
the mat (from which the bowls are delivered) may be placed anywhere between
    4 feet (1.22 metre) from the rear ditch and 27 yards (24.7 metres) from the
    ditch ahead (in practice giving a range of about 15 yards/13.7 metres for a
    competitor to choose where to begin from)
the jack is centred within the rink at the start of each 'end'
Points scoring:
only one player or side scores in every 'end' – when all the bowls of each
    player have been delivered – depending on final proximity to the smaller
    white ball, the jack; as many as that player or side has nearer the jack than
    any opponent's bowl ('wood') then count 1 each
in singles, the 2 competitors have 4 woods each, delivered alternately (thus 8
    woods to an end); first player to score 21 points wins
in pairs, the first two competitors deliver their 2, 3, or 4 woods (as previously
    decided) alternately before their partners deliver their woods similarly; the
    higher team score after 21 ends wins
in triples, the first two competitors deliver their 2 or 3 woods (as previously
    decided) alternately before the next two deliver their woods similarly, and
    the final two thereafter; the highest team score after 18 ends wins
in fours – the most common form of bowling – the first two competitors deliver
    their 2 woods alternately before the next two deliver their woods similarly,
    the third two follow, and the final two thereafter; the highest team score after
    21 ends wins
Dimensions of equipment:
the wood (which may in fact be rubber):
    diameter (wooden): maximum 5¾ inches (14.605 centimetres)
             (rubber): 4¾ – 5⅛ inches (12.065-13.018 centimetres)
    weight (wooden): maximum 3½ pounds (1.588 kilogram)
           (rubber): 3-3½ pounds (1.361-1.588 kilogram)

the jack:

diameter: 2¹⁵⁄₃₂ – 2¹⁷⁄₃₂ inches (6.271-6.429 centimetres)
weight: 8-10 ounces (226.8-283.5 grams)

the mat: 2 feet x 1 foot 2 inches (61 x 35.5 centimetres)

Crown green bowls is so called because the grass surface on which the game is played rises to a 'crown' or slight mound in the centre; it is not divided into rinks, and each game therefore ranges over considerably more area as a match progresses. It is unusual for more than individual players to play each other, generally with only two bowls each (thus four deliveries per end). The mat in crown green bowls is round, and called the 'footer'. The jack is most often black (rather than white, as in flat green bowls).

The name of the game *bowls*, like that of the cognate French *boules*, is probably ultimately akin to English *balls*.

boxing measurements and units [sport]

The dimensions of the ring:

maximum area: 20 x 20 feet (6.096 x 6.096 metres)

minimum area: professional – 14 x 14 feet (4.267 x 4.267 metres)
     amateur – 12 x 12 feet (3.658 x 3.658 metres)

minimum floor width outside ring: 18 inches (45.72 centimetres)

minimum height of ring floor off ground: 39 inches (1 metre)

height of ropes above floor:

bottom rope – 16 inches (40.64 centimetres)
middle rope – 32 inches (81.28 centimetres)
top rope – 52 inches (132.08 centimetres)

Times and timing:

generally 15, 12, 10, 8, 6, or 3 rounds each of three minutes' duration, depending on the experience of the contestants, on the boxing authority, and on
whether the contest is for individual boxers or teams, all rounds separated by
an interval of 60 seconds

Points scoring:

per round, winner: professional – 10 points; amateur – generally 20 points per
round,

loser: fewer points, in proportion to scoring blows additional points for good
style, initiative, attack, or defence

points deducted for foul, low, or late punches, careless use of the head,
wrestling, ducking below the opponent's waistline, and failing to 'break'
(separate) when ordered by the referee to do so

Methods of winning:

on points (higher total at end of final round)

by a knock-out (opponent fails to resume within 10 seconds)

on opponent's failure to leave corner at start of round

on a stoppage by the referee (to save opponent further punishment: a technical
knockout)

on the referee's disqualification of opponent

Dimensional rules on equipment:

glove weight:

standard glove weight: 8 ounces (226.8 grams)

professional welterweights and lighter may use 6-ounce (170-gram) gloves
professional heavyweights (and super-heavyweights) may use 10-ounce
(283.5-gram) gloves

taping the hands:

amateurs: up to 100 inches (2.54 metres) of 1.75-inch (4.445-centimetre) soft
dry bandage; up to 78 inches (1.98 metres) of same-width Velpeau dry
bandage

professionals: up to 18 feet (6 yards, 5.49 metres) of 2-inch (5.08-centimetre)
soft bandage; lighter than middleweight, up to 9 feet (3 yards, 0.9144
metre) zinc oxide tape; middleweight and heavier, up to 11 feet 1 inch
(133 inches, 3.378 metres) zinc oxide tape

knuckles must not be taped over

The rules of the modern form of international boxing are based on regulations first defined by the Marquis (Marquess) of Queensberry ('the Queensberry rules') in the 1860s. *See also* BOXING WEIGHTS.

boxing weights [sport] The classifications by weight of boxers for professional competition are virtually standardized around the world, although they may be expressed in the imperial units (pounds, or stones and pounds) on which they are based or in metric (kilograms and grams) units. Amateur boxers (in categories of the same names but under the aegis of such authorities as the Amateur Boxing Association) may have slightly different weight limits per classification. Not all countries observe strict demarcation in any case, but the following boxing weights are generally standard:

amateur boxers	professional boxers	weight up to		
		kilograms	pounds	stone lb
light flyweight		48.081	106	7 08
flyweight	flyweight	50.802	112	8 00
	bantamweight	53.524	118	8 06
bantamweight		53.978	119	8 07
featherweight	featherweight	57.153	126	9 00
	junior lightweight	58.967	130	9 04
lightweight		60.328	133	9 07
	lightweight	61.235	135	9 09
light welterweight	light welterweight	63.503	140	10 00
welterweight		63.956	141	10 01
	welterweight	66.678	147	10 07
light middleweight	light middleweight	70.760	156	11 02
	middleweight	72.574	160	11 06
middleweight		74.842	165	11 11
	light heavyweight	79.378	175	12 07
light heavyweight		80.739	178	12 10
heavyweight	heavyweight	more	more	more

Some national and international boxing authorities also recognize a category known as super-heavyweight. (*See also* BOXING MEASUREMENTS AND UNITS.)

box plot [maths] In statistics, a type of graph (bar chart) that represents the smallest, quartiles, median, and largest values of a set of data as blocks or boxes.

Boyle's law [physics] The volume of a perfect gas, at a given temperature, varies inversely in relation to the pressure it is subjected to.

Robert Boyle (1627-91) was an Irish physicist and philosopher.

braça, braccio, brasse, braza [linear measure] These measures, respectively Portuguese, Italian, French, and Spanish and Argentinian, must have derived from a single linear unit but are now all different. The original is undoubtedly the Latin *brachium* 'arm' – that is, the whole arm from shoulder to fingertips (as opposed to the north European ELL, based solely on the forearm) – used as a unit by extending both arms as widely as possible and taking the overall furthest distance between fingertips. In later Latin, however, the word was alternatively used just for the forearm, and the measure in some of these countries may in that case represent the ell used as a unit in exactly the characteristic fashion, as the distance between both elbows extended to the furthest possible. But it must be said that, far from measuring cloth, as is the major purpose of the ell, the 'arm' measure in at least two of these cases corresponds to the English FATHOM as a linear measure of vertical depth in liquid. (Perversely, in technical English the word *brachium* now refers to the upper arm, from shoulder to elbow.)

 1 Italian braccio (d'ara) = 70 centimetres, 0.70 metre
          (1 metre = 1.430 braccio d'ara)
        = 2.297 feet (2 feet 3.56 inches)
          (1 yard = 1.306 braccio d'ara
          1 fathom = 2.612 braccio d'ara)
   1 French brasse = 1.62 metres
          (2 metres = 1.234 brasse)
        = 5.315 feet (5 feet 3.78 inches)
          (1 fathom, 6 feet = 1.129 brasse)

```
        1 Spanish braza    =   8 (Spanish) PALMOS
                           =   1.67 metres
                                 (2 metres = 1.198 braza)
                           =   5.479 feet (5 feet 5.75 inches)
                                 (1 fathom, 6 feet = 1.095 braza)
     1 Argentinian braza   =   1.732 metres
                                 (2 metres = 1.154 Argentinian braza)
                           =   5.682 feet (5 feet 8.18 inches)
                                 (1 fathom, 6 feet = 1.056 Argentinian braza)
      1 Portuguese braça   =   10 (Portuguese) palmos
                           =   2.20 metres
                                 (2 metres = 0.910 braça)
                           =   7.218 feet (7 feet 2.62 inches)
                                 (1 fathom, 6 feet = 0.831 braça)
```

brace of partridges, of working dogs [collectives] A 'brace' is a pair because it corresponds to one for each arm (Norman French *bras*) of the hunter or master. Out of season, partridges come in COVEYS and foxhounds are quantified in PAIRS.

Bragg angle [physics: crystallography] For a maximum-intensity beam of X-rays diffracted by a crystal, the angle the incident and diffracted X-rays make with the crystal plane. It was named after the British physicist William Henry Bragg (1862-1942).

Bragg equation [physics: crystallography] In X-ray diffraction by a crystal, twice the spacing between atomic planes (in the crystal) times the sine of the angle between incident and refracted X-ray beams equals a whole number of X-ray wavelengths. The equation is alternatively known as *Bragg's law*.

Bragg rule [physics] The mass stopping power for alpha particles of an element is inversely proportional to the square root of the element's relative atomic mass (atomic weight).

Braille 'alphabet' *see* CELL [literary]

brain waves *see* ALPHA WAVES; BETA WAVES; DELTA WAVES

brake horsepower *see* HORSEPOWER

brass instruments' range [music]

	-2C	-1C	middle C	+1C	+2C
Bach (clarino) trumpet					
piccolo trumpet			E—(3 octaves)——		
Eb trumpet			A——(3 octaves)——		
D trumpet			G#——(3 octaves)——		
(non-transposing) C trumpet			F# ——(3 octaves)——		
(ordinary) Bb trumpet			E ——(3 octaves)——		
bass trumpet		F# ——(3 octaves)——			
Eb soprano cornet			A —(2½ octaves)—		
(ordinary) Bb/A cornet			Eb —(2½ octaves)—		
tenor cornet/melophone					
Eb tenor (sax)horn					
Bb baritone (sax)horn					
Bb alto flugelhorn			E —(2½ octaves)—		
(ordinary) F French horn	B ——(3½ octaves)——				
F/Bb double horn	B ———(4 octaves)———				
alto trombone					
tenor trombone		E —(2½ octaves)—			
tenor-bass trombone	B ——(3 octaves)——				
bass trombone	B —(2′ octaves)—				
double-bass (contrebass) trombone					

	-2C	-1C	middle C	+1C	+2C
B♭ euphonium ('tenor tuba')	B♭ ——(3 octaves)——				
F tuba	F ——(3 octaves)———				
E♭ tuba	E♭ ——(3 octaves)———				
B♭ double-bass tuba ('BB♭ bass')	B♭ ——(3 octaves)———				

(note: sousaphones are variant forms of tuba)

Wagner tenor tuba		B♭ —(2½ octaves)—		
Wagner bass tuba	B♭ —(2½ octaves)—-			

For saxophones, *see* WOODWIND INSTRUMENTS' RANGE

breadth [linear measure] In measuring the width of flags,

1 breadth	=	9 inches (22.86 centimetres)
4 breadths	=	36 inches, 3 feet, 1 yard (0.9144 metre)
		(1 metre = 4.374 breadths)

As a measure, it probably derives from using the thumb and little finger on one hand stretched the widest apart possible, bringing the thumb up to the little finger and, starting from there, stretching out the little finger again, counting the number of (hands') breadths off as necessary.

If this suggested derivation is correct, the unit corresponds exactly with the SPAN.

break [sporting term] In billiards, pool, and snooker, either the opening shot that disperses the arranged set-up of balls, or the total accumulated score over an uninterrupted sequence of shots by one player (without reference to the frame score).

Breathalyzer [medicine] An instrument for measuring the amount of alcohol on a person's breath, and thereby of the alcohol absorbed in the person's bloodstream.

breve [music] In musical notation, a breve – known in the United States as a double whole note – is represented by the symbol ⊫ and corresponds to

2 SEMIBREVES or whole notes

4 MINIMS or half-notes

8 CROTCHETS or quarter-notes

6 QUAVERS or eighth-notes

As the notated form representing the longest duration, it is ironic that the breve derives etymologically identically with the English word *brief*, but no doubt the symbol was much quicker to write than actually using 4 minims (half-notes).

The musical instruction *alla breve* indicates that a passage is to be played so fast as audibly to halve the nominal quantity of beats in each bar.

brewster [physics] Unit for the stress optical coefficient of a transparent medium in which, when plane polarized light passes perpendicular to the stress through a thickness of 1 millimetre, a stress of 10^5 newtons per square metre (10^5 pascals) results in a relative retardation of 1 angstrom.

1 brewster = 10^{-14} square metres per newton

The unit was named after the British physicist Sir David Brewster (1781-1868).

Brewster's law [physics] For a substance that polarizes light, the tangent of the angle of polarization (Brewster angle) equals the refractive index of the substance. It was named after the British physicist Sir David Brewster (1781-1868).

bridge scores *see* CONTRACT BRIDGE CALLS AND SCORES

brig [maths] Unit in which the ratio of two numbers is expressed as a logarithm to the base 10. It was named after the early British mathematician Henry Briggs (1561-1631) who, in 1616, drew up the first table of base 10 logarithms. *See also* LOGARITHMS.

brigade [military] Commanded by a major-general, a brigadier, or a colonel, a brigade is, in effect, a small army or task force, comprising two or more REGIMENTS or BATTALIONS often with differing strike or back-up capacities, together with headquarters, equipment, and support troops (such as engineers and air surveillance groups). Two or more brigades make up a division.

brigadier [military rank] A senior rank in the British army, ranking above a colonel and below a major-general. A brigadier is an equivalent of a commodore in the Royal Navy, and a senior group captain in the Royal Air Force.

There is no equivalent rank in the US armed forces, although the rank is techni-cally between those of colonel and brigadier-general in the US Army and Airforce, and captain and rear admiral in the US Navy.

brigadier-general [military rank] A senior rank in the US Army and Airforce, ranking between a colonel and a major-general, and the equivalent of a rear admiral (on the lower half of the list) in the US Navy.

There is no equivalent rank in the British army or Royal Navy, although the rank is technically between those of brigadier and major-general in the British army and commodore and rear admiral in the Royal Navy. The rank is, however, the equiva-lent in the Royal Air Force of air commodore.

Briggs logarithms [maths] Logarithms to the base 10. They were named after the early British mathematician Henry Briggs (1561-1631) who, in 1616, drew up the first table of base 10 logarithms. *See also* LOGARITHMS.

brightness [physics] An imprecise (and unquantifiable) term for *luminance*.

bril [physics] Unit on a scale of subjective brightness – and thus open to different interpretations by different observers.

The name of the unit derives from the first syllable of *brilliance*.

Brinell scale (of hardness) [physics; engineering] Scale of numbers representing the hardness of a material. Under standard conditions of loading, a hard steel ball is pressed into the material in question. The Brinell number is then calculated as the load on the ball (in kilograms weight) divided by the area of the indentation (in square millimetres).

The unit was named after the Swedish engineer Johann Brinell (1849-1925).

British thermal unit *see* THERMAL UNIT/THERM

Brix degrees [physics] Obsolete unit of specific gravity of a liquid, for use with a hydrometer. Devised in 1854 for measuring the specific gravity of sugar solutions (technically in 'degrees Brix'), the units corresponded to the concentration of sugar as a percentage either by mass or by volume (and which one it was had to be specified).

The unit was named after its inventor.

broad gauge [linear measure] In railway terminology, any measure between rails greater than the standard 56½ inches (4 ft 8½ in., 1.4351 metres). Finland and Russia have railway networks based on a gauge of 5 feet (1.524 metres); in Spain, Portugal, and parts of Argentina, Australia, India, and Pakistan, the gauge is 5 ft 6 in. (1.6764 metres).

broad jump, long jump *see* ATHLETICS FIELD EVENTS AREA, HEIGHT, AND DISTANCE MEASUREMENTS

Bronze Age [time] Period of civilization preceded by the Stone Age and followed by the Iron Age, referring not to any specific time but to the stage to which human culture had progressed. Some cultures even today have progressed little farther.

B series [paper] Series of paper sizes in the ISO system.

code	millimetres	inches
B0	1,000 x 1,414	39⅜ x 55⅝
B1	707 x 1,000	27⅞ x 39⅜
B2	500 x 707	19⅝ x 27⅞
B3	353 x 500	13⅞ x 19⅝
B4	250 x 353	9⅞ x 13⅞
B5	176 x 250	7 x 9⅞
B6	125 x 176	4⅞ x 7
B7	88 x 125	3½ x 4⅞
B8	62 x 88	2½ x 3½
B9	44 x 62	1¾ x 2½
B10	31 x 44	1¼ x 1¾

See also A SERIES; C SERIES.

BSI system of emulsion speed indicators [chemistry] A scale of film emulsion speeds, named after the British Standards Institution, which devised it. It has been largely superseded by the ASA SYSTEM.

Btu *see* THERMAL UNIT/THERM

bu [linear measure] Unit of length in Japan, the name of which is used also for a unit of area (*see below*).

$$1 \text{ bu} = 3.030 \text{ millimetres}, 0.003030 \text{ metre}$$
$$(1 \text{ centimetre} = 3.3003 \text{ bu})$$
$$= 0.119 \text{ inch}$$
$$(1 \text{ inch} = 8.382 \text{ bu})$$
$$100 \text{ bu} = 1 \text{ SHAKU}$$
$$600 \text{ bu, } 6 \text{ shaku} = 1 \text{ ken}$$
$$36{,}000 \text{ bu, } 360 \text{ shaku,}$$
$$60 \text{ ken} = 1 \text{ chô}$$

The term appears in origin to mean 'rate', 'ratio', 'proportion'.

bu [square measure] Unit of area in Japan, to be distinguished from the unit of length of the same name (*see above*).

$$1 \text{ bu} = 100 \text{ shaku}$$
$$= 3.306 \text{ square metres}$$
$$(1 \text{ square metre} = 0.302 \text{ bu})$$
$$= 35.585 \text{ square feet, } 3.954 \text{ square yards}$$
$$(36 \text{ square feet, } 4 \text{ square yards} = 1.0117 \text{ bu})$$
$$30 \text{ bu} = 1 \text{ SE}$$
$$3{,}000 \text{ bu, } 100 \text{ se} = 1 \text{ chô}$$

The square unit is alternatively known as a *tsubo*.

bulk [paper] For a sheet of paper, thickness divided by substance (grammage, or basis weight), equal to the reciprocal of density. The term is loosely (but incorrectly) used for thickness.

bulk modulus [physics] For an elastic substance, the ratio of the pressure (stress intensity) on the substance to its fractional decrease in volume. *See also* ELASTICITY.

bull's-eye *see* ARCHERY/DARTS MEASUREMENTS AND UNITS

bumper [quantitatives] Extraordinarily large, jumbo sized, mammoth.

The term originally (1600s-1800s) applied only to an extra-large cup or glass filled up right to the brim for drinking a toast, or to be passed around for each member of the company to drink three toasts from it three times over – 'three times three'. In the 1870s, however, the term was extended (as an adjective) to describe anything of unexpectedly great size or quantity. Most etymologists suggest that the word in this sense derives simply from the verb 'to bump' – 'to knock into' – on an analogy with 'thumping (great) size'. But, in the early 1600s, there was also a cylindrical leather-covered liquor jug called a *bombard*, and so called because it looked rather like a miniature cannon (Spanish *bombarda*, French *bombarde*); moreover, a frequent, regular, and notorious user of such a liquor jug could also be called a bombard. *See also* JUMBO.

bundle [paper; yarn] As a technical term in the paper industry,

$$1 \text{ bundle} = 2 \text{ REAMS}$$
$$= 960, 1{,}000, \text{ or } 1{,}032 \text{ sheets}$$

See also PAPER MEASURES.

As a technical term in the yarn industry,

$$1 \text{ bundle} = 20 \text{ HANKS}$$
$$= 16{,}800 \text{ yards of cotton yarn in skeins}$$
$$= 10{,}200 \text{ yards of worsted yarn in skeins}$$

See also YARN MEASUREMENT.

buqsha, bugsha [comparative values] Unit of currency in the Yemen.

$$40 \text{ buqsha} = 1 \text{ riyal}$$

See also COINS AND CURRENCIES OF THE WORLD.

burette, buret [chemistry; physics] A measuring tube, usually of glass, with a tap at the bottom, and calibrated in volumetric gradations down one side, used in laboratories for accurately measuring small quantities of liquids or gases, particularly in volumetric analysis.

The term derives from the French word that also means 'oil can' and 'cruet'.

burns [medicine] When a person's skin is seared by a flame or by caustic chemicals, the resultant wound is medically diagnosed as a first-, second-, or third-degree burn. First-degree burns affect only the top layer of the skin, the epidermis; they heal quickly, although there may be some peeling off of damaged skin. Second-degree burns are deeper and cause blisters to form, but some or all of the deep skin tissues are left, and healing usually occurs over time to leave little or no scar. Third-degree burns destroy all the skin tissues, leaving charred fragments and possibly exposing underlying muscles and organs; treatment must be by dermatological specialists and generally involves skin grafts.

When a person is burned by an electric current, diagnosis must be on a different basis, for the current may leave the top layers of skin virtually intact while injuring deeper tissues and organs.

bushel [volumetric measure] Old, imperial volumetric measure almost entirely of dry goods, especially of grain or fruit and vegetables. In some contexts, a bushel also represents the weight of the specified volume of the goods.

1 bushel	=	4 PECKS, 8 GALLONS, 32 quarts, 64 pints

1 UK bushel	= 36.370 litres	1 US bushel	= 35.238 litres
	= 36,370 cc		= 35,238 cc
	= 2,219.3 cu. in		= 2,150.2 cu. in
	= 1.284 cu. ft		= 1.244 cu. ft
	= 1.0321 US bushel		= 0.9689 UK bushel

1 cu.ft	=	0.7788 UK bushel, 0.8039 US bushel
1 m³	=	27.495 UK bushels, 28.378 US bushels

Despite conflicting etymologies proposed by several sources, the majority, however, suggesting a grain measure too small to be likely, the term probably derives from a diminutive of the medieval French for a wooden crate, as reflected in the present-day French words *bois* 'wood' and *boîôte* 'box' (cf. 'butt' below). Modern French for 'bushel' is *boisseau* (but *boisselage* is 'grain-measuring'); in Belgium, however, a *boisseau* corresponds to a unit measure of 15 litres, 0.53 cu. ft.

butt [cubic measure] A large barrel for wine or beer, but as a liquid measure different in size depending on the contents and depending also on the country of origin.

In the UK,

1 butt of ale, beer	=	108 UK gallons, 490.964 litres
	=	1.201 US butt of ale, beer
1 butt of wine	=	126 UK gallons, 572.791 litres
	=	1.201 US butt of wine

In North America,

1 butt of ale, beer	=	108 US gallons, 408.812 litres
	=	0.8327 UK butt of ale, beer
1 butt of wine	=	126 US gallons, 476.949 litres
	=	0.8327 UK butt of wine

By derivation, the English word *bottle* is a diminutive of *butt*, which corresponds neatly with the fact that the manservant in charge of the bottles in the cellar was the butler.

butut [comparative values] Unit of currency in the Gambia.

100 bututs	=	1 dalasi

See also COINS AND CURRENCIES OF THE WORLD.

bye [sporting term] In cricket, a bye is a run scored when the batsman has failed to hit the ball with the bat when making a definite stroke at it, and the ball has avoided the wicket and eluded the wicketkeeper. Although the ball may in fact have bounced off the batsman's body, he is not credited with the run as a personal score, and it contributes to the total score of runs under the heading 'Extras'.

bypass ratio [engineering] In a gas turbine (jet) engine, the amount of air that bypasses the combustion chambers divided by the amount that goes through them.

byte [electronics] In digital computers, a group of bits processed as 1 unit of data.

1 byte	=	8 bits
1,000 bytes	=	1 kilobyte

C

C (Roman numeral) [quantitatives] As a numeral in ancient Rome, the symbol C corresponded to 100 – yet, in this sense, it did not derive from the third letter of the Latin alphabet, the C that was in turn derived from the same source as the Greek letter *gamma* (which, to the Greeks as a numerical symbol, signified 3 or 3,000, and which only resembled anything like a C in its capital form). Instead, it comprised the single verticle stroke that was the numeral '1', with a horizontal dash at its head and at its foot (in virtual correspondence with the two noughts of our 'Arabic' 100).

This symbol was almost immediately assimilated to the capital letter C that neatly coincided with the first letter of the Latin *centum* '100' – so neatly that many authorities suggest that the symbol is in fact no more than the first letter of that word. *See also* L.

cab, kab [volumetric measure] In ancient Israel (and therefore the Bible), a volumetric measure of both liquid and dry capacity.

In liquid measure,

1 cab or kab	=	4 LOGS
	=	2.24 litres
		(1 litre = 0.4464 cab)
	=	3.942 UK pints, 4.734 US pints
		(1 UK gallon = 2.029 cab
		1 US gallon = 1.690 cab)
	=	78.848 UK fluid ounces, 75.736 US fluid ounces
3 cab	=	1 hin
6 cab, 2 hin	=	1 SEAH
18 cab, 6 hin, 3 seah	=	1 BATH

In dry measure,

1 cab or kab	=	4 logs
	=	2,240 cubic centimetres
	=	136.68 cubic inches
	=	3.942 UK dry pints, 4.064 US dry pints
1.8 cab	=	1 OMER or issaron
6 cab	=	1 seah
18 cab, 10 omers, 3 seah	=	1 EPHAH

caba [volumetric measure] In the Philippines, a volumetric measure of dry capacity.

1 caba	=	about 1 US BUSHEL (0.9689 UK bushel)
	=	about 35,238 cubic centimetres
	=	about 2,150.2 cubic inches, 1.244 cubic feet

The term and measure is apparently borrowed directly from Spanish.

Cabala, Cabbala [literary] System of interpreting the Hebrew scriptures by finding divine meanings in the numerical values of the words (in Hebrew, as in ancient Greek, the alphabetical characters also stood for numbers). The idea was put forward and popularized by rabbis during medieval times as a form of almost tantric mysticism – so much so that the word later became applicable to any sort of secret, esoteric, or even occult, doctrine.

caber-tossing [sport] A Scottish sport almost always included as an event in Highland Games meetings. Tossing the caber consists of holding a fir tree-trunk (the caber, which is about 17 feet/5.182 metres long) from under one end in both hands vertically against one side of the neck (it may initially be propped up there by assistants), moving gently forward while balancing it and, as it topples forward at increasing speed, bringing up the hands in a rapid motion to up-end the caber, releasing it so as to describe a loop in the air and come down as vertically as possible, and thereafter fall as parallel as possible to the original direction of movement. Marking is subjective but by experienced judges; points are given for apparent strength of lift, for verticality of the caber's landing position, and for how parallel its final resting position is in relation to the movement of the tosser. The name for the pole represents the Gaelic *cabar* 'beam', 'rafter'.

cable, cable's length [linear measure] A measure formerly used at sea by naval authorities, and differing between different navies.

1 US cable's length	=	120 fathoms, 720 feet (240 yards)
	=	219.456 metres
		(200 metres = 0.9113 US cable's length)
	=	1.185 UK cables
8.439 US cables' lengths	=	1 US nautical mile

$$
\begin{aligned}
\text{1 UK cable} \quad &= \quad \text{607.56 feet (101.26 fathoms)}\\
&= \quad \text{185.184 metres}\\
&\qquad \text{(200 metres = 1.0800 UK cables)}\\
&= \quad \text{0.8438 US cable's length}\\
\text{10 UK cables} \quad &= \quad \text{1 US nautical mile, 0.999 UK nautical mile}
\end{aligned}
$$

When there is no real need for total precision, the UK cable is often regarded as being one-tenth of a (UK) nautical mile.

The cable in this sense was the heavy rope or metal chain from which the anchor was suspended. Its actual length could not initially have been of real importance, except perhaps in relation to two or more kedging anchors – floating anchors that could be used to manoeuvre a ship that was not otherwise under way.

The term derives from Latin *cap(u)lum* 'halter', 'mooring-rope', a derivative in turn from the verb *capere* 'to hold fast', 'to seize'.

cade [volumetric measure] Obsolete unit of dry measure in relation to fish – generally sprats or herring – in barrels (French *cade* 'barrel', Latin *cadus* 'cask', 'jar').

$$
\begin{aligned}
\text{1 cade} \quad &= \quad \text{a barrel holding 500 (or 720) herring}\\
&= \quad \text{a barrel holding 1,000 sprats}
\end{aligned}
$$

See also KEG.

cadmium red line [physics] Line in the emission spectrum of the metal cadmium with a wavelength of 643.8496 nanometres, formerly used as a (reproducible) standard of length. *See also* METRE.

calando [music] Musical instruction: sing or play with diminishing speed and volume. The term derives directly from an Italian word meaning 'reducing'.

calculus *see* DIFFERENTIATION, DIFFERENTIAL COEFFICIENT; INTEGRATION

calendar *see* YEAR; ERA, CALENDRICAL; CALENDS

calends, kalends [time] In ancient Rome, the *calends* were originally the public proclamations (Latin *calendae* 'things for calling out', derivative of *calare* 'to call', and closely related to the English word *call*) that were made by the officers known as *pontifices* stating exactly when the NONES for the month were to be celebrated – on the fifth or seventh day of the month. But, because the proclamations always took place on the first day of the month, the term came itself to refer to the beginning of the month – an important day because, by Roman law, it was the day on which interest on loans became due . . . which is why many people's salaries and wages are still payable by the *calendar* month (the month that begins at the calends).

calibre, caliber [linear measure] The size of the barrel of a gun, rifle, or pistol. Originally, the barrel sizes of artillery pieces – large-calibre weapons – were stated in terms of the weight of shot they fired. If the shot was spherical and made of, say, iron, its weight defined also its diameter. Guns were given a designation such as '12-pounder', which specified its calibre.

This method of measurement was carried over to smaller-calibre weapons, such as muzzle-loading rifles and pistols but, for these, the calibre was defined by the lead balls (spherical bullets) they fired, in terms of the number of balls that could be cast from a pound weight of lead. Thus, the balls for a 16-bore rifle weighed 16 to the pound (that is, one ounce each), corresponding to a ball – and barrel – diameter of about 1.73 centimetres (0.68 inch). The system is still commonly used for shotguns, a popular size being 12-bore (called 12-gauge in the United States).

When bullets ceased to be spherical – becoming, appropriately, bullet shaped – the bore system was no longer suitable and, from that time, calibres were generally stated in terms of the diameter of the barrel. In Britain and the United States, calibres were originally in inches, although 'inch' was not stated. Common calibres for rifles and pistols are .22, .300, .303, .38, and .45, usually just called 'two-two' ('twenty-two' in the United States), 'three hundred', 'three-oh-three', 'thirty-eight', 'forty-five', etc. In Continental Europe, calibres were given in millimetres, such as 9 mm. Since the advent of NATO, millimetric calibres have become standard in the armed forces of nearly all Western nations. Larger-calibre weapons may still retain their various 'inches' (4-inch mortar) or 'millimetres' (80-millimetre canon) designations.

The term derives through French and Italian from Arabic *qalib* 'a bar on which to work (iron, leather, or other malleable materials)', from the time when barrels for firearms were made of iron wrought to shape by being beaten while being constantly turned over a rounded surface.

rifle and pistol calibres:			shotgun calibres:	
inches	*millimetres*		*bore*	*inches*
.177	4.50		8	.835-.860
.22	6.00		10	.775-.793
.25	6.35		12	.729-.740
.300	7.62, 7.63, 7.65		16	.662-.669
.303	7.70		18	.637
.32	8.13		20	.615
.38	9.00		24	.579
.45	11.35		28	.550
			32	.526

calliper, caliper [physics: engineering] An instrument resembling a pair of geometrical dividers with inwardly or outwardly curved ends, for measuring external or internal dimensions.

The term derives from much the same root as CALIBRE, but is more related to the corresponding Arabic verb *qalaba* 'to turn over', 'to measure by turning (the instrument) over'.

call [sporting term] In team games with one or more referees, umpires, or line judges, a decision announced by the official(s).

In card games, a bid relating to the number of tricks to be taken, a declaration of trumps, or a demand for a show of all hands still in the game.

In billiards, pool, or snooker, an announcement of the ball that a player intends to hit with the cue ball, especially when there might be some doubt.

Callier coefficient [physics: photography] The ratio of the density of an exposed photographic film or plate viewed in parallel light to its density viewed in diffused light (usually equal to approximately 1.5).

calorie [physics; medicine: nutrition] Unit of heat in the centimetre-gram-second (CGS) system, equal to the amount of heat needed to raise the temperature of 1 gram of water by 1°C at atmospheric pressure. For the calorific value of foods (a measure of their energy content), the usual unit is the kilocalorie (=1,000 calories), sometimes called the large calorie, which can also be written as the Calorie (with a capital C). But in this and other applications, the calorie/Calorie are being replaced by the corresponding SI units, the joule/kilojoule.

$$1 \text{ calorie} = 4.184 \text{ joules}$$
$$(4 \text{ joules} = 0.956 \text{ calorie}$$
$$1 \text{ joule} = 0.239 \text{ calorie})$$
$$1 \text{ kilocalorie or Calorie} = 1,000 \text{ calories}$$
$$= 4,184 \text{ joules}, 4.184 \text{ kilojoules}$$
$$(1 \text{ kilojoule} = 239 \text{ calories})$$

The term derives from the Latin *calor* 'warmth'.

Calorie [physics; medicine: diet] Unit of energy content of foods, equal to 1,000 calories (with a small c): *see* CALORIE.

$$1 \text{ Calorie (CGS units)} = 4.184 \text{ kilojoules (SI units)}$$
$$(1 \text{ kilojoule} = 0.239 \text{ Calorie})$$

calorific value [physics; medicine: diet] The heat liberated when a unit weight or unit volume of a substance (such as a fuel or a food) undergoes complete combustion. It is measured in the laboratory using a CALORIMETER. *See also* CALORIE.

calorimeter [physics] An apparatus for measuring the heat evolved or absorbed during chemical reactions, changes of state (such as melting), or solvation (dissolution in a solvent). It can also be used to measure the heat content of an object, to determine its specific heat capacity (specific heat). A type called a bomb calorimeter measures the heat evolved when a substance (such as a fuel or a food) is completely burned. *See also* CALORIE; CALORIFIC VALUE.

Cambrian period [time] A geological PERIOD during the Palaeozoic or Primary era. The first period of the era, immediately following the Precambrian or Archaeozoic era, it was followed by the Ordovician period and corresponded roughly to between

590 million years ago and 505 million years ago. The system of rocks dating from the period are the oldest in which fossils can be used for dating purposes, and contain the earliest forms of shelled life. By the end of the period, primitive invertebrate marine creatures were common, notably trilobites and bivalved brachiopods.

Era	Period		Stages
P r i m a r y	Permian		Thuringian Saxonian Autunien
	Carboniferous	Upper or Pennsylvanian	Stephanian Westphalian
		Lower or Mississippian	Dinantian
P r o t e r o z o i c	Devonian		Famennian Frasnian Givetian Eifelian Coblentzian Gedinnian
	Silurian		Ludlowian Wenlockian Llandoverian
	Ordovician		Ashgillian Caradocian Llandeilian Llanvirnian Arenigian Tremadocian
P a l a e o z o i c	Cambrian		Potsdamian Croixian Acadian Albertan Georgian Waucoban

The stages shown in this chart are meant as examples only: they represent predominantly north-west European or North American terms for which there are sometimes corresponding equivalents elsewhere in the world.

The period is named after the country of Wales, called by the Romans *Cambria* (cf. *Cymru*, the Welsh for Wales, found also in English place names and surnames as Cumbria, -gomery, and so forth), an area in which rocks of the period are prominent.

campimetry [physics] The measurement of a person's field of vision, using an instrument called a *campimeter*. Campimetry is known also as *perimetry*.

candela [physics] Unit of luminous intensity in the SI system, equal to the intensity of a source of light, frequency 540×10^{12} hertz, that gives a radiant intensity of $\frac{1}{683}$ watts per steradian in a given direction. It has superseded the INTERNATIONAL CANDLE and CANDLEPOWER (CGS units).

candle, international or standard *see* INTERNATIONAL CANDLE; STANDARD CANDLE

candlepower, candle power [physics] A former unit of luminous intensity, now superseded by the CANDELA.

canoeing disciplines and measurements [sport] Canoes used in competitions are:

single-seater canoes (C1)
   minimum length 4.0 metres (13 feet 1½ inches);
   minimum beam 70 centimetres (2 feet 3½ inches)
two-seater canoes (C2)
   minimum length 4.58 metres (15 feet);
   minimum beam 80 centimetres (2 feet 7 inches)
(single-seater) kayaks (K1)
   minimum length 4.0 metres (13 feet 1½ inches);
   minimum beam 60 centimetres (2 feet)

All of these canoes may be paddled sitting (most common in Europe) or kneeling ('Canadian-style', most common in North America – the 'C' in C1 and C2 is sometimes held to stand for 'Canadian'), and the chosen method should be authorized by an event's judges before participation.

The most common form of canoe event is the canoe slalom, in which competitors have to negotiate a complex course of between 25 and 30 'gates' (pairs of poles suspended just above the water) between which – and without touching them – the competitor has to make his or her way downstream, through at least four of them backwards (as indicated by the colours of the poles making up the gates, and the large number across the top of each gate). The maximum length of a course is 800 metres (880 yards), and courses are graded from I to VI according to difficulty, including a measure that takes into account the current (which must not be less than 2 metres/6 feet 6 inches per second, and is why the event is alternatively known as the 'white-water slalom') and natural hazards such as weirs, rapids, prominent rocks, and bridge piers. Contestants then compete according to their own standard. Team events are for teams of three boats.

The canoe slalom is scored on time: the total time taken to complete the course over two runs, plus any penalties incurred (each of which may add on 10, 20, or 50 seconds depending on severity).

An extension of this event is the 'wild water course' event that is at least 3 kilometres (1 mile 1,457 yards 10 inches) long and at least grade III in difficulty, including natural and artificial hazards (although gates are necessarily fairly widely scattered). For this event the canoes and kayaks are normally slightly larger:

single-seater canoes (C1)
   minimum length 4.3 metres (14 feet);
   minimum beam 70 centimetres (2 feet 3½ inches)
two-seater canoes (C2)
   minimum length 5 metres (16 feet);
   minimum beam 80 centimetres (2 feet 7 inches)
(single-seater) kayaks (K1)
   minimum length 4.5 metres (14 feet 9 inches);
   minimum beam 60 centimetres (2 feet)

The wild water course event is scored on total time (there is usually no need to include penalties because gates are so far apart).

canonical hours [time] In Catholic Christianity, the seven times of day reserved for prayer and worship, as decreed by canon law. Initially, the times were specific hours but, over the centuries, custom altered the actual timing of some (one or two by as much as three hours – *see* NONES), and now the terms involved refer more to specific liturgical offices for individual or congregational worship. The seven are: Mat(t)ins, also called Lauds; Prime; Tierce; Sext; Nones; Vespers; Compline (or Complin).

Matins or Mattins was originally celebrated before dawn (Latin *matutinus* 'of the morning'); Lauds is simply a version of a Latin word meaning 'prayers'. Prime, tierce, sext, and nones are French-altered variants of Latin expressions for the first, third, sixth, and ninth hours of the day. Vespers represented the onset of darkness (Latin *vespera* 'evening'). And Compline is an English rendition of Old French *complie*, the hour that 'completed' the day, the final -in(e) apparently adopted on analogy with Mat(t)ins.

cant [physics: engineering] On a curve in a road or railway track, the extent to which the road or track on the outer side of the curve is elevated above the inner side (so permitting higher cornering speeds). The cant is alternatively known as the *superelevation*.

canter *see* GAIT

capacitance [physics] The charge on one of the two conductors (plates) of a capacitor (condenser) divided by the potential difference (voltage) between them; there is an equal and opposite charge on the other conductor. It is measured in FARADS.

capacity [volumetric measure] There are various units for measuring the holding capacity – that is, the volume – of containers. Indeed, ordinary volume units can be used, especially for liquids: *see* VOLUME. Capacity units are more appropriate for powders and other solids.

The units of capacity most commonly used are:
> metric (litre, millilitre, etc.)
> imperial and US customary (pint, gallon, etc.)
> imperial dry (peck, bushel, etc.)
> household (teaspoon, cup, etc.)

capacity [physics; engineering] Of an electrical machine (such as an alternator or motor), a measure of the output, generally in KILOWATTS or HORSEPOWER.

Of an accumulator (battery), the amount of charge it can hold and deliver, measured in AMPERE-HOURS.

The term has also been used as a synonym for capacitance.

capital [comparative values] In business, the fund of money available to a company first to establish it and then to keep it running, with or without incoming revenue; the initial investment of the company's owners. Also, the net value of a company after the deduction of taxes, overheads, salaries and wages, and any outstanding debits.

capo, capodastro, capotasto [music] A device that consists primarily of a bar or strip which, when placed tightly across the neck of a stringed instrument (especially a guitar), raises the pitch of all the strings at once by an equal amount. In effect, this raises the 'home' or main key of the instrument – which is how it got its name: Italian *capo tasto* 'chief key'.

captain [military rank] Rank that differs widely in seniority between naval and other armed forces.

In the British army, a captain ranks above a lieutenant and below a major, but equally with a lieutenant of the Royal Navy and a flight lieutenant of the Royal Air Force, whereas in the Royal Navy a captain has a far senior position, ranking above a commander and below a rear admiral, but equally with a colonel or brigadier in the British army and a group captain in the Royal Air Force. Some senior naval captains may in fact be regarded as equal in status to a commodore.

In the US armed forces (and in some other armed forces of the world), the rank has a status identical to the British rank in relation both to the navy and to the army and airforce.

In the United States, the title is also accorded to officers of the police and fire departments likewise ranking above a lieutenant.

carat [minerals] Unit of mass used for gemstones, also known as the *metric carat*.

1 carat	=	200 milligrams, 20 centigrams, 2 decigrams
	=	3.0864 grains
	=	0.0070548 ounce avdp.
		(1 ounce avdp. = 141.7475 carats)

The term derives via Arabic *qīrāt* 'bean', 'weight of four grains' from the ancient Greek *keration* 'bean' (literally 'a little horny thing').

carat [minerals] Unit for stating the purity of gold – that is, the proportion of pure gold in an alloy – equal to one twenty-fourth part ($\frac{1}{24}$). Pure gold is thus 24 carats; 18-carat gold, for example, is $\frac{18}{24}$ pure gold alloyed with $\frac{6}{24}$ of another metal (such as silver or copper). *See also* FINENESS.

carbon dating *see* RADIOCARBON DATING

Carboniferous period [time] A geological PERIOD during the Palaeozoic or Primary era, occurring after the Devonian period but before the Permian, and corresponding roughly to between 360 million years ago and 286 million years ago. The period is generally divided into two unequal sections, the earlier lasting twenty million years, the later fifty-four million years: in Britain these are known as the Lower and Upper Carboniferous; in North America they are more commonly called the Mississippian and Pennsylvanian repsectively. Rocks dating from the earlier period are mostly marine shales representing the deep sediments formed by tiny, dead, shell-covered organisms and corals long turned to limestone; there are also some black shales that contain life forms such as trilobites. The later period was one in which club mosses and horsetails flourished amid warm freshwater lagoons, fossilizing where they fell to produce the world's present reserves of coal – after which the whole period is named. A time of considerable volcanic activity, glaciation took over towards the end, especially in the Southern Hemisphere near the modern equator.

For table of the Palaeozoic era *see* CAMBRIAN PERIOD.

carcel [physics] Obsolete unit of illuminating power in France, equal to 10 international candles. It was based on the light emitted by a standard Carcel lamp, a lamp devised by the French inventor Bertrand G. Carcel in the 1800s that relied on oil fed to a wick by means of a pump operated by clockwork, and that was primarily used in lighthouses.

cardinal numbers [maths] Ordinary counting numbers: 1, 2, 3, 4, and so on.

In mathematics (set theory), a cardinal number is the number of elements in a set; for example, all sets with seven elements have the cardinal number 7. *See also* ORDINAL NUMBERS.

cardiograph *see* ELECTROCARDIOGRAPHY

cardioid [maths] A heart-shaped curve that is the locus of a point on the circumference of a circle that rolls round another circle of the same radius.

The term derives through Latin and French ultimately from ancient Greek *kardia* 'heart' (or, surprisingly, 'stomach'). *See also* EPICYCLOID.

cardiotachometry [medicine] The timing and recording of the heart beat by a microphone pick-up and an electronic amplifier called a *cardiotachometer*, used to monitor the heart during surgery.

carene [comparative values] A religious penance involving fasting and the consumption only of bread and water for a period of forty days, imposed by a bishop on lesser clergy (or occasionally laity), or by an abbot on the monks in his charge.

The term began as a version of Latin *quadraginta* 'forty' but was quickly influenced by pious (or perhaps resentful) thoughts of Latin *carens* 'going without', 'missing'.

Carlovingian *see* CAROLINGIAN, CARLOVINGIAN

Carnegie unit [literary: education] Unit originally defined by the Carnegie Foundation for the Advancement of Teaching, in the United States, representing a year's study in a subject at high school (estimated as the equivalent of 120 hours), and used as a quantitative measure for college entrance standards.

Carolean [time] Descriptive of a style (in architecture, furniture design, or clothing) pertaining to the periods between 1625 and 1649, and 1660 and 1685, the times that Charles I and Charles II occupied the English throne.

Carolingian, Carlovingian [time] Period of time during which the second Frankish dynasty was in power in western and southern Europe –

in France, from AD 751 to 987

in Germany, from AD 751 to 911

in Italy, from AD 774 to 887

– and so called because many of the more important kings and emperors were called (the equivalent of) Charles. It began with the reign of Pepin the Short, who supplanted the last of the Merovingian kings (Childeric III) in 751, and who was son of Charles Martel, the Hammer of the Muslims at the Battle of Poitiers in 732. His own son was Charles the Great, or Charlemagne (ruled 800-14). The last of the Carolingians was Louis V ('*le Fainéant*'), whose kingdom in France was finally usurped by Hugh Capet.

carolus [comparative values] Name of several coins struck and issued during the reign of kings called Charles. Perhaps the most famous was the gold coin issued in

England during the reign of Charles I (ruled 1625-49) and worth twenty shillings (one pound).

carré [sporting term] In roulette, a bet on four numbers forming a square (French *carré*) on the board, made by placing the chips or money on the cross that represents the border of all four.

carrier wave [physics: telecommunications] While a radio station is on the air, the continuously transmitted radio signal that is modulated by another signal representing the sounds being broadcast. In AM broadcasting, the amplitude of the carrier wave is modulated (*see* AMPLITUDE MODULATION); in FM broadcasting, the frequency of the carrier wave is modulated (*see* FREQUENCY MODULATION).

Cartesian coordinates [maths] In coordinate geometry, a method of locating a point on a plane or in three-dimensional space. A point on a plane has two coordinates which specify the distance of its perpendicular along the (horizontal) *x*-axis – the *x* coordinate, or *abscissa* – and the distance of its perpendicular from the (vertical) *y*-axis – the *y* coordinate, or *ordinate*; the coordinates are thus (*x*, *y*).

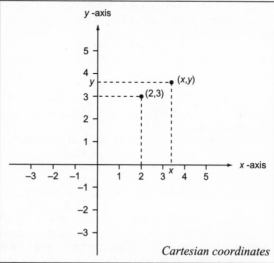

Cartesian coordinates

For example, the perpendiculars from the point with coordinates (2, 3) are located 2 units along the *x*-axis and 3 units along the *y*-axis. In three-dimensional coordinate geometry there are three axes at right angles, designated the *x*-axis, *y*-axis, and *z*-axis. A point in space has coordinates (*x*, *y*, *z*), representing the distances of its perpendiculars from these axes. For example, a point may have the coordinates (3, 4, 5).

This coordinate system was named after the French mathematician and philosopher René Descartes (1596-1650). *See also* POLAR COORDINATES.

Cartesian product [maths] Another name for CROSS PRODUCT.

cascade unit [astronomy] Unit of length applied to cosmic ray showers, equal to the shower unit (the mean length of path along which the energy of cosmic rays is reduced by 50 per cent as they pass through matter) divided by the natural logarithm of 2. It is alternatively known as the radiation length.

cash [comparative values] In India and China, a coin of minimal value. In China, the coin was visually distinct in having a square hole in the middle.

The derivation of this word (possibly from Malayalam) is quite different from the etymology of the English *cash* meaning 'ready money' (from Latin *capsa* 'money-box', 'container').

catacaustic curve *see* CAUSTIC CURVE

catchweight [sporting term] Descriptive of a contest (in horse racing, wrestling, and various other pursuits) in which there are no weight restrictions.

catenary [maths] A curve taken up by a heavy cable or chain supported at its ends. By extension, it also describes the supporting wire for the conductor wire in overhead electrification systems for trams and trains, which has the shape of a catenary and vertical wires hanging from it (of various lengths) to hold the conductor wire horizontal.

The term derives from the adjectival form of Latin *catena* 'chain'.

cathetometer [physics: surveying] An instrument for measuring comparatively small heights, consisting of a telescope that can slide up and down a calibrated vertical pillar. Readings are taken with the telescope focused on the top and then the bottom of the object whose height is being measured (and one reading is subtracted from the other).

cathode rays [physics] Streams of high-speed electrons emitted by a cathode (negatively charged electode) in a vacuum or vacuum tube (such as a cathode ray tube). They move at a speed of 595 kilometres per second for each volt difference in potential between the cathode and the anode.

The term 'cathode' is a relatively modern word that uses ancient Greek elements intended to suggest emission: *kath-hodos* 'the way out'. The opposite is *anode*, which corresponds to 'the way in'.

CAT scan *see* SCANNING SYSTEMS

catty, kati [volumetric measure] Unit of weight corresponding to one used all over the Far East during the colonial times of the mid-1800s, devised artificially as a measure for taxation purposes. What in China was called the *catty* or *chin* was in Korea known as the *catty* or *kon*, in Malaysia the *kati* or *gin*, in Hong Kong the *kan*, and in Japan the *kin*.

1 catty or kati	=	16 TAELS	
	=	1.333 pounds avdp.	
		(1 pound = 0.75 catty/kati)	
	=	604.775 grams, 0.604775 kilogram	
		(1 kilogram = 1.6535 catty/kati)	
3 catty/kati	=	4 pounds avdp. (1.8143 kilogram)	
In Malaysia, 100 kati	=	1 PICUL	

For some reason, the *catty* in Thailand was exactly twice the amount elsewhere:

1 Thai catty	=	20 Thai taels
	=	2.666 pounds avdp.
		(2 pounds = 0.75 Thai catty)
	=	1.20955 kilograms
		(1 kilogram = 0.82675 Thai catty)
3 Thai catty	=	8 pounds avdp. (3.6286 kilograms)

It was the Chinese catty that became the standard measure for weighing out tea exported to Europe from China; containers that held that quantity of tea very quickly became known as *caddies*.

caustic curve [physics; maths] In physics, after light rays are reflected (or re-fracted) by a curved mirror (or transparent medium), their tangents generate another curve. For example, the caustic curve produced by a semi-cylindrical mirror takes the form of two curves meeting at a cusp. The curve is described as 'caustic' probably because the light reflected from a mirror, or refracted by a lens, is concen-trated and can burn material on which it falls (Latin *causticus* 'capable of burning').

The mathematical definition is based on the optical phenomenon in physics: a caustic curve is the envelope of the rays from the radiant point of another curve after they have been reflected by it. A geometrically generated caustic curve is alternatively known as a *catacaustic curve*, in which 'catacaustic' literally means 'reflected caustic'.

cc *see* CENTIMETRE, CUBIC (CC)

cedi [comparative values] Unit of currency in Ghana.

1 cedi = 100 pesewas

The term derives from the local native (Fanta) name for the cowrie shell com-monly found on coastal shores. *See also* COINS AND CURRENCIES OF THE WORLD.

ceiling [physics: meteorology] In aviation, the distance between the surface of the Earth and the lowest clouds; or the greatest altitude above the Earth's surface that can be flown without oxygen masks or other life-support aids.

In a more specific sense, the ceiling is alternatively the greatest altitude at which an aircraft can still climb at a rate faster than 100 feet per minute (30 metres per minute).

The term is etymologically more appropriate to these meanings than to its common meaning 'roof of a room', deriving as it does from Latin *caelum* 'the skies', 'the heavens', the English form of the adjective from which is *celestial* (*see below*).

celestial coordinates [astronomy] Parameters that define the location of a celestial object; their distance referred to the celestial equator. They include DECLINATION and RIGHT ASCENSION, in one coordinate system, and celestial latitude and longitude (*see* LATITUDE; LONGITUDE).

celestial equator [astronomy] The great circle around the CELESTIAL SPHERE where the plane of the Earth's equator cuts it.

celestial sphere [astronomy] An imaginary sphere on to which the various celestial objects, such as stars, appear to be attached. The observer is at the centre, the celestial poles are immediately above and below the Earth's poles, and it is cut in half by the celestial equator.

cell [physics; biology: medicine] The basic unit of which living organisms are composed. It consists of a central nucleus surrounded by cytoplasm (containing various organelles). Some organisms, such as bacteria and protozoans, consist of merely a single cell, but most organisms are multicellular.

cell [literary] A 'character' of the Braille 'alphabet' used for reading by blind people. There are sixty-three in all, representing not only the ordinary letters of the English alphabet but also punctuation marks, numerals, consonantal combinations (such as 'ch' and 'th'), and a few common words.

cello *see* STRINGED INSTRUMENTS' RANGE

Celsius scale [physics] Temperature scale on which the two fixed points are the freezing point of pure water (0°C) and the boiling point of pure water (100°C). It is the same as the obsolescent centigrade scale.

A	F	K	P	U
B	G	L	Q	V
C	H	M	R	X
D	I	N	S	Y
E	J	O	T	Z

Braille 'alphabet'

To convert a Celsius (or centigrade) temperature to a fahrenheit temperature, multiply by ⅘ and add 32 to the product. Thus, for example,

     100°C   =   (100 x ⅘) + 32   =   180 + 32   =   212°F

To convert Celsius (or centigrade) to KELVIN, add 273.16 (there is no degree sign). Thus, for example,

     100°C   =   100 + 273.16   =   373.16 K

The original Celsius scale, as devised in 1742 by the Swedish astronomer Anders Celsius (1701-44), was inverted: it ran from 100° (the freezing point of water) to 0° (the boiling point). The scale was given its present form in 1850 by a scientist named Strömer.

Cenozoic era [time] Most recent geological ERA that includes the present day. Divided into the Tertiary and Quaternary periods – the latter differing from the former only in representing the time during which hominids have been on the Earth – and subdivided into seven epochs (from earliest to latest, the Palaeocene, the Eocene, the Oligocene, the Miocene, the Pliocene, all in the Tertiary period, and the Pleistocene and the Holocene of the Quaternary period), it began about 65 million years ago at the end of the Cretaceous period of the Mesozoic era. This was the time that mammals in general began to dominate the planetary surface, although some were greatly reduced in numbers and others became extinct during the various ice ages that took place in the Pleistocene epoch.

The name of the era corresponds to the progression inherent in the names of the previous eras: Palaeozoic 'ancient life', Mesozoic 'middle life', and thus Cenozoic 'new (or modern) life'.

See chart of the Cenozoic era *under* PALAEOCENE EPOCH.

census [quantitatives] An official count of the number of people in a territory and, most often, a recording of their ages, ethnic grouping, mode of livelihood, and any other factors in which the authority taking the census may be interested. In many countries a regular census permits an updating of statistical and demographic charts, and the correction of electoral registers.

census tract [square measure; quantitatives] In the United States, a unit area containing an average population of 4,000, as defined by the Bureau of the Census for the study of smaller metropolitan communities.

cent [comparative values] Unit of currency in the United States, Canada, and many other countries of the world that use a monetary system based on a 'DOLLAR' or a 'pound' divided into 100 smaller units.

100 cents = 1 dollar

The term is French in origin (*cent* '100', in turn derived from Latin *centum* '100'), and may or may not have been borrowed as an abbreviated form of the French for 'one-hundredth'. Certainly, in the United States, the terms 'dollar' and 'cent' were first proposed for a revised US currency system independent of the British system by the statesman and financial commentator Gouverneur Morris (1752-1816), in essays published in the *Pennsylvania Packet* in 1780.

cent [physics; music] A $\frac{1}{1,200}$ of the difference between two sound frequencies that are an octave apart (that is, are in the ratio 1:2). Each cent equals 3,968.31 times the logarithm of the ratio of the frequencies.

cent [physics] In nuclear physics, a unit of reactivity equal to one-hundredth of a DOLLAR.

cental [weight] Unit of weight in the AVOIRDUPOIS WEIGHT SYSTEM related to the 'short TON' of 2,000 POUNDS, and so used more in North America than in other English-speaking areas of the world (where the 'long ton' of 2,240 pounds is more in evidence, except where superseded by the TONNE of the metric system).

1 cental = 100 pounds
= 1 US QUINTAL
(1 UK quintal = 112 pounds, 1 HUNDREDWEIGHT, and in some countries 1 CENTNER or sentner)
= 0.8929 UK quintal
= 45.35924 kilograms
(50 kilograms = 1.10231 cental)
20 centals = 1 short ton, 2,000 pounds

centavo [comparative values] Unit of currency in Argentina, Bolivia, Brazil, Cape Verde, Colombia, Cuba, the Dominican Republic, Ecuador, El Salvador, Guatemala, Guinea-Bissau, Honduras, Mexico, Mozambique, Nicaragua, Peru, the Philippines, and Portugal.

In Argentina, Bolivia, Colombia, Cuba, the Dominican Republic, Guinea-Bissau, Mexico, and the Philippines:

100 centavos = 1 PESO

In Brazil:

100 centavos = 1 CRUZADO

In Cape Verde and Portugal:

100 centavos = 1 ESCUDO

In Ecuador:

100 centavos = 1 SUCRE

In El Salvador:

100 centavos = 1 COLON

In Guatemala:

100 centavos = 1 QUETZAL

In Honduras:

100 centavos = 1 LEMPIRA or peso

In Mozambique:

100 centavos = 1 METICAL

In Nicaragua:

100 centavos = 1 CÓRDOBA

In Peru:

100 centavos = 1 SOL

See also COINS AND CURRENCIES OF THE WORLD.

centenarian [time] Person who has passed his or her one-hundredth birthday.

centenary, centennial [time] Descriptive of an anniversary celebrated 100 years after the original event; also the one-hundredth anniversary itself.

center *see* CENTRE, CENTER

center of buoyancy *see* CENTRE OF BUOYANCY

center of curvature *see* CENTRE OF CURVATURE

center of gravity *see* CENTRE OF GRAVITY

center of mass *see* CENTRE OF MASS

centesimal [quantitatives] Of one-hundredth part; divided into 100 parts.

centesimo [comparative values] Unit of currency in Chile, Panama, and Uruguay, a notional unit of currency in Italy, and formerly a unit of currency in Somalia.

In Chile and Uruguay,
> 100 centesimos = 1 PESO

In Panama,
> 100 centesimos = 1 BALBOA

In Italy,
> 100 centesimi = 1 LIRA (but the unit is so small as to be meaningless in everyday terms)

Until 1961 in Somalia,
> 100 centesimi = 1 SOMALO

The term derives directly from Latin *centesimus* 'hundredth'. Some English-speaking authorities distinguish two plural forms as in the countries involved: centesimos and centesimi. *See also* COINS AND CURRENCIES OF THE WORLD.

centgener [quantitatives] Unit sample measure of domesticated animals or cultivated plants (comprising 100 individuals, or sometimes many more) regarded as true examples of their racial, generic, specific, or varietal type for biological comparison.

The term derives from the Latin elements *cent-* 'hundred', *gener-* 'type'.

centi- [quantitatives: prefix] Prefix that normally signifies 'one-hundredth' of the unit to which it is attached.

Example: centilitre – one-hundredth of a litre

But the prefix has in the past also been used to signify 'one hundred', and is occasionally still used in this sense today.

Examples: centigrade – comprising 100 degrees

centipede – animal with '100' legs (actually 280)

centiare [square measure] Unit of area in the metric system, more commonly known as a SQUARE METRE but, in this case, calculated as one-hundredth of an ARE.

centigrade heat unit (CHU) [physics] Unit of heat, equal to the quantity of heat required to raise the temperature of 1 pound of water by 1 degree Celsius. It is known also as the *pound-calorie*.

> 1 centigrade heat unit = 453.6 calories
> (500 calories = 1.1023 CHUs)
> = 1,898.5 joules
> (2,000 joules = 1.0535 CHUs)
> = 1.8 British thermal units
> (1 British thermal unit = 0.555 CHU)

centigrade scale [physics] Former name of the CELSIUS SCALE.

centigram, centigramme [weight] One-hundredth GRAM: a unit of weight in the metric system far too small for everyday use.

> 1 centigram = 0.01 gram, 0.10 DECIGRAM
> = 0.1543236 grain
> = 0.0003257396 ounce avdp.
> 6.5 centigrams = 1 GRAIN
> 100 centigrams = 1 gram
> 2,834.9523 centigrams = 1 ounce avdp.

centilitre, centiliter [volumetric measure] One-hundredth LITRE: a small volumetric unit of both liquid and dry capacity in the SI system. It is abbreviated as 'cl'.

> 1 centilitre = 0.01 litre, 0.10 DECILITRE, 10 MILLILITRES
> = 0.0176 UK pint, 0.0211 US pint
> = 0.3520 UK fluid ounce, 0.3376 US fluid ounce
> (1 UK fluid ounce = 2.841 centilitres
> 1 US fluid ounce = 2.962 centiliters)
> = 10 cubic centimetres (for most practical purposes)
> = 0.6102374 cubic inch
> (1 cubic inch = 1.6387 centilitre)
> 25 centilitres = about one cup(ful) in cooking

70-75 centilitres = 1 standard wine bottle

centime [comparative values] Unit of currency in Belgium, France, (and some French former colonies), Haiti, Luxembourg, and Switzerland.

In Belgium, France, Luxembourg, and Switzerland:

100 centimes = 1 FRANC

In Haiti:

100 centimes = 1 GOURDE

The term derives from a particularly French corruption of the Latin *centesimus* 'hundredth'. In Switzerland, centimes are alternatively known by the Germanic plural *rappen*, a traditional name deriving from a medieval coin on which there was a depiction of a raven (medieval German *rappe*). *See also* COINS AND CURRENCIES OF THE WORLD.

centimetre, centimeter [linear measure] One-hundredth METRE, a much-used small linear measure in the metric system. It is abbreviated as 'cm'.

1 centimetre = 0.01 metre, 0.10 DECIMETRE
= 0.3937008 inch, 0.03281 foot
(1 inch = 2.540 centimetres
1 foot = 30.48 centimetres)

100 centimetres = 1 metre
100,000 centimetres = 1 kilometre

centimetre, cubic (cc) [volumetric measure] Volumetric unit of liquid and of dry capacity in the metric system but in particular use in relation to the specifications of motor vehicle engines.

1 cubic centimetre = 0.000001 (one-millionth) cubic metre (10^{-6} m^3)
= 0.001 (one-thousandth) LITRE
= 1 MILLILITRE (1 ml), 0.10 centilitre (for most practical purposes)
= 0.00176 UK pint, 0.00211 US pint
= 0.06102374 cubic inch
(1 cubic inch = 16.387064 cc)

1,000 cc = 1 litre

centimetre, square [square measure] One ten-thousandth SQUARE METRE, a little-used unit of area in the metric system.

1 square centimetre (cm^2) = 0.0001 square metre (10^{-4} m^2)
= 0.1550003 square inch
(1 square inch = 6.4516 cm^2
1 square foot = 929.03 cm^2)

10,000 square centimetres = 1 square metre, 1 CENTIARE

centimetre-candle [physics] Unit of luminance (illumination) in the centimetre-gram-second (CGS) system, also known as a *phot*, and equal to 1 LUMEN per square centimetre; now superseded by the LUX.

centimetre-gram-second (CGS) system [quantitatives] System of units based on the centimetre (for length), gram (mass), and second (time). It was superseded first by the METRE-KILOGRAM-SECOND (MKS) SYSTEM, and then by the SI SYSTEM.

centimo [comparative values] Unit of currency in Costa Rica, Paraguay, Spain, and Venezuela.

In Costa Rica:

100 centimos = 1 COLON

In Paraguay:

100 centimos = 1 GUARANI

In Spain:

100 centimos = 1 PESETA

In Venezuela:

100 centimos = 1 BOLIVAR

The term derives in Spanish through a French corruption of the Latin *centesimus* 'hundredth'. *See also* COINS AND CURRENCIES OF THE WORLD.

centner, sentner [weight] European unit of weight differing between countries according to tradition and the completeness of assimilation to the metric system.

In the United Kingdom and some areas of Scandinavia:

$$
\begin{aligned}
\text{1 centner} &= \text{1 HUNDREDWEIGHT, 112 POUNDS avdp.} \\
&= \text{1 UK QUINTAL} \\
&= \text{1.12 US quintal or CENTAL} \\
&\quad \text{(1 US quintal/cental = 100 pounds avdp.} \\
&\quad \text{= 0.8929 UK centner)} \\
&= \text{50.80208 kilograms} \\
&\quad \text{(50 kilograms = 0.9842 UK centner)} \\
\text{20 centners} &= \text{1 (long) ton}
\end{aligned}
$$

Elsewhere in Europe, especially Germany:

$$
\begin{aligned}
\text{1 centner} &= \text{50 kilograms} \\
&= \text{110.23 pounds avdp.} \\
&\quad \text{(1 UK centner/quintal = 1.0161 Euro-centner} \\
&\quad \text{1 US quintal/cental = 0.9072 Euro-centner)} \\
\text{20 centners} &= \text{1 (metric) tonne}
\end{aligned}
$$

In some American dictionaries there is confusion between the US cental and the centner, but a cental is strictly related to North American use of the 'short' ton of 2,000 pounds, of which it is one-twentieth. The centner in all cases refers to one-twentieth of the 'long' ton of 2,240 pounds or of the closely equivalent metric tonne.

The spelling 'sentner' is used in Scandinavian countries, such as Sweden and Finland, which use a strictly phonetic orthography.

-cento [time: suffix] Suffix used to denote the century of origin of styles in art and other cultural achievements in Italy between AD 1200 and 1800. Thus:

duecento corresponds to	AD 1200s (thirteenth century)
trecento	1300s (fourteenth century)
quattrocento	1400s (fifteenth century)
cinquecento	1500s (sixteenth century)
seicento	1600s (seventeenth century)
settecento	1700s (eighteenth century)

The terms above represent the Italian for 'two hundred', 'three hundred', 'four hundred', and so on and, in each case, rely on an understood prior term *mille* 'one thousand'.

centrad [maths] Unit of angular measure equal to one-hundredth of a RADIAN.

$$
\begin{aligned}
\text{1 centrad} &= \text{0.01 radian} \\
&= \text{0.572958 degree} \\
&\quad \text{(1 degree = 1.7453 centrad)}
\end{aligned}
$$

centre, center [sporting term] In shooting sports, the part of the target that is nearest the middle and is most valuable in scoring.

In field team games apart from American football, a player whose positioning is mainly along a line between the two goals, and who may or may not begin play (with a kick-off, bully-off, ball-up, or equivalent in the centre circle). Also a pass from a player wide on one side of the field towards players of his or her own side near the goal.

In American football, the offensive player, in the line of scrimmage between the guards, who holds the ball and passes it through his legs to the quarterback (or other designated player) to begin play.

In baseball, the *center field* is the part of the outfield beyond second base, and the position of the player stationed there.

centre of buoyancy [physics: shipping] The CENTROID of the part of a ship that is in the water, at which the buoyant action (uplift) of the water appears to act.

centre of curvature [maths; physics] For a given point on a curve in geometry or coordinate geometry, the centre of a circle that touches the curve at that point. It is alternatively known simply as the curvature.

In optics, the geometric centre of a spherical mirror.

centre of gravity [physics] The point at which the gravitational forces experienced by the particles making up an object act. In a uniform gravitational field it is the same as the CENTRE OF MASS.

centre of mass [physics] The point at which the mass of an object can be regarded as being concentrated. *See also* CENTRE OF GRAVITY.

centrifugal force [physics] An apparent (but non-existent) outward force that seems to act on an object that is spinning or moving in a circular path. It is invoked as an opposite force to CENTRIPETAL FORCE.

centrifuge [physics; biology] An apparatus for separating particles in suspension by rotating them at high speeds. The rate of sedimentation depends on the speed of rotation and the size of the particles: at any given speed, the rate is quicker for larger particles than for smaller ones.

centripetal acceleration [physics] For an object moving at a constant speed in a curved path, the acceleration directed towards the CENTRE OF CURVATURE, equal to the square of its speed divided by the path's radius of curvature.

centripetal force [physics] The force (directed towards the centre) that makes an object move in a curved path (rather than a straight line), equal to its CENTRIPETAL ACCELERATION multiplied by its mass – that is, the product of the square of its speed and its mass divided by the radius of curvature. It is equal and opposite to the so-called CENTRIFUGAL FORCE.

centroid [physics] Within an irregularly shaped object, the point at which the CENTRE OF MASS would be if it had a uniform density. For a symmetrical object it is the same as the centre of mass.

centuple [quantitatives] A hundredfold, one hundred times; composed of 100 elements.

century [time; quantitatives; sporting term] One hundred years; the specific period of 100 years between two years divisible by 100, e.g. 1900 and 2000. Exactly when a century comes to an end (for example, on 31 December 1999 or 31 December 2000) remains controversial, depending on whether 'the first year of the century' begins on 1 January 00, or whether 'year 01 of the century' begins on 1 January 01. The latter understanding is favoured by traditionalists.

Also, any group of 100 elements (such as dollars) or people.

In sport, a score of 100 points (runs) and over. *See also* TON.

cephalic index [medicine] In comparative anatomy and anthropology, the ratio of the greatest breadth of the skull to its greatest length from back to front, expressed as a percentage (and therefore without units). A device to measure the dimensions involved is known as a *cephalometer*. Both terms derive ultimately from the ancient Greek *kephalos* 'head', which is cognate with Latin *caput*, German *Haupt*, Dutch *hoofd*, and English 'head' and 'hood'.

Cerenkov effect, Cherenkov effect [physics] The emission of radiation (such as light) when a particle travels through a medium (such as water) faster than the speed of light in that medium. Named after the Soviet physicist Pavel Cerenkov (1904-), it is sometimes alternatively called Cerenkov radiation.

cetane number, cetane rating [chemistry] A measure of the ignition value – anti-knock properties – of a diesel fuel, equal to the percentage of cetane (hexadecane) in a mixture of cetane and methylnaphthalene that has the same ignition factor as the fuel being tested. It is similar to the OCTANE NUMBER used for petrol fuels.

Cetane got its name because it is found in the oil of the sperm whale (and whales are cetaceans).

cete of badgers [collectives] Deservedly obsolescent (if not obsolete) collective for a group of badgers, first recorded in English in 1486. It would seem to be an attempt at a learned use of the Latin *co-etus* 'being together'.

CFA franc [comparative values] Unit of currency in Burkina Faso (formerly Upper Volta), Cameroon, Central African Republic, Chad, Congo, Equatorial Guinea, Gabon, Ivory Coast, Mali, Niger, Senegal, and Togo, all except Senegal and Togo originally French colonies or overseas departments. Intended as a common currency between mostly neighbouring countries, its value was standardized at what was initially a reasonably high value against the French franc (although never at parity with it). The value has since slipped to a very low percentage of that of the French franc. In all cases,

1 CFA franc = 100 centimes

The acronym stands for *Communauté Financière Africaine* (African Financial Community). *See also* COINS AND CURRENCIES OF THE WORLD.

CGS system, CGS units *see* CENTIMETRE-GRAM-SECOND (CGS) SYSTEM

chad [physics] A proposed – but seldom used – unit of neutron flux, equal to 1 neutron per square centimetre per second. It was named after the British physicist James Chadwick (1891-1974), who discovered the neutron.

chain [linear measure] A linear unit used mostly by engineers or surveyors. But the surveyors' (or Gunter's) chain is not the same as the engineers'.

1 surveyors' chain	=	100 surveyors' links
	=	66 feet, 22 yards
	=	20.117 metres
		(20 metres = 0.9942 surveyors' chain)
10 surveyors' chains	=	1 FURLONG, 220 yards
80 surveyors' chains,		
8 furlongs	=	1 mile
1 engineers' chain	=	100 engineers' links
	=	100 feet, 33.333 yards
	=	30.480 metres
		(30 metres = 0.9843 engineers' chain)

The surveyors' chain is clearly related to proportions of the mile; the engineers' chain (as used almost solely in the United States) evidently evolved as multiples of the standard foot. *See also* LINK.

chain, square [square measure] Unit of square measure used by surveyors, and thus based on the linear CHAIN (also called the Gunter's chain) used in surveying, as opposed to the engineers' chain.

1 square chain	=	1,000 SQUARE LINKS, 16 SQUARE RODS
	=	22 x 22 yards, 484 square yards
	=	404.694 square metres
		(400 square metres = 0.9884 square chain)
10 square chains	=	1 ACRE, 4,840 square yards
100 square chains	=	1 square furlong
6,400 square chains	=	1 square mile

Chalcolithic [time] Of the Bronze Age, a stage of human culture in which tools and weapons were made of copper or bronze. Referring to no specific period of human history, the word derives from two Greek elements meaning 'copper/bronze' and 'stone' respectively.

chaldron [volumetric measure] Obsolete measure of dry capacity in Britain, varying greatly in quantity during the different centuries between AD 1200 and 1878. Introduced by the Norman French (*chauderon*, a corruption of the Latin *calidaria* 'cooking pot'), it was evidently also originally a unit of liquid capacity – the same word as *ca(u)ldron*. But, in 1878, it was standardized by English law:

1 chaldron	=	36 (imperial) bushels
	=	46.224 cubic feet, 1.712 cubic yard
	=	1.681 cubic metre
		(= 37.1556 US bushels)

Its standardization, however, seems to have been tantamount to phasing it out altogether as a practical unit.

champagne glass *see* DRINKING-GLASS MEASURES

chang [linear measure] Linear unit used in China during the time in the 1700s to late 1800s when European colonists were fairly strident in their demands to trade, and not actually based on any Chinese measure. Moreover, the unit varied slightly in different Chinese ports.

1 chang	=	10 CH'IH, 2 PU
	=	about 11 feet 9 inches (141 inches)
	=	about 3.5814 metres
180 chang, 360 pu,		
1,800 ch'ih	=	1 LI

change ringing, changes [maths; music] Method of ringing four or more church bells or handbells that presents a different sequence each time the full number of bells is rung. For ease and practicality, ringers learn a relatively simple pattern of ringing that is followed, and that leads or contributes to the desired result, rather than having to try to keep in mind an individual, hideously complex and potentially irregular rota.

One such pattern is evident in this mathematical representation of four bells (here shown as 1, 2, 3, and 4). The number 1 bell (the 'treble' or highest-pitched bell) moves ('hunts') through the other bells from its primary position to the back, stays there one more change, and hunts forward again to the front. All the other bells follow the same pattern from different starting positions.

```
1   2   3   4
2   1   4   3
2   4   1   3
4   2   3   1
4   3   2   1
3   4   1   2
3   1   4   2
1   3   2   4
1   2   3   4
```

But it is evident that not all the possible changes have been rung by the time the four bells return to this intial position – the number of possible changes on four bells is factorial 4 (that is, 1 x 2 x 3 x 4) = 24, and only eight have here been achieved. It is necessary to put in some sort of 'blip' to change the initial order, so that the same pattern can be used to achieve eight more (different) changes, and again after that to use the same 'blip' once more to complete the full twenty-four possible changes.

The complexity of the overall pattern and the form of the 'blip' that recurs and makes possible the entire number of changes correspond to the differences between the various change-ringing methods in use (such as Plain Bob, Grandsire, or Stedman). The number of bells rung is most commonly four ('minimus'), six ('minor'), seven ('doubles', with an eighth 'tenor' bell rung as a permanent drone), eight ('major'), or nine ('triples', with a tenth 'tenor' bell rung as a permanent drone). The classic peal used for celebrations is of doubles (5,040 changes) and takes about three hours non-stop. The sample above is of Plain Bob minimus.

channel width [physics: telecommunications] The range of frequencies assigned to a particular purpose, such as transmissions by a radio or television station (requiring channels several megahertz wide), or speech signals or data on a coaxial cable or fibre-optic cable (requiring a few kilohertz).

character [literary; quantitatives: lighthouses] A symbol or mark (such as a letter of the alphabet, a punctuation mark, a mathematical sign, or any single meaningful device) in writing, printing, or on a video screen. Such symbols or marks on a video screen may be more technically described as *alphanumeric characters*.

Also the distinctive coding of flashes emitted by a lighthouse (or lightship). Each lighthouse has a timed pattern of duration and number of flashes per minute that is marked on marine charts, so that vessels out at sea can tell which lighthouse is visible (or, in fog, audible as sound signals).

The term derives directly from the ancient Greek for 'engraved mark'.

character code [electronics: computing] The BINARY NUMBERS that represent a CHARACTER (*see above*) in a computer. A commonly used standard character code is the American Standard Code for Information Interchange, or ASCII.

characteristic [maths] The non-fractional part of a logarithm. For example: in 4.3010 (the logarithm of 10,000 to the base 10), the characteristic is 4. The fractional part of a logarithm is the *mantissa*.

charge [physics] Property of an object because it has an excess or a deficiency of electrons. An electric charge may be termed merely positive or negative, or it may be quantified in COULOMBS.

charge-mass ratio [physics] For an ion or subatomic particle, its charge divided by its mass. *See also* ELECTRON CHARGE-MASS RATIO.

charka [volumetric measure] Former volumetric unit in Russia approximating to no unit in common modern use, except perhaps the DECILITRE.

1 charka   =   12.30 centilitres, 1.23 decilitre, 0.123 litre
(1 decilitre = 0.8130 charka)

	=	0.216 UK pint, 0.259 US pint
	=	4.33 UK fluid ounces, 4.15 US fluid ounces
6.25 charki	=	1 BOUTYLKA
10 charki	=	1 SCHTOFF
100 charki, 10 schtoffs,		
16 boutylki	=	1 (Russian) VEDRO

Charles's law [chemistry; physics] The volume of an ideal gas at a given pressure is proportional to its absolute temperature. Named after the French physicist Jacques Charles (1746-1823), it is alternatively known as Gay-Lussac's law.

charm of nightingales, of finches [collectives] It is easy to suppose that the collective noun *charm* refers to the pleasure derived by hearing the nightingale calling amid the trees in the dusk. The term is, however, a corruption of the medieval French *charme*, which ultimately derives from the Latin *carmen* 'song'. This, too, would seem utterly appropriate – but the French in fact later extended the meaning somewhat, to 'magical incantation', 'spell' (hence the English meaning of the verb 'to charm'). The nightingales are thus not so much charming as potentially dangerous, alluring innocent passers-by by singing at a time when no decent, upright, well-meaning bird should be awake.

Curious, then, that the word is used as a collective noun also for finches, that would seem to be fairly harmless, comparatively well-behaved members of avian society, neither particularly musical (although once the most popular cage bird in Europe) nor especially enchanting. It is not impossible that the usage in this case derives instead from Anglo-Saxon *cirm* 'squawking', 'tweeting'.

check, checkmate *see* CHESS PIECES AND MOVES

checkers *see* DRAUGHTS (CHECKERS) MOVES AND TERMS

cheme *see* KHEME, CHEME

chemical affinity [chemistry] A measure of how reactive a chemical compound or functional group is with a particular reagent.

chemical compound [chemistry] A substance that is made up of two or more elements combined together in definite proportions by weight. A pure compound always has the same composition no matter how it is made. *See also* LAW OF CONSTANT PROPORTIONS.

chemical energy [chemistry] The energy liberated by a chemical reaction, usually mainly in the form of heat (*see* EXOTHERMIC). In an explosion, great amounts of chemical energy are released.

chemical oxygen demand (COD) [biology; medicine: ecology] The amount of potassium dichromate (an oxidizing agent) needed to oxidize reducing material in a sample of water in the environment, expressed as oxygen concentration (in parts per million) chemically equivalent to the quantity of potassium dichromate used. *See also* BIOLOGICAL OXYGEN DEMAND.

chemical symbol *see* EQUATION; FORMULA

chenica [volumetric measure] In Iran, a unit of liquid capacity.

1 chenica	=	1.359 litres
		(1.5 litres = 1.1038 chenica)
	=	2.3918 UK pints, 2.8722 US pints
		(2 UK pints = 0.8362 chenica
		2 US pints = 0.6963 chenica)
6 chenica	=	1 COLLOTHUN
48 chenica	=	1 ARTABA

The name of the unit derives from the ancient Greek KHOINIX (plural: *khoinikes*), 48 of which amounted to 1 MEDIMNOS.

chereme [literary] In American Sign Language for the deaf, any unit signal.

The term is based on the ancient Greek word *cheir-* 'hand' with a suffix on analogy with 'morpheme', 'phoneme', etc.

Cherenkov effect *see* CERENKOV EFFECT, CHERENKOV EFFECT

chess pieces and moves [sport] The pieces ('men') on the chessboard, either black or white, in order of importance with the moves they are entitled to make, are: the *king*: one square in any direction (including diagonally); may not move to a square that places him in check; may move with a castle/rook to an intermediate square if not moved or placed in check earlier, and if room

the *queen*: any number of squares in any direction (including diagonally); the most versatile of the chess pieces

the two *rooks* or *castles*: any number of squares horizontally or vertically; may move with a king to an intermediate square if not moved or if king not placed in check earlier

the two *bishops*: any number of squares diagonally (one is on the white squares, the other on the black)

the two *knights* (figured as horses' heads): one square vertically (up or down) and two squares horizontally (left or right), or one square horizontally (left or right) and two squares vertically (up or down)

the eight *pawns*: first move, one square or two squares vertically up the board; thereafter one square vertically up the board, except to take an opponent, when one square diagonally; a pawn that reaches the opponent's first file (the opposite edge of the board) becomes any nominated piece

The object of the game is to 'capture' the opponent's king. Every time there is a threat of this happening on the next move, however, the word 'check' is said as a warning. (In amateur games and games with beginners, a similar warning may be given when the queen is in danger.) When the king cannot move out of danger at all, and may be captured at will, the situation is said to be 'checkmate', and the game is over. (If the king cannot move, and nor can any other piece, but the king cannot actually be taken at will, the situation is 'stalemate', and a draw.) It is the

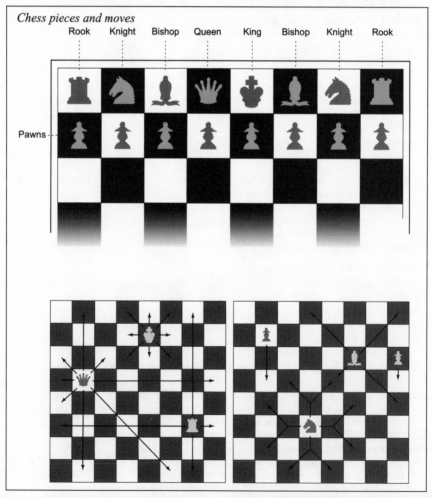

Chess pieces and moves

king (and his importance) that has given the game its name and that is responsible for some of these terms: 'checkmate' is thought to derive from Arabic *(al-)shah mat* '(the) king is dead', the first double-element of which became Old French *eschecs*, which in turn became English 'chess' and American 'checkers'.

A piece that is captured is removed from the board. A piece may return to the board, when a pawn reaches the opposite end file and that piece is nominated as its replacement.

In that the game is probably of Persian or Arabic derivation (in the form we have it in today), it is strange that so many Western courtly elements are included in the pieces. The original chess sets cannot have had bishops, at the very least, and the 'knights' are evidently more cavalry soldiers than members of the nobility (compare the German for 'knight', *Ritter* 'rider'), as demonstrated by their equine shape. Castles were certainly something completely different, as proved by their common name, 'rooks' – which is a form of the old Arabic/ Persian name for the fabulous bird, the *roc*.

chetrum [comparative values] Unit of currency in Bhutan.

100 chetrum = 1 ngultrum or tikehung.

See also COINS AND CURRENCIES OF THE WORLD.

cheval-vapeur [physics: engineering] The metric unit of horsepower used in France: *see* HORSEPOWER.

chief master sergeant [military rank] In the United States Air Force, a non-commissioned officer who ranks above a senior master sergeant. There is no equivalent in the US Army or Navy, nor in any of the British armed forces.

Chief of Staff [military rank] The most senior rank in the United States Army or Air Force. In Britain, the rank is represented not so much by military title as by elevation to the peerage: 'Field Marshal Lord — ', 'Marshal of the Royal Air Force Lord — '; the equivalent in the Royal Navy is first sea lord, or first lord of the admiralty.

chief petty officer [military rank] The most senior rank of non-commissioned officer in both the British (Royal) and United States Navies.

Chi'en Lung [time] In Chinese art and pottery, of the period between AD 1736 and 1796, the time when the Emperor Chi'en Lung ruled the country.

ch'ih [linear measure] Linear unit in China adopted by European colonial authorities during the 1700s to late 1800s when trading with China. The unit varied at different Chinese ports. In general, nonetheless, and for Customs purposes,

1 ch'ih	=	10 t'sun
1 ch'ih	=	about 14.1 inches, 35.814 centimetres
5 ch'ih	=	1 PU
10 ch'ih, 2 pu	=	1 CHANG
1,800 ch'ih, 360 pu,		
180 chang	=	1 LI
	=	about one-third of a mile

chiliad [quantitatives; time] Group of one thousand elements or people. Also, 1,000 years: a millennium.

The term derives from the ancient Greek *chilioi* 'thousand' – a word that is thought to represent a rather odd dialectal variant of the word from which the Romans independently derived their *mille* 'thousand'.

Chinese calendar *see* YEAR; ERA, CALENDRICAL

ching [square measure] Square unit in China adopted by European colonial authorities during the 1700s to late 1800s when trading with China, and used for Customs purposes.

1 ching	=	11 x 11 feet, 121 square feet
	=	13.444 square yards
	=	11.2413 square metres
15 ching	=	1 CHÜO
60 ching, 4 chüo	=	1 MOU
600 ching, 40 chüo,		
10 mou	=	1 CH'ING

ch'ing [square measure] Square unit in China adopted by European colonial authorities during the 1700s to late 1800s when trading with China, and used for Customs purposes.

$$
\begin{array}{rcl}
1\ \text{ch'ing} & = & 10\ \text{MOU},\ 40\ \text{CH\"UO},\ 600\ \text{CHING} \\
& = & 72,600\ \text{square feet} \\
& = & 8,066.666\ \text{square yards} \\
& = & 6,744.760\ \text{square metres}
\end{array}
$$

Ching dynasty, Ch'ing dynasty, Xing dynasty [time] In Chinese art and cultural works, the style of the period between AD 1644 and 1912 – the final dynasty of the Chinese Empire before the advent of the Chinese Republic.

chips [sporting term] In casinos and other places for gambling, counters of various colours or sizes and shapes that represent money with which to bet in roulette, blackjack, and other games. The casino's 'bank' charges a commission for changing real money into chips, and again at the end of the session for changing chips back into real money, thus ensuring the house a further percentage profit.

The chips are so called because they resemble discs 'chipped' or 'chopped' from a cylinder.

chi-square test [maths] In statistics, a test that measures how accurately the frequency distribution of real data corresponds to data calculated by statistical means. The test was devised in 1900 by the British mathematician Karl Pearson (1857-1936).

chô [linear measure] In Japan, a linear measure that is a multiple of the KEN, a unit approximating to the fathom.

$$
\begin{array}{rcl}
1\ \text{chô} & = & 60\ \text{ken},\ 360\ \text{shaku} \\
& = & 109.08\ \text{metres} \\
& & (100\ \text{metres} = 0.9168\ \text{chô}) \\
& = & 119.28\ \text{yards}\ (357\ \text{feet}\ 10.08\ \text{inches}) \\
& & (100\ \text{yards} = 0.8384\ \text{chô}) \\
& = & 59.64\ \text{fathoms}
\end{array}
$$

chô [square measure] In Japan, apart from the chô that is the linear measure (*see above*), a square unit used in the measure of land.

$$
\begin{array}{rcl}
1\ \text{chô} & = & 3,000\ \text{BU or tsubo},\ 100\ \text{se} \\
& = & 9,918.0\ \text{square metres} \\
& & (100\ \text{x}\ 100\ \text{metres},\ 10,000\ \text{square metres} = \\
& & 1.008\ \text{chô}) \\
& = & 11,862.0\ \text{square yards} \\
& & (100\ \text{x}\ 100\ \text{yards},\ 10,000\ \text{square yards} = \\
& & 0.843\ \text{chô})
\end{array}
$$

choenix *see* KHOINIX, CHOENIX

chon [comparative values] Unit of currency in South Korea.

$$
100\ \text{chon}\ =\ 1\ \text{won}
$$

See also COINS AND CURRENCIES OF THE WORLD.

chopin [volumetric measure] Obsolete unit of liquid capacity in north-western Europe. The measure was extremely variable according to location, but was based on an approximation to 'half a pint' in relation to old pints that were larger than today's. Its least and highest values are represented by the units of France and Scotland respectively.

In France:

$$
\begin{array}{rcl}
1\ \text{chopin(e)} & = & \text{one-half } pinte \\
& = & \text{about }0.4655\ \text{litre},\ 465.5\ \text{millilitres} \\
& = & \text{about }0.819\ \text{UK pint},\ 0.984\ \text{US pint}
\end{array}
$$

In Scotland:

$$
\begin{array}{rcl}
1\ \text{chopin} & = & \text{one-half Scotch pint} \\
& = & \text{about 1 UK QUART, 2 UK pints} \\
& = & \text{about }1.0567\ \text{US quarts},\ 2.1134\ \text{US pints} \\
& = & \text{about }1.1365\ \text{litres}
\end{array}
$$

chord [maths; aeronautics] In geometry, a straight line that connects two points on a curve. A chord that passes through the centre of a conic section is called a diameter.

In aeronautics, the width of a wing (a line joining the centres of curvature of the leading and trailing edges), which generally varies along the length of the wing.

The term derives from the ancient Greek *chorde* 'sinew', 'string'.

choriamb, choriambic foot *see* METRE IN VERSE

C horizon *see* SOIL HORIZONS, SOIL LAYERS

Chou dynasty [time] In Chinese art and cultural works, the style of the period between 1027 and 256 BC. This was the period during which the great Chinese philosophers Confucius and Mencius (proponents of what is now Confucianism) and Lao-Tzu (proponent of what is now Taoism) flourished.

chous *see* KHOUS, CHOUS

Christian Era *see* ERA, CALENDRICAL

chroma [physics] In the Munsell system of classifying colours, a number that represents the degree of saturation of a hue (that is, how much white, grey, or black there is in a colour). A chroma of 0 indicates neutral grey, and a chroma of 10 or more indicates total saturation (depending on the hue).

The term ultimately derives from ancient Greek *khrōma* 'surface', thus 'coloration'.

chromatic scale *see* SCALE; KEY, KEYNOTE

chromosome number [medicine] The number of different chromosomes in the nucleus of a body cell – a distinguishing factor between species. For example, human beings have forty-six chromosomes (in twenty-three pairs). Various congenital abnormalities occur in humans with fewer or more than the standard chromosome number (or chromosome set).

A chromosome is so called because, under the microscope, having previously been stained, it becomes a 'coloured body' (ancient Greek *khrōm-* 'coloured', *sōma* 'body').

chromosphere *see* ATMOSPHERE OF THE SUN

chronaxie, chronaxy [biology; medicine] The minimum time needed to excite a nerve when the stimulus (electric current) is twice the minimum threshold value required for the basic response. The unit is used to monitor changes in an individual's nervous or muscular response over time.

chronometer [physics: navigation] A very accurate clock kept on a ship as an aid to navigation, usually set to Greenwich Mean Time (GMT).

chronon [time] A very small unit of time, equal to the time a photon (at the speed of light) would take to cross the width of an electron; about 10^{-24} (one UK quadrillionth/one US septillionth) second.

The name of the unit is intended to reflect the ancient Greek *khronos* 'time'.

CHU *see* CENTIGRADE HEAT UNIT

chukka, chukkar, chukker *see* POLO MEASUREMENTS AND UNITS

chüo [square measure] Square unit in China adopted by European colonial authorities during the 1700s to late 1800s when trading with China, and used for Customs purposes.

1 chüo	=	15 CHING
	=	15 (11 x 11 feet), 1,815 square feet
	=	201.666 square yards
	=	168.619 square metres
4 chüo	=	1 Chinese MOU (as opposed to the Customs unit)
40 chüo, 10 mou	=	1 CH'ING

cicero [literary: printing] Unit of type size in the Didot point system (used in Continental Europe), equivalent to the PICA of the UK and USA.

1 cicero	=	12 Didot points
	=	12.8 (UK/USA) POINTS
	=	1.066 pica
		(1 pica = 0.9380 cicero)

cinque [quantitatives] The number 5 or any symbol representing the number. The term is a medieval (and modern) Italian form of Latin *quinque* 'five'.

cinquecento *see* -CENTO

circadian [time] Recurring every twenty-four hours or so, or recurring in a cyclic pattern completed in twenty-four hours or so – especially in relation to body rhythms such as regular digestive processes, peaks and troughs in mental alertness, and some consequent behavioural manifestations (such as defecation and sleep patterns).

The term derives, rather nastily, from Latin elements *circa* 'around' and *dies* 'day' (the adjectival compound form of which is actually *diurnus, diurnal-*, from which come the English words *diurnal* and *journal*).

circannian, circannual [time] Recurring once every year or so, or recurring in a cyclic pattern completed in a year or so – especially in relation to human body cycles dependent on the length of day or the amount of sunlight received, and especially in relation to body rhythms of animals that hibernate or aestivate on an annual basis.

The term derives from Latin elements *circa* 'around' and *annuus, annual-* 'year'; the dreadful English form *circannian* was coined on analogy with the unspeakable CIRCADIAN (*see above*).

circle [maths] A plane, closed, curved figure with an eccentricity of 0. It is the locus of a point that moves so that it is always at the same distance (the radius) from a fixed point (the centre). A circle is also generated as a conic section when a plane cuts through a cone parallel to its base.

A line that joins two points on a circle is a *chord*; a line that passes through the centre (and bisects the circle) is a *diameter*; and the distance round a circle (its perimeter) is the *circumference*. A 'slice' of a circle formed by two radii is a *sector*; a piece cut off by a chord is a *segment*; a length of circumference is an *arc*; and a straight line that touches the circumference at one point only is a *tangent*.

The area of a circle is equal to π times the square of its radius; the circumference is equal to π times the diameter.

See also CIRCULAR MEASURE.

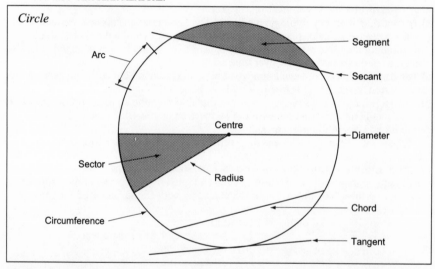

Circle

Arc — Segment — Secant — Centre — Diameter — Sector — Radius — Circumference — Chord — Tangent

circle of confusion [physics: optics] The blurry circular band that results when a point of light is incorrectly focused by a lens on to a plane surface. The inside of the band – the part of the image that is most in focus – is known as the circle of least confusion. The overall size of the circle of confusion is used in photography to gauge either focal sharpness or the aberration of a lens.

circle of curvature *see* RADIUS OF CURVATURE

circular inch [square measure] In the cross-sectional measurement of wire used in electrical circuits,

$$1 \text{ circular inch} = 0.785 \text{ square inch}$$
(a diameter of half an inch)
(1 square inch = 1.274 circular inch)
$$= 5.0645 \text{ square centimetres}$$
(a diameter of 1.27 centimetres)
(5 cm^2 = 0.9873 circular inch)

See also CIRCULAR MIL.

circular measure [maths] System of arc (angular) measure that uses *radians*, in which 1 radian equals the angle subtended at the centre of a circle by an arc (segment of circumference) equal in length to the radius. There are 2π radians in a complete revolution (360 degrees).

$$1 \text{ radian} = 57.2958 \text{ degrees}$$
$$1 \text{ degree} = 0.0174533 \text{ radian}$$
$$360 \text{ degrees} = 6.283 \text{ radians}$$

circular mil [square measure] In the cross-sectional measurement of wire used in electrical circuits,

$$1 \text{ circular mil} = \text{one-millionth CIRCULAR INCH}$$
(a diameter of 0.001 inch)
$$= 0.000000785 \text{ square inch}$$
$$= 0.000050645 \text{ square millimetre}$$
(a diameter of 0.025 millimetre)

circular number [maths] Numbers that, when multiplied by themselves, always end with themselves. For example: 5 is a circular number because 5 x 5 = 25, 5 x 5 x 5 = 125, 5 x 5 x 5 x 5 = 625, and so on, always ending in a 5. The only other true circular number is 6.

circular permutation [maths] The number of ways in which objects in a group can be arranged in a circle. For n objects, there are $(n − 1)!$ different circular permutations (! stands for factorial: *see* FACTORIAL). *See also* PERMUTATION.

circumcentre (of a triangle) [maths] The perpendicular bisectors of the sides of a triangle meet at the circumcentre, which is the centre of a circle that fits round the triangle, touching all three corners.

circumcircle *see* CIRCUMSCRIBED CIRCLE

circumference [maths] The boundary of a closed curve. For a circle, it equals π times the diameter.

circumscribed circle [maths] A circle that touches all the corners of a POLYGON (*see* CIRCUMCENTRE), also known as a *circumcircle*.

cissoid [maths] In coordinate geometry, a two-branched curve with a cusp at the origin and asymptotic to a line parallel to the y-axis (*see* ASYMPTOTE). It is the inverse of a PARABOLA (in respect of its vertex).

The term derives as a learned borrowing of ancient Greek *kisso-edes* 'ivy-shaped', in reference to the plant's leaves.

cistron [biology] A length of a chain of the DNA (deoxyribonucleic acid) molecule, which specifies one polypeptide unit and therefore influences the manufacture of proteins in a cell.

The term derives indirectly from the cis-trans ('inside-outside') test, a method of identifying genetic units.

civil twilight [physics; time] Part of evening during sunset, when the Sun's centre is optically 6 degrees below the horizon (as opposed to ASTRONOMICAL TWILIGHT, when the Sun's centre is optically 18 degrees below the horizon).

cl *see* CENTILITRE, CENTILITER

clamour of rooks, of starlings [collectives] Collective noun that is totally understandable in view of the unholy din attendant on a large assembly of those birds.

clarinet *see* WOODWIND INSTRUMENTS' RANGE

Clark degree [chemistry] Unit for expressing the hardness of water, equal to 1 part of calcium carbonate per 70,000 parts of water (equivalent to 1 grain per UK gallon). It was named after the British scientist Hosiah Clark (d.1898).

$$1 \text{ degree Clark} = 14 \text{ parts per million (approx.)}$$

Class, class [biology] In the taxonomic classification of life forms, the category between PHYLUM (animals) or DIVISION (plants) and ORDER. Of the five kingdoms generally specified in modern taxonomy, the full list of such categories and subcategories for the animals and plants is:

Animals (suffix)	*Plants (suffix)*
Kingdom	Kingdom
Subkingdom	
Phylum	Division (-phyta)
Subphylum	Subdivision (-phytina)
Superclass	
Class	Class (-opsida)
Subclass	Subclass (-idae)
Infraclass	

Animals (suffix)	*Plants (suffix)*
Cohort	
Superorder	
Order	Order (-ales)
Suborder	Suborder (-ineae)
Superfamily (-oidea)	
Family (-idae)	Family (-aceae)
Subfamily (-inae)	Subfamily (-oideae)
Tribe (-ini)	Tribe (-eae)
	Subtribe (-inae)
Genus	Genus
Subgenus	Subgenus
	Section
	Subsection
	Series
	Subseries
species	species
subspecies	subspecies
	variety
	subvariety
	form
	subform

In the plant world, suffixes beginning myc- refer to fungi, and suffixes beginning phyc- refer to algae. In the animal world, no real distinction is made after the classification of subspecies, although in some animals nonetheless there are recognizably and consistently different types defined as 'breeds'.

Organisms are categorized by as few taxonomic distinctions as possible to ensure precise identification. Few organisms, therefore, have been attributed names corresponding to every category listed above. But all living species are defined at least in Kingdom, Phylum/Division, Class, Order, Family, Genus, and species categories.

clause [literary] In grammar, a statement containing one verb. A short sentence may be made up simply of one (independent) clause; complex sentences may comprise two or more clauses.

The major difference between a PHRASE and a clause is the obligatory presence of a verb (and not simply a participle) in a clause.

clausius [physics] Unit of entropy equal to the energy of a system (in kilocalories) divided by its absolute temperature. It was named after the German physicist Rudolf Clausius (1822-88).

clean and jerk *see* WEIGHTLIFTING BODYWEIGHTS AND DISCIPLINES

clef [music] In musical notation, a way of showing the register of a piece of music – that is, the general pitch of the music as written for a specific instrument.

Most music – and especially keyboard and choral music – is written in a combination of two staves, the top one using the treble clef, and the bottom one using the bass clef.

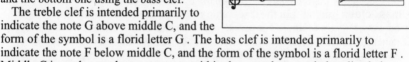

The treble clef is intended primarily to indicate the note G above middle C, and the form of the symbol is a florid letter G . The bass clef is intended primarily to indicate the note F below middle C, and the form of the symbol is a florid letter F . Middle C in each case does not appear within the stave but on a ledger line below or above it.

Orchestral scores may comprise a number of staves in which those at the top are in the treble clef and those underneath are in the bass clef. But more often they include some staves that are in clefs more appropriate to instruments with registers significantly above the treble clef or between treble and bass clefs.

Such clefs all primarily indicate the note middle C (although it is hard to believe that the symbol for these clefs represents a florid letter C), and comprise the soprano clef (above the treble clef in register), and the alto and tenor clefs (below the treble clef in register). In two out of three cases, the symbol itself is obliged to overhang the stave.

C clefs

Soprano   Alto   Tenor

All three show middle C

The soprano clef may be used for certain high-pitched instruments like the piccolo or clarino trumpet; the alto clef is quite ordinarily used in the musical parts played by violas and cellos; the tenor clef is fairly commonly used for the parts played by horns and bassoons.

The word *clef* is the French for 'key'. *See also* KEY SIGNATURES.

clepsydra [time] A water clock: a device for measuring the passage of time that relies on the constant flow of water (or occasionally some other liquid, rarely mercury) through a stricture.

Even the ancient Egyptians had some device of this kind. The term is the Latin form of the Greek for 'water clock'.

clerihew *see* VERSE FORMS

clinical thermometer [medicine] A narrow-range thermometer used for measuring body temperature (the 'normal' average value of which is 37°C/98.6°F).

clinometry [physics: geology] Measurement of dip (inclination) within rock strata, generally using a *clinometer*: a device that measures deviation from the horizontal. The first element of both terms derives from ancient Greek *klinein* 'to slope'.

Clinton [time] A geological stage during the SILURIAN PERIOD of the Palaeozoic era in what is now the United States.

clo [physics; textiles] Unit of thermal insulation (for cloth), arbitrarily defined as the insulation needed to keep a resting person comfortable indoors.

$$1 \text{ clo} \quad = \quad 1.5504 \text{ tog}$$
$$(1 \text{ tog} = 0.645 \text{ clo})$$
$$= \quad 0.875 \text{ foot hour degree (Fahrenheit) per British thermal unit}$$
$$= \quad 0.506 \text{ metre degree (Celsius) per watt}$$

The name of the unit is evidently taken from the first syllable of *cloth* or *clothing*. *See also* TOG.

clock *see* O'CLOCK.

closed curve [maths] Any curve that is a version of a loop, a curve with no ends (that is, a line with one 'end' joined to the other) – for example a circle, ellipse, oval.

closed set [maths] In set theory, a set in which the combination of any two of its members under an operation always gives a member of the original set. For example, for the set of positive whole numbers {1, 2, 3, 4, 5, ...}, the operation of addition always results in a member of the set – add together any two numbers and the sum is always another number in the set – and so it is closed under the operation of addition (as it is also under multiplication; it is an open set under subtraction and division).

close harmony [music] Form of musical writing for three or more voices – generally all-male or all-female – in which the harmonies never extend over an interval of more than a tenth (that is, an octave and a third). The type of singing known as Barbershop or performed by Welsh male-voice choirs is mostly written in close harmony.

clotting time [medicine] The time it takes a small sample of blood to coagulate (clot). It can be determined by collecting a sample of fresh blood in a capillary tube and breaking off small sections of tube every thirty seconds until fine strands of fibrin are seen at the break (usually six to seventeen minutes). More precise (and convenient) values are obtained using a *coagulometer*.

cloud point *see* DEW POINT

cloze reading test [literary; medicine] In education, a test to measure a student's comprehension of what is read, involving a passage of text in which words have been omitted and a space left for them. The student has to supply the missing words, the spaces for which become gradually more frequent, the sense gradually more complex.

The term *cloze* is an abbreviated form of *closure* which, as a technical term in psychology, means the mental process by which incomplete information is added to (even at the risk of inaccuracy) by the brain to resolve incomprehensibility.

clubs *see* GOLF; SUITS OF CARDS

clutch of eggs, of chicks [collectives] But hens do not actually 'clutch' either eggs or chicks, so this collective has nothing to do with the word that is a frequentive of *cling*. Instead it is a dialectal form of Scandinavian Germanic and Scots *cleck* 'to hatch'.

cm *see* CENTIMETRE, CENTIMETER

coagulation of blood *see* CLOTTING TIME

Coal Age *see* CARBONIFEROUS PERIOD

Coal Measures [time] A geological stage during the Upper or Pennsylvanian division of the CARBONIFEROUS PERIOD of the Palaeozoic era in what is now the United States. It is named after the alternating strata of coal and clayey sandstone that are characteristic of it.

coal sizes [minerals: geology] Coal for industrial use is described in terms of its thermal properties and its particle size. The sizes used in Britain are:

	inches	*millimetres*
grains	⅛–¼	3.18–6.35
peas	¼–½	6.35–12.7
singles	½–1	12.7–25.4
doubles	1–2	25.4–50.8
trebles	2–3	50.8–76.2
cobbles	2–4	50.8–101.6
large cobbles	3–6	76.2–152.4
large coal	over 6	over 152.4

co-altitude [astronomy] The complement of the altitude of a celestial object, equal to its ANGULAR DISTANCE from the zenith, and so alternatively called the *zenith distance*.

coat sizes [quantitatives; linear measure] There are three major world systems for classifying the size of coats and overcoats: the systems of Britain (the UK), of the United States, and of continental Europe. All the systems render men's and women's sizes differently.

In women's coats,

UK size	*US size*	*European size*
30	8	36
32	10	38
34	12	40
36	14	42
38	16	44
40	18	46
42	20	48

In men's coats,

UK/US size	*European size*
34	44
36	46
38	48
40	50
42	52
44	54
46	56
48	58

coaxial [maths] Describing two or more circles with a common centre, or two or more cylinders with a common axis.

codeclination [astronomy] The complement of the declination of a celestial object, equal to its ANGULAR DISTANCE from the celestial pole, and so known alternatively as the *polar distance*.

codon [biology] In a molecule of DNA (deoxyribonucleic acid), three consecutive bases which code for (specify) a particular amino acid; by interpreting the code, the components of a cell join together various amino acids to form polypeptides and, eventually, proteins.

The term is based on the word *code*. *See also* CISTRON.

coefficient [maths] The number (factor) by which a variable is multiplied in an algebraic expression or equation. So, for example, in $3x^2 - 5xy + y^2$ the coefficients of x^2, xy, and y^2 are 3, –5, and 1, respectively (the coefficient 1 is not written but assumed).

coefficient of compressibility [physics] A number that indicates the extent to which a real gas deviates from BOYLE'S LAW.

coefficient of expansion [physics] For a material that expands on heating, the expansion per unit length, area, or volume per degree rise in temperature – termed the coefficients of linear, superficial, and cubical expansion, respectively.

coefficient of friction *see* FRICTION, COEFFICIENT OF

coefficient of restitution [physics] For two elastic spheres that collide, their relative velocity after impact divided by their relative velocity before impact. It is a measure of a material's ability to bounce. For a sphere dropped on to a plane surface, it equals the square root of the rebound height divided by the square root of the drop height. *See also* ELASTICITY.

coefficient of viscosity [physics] For a fluid, the tangential force per unit area needed to maintain unit relative velocity between a pair of parallel planes unit distance apart, measured in POISE (SI units).

coercive force [physics] For a magnetized ferromagnetic material, the magnetizing force needed to return its magnetization to zero.

coherent units [quantitatives] A system of units in which the base units are divided or multiplied to produce all the necessary practical units, without the need to introduce any numerical constants; for example, the SI SYSTEM of units.

cohort [military] One-tenth of a Roman imperial legion, comprising a body of between 300 and 600 men under the command of three maniples or six centurions. By extension, any group of soldiers or men with a single purpose.

The word is identical in derivation to the English word *court*, originally referring to an enclosed square.

Cohort [biology] In the taxonomic classification of life forms, the relatively rare distinction in categorization of animal life between a Class and an order; it has no equivalent in the plant world. *See* full list of taxonomic categories *under* CLASS, CLASS.

coins and currencies of the world [comparative values] Below is an alphabetical list of countries and beside each is the form of currency it uses. In a few of the newer countries – such as those of the Commonwealth of Independent States – the form of currency is still in the process of establishing itself.

For the reverse information – distinguishing the country of origin of a known coin or standard currency – *see* individual coins and currencies as included within the main text. So, for example, to find out which countries use the peso, *see* PESO.

country	currency
Afghanistan	100 puls = 1 afghani
Albania	100 qintar = 1 lek
Algeria	100 centimes = 1 (Algerian) dinar
Andorra	(Spanish) peseta/(French) franc
Angola	100 lwei = 1 kwanza
Anguilla	100 cents = 1 (Caribbean) dollar
Antigua/Barbuda	100 cents = 1 (Caribbean) dollar
Argentina	100 centavos = 1 (Argentine) peso
Armenia	100 kopecks = 1 (Russian) r(o)uble
Aruba	100 cents = 1 (Dutch) gulden
Ascension Island – *see* St Helena-Ascension	
Australia	100 cents = 1 (Australian) dollar
Austria	100 groschen = 1 schilling
Azerbaijan	100 kopecks = 1 (Azerbaijani) r(o)uble
Bahamas	100 cents = 1 (Bahamian) dollar
Bahrain	1,000 fils = 1 (Bahrain) dinar
Bangladesh	100 paisa = 1 taka
Barbados	100 cents = 1 (Barbados) dollar

Barbuda – *see* Antigua/Barbuda
Belarus – *see* Byelorussia
Belau/Palau   100 cents = 1 (US) dollar
Belgium   100 centimes = 1 (Belgian) franc
Belize   100 cents = 1 (Belize) dollar
Belorus – *see* Byelorussia
Benin   100 centimes = 1 (CFA) franc
Bhutan   100 chetrum = 1 ngultrum, tikehung/Indian rupee
Bolivia   100 centavos = 1 (Bolivian) peso
Bosnia-Herzegovina   [not yet established]
Botswana   100 thebe = 1 pula
Brazil   100 centavos = 1 cruzado
British Virgin Islands – *see* Virgin Islands
Brunei   100 cents = 1 (Brunei) dollar
Bulgaria   100 stotinki = 1 lev
Burkina Faso   100 centimes = 1 (CFA) franc
Burundi   100 centimes = 1 (Burundi) franc
Burma   100 pyas = 1 kyat
Byelorussia   100 kopecks = 1 (Russian) r(o)uble
Cambodia   100 sen = 1 riel
Cameroon   100 centimes = 1 (CFA) franc
Canada   100 cents = 1 (Canadian) dollar
Cape Verde   100 centavos = 1 (Cape Verde) escudo
Cayman Islands   100 cents = 1 (Jamaican) dollar
Central African R.   100 centimes = 1 (CFA) franc
Chad   100 centimes = 1 (CFA) franc
Chile   1,000 escudos = 1 (Chilean) peso
China   100 fen = 1 yüan
Colombia   100 centavos = 1 (Colombian) peso
Comoros Islands   100 centimes = 1 (CFA) franc
Congo   100 centimes = 1 (CFA) franc
Cook Islands   100 cents = 1 (Cook Island/New Zealand) dollar
Costa Rica   100 centimos = 1 (Costa Rican) colon
Côte d'Ivoire – *see* Ivory Coast
Croatia   (Croatian) dinar
Cuba   100 centavos = 1 (Cuban) peso
Curaçao   100 cents = 1 (Dutch) gulden
Cyprus   100 cents = 1 (Cyprus) pound
Czech Republic   100 haleru = 1 koruna (crown)
Denmark   100 øre = 1 (Danish) krone
Djibouti   100 centimes = 1 (Djibouti) franc
Dominica   100 cents = 1 (East Caribbean) dollar
Dominican Republic   100 centavos = 1 (Dominican) peso
Dubai   100 dirhams = 1 (Dubai) riyal
Ecuador   100 centavos = 1 sucre
Egypt   100 piastres = 1 (Egyptian) pound
Eire – *see* Irish Republic
El Salvador   100 centavos = 1 (El Salvador) colon
Equatorial Guinea   100 centimes = 1 (CFA) franc
Estonia   100 senti = 1 kroon
Ethiopia   100 cents = 1 birr
Falkland Islands   100 pence = 1 pound (sterling)
Faeroe Islands   100 øre = 1 (Danish) krone
Fiji   100 cents = 1 (Fiji) dollar
Finland   100 penniä = 1 markka
France   100 centimes = 1 (French) franc
Gabon   100 centimes = 1 (CFA) franc
Gambia   100 bututs = 1 dalasi
Georgian Republic   [not yet established]

Germany	100 pfennigs = 1 (Deutsche) Mark
Ghana	100 pesewas = 1 cedi
Gibraltar	100 pence = 1 (Gibraltar) pound/Spanish peseta
Great Britain – *see* United Kingdom	
Greece	100 lepta = 1 drachma
Greenland	100 øre = 1 (Danish) krone
Grenada	100 cents = 1 (East Caribbean) dollar
Guatemala	100 centavos = 1 quetzal
Guinea	100 centimes = 1 (Guinean) franc
Guinea-Bissau	100 centavos = 1 (Guinea-Bissau) peso
Guyana	100 cents = 1 (Guyana) dollar
Haiti	100 centimes = 1 gourde
Holland – *see* Netherlands	
Honduras	100 centavos = 1 lempira (or peso)
Hong Kong	until 1997: 100 cents = 1 (Hong Kong) dollar
Hungary	100 filler = 1 forint
Iceland	100 aurar = 1 króna
India	100 paisa = 1 (Indian) rupee
Indonesia	100 sen = 1 rupiah
Iran	100 dinar = 1 (Iranian) rial
Iraq	1,000 fils = 20 dirhams = 1 (Iraqi) dinar
Irish Republic	100 pence = 1 punt
Israel	100 agorot = 1 shekel
Italy	(Italian) lira
Ivory Coast	100 centimes = 1 (CFA) franc
Jamaica	100 cents = 1 (Jamaican) dollar
Japan	(100 sen =) 1 yen
Jordan	1,000 fils = 1 (Jordanian) dinar
Kazakhstan	[not yet established]
Kenya	100 cents = 1 (Kenya) shilling
Kirghizia – *see* Kyrgyzstan	
Kiribati	100 cents = 1 (Australian) dollar
Korea – *see* North Korea, South Korea	
Kuwait	1,000 fils = 1 (Kuwaiti) dinar
Kyrgyzstan	[not yet established]
Laos	100 at(t) or cents = 1 kip
Latvia	100 kopecks = 1 (Latvian) r(o)uble
Lebanon	100 piastres = 1 (Lebanese) pound
Lesotho	100 lisente = 1 loti
Liberia	100 cents = 1 (Liberian) dollar/US dollar
Libya	1,000 dirhams = 1 (Libyan) dinar
Liechtenstein	100 centimes = 1 (Swiss) franc
Lithuania	100 kopecks = 1 (Lithuanian) r(o)uble
Luxembourg	100 centimes = 1 (Luxembourg/Belgian) franc
Macao	100 avos = 1 pataca
Madagascar	100 centimes = 1 (Madagascar) franc
Malawi	100 tambala = 1 kwacha
Malaysia	100 sen = 1 ringgit (or Malaysian dollar)
Maldive Islands	100 laari = 1 rufiyaa (rupee)
Mali	100 centimes = 1 (CFA) franc
Malta	1,000 mils = 100 cents = 1 (Maltese) lira
Marshall Islands	100 cents = 1 (US) dollar
Mauritania	5 khoums = 1 ouguiya
Mauritius	100 cents = 1 (Mauritius) rupee
Mexico	100 centavos = 1 (Mexican) peso
Micronesia	100 cents = 1 (US) dollar
Moldavia	[not yet established]
Monaco	100 centimes = 1 (Monégasque/French) franc
Mongolia	100 mung or mongo = 1 tugrik or togrog

Monserrat	100 cents = 1 (Caribbean) dollar
Morocco	100 francs = 1 (Moroccan) dirham
Mozambique	100 centavos = 1 metical
Namibia	100 cents = 1 (South African) rand
Nauru	100 cents = 1 (Australian) dollar
Nepal	100 paisa = 1 (Nepalese) rupee
Netherlands	100 cents = 1 (Dutch) gulden (guilder)
Nevis	100 cents = 1 (East Caribbean) dollar
New Caledonia	100 centimes = 1 (French) franc
New Zealand	100 cents = 1 (New Zealand) dollar
Nicaragua	100 centavos = 1 córdoba
Niger	100 centimes = 1 (CFA) franc
Nigeria	100 kobo = 1 naira
Niue	100 cents = 1 (New Zealand) dollar
North Korea	100 jun (or chon) = 1 won
Norway	100 öre = 1 (Norwegian) krone
Oman	1,000 baiza = 1 rial saidi (Omani rial)
Pakistan	100 paisa = 1 (Pakistani) rupee
Palau – *see* Belau/Palau	
Panama	100 centesimos = 1 balboa
Papua New Guinea	100 toae = 1 kina
Paraguay	100 centimos = 1 guarani
Peru	100 centavos = 1 sol
Philippines	100 centavos = 1 (Philippines) peso
Pitcairn Islands	100 cents = 1 (New Zealand) dollar
Poland	100 groszy = 1 zloty
Portugal	100 centavos = 1 escudo
Puerto Rico	100 cents = 1 (US) dollar
Qatar	100 dirhams = 1 (Qatari) riyal
Romania	100 bani = 1 leu
Russia	100 kopecks = 1 (Russian) r(o)uble
Rwanda	100 centimes = 1 (Rwanda) franc
St Helena-Ascension	100 pence = 1 pound (sterling)
St Kitts	100 cents = 1 (East Caribbean) dollar
St Lucia	100 cents = 1 (East Caribbean) dollar
St Vincent Islands	100 cents = 1 (East Caribbean) dollar
Samoa, Western – *see* Western Samoa	
San Marino	(Italian) lira
São Tomé-Principe	100 centimes = 1 dobra
Saudi Arabia	100 hallalas = 20 guersh = 1 (Saudi) rial
Senegal	100 centimes = 1 (CFA) franc
Serbia	(Serbian) dinar
Seychelles	100 cents = 1 (Seychelles) rupee
Sierra Leone	100 cents = 1 leone
Singapore	100 cents = 1 (Singapore) dollar
Slovak Republic	100 haleru = 1 koruna (crown)
Slovenia	(Slovene) dinar
Solomon Islands	100 cents = 1 (Solomon Islands) dollar
Somalia	100 cents = 1 (Somali) shilling
South Africa	100 cents = 1 rand
South Korea	100 chon = 1 won
Spain	100 centimos = 1 peseta
Sri Lanka	100 cents = 1 (Sri Lankan) rupee
Sudan	1,000 millièmes, 100 piastres = 1 (Sudan) pound
Surinam	100 cents = 1 (Surinam) gulden (guilder)
Swaziland	100 cents = 1 lilangeni
Sweden	100 öre = 1 krona
Switzerland	100 centimes (rappen) = 1 (Swiss) franc/franken
Syria	100 centimes = 1 (Syrian) piastre (or pound)

Tadjikistan	[not yet established]
Tahiti	100 centimes = 1 (French) franc
Taiwan	100 cents = 1 (Taiwan) yüan (or dollar)
Tanzania	100 cents = 1 (Tanzanian) shilling
Thailand	100 satang = 1 baht
Tibet	100 fen = 1 (Chinese) yüan
Togo	100 centimes = 1 (CFA) franc
Tonga	100 seniti = 1 pa'anga
Trinidad & Tobago	100 cents = 1 (Trinidad & Tobago) dollar
Tunisia	1,000 millimes = 1 (Tunisian) dinar
Turkey	100 kurus (piastres) = 1 (Turkish) lira
Turkmenistan	[not yet established]
Turks & Caicos Is.	100 cents = 1 (US) dollar
Tuvalu	100 cents = 1 (Tuvalu/Australian) dollar
Uganda	100 cents = 1 (Uganda) shilling
Ukraine	(Ukrainian) grivna
United Arab Emir.	100 fils = 1 (UAE) dirham
United Kingdom	100 pence = 1 pound (sterling)
United States of A.	100 cents = 1 (US) dollar
Upper Volta – *see* Burkina Faso	
Uruguay	100 centesimos = 1 (Uruguayan) peso
US Virgin Islands – *see* Virgin Islands	
Uzbekistan	[not yet established]
Vanuatu	(Vanuatu) vatu
Vatican City	100 centesimi = 1 (Vatican) lira/Italian lira
Venezuela	100 centimos = 1 bolivar
Vietnam	100 xu or sau = 1 dong
Virgin Islands	100 cents = 1 (US) dollar
Western Sahara	100 centimes = 1 (Moroccan) dirham
Western Samoa	100 sene = 1 tala
Xizang – *see* Tibet	
Yemen	100 fils = 1 (Yemeni) rial
Zaire	100 makuta = 1 zaire
Zambia	100 ngwee = 1 kwacha
Zimbabwe	100 cents = 1 (Zimbabwe) dollar

colatitude [physics: geology] The complement of a place's latitude, equal to its ANGULAR DISTANCE from the pole (that is, 90 minus its latitude, in degrees).

colla parte [music] Musical instruction to an accompanying player or players to take the tempo from the soloist. The expression is the Italian for 'with the part'.

collateral [comparative values] An amount of property – especially property readily convertible to cash (such as stocks and bonds) – that, in value, is equal to or more than the amount required as a loan, that is owned by the person who requires the loan, and that can therefore stand as a guarantee for the loan's repayment over time.
    The term derives from the Latin for 'alongside', 'back-up'.

colla voce [music] Musical instruction for a singer to take the tempo from the leading singer. The expression is the Italian for 'with the voice'.

collective dose equivalent *see* X-RAYS

collective nouns [collectives] Below is a list of items which, in groups or assemblies, have a special term to describe them. For the reverse information, about selected collective nouns, *see* individual collective nouns as listed within the main text. So, for example, to see what items are grouped in 'charms', *see* CHARM OF NIGHTINGALES, OF FINCHES.

items	*collective*
aircraft	fleet, flight, squadron, wing
angels	band, flight, host
antelope	herd, troop
ants	colony, nest
apes	shrewdness
arrows	flight, quiver

asses	herd, pace
baboons	troop
badgers	cete
bears	sleuth, sloth
bees	grist, hive, swarm
birds	flock
bitterns	sedge
boars	singular
boats	armada, flotilla
buffalo	herd
buses	fleet, line
camels	caravan, herd, string
cars	fleet
cats	clouder, clowder, cluster
cattle	drove, herd
chickens	brood
choughs	chattering
colts	rag
coots	covert
cranes	herd, sedge, siege
crows	murder
cubs	litter
deer	herd
dogs	pack
dolphins	school
donkeys	herd, pace
doves	dule, flight
draught animals	team, yoke
ducklings	brood, team
ducks	badling (badelynge), flight, paddling
eggs	clutch
elephants	herd
elk	gang, herd
ferrets	fesnying
finches	charm
fish	catch, draught, haul, school, shoal
flamingos (flying)	skein
flies	swarm
foxes	skulk
giraffes	herd
geese	flock, gaggle, skein
gnats	cloud
goats	flock, herd, tribe
grouse	covey, pack
gulls	colony
hares	down, husk
hawks	cast
hens	brood, clutch
herons	colony, sedge, siege
hogs	drift, herd, sounder
hounds	kennel, mute, pack
horses	herd, string
hyenas	pack
jellyfish	smuck
kangaroos	mob, troop
kittens	kindle, litter
lapwings	desert
larks	bevy, exaltation
leopards	leap, lepe

lies	pack
lions	pride
lorries, trucks	fleet
mallards	flush, sord, sute
martens	richesse
mice	nest
moles	labour
monkeys	troop
nightingales	charm, watch
oxen	drove, herd
partridges	brace, covey
peacocks	ostentation
peafowl	muster
penguins	rookery
people	army, crowd, horde, host, queue, throng
pheasants	nide, nye
pigeons	flight, flock, loft
piglets	farrow, litter
pigs	drift, herd, sounder
pigs, tame	doylt
pine martens	richesse
plovers	congregation, wing
pochards	knob, rush
porpoises	school
pup(pie)s	litter
quails	bevy, covey
rabbits	nest, warren
ravens	unkindness
rhinos	crash
rooks	building, clamour, rookery
seals	pod, rookery
sheep	flock, herd
sheldrake	dopping
ships	armada, fleet
snipe	walk, wisp
soldiers	battalion, brigade, troop
sparrows	host
spiders	purse
squirrels	drey, dray
starlings	clamour, murmuration
storks	mustering
swallows	flight
swans	bevy, wedge
swine	herd
taxis	fleet
teal	knob, spring
thieves	den, pack
threats	plethora
toads	knob
trucks, lorries	fleet
turtles	bale, dule
wasps	nest
waterfowl	plump
whales	gam, pod, school
whiting	pod
widgeon	company, knob
wildfowl	sute
wolves	pack, rout
woodcock	fall

yaks	herd
zebra	herd

collision [physics] Strictly, in physics, interaction between objects in which momentum is conserved – that is, the algebraic sum of momentums before impact equals the sum of momentums after impact. Interaction does not necessarily involve contact (as, for example, in the repulsion of one subatomic particle by another). If kinetic energy is also conserved, it is termed an *elastic collision*.

collision number [chemistry; physics] For molecules of gas (which continually collide with one another and with the walls of their container), the frequency (number per unit time) of collisions per unit concentration.

colloid [chemistry] A form of matter in which a substance exists as very small particles between one ten-thousandth and one-millionth of a millimetre (1-100 nanometres) across dispersed in a fluid. The fluid may be a gas, as with fog and mist (which are termed aerosols), or a liquid, as with gelatine and rubber (termed gels). The colloidal state is particularly important in biological systems, many of which involve particles (for example, MACROMOLECULES) of colloidal size.

The term derives from an ancient Greek adjective meaning 'glue-like'. *See also* SOLUTION.

collothun [volumetric measure] In Iran, a unit of liquid capacity.

1 collothun	=	6 CHENICA
	=	8.22 litres
		(8 litres = 0.9732 collothun)
	=	1.8082 UK gallons, 2.1716 US gallons
		(2 UK gallons = 1.1061 collothun
		2 US gallons = 0.9210 collothun)
8 collothun	=	1 ARTABA

cologarithms [maths] Cologarithms are effectively the LOGARITHMS of a reciprocal of a number. The logarithm of any fraction therefore corresponds to the logarithm of the numerator (the upper figure) added to the cologarithm of the denominator (the lower figure).

colon [comparative values] Unit of currency in Costa Rica and El Salvador.

In Costa Rica:

1 colon	=	100 centimos

In El Salvador:

1 colon	=	100 centavos

The term is derived from the Spanish form of the name (Christopher) Columbus: (Cristobal) Colón. *See also* COINS AND CURRENCIES OF THE WORLD.

colonel [military rank] Senior commissioned rank in the army and airforce.

In the British army, a colonel ranks above a lieutenant colonel and below a major general, but has a rank equal to that of a brigadier – the difference between colonel and brigadier being largely a matter of whether the officer is directly concerned with combat-readiness (brigadier) or with regimental authority (colonel). In the latter sense, the colonels of some regiments are actually retired officers. The equivalent of a colonel in the Royal Navy is a captain, in the Royal Air Force is a group captain.

In the United States Army, Marines, and Air Force a colonel ranks between a lieutenant colonel and a brigadier general, and is equivalent in rank to a captain in the US Navy.

colonel-in-chief [military rank] By tradition, many British army regiments have a colonel-in-chief who is not specifically concerned with the everyday running of the regiment, who is little more than a figurehead for public occasions, and who is a member of the British royal family or the royal family of some other (friendly) country.

colorimetry [chemistry; physics] In analytical chemistry, the measurement of the intensity of colour of a solution (by comparing it with a standard). The colour can then be related to concentration and the relationship provide a method of quantitative analysis.

In physics, the measurement of the brightness, hue, and purity of a colour.

Both techniques use (different) instruments called colorimeters.

color index *see* COLOUR INDEX, COLOR INDEX

color sergeant *see* COLOUR SERGEANT

color temperature *see* COLOUR TEMPERATURE, COLOR TEMPERATURE

colour index, color index [physics: astronomy] For a star, the difference between its photographic magnitude and visual magnitude, which provides a way of estimating its temperature.

colour sergeant [military rank] In the armed forces, the SERGEANT responsible for carrying and maintaining his or her regiment's flag (or 'colours') on parades and in combat.

colour temperature, color temperature [physics: photography] For a particular source of light, the temperature of a black body (an ideal radiator) that radiates the same wavelength(s). It is used mainly in colour photography because colour films are balanced for light of a particular temperature (which is why 'daylight' colour film gives an incorrect rendering of colours illuminated with artificial light, for which film balanced for tungsten light should be used).

Some 'standard' colour temperatures are:

incandescent (tungsten) lamp	2854 K
direct sunlight	4810 K
overcast sky	6770 K

colure [astronomy] Either of two great circles on the CELESTIAL SPHERE. The *solstitial colure* passes through both SOLSTICES (on the celestial equator) as well as the celestial poles. The *equinoctial colure* passes through both EQUINOXES (on the equator) and the celestial poles.

Latin *colurus*, a version of Greek *kolo-* 'cut off' *oura* 'tail', is said to be the derivation of the English term 'because in Greece part of each circle is cut off by the horizon'.

combination [maths] A group of items, irrespective of their order, selected from a larger number of items. The number of combinations of r items that can be made from a set of n items is $n!$ divided by $r!(n-r)!$ (The symbol ! stands for 'factorial': *see* FACTORIAL.)

For example: the number of ways of choosing 3 different objects from a group of 8 objects is

$$
\begin{aligned}
&8!/3!(8-3)! \\
=\ &8!/(3! \times 5!) \\
=\ &40{,}320/(6 \times 120) \\
=\ &40{,}320/720 \\
=\ &56
\end{aligned}
$$

[A quicker method of calculation in this instance is to start at the expression $8!/(3! \times 5!)$ and note that factorial 5 occurs in both halves of the fraction and can thus be discarded. The resultant fraction is $(8 \times 7 \times 6)/(3 \times 2 \times 1) = 56$.]

combining weight [chemistry] Another name for EQUIVALENT WEIGHT.

come prima [music] Musical instruction to play or sing at the speed and in the manner of earlier in the piece, following an intervening element at a different speed or in a different manner. The expression is the Italian for 'as at first'.

commandant, kommandant [military rank] In the French and some other armies, a rank between captain and lieutenant colonel, the equivalent of a major in the British and US armies.

In German, the term also means 'commanding officer', a title that may be applied to an officer more senior in rank (equivalent perhaps to a colonel or brigadier), in charge of a brigade headquarters or garrison, camp, or military school.

commander [military rank] Senior naval rank.

In the Royal Navy, a commander ranks above a lieutenant commander but below a captain, and is equivalent in rank to a lieutenant colonel in the British army and a wing commander in the Royal Air Force.

In the US Navy, a commander ranks above a lieutenant commander but below a captain, and is equivalent in rank to a lieutenant commander in the US Army and Air Force.

commander-in-chief [military rank] Very senior member of the armed forces – appointed from army, navy, or airforce – or the diplomatic leader of the country (such as the President of the United States) in charge of all national troops and back-up measures (including engineers, signallers, ordnance, and supply lines) in a specified area defined as a theatre of combat, although generally operating from a base within the home country. In a war situation, he or she may additionally be in charge of similar troops and back-up measures of other, allied, countries, and may visit the war arena for briefings.

commensurable numbers [maths] Two or more numbers that are each whole number multiples of another one (common to them all) – that is, they all share the same unique (common) factor. For example, 8, 20, and 32 are commensurable (because they are all multiples of 4); 8, 20, and 33 are not commensurable.

commodities [comparative values] In share dealings, shares that have to do with manufactured goods or perishable produce, both of which are already in existence (and thus not the subject of dealing in 'futures') and have to be sold to maintain an annual profit and turnover.

commodore [military rank] Senior commissioned officer in the Royal Navy, whose rank is technically equal to that of a captain but who has different responsibilities more to do with combat action (in just the same way as a brigadier has different responsibilities from a colonel in the British army). The equivalent rank in the Royal Air Force is that of group captain.

There is no real equivalent in the US armed forces, although the title was formerly applied to a naval rank between a rear admiral and a captain.

common denominator [maths] For two or more fractions, the same denominator (number below the line) assigned to all of them, so that they can be added or subtracted.

For example: to subtract $\frac{2}{3}$ from $\frac{3}{4}$, first express them all as twelfths – giving them both the common denominator of 12. They are restated as $\frac{8}{12}$ and $\frac{9}{12}$, and the subtraction becomes $\frac{9}{12} - \frac{8}{12} = \frac{1}{12}$. The corresponding addition is $\frac{8}{12} + \frac{9}{12} = \frac{17}{12} = 1\frac{5}{12}$.

common difference [maths] The difference between any two consecutive terms in an arithmetic progression/series.

common factor, common divisor [maths] A number that divides into two other given numbers. For example: 4 is a common factor (or divisor) of 12 and 20.

common fraction *see* VULGAR FRACTION

common measure, common meter [literary; music] In verse, and especially in hymns and the music set to hymns, a quatrain (group of four lines) the first and third lines of which are of four iambic feet, and the second and fourth of which are of three iambic feet. A well-known example is the hymn *O God, our help in ages past* (and the tune most commonly sung to it, *St Anne*).

In relation to music, common measure is alternatively known sometimes as common time, also known as $\frac{4}{4}$ time.

common ratio [maths] The ratio between any two consecutive terms in a GEOMETRIC PROGRESSION/SERIES.

common time *see* COMMON MEASURE

Commonwealth [time] Period in England between 1649 and 1660 following the execution of King Charles I and ending with the accession to the throne of Charles II. During this time the country was ruled by the Lord Protector, Oliver Cromwell and by his son and successor, Richard.

commutative [maths] Describing a mathematical operation that is independent of the order in which it is done. For example, the addition $2 + 3 + 4$ has the same sum as $3 + 4 + 2$ (or any other order); addition is therefore commutative. (Subtraction is not commutative; e.g. 3 - 2 is not the same as 2 - 3). *See also* ASSOCIATIVE.

company [military] In the army, a troop of soldiers made up of two or more PLATOONS, most often under the command of a captain. A number of companies together (in the British army, between two and five; in the United States Army, ordinarily five) make up a BATTALION.

comparison of adjectives, of adverbs *see* ADJECTIVE; ADVERB

compass points *see* POINTS OF THE COMPASS

complanation [maths] The mathematical calculation of the area of a curved surface in terms of an equivalent flat (plane) surface area.

complement [maths] In pure mathematics (and computers), the result of subtracting each of a number's digits from one less than the number's base (radix) and adding 1. For example: the complement of the number 24 in the base 10 (ordinary numbers) is $(9 - 2)(9 - 4) + 1 = 75 + 1 = 76$. Complements are particularly important in computers because (the mathematical operation of) subtraction can be achieved by addition of a complement (and ignoring the highest digit) – and addition is the only basic operation that a computer can perform. As an illustration, consider the addition $47 - 24$. This can be restated as the addition of 47 and the complement of 24 (which is 76, *see earlier*). The operation therefore becomes $47 + 76 = 123$; ignoring the highest digit gives 23. In computers, this method is applied to binary numbers. By extension, division (as multiple subtraction) can also be achieved.

In set theory, the set of all members of the universal set that are not members of a given set (whose complement it is).

complement, complementary angle [maths] In geometry, one of any two angles that together add to 90 degrees (such as the angles other than the right-angle in a right-angled triangle). Each angle is the complement of the other.

complementary colours [physics] Any two colours that combine to give white. *See also* PRIMARY COLOURS; SECONDARY COLOURS; TERTIARY COLOURS.

complex number *see* ARGAND DIAGRAM; NUMBER

composite number [maths] Any whole number into which another whole number – other than 1 – can be divided; the opposite of a PRIME NUMBER.

compound [chemistry] A substance that is made up of two or more chemical elements combined in definite proportions by weight: *see* LAW OF CONSTANT PROPORTIONS.

compound interest *see* INTEREST

compound pendulum [physics] A bar or rod with a weight at its lower end and pivoted near the top so that it hangs vertically: the type in most pendulum clocks. When it is displaced through a small angle, it swings in much the same way as a simple pendulum. If its centre of mass is a distance d from the pivot, and the pendulum's radius of gyration is r, the period (time for one complete swing, there and back) is 2π times the square root of the ratio of d^2 plus r^2 to d times the acceleration of free fall (g, or the acceleration due to gravity). For the mathematically minded:

$$t = 2\pi[(d^2 + r^2)/dg]^{1/2}$$

See also PENDULUM.

compressibility [physics] For an elastic material, the reciprocal of its BULK MODULUS: *see* COEFFICIENT OF COMPRESSIBILITY.

compression ratio [physics] For a petrol or diesel engine, the total volume inside the cylinder (at outer dead centre) divided by the volume at the top end of the compression stroke. Other things being equal, high-compression engines produce more power than low-compression ones of the same overall capacity.

Compton wavelength [physics] The wavelength ascribed to a moving subatomic particle, equal to Planck's constant (6.626×10^{-24}) divided by the product of its rest mass and the speed of light – that is, it depends only on the particle's mass. For example: an electron has a Compton wavelength of 2.426321×10^{-12} metres. It was named after the American physicist Arthur Compton (1892-1962). *See also* DE BROGLIE WAVELENGTH.

computerized axial tomography (CAT/CT scan) *see* SCANNING SYSTEMS

concentration [physics; chemistry] The amount of a particular substance in a mixture or a solution. It can be expressed in various ways, such as the mass per unit volume (for example, grams per litre of solvent or litre of solution, pounds per gallon, etc.), percentage (parts per hundred by weight or volume), parts per million (ppm), molality (moles per kilogram of solvent), molarity (moles per litre of solution) or normality (gram-equivalents per litre of solution). *See also* MOLALITY; MOLARITY; MOLE; NORMALITY.

concentric [maths] Describing two or more circles with a common centre, or two or more cylinders with a common axis. Two circles, one inside the other, that do not share a common centre are said to be *eccentric*.

concertina *see* HARMONICAS' AND ACCORDIANS' RANGE

concert pitch [music] A standard pitch (frequency) of a note to which instruments are tuned so that they can be played together harmoniously. In this sense, it is arbitrary but, nowadays, it is the internationally agreed frequency of 440 hertz for the A above middle C.

condensation [chemistry] A chemical reaction – also called a condensation reaction – in which small molecules combine to form a larger one. Usually a simple substance, such as ammonia or water, is also formed. *See also* POLYMERIZATION.

condensation number [physics] The number of molecules condensing on a surface divided by the total number of molecules striking it.

conditional probability [maths] The probability that an event will occur, given that another (conditioning) event will definitely happen: *see* PROBABILITY.

conductance [physics] The ability of a material to conduct heat or electricity. Heat conductance is better known as THERMAL CONDUCTIVITY. Electrical conductance is the current flowing in a conductor divided by the potential difference (voltage) across it. It is the reciprocal of resistance and is measured in SIEMENS. *See also* ELECTRICAL CONDUCTIVITY.

conductance ratio [physics] The equivalent conductance of a solution divided by its conductance at infinite dilution.

conductimetric analysis [chemistry] A type of VOLUMETRIC ANALYSIS in which the end-point is detected by a significant change in a solution's CONDUCTANCE.

conductivity (electric and thermal) *see* ELECTRICAL CONDUCTIVITY; THERMAL CONDUCTIVITY

cone [maths] A solid figure with a flat circular base and sloping curved sides that come to a point (the apex). If the apex is directly over the centre of the base, it is described as a right cone.

The volume of a cone is $\frac{1}{3}\pi$ times the product of the square of the base's radius and the perpendicular height.

The area of the curved surface of a cone is π times the product of the base's radius and the slant height.

conformal projection [geography: cartography] Method of cartographic projection, a variety of conical projection, that allows for the correct measurement of angles on a map. Most useful for detailed maps in large scale, the projection nonetheless may distort the contours of large areas. Mercator's cylindrical projection is a well-known example of this type, used specifically because the shortest route between any two points as defined on the map really is a straight line.

A special case of the conformal projection is provided by Lambert's projection which combines the conformal with EQUAL-AREA PROJECTION.

conformational analysis [chemistry] The visualization of the shape of a molecule of a substance, particularly an organic compound with a single carbon-carbon bond that can be rotated to alter the relative positions of the other atoms.

cong [square measure] In Vietnam, a unit measure of land based on a unit within the metric system.

1 cong	=	1,000 square metres, 10 ARES
	=	1,196 square yards, 10,764 square feet
		(30.25 x 30.25 yards, 1 SQUARE ROD = 0.765 cong
		34 x 34 yards, 1,156 square yards = 0.966 cong)
	=	0.2471 ACRE (just under one-quarter of an acre)
		(¼ acre, 1,210 square yards = 1.0117 cong)

congius [volumetric measure] In ancient Rome, a measure of liquid capacity, especially in the preparation or preservation of wine and oil.

1 congius	=	6 SEXTARII ('sixths', or Roman pints)
	=	3.1875 litres
	=	0.701 UK gallon, 5.61 UK pints
	=	0.842 US gallon, 6.74 US pints
4 congii	=	1 URNA
8 congii	=	1 AMPHORA

In more recent pharmaceutical use, a congius has been taken to correspond to 1 UK GALLON.

congruent [maths: geometry] Describing plane or solid figures that have exactly the same size and shape. A pair of congruent plane figures can be made to overlap exactly (one figure may require rotation or reflection first).

conical projection [physics: cartography] Type of map projection in which lines on the Earth – lines of longitude (meridians) and lines of latitude (parallels) – are projected on to a cone, usually located so that its axis is parallel to the Earth's axis. On the resulting map, the parallels are depicted as concentric circular arcs, and the meridians are radiating straight lines. Scale varies across the map.

conic section [maths] A curve produced when a plane cuts through a circular cone. If the cutting plane is parallel to the cone's base, the conic section is a circle. If the cutting plane slopes, the section is an ellipse. If the cutting plane cuts through the base of the cone, the section is a parabola. If the cutting plane is parallel to the cone's sloping side, the section is a hyperbola.

Each conic section can also be defined as the locus of a point that moves in such a way that its distance from a fixed point (the focus) divided by its distance from a fixed line (the directrix) is a constant, called the ECCENTRICITY. *See also* CIRCLE; ELLIPSE; HYPERBOLA; PARABOLA.

conjugate [maths] Describing (1) two angles that add to 360 degrees; (2) two arcs that join to form a complete circle; or (3), of the complex number $x + iy$, the complex number $x - iy$ (that is, a reflection on the opposite side of the real axis on an ARGAND DIAGRAM). All of these meanings reflect the term's derivation from Latin words that originally meant 'yoked together'.

conjunction [literary] In English grammar, a word that *conjoins* two other words, two phrases, two clauses, or two sentences: the classic example is the word 'and'.

conjunction [astronomy] The lining up of two celestial objects so that they have the same apparent longitude as viewed from Earth. For the planets Mercury and Venus, *inferior conjunction* occurs when either planet is between the Earth and the Sun; *superior conjunction* occurs when the Sun is between the Earth and the planet.

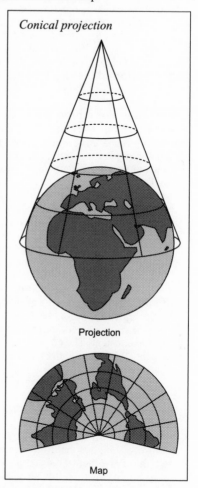

Conical projection

Projection

Map

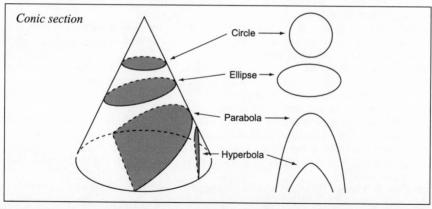

Conic section

Circle →

Ellipse →

Parabola →

Hyperbola →

conoid [maths] A solid figure formed by revolving a conic section about an axis. For example: a circle rotates to form a sphere, an ellipse forms an ellipsoid, a parabola a paraboloid, and a hyperbola a hyperboloid.

The term reflects its derivation from ancient Greek elements meaning 'of a cone'.

consequent [maths] The second term in a ratio. For example: in the ratio 3:4, 4 is the consequent (and 3 the antecedent).

conservation of charge [physics] A basic law of physics that states that, for all electromagnetic processes occurring in an isolated system (or the Universe), the total electric charge remains constant.

conservation of energy [physics; chemistry] A basic law of chemistry and physics that states that, for all processes occurring in an isolated system, the total energy remains constant. It is a restatement of the first law of thermodynamics: energy can be neither created nor destroyed by ordinary chemical or physical processes.

conservation of mass [chemistry] A basic law of chemistry that states that the combined mass of the products of a chemical reaction is the same as that of the reactants.

conservation of momentum [physics] A law of physics that states that the combined momentum of two colliding objects is the same after impact as before it.

consols [comparative values] Financial bonds issued by the government of the United Kingdom relating to its permanent funds, as opposed to funds raised by annual or special parliamentary voting. The term is an abbreviation for 'consolidated annuities'.

consonants [literary: phonetics] Letters of the alphabet and sounds made in speech that are not VOWELS (pure sounds modified only by the position of the jaws or the muscles of the mouth), and thus require the use of teeth, tongue, palate, lips, or breath to pronounce them.

In phonetics, consonants may be described as dental, fricative, labial, palatal, plosive, velar, or any combination of these types.

See also DENTAL CONSONANTS; FRICATIVE CONSONANTS; LABIAL CONSONANTS; PALATAL CONSONANTS; PLOSIVE CONSONANTS; VELAR CONSONANTS.

constant [maths; physics] A quantity that always remains the same. For example: in the algebraic expression $3x^2 = 7xy$, 3 and 7 are constants (x and y are variables). The ratio of a circle's circumference to its diameter, the irrational number π, is an absolute constant. The constant resulting from indefinite integration is an arbitrary constant (for example: the integral of x^4dx is $\frac{1}{5}x^5$ + K, where K is an arbitrary constant).

In physics, various standardized fundamental quantities are known as physical constants. Examples include the acceleration of free fall (acceleration due to gravity), Avogadro's constant, the charge on an electron, Planck's constant, the speed of light, and so on.

constant-boiling mixture [chemistry] A mixture – also known as an *azeotropic mixture* – of liquids that form a loose compound whose boiling point (different from those of either of the components) remains the same as vapour is boiled off. For example ethanol (ethyl alcohol) and water form a constant-boiling mixture containing 95.57 per cent ethanol, which boils at 78°C.

constant composition *see* LAW OF CONSTANT PROPORTIONS

constringence [physics] For a transparent material, its mean refractive index minus 1, divided by the difference in its refractive indices for red and violet light. It is the reciprocal of the dispersive power.

consumer price index [comparative values] An example of an index that indicates the comparative values of household commodities and services, and that is close to being identical with a cost-of-living index.

contact angle [physics] *see* ANGLE OF CONTACT

continental (Morse) code [literary; physics] English-speaking term for the variant of Morse code used on the continental mainland of Europe. Known also as the 'international code', it has eleven symbols that differ from the ordinarily accepted English version of the code. Neither form of the code is much used today, however. *See also* BAUD.

continuous wave [physics] An electromagnetic wave with a constant AMPLITUDE.

conto [comparative values] Unit of monetary value in Portugal and Brazil. The term derives from a Portuguese word originally meaning '1 million', and initially referred to a Portuguese monetary unit of a value of 1 million *reis*. Time devalued the unit, however, and now

1 Portuguese conto	=	1,000 ESCUDOS
1 Brazilian conto	=	1,000 CRUZADOS

contour, contour line [physics: cartography; geology] A line on a map that joins places of the same height above sea level (that is, the same altitude).

contrabass, contrabasso [music] A bass voice of the lowest pitch. The term is also used for the deep-pitched stringed instrument the double bass.

contract bridge calls (bids) and scores [sport] In (contract) bridge, the bidding is to establish that one side is to win a specified number of tricks with a specified suit as trumps or with no trumps. It is understood from the outset that six tricks will be the minimum to be won, and so a bid of 1 trick actually means that 7 tricks are to be taken, and a bid of 7 tricks means a slam of 13 tricks altogether must be taken. The suit nominated becomes the trumps. The order of value of the suits (from lowest to highest) is clubs, diamonds, hearts, and spades; no-trumps is of higher value still. The scoring value of a bid may be doubled (by an opponent) or re-doubled (by a partner thereafter) without changing the actual bid.

The highest bidder ('declarer') succeeds in nominating trumps or no trumps and must thereon make good the bid. His or her partner lays down the cards in his or her hand as the dummy for the round.

The score is kept each side of a horizontal line on the score sheet or bridge block. Above the line is the hono(u)r score; below it is the trick score.

Points for tricks (over the non-bid 6):

	trumps		*no trumps*
	spades/hearts	*diamonds/clubs*	
first trick (= 7th)	30 points	20 points	40 points
subsequent tricks	30 points	20 points	30 points

The tricks that correspond to the contract are included in the trick score; points for extra tricks are included in the hono(u)r score (above the line). If the contract was doubled, the trick score is also doubled, and extra tricks count 100 points each in the hono(u)r score if the declarer's side has not already won a game (is 'not vulnerable'), and 200 points each if the side has already won a game (is 'vulnerable'). If the contract was redoubled, these scores are multiplied by 2 . A fulfilled contract that has been doubled or redoubled receives a bonus of 50 points in the hono(u)r score.

If, however, the contract is not fulfilled, the opponents score for every trick that the declarer fails to complete the contract by. The score involved also depends on whether the declarer was vulnerable or not vulnerable (as above).

	vulnerable			*not vulnerable*		
	standard	doubled	redoubled	standard	doubled	redoubled
first undertrick	50 pts	100 pts	200 pts	100 pts	200 pts	400 pts
subsq. undertricks	50 pts	200 pts	400 pts	100 pts	300 pts	600 pts

When a score of 100 or more points is reached in the trick score, the scorers are held to have won a game. A new horizontal line is drawn on the score sheet/bridge block, and another game begins. (Only trick scores count towards the winning of games; hono(u)r scores are maintained separately.) Two games win a 'rubber', and collect a bonus – of 500 points if the opponents have won a game, of 700 if the opponents have not won a game. At the end of a rubber, all the trick scores and hono(u)r scores are totalled for both sides, and the side with the higher total receives the difference to the nearest 100 up or down.

Hono(u)r scores include bonuses for any player who holds the four top-value trump cards in the hand (100 points in the hono(u)r score above the line) or five top-value trumps or four aces (150 points above the line). For a fulfilled contract of 6 (a 'little slam'), a bonus of 500 points is scored by a side that is not vulnerable, of 750 points if the side is vulnerable. For a fulfilled contract of 7 (a 'grand slam'), a bonus of 1,000 points is scored by a side that is not vulnerable, of 1,500 points if the side is vulnerable. (A side declaring 6 but making 7 has fulfilled a contract for a little slam and has one additional overtrick.)

The original form of the game of bridge is now known as bridge whist, in acknowledgement that it began as an extension of the game of whist in the early 1890s. Contract bridge replaced bridge whist as the most popular serious (and competition) card game in about 1930.

conversion [quantitatives; maths; comparative values] The rendering of a number of units in one unitary system into a corresponding number of units in another unitary system. In this book, such conversions are presented as part of the information under every headword that represents a quantitative unit.

conversion factor [quantitatives; maths; comparative values] A multiplication factor that converts quantities in one set of units into another set of units. For example: the conversion factor for changing inches into millimetres is 25.4 (1 inch = 25.4 millimetres). Some conversions use operations other than mere multiplication (*see*, for example, CELSIUS SCALE for the conversion of Celsius temperatures to Fahrenheit).

coordinates [maths] A pair of numbers that specify the position of a point on a plane (or a set of three numbers that define a point in space): *see*, for example, CARTESIAN COORDINATES; POLAR COORDINATES.

coordination number [chemistry] For an atom or ion in a chemical compound, its number of nearest neighbours. The term is generally applied to a coordination compound (also called a complex compound), in which a central atom is bonded to several surrounding groups, called *ligands*, by coordinate bonds. (A coordinate bond is a type of covalent bond in which both of the bonding electrons – often a lone pair – are contributed by one of the combining atoms; it is also called a dative bond.)

copeck *see* KOPECK

coplanar [maths] Describing two or more points or lines that lie in the same plane.

Copper Age [time] A less well-attested name for the Chalcolithic or Bronze Age, the stage in human culture in which tools and weapons were made of copper or bronze, but referring to no specific period of human history.

cor *see* KOR, HOMER

cor anglais *see* WOODWIND INSTRUMENTS' RANGE

cord [cubic measure] Cubic unit in the measurement of rough-cut timber (logs).

1 cord	=	128 cubic feet, 4.7407 cubic yards
	=	a pile 4 feet high, 4 feet deep, and 8 feet wide (that is, a stack of 8-foot lengths)
	=	8 CORD FEET
	=	3.62455 cubic metres (4 cubic metres = 1.1036 cord)

See also FACE CORD.

cord foot [cubic measure] Cubic unit in the measurement of rough-cut timber (logs) in North America.

1 cord foot	=	16 cubic feet, 0.5926 square yard
	=	a pile 4 feet high, 4 feet deep, and 1 foot wide
	=	one-eighth of a CORD
	=	0.4531 cubic metre (1 cubic metre = 2.2072 cord feet)

See also FACE CORD.

córdoba [comparative values] Unit of currency in Nicaragua.

1 córdoba	=	100 centavos

The unit is named after the Spanish explorer Francisco Hernández de Córdoba (*c*.1475-1526) who travelled the country during the mid-1520s on behalf of the governor of Panama – who then executed him. *See also* COINS AND CURRENCIES OF THE WORLD.

Coriolis effect, Coriolis force [physics: meteorology] A hypothetical force used to explain certain movements in systems that are rotating. For example, a location at a high northern latitude on Earth is travelling more slowly eastwards (with the Earth's rotation) than a place on the equator; as a result, a wind blowing northwards from the equator appears to drift eastwards, apparently driven by the Coriolis 'force'. A similar effect applies to water in the oceans, and even to bath water spiralling down the plug-hole (clockwise in the Northern Hemisphere, anticlockwise in the Southern). It was named after the French mathematician and engineer Gaspard Gustave de Coriolis (1792-1843), who first described it.

cornet [military rank] Obsolete rank in cavalry regiments of the British army: a junior officer whose duty it was to carry the regimental flag on parade or into battle. In other, infantry, regiments this duty was performed by the ENSIGN. *See also* COLOUR SERGEANT.

cornet [music] *see* BRASS INSTRUMENTS' RANGE

corn-hog ratio [comparative values] In the United States' pig-farming industry, the ratio of the market price of one hog weighing 100 pounds (45.359 kilograms) to that of one US bushel (0.9689 UK bushel, 35.238 litres, 1.244 cubic feet) of corn (maize), used as an indication of profit margins.

corona [astronomy] The layer of luminous gases that surrounds the Sun, corresponding to the Sun's outer atmosphere. A strong X-ray source, it has a temperature of 500,000 to 1,000,000 K. It is visible during a total eclipse of the Sun, but can be studied at any time using a coronagraph (a type of telescope with a disc that blocks off the Sun's disc).

The term derives directly from the Latin for 'crown', which in turn is derived from or akin to the ancient Greek *korone* 'curved beak', 'rounded tip'. *See also* ATMOSPHERE OF THE SUN.

corporal [military rank] Low-ranking non-commissioned officer. In the United States' armed forces, a corporal ranks above a lance corporal but below a sergeant. In the British armed forces (and the armed forces of many other countries), a corporal is the first grade up from the unranked private.

The title derives in English as a French corruption of the Italian *caporale* 'of the head', 'of the leader'. *See also* LANCE CORPORAL.

correlation coefficient [maths] In statistics, a number between -1 and +1 that is a measure of the correlation between two variables (items of data), such that changes in one variable lead to changes in the other. If an increase (or decrease) in one leads to an increase (or decrease) in the other, correlation is positive; if an increase in one leads to a decrease in the other (or vice versa), correlation is negative. For example, there is a positive correlation between the amount of a person's suntan and his or her exposure to sunlight.

correspondence [maths] In set theory, one of four ways in which the elements in one set can be matched with those in another. The possible types of correspondence are one-to-one, one-to-many, many-to-one, and many-to-many.

corresponding angle [maths] One of two angles on the same side of a transversal (line) that cuts a pair of lines; corresponding angles are those between the transversal and the other lines. If the pair of intersected lines are parallel, the corresponding angles are equal.

cosecant (cosec) [maths] For an angle in a right-angled triangle, the length of the hypotenuse (longest side) divided by the length of the side opposite the angle. It is the reciprocal of the SINE.

cosech *see* HYPERBOLIC FUNCTION

cosh *see* HYPERBOLIC FUNCTION

cosine (cos) [maths] For an angle in a right-angled triangle, the length of the side adjacent to the angle divided by the length of the hypotenuse (longest side).

cosmic radiation [physics: astronomy] Very short-wavelength, penetrating radiation (about 10^{-15} metres wavelength) that originates in outer space. There are two types of cosmic rays. Protons, other nuclei, and subatomic particles make up primary rays. Collisions between these and gas atoms (nitrogen, oxygen) in the Earth's upper atmosphere generate secondary cosmic rays, which consist of elementary particles and gamma-rays.

cosmic speed *see* ESCAPE VELOCITY

cosmic year [time; physics; astronomy] The time the Sun takes to complete one orbit within the (revolving) Galaxy, thought to be about 200 million years. There have been only about twenty-three cosmic years, or solar orbits, since the Solar system came into existence, and only about seventy cosmic years, or galactic revolutions, since the Galaxy itself was created. *See also* AGE.

cost unit *see* UNIT COST

cotangent (cot) [maths] For an angle in a right-angled triangle, the length of the side adjacent to the angle divided by the length of the side opposite it. It is the reciprocal of the TANGENT.

coth *see* HYPERBOLIC FUNCTION

cotula *see* COTYLA, KOTYLE

cotyla, kotyle [volumetric measure] Unit of liquid and dry capacity in ancient Greece, closely approximating to half of a modern pint.

1 cotyla	=	72 kheme, 36 setier, 24 mystra, 6 KYATHOI, 4 oxybatha
	=	26.9695 centilitres, 0.269695 litre
	=	0.4747 UK pint, 0.5700 US pint
	=	0.4747 UK dry pint, 0.4894 US dry pint
	=	296.695 cubic centimetres
	=	16.4585 cubic inches
2 cotylai	=	1 SEXTE or xestes
4 cotylai, 2 sexte	=	1 KHOINIX (dry measure only)
12 cotylai, 6 sexte	=	1 KHOUS
144 cotylai, 72 sexte, 12 khoes	=	1 (Greek) AMPHORA

The *kotyle* (Latinized later to *cotyla* or *cotula*) was the equivalent of the drinking glass in ancient Greece, although made of pottery. Rather than cylindrical in shape, however, it was as wide as it was deep, like a soup bowl. Because of this shape, the word was in ancient Greek applied also the socket of the hip joint. *See also* ACETABULUM; OXYBATHON.

coulomb [physics] Unit of electric charge, equal to the quantity of electricity carried by 1 AMPERE of current in 1 second. It was named after the French physicist Charles Coulomb (1736-1806).

coulomb force [physics] The force of attraction between two dissimilar electric charges, or the force of repulsion between two similar ones. Its dimension is defined by COULOMB'S LAW.

Coulomb's law [physics] The force of attraction (or repulsion) between two point electric charges is proportional to the product of the charges and inversely proportional to the square of the distance between them. The constant of proportionality equals the reciprocal of 4π times the permittivity of the medium between the charges (in SI units). It was named after Charles Coulomb (*see above*).

Coulomb's law for magnets [physics] COULOMB'S LAW (*see above*) can be applied to hypothetical magnetic poles: the force of attraction (or repulsion) between two magnetic poles is proportional to the product of their pole strengths and inversely proportional to the square of their distance apart times the permeability of the medium separating them.

coulometer [physics] An instrument for measuring (small) amounts of electricity in terms of the quantities of substances released at the electrodes of an electrolytic cell. *See also* VOLTAMETER.

counter-tenor *see* ALTO RANGE

couple [physics] A pair of equal forces that act (in parallel but opposite directions) on an object. The result is a torque (turning effect), equal to the product of one of the forces and the distance between them; this is also termed the moment of the couple.

couplet (in verse) [literary] Two lines of verse that rhyme and have the same number of feet.

The term is an originally French diminutive form of *couple*.

cousin *see* RELATIONS AND RELATIVES

covado [linear measure] Obsolete unit of length in Portugal, probably in origin an assimilation to the metric system of a Portuguese form of CUBIT.

1 covado	=	3 (Portuguese) palmos
	=	66 centimetres, 0.66 metre (1 metre = 1.5152 covado)
	=	25.98 inches (2 feet 1.98 inches) (24 inches, 2 feet = 0.9238 covado)

covalence, covalency [chemistry] A type of chemical bonding in which each combining atom contributes an electron to the bond; it is the typical sort of valence in organic compounds. *See also* COORDINATION NUMBER.

covalent radius [chemistry] For two similar atoms joined by a covalent bond in a molecule of a compound, half the distance between their atomic nuclei. When dissimilar atoms are covalently bonded, the sum of their covalent radii equals the bond length.

coversed sine [maths] For a given angle, a trigonometrical function equal to 1 minus the sine of the angle: *see* SINE.

covert of coots [collectives] A collective for coots that suggests the birds' relatively secretive nesting habits, covert (or covered) among reeds or bushes at the water's edge. The word in this sense, however, has nothing to do with being hidden but is instead a slightly less Anglicized variant of the collective noun COVEY (*see below*).

covey of partridge, of quail (out of season) [collective noun] A covey describes a small flock of several different game birds. The term derives from the past participle of the medieval French *cover* 'to brood', 'to incubate eggs': *covée* 'brood', 'clutch', but is related through the Germanic to other English collective nouns such as *coop* and *hive*.

covido [linear measure] In Arabia, a unit of length probably deriving as a borrowing from one or other European country's version of the CUBIT.

$$
\begin{aligned}
1 \text{ covido} \ &= \ 19 \text{ inches, 1 foot 7 inches} \\
&\quad (18 \text{ inches, } 1\tfrac{1}{2} \text{ foot} = 0.9474 \text{ covido}) \\
&= \ 48.26 \text{ centimetres, } 0.4826 \text{ metre} \\
&\quad (50 \text{ centimetres, } 0.5 \text{ metre} = 1.0361 \text{ covido})
\end{aligned}
$$

cran [volumetric measure] In Britain, a unit volume of fresh-caught herrings.

$$
\begin{aligned}
1 \text{ cran of herrings} \ &= \ 37.50 \text{ UK gallons of fish} \\
&= \ 45.0356 \text{ US gallons of fish} \\
&= \ 170.474 \text{ litres of fish} \\
&= \ \text{about 750 herrings}
\end{aligned}
$$

The term evidently derives from (Scottish) Gaelic, although the exact meaning is not clear. The Gaelic word *crann* means 'portion', 'allocation', however – which is of course highly relevant to a volumetric unit.

cranial index *see* CEPHALIC INDEX

crash of rhinos [collectives] A strange collective for the generally mild, inoffensive, and herbivorous rhinoceros. The African black rhino, however, has a highly nervous disposition and is liable to panic and charge blindly at any unexpected object (such as a human) when disturbed. In doing so, it may well career into intervening trees and tall scrub, causing a characteristic crashing noise.

crease *see* CRICKET MEASUREMENTS, UNITS, AND POSITIONS

credit rating [comparative values] Valuation set by a bank or finance company on a person's ability to pay back a loan with interest by means of regular repayment instalments, usually per month. The higher the credit rating, the higher the initial loan may be, because the finance house believes the higher repayment instalments can be met.

It is perhaps ironic that only by owing money can a person get a 'credit' rating yet, without a credit rating, a person may not be able to get a loan.

creep [metallurgy] Creep is the very slow but continuous deformation (usually extension) of a metal under the effect of a steady load (for example, lead on a sloping roof gradually creeps downwards). A metal's ability to resist such movement is called *creep strength*, which is measured by the amount of movement produced by a given stress (for a certain time and at a particular temperature).

crescendo [music] Musical instruction to play or sing with gradually but consistently increasing volume. Italian 'increasing'. The opposite is DIMINUENDO.

crescent moon *see* LUNAR PHASES

Cretaceous period, Upper and Lower [time] A geological PERIOD at the end of the Mesozoic or Secondary era, after the Jurassic period and before the comparatively sudden and dramatic onset of the Pleistocene period of the Cenozoic era, corresponding roughly to between 144 million years ago and 65 million years ago. The period is generally considered to comprise two divisions: first the Lower Cretaceous, and then the Upper. The Lower Cretaceous lasted from roughly 144 million years ago to about 102 million years ago; the Upper from roughly 102 million years ago to about 65 million years ago. During the Cretaceous, dinosaurs roamed the land; mammals were comparatively few and insignificant, feeding on the new form of vegetation, flowering plants; in the gradually increasing waters, molluscs were at first abundant but declined seriously thereafter. The distinction between the Lower and Upper Cretaceous was marked by severe intrusion by the sea on to land: not since Palaeozoic times had so much seawater covered the Earth. The Upper Cretaceous was the time during which many of the chalk strata evident in northern Europe and the western United States were laid down, the strata that gave the period its name (Latin *cretaceus* 'chalky'). It was at the end of the Cretaceous period that the still controversial cataclysm occurred that rendered the dinosaurs (and various other types of creatures, including molluscs such as ammonites) extinct.

For table of the Mesozoic era *see* TRIASSIC PERIOD.

cricket measurements, units, and positions [sport] Cricket is a highly evolved game, with complex rules, multiple methods of scoring, multiple methods of being 'out' when batting, and frequent changes of position by the fielding team. The object of the game is for a team of eleven batsmen to make as many runs as possible over one or two innings (depending on the formality of the match), two batsmen batting ('in') at a time but only one receiving the bowling, and each bowler bowling in overs of six balls (in some Australian games eight balls) from one direction before 'over' is called (by one of two umpires) and the next bowler begins his over from the other direction. In some professional games, the number of overs is limited to ensure that a game finishes within a day's play; in top-class international cricket, a game (Test match) may last for up to five playing days. Matches not completed are drawn.

The dimensions of the pitch:
> playing area: as large as the (generally oval) boundary line
> pitch or wicket (close-mown area between sets of stumps):
>> length: 22 yards (66 feet, 20.12 metres) between the stumps (wickets); the wicket may be as close mown behind each set of stumps for a farther 4 feet (1.22 metres)
>> width: 8 feet 8 inches (2.64 metres); the line marking the popping crease each end (at which the batsman stands and from which the bowler bowls at the other end) is usually extended a farther 20 inches (50.8 centimetres) each side
>> the popping crease: 4 feet (1.22 metres) inside the stumps at each end
>> the three stumps/the wicket:
>>> overall height – 2 feet 4 inches (81.5 centimetres)
>>> overall width – 9 inches (22.8 centimetres)

Positions of fielding players: as illustrated *opposite* (note these are the names of the positions, but only eleven fielders are on the field at any one time, including the bowler and the wicket-keeper)

Times and timing: according to the formality of the match, and how long the innings last
> intervals: 2 minutes for an incoming batsman to take his place (in professional cricket: 40 minutes for lunch, 20 minutes for tea; 10 minutes between innings)

Runs scoring:
> as many runs between wickets as can be made without the ball hitting the stumps when a batsman is not in the crease
> four (4; when ball reaches boundary having touched the field)
> six (6; when ball is hit directly over the boundary)

Extras (runs scored without running, as called by umpire):
> no ball (e.g. improper bowling, misplaced fielders)
> wide
> bye, leg-bye
> improper fielding (e.g. ball stopped with cap) 5 runs

Methods of being out:
> bowled
> leg-before-wicket (as judged by umpire)

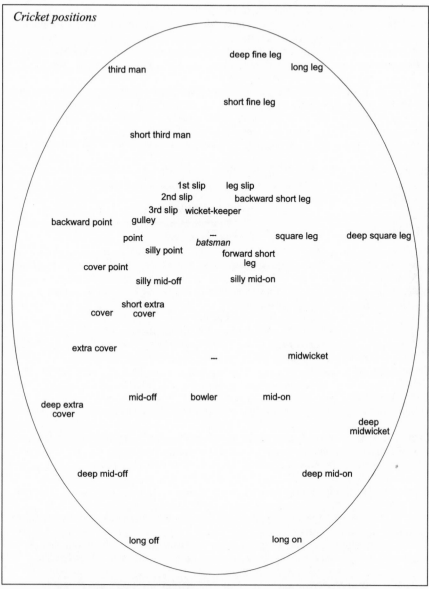

Cricket positions

deep fine leg
third man
long leg
short fine leg
short third man
1st slip   leg slip
2nd slip   backward short leg
3rd slip wicket-keeper
backward point gulley
point   *batsman*   square leg   deep square leg
silly point   forward short
cover point   leg
silly mid-off   silly mid-on
short extra
cover cover
extra cover
midwicket
mid-off   bowler   mid-on
deep extra
cover
deep
midwicket
deep mid-off   deep mid-on
long off   long on

stumped (as judged by umpire)
caught
run out (as judged by umpire)
hit wicket (so as to dislodge bails)
handled ball (including to protect wicket)
obstructed fielder
hit ball twice (unless protecting wicket)
retired hurt (may return, with opponents' consent)
Dimensions of equipment:
the bat:
maximum overall length: 3 feet 2 inches (96.5 centimetres)
maximum overall width: 4¼ inches (10.8 centimetres)
weight: (usually) 2 pounds 4 ounces (1.021 kilograms)
the ball:
circumference: 8¹³⁄₁₆–9 inches (22.4–22.9 centimetres)

weight: 5½–5¾ ounces (156–163 grams)

Although cricket is played on an international basis, it remains largely popular only in England and some ex-colonial states. Following the establishment of a World Cup cricket competition for women during the early 1990s, women's cricket then also achieved a high standing in those countries. The rules of the women's game are the same as those of the men's game.

The name of the game reflects the shape of a stump, which has a notch in the top on which the bail is lodged. Originally there were only two stumps comprising a wicket, and each stump had a deeper notch to carry a crossbar: they were distinctly sticks with hooks on top – and *cricket* is a sort of diminutive of a word that is found in English also as *crutch* and *crook*, a name thus also essentially the same as *croquet*.

crinal [physics] Unit of force in the defunct decimetre-kilogram-second system, equal to the force that provides a mass of 1 kilogram with an acceleration of 1 decimetre per second.

$$
\begin{aligned}
1 \text{ crinal} \quad &= \quad 0.1 \text{ (one-tenth) newton} \\
&= \quad 1.0197 \text{ grams force} \\
&= \quad 0.35984 \text{ ounce force}
\end{aligned}
$$

The unit's name derives from the (adjectival form of the) Latin *crinis* 'hair', because the force (about 10 grams weight) was thought to be that required just to break a human hair.

crith [physics] Unit of mass of a gas, equal to the mass of 1 litre of hydrogen at standard temperature and pressure.

$$
\begin{aligned}
1 \text{ crith} \quad &= \quad 89.88 \text{ milligrams} \\
&= \quad 8.988 \times 10^{-7} \text{ kilogram}
\end{aligned}
$$

The unit's name derives from the ancient Greek *krithe* 'barleycorn', regarded as a thing of virtually no weight at all.

critical angle [physics] For a ray of light incident on a block of transparent material, the smallest angle of incidence at which there is total internal reflection of the ray – that is, the angle of refraction is 90 degrees.

critical mass [physics] The smallest amount of fissile material that is needed to maintain a nuclear chain reaction.

critical point [physics] On the curve of pressure against temperature for a gas (its isothermal), the point at which the gas and its liquid form coexist – that is, gas and liquid densities are the same. The gas is at its CRITICAL PRESSURE and CRITICAL TEMPERATURE.

critical potential [physics] The energy needed to turn an atom into an ion, measured in ELECTRON-VOLTS.

critical pressure [physics] The minimum pressure needed to condense a gas into a liquid at its CRITICAL TEMPERATURE.

critical state [physics] The temperature and pressure at which the liquid and gas phases of a substance become a single phase (and have the same density). The substance is then at its CRITICAL POINT.

critical temperature [physics] The temperature above which a gas cannot be liquefied by pressure alone.

Also, the temperature at which a magnet loses its magnetism. In this sense the critical temperature is known alternatively as the curie point, after the Polish-born French physicist Marie Curie (1867-1934).

critical velocity [physics] For a flowing fluid, the speed at which laminar flow changes to turbulent flow: *see* REYNOLDS NUMBER.

critical volume [physics] The volume of 1 mole of a substance at its critical temperature and pressure.

cron [time] A suggested – but unadopted – unit of time equal to 1,000,000 years. One can only be grateful that such a hair-raising example of extreme illiteracy (the Latin/Greek for 'time' is *chron-*; the word element *cron-* – as in the name of the deity Cronus – is a variant of the English words *corn* and *grain*, and nothing whatever to do with time) never made it into the conversion charts.

croquet scoring [sport] The object of the game in croquet is to score points by hitting balls with a mallet through six hoops on each of two courses (twelve hoops in all), and ending by hitting a wooden peg. Four balls are used – blue, black, red, and yellow – whether two or four players (in two teams) are contesting the match. A player's turn is normally one stroke at a time, but *continuation strokes* may be earned by getting a ball through a hoop, and a *croquet shot* is gained by making the ball strike an opponent's ball (roquet) without either ball's leaving the playing area, after which the player's ball may be resited behind the roqueted ball so that the player's ball cannons through a hoop and the opponent's ball hurtles off somewhere else.

The dimensions of the court:

> classic length: 35 yards (32 metres)
> classic width: 28 yards (25.6 metres)
> outer hoops (1, 2, 3, 4): 7 yards (6.4 metres) in from edges
> inner hoops (5,6): down the centre, 7 yards (6.4 metres) ahead of outer hoops and in front of the peg

Points scoring:

> for each hoop gone through: 1 point
> the player or side who completes the two courses first, or with the more points when the game is concluded, wins

Dimensions of equipment:

> mallet: any length, any weight, provided faces flat and identical
> balls: weight – 15¾–16¼ ounces (446.5–460.7 grams); all four used must be identical
> hoops: interior height – 11½ inches (30 centimetres)
> interior width – 3¾ inches (9.5 centimetres)
> peg width: 1¼ inch (3.1 centimetres)

The game was originally played with a hooked stick much like a hockey stick instead of a mallet, and it is as a diminutive of the Old French word for 'crook' (as in the shepherd's variety, with which English word it is also cognate) in reference to that stick that the game gets its name.

crore [quantitatives] In India, 10 million (10^7).

> 1 crore = 10,000,000
> = 100 LAKHS (100 x 100,000)

The Hindi word is more accurately transliterated *karor*.

cross-country skiing *see* SKIING DISCIPLINES

cross-over frequency [physics] In the design of a loudspeaker system with two speakers in one enclosure, the frequency at which the input is split into two signals (one for the high-frequency tweeter and one for the low-frequency woofer); it is usually chosen so that the high- and low-frequency speakers deliver equal power.

cross product [maths] Another name for VECTOR PRODUCT.

cross-section [maths; physics] In geometry and engineering, the shape produced by cutting (or visualizing a cut through) an object at right-angles to its length or its main axis.

In nuclear physics, the likelihood that two particles will interact, visualized as an area surrounding the target particle, and measured in BARNS.

crotchet [musical measure] In musical notation, a crotchet – known in the United States as a quarter-note – is represented by the symbol ♩ and corresponds to

> ¼ semibreve or whole note
> ½ minim or half-note
> 2 quavers or eighth-notes
> 4 semiquavers or sixteenth-notes

The term derives as a description in late medieval French of its contemporary written shape: *crochet* 'little hooked thing', and is in effect the same word in English as both *crochet* and *croquet*.

crown [comparative values] Coin issued in Britain (and in some British colonies) but, by the twentieth century, only in commemoration of royal or state celebrations. It remains nonetheless legal tender.

> 1 crown = 5 shillings (60 old pence)
> = 25 (new) pence
> 4 crowns = 1 pound (sterling)

For many decades it was the coin of the highest denomination in the British monetary system; fiscal units of higher denomination were issued only in the form of notes. Inflation and the introduction of decimal currency in 1971 – both of which led rapidly to the phasing in of coins of higher values, specifically the fifty-pence coin and the one-pound coin – rendered the coin obsolete in terms of value, but rare examples of commemorative issues (at the modern value of twenty-five pence) still occur.

The equivalent of the word *crown* is used for major units of currency in a number of other countries, notably four out of five Scandinavian countries, Estonia, and the Czech and Slovak Republics.

crown [paper size] *see* PAPER SIZES

cru [time] In viticulture, a term denoting the year in which a wine was made, or even whether the grapes involved were those of the first harvest ('premier cru') or a later harvest.

The word represents the past participle of the French verb *croire* 'to grow'.

cru [comparative values] Monetary unit as yet purely notional, but suggested for use with other currencies in the reserves of international central banks around the world for transactions involving payments between countries and governments.

The term derives as an acronym from Collective Reserve Unit.

crunode [maths] A point where two branches of a curve intersect but each branch has its own tangent. *See also* SPINODE.

crusado *see* CRUZADO, CRUZEIRO

cruzado, cruzeiro [comparative values] Units of currency in Brazil, one now obsolete.

$$100 \text{ centavos} = 1 \text{ cruzado (until 1986, 1 cruzeiro)}$$

Both units correspond to earlier Portuguese coins that were marked by a cross (Portuguese *cruz*), although the Portuguese form of 'cruzado' was in fact *crusado* 'crossed'.

cryometer [physics] An instrument for measuring extremely low temperatures.

crysoscope [physics] An instrument for measuring the freezing points of a liquid (equivalent to the melting point of the same substance in the solid state).

crysoscopic method [chemistry] A way of finding the relative molecular mass (molecular weight) of a substance by measuring the extent to which it lowers the freezing point of a solvent in which it is dissolved: *see* DEPRESSION OF FREEZING POINT.

crystal axes [physics: crystallography] The axes of a crystal, usually the lines at right-angles to their faces which thereby form a coordinate system for the CRYSTAL LATTICE.

crystal forms *see* CRYSTAL LATTICE

crystal goniometer [physics: crystallography] An instrument for measuring the angles between the faces of a crystal, a characteristic useful in identification.

crystal lattice [physics: crystallography] The repeating three-dimensional array of atoms (or ions) that make up a crystal. There are various ways of classifying crystal types (called systems), depending on the orientation and lengths of their axes; types may in turn be subdivided into classes and forms.

Common systems include:

cubic (based on a cube – simple, body-centred or face-centred – with three equal axes at right-angles)

hexagonal (based on a hexagon-based prism, with three equal axes at 60 degrees to each other and a fourth axis at right-angles)

monoclinic (based on a parallelogram-based prism, with three unequal axes, two at right-angles)

orthorhombic (based on rectangle-based prism, with three different axes all at right-angles)

tetragonal (based on a square-based prism, with three axes at right-angles, two equal and the third longer or shorter)

triclinic (based on a shape with no right-angles, only a centre of symmetry)

trigonal (based on a cube stretched along a diagonal, giving threefold symmetry)

C series [paper] A series of envelope sizes in the ISO system, designed to hold paper from the A series of paper sizes, or A sizes folded (with adequate allowance for inserting into the envelopes). For example:

code	millimetres	inches	description
C4	229 x 324	9 x 12¾	holds A4 flat
C5	162 x 229	6⅜ x 9	holds A5 flat or A4 folded in half
C6	114 x 162	4½ x 6⅜	holds A6 flat, A5 folded in half or A4 folded in four
DL	108 x 219	4¼ x 8⅝	holds A4 folded twice, into thirds

DL was not an original ISO designation, but had to be introduced because people insisted on folding their A4 letters in three – and why not! *See also* A SERIES.

CT scan, CAT scan *see* SCANNING SYSTEMS

cuartillo [volumetric measure] In Spain, a unit in both liquid and dry measure.

In dry measure:

$$
\begin{aligned}
1 \text{ cuartillo} &= 4 \text{ OCTAVILLOS} \\
&= 1{,}156.25 \text{ cubic centimetres} \\
&= 2.035 \text{ UK dry pints, } 2.098 \text{ US dry pints} \\
&\quad (2 \text{ UK dry pints} = 0.9828 \text{ dry cuartillo} \\
&\quad 2 \text{ US dry pints} = 0.9533 \text{ dry cuartillo}) \\
&= 70.462 \text{ cubic inches}
\end{aligned}
$$

4 cuartillos = 1 ALMUDE
48 cuartillos, 12 almudes = 1 FANEGA

In liquid measure:

$$
\begin{aligned}
1 \text{ cuartillo} &= 0.504 \text{ litre} \\
&= 0.887 \text{ UK pint, } 1.065 \text{ US pint} \\
&\quad (1 \text{ UK wet pint} = 1.1274 \text{ wet cuartillo} \\
&\quad 1 \text{ US wet pint} = 0.9390 \text{ wet cuartillo})
\end{aligned}
$$

The term evidently derives from the Spanish for 'one-quarter' and presumably relates to the dry measure's proportion of an almude.

cuathys, cuathus *see* KYATHYS, CYATHYS

cube [maths] In geometry, a solid figure with six squares as its faces. Its area equals six times the square of the length (of any side), and its volume equals the length cubed.

In arithmetic, to cube a number is to raise it to the power 3.

For example: $4^3 = 4 \times 4 \times 4 = 64$.

cube root [maths] Of a given number, another number which, when multiplied by itself twice, equals the given number. For example: the cube root of 8 (written $\sqrt[3]{8}$) is 2 because $2 \times 2 \times 2 = 8$.

The power or exponent for cube root is $\frac{1}{3}$. For example: $8^{\frac{1}{3}} = 2$.

cubic [maths] In algebra, an expression (polynomial) of degree 3, generalized as:

$$ax^3 + bx^2 + cx + d$$

cubic centimetre, cubic centimeter *see* CENTIMETRE, CUBIC (CC)

cubic inch *see* INCH, CUBIC

cubic foot *see* FOOT, CUBIC

cubic kilometre, cubic kilometer *see* KILOMETRE, CUBIC

cubic metre, cubic meter *see* METRE, CUBIC

cubic mile *see* MILE, CUBIC

cubic yard *see* YARD, CUBIC

cubit [linear measure] Ancient measure, in tradition at least as old as ancient Egypt and Babylon, and almost everywhere deriving as the length from the point of an elbow to the tip of the middle finger on the same arm. The name of the unit itself suggests this, although it is as late as Latin: Latin *cubitus* 'elbow', but also 'cubit', and related to the Roman habit of reclining on one arm (*cubitans*).

In ancient Egypt:

$$
\begin{aligned}
1 \text{ cubit} &= 6 \text{ palms, } 24 \text{ digits} \\
&= 45.110 \text{ centimetres, } 17.76 \text{ inches} \\
1 \text{ royal cubit} &= 1\frac{1}{2} \text{ royal feet, } 7 \text{ palms, } 28 \text{ digits} \\
&= 52.416 \text{ centimetres, } 20.72 \text{ inches}
\end{aligned}
$$

In ancient Babylon:

$$
\begin{aligned}
1 \text{ cubit} &= 30 \text{ shusi} \\
&= \text{about } 52.8 \text{ centimetres, } 20.787 \text{ inches}
\end{aligned}
$$

In ancient Israel:

1 ordinary cubit (*ammah*)	=	2 spans, 6 palms, 24 digits
	=	44.96 centimetres, 17.7 inches
4 ordinary cubits	=	1 fathom
1 royal cubit	=	one and one-sixth ordinary cubit
	=	7 palms, 28 digits
	=	52.416 centimetres, 20.72 inches

In ancient Greece:

1 cubit	=	1½ 'feet', 2 spans, 6 palms, 24 digits
	=	46.20 centimetres, 18.24 inches
4 cubits	=	1 fathom
100 cubits	=	1 ride

In ancient Rome:

1 cubit	=	1½ 'feet', 18 'inches' (*unciae*), 24 digits
	=	44.40 centimetres, 17.52 inches

In Britain (and ex-British colonies) and North America, the cubit was never a specific measure. Used mainly as a unit of length in quantifying short distances over terrain, it was purely subjective – according to the individual measurer's own arm length – and could be anything between 18 and 22 inches (45.72 and 55.88 centimetres). *See also* DIGIT; ELL; SPAN.

cuboid [maths] A parallelepiped with rectangular faces – the shape of a typical brick – alternatively known as a rectangular prism.

Cuisenaire rods [maths] Set of ten short sticks representing the numerals 1 to 10, each one also represented as its number's unit length (so that the rods representing 1 and 9 when placed end-to-end are together the length of the rod representing 10, and so on). The rods were designed as a means to teach elementary addition and subtraction to children by the Belgian schoolmaster Georges Cuisenaire.

cum laude *see* SUMMA CUM LAUDE

cup [volumetric measure: cookery, cocktails] In cooking and in making some cocktails, a cup or cupful is not a specific quantity – but in general,

1 cup	=	about 2½ DECILITRES, 25 CENTILITRES,
		250 MILLILITRES
	=	about 0.44 UK pint, 0.53 US pint
	=	8.80 UK fluid ounces, 8.48 US fluid ounces

curie [physics] Unit of radioactivity equal to 3.7 disintegrations per second (similar to the activity of 1 gram of radium-226). It has been superseded by the BECQUEREL. The curie was named after the Polish-born French physicist Marie Curie (1867-1934).

$$1 \text{ curie} = 3.7037 \times 10^{10} \text{ becquerel}$$

Curie point, Curie temperature [physics] For a magnetic material, another name for CRITICAL TEMPERATURE.

Curie's law [physics] For a paramagnetic material (which is magnetic only in the presence of a magnetizing field), its magnetic susceptibility is inversely proportional to its absolute temperature. The constant of proportionality is called the Curie constant.

curling measurements and units [sport] Curling is a game played mainly in Scotland and Canada by teams of four called rinks. Each rink plays another rink under the direction of a skipper, in a consistent order decided by the skipper. Players deliver their two stones thus alternately always against the same opponent. One full turn (of two stones) for each player of each rink completes a 'head' or 'end' (of eight stones played).

The dimensions of the playing area:

width: (usually) 14 feet (4.27 metres)

distance from the footscore line to the hog score line (the farthest from where the stone must be delivered): 11 yards (10.06 metres)

distance from the hog score line to the centre of the house (the circular target) or tee (up to which the ice may be swept by player's team-mates): 31 yards (28.36 metres)

distance from the tee to the far end of the house (= radius of target circle, in

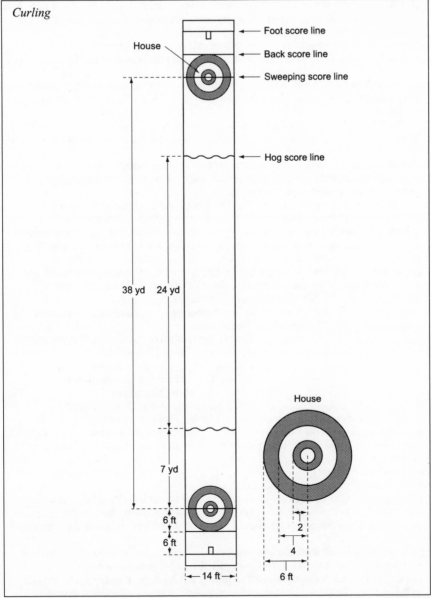

Curling

House

Foot score line

Back score line

Sweeping score line

Hog score line

38 yd   24 yd

House

7 yd

6 ft

6 ft

14 ft

2

4

6 ft

which ice may be swept by player's opponents): 2 yards (6 feet, 1.83 metres)
  (the ice is swept to increase the stone's gliding capacity)
Timing: as long as it takes to complete the number of heads or ends decided upon
  by the skippers
Points scoring:
  only one side scores in every end – when both stones of each player have been
    delivered – depending on final proximity to the centre of the house, the
    tee; as many as that side has inside the house and nearer the tee than any
    opponent's stone then count 1 each
Dimensions of the stone:
  maximum circumference: 36 inches (3 feet, 91.44 centimetres)
  maximum height: 4½ inches (11.45 centimetres)
  maximum weight: 44 pounds (3 stone 2 pounds, 19.96 kilograms)
The game is first recorded under its present name in the early 1600s – but why it

should have been given the name remains conjectural. It is probable that it derives from a Scottish dialectal variant of a Germanic root meaning 'to roll', referring to the gliding of the stones.

currencies of the world *see* COINS AND CURRENCIES OF THE WORLD

current [physics] The rate of flow of charge passing through a solid, liquid, or gas. In a solid, the charge carriers are electrons, and, by convention, the direction of an electric current is opposite to that of electron flow (that is, in a circuit, current flows from positive to negative; electrons flow the other way). Current is measured in AMPERES.

The rate of flow of a fluid, particularly a liquid, is also a current, measured in units of volume per unit time (for example: litres per minute, gallons per hour). *See also* CUSEC; FLOW.

current density [physics] The electric current flowing per unit area of conductor, measured in amperes per square metre (SI units) or amperes per square foot (FPS units, commonly used in electroplating).

1 ampere per square metre =   10.7636 amperes per square foot
1 ampere per square foot   =   0.09291 ampere per square metre

current efficiency [physics] The mass of a substance liberated during electrolysis divided by the mass predicted by Faraday's second law of electrolysis: *see* FARADAY'S LAWS OF ELECTROLYSIS.

curvature [maths] The rate of change in direction of a curve at a given point. The circle that touches the curve there (the circle of curvature) has its centre at the *centre of curvature*. *See also* RADIUS OF CURVATURE.

cusec [speed/flow] Unit rate of flow in engineering or in measuring currents in rivers and oceans. The pressure is unspecified.

1 cusec =   1 cubic foot per second, 60 cubic feet per minute
         =   1 cubic yard every 27 seconds
             (2.222 cubic yards per minute)
         =   28,316.85 cubic centimetres per second
         =   1.699 cubic metre per minute
             (1 cubic metre every 35.315 seconds)

The term derives as an abbreviation for 'cubic foot per second'.

cusp [maths] A point on a curve at which it crosses itself, or at which two curved branches of a curve meet. Examples of curves with cusps include cardioid, cissoid, and cycloid.

cwt *see* HUNDREDWEIGHT

cyathys, cyathus *see* KYATHYS, CYATHYS

cycle [physics] One of a (usually) recurrent series of similar changes, such as a vibration or wave motion. One cycle is the period of motion; the number of cycles per second is the *frequency* (measured in HERTZ, which has superseded cycles per second).

cycling disciplines and distances [sport] Cycle racing is either on roads or on (indoor or outdoor) tracks (sometimes called velodromes), or as a sort of cross-country event that includes hazards requiring a rider to pick up the cycle and run.

Road races comprise either *tours* (over several or many stages and generally a number of days), *circuit races* ('criteriums', often around city centres, which generally last for a set number of hours or fractions of hours), or *time trials* (in which competitors set off at 1-minute intervals over a course specifically racing against the clock). The larger tours include criterium stages and time-trial stages, and individual and team events in both. All road races are scored on time, including bonuses for being in the lead or among the leaders at specific points within each stage or course, or after a set number of circuits.

Track races are:

sprints: 1 to 4 laps of the track, only the last 200 metres (220 yards) measured
individual pursuit: a race between two riders over 3,000 metres (1.864 miles), 4,000 metres (2.485 miles) or 5,000 metres (3.107 miles), in which the riders start on opposite sides of the track; first one home wins, unless one rider overtakes the other first
team pursuit: like individual pursuit but in teams of four; first three riders of

each team home count for timing, unless entire team overtaken; *Australian pursuit* is a team pursuit event with teams of eight riders; *Italian pursuit* is a team event with teams of up to five riders, four of which drop out at successive intervals to leave only two finalists on the track

Madison racing is relay racing in teams of two, only one of each pair on the track at a time, the other rejoining to be pushed ahead by the temporarily retiring rider

Longer track events may be paced by a moped or motorcycle; in some events longer than 10 kilometres (6.214 miles), each rider is allowed a motorcycle pacer.

Cross-country cycling is *cyclo-cross*, in which riders race around a course over natural countryside (involving woodland, hillsides, fieldside paths, and a short distance on roads) negotiating hazards that may require the bike to be carried. Maximum recommended distance is 24 kilometres (14.913 miles), which may be made up of several circuits.

Even more strenuous is a *hill-climb*, which may be a cross-country climb of one hill (maximum recommended distance 5 kilometres/3.107 miles), or may be a combination of ups and downs in a carefully marked out quarry.

The popularity of mountain bikes from the late 1980s has also seen an increase in *cycle obstacle-course events*, in which riders have to negotiate slopes, pedal along narrow surfaces (such as logs or seesaw planks), ride through streams, lift the front wheel while swinging the bike round, and other feats, all without putting a foot on the ground.

cycloid [maths] A curve traced by a point on the circumference of a circle that rolls along a straight line. *See also* EPICYCLOID.

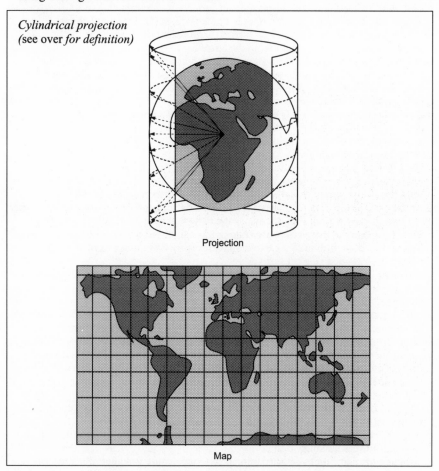

Cylindrical projection
(see over *for definition)*

Projection

Map

cylinder [maths] A symmetrical solid figure with a circular cross-section; it is equivalent to a prism with a circular end. If the ends are perpendicular to the central axis, it is a right cylinder.

The volume of a cylinder equals the product of π, the square of the radius of the base, and the height. The area of the curved surface is 2π times the base's radius, times the height.

cylindrical coordinates [maths] A set of three coordinates that represent the position of a point in space. Two of the coordinates are its distance from a fixed point (the origin) and its angle from a fixed line (exactly like POLAR COORDINATES), and the third coordinate is its height above a fixed plane.

cylindrical projection [physics: cartography] Type of map projection in which lines on the Earth – lines of longitude (meridians) and lines of latitude (parallels) – are projected on to a cylinder, usually with the axis of the cylinder parallel to the Earth's axis. On the resulting map, both meridians and parallels appear as straight lines. The simple cylindrical projection can be modified to produce an equal-area map or Mercator's projection, on which straight lines represent true bearings. *See previous page* for illustration.

D

D (Roman numeral) [quantitatives] As a numeral in ancient Rome, the symbol D corresponded to 500 – yet in this sense it did not derive from the fourth letter of the Latin alphabet, a variant on the Greek *delta* (which, to the Greeks, signified either 4 or 4,000). Instead, it derived from the old Tuscan numeral much like a Greek capital *phi*, or a capital I within rounded parentheses: (I). This symbol denoted '1,000' – so the Romans took half of it, the equivalent of I), for their '500', and from this form it was assimilated to the letter D.

Dachstein [time] Geological division or stage during the TRIASSIC PERIOD of the Mesozoic era, between 248 and 213 million years ago. It is named after the mountains in the Austrian Alps that were formed at this time.

dactyl, dactylic foot *see* FOOT [literary]

dalasi [comparative values] Unit of currency in the Gambia.

$$1 \text{ dalasi} = 100 \text{ bututs}$$

See also COINS AND CURRENCIES OF THE WORLD.

dalton *see* ATOMIC MASS UNIT/DALTON

Dalton's law [chemistry; physics] In a mixture of gases, the pressure exerted by any one of them is the same as if that gas alone occupied the total volume. It was named after the British scientist and atomic theorist John Dalton (1766-1844).

dan, Dan [sporting term] In martial arts, the grades of proficiency and expertise after the initial student grades are known as the dan (Japanese for 'leader') grades. In judo, for example, there are twelve dan grades, signified by the wearing of a belt of a specific colour, although the eleventh and twelfth dan grades have by tradition never been attained (*see* BELTS).

dandiprat [comparative values] Coin issued in early sixteenth-century England, then worth 1½ pence or one-quarter of a sixpence (there were twelve pence to the shilling, 240 to the pound).

The derivation of the term is unknown, but might well relate to a combination of the elements 'dandy' (something fine to look at) and 'prat' (but worthless).

Danian [time] Geological division or stage during the early PALAEOCENE EPOCH of the Cenozoic era, but sometimes held instead to be an extremely late stage of the Upper CRETACEOUS PERIOD just inside the Mesozoic era. Named after the country Denmark, the resultant rock formation is nonetheless geologically attested mostly in France and Belgium.

daraf [physics] Unit of elastance of a component in an electrical circuit, equal to the reciprocal of CAPACITANCE. The term is the word FARAD backwards.

darcy [geology] Unit of porosity (permeability coefficient) of a rock, providing an indicator of its resistance to the flow through it of a liquid or gas. It is defined as the volume of fluid of unit viscosity, subjected to a pressure gradient of 1 atmosphere per unit distance, that passes through unit area of porous material in unit time. In SI units, it is ¹⁄₁,₀₀₀ of the porosity that allows a flow of 1 cubic metre per second through a metre cube of material under a pressure of 1 newton per square metre.

The unit was named after the French physicist H. Darcy (1803-58). For everyday use, the most practical form of the unit is the millidarcy: 0.001 (one-thousandth) darcy.

Dark Ages [time] Period in Europe between the end of the ancient Roman civilization (around AD 450–500) and about AD 1000 or, according to some authorities, until the beginning of the European Renaissance (around 1400–1450), during which scientific discoveries were few and far between, and cultural advances equally scanty, probably as a result largely of religious intolerance.

darts measurements [sport] Competition darts is popular in Europe, but not well known outside that continent. The dartboard is the classic English type divided into 20 equal sectors (worth 1 to 20 points, plus doubles and triples areas in each sector) around two small concentric rings (worth 50 and 25 points).

sector subdivision widths:
doubles area: ⅜ inch (9 millimetres)
outer singles area: 2⅛ inches (54 millimetres)
triples area: ⅜ inch (9 millimetres)
inner singles area: 2⅞ inches (73 millimetres)
25-point area: ⅝ inch (15 millimetres)
50-point area (bull): ¼ inch (6 millimetres)

The name of the game is apparently taken from early Celtic roots borrowed both by Norman French and medieval Germanic languages.

dasymeter [physics] Instrument that measures and displays the density of a gas.

date [literary] There are several conventions that relate to setting out a date including day, month, and year.

In British usage, the preferred forms are:
26 April 1999   26th April 1999   26/4/99   26.04.99
In US usage, the preferred forms are:
April 26, 1999   April 26th, 1999   4/26/99   04.26.99
For the different calendars in which dates occur, *see* YEAR.

datum line, datum plane, datum point [physics] In surveying and engineering, the line, plane, or point that is the basis from which measurements of height or depth are taken.

day [time] In normal parlance,
1 day = 24 hours = 1,440 minutes = 86,400 seconds
and begins and ends at midnight, representing one complete revolution of the Earth on its axis, in relation to the Sun. Lunar days (measured in relation to the Moon) and sidereal days (measured in relation to the stars) are not used for scientific purposes although, in accumulation over a year, the differences may be substantial (*see* EQUATION OF TIME; YEAR).

Some cultures and religions reckon a day to begin and end at times other than midnight: in Israel, for example, the day traditionally begins at sunset.

The division of the day into twenty-four hours is of comparatively late cultural development, arising initially as a necessity for making trading appointments and deals during daylight. At first only daylight time was divided into hours, and the hours themselves varied in length depending on the time of year and the amount of daylight over which they then had to stretch.

That seven days makes one week probably derives from the use in the first millennium BC by several Mediterranean cultures (notably the ancient Israelites) of the mystic number seven in this context. Again, it is thought to result primarily from trading requirements.

db, dB *see* DECIBEL

dead heat *see* HEAT

deadweight tonnage [weight] The overall weight of a merchant ship including the weight of crew, cargo, supplies for the voyage, spare parts, and passengers, reckoned in (long) TONS. The weight of the ship by itself is the *gross tonnage*.

deal [sporting term] In card games, to distribute some or all of the cards proportionately to each player; the responsibility or action of doing so. Most card games rely on a dealer's distributing the cards one at a time to each player in turn. Some games, however, such as contract bridge and solo whist, require the dealer to deal the cards out three at a time to all but one player and four to the other player, the player receiving four being a different player successively each time round.

In the original Anglo-Saxon, the *deal* (a noun only) was what was *doled* out (verb).

death rate [medicine] The number of deaths in one year within a specified area or territory, as compared with the total population (or per thousand thereof), with previous years' figures, with the area's BIRTH RATE, or with any or all of these figures of another area.

deben [weight] Unit of weight in ancient Egypt. The modern equivalent is not known and, in any case, the unit seems to have varied over time.

$$1 \text{ deben} = 10 \text{ KITE}$$
$$10 \text{ deben} = 1 \text{ sep}$$

de Broglie wavelength [physics] One of the postulates of quantum physics is the dual particle and wave behaviour of subatomic particles. Any fast-moving particle can thus be assigned a wavelength, which equals Planck's constant (h) divided by the product of the particle's mass and velocity. It was named after the French physicist Louis de Broglie (1892-1987).

debye [physics; chemistry] Unit of electric DIPOLE MOMENT, usually applied to polar (ionized) molecules, equal to 3.34×10^{-30} coulomb metre (= 10^{-18} electrostatic unit). It was named after the Dutch physical chemist Peter Debye (1884-1966).

deca-, deka- [quantitatives: prefix] Prefix which, when it precedes a unit, multiplies the unit by 10 times its standard size or quantity.

Example: decastere – 10 steres

The term derives directly from ancient Greek *deka* 'ten', and should not be confused with the prefix DECI-, meaning 'one-tenth'.

decad [quantitatives] A group of ten. The adjective is *decadal* or *decadic* ('of a group of ten').

decade [time; quantitatives; physics] Ten years; the specific period of ten years between two years divisible by 10 – for example: the period between 1 January 1980 and 31 December 1989 – 'the eighties'. Also *decenniad* or *decennium*.

Also, a group of ten items, especially beads on a rosary (some of which are arranged in five 'decades') or on a decade ring, also used in co-ordinating prayers.

Also, in the context of physics, a ratio of 10:1, particularly as applied to frequencies (for example: the frequencies 20 kilohertz and 2 kilohertz are a decade apart).

decagon [maths: geometry] A POLYGON that has ten sides (and ten angles).

decagram [weight] Ten GRAMS. A small and not very useful weight.

$$1 \text{ decagram} = 0.3527396 \text{ ounce}$$
$$= 5.4268 \text{ drams avdp., } 148.39 \text{ grains}$$

decahedron [maths] A solid figure (POLYHEDRON) with ten plane faces.

decalescent temperature [physics: metallurgy] For a ferromagnetic material, such as iron or steel, decalescence is the absorption of heat that occurs as it is heated through a temperature at which a change takes place in its crystal structure (when there is often also a change in magnetic properties). That temperature is the decalescent temperature.

The term is based on the Latin elements *de-calescens* 'warming down'.

decalitre, decaliter [volumetric measure] Ten LITRES. As a unit, its most meaningful approximation in conversion is to the PECK (which differs between UK and US usages).

$$1 \text{ decalitre} = 10,000 \text{ cubic centimetres}$$
$$= 2.2 \text{ UK gallons, } 2.64179 \text{ US gallons}$$
$$= 1.0999 \text{ UK pecks, } 1.135 \text{ US pecks;}$$
$$(1 \text{ UK peck} = 0.9092 \text{ decalitre}$$
$$1 \text{ US peck} = 0.8811 \text{ decaliter})$$
$$= 0.27495 \text{ UK bushel, } 0.28378 \text{ US bushel}$$

decametre, decameter [linear measure] Ten METRES. Another fairly arbitrary measure as a unit.

decametric frequency [physics: telecommunications] Any short- to medium-wave radio frequency in the range 3–30 megahertz, corresponding to wavelengths of between 10 and 100 metres.

decare [square measure] Ten ARES, a unit not much used.

decathlon events, measures, and points [sport] The events of the decathlon are:

first day:	second day:
100-metre sprint	110-metre hurdles
long (broad) jump	discus
shot-put	pole-vault
high jump	javelin
400-metre sprint	1,500-metre run

Points scoring:

according to tables issued by the International Amateur Athletic Federation (IAAF), relating to ideal times, heights and distances, in thousands of points

The decathlon is an event for men only; the equivalent for women is the PENTATHLON.

decay constant *see* DISINTEGRATION CONSTANT

decay time [physics] For an exponentially decaying quantity (such as radioactivity), the time in which its value reduces to a fraction $1/e$ (= 0.3678, or about 37 per cent) of its original value.

deceleration [physics] Slowing down, or the rate of decrease in speed; it is equivalent to negative acceleration.

decenniad, decennium *see* DECADE

decennial [time] Of ten years; occurring every ten years; lasting for ten years.

deci- [quantitatives: prefix] Prefix which, when it precedes a unit, divides it by ten, reducing it to one-tenth (0.1) of its standard size or quantity.

Example: decistere – one-tenth of a STERE

The prefix derives from Latin *decimus* 'tenth'; the medieval French are responsible for the elision by which the -m- has disappeared from the prefix.

deciare [square measure] One-tenth ARE, a unit not much used.

decibel (dB) [physics] Unit of sound pressure level, equal to ten times the logarithm of the ratio of the square of the RMS value of the sound pressure to the square of a reference pressure (usually the threshold of hearing). By extension, the decibel is used for other power ratios (such as the input and output voltages of an amplifier).

1 decibel　=　0.1 (one-tenth) bel

See also BEL.

decigram [weight] One-tenth GRAM, an extremely small weight.

1 decigram	=	0.003527396 ounce
	=	1.5432 grains

decile [maths] In statistics, one of a series of lines or columns that divide frequency distribution data into ten equal parts.

decilitre, deciliter [volumetric measure] One-tenth LITRE: a unit of volume that is in fairly common household use in parts of Europe.

1 decilitre	=	0.1 litre, 100 cubic centimetres
	=	0.176 UK pint, 0.211 US pint
	=	3.520 UK fluid ounces, 3.376 US fluid ounces
		(5 UK fl. oz, ¼ UK pint = 1.420 decilitres
		4 US fl. oz, ¼ US pint = 1.185 deciliters)
	=	6.102374 cubic inches
2½ decilitres	=	about 1 cup(ful) in cooking

decilog [physics] Unit of (low) pressure equal to the negative logarithm (to the base 10) of ten times the pressure in millimetres of mercury. Decilog was once also suggested as an alternative term for *decibel*, but has not been adopted as such.

decimal, decimal point [maths] The number system that uses the base 10 – that is, the numerals 1 to 9 and 0 – is called the decimal system. Fractions in this system are represented by putting a decimal point after the units figure (equivalent to an ordinary fraction in which the denominator is a power of 10). Thus $2\frac{3}{10} = 2.3$, $15\frac{27}{100} = 15.27$ and $\frac{37}{1,000} = 0.037$. The decimal system greatly simplifies arithmetical operations on fractions. The decimal point may be written in line with the centre of the numerals (as in 2·3), but increasingly it is written (and printed) on the base line (as in 4.5).

In continental Europe, a comma is generally used in place of a written or printed point or stop (as in 6,7), and the word *comma* (or equivalent) is accordingly used when speaking of decimal fractions.

decimate [quantitatives] Although, in ordinary parlance, to 'decimate' is to reduce by a large quantity, and is thus virtually identical with 'devastate', the verb derives from the imperial Roman army practice of punishing mutinous legions by executing every tenth (Latin *decimus*) man, as selected by lot. This left 90 per cent of a legion intact but badly shocked and in a much more docile frame of mind.

decimetre, decimeter [linear measure] One-tenth METRE, a unit that is used relatively scarcely, in contrast to CENTIMETRES and MILLIMETRES.

$$1 \text{ decimetre} = 10 \text{ centimetres; } 100 \text{ millimetres}$$
$$= 3.937008 \text{ inches, } 0.3281 \text{ foot}$$
$$(1 \text{ foot} = 3.048 \text{ decimetres})$$

decimetre-kilogram-second (DKS) system [quantitatives] An obsolete system of units based on the decimetre (for length), kilogram (for mass), and second (for time). It was superseded by the metre-kilogram-second (MKS) system, which became the basis of SI units.

decimetric frequency [physics: telecommunications] Any very short-wave radio frequency in the range 300–3,000 megahertz, corresponding to wavelengths of between 0.1 and 1 metre. The shortest of these (less than 0.3 metre) come within the range normally called MICROWAVES.

decistere [cubic measure] One-tenth STERE, the equivalent in cubic, dry measure to 100 litres in volumetric liquid measure.

$$1 \text{ decistere} = \frac{1}{10} \text{ cubic metre, } 100,000 \text{ cubic centimetres}$$
$$= 3.5315 \text{ cubic feet, } 6,102.374 \text{ cubic inches}$$

deck [sporting term] In the United States a deck of cards is what is in Britain called a pack of cards. (In US street slang, a 'deck' may also be a three-grain package of heroin.)

The term seems to derive (possibly through Dutch) from the idea of the set of cards' being a collection of flat layers, as each deck of a ship lies upon another. (The ultimate derivation would then be an element akin to Latin *tegere* 'to cover', 'to roof', Dutch *dekken* 'to cover'.)

declination [astronomy; physics; geology] In astronomy, declination is one of the coordinates that define the position of a heavenly body (the other is its AZIMUTH). It is its ANGULAR DISTANCE from the celestial equator along the meridian that passes through it. Northern declinations are regarded as being positive; southern ones are negative.

In geology, declination is another term for magnetic declination (the angle by which magnetic north differs from true north).

declination circle [astronomy] A calibrated circular scale on the declination (horizontal) axis of an astronomical telescope for setting or reading declinations.

defective year [time] In the Jewish calendar a year of 353 days, representing the shortest form of the three common types of year: the other two types are the ABUNDANT YEAR and the REGULAR YEAR.

deficient number [maths] A whole, positive number of which the numbers by which it can be divided add up to less than itself.

Example: 8 (divisors $1 + 2 + 4 = 7$)

definite article *see* ARTICLE

degeneracy [physics] For an atom or subatomic particle, the property of two or more quantum states that have the same energy (stated as a degeneracy of 2, 3, or whatever).

degree [maths] In geometry, a unit of angle (arc) derived by dividing a complete revolution (circle) into 360 segments; symbol $°$. Degrees may be subdivided into minutes (symbol ') and seconds (symbol ") of arc.

$$1 \text{ degree} = 60 \text{ minutes} = 3,600 \text{ seconds}$$

In algebra, the degree of an expression or equation in a single variable is the highest power (exponent) to which the variable is raised. For example: $3x^4 + 4x^2 - 7x + 2$ is of degree 4. In a term with more than one variable, the degree is the sum of the powers of all the variables. For example: $3x^3y^4z^5$ is of degree 12.

degree [physics] In thermometry, a unit of difference in temperature on a temperature scale, represented by the symbol ° . Common units are degrees Celsius (or centigrade, both written °C) and degrees Fahrenheit (°F). Note that the SI unit of temperature, the kelvin, does not have a degree sign (0°C = 273 K). *See also* TEMPERATURE.

In hydrometry, there are various types of degrees used to express densities (specific gravities) of liquids, such as BAILLING, BAUMÉ, and BRIX degrees (*see also* next article). Likewise, there are various types of degrees, also named after their inventors, for expressing viscosities of fluids (such as ENGLER DEGREES).

degree [chemistry: industrial chemistry] API gravity, used for expressing the density of petroleum fractions and named after the American Petroleum Institute, is expressed in degrees. Degrees API equal the ratio of 141.5 to the specific gravity of the liquid at 60°F (15.56°C), minus 131.5. Special hydrometers may be calibrated directly in degrees API.

Hardness of water was formerly expressed in degrees which varied in definition from country to country. For example, an English (or Clark) degree equalled 1 part of calcium carbonate in 70,000 parts of water (= 1 grain per UK gallon); a French degree equalled 1 part of calcium carbonate per 100,000 parts water; and a German degree equalled 1 part of calcium oxide in 100,000 parts water (equivalent to 17.8 parts of calcium carbonate per million). Today, hardness of water is generally expressed in parts of calcium carbonate per million parts of water.

degree of freedom [physics; chemistry] A variable factor (such as concentration, pressure, or temperature) that has to be specified for a system's equilibrium condition to be defined.

Also, the rotational, translational, or vibrational energy of atoms in a molecule, resulting in every molecule's having $3n$ degrees of freedom (where n is the number of atoms in the molecule).

delta-particle [physics] A very short-lived particle (a hyperon) which decays almost instantaneously (because of the strong interaction).

delta ray [physics] A beam of electrons that is ejected by atoms struck by high-energy particles.

delta waves [medicine] Low-frequency electrical waves from the human brain (frequencies in the range 1 to 8 hertz); they can be recorded by electroencephalography.

demi- [quantitatives: prefix] A comparatively uncommon prefix denoting 'half'.
Example: demi-semiquaver – half a semiquaver
It derives through French from Latin *di(s)-midius* '(cut) in two at the middle'.

demijohn [volumetric measure] Large glass (or sometimes ceramic) bottle half encased in wickerwork or interlaced rushes. Of no specified capacity but always sizable, the bottle had a characteristic shape – bulging at the base, narrow at the neck, and with one or two rounded, arm-like handles – which probably gave it its name: in medieval French originally *Dame Jeanne* 'Lady Jane'.

demi-octavo, demy octavo *see* DEMY OCTAVO
demi-quarto, demy quarto *see* DEMY QUARTO
demi-semiquaver [music] In musical notation, a demi-semiquaver – known in the United States as a thirty-second note – is represented by the symbol ♫ and corresponds to
½ SEMIQUAVER or sixteenth-note
¼ quaver or eighth-note
⅛ crotchet or quarter-note
See also QUAVER.

demography [medicine] Science of the statistics that relate to human populations and their distribution.

demy [paper: printing] Size of paper sheet, different in UK and US usages.
Writing paper:
UK demy – 15½ x 20 inches (393.7 x 508 millimetres)
US demy – 16 x 21 inches (406.4 x 533.4 millimetres)
Printing sheet for use in books:
demy – 17½ x 22½ inches (444.5 x 571.5 millimetres)

demy octavo [paper sizes] A size of paper, and books, equal to 8½ x 5½ inches (216 x 140 millimetres).

demy quarto [paper sizes] A size of paper, and books, equal to 11 x 8⅝ inches (279 x 219 millimetres).

denarius [comparative values] Two different coins at different times in ancient Rome, one silver and the other gold. As a silver coin it was first issued in 269 BC and rapidly became the standard unit of currency for the Roman Empire. Its value was not static, however: it was originally equal to 10 *asses* but, in times of national financial stress, occasionally devalued to 16 asses. Even at such times, by tradition, the imperial army always received its pay at the scale of 10 asses to the denarius. Indeed, it was as this tenfold unity that it received its name, *denarius* 'group of ten'.

The name of the coin has historically been borrowed by many succeeding civilizations and cultures, especially Arabic states (*see* DINAR), but one of the few European nations to retain any vestige of it was Britain which, until 1971 and the introduction of decimal coinage, customarily abbreviated 'pence' as 'd.' for 'denarii'.

denary notation [maths] In computing (mainly), the ordinary numbers to the base 10 (0, 1, 2, 3, 4, 5, and so on). *See also* DECIMAL.

dendrochronology [time] The scientific study of time past as recorded in tree rings, each of which corresponds to information on a tree's growth and environmental conditions during one solar year. By comparison of the rings from many trees, some of them preserved by freak of nature, it has been possible to build up a profile of geophysical, climatic, and atmospheric conditions in certain localities year by year backwards for at least 13,000 years.

denier [textiles] A measure of the fineness of silk, nylon, or other yarn, based on the weight of 9,000 metres of the yarn in grams.

$$
\begin{aligned}
15 \text{ denier} \quad &= \quad 15 \text{ grams per } 9{,}000 \text{ metres} \\
&= \quad 0.02688 \text{ ounce per } 1{,}500 \text{ feet} \\
&\qquad (1 \text{ ounce per } 10.5689 \text{ miles}) \\
30 \text{ denier} \quad &= \quad 30 \text{ grams per } 9{,}000 \text{ metres} \\
&= \quad 0.05377 \text{ ounce per } 1{,}500 \text{ feet} \\
&\qquad (1 \text{ ounce per } 5.2844 \text{ miles})
\end{aligned}
$$

See also DREX; TEX.

denominator [maths] In a fraction, the number beneath the line – denoting the number of parts into which unity has been divided. *See also* LOWEST COMMON DENOMINATOR.

density [physics] Density of a material is its mass per unit volume, and so is expressed in such units as grams per cubic centimetre (CGS units), kilograms per cubic metre (SI units), and pounds per cubic foot. For gases, which have very low densities, the usual units are grams per cubic decimetre (equivalent to grams per litre) at standard temperature and pressure (STP, = 0°C and 101,325 newtons per square metre). Fluid densities may be measured using a densimeter which, at its simplest, takes the form of a density bottle (an accurately calibrated vessel for weighing a known volume of liquid).

Relative density, which is exactly the same as SPECIFIC GRAVITY, is the mass of a given volume of a material divided by the mass of an equal volume of water at its maximum density (that is, at 4°C). Because it is a ratio of two masses, it has no units.

Density is also used to describe the number of items in a given area (for example, population density – the number of people per square kilometre or per square mile – and charge density). *See also* OPTICAL DENSITY.

dental consonants [literary: phonetics] Consonants that are produced mostly or entirely by placing the tip of the tongue against or adjacent to the backs of the teeth on the upper jaw (Latin *dens, dent-* 'tooth'). In English these correspond to:

d,   f,   n,   t,   th,   v

dental formula [medicine] Notation used in dentistry to record the number and kind of teeth in a person's (or indeed other mammal's) mouth. In the order incisors (front teeth), canines ('eye' teeth), premolars (or carnassials for a carnivore), and molars, the formula gives the number of teeth in one side of the jaw, with the number in the upper jaw over the number in the lower jaw written

like a fraction. For example: an adult human being who has all of his or her teeth
has a total of 4 incisors, 2 canines, 4 premolars, and 6 molars in each jaw. The
human dental formula is therefore ⅔ ⅟₁ ⅔ ⅜.

deoxyriboneucleic acid (DNA) *see* GENE

depression [astronomy; surveying; meteorology] In astronomy, the ANGULAR
DISTANCE of a celestial body below the visible horizon.

In surveying, the equivalent: the angular distance of an object below the horizontal plane at the point of observation.

In meteorology and weather forecasting, a region of low atmospheric pressure
(cyclone): *see* WEATHER MAP.

depression of freezing point [chemistry; physics] When a substance (solute) is
dissolved in a liquid (solvent), the freezing point of the solvent is lowered. For
dilute solutions in a non-volatile solvent, and at constant pressure, the depression of
freezing point depends directly on the concentration of the dissolved substance (and
thus can be used as a method of determining concentration).

depth [linear measure] Depth is a distance, measured downwards, and therefore
expressed in the appropriate unit of length, from millimetres for the depth of a small
dent to fathoms, kilometres, or miles for the depth of the ocean. A wide variety of
instruments are used for measuring depth, such as a depth gauge in woodworking
and metalworking, a simple lead weight on a length of line for taking soundings in
shallow water, and an echo sounder for measuring depths in deep water.

depth of field [physics: photography] The range of distances behind and in front of
an object focused by an optical instrument (such as a camera or microscope) within
which any other objects are acceptably in focus. With a camera, it depends on the
focal length of the lens and its aperture (F-NUMBER), smaller apertures (large
f-numbers) giving the greatest depth of field. *See also* DEPTH OF FOCUS.

depth of focus [photography] In a camera, the range of distances from the back of
the lens to the surface of the film within which the image is acceptably sharp. *See
also* DEPTH OF FIELD.

dessertspoonful [volumetric measure] Imprecise measure in cooking and cocktail
mixing, sometimes held to be equivalent to around 2½ fluid drams (around 9
millilitres). In more general terms:

1 dessertspoonful	=	2 teaspoonfuls
2 dessertspoonfuls	=	1 tablespoonful

dessiatina [square measure] In Russia, an old measure approximating to a HECTARE.

1 dessiatina	=	2,400 square SADZHEN'
	=	1.09254 hectare, 109.254 ares
		(1 hectare = 0.913 dessiatina)
	=	10,925.4 square metres
	=	2.6996 acres, 13,066.1 square yards
		(1 acre = 0.3704 dessiatina)
	=	26.996 square chains

The term appears to mean 'tenfold' in Russian.

desyatina *see* DESSIATINA

determinant [maths] A quantity found, following certain rules, by adding the
products of a square MATRIX.

deuce [sporting term] In cards and dice, the 'two'. In the game of canasta, the deuces
are wild cards.

In tennis, the score 40-40, following which the game is won by the player who
obtains two further clear points in a row.

The term derives from Old French *deus*, a variant of the earlier Latin *duos* 'two'.

deunam *see* DONUM, DEUNAM

Deutsche mark, Deutschemark [comparative values] Unit of currency in
Germany.

> 1 Deutsche mark   =   100 pfennigs

The 'mark' is a very ancient term for a coin that has been stamped out (Latin
marcatus) for use at the market by merchants in commerce. 'Deutsche' is
(etymologically) the same word as 'Dutch', and reflects the early name for the
Germans, the Teutons.

devaluation [comparative values] Official reduction by a nation's fiscal authorities in the rate at which the national currency is valued against foreign currencies. The result in overall terms is that the national currency is worth less than it was: for the nation's own people it costs more to convert the national currency into foreign currency; and, at the same time, a foreigner receives more than previously in the national currency for his or her foreign currency (but can only do the same or less with it). In the short term, imports may be hit but exports may increase as a result of the changed rate.

deviance [maths] In statistics, the extent to which a (simple) statistical model fits a more (complex and) complete one.

deviation [maths] In statistics, the difference between an observed value and a fixed (typically the mean) value.

Devonian period [time] A geological PERIOD during the Palaeozoic or Primary era, after the Silurian period but before the Carboniferous period, corresponding roughly to between 408 million years ago and 360 million years ago. The period is sometimes known as the Age of Fishes, because salt or brackish water covered much of the land surface, and fishes of all kinds were the dominant life form; corals and seaweeds also flourished. Plants on the rare outcrops of swampy marshland hardly ever grew to more than a couple of feet (60 cm) high, but eventually evolved into tree ferns which were common by the end of the period. By that time, too, some fishes had begun to develop lungs and an amphibian mode of existence was becoming feasible.
For table of the Palaeozoic era *see* CAMBRIAN PERIOD.

Dewey decimal system [maths; literary] A system of classifying books in a general library by field and by subject. It was devised in 1872 by Melvil Dewey (1851-1931) specifically for use in the library of Amherst College, Massachusetts, and was published in 1876. The original plan – revised slightly thereafter by Dewey and others – divided knowledge into 10 fields, subdivided decimally again and again to provide subject areas. From it was developed the Universal Decimal Classification (UDC) system (first published in France in 1899) now used internationally, which is capable of taking into account not only the rapid expansion of whole branches of knowledge but can also be used to classify highly specialized material as found in academic reports and articles in periodicals.

dew point, cloud point [physics: meteorology] For air in contact with the ground, the temperature at which it becomes saturated with moisture, and water droplets condense out of the air and are deposited as dew. Above the ground, the water droplets form mist or fog; high in the air they form clouds – hence the alternative name, cloud point.

dextrorotatory [chemistry; physics] Describing an optically active substance that rotates the plane of polarized light clockwise – to the right – as viewed from against the incoming light.

di- [quantitatives: prefix] Prefix meaning 'double', 'twofold', 'twice over'.
Example: dihedral – having two plane faces (*but see* DIHEDRAL)
The term derives from ancient Greek *dis* 'twice'.

diagonal [maths] A line that joins any two non-adjacent corners of a polygon. In the special cases of a square, rectangle, and rhombus, either of the two diagonals bisects the figure.
In general, a polygon with n sides has $\frac{1}{2}n(n-3)$ diagonals.

diamagnetic [physics] Describing a material that becomes weakly magnetized in a direction opposite to that of an applied magnetizing field – a property of all substances. Frequently, diamagnetism is masked by stronger effects such as ferromagnetism and paramagnetism.

diameter [maths] A straight line that passes through the centre of a circle (or other conic section) and bisects it. It is twice the radius in length, and is the longest chord of a circle.

diametral pitch [physics; engineering] In a gearing system, the number of teeth per inch diameter of a gear cog.

diamond anniversary, diamond jubilee *see* ANNIVERSARIES

diamonds *see* SUITS OF CARDS

diastole *see* BLOOD PRESSURE

diatomic [chemistry] Describing a compound that has two identical atoms in its
molecules: for example, hydrogen (H_2), chlorine (Cl_2).

diatonic scale [musical] The ordinary major or minor scale of a key, with no
additional accidentals (sharps, flats, or naturals). On a piano, for example, the
diatonic scale of C is represented by the white notes between one C and another.

dice, die [sport] Small cubic object(s) thrown in gambling games, on each face of
which is marked – usually in dots – a number between 1 and 6 (inclusive) in such a
way that numbers on opposite sides of the cube add up to 7. When thrown or rolled,
the number that is uppermost when the cube comes to rest is the 'result' for that
dice (or, for the more traditional, die).

Many games are played with two or more dice. In special games, dice may have
markings other than numeric dots on them: poker dice, for example, have symbols
corresponding to the playing cards 9, 10, knave (jack), queen, king, and ace, with
which players attempt to make the combinations that score in the game of poker.

Dice games are probably the oldest form of gambling, known in ancient Egypt
well before 2000 BC, and worldwide in one form or another – occasionally with
only two scoring faces, sometimes with four scoring faces – since then.

The word *dice*, as a plural of *die*, corresponds precisely to an English variant of
an Old French corruption of the Latin *data* – which, even in English, can appropriately mean 'figures', 'information presented', 'things given'.

Didot point system [linear measure: typography] The usual system of type
measure in continental Europe. It is based on the 12-point CICERO, which takes the
role of the British pica, but is slightly larger.

$$\begin{aligned}
\text{1 Didot point} \quad &= \quad 0.376 \text{ millimetre, } 0.0148 \text{ inch} \\
&= \quad 1.0712 \text{ UK/US points} \\
&\quad\;\; (1 \text{ UK/US point} = 0.351 \text{ millimetre} \\
&\qquad\qquad\qquad\qquad\; = 0.9335 \text{ Didot point}) \\
\text{12 Didot points} \quad &= \quad 1 \text{ cicero} \\
&= \quad 12.8 \text{ UK/US points} \\
&\quad\;\; (12 \text{ UK/US points} = 1 \text{ pica})
\end{aligned}$$

See also CICERO; PICA; POINT.

dielectric strength [physics] The property of a non-conductor (insulator) that
enables it to withstand electric stress. The breakdown stress (that is, the maximum
dielectric strength) is usually expressed in kilovolts per millimetre of thickness.

differentiation, differential coefficient [maths] In calculus, a method used to
find the *derivative* of a function – the derivative (also called *differential coefficient*
and *derived function*) is the rate of change of the function. If the function is $y = f(x)$,
the differential coefficient $f' = dy/dx$.

For a function representing a curve (in coordinate geometry), differentiation gives
the slope of the curve at a given point.

diffraction [physics] Diffraction is the bending of a beam of light (or sound, electrons, X-rays, or other electromagnetic radiation) at the edge of an object or
aperture. With light from a point source, it results in fringes (alternate light and dark
bands) at the edge of a shadow; with sound, it allows sound to travel around
corners. The diffraction of X-rays or electrons by the atoms in crystals, in a
diffractometer, provides information about crystal structure. The diffraction of light
by a diffraction grating (a glass plate engraved with very closely spaced lines)
produces one or more spectra, an important technique in spectroscopy.

diffusivity [physics] A measure of how fast heat diffuses through a material, equal to
the ratio of the material's thermal conductivity to the product of its specific heat and
density (expressed in square metres per second).

digit [linear measure] In the ancient world (notably ancient Egypt, ancient Palestine,
ancient Greece, and ancient Rome), a short unit of length, remarkably consistent in
value over centuries and over thousands of miles, although representing a factor of
different units.

In ancient Egypt:

$$\begin{aligned}
\text{1 digit (or } zebo) \quad &= \quad 0.74 \text{ inch, } 18.72 \text{ millimetres} \\
\text{4 digits} \quad &= \quad 1 \text{ palm (or } shep) \\
\text{28 digits} \quad &= \quad 1 \text{ royal CUBIT}
\end{aligned}$$

In ancient Israel:

1 digit or finger (or *azba*)	=	0.74 inch, 18.72 millimetres
4 digits	=	1 palm (or *tefah*)
12 digits, 3 palms	=	1 SPAN (or *zeret*)
24 digits, 2 spans	=	1 ordinary cubit
28 digits	=	1 royal cubit

In ancient Greece:

1 digit or finger	=	0.76 inch, 19.25 millimetres
4 digits	=	1 palm
12 digits, 3 palms	=	1 span
6 digits, 4 palms	=	1 'FOOT'
24 digits, 6 palms, 2 spans	=	1 cubit

In ancient Rome:

1 digit (*digitus*)	=	0.73 inch, 18.50 millimetres
16 digits	=	12 'inches' (*uncia*), 1 'foot' (*pes*)
24 digits, 18 inches, 1½ 'feet'	=	1 cubit

The Latin word *digitus* 'finger' has the root meaning 'the pointer', 'the shaper', 'the indicator of correctness', as found in many Latin and English words that include the elements dig-, dic-, -dex. But the element's etymon *dh + k is akin to, if not identical with, that of the Latin *fingere* 'to shape', 'to form', from which English derives the word *figure* – which is probably the same word (though through Germanic sources) as *finger* anyway. But that connection shows why the word *digit* in English has come also to mean 'a numeral', 'a symbol' (especially on video screens).

digit [maths: computing] A single numeral, any integer (whole number) under 10. In computers that utilize the BINARY SYSTEM (as nearly all do) a digit is 1 or 0, corresponding to a BIT. *See also* NUMBER.

dihedral [maths; physics: aeronautics] In maths, the angle between two planes (measured in the plane at right-angles to the line of intersection).

For a fixed-wing aircraft, the (upward) angle between the wing (or tailplane) and the horizontal. If the wing slopes downwards, the angle is termed negative dihedral, or anhedral.

diheptal [physics: engineering] Describing something with fourteen parts, specifically the pins on a valve base or valve (electron tube).

dilatometer [physics] Some substances can exist in more than one crystal form, and the temperature at which they change from one to another is called the transition point (at which both forms can co-exist). A dilatometer is an instrument for measuring the transition point of a solid, by detecting the accompanying marked change in its volume.

dime [comparative values] Unit of currency in the United States and Canada.

1 dime	=	10 cents
10 dimes	=	1 dollar

The term derives via Norman French *disme* ultimately from Latin *decima* (*pars*) 'one-tenth (part)'. *See also* COINS AND CURRENCIES OF THE WORLD.

dimensions, dimensional analysis [quantitatives; physics] Each unit system is based on what are called fundamental units – in the internationally used SI SYSTEM they are the metre (for length), kilogram (for mass), and second (for time). These three basic dimensions (length, mass, and time) are denoted by L, M, and T. Any derived unit can be expressed in terms of two or more of the basic ones and dimensions ascribed to it. For example, velocity (speed) is expressed as a distance divided by a time, such as kilometres per hour or metres per second, and therefore has dimensions LT^{-1}. Force, which can be expressed as the product of mass and acceleration, has dimensions MLT^{-2}.

Noting the dimensions of the quantities in an equation in physics (*dimensional analysis*) is an important way of checking the validity of the relationship, because the combined dimensions of the terms on the left of the equation must be the same as those on the right of the equation.

For example: one of the equations of motion is
$$s = vt + \tfrac{1}{2}at^2$$
where s is the distance travelled in time t by an object with initial velocity v accelerating with an acceleration a. Writing this equation in the form of dimensions gives the following (numbers have no dimensions):
$$[L] = [LT^{-1}][T] + [LT^{-2}][T^2]$$
$$= [L] + [L]$$
$$= [L]$$
– that is, both sides of the equation simplify to the same dimension, [L], and the equation is therefore dimensionally consistent.

dimer [chemistry] A molecule that is formed by the joining together of two other similar molecules. *See also* POLYMER; TRIMER.

diminuendo [music] Musical instruction: decrease volume. Italian: 'diminishing'. The opposite is CRESCENDO.

dinar [comparative values] Unit of currency in Algeria and Tunisia, Kuwait, Iraq and Jordan, the Yemen, and the former Yugoslavia.

1 Algerian dinar	=	100 centimes
1 Tunisian dinar	=	1,000 millimes
1 Kuwaiti, Iraqi, Jordanian or Yemeni dinar	=	1,000 fils
1 former Yugoslavian dinar	=	100 paras

The term was, in past centuries, used in several Arabic countries as the name of a gold coin. It derives from the Latin *denarius*. *See also* COINS AND CURRENCIES OF THE WORLD.

DIN system of emulsion speed indicators [chemistry: photography] A logarithmic system of film speed ratings, often specified in degrees, named after the German standards organization, Deutsche Institut für Normung (DIN). *See also* ASA SYSTEM.

dioptre, diopter [physics] Unit of lens power, equal to the reciprocal of a lens's focal length in metres. Thus, a positive (converging) lens with a focal length of 50 centimetres (0.5 metre) is +2 dioptres. Negative (diverging) lenses are ascribed negative dioptre values. The system is most commonly used in specifying the power ('strength') of lenses in spectacles and contact lenses.

The name of the unit derives (via Latin and French) from ancient Greek elements meaning 'seeing through'.

dip [physics: geology] At any point on the Earth's surface, the vertical angle between the horizontal and the Earth's magnetic field. Measured using a dip circle or dip needle (also called an *inclinometer*), it is sometimes known alternatively as the inclination or magnetic dip.

Also, the angle at which a planar feature, such as a bed of rocks, is measured from the horizontal.

diphthong [literary: phonetics] Technically, in phonetics, a double vowel sound which, when written, phonetically combines two vowels but which may or may not be represented by two vowels in the conventionally written word.

the ou in 'mouse' [maus] is a diphthong
the ay in 'way' [wei] is a diphthong
the i in 'find' [faind] is a diphthong
the o in 'rogue' [roug] is a diphthong

The term derives from ancient Greek *di-phthongos* 'two-syllabled'.

diplo- [quantitatives: prefix] Double; two combined. From ancient Greek *diploos* 'double'. The corresponding prefix for 'single' is HAPLO-.

dipole [physics; chemistry] An electric dipole consists of two equal and opposite electric charges a short distance apart (as in some diatomic molecules).

A magnetic dipole consists of two separated magnetic poles (as in a bar magnet or a coil carrying an electric current).

dipole moment [chemistry; physics] For two opposite electric charges (or magnetic poles), the product of the charges (or pole strengths) and the distance between them: *see* DIPOLE.

dipstick, dip stick [physics: engineering] A rod, often calibrated in volumetric units, inserted vertically into a tank or sump of liquid (such as fuel or oil) to measure its depth and hence its volume.

diraa *see* PIK

Dirac's constant [physics] Planck's constant (6.626 x 10^{-34} joule-seconds) divided by 2π. It is the unit in which the spin of subatomic particles is expressed. It was named after the British physicist Paul Dirac (1902-).

dirham [comparative values] Unit of currency in Dubai, Iraq, Libya, Morocco, and Qatar.

In Dubai and Qatar:

$$100 \text{ dirhams} = 1 \text{ RIYAL}$$

In Iraq:

$$20 \text{ dirhams} = 1{,}000 \text{ fils} = 1 \text{ DINAR}$$

In Libya:

$$1{,}000 \text{ dirhams} = 1 \text{ dinar}$$

In Morocco:

$$1 \text{ dirham} = 100 \text{ FRANCS}$$

The term derives from the Latin *drachma*, a small coin, the name of which was in turn borrowed from ancient Greek, in which it was the name both of a small coin and of a small weight. *See also* COINS AND CURRENCIES OF THE WORLD.

dirhem [weight] In Egypt, a small unit of weight.

$$1 \text{ dirhem} = 3.12 \text{ grams, } 0.110 \text{ ounce}$$
$$(5 \text{ grams} = 1.6026 \text{ dirhems}$$
$$1 \text{ ounce} = 9.0909 \text{ dirhems})$$

The term derives from the ancient Greek *drachme*, the name both of a small weight and of a small coin.

discus throwing *see* ATHLETICS FIELD EVENTS AREA, HEIGHT, AND DISTANCE MEASUREMENTS

disintegration constant [chemistry; physics] For an unstable atomic nucleus, the probability of radioactive decay per unit time, which is a measure of the exponential decay (of radioactivity for a particular isotope) over a period of time. It is also known as *decay constant* and *transformation constant*. *See also* HALF-LIFE.

dispersive power [physics] When an electromagnetic radiation, such as light, passes through a transparent medium, it is diffracted or refracted (*see* DIFFRACTION; REFRACTION). Different wavelengths are affected to different extents, and dispersion is the splitting of an electromagnetic radiation (as it passes through a medium) into its component wavelengths as a result of this effect. The wavelengths emerging from the medium generally take the form of a spectrum. For light, the dispersive power of a medium is the difference in refractive indices for red and violet light divided by the mean refractive index minus 1. It has no units.

displacement [maths; engineering] In mathematics, the position of a point in relation to another, stated in terms of the distance between them and the direction of the first point from the second (it is thus a vector quantity).

In shipping, the weight of water displaced by a floating vessel (equal to its total weight: vessel and contents).

displacement ton [volumetric measure] Unit measure of stowage capacity aboard a ship.

$$1 \text{ displacement ton} = 7 \text{ BARRELS BULK}$$
$$= 35 \text{ cubic feet, } 1.2963 \text{ cubic yards}$$
$$(1 \text{ cubic yard} = 0.7714 \text{ displacement ton})$$
$$= 0.9905 \text{ cubic metre}$$
$$(1 \text{ cubic metre} = 1.0096 \text{ displacement tons})$$

The unit is based on an approximation of the volume of salt water displaced by a weight of 2,240 pounds, 1 (long) TON.

The *displacement tonnage* of a ship is its gross weight expressed as the number of (long) tons of water displaced by it. *See also* GROSS TON.

distance ratio, velocity ratio [physics; engineering] For a simple machine, such as a lever or a pulley, the distance moved by the load divided by the distance moved by the effort. *See also* FORCE RATIO, MECHANICAL ADVANTAGE.

distribution *see* FREQUENCY DISTRIBUTION

distributive [maths] Describing a mathematical operation in which the sum of the results of applying it individually to several terms is the same as applying it just once to the sum of the terms. For example:

$$3 \times (4 + 5) \quad \text{equals} \quad (3 \times 4) + (3 \times 5)$$
$$27 \quad = \quad 27$$

so that multiplication distributes over addition. The opposite is not true, however:

$$3 + (4 \times 5) \quad \text{does not equal} \quad (3 + 4) \times (3 + 5)$$
$$23 \quad \neq \quad 56$$

– that is, addition does not distribute over multiplication.

divalence, divalency [chemistry] A valence (valency) of 2, possessed by an atom that is able to combine with two atoms of hydrogen (or their equivalent); formerly called bivalence (bivalency). For example: oxygen in water (H_2O) is divalent.

diversity factor [physics: electrical engineering] In an electricity supply system, the sum of the individual maximum demands of several consumers divided by maximum demand of all the consumers.

dividend [maths; comparative values] A number or quantity that is divided mathematically by another (which is the *divisor*; the result of the division is the *quotient*).

By extension, a sum of money representing profits or a proportion of received premiums that is divided mathematically between owners, stockholders, shareholders, and/or members of a company.

division [maths] Mathematical operation in which a *dividend* is divided by a *divisor* to produce a *quotient*, indicated by the division sign ÷ . In the sum $27 \div 9 = 3$, 27 is the dividend, 9 is the divisor and 3 is the quotient. Division – the inverse of multiplication – has the effect of sharing a quantity into an equal number of parts. If an exact number does not result from the division, any part that is left over is called the remainder. *See also* FRACTION.

Division [biology] In the taxonomic classification of life forms, the category of plant life between Kingdom and Class, equivalent in the animal world to the category Phylum.

See full list of taxonomic categories under CLASS, CLASS.

divisor [maths] A number or quantity by which another (the *dividend*) is mathematically divided. The result of the division is the *quotient*.

dix [sporting term] In some card games, the lowest trump. The term derives from the French game of pinochle, in which the lowest trump scores ten points: French *dix* 'ten'.

djerib *see* DONUM, DEUNAM

DKS system *see* DECIMETRE-KILOGRAM-SECOND (DKS) SYSTEM

dl *see* DECILITRE, DECILITER

D layer, D region [physics] In the ionosphere about 80 kilometres (50 miles) above the surface of the Earth, the lowest layer of ionized particles. It is the D layer that absorbs the energy of short-wave radio waves reflected by other layers.

DNA *see* GENE

do, doh *see* SCALE

dobra [comparative values] Unit of currency in São Tomé and Principe. Its name is taken from that of several former Portuguese coins: Portuguese *dobra* 'double'. *See also* COINS AND CURRENCIES OF THE WORLD.

dodeca- [quantitatives: prefix] Prefix signifying 'twelve'. For examples, *see below*. *See also* DUODEC-.

dodecagon [maths] A POLYGON that has twelve sides (and twelve angles), known also as a *duodecagon*.

dodecahedron [maths] A solid figure (POLYHEDRON) with twelve plane faces.

doh *see* SCALE

dol [medicine] Unit of pain and pain intensity, measured on a scale of 1 up to 10 on an instrument called a dolorimeter, and based on the application of heat produced by a lamp on the skin.

The name of the unit derives from Latin *dolor* '(mental or physical) pain'.

doldrums [physics; geography] Area of calm in oceans astride the equator. At the approach of the summer solstice (mid-June in the Northern Hemisphere), the area moves about 5° northwards; at the winter solstice (mid-December) it moves about 5° southwards.

The frustration and misery caused by the preternatural calm to mariners in solely wind-powered sailing vessels of former centuries gave rise to the English expression '(down) in the doldrums'. But the word itself is a combination of elements, the first of which corresponds to English *dull*, and the second to a suffixial ending that apparently represents an analogue of the -trum in *tantrum*.

dollar [comparative values] Unit of currency in the United States and Canada, Australia and New Zealand, Hong Kong and Taiwan, Ethiopia and Liberia, Guyana and Belize, Mexico, Malaysia, and many of the islands of the Pacific Ocean and the Caribbean Sea.

$$1 \text{ dollar} = 100 \text{ cents}$$

The term derives from a coin made of silver mined close to the town of Sankt Joachimsthal (now Jachymov, a short distance north of Prague, in the Czech Republic) and issued from 1519. The coins soon became known as 'Joachimsthalers', and this mouthful was then abbreviated just to 'thalers' or dollars.

Fifty or so years later, this term was applied mostly to the Spanish PESO ('piece of eight', worth eight *reales*), a coin that came to be much used in the British colonies in North America. Its name was then borrowed for official use during the 1780s, at which time the dollar sign or dollar mark ($) seems to have been devised. *See also* CENT; COINS AND CURRENCIES OF THE WORLD.

dollar [physics] In nuclear physics, a unit of reactivity equal to that contributed by delayed neutrons.

$$1 \text{ dollar} = 100 \text{ cents}$$

dollar gap [comparative values] A trade balance deficit between a country of which the currency is linked to the US dollar and another country of which the currency is not linked to the US dollar, resulting in the need to settle up by a dollar exchange or in bullion.

dollars per week/per month *see* REPAYMENT RATES

dolorimeter *see* DOL

domain [maths] In an equation involving the variables x and y, the set of all numbers that can be assigned as values of x, also known as a *replacement set* (and so called because the members of such a set may take the place of the variable in a given relation).

The term derives through French ultimately from Latin *dominium* 'property of a lord'.

dominoes [sport] There are no set dimensions for dominoes, although the classic shape is approximately 2 x 1 x ⅙ inches (50.8 x 25.4 x 4.23 millimetres). The object of the game is to use all the dominoes in one's hand, placing them face-up one at a time in one's turn so that a number appears adjacent to an identical number and, if possible, so that each time the total sum of the numbers at the end of the string of dominoes on the table scores.

Various methods of scoring are used around the world, mostly reliant on making multiples of 5 (so that a total of 15 at the ends of the string of dominoes scores 3) or 3 (so that a total of 15 at the ends of the string of dominoes scores 5), or both.

They are called dominoes after the black domino mask used in masqued balls in the eighteenth to nineteenth centuries (presumably in reference to the whites of the eyes appearing on a black background much like the classic white dots on the dominoes' black background). But the name of the mask itself is an extension of the name of the black ceremonial hood worn by schoolmasters from the 1600s (until modern times in some more eclectic establishments) – a schoolmaster in Scotland and in some parts of England being known as a *dominie* (from Latin *dominus* 'lord of the household').

dong [comparative values] Unit of currency in Vietnam.

$$1 \text{ dong} = 100 \text{ xu or sau}$$

See also COINS AND CURRENCIES OF THE WORLD.

donum, deunam [square measure] A former measure of land area commonly used in Yugoslavia, Turkey, and the Near East, but now standardized to a metric unit.

$$
\begin{aligned}
1 \text{ donum} &= 50 \times 50 \text{ metres} \\
&= 2{,}500 \text{ square metres, } 25 \text{ ARES, } 0.25 \text{ hectare} \\
&= 2{,}990 \text{ square yards, } 26{,}910 \text{ square feet}
\end{aligned}
$$

$$= \quad 0.6178 \text{ ACRE}$$
$$(1 \text{ acre} = 1.6186 \text{ donum})$$
$$4 \text{ donums} \quad = \quad 1 \text{ djerib (hectare)}$$

dopping of sheldrake [collectives] Collective term for a flock of shelducks on the water or beneath the surface – for the word is cognate with the more familiar participle *dipping* – that is, diving. Why in dictionaries the collective seems normally to apply only to the male of the species, no one knows.

Doppler effect [physics] Apparent change in the wavelength (and frequency) of an electromagnetic wave or sound caused because the source is moving in relation to the observer. If the source is approaching the observer, the wavelength decreases (and frequency increases – so sound, for example, rises in pitch); if the source is receding, the wavelength increases (and frequency decreases – so sound falls in pitch).

In astronomy, the phenomenon is responsible for the red shift – the movement of a receding object's spectral lines towards the red end of the spectrum. The effect was named after the Austrian physicist Christian Doppler (1803-53).

Dorian mode [music] Medieval form of KEY, the SCALE of which is characterized on a modern piano by the white notes between one D and the next. In technical terms, its main effect is that of a minor key with a major subdominant chord and a minor dominant chord. (Take, for example, the key of A minor, which normally has a subdominant chord of D minor and a dominant chord of E major: in the Dorian mode, the key would then have a subdominant chord of D major and a dominant chord of E minor.)

Its name derives from one of the musical modes used in ancient Greece, and held to be particularly solemn but simple – but the ancient Greeks had completely different musical referents.

dosemeter, dosimeter [physics] An instrument or device for measuring the dose of radiation received in a given time by a person or by objects within a particular area. The simplest type is a film badge; the film is developed and the degree to which it is fogged gives an indication of radiation dosage.

dot product [maths] In vector analysis, the product of the magnitudes of two vectors and the cosine of the angle between their directions. Given its name because it is denoted by a dot sited between the vectors (as in v·w), it is a scalar (that is, a non-vector) quantity, and is also known as the *scalar product* or the *inner product*. *See also* CROSS PRODUCT.

dotted notes [music] In musical notation, a dot next to a note extends the note to 1½ times its standard duration. Thus:

♩. has a duration of ♪♪♪ (usual ♪ plus half again)
♪. has a duration of ♪♪♪ (usual ♪ plus half again)

The symbol signifying a rest can also be dotted in this fashion.

A dot sited *above* a note indicates that it is to be played or sung *staccato*, as a short individual note.

double [quantitatives; sport] Two times the size, quantity, or degree; twice over; twofold, in two segments. Deriving through French from Latin (*duplus, duplex*) and ancient Greek (*diploos*), variants in English include 'duple', 'duplex', and 'diplo-'.

double [square measure] Standard size of roofing slate, 13 x 6 inches (330 x 152 millimetres): *see* SLATE SIZES.

double amplitude [physics] For a wave motion, the peak-to-peak distance – that is, twice the normal amplitude.

double bass *see* STRINGED INSTRUMENTS' RANGE

double bed [square measure] The standard size for a double bed is 74 x 54 inches (6 ft 2 in. x 4 ft 6 in.; 187.96 x 137.16 centimetres), although smaller and larger sizes are, of course, available.

double entry [comparative values] In old-fashioned ledger bookkeeping, a system by which each transaction is recorded twice over, once on the credit side and again on the debit side.

double flat, double sharp [music] Rare form of musical notation required only when a natural (a note that is neither sharp nor flat, as written) has already been made sharp or flat for the purposes of a melody or of creating a harmony, and has then to be made sharper or flatter still for the same purposes within a bar. Alternatively, but only technically, as written, a flat note in a key that is one of the flats (such as B♭ in the key of F, one flat) that has then to be made flatter still (B♭♭, =effectively A) for the purposes of a melody or of creating a harmony, or a sharp note in a key that is one of the sharps (such as C# in the key of A, three sharps) that has then to be made sharper still (C##, = effectively D) for the purposes of a melody or of creating a harmony. *See also* FLAT; SHARP.

double time [speed; music] A rate of military marching intended to approximate to twice as fast as QUICK TIME. In the United States, for example, quick time corresponds to a regulated 120 steps of 30 inches per minute (3.4091 mph, 5.4864 km/h), but double time corresponds to 180 steps of 36 inches per minute (6.1364 mph, 9.8755 km/h). Double time in the British army corresponds officially to 165 steps of 33 inches per minute (5.1563 mph, 8.2979 km/h). *See also* HALF-STEP.

'Double time' also describes music that has a beat and tempo suitable for marching at this speed – most often a fastish 4/4 with a strong mid-bar beat that makes it sound like 2/2.

doubloon [comparative values] Old Spanish coin widely used in South America from the 1600s to the 1800s, and for some of that time also in North America.

$$1 \text{ doubloon} = 2 \text{ pistoles, } 8 \text{ pesos ('dollars'), } 64 \text{ reales}$$

It was a 'doubloon' because it was originally double the value of a pistole.

dovap, DoVAP [physics] Measurement of the DOPPLER EFFECT on radio signals transmitted and returned to gauge the speed and position of an approaching or passing object (such as an aircraft or missile). The term derives as an acronym of Doppler Velocity And Position.

Dow-Jones Index, Dow-Jones Industrial Average [comparative values] Authoritative INDEX of New York and US stock-market quotations of the current values of industries, utilities, and railways/railroads published daily for comparison and analysis in relation to the national economy.

Based largely on a mathematical theory propounded first by the US economist Charles H. Dow (1851-1902), the Index/Average is named after him and after his compatriot Edward D. Jones (? -1920).

downhill skiing *see* SKIING DISCIPLINES

downwash [aeronautics] The continuous downward rush of air directed by a helicopter or a vertical-take-off jet plane or rocket, measured in terms of speed (in miles per hour, kilometres per hour, or knots).

doylt of tame pigs, of tame swine [collectives] This collective applies only to tame pigs, domesticated to the stage of expecting to be housed and fed at all times. The word appears in very few dictionaries, however, and would seem to be fairly locally dialectal in origin – possibly northern English, southern Scots. It may or may not be the variant of English *docility* represented in Scots *doilt* 'tame to the point of stupidity', 'dumb'.

dozen [quantitatives] Twelve, a group of twelve. The term derives from French *douzaine* 'group of twelve', 'twelvefold', from *douze* 'twelve', ultimately from Latin *duodecem* 'twelve' (literally '2, 10'). *See also* BAKER'S DOZEN.

drachm, dram [weight] Unit of weight in the APOTHECARIES' WEIGHT SYSTEM used during the seventeenth to nineteenth centuries, and fundamentally based on the weight of the ounce in the TROY WEIGHT SYSTEM. The unit is spelled *drachm* in the United Kingdom and in some other English-speaking countries, and spelled *dram* in the United States. The latter spelling – although reflecting the normal pronunciation of the word – is potentially very confusing, in that drams also feature in the AVOIRDUPOIS WEIGHT SYSTEM, and at a different number to the (avoirdupois) ounce. On the other hand, in fluid measure both the United Kingdom and the United States not only normally calculate in fluid *drams*, but there are in both cases only 8 fluid drams to the fluid ounce, as in the apothecaries' system.

According to the apothecaries' weight system:

$$
\begin{aligned}
1 \text{ dra(ch)m} &= 3 \text{ scruples, } 60 \text{ grains, } 0.125 \text{ troy ounce} \\
&= 2.1943 \text{ drams avoirdupois} \\
&= 3.887959 \text{ grams} \\
&\quad (1 \text{ dram avdp.} = 0.4557 \text{ dra(ch)m apoth.}
\end{aligned}
$$

```
        8 dra(ch)ms  =   1 troy ounce (1 oz apoth.), 480 grains
                         (1 troy ounce = 1.09709 ounce avoirdupois
                          1 oz avdp., 16 drams = 0.9115 troy ounce
                                              = 28.349523125 grams)
```

drachm, fluid *see* DRAM, FLUID

drachma [weight] Small units of weight in the Netherlands, Greece, and Turkey, based upon ancient apothecary measures.

```
    1 Dutch drachma  =   3.906 grams
                         (5 grams = 1.2801 Dutch drachma)
                     =   1.005 dra(ch)m apoth., 2.2043 drams avdp.
                     =   0.13777 ounce avdp.
    1 Greek drachma  =   3.20 grams
                         (5 grams = 1.5625 Greek drachma)
                     =   0.823 dra(ch)m apoth., 1.8060 dram avdp.
                     =   0.11288 ounce avdp.
    1 Turkish drachma =  3.21 grams
                         (5 grams = 1.5576 Turkish drachma)
                     =   0.826 dra(ch)m apoth., 1.8117 dram avdp.
                     =   0.11323 ounce avdp.
```

The term derives from the ancient Greek *drachme*, the name both of a small weight and of a small coin.

drachma [comparative values] Unit of currency in Greece.

```
        1 drachma  =   100 lepta (or leptons)
```

The term derives from the ancient Greek *drachme*, the name both of a small weight and of a small coin. *See also* COINS AND CURRENCIES OF THE WORLD.

drachme [comparative values] Unit of currency in ancient Greece, deriving initially as a unit of weight in silver.

```
        1 drachme  =   6 oboloi, 48 khalkoi
        4 drachme  =   1 tetradrachm
```

The name of the unit is still current as the name both of a coin and of a weight (*see* DRACHMA *above*).

draft, draught *see* DRAUGHT, DRAFT

drag [physics: aeronautics] The resistance to movement through a fluid, including gases such as air. For an aircraft in flight, it is the component of the force measured parallel to the direction of motion (the drag axis). It is reduced by streamlining, and is deliberately induced by parachute braking systems and air brakes (which stick out into the airflow).

dram [weight] For the purpose of avoiding confusion, in this book the unit of weight within the APOTHECARIES' WEIGHT SYSTEM is defined under DRACHM, DRAM, and it is the dram that belongs to the AVOIRDUPOIS WEIGHT SYSTEM that is defined here.

```
        1 dram  =   27.344 grains
                =   1.772 gram
                    (1 gram = 0.5643 dram
                     = 0.03527396 ounce)
      16 drams  =   1 ounce avoirdupois (1 oz av. or avdp.)
                =   28.349523125 grams
```

dram, fluid [volumetric measure] A very small unit of liquid volume, different in UK and US usages, and deriving from the APOTHECARIES' WEIGHT SYSTEM (*see* DRACHM, DRAM) – which is why there are 8 (fluid) drams to the (fluid) ounce, and not 16.

In the United Kingdom:

```
    1 fluid dram  =   one-eighth of a FLUID OUNCE
                  =   ¹⁄₁₆₀ of a PINT, 0.006250 pint
                  =   0.0035512 litre, 3.5512 millilitres
                      (4 millilitres = 1.1264 fluid dram)
                  =   0.9606 US fluid dram
    8 fluid drams =   1 fluid ounce (28.4 millilitres)
   40 fluid drams,
    5 fluid ounces =  1 GILL or quartern
```

In the United States:

1 fluid dram	=	one-eighth of a fluid ounce
	=	$\frac{1}{128}$ of a pint, 0.007812 pint
	=	0.0036968 liter, 3.6968 milliliters (4 milliliters = 1.0820 fluid drams)
	=	1.0410 UK fluid dram
8 fluid drams	=	1 fluid ounce (29.6 milliliters)
32 fluid drams, 4 fluid ounces	=	1 gill

draught, draft [weight; comparative values] In shipping, the depth to which a ship sinks in water ('draws'), equal to the minimum water depth a ship needs to float. This may be calculated mathematically in relation to the hull shape and the DEADWEIGHT TONNAGE.

In commerce, an allowance made for the natural wastage of some specific consumer goods that are sold by weight.

draughts (checkers) moves and terms [sport] The pieces ('draughts' in Europe and elsewhere, 'checkers' in North America) play only on the black or the white squares of a chessboard.

The moves are:

one square diagonally forward

two squares diagonally forward if 'jumping' an opposing piece (in which case the piece is removed from the board)

Draughts (checkers)

If another opposing piece can immediately be 'jumped' (by the same 'jumper'), the turn continues, and the second opposing piece is also removed from the board. It is not possible to 'jump' to a square already occupied. A piece that reaches the opponent's far file becomes a 'king'; a piece that has already been taken and is off the board is then added to the king (on top of it), signifying that the king can now move in any direction, backward or forward, in the usual moves.

The European name for the game refers to the fact that the pieces are 'drawn' by hand across the board in a series thus of 'draughts'. The North American name refers to the board: *see* CHESS PIECES AND MOVES.

draw [sporting term] In games, a contest that remains undecided at the end – either because the score of each competitor is identical, or because no result is possible in the time available – a definition that distinguishes a draw from a TIE.

The term originates as a nineteenth-century abbreviation for 'a drawn game', a seventeenth-century expression that apparently refers to the competitors' being drawn (pulled, stretched) together to the identical score or attainment.

dray of squirrels *see* DREY OF SQUIRRELS

D region *see* D LAYER, D REGION

dressage [sport] Event on the first day of the three-day event in equestrianism, in which the rider requires the horse to show extreme skill and obedience in fulfilling subtle commands at the walk, at the canter, and at the trot, within the fairly confined space of a show-ring.

drex [textiles] Unit of thickness of yarn, equal to the mass (in grams) of a 10,000-metre length of yarn.

1 drex is equivalent to 0.9 denier
(1 denier is equivalent to 1.1111 drex)

See also DENIER; TEX.

drey of squirrels [collectives] The *drey* or *dray* is just another word for the squirrels' nest, and is not a particularly satisfactory collective, therefore. It would seem to derive from the squirrels' *dragging* grasses, leaves, and other soft materials together to line the nest.

drift [meteorology; navigation] The distance a ship or an aircraft travels sideways off its straight course because of prevailing cross-currents. Expressed generally as the angle (called the drift angle) between the course and the actual heading (and in aircraft registered on an instrument known as a drift meter), it can be used to plot a new course by which the destination can be reached in a direct line.

drill sizes [physics: engineering] The diameters of small drills for boring metals are often specified by a series of numbers, from 1 = 0.228 inches (5.77 millimetres) downwards to 80 = 0.0135 inch (0.343 millimetres). Larger sizes are specified using letters, ranging upwards from A = 0.234 inch (5.94 millimetres) to Z = 0.413 inch (10.49 millimetres).

The diameter sizes of drills for boring holes in wood are generally given in fractions of an inch or millimetres.

drinking glass measures [volumetric measure] Standard drinking-glasses are:

	capacity		
	UK fl. oz	*US fl. oz*	*centilitres*
litre mug (stein)	(35.2)	(33.8)	100
'pint' mug	17.6	(16.9)	(50)
highball glass	14.5	13.9	41.2
white wine glass			
(tall stem, high sides)	12.0	11.5	34.1
red wine glass			
(short stem, rounded sides)	10.0	9.6	28.4
half-pint mug	10.0	9.6	28.4
old-fashioned glass	10.0	9.6	28.4
short glass	8.0	7.7	22.7
sherry schooner	6-8	5.75-7.7	17.0-22.7
port glass	3-6	2.875-5.75	8.5-17.0
liqueur glass	2-4	1.92-3.8	5.7-11.4

There are many variants on the wine glass, of which the tulip glass is probably the most widespread of those that have no specific capacity and has the greatest range of sizes. Liqueur glasses also differ in capacity around the world. The highball glass is possibly the most common form of glass in use today, as popular in households as it is in bars: the tall upright tumbler from which children drink milk and adults sip their cooling cocktails while avoiding the slice of lemon and the paper umbrella.

driving clock [astronomy; telecommunications] A clock linked to a synchronous electric motor that drives an apparatus or machine consistently for twenty-four hours. Such a clock may be particularly important in astronomy or in satellite broadcast reception, where it may be necessary to aim a telescope or dish antenna consistently at the same point in the sky.

drop goal, dropped goal *see* RUGBY LEAGUE/UNION MEASUREMENTS, UNITS, AND POSITIONS

drosometer [physics: meteorology] An instrument that measures the amount of dew deposited.

The first element of the term comes from ancient Greek *drosos* 'dew'.

dual *see* DUO, DUO-

ducat, ducatoon [comparative values] Various units of currency in Europe between medieval times and the nineteenth century. The best known and most used was probably an Italian silver coin, although the earliest example is thought to have been the gold coin issued in Sicily by Duke Roger II of Apulia. It was thus a coin that was minted on behalf of the duchy (medieval Italian *ducato*). In time, most of the countries of Europe came to have a unit of currency with this name or similar. The variant *ducatoon* was originally French.

duck [sporting term] In cricket, a batsman's score of zero for an innings. The term originated as an abbreviation for 'duck-egg', the oval shape on the scoreboard that corresponds to the humiliating zero.

duet [quantitatives: musical] Group of two, contributing to a unity, especially in musical performances and similar contexts.

dule of doves, of turtles [collectives] Obscure collective for doves that most probably reflects the mournful nature of their cooing, akin thus to *dolour* and *doleful*. The turtles here are in that case almost definitely turtle doves, not the marine and riverine reptiles. This might also explain the other extraordinary collective used for turtles, *bale* – for if they are doleful, they may equally sound *baleful*.

Dulong and Petit's law [chemistry; physics] The specific heat capacity of a solid chemical element is inversely proportional to its relative atomic mass (atomic weight). In other words, the product of the specific heat and atomic mass (a quantity termed the atomic heat) is a constant, equal to 26.4 joules per kelvin per mole. The law was named after the French chemist Pierre Dulong (1785-1838) and the French physicist Alexis Petit (1791-1820).

duo, duo- [quantitatives: prefix] Group of two, twofold unity; of two parts.

duodec- [quantitatives: prefix] Prefix signifying 'twelvefold', 'of twelve'.
  Example: duodecimal – of twelve parts, in twelfths
  The term derives from Latin *duodecem* 'twelve' (literally '2, 10').

duodecagon [maths] A POLYGON that has twelve sides (and twelve angles). Also *dodecagon*.

duodecal [physics; engineering] Describing something with twelve parts, specifically the number of pins on a valve base or valve (electron tube).

duodecimal system [maths] A number system with the base 12. Its 'numerals' are 0, 1, 2, 3, 4, 5, 6, 7, 8, 9, T, E, and 10 (duodecimal T, E, and 10 = 10, 11, and 12 in the decimal system). There are remnants of this form of numeral system in such English terms as *dozen* and *gross*.

duodecimo [paper] One-twelfth of a sheet of paper, or a sheet folded to make twelve leaves (= twenty-four pages). It is often abbreviated to '12mo'.

duple, duplex *see* DOUBLE

dynameter [physics] An instrument for measuring the magnifying power of a telescope.

dynamometer [physics] Any of various instruments for measuring force or power, in the mechanical sense.

dyne [physics] Unit of force in the centimetre-gram-second (CGS) system, equal to the force that gives an object of mass 1 gram an acceleration of 1 centimetre per second per second.

$$1 \text{ dyne} = 7.233 \times 10^{-5} \text{ POUNDAL}$$
$$10^5 \text{ dynes} = 1 \text{ NEWTON (the SI unit of force)}$$
$$981 \text{ dynes} = 1 \text{ gram weight}$$

  The name of the unit is taken from the first syllable of the ancient Greek *dynamis* 'power'.

dyne-centimetre, dyne-centimeter [physics] Unit of work or energy in the centimetre-gram-second (CGS) system, equal to the amount of work done when a force of 1 DYNE moves though 1 centimetre. It is much better known as the ERG (= 10^{-7} JOULES).

dystectic [physics] A mixture that has the maximum possible melting point: *see* EUTECTIC POINT.

E

E *see* ENERGY; RELATIVITY: THE MASS-ENERGY EQUATION

e [maths] A term for the IRRATIONAL NUMBER that corresponds to the base of NATURAL LOGARITHMS (also called Napierian logarithms): 2.71828183 . . .

each-way bet [sporting term] In betting on horses and greyhounds, a bet on one animal either to come first (at the full, quoted, starting odds) or to be in the first three past the post (at considerably reduced odds). The amount of money staked is thus double the amount expressed: 'two pounds each way' on a horse implies a total stake of four pounds. If the horse duly comes first, the better receives the full odds (plus premium) for the first half of the stake, and the appropriate reduced odds (plus premium) on the second half of the stake. If the horse comes second or third, the better forfeits the first half of the stake altogether and receives only the reduced odds (plus premium) on the second half – but may yet more than cover the initial outlay.

It is more common for accustomed betters to place bets for a win only. Nonetheless, it is possible instead to put bets on a horse or dog only 'for a place' – that is, to be in the first three – at reduced odds.

eagle [sporting term] In golf, a birdie is one under par (one shot less than the standard number for the hole); an eagle is two under par; and an albatross is no fewer than three under par. Presumably the terms 'eagle' and 'albatross' represent gradations in size (and status) based on the initial term 'birdie' – but why that first term was chosen, no one knows.

eagle [comparative values] The name of several coins current at various times in the history of English-speaking peoples. The earliest was a coin of little value issued in England during the second half of the thirteenth century, so called because it had a representation of an eagle on the obverse.

The best-known eagle was the gold coin issued in the United States from the 1780s for well over a century, and worth 10 dollars. For a time there was, in addition, a US coin known as the double eagle, worth 20 dollars.

earned run average [sporting term] In baseball, the average number of runs actively scored by a batter (and not the result of defensive errors) per game, calculated over a series or season of 9-inning games.

Earth: planetary statistics *see* PLANETS OF THE SOLAR SYSTEM

Easter day, date of [time; astronomy] The Christian feast of Easter – technically the most important Christian festival of the year – is a movable feast for which the timing depends on a particular full Moon (the Paschal full Moon). Rather than need always to have astronomical calendars to hand, however, a table has been available for at least three centuries from which the date of Easter in the Western Church may be determined at least up to the year 2199. It depends on first determining the GOLDEN NUMBER and the SUNDAY LETTER of the year for which the date of Easter is required. With them, it is a simple matter to read off the appropriate date.

golden number	*Sunday letters*						
	A	B	C	D	E	F	G
1	Apr 16	Apr 17	Apr 18	Apr 19	Apr 20	Apr 21	Apr 15
2	Apr 9	Apr 10	Apr 4	Apr 5	Apr 6	Apr 7	Apr 8
3	Mar 26	Mar 27	Mar 28	Mar 29	Mar 30	Mar 24	Mar 25
4	Apr 16	Apr 17	Apr 18	Apr 12	Apr 13	Apr 14	Apr 15
5	Apr 2	Apr 3	Apr 4	Apr 5	Apr 6	Apr 7	Apr 1
6	Apr 23	Apr 24	Apr 25	Apr 19	Apr 20	Apr 21	Apr 22
7	Apr 9	Apr 10	Apr 11	Apr 12	Apr 13	Apr 14	Apr 15
8	Apr 2	Apr 3	Apr 4	Mar 29	Mar 30	Mar 31	Apr 1
9	Apr 23	Apr 17	Apr 18	Apr 19	Apr 20	Apr 21	Apr 22
10	Apr 9	Apr 10	Apr 11	Apr 12	Apr 6	Apr 7	Apr 8
11	Mar 26	Mar 27	Mar 28	Mar 29	Mar 30	Mar 31	Apr 1
12	Apr 16	Apr 17	Apr 18	Apr 19	Apr 20	Apr 14	Apr 15
13	Apr 9	Apr 3	Apr 4	Apr 5	Apr 6	Apr 7	Apr 8
14	Mar 26	Mar 27	Mar 28	Mar 29	Mar 23	Mar 24	Mar 25
15	Apr 16	Apr 17	Apr 11	Apr 12	Apr 13	Apr 14	Apr 15
16	Apr 2	Apr 3	Apr 4	Apr 5	Apr 6	Mar 31	Apr 1
17	Apr 23	Apr 24	Apr 18	Apr 19	Apr 20	Apr 21	Apr 22
18	Apr 9	Apr 10	Apr 11	Apr 12	Apr 13	Apr 14	Apr 8
19	Apr 2	Apr 3	Mar 28	Mar 29	Mar 30	Mar 31	Apr 1

The examples of the golden number and Sunday letter given in those entries related to the year 2000, for which the golden number is 6 and the Sunday letter is A. Unless otherwise decreed by Church and/or state authorities, Easter Day in 2000 should therefore be celebrated on 23 April.

ebulliometer [physics] An instrument for measuring and displaying the BOILING POINTS of liquids, especially solutions. The first element of the word derives through French from Latin *ebullire* 'to boil away'.

ebullioscopy [physics] A method of finding the relative molecular mass (molecular weight) of a substance by measuring the extent to which it raises the boiling point of a solvent, using an EBULLIOMETER. *See also* ELEVATION OF BOILING POINT.

eccentricity [maths; astronomy] The extent by which a closed path or curve differs from being precisely circular – the eccentricity of a circle is 0.

More specifically, for any conic section (ellipse, parabola, hyperbola), it is the ratio of the distance between the curve and the focus to its distance from the directrix. For an ellipse it is less than 1; for a parabola it equals 1; and for a hyperbola it is greater than 1. In the case of a planet orbiting the Sun in an elliptical orbit, for example, for its orbit to be described as 'highly eccentric' indicates that there is a large difference between the planet's nearest and farthest distances from the Sun.

The term derives ultimately from the ancient Greek elements *ex-kentron* 'away from the centre'.

echolocation [physics] A technique for measuring the range and bearing of an object by detecting echoes of high-frequency (ultrasonic) sounds reflected by it. Employed in nature by bats, dolphins, and whales, and by some cave-dwelling birds, it is the basis of sonar, and of echo sounders that measure the depth of water beneath the hull of a ship. Such soundings are displayed on a graph or chart called an echogram, which is calibrated directly in feet or metres.

eclipse [astronomy] The two dimensions in which eclipses of the Sun and Moon are measured are the duration (how long the eclipse lasts) and the extent (whether an eclipse is total or partial).

The duration of a partial eclipse may be no more than a few seconds, but a total eclipse may last for a maximum of 7 minutes 40 seconds in totality and more than 28 minutes from onset to separation.

The extent of occultation, if not total, may be measured in units known as DIGITS, 1 digit corresponding to half the apparent diameter of the Sun or Moon (which from the Earth have virtually identical apparent diameters).

eclipse year [astronomy] The time that elapses between two apparent passages of the Sun through the same point on the Moon's orbit, equal to 346 days 14 hours 52 minutes and 48 seconds (346.62 days). *See also* YEAR.

ecliptic [astronomy] Apparent annual path of the Sun as it passes through the 12 constellations of the zodiac on its way around the celestial sphere. The ecliptic crosses the CELESTIAL EQUATOR at the two EQUINOXES.

ectomorph [medicine] In HUMAN MORPHOLOGY, the human body shape supposedly corresponding to the type that is thin, not to say skinny, and of a rather nervous and introverted disposition. The other two types so classified are MESOMORPHS and ENDOMORPHS.

The term derives from scientific Greek *ekto-* 'of outer structures' (implying the skin and the nerves) + *morph-* 'form', 'shape'.

ecu, ECU [comparative values] Coin and currency created in the 1980s to be a standard in the countries of the European Community.

The term was devised partly to represent the initials of 'European Community (or Currency) unit', but partly because in France – in the language of which those words would require the initials to be reversed – there had already been a number of coins issued from the thirteenth century on called *écus*, featuring an embossed shield (French *écu* 'shield', cf. Portuguese *escudo*, English *escutcheon*; cf. also English *shilling*, Austrian *schilling*, both from a diminutive of what is in present-day English *shield*). *See also* COINS AND CURRENCIES OF THE WORLD.

Eddington limit [astronomy] In measuring the brightness of a star or other stellar object, the maximum brightness that can be attained, as calculated in proportion to the star or object's mass.

edge numbers [photography] A series of letters and numbers along the edge of
35-mm photographic film that enables individual frames to be identified and found.
Edwardian age/era [time] Period from 1901 to 1911, during which King Edward
VII occupied the British throne.
effective dose equivalent *see* X-RAYS
effective value [physics: engineering] The value of a simple measurement that is
effectively the same as a complex measurement in most applications. For example:
the DC value of an alternating voltage is effectively the same as the ROOT MEAN
SQUARE value of the AC.
efficiency (of a machine) [physics] The usable energy output of a machine
divided by the energy input, usually expressed as a percentage. For simple
machines, efficiency is calculated as the FORCE RATIO (or mechanical advantage)
divided by the DISTANCE RATIO (or velocity ratio).
effusiometer [physics] An apparatus for comparing the relative molecular masses
(molecular weights) of gases by measuring the time they take to pass (effuse)
through a small aperture: *see* EFFUSION.
effusion [physics] The flow of gas through a small orifice, at a rate that is approximately proportional to the square root of the difference in pressures on each side of
the orifice. It is more difficult to quantify than diffusion (*see* GRAHAM'S LAW).
    The term derives from the Latin past participle *effusus* 'poured out'.
egg size and quality measurement [nutrition] Through careful selection and
breeding of chickens, the size of hens' eggs has increased markedly over the
thousands of years that the birds have been kept domestically. The quantity of the
contents has also increased, but the amount of shell enclosing that quantity has, in
fact, remained much the same. As a result, the shell has had to 'stretch' and become
thin still to cover the egg. This is why domestic hens' eggs break much more easily
than the eggs of other birds (it is possible for an adult human to stand on an ostrich
egg without breaking it).
    In most countries, eggs for sale are generally sorted into sizes specified by the
relevant marketing or governmental authority. In the UK, for example, Size 1 is the
largest size, Size 6 is the smallest. Elsewhere in the world, names are given to the
sizes instead of numbers. In the United States until the 1970s, for example, the
largest egg size was 'jumbo', the smallest 'peewee'. All these apparent size
classifications in fact correspond to a measurement by weight (not dimensions) as
the standard per dozen eggs.
    Egg quality measurement is more often alphabetical. Judged by a process known
as candling (by which eggs are examined under strong light in a darkened room to
record details of the white, the yolk and the air cell), the best eggs are given a rating
of AA; in decreasing quality, eggs may alternatively be classified as A, B, or C.
egg-timer *see* HOURGLASS, SANDGLASS, EGG-TIMER
Egyptian pound [comparative values] Unit of currency in Egypt.
            1 Egyptian pound   =   100 piastres
    See also COINS AND CURRENCIES OF THE WORLD.
eigenfrequency [physics] In acoustics, the vibrational frequency of something that
can vibrate without restriction.
    The prefix *eigen-* is the German for 'own', in the sense of 'characteristic'.
eigenfunction [physics; maths] In quantum mechanics, one of the possible solutions
of the Schrödinger wave equation, or the solution of any wave equation that
satisfies certain conditions.
    In differential calculus, a solution that satisfies specified conditions for only
certain values (EIGENVALUES) of a parameter.
    The prefix *eigen-* is the German for 'own', in the sense of 'characteristic'.
eigenvalue [physics; maths] A possible value of a quantity derived from an equation
of which the solutions comply with certain conditions. For example: eigenvalues of
energy from the Schrödinger wave equation in quantum mechanics correspond to
possible energy levels.
    The prefix *eigen-* is the German for 'own', in the sense of 'characteristic'. *See
also* EIGENFUNCTION.
eight [quantitatives] 8.

$$^8/_{10} = 0.8 = 80\%$$
$$8 \text{ pints} = 1 \text{ gallon}$$
$$8 \text{ furlongs} = 1 \text{ mile}$$

A group of eight is an octad, an ogdoad, an octave, an eighthsome, or an octet.

A POLYGON with eight sides (and eight angles) is an octagon.

A solid figure with eight plane faces is an octahedron.

As is evident from the above, the prefix meaning 'eight-' in English corresponds to 'oct-', 'octa-', or 'octo-'.

The name for the number derives from common Indo-European, attested over the entire range of languages from Sanskrit to Welsh. It has been suggested that in some way it derives from an expression meaning 'two times four' (after which 9 would be the *new* or *next* number: *see* NINE), but this is unlikely ever to be proved.

See also EIGHTH; OCTAL; OCTANT; OCTONAL; OCTONARY; OCTUPLE.

eighteen inches [linear measure] Imperial measure of a length with many practical applications.

$$
\begin{aligned}
18 \text{ inches} \ &= \ 1 \text{ ft } 6 \text{ in } (1\tfrac{1}{2} \text{ feet}) \\
&= \ \text{one-quarter of a FATHOM, half a YARD} \\
&= \ \text{(approximately) 1 CUBIT} \\
&= \ 2 \text{ SPANS or BREADTHS} \\
&= \ 47.72 \text{ centimetres, } 0.4572 \text{ metre} \\
&\quad\ \ (50 \text{ centimetres, half a metre} = 19.685 \text{ inches})
\end{aligned}
$$

eighth [quantitatives; music] $\frac{1}{8}$ (one-eighth); the element in a series between the seventh and the ninth, or the last in a series of eight.

$$
\begin{aligned}
\tfrac{1}{8} \ &= \ 0.125 \ = \ 12\tfrac{1}{2}\% \\
\tfrac{2}{8} \ = \ \tfrac{1}{4} \ &= \ 0.25 \ = \ 25\% \\
\tfrac{3}{8} \ &= \ 0.375 \ = \ 37\tfrac{1}{2}\% \\
\tfrac{4}{8} \ = \ \tfrac{1}{2} \ &= \ 0.5 \ = \ 50\% \\
\tfrac{5}{8} \ &= \ 0.625 \ = \ 62\tfrac{1}{2}\% \\
\tfrac{6}{8} \ = \ \tfrac{3}{4} \ &= \ 0.75 \ = \ 75\% \\
\tfrac{7}{8} \ &= \ 0.875 \ = \ 87\tfrac{1}{2}\%
\end{aligned}
$$

In music, an eighth is the interval of an OCTAVE.

eighth-note *see* QUAVER

eighth of an inch [linear measure] A small imperial measure.

$$
\begin{aligned}
\tfrac{1}{8} \text{ inch} \ &= \ 0.3175 \text{ centimetre, } 3.175 \text{ millimetres} \\
&= \ 9 \text{ POINTS (in typography)} \\
&= \ 125 \text{ MILS}
\end{aligned}
$$

eight millimetre, 8 mm [photography] The smallest gauge of cine film, originally obtained by splitting 16-millimetre film lengthways after processing, but later – as Super-8 – purpose made.

Compact camcorders use 8-mm videotape, the narrowest commercially available.

eights *see* ROWING MEASUREMENTS AND TERMS

Einstein equation *see* RELATIVITY: THE MASS-ENERGY EQUATION

Einstein unit, einstein [physics] Unit of electromagnetic energy (usually light or ultraviolet light) involved in a mole of a substance undergoing a photochemical reaction, equal to $Nh\nu$ where N is AVOGADRO'S NUMBER (Avogadro's constant), h is Planck's constant, and ν (the Greek letter nu) is the frequency of the initiating radiation.

The unit was named after the German-born US physicist Albert Einstein (1879-1955).

Einstein shift [physics; astronomy] A shift of the spectral lines of a celestial object towards the red end of the spectrum that occurs when an object with a significant gravitational field emits electromagnetic radiation.

eka- [quantitatives: prefix] Prefix meaning 'one immediately after', 'the next one to', and once applied to hypothetical chemical elements additional to those in the incomplete PERIODIC TABLE OF ELEMENTS, but with properties adduced from that table.

Example: ekasilicon – name posited by Dmitri Mendeleev for the then unknown element one above silicon in the Periodic Table, later isolated and named germanium.

The term derives from the ancient Greek *ek-* 'from', 'beyond', although some commentators have instead suggested it originates from the Sanskrit *ekas* 'single-', 'mono-', presumably in the sense '(plus) one'.

ekuele, ekpwele [comparative values] Unit of currency in Equatorial Guinea, from 1973 replacing the peseta, corresponding to a local name for what is better known as the CFA franc.

      1 ekuele/CFA franc   =   100 centimes

    See also COINS AND CURRENCIES OF THE WORLD.

elastance [physics] Elastance – the reciprocal of capacitance – is measured in DARAF (a term representing *farad* backwards). The unit got its name by mechanical analogy with the elasticity of a spring.

elastic constant [physics] Constant that describes the elastic properties of a material, particularly its behaviour when subjected to compression, longitudinal stress, or shear stress. The constant, also known as the elastic modulus, is generally the ratio of stress to strain under certain specified conditions.

    See also BULK MODULUS; POISSON'S RATIO; RIGIDITY MODULUS; YOUNG'S MODULUS.

elasticity [physics] Elasticity is that property of a material being deformed that causes it to take on its original shape again after the deforming force is removed (assuming that the material's elastic limit is not exceeded, when permanent deformation results). A material with this property is called an elastomer.

    Elasticity is measured in terms of the various elastic constants, such as BULK MODULUS, RIGIDITY MODULUS, and YOUNG'S MODULUS, which are essentially the ratio of stress to strain.

elastic limit [physics] Up to the elastic limit, the stress applied to an elastic material is proportional to the strain (*Hooke's law*). Beyond the elastic limit, a small increase in stress produces a large strain (the material may become plastic and stretch markedly) and a permanent deformation results.

    See also ELASTICITY.

E layer, E region [physics: telecommunications] A layer of ionized gases in the atmosphere at a height of 110 to 120 kilometres (68.4–74.6 miles) which, during the day, reflects radio waves. It is also known as the Heaviside layer, after the British physicist Oliver Heaviside (1850-1935).

electrical conductivity [physics] A measure of a material's ability to conduct electricity, equal to the current density divided by the applied electric field. It is the reciprocal of RESISTIVITY, and is expressed in units of ohm^{-1} metre^{-1} (SI units) or siemens per metre. At high temperatures, the conductivity of a metal is inversely proportional to its absolute temperature.

electrical resistance *see* RESISTANCE

electric constant [physics] Alternative name for the PERMITTIVITY of free space.

electric current [physics] The movement of electrons along a conductor caused by a potential difference (voltage) across it, measured in AMPERES or their submultiples.

electric field strength [physics] The strength of an electric field – the region around an electric charge in which a force acts on a charged particle – measured in terms of the force that acts on a unit charge (at a given point), and is expressed in VOLTS per metre.

electric flux [physics] The total intensity of the electric field at right-angles to a surface over a given area, generally measured in VOLT-METRES.

electric polarization [physics] The charge separation of a given dielectric, measured in terms of the DIPOLE MOMENT per unit volume.

electric susceptibility [physics] The amount by which a dielectric's relative permittivity (dielectric constant) is greater than 1, expressed in FARADS per metre.

electrocardiography [medicine] The measurement of the electrical activity of the heart muscle, as recorded on an electrocardiograph (ECG or EKG).

electrochemical equivalent [physics] In electrolysis, the mass of substance deposited or liberated at the cathode per COULOMB of electric charge passing through it.

electrochemical series, electromotive series [chemistry; physics] A list of metals in order of their ELECTRODE POTENTIALS. Hydrogen is also included in the list, and ascribed the (arbitrary) potential of zero. A given metal will spontaneously displace from solutions of their salts any metals below it in the series.

electrode potential [chemistry; physics] The equilibrium potential (voltage) that

develops between a metal and a solution of its ions. *See also* STANDARD ELECTRODE
POTENTIAL.

electroencephalography [medical] The measurement of the electrical activity of
the brain, using an electroencephalograph (EEG). *See also* ALPHA WAVES; BETA
WAVES; DELTA WAVES.

electromagnetic moment *see* MAGNETIC MOMENT

electromagnetic spectrum [physics] The range of electromagnetic radiations
(waves) from long radio waves (wavelengths down to 10^3 metres), through me-
dium- and short-wave radio, microwaves, infra-red radiation, visible light, ultraviolet radiation and X-rays to gamma-rays (at wavelengths down to 10^{-15} metre).
All move at the velocity of light, and consist of electric and magnetic fields moving
transversely at right-angles.

electromagnetic units, EMU [physics] A system of electrical units based on the
unit magnetic pole and a given value of the permittivity of free space (electric
constant).

Based within the CGS system, 1 EMU is in the SI system assigned a value of
8.854×10^{-12} FARADS per metre.

electromagnetic wave [physics] A wave that consists of electric and magnetic
fields at right-angles, moving at a speed of 2.997924×10^8 metres per second in
vacuum (the velocity of light).

For a list of the various types of electromagnetic wave, *see* ELECTROMAGNETIC
SPECTRUM.

electrometer [physics] An instrument for measuring potential difference (voltage)
by means of the effects of accumulated charges on metal plates.

electromotive force (EMF, emf) [physics] Electromotive force corresponds to the
potential difference (voltage) produced by electrical energy sources that can drive
currents along conductors, and is measured in VOLTS.

electromotive series *see* ELECTROCHEMICAL SERIES, ELECTROMOTIVE SERIES

electron [physics] A subatomic particle with a negative charge of 1.602102×10^{-19}
coulombs, rest mass of 9.10908×10^{-31} kilogram, and radius of 2.81777×10^{-15}
metre. The electron charge (e) is a fundamental physical constant. The electron's
antiparticle is the positron (so called because it is positively charged) and, together
with muons and neutrinos, electrons make up the group of subatomic particles
known as leptons.

All neutral (unionized) atoms contain one or more electrons; the number of
electrons in an atom – equal to the number of protons in its nucleus – is an
element's ATOMIC NUMBER.

The term derives ultimately from the ancient Greek *elektron* 'amber', so called
because light shines through it (the *-lect-* would seem to be cognate with the
English word *light*), and – possibly coincidentally – a substance that when rubbed
acquires an electrostatic charge.

electron affinity [chemistry; physics] Electron affinity is the energy required to
neutralize a negative ion by removing an electron from it, or the energy released
when a negative ion acquires an electron. It is measured in JOULES. *See also* WORK
FUNCTION.

electron charge/mass ratio [physics] A fundamental constant quantity in physics,
the ratio of an electron's charge to its mass equals 1.759×10^{11} coulombs per
kilogram.

electron density [physics] The number of electrons in a gram of a substance. For
most light elements it is of the order of 10^{-23} (Avogadro's constant).

electron octet [chemistry] The eight electrons in the outer (valence) shell of an
atom or molecule, the presence of which confers great stability. For example,
helium – and all the other noble gases – have eight outer electrons; sodium and
chlorine as ions in sodium chloride also each achieve an electron octet (by sharing),
forming a stable molecule from two highly reactive atoms.

electron pair [chemistry] Two (valence) electrons that are shared by a neighbouring
atomic nucleus, so forming a covalent chemical bond.

electron spin resonance, ESR [physics] A type of microwave spectroscopy in
which a paramagnetic substance (which has unpaired electrons) resonates after

absorbing radiation. The application of a strong magnetic field splits the energy levels of the electrons, and the energy difference provides information about the substance.

electron volt, eV [physics] Unit of (kinetic) energy equal to the work done on an electron when it passes through a potential gradient of 1 VOLT, equal to 160.206 x 10^{-21} JOULE. Used to express the energy of moving particles, it is such a small unit that the most practical multiples in ordinary measurements are mega-electron volts (MeV, $= 10^6$ eV) and giga-electron volts (GeV, $= 10^9$ eV).

electroscope [physics] An instrument for detecting the presence of electric charge, usually by means of a pair of thin gold leaves that diverge as they become charged.

electrostatic units, ESU [physics] A system of electrical units defined in terms of the force of attraction or repulsion between two electric charges and a given value of the permittivity of free space (electric constant). Based within the CGS system, 1 ESU is in the SI system assigned a value of 8.854 x 10^{-12} FARAD per metre.

electrovalence, electrovalency [physics] An electrovalent, or ionic, bond is a type of chemical bond in which one or more ELECTRONS move from one atom to another, and there is electrostatic attraction between the resulting pair of ions. The electrovalence of an atom is equal to the number of electrons lost or gained to form such a bond.

elegiac verse *see* VERSE FORMS

element [maths] In set theory, an individual member of a set. For example, for the set {even numbers}, 8 is an element. This is written 8 ϵ {even numbers}.

elementary particles [physics] Subatomic particles that are not made up of smaller particles; also called fundamental particles. There are three main kinds:
> leptons (electrons, negative muons, tau-minus particles and their neutrinos);
> quarks (six kinds, and their antiquarks);
> and gauge bosons (photons, gluons, gravitons, and intermediate gauge bosons).

elements, chemical *see* PERIODIC TABLE OF ELEMENTS

elevation [physics; astronomy; surveying] In astronomy and in surveying, the angular distance (that is, the measured angle) of an object above the horizontal plane (in astronomy, the horizon) in relation to the point of observation.

The term is also used to describe a place's height above sea-level (*see* ALTITUDE).

elevation of boiling point [physics; chemistry] When a substance (solute) is dissolved in a liquid (solvent), it raises the boiling point of the solvent. For a non-volatile solvent, the increase is proportional to the relative molecular mass (molecular weight) of the solute, which can be determined in this way.

eleven [quantitatives] 11. The number of members in a soccer, netball, field hockey, or cricket team.
$$11\% = 0.11$$
A POLYGON with eleven sides (and eleven angles) is an undecagon.

A solid figure (POLYHEDRON) with eleven plane faces is an undecahedron.

The name of the number derives solely through Germanic elements (unlike the names of all the numbers up to it, which are common Indo-European in origin). In Old English it was *endleofan*, which corresponds in present-day words to 'one leave' – that is, the number is '(ten plus) one leave (over)' . . . just as the next number is 'two leave', or twelve.

eleventh [quantitatives; music] $\frac{1}{11}$ (one-eleventh); the element in a series between the tenth and the twelfth, or the last in a series of eleven.
$$\frac{1}{11} = 0.90909 = 9.1\%$$
In music, an eleventh is the interval of an OCTAVE plus a perfect FOURTH: example – C to F above the next C. It is not in itself a harmony, but may contribute to a harmonic chord.

Elizabethan age/era [time] Almost always, the period between 1558 and 1603 during which Queen Elizabeth I occupied the English throne.

But occasionally, in contrast to the preceding Edwardian and late Georgian eras, the period following 1953 during which Queen Elizabeth II has occupied the British throne.

ell [linear measure] A linear measure well established in northern Europe during former centuries, particularly in relation to the manufacture and sale of cloth. In Old English it was an *eln*, in Old French an *alne* (present-day French *aune*), and both these variants display very neatly the etymological link with the Latin *ulna* (which has been adopted in medical English as the name for) the bone of the forearm. Indeed, the word *elbow* itself represents the 'bow' or bend at (the end of) the 'ell' or ulna. The link is significant, for the measure seems to have originated in English- and French-speaking countries as the average length between the two elbows with the bent arms stretched out as far apart as possible. This fairly rough-and-ready method accounts presumably for the differences in the unit's length between even close countries.

1 Jersey (Channel Island) ell	=	48 inches (1.22 metres)
1 French aune	=	1.188 metres (46.77 inches)
1 old Belgian aune	=	1.2 metres (47.24 inches)
1 English ell	=	45 inches (1.143 metres)
	=	5 SPANS, 20 NAILS

Other predominantly Germanic-speaking countries seem instead to have taken the measure to be between the elbows with both hands together as tightly clenched fists or perhaps even on top of each other.

1 Scottish ell	=	32.7 inches (83.06 centimetres)
1 old Dutch el	=	27.08 inches (68.78 centimetres)
1 Flemish ell	=	27 inches (68.58 centimetres)

Elsewhere in Europe, other nations seem to have taken the name but used it in relation to the CUBIT (a measure also related to the forearm) or to their own special units.

1 Swiss elle	=	60 centimetres (23.622 inches)
1 Latvian elle	=	53.7 centimetres (21.142 inches)

See also PIK.

ellipse [maths] A plane, closed curved figure with an ECCENTRICITY of less than 1. It is the locus of a point that moves around two fixed points (the *foci*) so that the sum of its distances from the foci is constant.

An ellipse is also generated as a CONIC SECTION when a plane cuts through a cone at an angle to the base.

elongation [astronomy] The angular distance between a planet (or the Moon) and the Sun.

em [linear measure: type] In any typeface, the width of the capital letter M (initially the widest letter), used in multiples to form a long dash (an em-dash), as opposed to the shorter en-dash based on the letter N . The em-dash is usually printed tight up against the preceding and following words; the en-dash is preceded and followed by a thin space.

By extension, and more commonly, a unit of pica type used to measure the width of book or newspaper columns, and giving some idea of the average number of characters per line, per typeface.

In this sense, 1 pica em = ⅙ inch, 12 POINTS
          = 4.233 millimetres

emalangeni *see* LILANGENI

embolismic [time] Describing a year in which a period of time has been intercalated (inserted additionally). Leap years are embolismic years, as are those years in the Jewish calendar in which an extra lunar month is intercalated. *See also* YEAR.

emergent year [time] The year or date at which an era begins: the year 0 in a calendar. For most Western people (using the Gregorian calendar, an extension of the Julian), the emergent year is the non-existent year between 1 BC and AD 1.
See also ERA, CALENDRICAL; YEAR.

EMF, emf *see* ELECTROMOTIVE FORCE

emissivity [physics] At a given temperature, the emissivity of a surface is the emissive power (energy radiated at all wavelengths per unit area per unit time) divided by the emissive power of a perfect black body at the same temperature. It is highest for matt black surfaces and lowest for polished metallic ones. As a ratio, it has no units.

empirical formula [chemistry] A chemical formula giving the simplest ratio between the atoms in a molecule, usually determined by analysis. For example:

acetic acid (ethanoic acid) and glucose have the same empirical formula of CH_2O (their molecular formulae are $C_2H_4O_2$ and $C_6H_{12}O_6$, respectively).

Such formulae may be worked out by trial and error – as indicated by the derivation of the term *empirical* from the Greek *empeirikos* 'experienced', itself deriving from the elements *en-peira-* 'in experiment'.

empty set [maths] In set theory, a set that has no members, denoted by the symbol $\emptyset$. For example, the set {moons of the planet Venus} $= \emptyset$.

EMU, emu *see* ELECTROMAGNETIC UNITS

en [linear measure: type] In any typeface, the width of the capital letter N (traditionally half that of the widest letter, M). As a unit of length it is used primarily in the form of an en-dash, a dash of the N's width, as opposed to the longer em-dash which corresponds to multiples of a dash the width of the M. The em-dash is usually printed tight up against the preceding and following words; the en-dash is preceded and followed by a space.

endomorph [medicine] In HUMAN MORPHOLOGY, the human body shape supposedly corresponding to the type that is well built, not to say fat, and of a rather extrovert disposition. The other two types so classified are MESOMORPHS and ECTOMORPHS.

The term derives from scientific Greek *endo-* 'of inner structures' (implying fatty layers) + *morph-* 'form', 'shape'.

endothermic [chemistry] Describing a chemical reaction that requires a supply of heat to make it happen (which is the case for most reactions). *See also* EXOTHERMIC.

end-point [chemistry] For a titration in VOLUMETRIC ANALYSIS, the stage at which reaction is complete (that is, the reacting volumes of the solutions are exactly equivalent). It is often detected by using an indicator (such as litmus or methyl orange for an acid-alkali titration).

endurance [aeronautics] The longest time an aircraft can keep flying before running out of fuel, sometimes stated in terms of distance (maximum range) rather than actual time.

energid [biology] A 'unit' of life consisting of the nucleus and cytoplasm of a cell (but no outer membrane, and not therefore a complete cell).

energy [physics] Energy – the capacity for doing work – is measured in JOULES in the SI system, and in ERGS in the centimetre-gram-second (CGS) units.

$$1 \text{ erg} = 10^{-7} \text{ joule}$$
$$1 \text{ joule} = 10 \text{ million ergs}$$

There are various kinds of energy – such as chemical energy, electricity, heat, kinetic energy, light, potential energy, and sound. It is a tenet of modern physics that they are interconvertible, in the way, for example, a dynamo converts mechanical (kinetic) energy into electrical energy, and, as in an explosion, chemical energy is converted into heat, light, and sound. It can be argued that all types of energy are manifestations of kinetic energy (that, for example, heat in a material results from the kinetic energy of its vibrating atoms or molecules). According to the mass-energy equation in the special theory of relativity, energy and mass are equivalent:

$$E = mc^2$$

where c is the velocity of light (*see* SPEED OF LIGHT).

See also KINETIC ENERGY; POTENTIAL ENERGY.

energy barrier [chemistry] The amount of FREE ENERGY that a chemical must have before it will undergo a particular chemical reaction. *See also* ACTIVATION ENERGY.

engine speed [physics; engineering] The speed of an engine is usually stated in revolutions per minute (rpm) of the main rotor (for a turbine) or crankshaft (for a reciprocating engine).

Engler degree [physics] Unit of viscosity of a liquid, defined as the time needed for a given volume of the liquid to flow through a standard orifice divided by the time taken for an identical volume of water to flow through the same orifice (at 20°C). One Engler degree equals a dynamic viscosity of one hundred-thousandth square metre per second (10^{-5} m^2s^{-1}).

English system (of measurements) *see* IMPERIAL SYSTEM

ennead [quantitatives] A group of nine. The adjective is *enneadic* ('of a group of nine').

enneagon [maths] A POLYGON with nine sides (and nine angles), more often called a nonagon.

enneahedron [maths] A solid figure (POLYHEDRON) with nine plane faces.

ensign [military rank] In the United States Navy, the lowest rank of commisioned officer, just above the (non-commissioned) warrant officer and below a lieutenant junior grade.

In former centuries the title was also that of a junior officer in the British army, whose duties included carrying the regimental colours or flag (or ensign) on parade or into battle.

enthalpy (heat content) [physics] Enthalpy – the amount of heat energy in a substance – is measured in JOULES in terms of the change in heat accompanying a chemical reaction that takes place at constant pressure.

For a system of internal energy U, pressure p and volume V,
$$\text{enthalpy } (H) \; = \; U + pV$$

entropy [physics] A difficult concept in thermodynamics, entropy is a measure of the unavailability of a system's energy to do work – that is, a measure of its disorder.

In a reversible process, the ratio of the energy (usually heat) absorbed to the absolute temperature is equal to the change in entropy, and hence is measured in units of JOULES per KELVIN.

E-numbers [medicine: nutrition] A set of numbers allocated to food additives, the E prefix indicating that they are recognized throughout the European Community. They include colourings (numbers E100 to E180), preservatives (E200 to E297), antioxidants (E300 to E321), emulsifiers and stabilizers (E322 to E499), acids, anticaking agents, and others (500 to 578, with no E), flavour enhancers and sweeteners (620 to 637), and miscellaneous additives (900 to 927).

envelope [maths] A plane curve that touches all members of a family of curves; the family is also known as an envelope.

enzyme unit [medicine] The amount of an enzyme that will catalyse the transformation of 1 micromole (one-thousandth of a gram-molecular weight) of substrate per minute; temperature, pH (acidity), and concentration of substrate have to be specified. Activity is expressed in units per milligram. The second element of the word *enzyme* derives from ancient Greek *zume* 'leaven', '(cause of) fermentation', referring to the fact that enzymes do indeed cause digestive fermentation in the human stomach and duodenum (cf. ZYMOMETRY).

Eocene epoch [time] A geological EPOCH during the Tertiary period of the Cenozoic era, corresponding roughly to between 55 million years ago and 38 million years ago. During this epoch there was a general warming of the Earth's surface and atmosphere. By the end, tropical and subtropical vegetation abounded in many forms still visible today.

The name of the epoch derives through French from ancient Greek *eos* 'dawn', *kainos* 'new', 'recent'.

For table of the Cenozoic era *see* PALAEOZOIC EPOCH.

Eogene [time] In geological time, a collective name for the Palaeocene, Eocene, and Oligocene EPOCHS of the Tertiary period in the Cenozoic era, corresponding roughly to between 65 million years ago and 25 million years ago (when the Neogene began with the Miocene epoch). The Eogene is sometimes alternatively called the Nummulitic or the Pal(a)eogene.

Eolithic [time] From prehistoric times, the beginning of the Stone Age (Greek *eos* 'dawn', *lith-* 'stone'). At this stage, early people made and used crude stone tools now known as eoliths.

eon *see* AEON, EON

Eotvos unit [physics; geology] In geophysics, a unit for measuring the change in gravitational field intensity with distance.
$$1 \text{ Eotvos unit} \; = \; 10^{-9} \text{ galileo per centimetre}$$
(where 1 galileo is an acceleration of 1 centimetre per second per second, or 10^{-2} ms^{-2}). The unit was named after the Hungarian physicist Baron Roland von Eotvos (1848-1919), better known as the inventor of the torsion balance.

Eozoic [time] An imprecise term intended to represent the geological time at which life forms of any kind first appeared on the planet Earth [and derived suitably from ancient Greek: *eos* 'dawn', *zoion* 'life (form)']. That time for most authorities is subsumed within the PRECAMBRIAN ERA, sometimes alternatively called the Archaeozoic era.

epact [time] The 11 days (12 in a leap year) by which the solar year is longer than the lunar year: 365 (or 366) as opposed to 354. The term is occasionally applied instead to the number of days in the lunar month that have already passed at the beginning of the solar year (1 January). The term derives from an ancient Greek participle meaning 'brought in' – that is, intercalated or inserted additionally. *See also* YEAR.

epagomenal [time] Describing a period of a day or days inserted into the calendar. In leap years, 29 February is an epagomenal day.

epee, épée *see* FENCING MEASUREMENTS AND POSITIONS

ephah [volumetric measure] In ancient Israel (and therefore in the Bible), a measure of dry capacity.

1 ephah	=	3 SEAH, 10 OMERS, 18 CAB or kab, 72 LOGS
	=	40,320 cubic centimetres, 0.04032 cubic metre (1 cubic metre = 24.802 ephahs)
	=	2,460.24 cubic inches, 1.4238 cubic foot (1 cubic foot = 0.7023 ephah)
	=	1.1159 UK bushel, 1.1445 US bushel (1 UK bushel = 0.8961 ephah 1 US bushel = 0.8737 ephah)
5 ephahs	=	1 lethech
10 ephahs	=	1 KOR or homer

ephemeris [astronomy; navigation] A set of tables that give the daily positions of the Sun, Moon, planets, and some stars, used as an aid to navigation and astronomical observation. The term derives from ancient Greek elements *epi-hemera* 'upon (that is, relevant only for) a day'.

ephemeris second [physics: astronomy] Fundamental unit of time equal to $\frac{1}{31556925.9747}$ of the tropical year for the beginning of January in 1900: *see* EPHEMERIS TIME.

ephemeris time [time: astronomy] A standardized measure of time adopted in 1956 by the International Bureau of Weights and Measures, and based upon units (ephemeris seconds) derived from the precise time taken by the Moon and the planet Earth to accomplish one complete orbit each during the (solar) year 1900. It differs by several seconds from universal time – which varies because of variations in the Earth's rate of rotation – and is sometimes alternatively known as Newtonian time. *See also* EPOCH.

epicentre [seismology] The point at the Earth's surface directly above the centre of an earthquake, and from which (from the viewpont of surface dwellers) the earthquake itself seems to radiate.

epicycle, epicycloid [maths] A cusped curve generated by a point on the circumference of a circle that rolls around the circumference of another circle. The epicycle that results when both circles are of the same diameter is known as a *cardioid*.

epoch [physics: time] Outside geological contexts, and in a general sense, an epoch is any lengthy period of time (over at least ten years) during which something is constant or recurrent. But in a technical sense in astronomy, an epoch is a precise instant of recorded time to which astronomical observations or calculations are referred. In this sense, for example, EPHEMERIS TIME is defined in relation to an epoch at the beginning of January in the year 1900.

The word derives as a noun from the ancient Greek verb *ep-echein* 'to hold on' (in all senses – including thus both 'to keep going' and 'to stop').

epoch, geological [time] Division of geological time within the Cenozoic era (the most recent ERA) during either of the geological PERIODS known as the Tertiary and the Quaternary. Similar divisions occur also in two of the other three eras, but are not generally called epochs (and are for the most part divisions of relevant periods into 'Upper', 'Middle', and 'Lower' parts).

According to most authorities there are seven epochs. From the oldest to the most modern they are:

Palaeocene	
Eocene	Tertiary period
Oligocene	(65 million years ago
Miocene	to 2-3 million)
Pliocene	

Pleistocene	Quaternary period
Holocene	(from 2-3 million years ago)

See information *under each* individual epoch.

equal area projection [geography: cartography] Method of cartographic projection, a variety of conical projection. Most useful for charting larger areas on a small scale, the projection has the disadvantage of being less accurate in representing angles and compass bearings.

equal sets [maths] Sets that contain the same elements.
Example: $\{2, 4, 6, 8\}$ and $\{8, 6, 4, 2\}$

equation [maths] A mathematical statement of equality between quantities or algebraic expressions.
For example:

$$3 + 2 = 5,$$
$$4x - 3 = 13,$$
$$\text{and} \quad x^2 - 3x + 2 = 0$$

are all simple equations. The second example, $4x - 3 = 13$, is a *linear equation* (in which $x = 4$); $x^2 - 3x + 2 = 0$ is an example of a *quadratic equation* (in which $x = 1$ or 2).

equation [chemistry] Shorthand method of describing a chemical reaction using chemical formulae. The formulae use the symbols of the chemical elements to represent atoms, with suffixes (inferior numbers) to indicate the number of those atoms in a particular molecule. For example: each molecule of water – formula H_2O – consists of two hydrogen atoms combined with one oxygen atom; sulphuric acid – formula H_2SO_4 – has two hydrogen atoms, one sulphur atom, and four oxygen atoms.

In a chemical equation, the reactants (the molecules that react together, represented by their formulae) are written on the left, and the products of the reaction are written on the right, with a directional arrow or equals sign between them. The equation must balance – that is, the total number of atoms of an element on the left of the equation must equal the number of those atoms on the right. Thus, the equation for the formation of water by burning hydrogen in air or oxygen is written

$$2H_2 + O_2 \rightarrow 2H_2O$$

which indicates that two molecules of hydrogen combine with one molecule of oxygen to form two molecules of water. (Note that O_2 on the left side of the equation is balanced by 2 ... O on the right.) The equation for the reaction between sulphur trioxide and water to form sulphuric acid is

$$H_2O + SO_3 \rightarrow H_2SO_4$$

equation of state [physics/chemistry] An equation that describes a system in terms of its volume, pressure, and temperature; for example: VAN DER WAALS' EQUATION.

equation of time [astronomy; time] The difference between mean solar time (as shown on a sundial) and apparent time (as shown on a clock). Measured in minutes, it has a positive maximum value of about 14.5 minutes in February and a negative maximum value of about 16.5 minutes in November. Four times a year – at the solstices and equinoxes – it is equal to zero.

equations of motion [physics] The motion of an object that takes time t to move a distance s in a straight line, with an initial velocity v_1 and final velocity v_2 that involves an acceleration a, can be described in terms of five equations, each containing only four of these parameters.

$$s = v_2 t - \tfrac{1}{2} at^2 \qquad\qquad v_2 = v_1 + at$$
$$s = v_1 t + \tfrac{1}{2} at^2$$
$$s = \tfrac{1}{2} t(v_1 + v_2) \qquad\qquad v_2^2 = v_1^2 + 2as$$

equator [physics; geography; maths] A circle that divides a sphere into two equal parts. The Earth's equator – latitude $0°$ – runs around the greatest girth of the planet equidistant from the poles, dividing the Earth into Northern and Southern Hemispheres. At the equator, day and night are always of equal length. The terrestrial equator is 40,075.03 kilometres (24,901.47 miles) long. Because the planet is not an exact sphere, however – the globe is very slightly flattened at the poles – the circumference of the Earth through the poles is only 40,065.58 kilometres (24,895.75 miles). The term derives directly from a Latin word meaning precisely 'equator' – that is, equalizer (of day and night).

equilateral [maths] Of a plane figure, having sides of equal length (so that one side is equal in length to all others). In an equilateral triangle, for example, the sides are all the same length, and the angles are all equal at 60 degrees.

equilibrium constant [chemistry] For a reversible chemical reaction at a given temperature, the ratio of the product of the concentrations (active masses) of the products of the reaction to the product of the concentrations of the reactants. The concentrations are those of the substances at equilibrium, as indicated by the chemical EQUATION for the reaction.

equimolecular [chemistry] Description of a mixture in which the constituents are in equal molecular proportions.

equinox [astronomy] The time of year at which day and night are of equal length, as the Sun appears to cross the CELESTIAL EQUATOR. There are two equinoxes every year, in the spring (the vernal equinox) and autumn (autumnal equinox). The vernal equinox in the Northern Hemisphere corresponds exactly to the autumnal equinox in the Southern, and vice versa. The term derives from Latin words meaning 'equal night (and day)'. The two times of year at which the lengths of day and night are conversely least similar are the SOLSTICES.

equity capital [comparative values] The finance put into a business or property as personal investment by the owner or owners, and not loaned by or borrowed from others.

equivalent conductance [physics; chemistry] For a solution containing 1 gram equivalent weight of solute (dissolved substance) per litre (a normal solution), the electrical conductance of the solution per millimetre thickness. *See also* MOLAR CONDUCTANCE.

equivalent sets [maths] Sets that have the same cardinality – that is, sets in which each element in one set has a one-to-one correspondence with an element in the other set or sets, denoted by the symbol R .

   Example: {Monday, Tuesday, Wednesday} R {A, B, C}

equivalent weight [chemistry] The mass of one substance that reacts with a given mass of a standard specified substance. This is usually taken to mean the mass of a substance (element or compound) that will combine with or displace 1.00797 grams of hydrogen or 8 grams of oxygen.

era, calendrical [time] Historical period, the beginning of which is represented by an EMERGENT YEAR, a year from which the calendar is dated thereafter. In ordinary Western terms, BC ('before Christ') and AD (*anno Domini* 'in the year of the Lord') are the calendrical eras mostly encountered – BC after the year number, AD before the year number. In academic circles, however, the preference in relation even to these periods of historical time is to describe years BC as years BCE ('before the common era') and AD years as years CE ['(of the) common era' or '(of the) Christian era'] – both BCE and CE after the year number.

   Other calendars with different emergent years naturally relate to their own era. The Jewish calendar, for example, is calculated to date from the year of the Creation, and history is reckoned therefore in 'years of the world', *anno mundi*, or AM (or in the Hebrew equivalent), after the year number. Similarly, the emergent year of the Hindu calendar corresponds to AD 78. Many Muslim communities use a calendar that celebrates the migration of the Prophet Muhammad from Mecca to Medina, the Hegira (or Hijrah), and therefore date their years from that time (AD 622), in years 'after the Hegira', or AH, after the year number. *See also* YEAR.

era, geological [time] Major division of the geological history of the planet Earth. According to most authorities, there are four eras, the Precambrian, the Palaeozoic, the Mesozoic, and the present era, the Cenozoic. In turn, the eras are divided into PERIODS of various durations corresponding to the different geological events that took place.

Era	Period	millions of years ago
Cenozoic	Quaternary	from 2-3 on
	Tertiary	about 65 to about 2-3
Mesozoic	Cretaceous	about 144 to about 65
	Jurassic	about 213 to about 144
	Triassic	about 248 to about 213
Palaeozoic	Permian	about 286 to about 248
	Carboniferous	about 360 to about 286
	Devonian	about 408 to about 360
	Silurian	about 438 to about 408
	Ordovician	about 505 to about 438
	Cambrian	about 590 to about 505
Precambrian		about 4,600 to about 590

The Precambrian is sometimes alternatively known as the Archaeozoic, and the term 'Precambrian' is in that case used as a name for the single period within the era. The periods of the Cenozoic era are also divided into EPOCHS of various durations. All the periods of every era are also subdivided again into 'stages' that refer to particular geological formations in specific parts of the world.

erg [physics] Unit of energy in the centimetre-gram-second (CGS) system of measurement. It is defined as the work done when a force of 1 DYNE moves a distance of 1 centimetre. In terms of SI units,

$$1 \text{ erg} = 10^{-7} \text{ JOULE}$$
(1 joule = 10 million ergs)

The unit gets its name from the ancient Greek *ergon*, which not only means 'work' but is etymologically the same word. *See also* ENERGY.

ergometer [physics] Alternative term for a DYNAMOMETER.

eric [comparative values] In medieval Ireland, the blood money paid by a murderer or accidental killer (or his family) to the victim's family in full and complete satisfaction for the death, so that no further punishment or obligation would be imposed or sought.

The Irish word *éiric*, from which 'eric' derives, in present-day Irish Gaelic means no more than 'forfeit', 'retributive loss'.

eriometer [physics] An optical instrument used in the measurement of the diameters of fibres or wool particles. It works by viewing the fibres or particles under conditions of light diffraction that produce coloured rings of proportionately differing sizes.

erlang [physics: telecommunications] Unit of traffic flow in telephony, equal to the product of the number of telephone calls per hour and the average duration of each call. The unit was named after the Danish mathematician A. K. Erlang (1878-1929).

escadrille [military unit] During the first half of the twentieth century, a small squadron most commonly made up of six aircraft or six warships (with all their men and equipment). The term derives, like *squadron*, ultimately from a diminutive of the noun formed from the past participle of the late Latin (soldiers' slang) verb *exquadrare* 'to form a square'.

escape velocity [physics; astronomy] The velocity that a rocket or satellite has to attain immediately after launching to overcome the gravitational attraction of the planet or moon it is leaving and go on out into space. Gravity will still affect it but, at this speed and above, it will not have obligatorily to fall back as it proceeds along its parabolic trajectory. Also known as the cosmic speed, it may be calculated by means of the following equation:

$$\text{Escape velocity} = (2GM/R)^{½}$$

where G is the gravitational constant, M is the mass of the body the projectile is leaving, and R is the distance from the centre of that body to the projectile at launch.

The escape velocity at the surface of the planet Earth is 11.2 kilometres per second, 6.959 miles per second.

The escape velocity at the surface of the Moon is 2.4 kilometres per second, 1.491 miles per second.

The escape velocity at the 'surface' of the Sun is 617.7 kilometres per second, 383.821 miles per second.

The escape velocity, or cosmic speed, is also the velocity at which a body arriving from an infinite distance hits the surface of a massive celestial body.

escudo [comparative values] Unit of currency in Portugal, formerly also the unit of currency in Angola, Chile, Guinea-Bissau, and Mozambique.

$$1 \text{ escudo } = 100 \text{ centavos}$$

The name derives from the embossed shield (Portuguese *escudo*, cf. English *escutcheon*) on one side of the coin.

ester value [chemistry] An ester is a chemical compound formed between an alcohol and an acid, and esters are the basic constituents of fats and oils. The breaking up of an ester by strong alkali (as in soap making) is called saponification. The ester value is the amount of potassium hydroxide needed to saponify 1 gram of a fat; it equals the difference between the saponification value and the acid value.

ESU *see* ELECTROSTATIC UNITS (ESU)

etalon [physics: interferometry] A type of interferometer used to measure spectral wavelengths by means of a thin film of air between semi-silvered flat glass or quartz plates: multiple reflections of light between them produce interference fringes. The term derives from French *étalon* 'standard (measure)'.

etesian [time] Annual, yearly. (From ancient Greek *etesios* 'yearly'.)

eudiometer [chemistry; physics] An apparatus for determining the quantity of electricity passing by measuring the volumes of gases produced at electrodes during electrolysis; it is a type of VOLTAMETER.

Also, a similar apparatus for measuring the change in volume when a mixture of gases react together (for example, by combustion). In former times, however, a eudiometer was a simple meteorological instrument used to measure the purity of air, and especially the volume of oxygen in it. That is how it got its name, the first element of which corresponds to ancient Greek *eudios* 'clear (air)', 'fine (weather)'.

Euler number [physics; flow] Unit of fluid flow equal to the pressure in a liquid divided by the product of its density and the square of its velocity (p/dv^2).

The unit was named after the Swiss mathematician Leonhard Euler (1707-83), who devised Euler numbers as part of a mathematical series.

euploid [medicine: genetics] Description of a cell in which the nucleus has one or more whole sets of chromosomes – that is, is haploid, diploid, or polyploid.

Eurodollar [comparative values] A US dollar owned by a (non-US) European national and used as a medium of credit in financial affairs in Europe.

euryon [medicine] In the anatomy of the skull, either of the points at each end of the greatest diameter across the top of the skull from one side to the other. The term derives as a learned borrowing based on the ancient Greek *eurys* 'wide'.

eutectic point [physics] A eutectic is a mixture of substances of such a composition that the melting point of the mixture is the lowest possible: any change to the composition raises the melting point. The range of compositions of a two- or three-constituent mixture can be indicated on a phase diagram (a graph of temperature in relation to composition), and the eutectic point is the temperature on the diagram that corresponds to the formation of a eutectic. The term derives as a learned borrowing from ancient Greek *eu-tektos* 'readily melted', 'easily soluble'.

eV *see* ELECTRON VOLT

evection [physics; astronomy] The orbit of the Moon around the Earth is not regular – it is not a perfect ellipse – but varies periodically. The variation can be defined by a mathematical expression, and the largest of the variations in it is called the evection. Its maximum value is 1 degree 16 minutes, with a periodicity of 31 days 19 hours 26 minutes 24 seconds (31.81 days).

even-even nucleus [physics] An atomic nucleus that has even numbers of both protons and neutrons (a property associated with stability). *See also* EVEN-ODD NUCLEUS; ODD-ODD NUCLEUS.

even money [comparative values] The result of a bet in which the better profits by precisely the amount put up as the initial stake.

even-odd nucleus [physics] An atomic nucleus that has an even number of protons but an odd number of neutrons. *See also* EVEN-EVEN NUCLEUS; ODD-ODD NUCLEUS.

evolute [maths] A curve formed by linking the centres of curvature of points on another curve (which is called the involute).

exa- [quantitatives: prefix] Prefix which, when it precedes a unit, multiplies the unit by 1 UK trillion/1 US quintillion (10^{18}) times its standard size or quantity.

Example: 1 exavolt = 1,000,000,000,000,000,000 volts

In contrast to virtually all other quantitative prefixes, not only is the derivation of this prefix not readily evident, it is unknown.

exacta [sporting term] In betting on horses or greyhounds in North America, a system by which the better must nominate the first and second animals in the correct order to win the bet. There is also the quiniela system by which the better correctly nominates the first two animals home, with no reference to which comes first and which second, to win the bet.

exaltation of larks [collectives] It is only outside the breeding season that a person might see an exaltation of larks – a group of larks in flight – for groups on the ground are merely 'flocks', and, during the breeding season, larks tend to move only in pairs. In any case, larks are more often heard than seen in Britain (two native species), in North America (one native species, one introduced species), and in Australasia (similarly one native species, one introduced species).

The collective noun refers by derivation to the fact that larks in summer tend to fly up vertically to a great height (Latin *ex-altus* 'up high') and not to any rapturous emotion on the part either of the larks or of an observer.

excess loading [comparative values] In insurance, the sum that the person insured agrees to pay as the first part of any substantiated claim by a third party or of any claim that results from other specified circumstances: the insurance company agrees to indemnify the insured up to a stipulated amount 'in excess' of this primary sum. It is a means by which insurers try to avoid having to pay out on minor or frivolous claims.

exchange rates [comparative values] The rates or values set by banks and other monetary authorities in one country at which, until further notice, they will give out the currencies of other countries in exchange for that of their own country (the selling rate), and at which they will give out their own country's currency in exchange for those of other countries (the buying rate). The rates as displayed outside individual banks or exchange organizations often include a percentage corresponding to the commission chargeable to the customer for a transaction in either direction. *See also* COINS AND CURRENCIES OF THE WORLD; VALUTA.

Exchequer bills and bonds [comparative values] Exchequer bills were formerly issued by the British government to raise funds for some specified event or national activity (such as a campaign of war). Such bills were negotiable bills of credit on which interest was payable at a rate subject to variation. Exchequer bonds are issued by the British government, representing receipts for the loan of sums of money over a specified time at a fixed rate of interest. The fiscal authority by which such bills and bonds were made available to the public used tables or desks, the tops of which were inlaid with squares of black and white veneer. Coins or notes stacked in set amounts on the squares could then readily be counted by the number of squares covered. The appearance of these tables gave the fiscal authority its name (medieval French *eschequier* 'chessboard', 'checkerboard'), the Exchequer.

exclusion principle *see* PAULI EXCLUSION PRINCIPLE

exhaust velocity [physics: rocketry] For a rocket motor, the speed at which the propellant gases leave it, equal to the product of the specific impulse and the acceleration of free fall.

exon [biology: genetics] In all animal and plant cells, DNA (deoxyribonucleic acid) carries the genetic code that represents the instructions on how cells are to replicate, and RNA (ribonucleic acid) helps to transmit and decode those instructions. The exon is the part of the transcribed RNA of nucleated cells that becomes messenger RNA after removal of its introns.

exosphere *see* ATMOSPHERE, COMPOSITION OF

exothermic [chemistry] Describing a chemical reaction that evolves heat.

expansion [maths] In algebra, the extended or expanded form of an expression. For example: the expansion of $(2 + x)^2$ is $4 + 4x + x^2$.

expansion, coefficient of *see* COEFFICIENT OF EXPANSION

expansion of gases [physics] An ideal (perfect) gas increases in volume by $\frac{1}{273}$ of its volume at 0°C (273 K) for every degree rise in temperature (a statement of Charles's law). The actual expansion can be measured in any convenient volumetric units, such as millilitres or litres.

expansion of the universe [astronomy] In the expanding universe, galaxies recede from us at a speed that is proportionate to their distance from us: the most distant galaxies are thus receding the fastest. A measure of the rate at which the universe is expanding – in the range 50 to 100 kilometres per second per megaparsec – is HUBBLE'S CONSTANT, which was named after the American astronomer Edwin Hubble (1889-1953).

expansivity [physics] A measure of the increase in size of a substance (caused by thermal expansion) for every degree rise in temperature. It applies to length (linear expansivity), area (superficial expansivity), or volume (volumetric expansivity), and is expressed in such units as millimetres per degree, square centimetres per degree, and millilitres per degree. *See also* EXPANSION OF GASES.

expectancy/expectation of life *see* LIFE EXPECTANCY

expectation [maths] In statistics, the average (mean) value from an infinite series of observations of a random variable.

explicit function [maths] When a variable x can be expressed directly in terms of y, x is said to be an explicit function of y.

exponent, power [maths] A number indicating the power to which the BASE (another number) is to be raised (multiplied by itself), written as a superior number or symbol. For example: in the expressions 3^5, y^3, 2^x, the exponents are 5, 3, and x.

The exponent/power 2 is known as square ($x^2 = x$ squared $= x \times x$); and the exponent/power 3 is called cube ($5^3 = 5$ cubed $= 5 \times 5 \times 5$). The exponent/power $\frac{1}{2}$ indicates a square root; $\frac{1}{3}$ indicates a cube root. Exponents/powers expressed as minus numbers represent reciprocals (whereas 10^6 represents 1 million, 10^{-6} represents one-millionth). Any number to the exponent/power 0 is 1 ($10^0 = 1$, $256^0 = 1$, $x^0 = 1$).

exponential function [maths] The function e^x, often associated with systems undergoing natural growth or decay. *See also* DISINTEGRATION CONSTANT; E.

exposure meter [physics: photography] Instrument used in photography for measuring the light incident on or reflected from an object to determine the correct exposure (combination of shutter speed and aperture) for a given speed of film. Also called light meters, most consist of some sort of photoelectric cell of which the output is indicated by a voltmeter calibrated directly in shutter speed (in seconds) or aperture (in f-numbers). Light meters are also used to assess the brightness of daylight in some sports (for example in cricket, in which bad light can – and frequently does – stop play).

exposure value [physics: photography] In photography, a number representing any correct combination of aperture (f-number) and shutter speed for correct exposure (in particular lighting conditions). For example, changing the aperture leads to a complementary change in shutter speed to maintain the same exposure value.

extensometer [physics] An instrument that detects and displays the minute changes in length or shape in a material (especially a metal) that result from an increase or decrease in temperature. One type consists of a capacitor with one plate fixed and the other attached to the object whose dimensions are being monitored.

exterior angles [maths] In geometry, when two parallel lines are crossed by a single line, any of the four angles subtended outside the parallel lines; exterior angles on opposite sides of the crossing line are equal, exterior angles on the same side of the crossing line are supplementary (add to 180°).

extinction coefficient [chemistry; physics] A characteristic of a dissolved substance in terms of the amount of light (infra-red, visible, or ultraviolet) it absorbs. Also called molar absorbance, it is defined as the ABSORBANCE of a solution, concentration 1 mole per cubic decimetre, measured in a cell 1 centimetre thick.

extra- [quantitatives: prefix] Prefix meaning 'outside' and thus sometimes 'more than normal', 'more than necessary'.

extrapolation [maths] Adducing a value or values outside the range of known values, usually by extending a line on a graph.

extras *see* CRICKET MEASUREMENTS, UNITS, AND POSITIONS

extremely high frequency [physics: telecommunications] Imprecise term for radio frequencies in the MEGAHERTZ range.

extremely low frequency [physics: telecommunications] Imprecise term for radio frequencies in the range below about 100 KILOHERTZ.

extremum [maths] The maximum or the minimum value of a function.

eye tests *see* OPHTHALMIC TESTING

F

F, F. *see* FAHRENHEIT SCALE OF TEMPERATURE; FRANC

f, f. *see* FARAD; FEMTO-; FORTE, FORTISSIMO; FRANC; FREQUENCY

F1 layer *see* F LAYER, F REGION

F2 layer *see* F LAYER, F REGION

face card [sporting term] In card games in the United States, the jack (knave), queen, or king of any suit. In Britain they are more commonly known as court cards.

face cord [cubic measure] In the measurement of cut firewood in the United States, a unit evidently meant to relate to the CORD and CORD FOOT, but varying between different States. In general, however,

1 face cord	=	64 cubic feet, 2.3704 cubic yards
	=	a pile 4 feet high, 2 feet deep, and 8 feet wide
	=	half a cord, 4 cord feet
	=	1.8123 cubic metre
		(2 cubic metres = 1.1036 face cord)

facial angle [biology] In comparative anatomy, the angle between a line from the nostril to the centre of the ear and a line from the nostril to the centre of the forehead.

facial index [biology] In comparative anatomy, the ratio of the length of a person's face to its width, usually expressed as a single figure corresponding to how much the former is greater than the latter. An alternative method of calculation is to express the ratio of length to width as a percentage.

faciend [maths] An alternative name for MULTIPLICAND.

factor [maths] A number that will divide exactly into another number, or a polynomial that will divide into another polynomial. For example, 1, 2, 3, 6, 12, 24, and 48 are all factors of 48; $(x + 1)$ and $(x - 3)$ are factors of $x^2 - 2x - 3$. The term derives immediately from Latin, meaning 'doer', 'maker'.

factorial, factorial number [maths] The product of all the integers (whole numbers) from any particular integer down to 1. If the particular integer is n, its factorial is written $n!$ and it equals $n \times (n - 1) \times (n - 2) \times (n - 3)... 3 \times 2 \times 1$.

For example: factorial 7 (7!) equals $7 \times 6 \times 5 \times 4 \times 3 \times 2 \times 1 = 5,040$.

Factorial n ($n!$) equals the number of different ways of arranging n objects: *see* PERMUTATION.

Fahrenheit scale of temperature [physics] A temperature scale on which the fixed points are the freezing point of pure water ($= 32°F$) and the boiling point of pure water ($= 212°F$). To convert a Fahrenheit temperature to a Celsius (formerly centigrade) temperature, subtract 32 and multiply the remainder by 5/9 . Thus, for example:

$$212°F = (212 - 32) \times 5/9 = 180 \times 5/9 = 100°C$$

It was named after the German physicist Gabriel Fahrenheit (1686-1736). *See also* CELSIUS; KELVIN.

Fajan's rules [chemistry] A set of conditions that expresses the likelihood that ionic (polar) bonds will be formed between atoms or ions in a chemical compound (rather than covalent bonds). The conditions include the presence of a large cation (positive ion), a small anion (negative ion), and small ionic charge. If the cation has a stable electronic structure (such as an outer octet of electrons, like those of an inert gas), ionic bonding is also favoured. The rules were named after the Polish chemist Kasimir Fajans (1887-1975).

Fajans-Soddy law [physics] For an element undergoing radioactive decay, its atomic number increases by 1 if it emits BETA PARTICLES, or decreases by 2 if it emits ALPHA PARTICLES. The law was named after Kasimir Fajans (*see above*) and the British physicist and chemist Frederick Soddy (1877-1956).

fall line [geology] A contour, generally at the inland edge of a coastal plain, that represents the border between harder rock (often forming a ridge or mountain) and softer rock (of which the coastal plain is made). The best-known example is the Fall Line that runs north-south to the east of the Appalachian Mountains, on which the cities of Richmond, Baltimore, and Philadelphia were built. The term derives from the fact that the line corresponds to where streams and rivers that have fallen fast down the ridge or mountain slope take on a more peaceful, languid flow.

Fallot's tetralogy [medicine] A four-fold malformation of the heart in babies and children that may lead to severe under-oxygenation of the blood, but that may be corrected surgically. The four defects are:

> a hole between left and right ventricles
> increased thickness in the wall of the right ventricle
> a narrowed valve into the pulmonary circulation
> displacement of the aorta

The condition was named after the French physician Etienne-Louis Fallot (1850-1911).

Family, family [biology: taxonomy] In the taxonomic classification of life-forms, the major category between Order and (Tribe or) genus.

For full list of taxonomic categories *see* CLASS, CLASS.

fanega [volumetric measure] A unit used in Spain and Portugal (and therefore sometimes in South American countries that were once politically linked with them) in both liquid and dry measure, although the Spanish variant is applied predominantly to dry goods. The unit differs by a small margin, however, between the two traditions, and differs again in derivatives.

1 Spanish fanega	=	48 (dry) CUARTILLOS, 12 Spanish ALMUDES
	=	55.50 litres
		(50 litres = 0.9009 Spanish fanega)
	=	12.21 UK gallons, 14.66 US gallons
		(12 UK gallons = 0.9828 Spanish fanega
		15 US gallons = 1.0232 Spanish fanega)
	=	1.526 UK BUSHEL, 1.575 US bushel
		(1 UK bushel = 0.6553 Spanish fanega
		1 US bushel = 0.6349 Spanish fanega)
	=	1.9594 cubic feet
		(2 cubic feet = 1.0207 Spanish fanega)
1 Portuguese fanega	=	16 QUARTOS, 3.315 Portuguese almudes
	=	55.364 litres
		(50 litres = 0.9031 Portuguese fanega)
	=	12.180 UK gallons, 14.626 US gallons
		(12 UK gallons = 0.9852 Portuguese fanega
		15 US gallons = 1.0256 Portuguese fanega)
7.75 Portuguese fanegas	=	1 Portuguese (old) PIPA
	=	429 litres
	=	94.38 UK gallons, 113.33 US gallons
1 Argentinian fanega	=	137 litres
	=	30.1366 UK gallons, 36.1926 US gallons
		(30 UK gallons = 0.996 Argentinian fanega
		36 US gallons = 0.995 Argentinian fanega)

The term derives as an Arabic expression – *faniqa* 'big sack' – that would appear originally to have referred only to dry capacity.

fanega, fanegada [square measure] A measure of land in Spain, Peru, and Mexico, presumably originally the same but now very different.

In Spain,	1 fanegada	=	about 80 x 80 metres, 6,400 square metres
		=	about 1.58 acre, 7,647 square yards
In Peru,	1 fanega	=	about 6,536 square metres
		=	about 1.615 acre, 7,816.6 square yards
In Mexico,	1 fanega	=	about 3.56 hectares, 35,600 square metres
		=	about 8.80 acres, 42,592 square yards

Again the term derives ultimately from Arabic *faniqa* 'big sack' (*see above*).

farad [physics] Unit of electrical capacitance, equal to the capacitance that carries a charge of 1 coulomb when it is charged by a potential difference of 1 volt. It is a large unit and, in practice, submultiples such as microfarads (10^{-6} farad) or picofarads (10^{-12} farad) are generally used.

$$1 \text{ farad} = 10^{-9} \text{ electromagnetic unit (EMU)}$$
$$= 9 \times 10^{11} \text{ electrostatic units (ESU)}$$

The unit was named after the British physicist and chemist Michael Faraday (1791-1867). *See also* FARADAY.

faraday [physics] Unit of electric charge, equal to the amount of charge that will liberate 1 mole (1 gram-equivalent weight) of an element during electrolysis. It equals 9.6487×10^4 coulombs per mole (gram-equivalent), and is the product of the electronic charge and AVOGADRO'S NUMBER. The unit is alternatively known as the Faraday constant, and was named after the British physicist and chemist Michael Faraday (1791-1867). *See also* FARAD.

Faraday's laws of electrolysis [chemistry; physics]

Two laws:

During electrolysis, the amount of chemical decomposition is proportional to the electric current passed.

The amounts of substances liberated (by a given quantity of electricity) during electrolysis are proportional to their chemical equivalent weights.

Faraday's laws of (electromagnetic) induction [physics]

Two laws:

Whenever the magnetic field linking an electric circuit changes, an induced electromotive force (EMF) is set up in it.

The size of the induced EMF in a circuit is proportional to the rate of change of the magnetic flux linking it.

farsang *see* PARASANGES, PARASANG, FARSANG

farthing [comparative values] Coin formerly used in Britain, and worth one-quarter of one penny. In fact, it was this proportion that gave it its name ('fourth-ing'). In the time of Geoffrey Chaucer (late 1300s), however, when few had detailed knowledge of arithmetic, the word was used for any small measure or quantity (of land, of a substance, etc.). The sixteenth-century translators of the New Testament of the Bible used 'farthing' to represent the Roman coins *as* and *quadrans*.

fathom [linear measure; volumetric measure] In shipping and on nautical charts, a unit used to measure the depth of water. Much less commonly, the length of nautical cables and hawsers may also be expressed in fathoms.

$$1 \text{ fathom} = 6 \text{ feet, 2 yards}$$
$$= 1.8288 \text{ metres}$$
$$(2 \text{ metres} = 1.0936 \text{ fathom})$$

In forestry and mining, a unit of volume of related size.

$$1 \text{ fathom} = 6 \times 6 \times 6 \text{ feet, 216 cubic feet}$$
$$= 6.1128 \text{ cubic metres}$$

Like various other measures (such as the ELL and the FOOT), the fathom seems to have originated as a unit that depended on human body proportions. The Old English verb *faethmian* meant 'to use both arms (to hold)', thus 'to embrace' and only thence 'to fathom', 'to comprehend'; the corresponding noun therefore meant 'the use of both arms', and the linear measure in this way derives as the distance between the tips of the middle fingers with both arms outspread. But the volumetric measure is very unlikely to have originated as an amount that could be picked up in both arms. *See also* BRAÇA, BRACCIO, BRASSE, BRAZA.

fatigue strength [metallurgy] For a metal, the maximum stress that it can tolerate without weakening or snapping. The fatigue strength of some metal bars can be increased by up to one-quarter by machining a thin surface layer off the bar.

fault [sporting term] In tennis, an error that negates a serve (so that a second service has to take place, or if the fault occurs on a second service, so that the point is lost). Such faults are:

hitting the ball into the net

hitting the ball so it does not land within the service court

serving with one or both feet across the base line

In show jumping and in three-day events cross-country courses, faults are the penalty points awarded for misdemeanours, such as knocking a rail or pole off a fence, refusing in front of a fence, unseating the rider, and (in some competitions) taking too much time and so transgressing a time restriction.

fault line, fault plane [geology] In rock strata, the fault plane is the plane along which a geological fault (sectional movement) takes place. The fault line is a contour that corresponds to the intersection of a fault plane with the ground surface.

featherweight *see* BOXING WEIGHTS

feddan [square measure] A measure of land in Egypt and the Sudan, approximating closely to the ACRE.

$$
\begin{array}{rcl}
1 \text{ feddan} & = & 4{,}200 \text{ square metres, } 0.42 \text{ hectare} \\
 & & (1 \text{ hectare} = 2.3809 \text{ feddans}) \\
 & = & 1.038 \text{ acre, } 5{,}023.92 \text{ square yards} \\
 & & (1 \text{ acre} = 0.9634 \text{ feddan})
\end{array}
$$

feeler gauge [engineering] A thin piece of hard metal (usually spring steel) of accurately known thickness for measuring or setting small gaps (such as the gap between the electrodes of a spark plug). Feeler gauges often come in sets of various thicknesses, labelled in millimetres or thousandths of an inch.

feet *see* foot

feet per minute [speed] Unit for measuring speed or velocity.

$$
\begin{array}{rcl}
1 \text{ foot per minute} & = & 0.2 \text{ inche per second} \\
 & & (1 \text{ inch every 5 seconds}) \\
 & = & 60 \text{ feet per hour, } 0.011364 \text{ mile per hour} \\
 & & (1 \text{ mile every 88 hours 7.2 seconds}) \\
 & = & 0.3048 \text{ metre per minute} \\
 & & (1 \text{ metre every 3 minutes 16.85 seconds}) \\
 & = & 18.288 \text{ metres per hour} \\
 & = & 0.018288 \text{ kilometre per hour} \\
 & & (1 \text{ kilometre every 54 hours 38.7 minutes})
\end{array}
$$

feet per second [speed] Unit for measuring speed or velocity.

$$
\begin{array}{rcl}
1 \text{ foot per second} & = & 60 \text{ feet per minute, } 20 \text{ yards per minute} \\
 & = & 1{,}200 \text{ yards per hour} \\
 & = & 0.6818 \text{ mile per hour} \\
 & & (1 \text{ mile every 1 hour 28 minutes}) \\
 & = & 0.3048 \text{ metre per second} \\
 & & (1 \text{ metre every 3.2808 seconds}) \\
 & = & 18.29 \text{ metres per minute} \\
 & = & 1.0974 \text{ kilometres per hour (km/h)} \\
 & & (1 \text{ kilometre every 54 minutes 40.5 seconds})
\end{array}
$$

femorotibial index [biology] In comparative anatomy, the length of the tibia (shinbone) expressed as a percentage of the length of the femur (thighbone).

femto- [quantitatives: prefix] A metric prefix which, when it precedes a unit, reduces the unit to 1 UK thousand billionth/1 US quadrillionth (10^{-15}) of its standard size or quantity.

Example: 1 femtometre (or fermi) = 0.000000000000001 metre

The prefix derives from the Danish *femten* 'fifteen'. (The only other internationally accepted quantitative prefix deriving from Danish is ATTO-.)

fen [comparative values] Unit of currency in (the People's Republic of) China.

$$
100 \text{ fen} \quad = \quad 10 \text{ chiao} \quad = \quad 1 \text{ yüan}
$$

The unit derives from an old Chinese measure of weight in silver. *See also* COINS AND CURRENCIES OF THE WORLD.

fencing measurements and positions [sport] Fencing is a highly evolved form of swordsmanship in which both electronics and strong protective clothing may play an essential part. There are three events, each using a different weapon.

The three weapons are:

foil

épée

sabre/saber

The dimensions of the piste:

width: 2 metres (6 feet 6 inches)

minimum length: 13 metres (39 feet 6 inches)

on guard line (each side of centre line): 2 metres (6 feet 6 inches) from centre line

warning line for épée and sabre events: 5 metres (16 feet 5 inches) from centre line

warning line for foil event: 6 metres (19 feet 8 inches) from centre line

Timing:

bouts for men – maximum 6 minutes

bouts for women – maximum 5 minutes

timing stops at each halt in the bout

bouts otherwise end when one fencer has scored 5 hits (men) or 4 hits (women)

Scoring:

hits with the point in the target area count 1 each in sabre – cuts with the edge or top of the back edge also count

off-target hits in foil and sabre may count if the defendant has greatly distorted the body to avoid the hit

electronic equipment is ordinarily used in foil and épée; in all events, a president and four judges adjudicate on hits

there are various sequences ('phrases') of attack and defence that must be observed

Dimensions of weapons:

foil: maximum weight – 500 grams (1 pound 1.637 ounces)

blade length – 90 centimetres (2 feet 11½ inches)

minimum pressure to register hit electronically – 500 grams (1 pound 1.637 ounces)

épée: maximum weight – 770 grams (1 pound 11.161 ounces)

blade length – 90 centimetres (2 feet 11½ inches)

minimum pressure to register hit electronically – 750 grams (1 pound 10.456 ounces)

sabre: maximum weight – 500 grams (1 pound 1.637 ounces)

blade length – 88 centimetres (2 feet 10½ inches)

The word *fencing* derives from the same (Latin) root that in American English is represented by 'offense' and 'defense'.

fermentation, measurement of *see* ZYMOMETRY

fermi [linear measure] A very small unit of length equal to 1 UK thousand billionth/ 1 US quadrillionth metre (10^{-15} metre), also called a femtometre, used for expressing the sizes of subatomic particles. For example: a proton is 2.4 fermi across. It was named after the Italian-born US nuclear physicist Enrico Fermi (1901-54).

Fermi-Dirac statistics [maths; physics] The statistical mechanics that are applicable to systems of particles in which the WAVE FUNCTIONS change sign when any two particles are interchanged (thus complying also with the Pauli exclusion principle). They were named after Enrico Fermi (*see* FERMI *above*) and the English physicist Paul Dirac (1902-).

fermion [physics] A particle that obeys FERMI-DIRAC STATISTICS and complies with the Pauli exclusion principle; for example, baryons and leptons.

ferromagnetic [physics] Describing a material (such as iron, cobalt, or nickel) that becomes permanently magnetized when placed in a magnetic field – that is, the magnetism remains even after the magnetizing field has been removed.

fertilizers *see* N-P-K SYSTEM

fff *see* FORTE, FORTISSIMO

Fibonacci series [maths] A sequence of integers (whole numbers) in which each is the sum of the two preceding ones – also known as Fibonacci numbers and the Fibonacci sequence. The series begins 1, 1, 2, 3, 5, 8, 13, 21, 34, 55, ... It was named after the Italian mathematician Leonardo Fibonacci (*c*.1175-*c*.1240). *See also* SERIES AND SEQUENCES.

field density [physics] For an electric or magnetic field, the number of lines of force passing at right-angles through unit area of it.

field events *see* ATHLETICS FIELD EVENTS AREA, HEIGHT, AND DISTANCE MEASUREMENTS

field frequency [physics; telecommunications] In television and video, the number of frames (complete screenfuls of picture) scanned per second, expressed in hertz. In Britain and most of continental Europe, the usual field frequency is 50 hertz (50 pictures per second); in the United States it is usually 60 hertz.

field goal [sporting term] In American football, a goal kicked from a place-kick (generally instead of a last down when there is little likelihood of a touchdown from the current drive) or, very rarely, from a drop-kick. It scores three points.

In rugby, an Australasian term for a goal scored from a drop-kick: a dropped goal: *see* RUGBY LEAGUE/UNION MEASUREMENTS, UNITS, AND POSITIONS.

In basketball, a goal scored in open play, not as a result of a free throw. It scores two points.

field hockey *see* HOCKEY MEASUREMENTS, UNITS, AND POSITIONS

fielding average [sporting term] For a baseball player, the number of put-outs and assists while fielding divided by the total number of chances, expressed as a decimal fraction to three decimal places.

field intensity, field strength [physics] In an electric or magnetic field, the electric (or magnetic) force divided by the charge (or pole); the direction is specified, and so it is a vector quantity.

field marshal [military rank] Highest commissioned rank in the British Army and the armies of several European countries, subordinate only to the commander-in-chief and ranking above a general. In Britain, a field marshal is of a rank equivalent to that of an admiral of the fleet in the Royal Navy, and to that of a marshal of the royal air force.

The equivalent rank in the United States Army is that of GENERAL OF THE ARMY.

field officers [military rank] Ranks in the army above that of captain and, in the United States, below that of brigadier-general, in Britain below that of major-general. These are the officers directly responsible for deploying troops on the field of battle.

field of force [physics] Gravity, electric charges, and magnetic poles exert their influence at a distance, and the field of force is the region in which their influence can be felt or measured. *See also* INVERSE SQUARE LAW.

field strength *see* FIELD INTENSITY, FIELD STRENGTH

fifteen [quantitatives] 15.

$$15\% = 0.15$$

The number of players in a rugby union team, sometimes also called 'a fifteen'. A polygon that has 15 sides (and 15 angles) is a quindecagon.

The name of the number derives through Germanic sources and is literally '5, 10'. *See also* FEMTO-.

fifth [quantitatives] ⅕ (one-fifth); the element in a series between the fourth and the sixth, or the last in a series of five.

$$⅕ = 0.2 \quad ⅖ = 0.4 \quad ⅗ = 0.6 \quad ⅘ = 0.8$$

See also ANNIVERSARIES; FIVE.

fifth [volumetric measure] In the United States, a unit of liquid measure for alcoholic drink closely approximating to the capacity of a standard wine bottle.

$$\begin{aligned} 1 \text{ fifth } &= 0.2 \text{ (one-fifth) US gallon, } 1.6 \text{ US pint} \\ &= 0.757 \text{ liter, } 75.706 \text{ centiliters} \\ &= 0.1669 \text{ UK gallon (1.3352 UK pint)} \end{aligned}$$

fifth [music] In music, the interval between a keynote (tonic) up to the dominant of the key (*see* SCALE). For example: the interval between C and the G above it (and the sound of those notes when played). Unlike most other intervals that contribute to harmonies or discords, a fifth is neither harmonious nor discordant, and can contribute to a chord that is in either a major or a minor mode. It is therefore often described as a 'perfect' fifth. Its inverse (for example, the interval between C and the G below it) is accordingly a perfect fourth.

A fifth in which the upper note is flattened by a semitone (half-step) is known as a diminished fifth or an augmented fourth, and constitutes a harmony that requires resolution with another.

fifty [quantitatives] 50.

$$50\% = 0.5 = \tfrac{1}{2}$$

And because 50% is half of 100%, 'fifty-fifty' means 'half-and-half' or 1:1.

The name of the number derives from Germanic variants of the common Indo-European term, again combining elements that mean '5 (times) 10'. *See also* FIVE.

figure [maths] In arithmetic, an alternative word for (whole) NUMBER; a digit.

figured [music] Describing keyboard music (mostly eighteenth-century music for harpsichord) in which only the bass part is written out in full, the harmonies being indicated by numbers above, the melody to be improvised by the player.

figure-skating *see* ICE-SKATING DISCIPLINES AND EVENTS

filler [comparative values] Unit of currency in Hungary.

$$100 \text{ filler} = 1 \text{ forint}$$

The name of the unit probably derives from the old Austro-German *heller* that was one-hundredth of a *krone* 'crown'. *See also* COINS AND CURRENCIES OF THE WORLD.

film badge *see* DOSEMETER, DOSIMETER

film speed [photography] A measure of the sensitivity (to light) of a photographic emulsion, which can be used in working out the correct exposure for taking a photograph: *see* ASA SYSTEM; DIN SYSTEM.

fils [comparative value] Unit of currency in Bahrain, Iraq, Jordan, Kuwait, and the Yemen.

In Bahrain, Iraq, Jordan, and Kuwait:

$$1,000 \text{ fils} = 1 \text{ (national)} \text{ DINAR}$$

In the Yemen:

$$100 \text{ fils} = 1 \text{ RIAL}$$

See also COINS AND CURRENCIES OF THE WORLD.

filter factor [photography] A multiplying factor by which a photographic exposure must be increased to take account of the attenuation of light by a filter fitted in front of the camera lens.

Financial Times index *see* FTSE INDEX

fin-de-siècle [time] Describing a work of art, of architecture, or of clothing that is characteristic of the mid- to late 1890s: the term is French for 'end of (the) century'.

fineness [metallurgy] The purity of gold or silver in an alloy, expressed as the number of parts per thousand that are precious metal. For example: a gold alloy described as 950 fine is 95 per cent gold (and 5 per cent of another metal, such as copper). *See also* CARAT.

finger [linear measure] In ancient Israel and ancient Greece, a small linear measure.

To the ancient Hebrews,

1 finger (*azba*)	=	0.74 inch, 18.72 millimetres
4 fingers	=	1 palm (*tefah*)
12 fingers	=	1 span (*zeret*)
24 fingers	=	1 ordinary cubit (*ammah*)
28 fingers	=	1 royal cubit

This system strongly resembled that of the ancient Egyptians, whose equivalent measure is more often known now as a DIGIT.

To the ancient Greeks,

1 finger	=	0.76 inch, 19.25 millimetres
4 fingers	=	1 palm
12 fingers	=	1 span
16 fingers	=	1 'foot'
24 fingers	=	1 cubit

The Greek measure is also known as often as a digit.

In the United States, a finger is occasionally taken to refer instead to a linear measure of about 4½ inches (11.4 centimetres), intended to represent the length of a man's finger – although a finger of that length is comparatively unusual.

finger [volumetric measure] A fairly rough-and-ready measurement in pouring alcoholic spirits into a small whisky glass, based on the width of a man's finger.

           1 finger   =   about ¾ inch, 19 millimetres
           2 fingers =   about 1½ inch, 38 millimetres

The unit as here presented is remarkably similar to the equivalently named measurements in ancient Egypt, ancient Israel, and ancient Greece (*see above*).

fingerprinting [physics; biology] A method of forensic identification that relates to the amazing individuality of the sensitive surfaces of human fingertips. It is possible, even common, to be able to discern who was at the scene of a crime from a single fingerprint (provided he or she already has a criminal record, and the fingerprints are thus on record too) – although it is much easier to be certain if all the fingers of one hand (a 'full set') are present for study. Significant factors in identification are the loops, whorls. and radial lines present in the fingerprint, and their number and direction, the overall shape of the finger (if enough is shown), and, of course, any permanent scarring on the finger. Computerization of the fingerprint files of most modern police forces, and of a methodology for checking them, has vastly increased the speed at which fingerprints can now be identified.

Finsen unit [physics; medicine] Unit of ultraviolet intensity equal to that of UV radiation (of wavelength 296.7 nanometres) with an energy density of 10^5 watts per square metre, formerly used in expressing the 'power' of UV rays used in medical treatment. The unit was named after the Danish physician Niels Finsen (1860-1904).

firkin [volumetric measure] Originally a small cask, intended to be one-quarter the size of a BARREL, or half a KILDERKIN, and thus varying a little in actual capacity according to the goods within. But in general,

           1 firkin =   9 UK gallons, 10.81 US gallons
                    =   40.9137 litres
                    =   one-quarter of an international beer barrel
                    =   1.125 UK BUSHEL, 1.161 US bushel
                    =   1.4445 cubic foot

The term derives from medieval Dutch, in which it corresponds to a 'fourth-kin'.

firkin [weight] In the United States, a measure of weight that presumably corresponds to the weight of some specific goods as contained in the cask that became the volumetric measure of the same name (*see above*). Ironically, the weight relates to a unit that has never been popular in North America, and is almost solely used in Britain.

           1 firkin =   4 STONE (2 UK QUARTERS)
                    =   56 pounds, half a (long) HUNDREDWEIGHT
                    =   25.401 kilograms
                     (25 kilograms = 0.9842 firkin)

first [quantitatives] The earliest of a sequence; the leader, the beginner, the original; the most significant, the chief. The word is used as the ordinal for 'one', and is therefore often abbreviated to '1st'.

By derivation, the term corresponds to 'fore(mo)st'.

first cousin *see* RELATIONS AND RELATIVES

first-degree burn *see* BURNS

first-order reaction [chemistry] A chemical reaction the rate of which is proportional to the concentration (active mass) of only one reactant.

first person singular/plural *see* PERSON

First Point of Aries [astronomy] The point on the CELESTIAL EQUATOR at which it is intersected by the ECLIPTIC, from which right ascension and celestial longitude are measured.

first quarter *see* LUNAR PHASES

fiscal year [comparative values; time] The tax year, as established by the government.

fission parameter [physics] For an element that can undergo nuclear fission, the ratio of its atomic number to its relative atomic mass (atomic weight).

Fitzgerald-Lorentz contraction [physics] For an object moving at near the velocity of light, the reduction in length (or time scale) relative to the frame of reference from which the measurements are made. It is a consequence of relativity, and was named after the Irish physicist George FitzGerald (1851-1901) and the Dutch physicist Hendrik Lorentz (1853-1928).

five [quantitatives] 5.

$$^5/_{10} = 0.5 = ^1/_2 \qquad\qquad 5\% = 0.05$$

A group of five is a quintet or pentad.

A group of five arranged like the 5 on a dice is a quincunx.

A sequence of five is sometimes called a quint.

A polygon with 5 sides (and 5 angles) is a pentagon.

A solid figure with five plane faces is a pentahedron.

There are five players on court in a basketball team.

As is evident from the above, the prefixes meaning 'five-' in English are 'quint-', 'quin-' (from Latin) and 'penta-', 'pent-' (from Greek).

The name for the number derives from common Indo-European, attested over the entire range of languages from Sanskrit to Welsh. It is, however, the number that has probably undergone the greatest variation through the centuries and tongues. The etymon is thought to be something like *penkwe leading thus to Sanskrit pañca, Greek pente and (via a possible *pinque) to Latin quinque; Germanic forms correspond to a variant *finhw-, hence German fünf, denasalized in English to five.

It is possible that the word five is related to fist, the root meaning evidently a collection of five fingers; it is also possible that five is somehow akin to finger (see DIGIT) although, in that case, the number is the derivative. It should be remembered that the very first forms of arithmetical addition most probably involved not five but four fingers counted off against the thumb (see EIGHT; NINE).

five-star [military rank] Describing a most senior admiral or general in the United States armed forces, entitled to wear five stars as part of his or her insignia.

fixed point [physics] One of the two accurately reproducible temperatures used in calibrating a thermometer and defining its scale. For example: the Fahrenheit and Celsius scales use the freezing point and boiling point of pure water as their fixed points. They are just two of the ten internationally recognized fixed points. See also FUNDAMENTAL INTERVAL.

fjerding [volumetric measure] A volumetric unit of dry capacity in Denmark, closely approximating to the BUSHEL.

$$
\begin{aligned}
1 \text{ fjerding} &= 36 \text{ (Danish) pots, 2 SKAEPPE} \\
&= 34.815 \text{ litres, } 34{,}815 \text{ cubic centimetres} \\
&= 0.988 \text{ US bushel, } 0.95725 \text{ UK bushel} \\
&\quad (1 \text{ US bushel} = 1.01215 \text{ fjerding} \\
&\quad\ \ 1 \text{ UK bushel} = 1.04466 \text{ fjerding}) \\
4 \text{ fjerdings} &= 1 \text{ tønde}
\end{aligned}
$$

A similar unit of capacity also deriving its name as a form of 'fourth(-unit)' is the FIRKIN.

$$
\begin{aligned}
1 \text{ fjerding} &= 0.8509 \text{ firkin} \\
&\quad (1 \text{ firkin} = 1.1752 \text{ fjerding})
\end{aligned}
$$

flag captain [military rank] In the navy, the CAPTAIN of a flagship, responsible for the crew and for the safety of the ship, but under the authority of the admiral or commodore on board.

flag lieutenant [military rank] In the United States Navy, a LIEUTENANT who acts as personal assistant to an admiral on board his flagship.

flame photometry [physics; medicine] The use of an instrument known as a flame photometer to spray a solution into a hot flame to record the relative intensities of the resultant emission spectra of the elements present in the solution, and so measure the concentration. In this way, for example, the quantity and concentration of potassium and sodium salts in blood can be determined.

flank speed [speed] For a ship, the maximum speed permitted by a marine or waterway authority. Also, sometimes, a ship's maximum possible speed.

flash point [physics; chemistry] For a liquid fuel, the temperature at which it gives off enough vapour to cause a short flash when a naked flame is applied to the vapour. Volatile flammable liquids present a high fire risk because they have low flash points.

flash spectrum [astronomy; physics] In a total eclipse of the Sun, just as all direct light is cut off at the beginning of the totality and just as direct light is about to be restored at the end of the totality, a bright-line spectrum from the edge of the solar disc, lasting for only a couple of seconds at a time.

flask [volumetric measure] In the transportation of liquid mercury, a flask is a bottle-shaped iron container that holds 76 pounds (34.5 kilograms) of the metal.

Other flasks are of no specific capacity, although hip-flasks for spirits may approximate closely to half-bottle and quarter-bottle sizes.

flat [music] A note that is technically a semitone (or half-step) lower in pitch than a given note: the note A♭ is thus one semitone (half-step) lower than the note A . On a piano keyboard, such notes are not always represented by black keys: the note C♭, for example, corresponds to the note B (because the note B is only a semitone or half-step lower than C anyway).

Many keys incorporate flats as part of their tonic scales: for a list of the flat keys *see* KEY, KEYNOTE.

But the term as an adjective or adverb in less technical musical terminology means merely at a slightly lower pitch. So to sing flat is to sing out of tune by not reaching up to the desired note. Unaccompanied choirs in hot halls quite often sing progressively flatter and flatter. *See also* DOUBLE FLAT, DOUBLE SHARP.

flavour, flavor [physics] One of six parameters that characterize different kinds of quarks. They are: up, down, strangeness, charm, beauty (or bottom), and truth (or top), abbreviated to u, d, s, c, b and t, respectively. For antiquarks, the parameters are anti-up, anti-down, and so on.

flavour enhancers *see* E-NUMBERS

F layer, F region [physics; telecommunications] A layer of ionized gases in the atmosphere at a height of 200 to 300 kilometres (124.3-186.4 miles) which reflects radio waves of up to 50 megahertz frequency. The outermost layer of the iono-sphere, sometimes, in fact, regarded as two layers (known respectively as the F1 and F2 layers), it is alternatively known as the Appleton layer [after the British physicist Edward Appleton (1892-1965)].

It is called the F layer because it is outside the E layer: *see* E LAYER, E REGION.

fleet admiral [military rank] The most senior rank in the United States Navy, ranking higher than an admiral, and equivalent to general of the army and general of the airforce.

The title of the equivalent rank in the (British) Royal Navy is ADMIRAL OF THE FLEET.

fleet of ships, of aircraft, of buses, of cars, of trucks [collectives] The word *fleet* should technically be applied only to boats and ships, for it is a derivative of the Old English form of the verb 'to float' (and the adjective *fleet*, 'speedy', similarly originally meant 'fast-flowing', as used in the names of many inland streams and rivers). But, by extension, the noun has in modern times been much used in relation to large collections of all types of vehicles, especially when owned by one country, one authority, or one person.

A little fleet is a *flotilla* (via Spanish).

flesh-printing [biology] The use of an electronic device to measure and display the protein patterns in the flesh of caught fish, to relate each fish to the waters in which it grew to maturity and so determine migratory routes (and, if necessary, take measures to protect sea areas in which the fish are vulnerable).

flight lieutenant [military rank] Commissioned rank in the (British) Royal Air Force and the airforces of some other English-speaking countries (such as Australia and Canada).

A flight lieutenant ranks above a flying officer but below a squadron leader, and is equivalent in rank to a captain in the British army and a lieutenant in the Royal Navy.

The equivalent rank in the US Airforce is CAPTAIN.

flight of aircraft, of angels, of arrows, of steps [collectives] Collective noun for assemblies of several different items, most of which have one thing in common: they fly. And, of course, the noun is the (past-participular) abstract noun from the verb 'to fly'.

A flight of stairs or steps, however – which is not a true collective – may or may not derive from the same source. It is possible instead that the noun in this event comes from Germanic equivalents to Latin *flect-*, *plect-* 'curving inwards', 'bending', in reference to the spiral nature of staircases in former centuries. At the same time, it is interesting that a flight of steps does indeed rise vertically, and may descend to a 'landing' . . .

    See also FLOCK OF BIRDS

flight path [aeronautics] The true three-dimensional path of an aircraft or rocket through the air (as opposed to its track, which is its path projected on to the ground).

flight recorder, black box [aeronautics] The flight recorder in every commercial aircraft – also known as the black box, although only one element of the equipment is box-like and it is more often bright orange or red in colour – is a multiple device that makes a record of the mechanical operations registered on specific dials and meters on the flight deck panels, and also records the voices of the flight crew, during every flight. Its purpose is to provide clues to what actually happened in the event of a crash, especially if the air crew are unable themselves to give evidence. The equipment is therefore made as crash-resistant as possible.

    From 1989, commercial shipping has by maritime law been required to carry similar recording devices monitoring the operation and navigation of ocean-going ships, known as a *voyage recorder*.

flight shooting [sporting term] In archery, shooting in competition to find who can attain the greatest distance. Specially light and well-feathered arrows are used for the purpose, known as flight arrows.

flock of birds, of sheep, of goats, etc. [collectives] It is likely that, by derivation, the sort of flock that birds are seen in is not the same sort of flock that sheep and goats roam in. The *flock* of birds is probably a variant of the appropriate word *flight*, itself a collective for certain specific birds (doves, geese, pigeons, swallows), other flying items (angels, arrows), and objects that rise vertically (stairs).

    Most commentators suggest that the flock representing a group of sheep or goats etymologically derives from a northern Germanic variant of the English word *folk* that meant 'crowd', 'gathering'. (Some authorities prefer this derivation for the flock of birds too.) But it is interesting that the *flock* that is used to stuff mattresses or is sprayed on wallpaper to form a velvety surface is itself etymologically akin to such English words as *fleece*, *villus* (medical term), and even *wool* and *felt*.

floral formula [biology; botany] A method of representing the structure of a flower in terms of the number of parts in its perianth (denoted by the letter P followed by the number), calyx (K), corolla (C), androecium (A), and gynoecium (G). Numbers in parentheses indicate the connation (fusing of parts), and a bar over or under the G indicates whether the ovary is superior or inferior.

florin [comparative values] Unit of currency first issued in the form of a gold coin at Florence, Italy, in 1252, with a representation of a lily (medieval Italian *fiore*, French *fleur-de-lis*) on one side and the name of the city (Italian *Firenze*, French *Florence*) on the other, and for the one or the other reason therefore called a *fiorino* in Italian (and *florin* in French). Thereafter, most countries of Europe sooner or later had a coin named after it.

    In Britain, the first florin was that minted during the reign of Edward IV in 1480, and worth 6 shillings and 8 pence (one-third of a pound). After that coin became obsolete, it apparently became the English custom by the early 1600s to call various other European coins by the name, even though most or all of them had their own national names. It was in this way that the Dutch guilder (*gulden*) took on the ascription, and in Britain may even today be abbreviated in the form 'Dfl.' for Dutch florins.

    It was not until 1849 that Britain had another florin of its own: a silver coin worth 2 shillings (24 pence, one-tenth of a pound), of a shape and size that survived right up until 1993 – albeit by then, since 1971, in decimal currency as a 10-pence piece.

    See also FORINT.

flow [engineering] In engineering and other industrial processes, the rate of flow of fluids (liquids or gases) is usually stated as volume per unit time (such as pints per minute or litres per second), as measured by a flowmeter. In scientific applications, the rate of flow (of mass, volume, or energy) is usually called FLUX. *See also* FLUIDITY; VISCOSITY.

fluid dram *see* DRAM, FLUID

fluidity [physics] A measure of the ease with which a fluid flows; the reciprocal of VISCOSITY.

fluid ounce *see* OUNCE, FLUID

fluid scruple *see* SCRUPLE, FLUID

fluorometry, fluorimetry [physics] The use of a device called a fluorometer to measure fluorescent radiation – visible light given off by a substance exposed to rays outside the visible range (such as X-rays and ultraviolet light).

flush *see* POKER SCORES

flute *see* WOODWIND INSTRUMENTS' RANGE

fluviometer [physics; meteorology] Any apparatus – often simply a board or plank calibrated in feet or metres – from which the depth of water in a river can be read.

flux [physics] The rate of flow of mass, volume, or energy per unit cross-sectional area (at right-angles to the direction of flow): *see* ELECTRIC FLUX; LUMINOUS FLUX; MAGNETIC FLUX.

flux density [physics] The amount of radiant energy (such as light) passing through unit cross-sectional area (at right-angles to the beam of radiation).

fluxmeter [physics] An instrument for measuring magnetic flux (by means of a probe that is placed in the magnetic field).

flux unit [astronomy] In radioastronomy, the flux of incident radiant energy; 1 flux unit equals 10^{-26} watt per square metre per hertz.

flying officer [military rank] Commissioned rank in the (British) Royal Air Force and the airforces of some other English-speaking countries (such as Australia and Canada).

A flying officer ranks above a pilot officer but below a flight lieutenant, and is equivalent in rank to a lieutenant in the British army and a sublieutenant in the Royal Navy.

The equivalent rank in the US Airforce is FIRST LIEUTENANT.

flyweight *see* BOXING WEIGHTS

f-number [photography] For a camera lens, a measure of the amount of light it will pass, equal to the focal length divided by the diameter of the aperture. Small *f*-numbers indicate large apertures (and vice versa). For example: at *f*8 the diameter of the aperture is effectively one-eighth of its focal length.

In the usual sequence – *f*22, *f*11, *f*8, *f*5.6, *f*4, and so on – each aperture has twice the area of the one preceding it (and therefore lets through twice as much light).

focal length [physics; photography] For a lens or lens system, the distance from the centre of the lens (or its equivalent in a compound lens) to the point at which an image is brought to a focus. *See also* BACK FOCUS.

focus [maths] For a CONIC SECTION, a fixed point on the concave (inner) side which, together with the DIRECTRIX and ECCENTRICITY, defines the curve. Ellipses and hyperbolas have two real foci; other conics have one.

focus [physics] For a lens or curved mirror, the point at which refracted or reflected light rays meet.

fod [linear measure] An old Danish linear measure corresponding in its history to the closely equivalent statute FOOT of English-speaking countries: singular fod, plural fødder.

$$
\begin{aligned}
1 \text{ fod} \quad &= \quad 12.365 \text{ inches}, 1.0297 \text{ foot} \\
&\qquad (1 \text{ foot} = 0.9712 \text{ fod}) \\
&= \quad 31.407 \text{ centimetres}, 0.31407 \text{ metre} \\
&\qquad (1 \text{ metre} = 3.1840 \text{ fødder})
\end{aligned}
$$

See also FOOT; FOT.

foghorn *see* CHARACTER

foil *see* FENCING MEASUREMENTS AND POSITIONS

folio [literary: legal] In law, a specified number of words used as a unit for defining the overall length of a document. In the United States, the customary number of words is 100; in Britain, the number is either 72 or 90. The method of quantification derives from a time when the average number of words on a single sheet or leaf (Latin *folio* 'on the leaf') of a legal document could be readily estimated in this way.

The folio is also a technical term for the page number on a sheet of a book or newspaper.

food additives *see* E-NUMBERS

foolscap [paper size] An obsolete paper size of 13½ x 17 inches (342.9 x 431.8 millimetres), usually taken as 13 x 16 inches (330.2 x 406.4 millimetres) in the United States.

Paper of this size originally had a watermark in the form of a court jester's tricorn cap with bells – hence the name.

foot [linear measure] 'And did those feet in ancient time . . . ' – that is, in ancient Babylon, ancient Greece, and ancient Rome – bear any resemblance to the modern statute foot of English-speaking countries? Well, surprisingly, they were quite close. Even more surprisingly, one or two of them were actually slightly larger – and that is truly remarkable, for if the measure is based upon the bodily dimension of a human foot, then those ancient peoples had proportionately much larger feet than those of the average person today. Today's unit measure of 1 foot (12 inches), after all, corresponds to a (adult man's) shoe size in Britain of 13½-14, in the United States of 14-14½, and in continental Europe of 47-48 – a size that is too large to be normally available.

The answer is that the foot in ancient times was more often an exact number of smaller units ('fingers', 'palms', and so forth), and was called a foot simply because that was a part of the body that was a tolerably close approximation in length, and moreover could be linked directly in the mind with the next greater measure of length, the step or pace.

In ancient Babylon,

1 'foot'	=	20 shusi
	=	13.93 (modern) inches, 35.39 centimetres
1.5 'foot', 30 shusi	=	1 CUBIT (*kus*)

In ancient Greece,

1 'foot'	=	4 'palms', 16 DIGITS or 'fingers'
	=	12.16 (modern) inches, 30.89 centimetres
1.5 'foot', 6 'palms', 24 digits	=	1 cubit
2.5 'feet', 15 'palms', 40 digits	=	1 'step'
60 'feet'	=	1 'cable'
600 'feet'	=	1 STADION

In ancient Rome,

1 'foot' (*pes*)	=	16 digits, 12 UNCIAE ('inches'), 4 palmae
	=	11.68 (modern) inches, 29.67 centimetres
1.5 'feet', 24 digits, 18 unciae, 6 palmae	=	1 cubit
2.5 'feet', 40 digits, 30 unciae, 10 palmae	=	1 'step' (*gradus*)
5 'feet'	=	1 pace (*passus*)
625 'feet', 125 paces	=	1 stadium
5,000 'feet'	=	1 mile (*milia passuum* 'thousand paces')

In modern English-speaking countries,

1 foot	=	12 INCHES
	=	72 pica ems (in typography)
	=	144 lines, 1,440 gries
	=	12,000 mils
	=	30.48 centimetres (exactly), 0.3048 metre (1 metre = 3.2808398 feet: 3 feet 3.37 inches)
3 feet	=	1 YARD (91.44 centimetres, 0.9144 metre)
6 feet, 2 yards	=	1 FATHOM
16½ feet, 5½ yards	=	1 ROD or pole
66 feet, 22 yards, 4 rods	=	1 CHAIN
660 feet, 220 yards, 10 chains	=	1 FURLONG
5,280 feet, 1,760 yards, 80 chains, 8 furlongs	=	1 MILE

For similar historical reasons, many non-English-speaking European countries also retained variants of the ancient foot until use of the metric system became virtually universal.

Representing Germanic traditions:

$$1 \text{ Austrian fusz} = 12.444 \text{ inches, } 31.608 \text{ centimetres}$$
$$(1 \text{ foot} = 0.9643 \text{ Austrian fusz})$$
$$1 \text{ Danish fod} = 12.365 \text{ inches, } 31.407 \text{ centimetres}$$
$$(1 \text{ foot} = 0.9712 \text{ Danish fod})$$
$$1 \text{ Norwegian fot} = 12.350 \text{ inches, } 31.370 \text{ centimetres}$$
$$(1 \text{ foot} = 0.9717 \text{ Norwegian fot})$$
$$1 \text{ Swedish fot} = 11.689 \text{ inches, } 29.690 \text{ centimetres}$$
$$(1 \text{ foot} = 1.0266 \text{ Swedish fot})$$

Representing Latinate traditions:

$$1 \text{ Portuguese pé} = 13.119 \text{ inches, } 33.324 \text{ centimetres}$$
$$(1 \text{ foot} = 0.9147 \text{ Portuguese pé})$$
$$1 \text{ French pied de roi} = 12.790 \text{ inches, } 32.487 \text{ centimetres}$$
$$(1 \text{ foot} = 0.9382 \text{ French pied de roi})$$
$$1 \text{ Belgian pied} = 11.81 \text{ inches, } 29.997 \text{ centimetres}$$
$$(1 \text{ foot} = 1.0161 \text{ Belgian pied})$$
$$1 \text{ Italian pie} = 11.73 \text{ inches, } 29.794 \text{ centimetres}$$
$$(1 \text{ foot} = 1.0230 \text{ Italian pie})$$
$$1 \text{ Spanish pie} = 10.97 \text{ inches, } 27.864 \text{ centimetres}$$
$$(1 \text{ foot} = 1.0939 \text{ Spanish pie})$$

See also FOOT, CUBIC; FOOT, SQUARE; FOOT-POUND-SECOND SYSTEM (FPS); BOARD FOOT; CORD; CORD FOOT.

foot [literary: poetic metre] The basic unit of verse, comprising two or more syllables of specified rhythm and length which with other similar feet make up a line. The 'rhythm' of the foot depends on how its syllables are stressed and/or unstressed; the length of each syllable depends on the duration of the vowel sound or sounds it contains.

In English verse, the most common form of foot is the *iamb*, a foot of two syllables, the first unstressed and short, the second stressed and (generally) longer. Individual words that correspond to this type of foot are, for example, 'between', 'around', 'surprise', 'delight' although, in verse, an iamb frequently instead contains either two words or parts of two words, or alternatively represents part of only a single word.

Example:

Thus un-/lamen-/ted pass/the proud/away,

The gaze/of fools,/and page-/ant of/a day!

(Alexander Pope: *Elegy to the Memory of an Unfortunate Lady*)

This type of verse, highly popular in the first decades of the 1700s, comprises lines of five iambs (iambic feet) – lines that are accordingly known as iambic pentameters.

Such lines can be represented as unstressed (u) followed by stressed (—) syllables:

$$u — / u — / u — / u — / u —$$

and one iambic foot can thus be represented as

$$u —$$

Other types of feet commonly represented in English poetry are:

	rhythm	sample words
trochee (trochaic foot)	— u	common, simple, budget
spondee (spondaic foot)	— —	salt-caked smoke-stack
anapest, anapaest	u u —	entertain, correspond
dactyl (dactylic foot)	— u u	fallacy, thunderer
amphibrach	u — u	tomorrow, example

and there are many other forms of poetic feet, each with a specific name and some of them combining the above forms, as classified in the works of ancient Greek and Latin poets. *See also* METRE IN VERSE; VERSE FORMS.

foot, cubic [cubic/volumetric measure] Volumetric measure of dry and liquid capacity, now in Europe largely superseded by units of the metric system.

1 cubic foot	=	12 x 12 x 12 inches, 1,728 cubic inches
	=	0.0370 cubic yard
	=	28,316.85 cubic centimetres, 0.0283 cubic metre
		(1 cubic metre = 31.3145 cubic feet)
	=	0.7788 UK bushel, 0.8039 US bushel
		(1 UK bushel = 1.284 cubic feet
		1 US bushel = 1.244 cubic feet)
27 cubic feet	=	1 cubic yard

See also BOARD FOOT; CORD FOOT.

foot, cubic, per second *see* FOOT-SECOND

foot, square [square measure] Square measure in the imperial (English) unitary system now in Europe largely superseded by units of the metric system.

1 square foot	=	12 x 12 inches, 144 square inches
	=	0.1111 square yard
	=	929.0304 square centimetres, 0.0929 square metre
		(1 square metre = 10.7643 square feet)
9 square feet	=	3 x 3 feet, 1 square yard

See also YARD, SQUARE.

football *see* AMERICAN FOOTBALL MEASUREMENTS, UNITS, AND POSITIONS; AUSTRALIAN RULES MEASUREMENTS, UNITS, AND POSITIONS; GAELIC FOOTBALL MEASUREMENTS, UNITS, AND POSITIONS; SOCCER MEASUREMENTS, UNITS, AND POSITIONS; RUGBY LEAGUE/UNION MEASUREMENTS, UNITS, AND POSITIONS.

foot-candle [physics] A former unit of illumination (brightness) equal to that at a distance of 1 foot from a light source of luminous intensity 1 international candle. It was superseded by the lumen per square foot or candela per square metre.

1 foot-candle	=	10.764 lux
	=	1.0764 milliphot
		(1 milliphot = 0.9290 foot-candle)

See also CANDLE; LUX.

foot-fault *see* FAULT

foot-lambert [physics] A former unit of luminance equal to that of a surface that emits or reflects 1 lumen per square foot (and the theoretical average luminance of a perfectly reflecting surface with an illumination of 1 foot-candle).

1 foot-lambert	=	1.076 millilambert
		(1 millilambert = 0.9290 foot-lambert)
	=	$\frac{1}{144}\pi$ candle per square inch

See also LAMBERT; LUX.

foot per second *see* FEET PER SECOND; FOOT-SECOND

foot-pound [physics] Unit of energy or work in the foot-pound-second (fps) system, equal to the work done when a mass of 1 pound is lifted 1 foot against the force of gravity – that is, when a force of 1 pound weight (32 poundals) acts over a distance of 1 foot, and therefore alternatively known as a foot-pound force.

1 foot-pound	=	1.3558 joule
		(1 joule = 0.7376 foot-pound)

See also POUND; POUNDAL.

foot-poundal [physics] Unit of energy or work in the foot-pound-second (fps) system, equal to the work done when a force of 1 poundal moves through a distance of 1 foot.

1 foot-poundal	=	0.04214 joule
		(1 joule = 23.73 foot-poundals)

See also POUND; POUNDAL.

foot-pound-second (FPS) system [quantitatives: unitary system] System of units based on the imperial system, using as its base units the foot (length), pound (mass), and second (time). In scientific usage, it has been almost completely superseded by the SI SYSTEM, although elements of it are retained in trade and industry (especially in the United States).

foot-second [speed] A misleadingly named unit of velocity (better expressed as feet per second) sometimes used to describe linear and volumetric flow.

As a unit of ordinary linear velocity,

> 1 foot-second = 1 foot per second: *see* FEET PER SECOND.

As a measure of the flow of a liquid,

> 1 foot-second = 1 cubic foot per second
> = 60 cu. ft per minute, 3,600 cu. ft per hour
> (1 cubic yard every 27 seconds)
> = 28,317 cubic centimetres per second
> = 1.6990 cubic metres per minute
> (1 cubic metre every 35.3 seconds)

foot-ton [physics] Unit of work equal to the energy required to lift a mass of 1 (long) ton (2,240 pounds, 1,016.0416 kilograms) a distance of 1 foot against the force of gravity.

> 1 foot-ton = 2,240 FOOT-POUNDS
> = 3,036.992 joules

force [physics] The influence that can cause a moving object to change speed or direction, or a stationary object to begin moving – that is, it brings about a change in the momentum of the object. For such objects, it equals the product of mass and acceleration. Its principal units are the newton (SI units), dyne (CGS units), and poundal (FPS units) – and their multiples and submultiples.

force ratio, mechanical advantage [physics; engineering] For a simple machine, the load (output force) divided by the effort (input force) – now generally called the force ratio but formerly better known as the mechanical advantage.

For a system of pulleys it is equal to the number of pulleys, or to the number of supporting ropes. *See also* DISTANCE RATIO, VELOCITY RATIO.

forint [comparative values] Unit of currency in Hungary.

> 1 forint = 100 FILLER

It is thought that the name of the unit derives from the medieval Italian *fiorino*: *see* FLORIN.

See also COINS AND CURRENCIES OF THE WORLD.

form [biology] In the classification of plants, the most minutely characteristic category. Most SPECIES are subdivided into SUBSPECIES and perhaps VARIETIES but, in even more detail, the categorization may continue down to subvariety, form, and subform. *See* full list of taxonomic categories *under* CLASS, CLASS.

formant [literary: phonetics] The normal relative pitch within the register of a voice when producing specific vowel sounds: back vowels (such as 'ah' and 'oo') tend to be pronounced at a deeper pitch (a lower formant) than front vowels (such as 'ay' and 'ee').

form factor [biology; maths] In forestry, the mathematical relationship between the overall volume of a tree and the volume of a cylinder (or other regularly shaped solid figure) that has the same base dimensions and height as the tree.

formula [maths; physics; chemistry] In mathematics, an expression that shows the relationship between quantities (*see*, for example, the formulae for AREAS and VOLUMES in the articles of those names).

In chemistry, a formula represents the chemical composition of a substance in terms of the type and number of atoms in its molecules. The type of atom – the chemical element concerned – is denoted by its symbol (*see* PERIODIC TABLE OF ELEMENTS) and the number of atoms (in excess of 1) is denoted by an inferior numeral (subscript). For example: the chemical formulae of water and sulphuric acid are H_2O and H_2SO_4.

See also DENTAL FORMULA; EQUATION; FLORAL FORMULA.

Formula motor racing *see* MOTOR VEHICLE RACING

forte, fortissimo [music] Musical instruction: play or sing loudly (Italian *forte*; symbol f) or as loudly as possible (*fortissimo*; symbol ff or fff).

fortnight [time] Two weeks. The term is a contraction from the Old English for 'fourteen nights'. In Shakespeare's time, a week was known also as a *sennight* – a contraction from the Old English for 'seven nights'. Both these expressions are so ancient as to refer to a time when it was easier to talk about nights than to talk about days which had no set hours.

forty days, forty years [time] In ancient Israel (and therefore the Bible), a mystical period regarded (in both cases) as a long time and one with overtones of greatness –

although the number 40 had no such particular meaning.

forty-eightmo, 48mo [paper sizes] A 48th of a sheet of paper, formed by folding a single sheet to make 48 leaves (96 pages).

fot [linear measure] An obsolescent linear measure in Norway and Sweden approximating closely to the statute FOOT, with which it is strongly linked historically.

In Norway,

$$1 \text{ fot} = 12.350 \text{ inches}$$
$$(1 \text{ foot} = 0.9717 \text{ Norwegian fot})$$
$$= 31.370 \text{ centimetres}$$
$$(1 \text{ metre} = 3.1878 \text{ Norwegian føtter})$$

In Sweden,

$$1 \text{ fot} = 11.689 \text{ inches}$$
$$(1 \text{ foot} = 1.0266 \text{ Swedish fot})$$
$$= 29.690 \text{ centimetres}$$
$$(1 \text{ metre} = 3.3681 \text{ Swedish fötter})$$

$$10 \text{ fötter} = 1 \text{ stong ('rod' or 'pole')}$$
$$100 \text{ fötter, 10 stänger} = 1 \text{ ref}$$
$$36,000 \text{ fötter, 3,600}$$
$$\text{stänger, 360 revar} = 1 \text{ mil ('mile')}$$

See also FOD.

four [quantitatives] 4.

$$\tfrac{4}{10} = 0.4 = \tfrac{2}{5} = 40\% \qquad 4\% = 0.04$$

A group of four is a quartet, tetrad, or foursome.

A polygon with 4 sides (and 4 angles) is a tetragon or quadrilateral; if the angles are right-angles, the figure is a rectangle (oblong) or a square.

A solid figure with four plane sides is a tetrahedron; if the four sides are equilateral triangles, the figure is a regular tetrahedron.

The prefixes meaning 'four-' in English are '*quadr-*' (from Latin) and '*tetra-*' (from Greek).

The name for the number derives from common Indo-European, attested over the entire range of languages from Sanskrit to Welsh. The etymon is thought to be something like **kwetwor*, most evident in the Latin form *quattuor*, plain in Welsh *pedwar*, and still visible in Gothic *fidwor*, Old High German *fior*, and thus English *four*. Some Indo-European groups have tended toward a different direction: the Sanskrit for 'four', for example, is *catvaras*, and most of the Slavonic tongues have a number that sounds like 'cheteeri'. Perhaps it was via this sort of variance that the (ancient and modern) Greek form eventually settled on *tessares* (compound form: *tetra-*), in which the first consonant has undergone an unusual (but not unknown) mutation.

Some authorities have suggested that, within the name of the number, there are possibly elements meaning 'twice two', and it is certainly clear that there is a good case at least for the 'two' – but the first syllable should then be close in form, and in virtually every language it is not.

As a number, four was probably originally more important than five, despite the modern predilection for fives and tens. It should be remembered that the very first forms of arithmetical addition most probably involved not five but four fingers counted off against the thumb (*see* EIGHT; NINE).

four-ball [sporting term] Describing a game of golf in which two teams of two players compete, each player having his or her own ball, but only the better ball of each team counting towards the score at each hole.

four-dimensional *see* FOURTH DIMENSION

four-four time (4/4 time) [music] In music, a style that stresses the first of every four beats (and that therefore has four beats to the bar, each of a duration of a crotchet or quarter-note). A classic bar in 4/4 time thus comprises

although, of course, it may be broken up into different units:

Four-four time is by far the most common style of music in the English-speaking world. Its derivative eight-eight time (eight beats in the bar, each of a duration of a quaver or eighth-note) is fairly common also in Latin American countries. *See also* THREE-FOUR TIME (3/4 TIME); SIX-EIGHT TIME (6/8 TIME).

Fourier analysis [maths] A method of representing a complex function that stands for a wave motion as a series of simpler sine waves. It was named after the French mathematician and physicist Jean Fourier (1768-1830).

Fourier number [physics] A dimensionless number employed in calculations involving the flow of heat. For a particular material of given length, it equals the product of thermal conductivity and time, divided by the product of specific heat, density, and length squared.

fours, four of a kind *see* POKER SCORES

four score, four-score [quantitatives] Four times twenty, 80: *see* SCORE.

four-star [military rank] Describing a senior admiral or general in the United States armed forces, who is entitled to wear four stars as part of his or her insignia.

fourth [quantitatives] ¼ (one-fourth, one-quarter); the element in a series between the third and the fifth, or the last in a series of four. In spoken English, the term 'fourth' is used most commonly as the ordinal number between third and fifth; for the fraction, 'one-quarter' is the more usual expression. By derivation, however, *fourth* and *quarte(r)* are the same word anyway: *see* FOUR.

$$\tfrac{1}{4} \;=\; 0.25 \;=\; 25\%$$

See also QUARTER.

fourth [music] In music, the interval between a keynote (tonic) down to the dominant of the key (*see* SCALE). For example: the interval between C and the G below it (and the sound of those notes when played).

Unlike most other intervals which contribute to harmonies or discords, a fourth is neither harmonious nor discordant, and can contribute to a chord that is in either a minor or a major mode. It is therefore often described as a 'perfect' fourth. Its inverse (for example, the interval between C and the G above it) is accordingly a perfect fifth.

A fourth in which the lower note is flattened by a semitone (half-step) is known as an augmented fourth or a diminished fifth, and constitutes a harmony that requires resolution with another.

fourth dimension [maths; physics] A solid object can be described in terms of the three dimensions of height, length, and width. In mathematical terms, any point on a solid (that is, the location of any point in space) can be described in terms of three coordinates, such as, for example, the three Cartesian coordinates x, y, and z related to a set of three axes at right-angles to each other. Mathematically, additional dimensions can be introduced, and one more dimension would, therefore, define a point in four-coordinate or four-dimensional space.

In physics, the fourth dimension is often ascribed to time; the combination of three linear dimensions and a time dimension can thus define something in terms of its location in time and space, and relativity theory describes phenomena in terms of such a space-time continuum.

fpm *see* FEET PER MINUTE

fps *see* FEET PER SECOND

FPS system Abbreviation for the FOOT-POUND-SECOND SYSTEM.

fractal [maths; physics] Type of mathematical curve created by replacing the sides of a regular polygon by a (longer) generator curve, then replacing the segments of the resulting shape with the generator, then replacing those new segments with the generator, and so on for as many generations as required. Many natural phenomena exhibit fractals (such as parts of the path of a particle undergoing Brownian motion), and, because of their resemblance to one well-known natural shape, they are sometimes alternatively called snowflakes.

fraction [maths] A part of a whole, usually represented in mathematics by a pair of numbers, one over the other (but *see* DECIMAL). The two numbers in a fraction are separated by a horizontal or diagonal line (for example, ⅔, ¾, and ⁹⁄₁₀, which represent two-thirds, three-quarters, and nine-tenths). The upper number is called the *numerator*, and the lower number is the *denominator*. Fractions of a hundred may be stated as percentages; for example: $^{45}/_{100}$ equals 45 per cent (or 45%).

frame [sporting term] In snooker and pool, the wooden triangle used to position the balls at the beginning of each game. Also, one game in a snooker or pool match.

In ten-pin bowling, one round of the ten rounds in a match. Also, a square on the score-sheet used for ten-pin bowling.

franc [comparative values] Unit of currency in many countries in which French is the first or second language.

In Andorra, Belgium, Benin, Burkina Faso, Burundi, Cameroon, Central African Republic, Chad, Comoros Islands, Congo, Djibouti, Equatorial Guinea, France, Gabon, Guinea, Ivory Coast, Liechtenstein, Luxembourg, Mali, Monaco, New Caledonia, Niger, Rwanda, Switzerland, Tahiti, Togo,

$$1 \text{ franc} = 100 \text{ centimes}$$

In Morocco,

$$100 \text{ francs} = 1 \text{ dirham}$$

The first coin of the name was struck in 1360; made of gold, it celebrated the contemporary King of the Franks, Jean le Bon (ruled 1350-64). Two hundred years later, another coin was issued in France and given the same name – this one was of silver. The present French franc (after which all others around the world are named) was first minted amid revolutionary zeal in 1795. *See also* CFA FRANC.

franklin [physics] Obsolete unit of electrostatic charge in the centimetre-gram-second (CGS) system, equal to a charge that exerts a force of 1 dyne on an equal charge 1 centimetre away in a vacuum. It was named after the American scientist Benjamin Franklin (1706-90).

Fraunhofer lines [physics; astronomy] A series of about 25,000 fine dark lines in the continuous emission spectrum of the Sun. The lines correspond to the absorption (in the cooler, outer layers of the Sun's atmosphere) of light wavelengths characteristic of chemical elements such as calcium, hydrogen, magnesium, and sodium within the hotter parts of the Sun. Some of the more prominent lines are designated by letters; for example: the [C] line is due to hydrogen. A few Fraunhofer lines are due to absorption by gases in the Earth's atmosphere; for example: the [A] line is caused by oxygen. They were named after the German (Bavarian) physicist and optician Joseph von Fraunhofer (1787-1826).

freeboard [linear measure] The distance between the sea surface and the main outward deck of a ship. Also the outer area of the ship's hull between sea surface and the main outward deck.

By extension, the gap between the road surface and the underframe of a car, beach buggy, go-kart, jeep, or truck.

free energy [physics] A measure of the capacity of any system to perform work. *See also* GIBBS FREE ENERGY.

free fall [physics] Movement of an object under the force of gravity (that is, in a gravitational field) when there are no other forces acting on it; it will then appear to be weightless. No matter what their masses, all objects falling freely in a vacuum have the same constant acceleration, called the ACCELERATION OF FREE FALL (acceleration due to gravity). *See also* TERMINAL VELOCITY.

free reserve [comparative values] In banking in the United States, the difference between a bank's total reserve funds and its minimum permitted reserve plus current liabilities.

free skating *see* ICE-SKATING DISCIPLINES AND EVENTS

free-style *see* ICE-SKATING DISCIPLINES AND EVENTS; SKIING DISCIPLINES AND EVENTS; SWIMMING DISCIPLINES AND COMPETITIVE DISTANCES

freezing point [physics] The temperature at which a substance in the liquid state solidifies, known also as the solidification point. It is the same as its melting point.

F region *see* F LAYER, F REGION

freight ton [volumetric measure] Unit of volume (not weight) in the transportation of goods by ship, corresponding therefore to a measure of stowage capacity.

$$\begin{aligned} 1 \text{ freight ton} &= 8 \text{ BARRELS BULK} \\ &= 40 \text{ cubic feet, } 1.4815 \text{ cubic yard} \end{aligned}$$

$$(1 \text{ cubic yard} = 0.6750 \text{ freight ton})$$
$$= \quad 1.1320 \text{ cubic metre}$$
$$(1 \text{ cubic metre} = 0.8834 \text{ freight ton})$$
$$= \quad 31.1243 \text{ UK bushels, } 32.1238 \text{ US bushels}$$
$$2\frac{1}{2} \text{ freight tons} \quad = \quad 1 \text{ REGISTER TON}$$

The freight ton is known alternatively as a *measurement ton* or a *shipping ton*.

French horn *see* BRASS INSTRUMENTS' RANGE

French pitch [music] Standard of musical pitch (frequency) in which the A above middle C is defined as 435 hertz, slightly lower than CONCERT PITCH. *See also* INTERNATIONAL PITCH.

French Revolutionary (Republican) calendar [time] Between the (Gregorian) years 1793 and 1805 inclusive, the French Revolutionary authorities adopted an annual calendar of 12 months of 30 days each, with five supplementary days added in ordinary years and six in leap years, known as *sans-culottides*, added at the end of the year as the final days of Fructidor, and celebrated then as festivals. The new names for the months were proposed by one Fabre d'Eglantine, and were deliberately based on words and verbal rhythms that were considered appropriate to the months involved. The autumn months, at the beginning of the new year, sounded dignified and of medium duration; winter months sounded long and heavy; spring months sounded lively but brief; and summer months sounded open and broad.

month	meaning	Gregorian months
Vendémiaire	'vintage'	mid-Sep to mid-Oct
Brumaire	'misty'	mid-Oct to mid-Nov
Frimaire	'frosty'	mid-Nov to mid-Dec
Nivôse	'snowy'	mid-Dec to mid-Jan
Pluviôse	'rainy'	mid-Jan to mid-Feb
Ventôse	'windy'	mid-Feb to mid-Mar
Germinal	'seed-growth'	mid-Mar to mid-Apr
Floréal	'flowering'	mid-Apr to mid-May
Prairial	'meadowy'	mid-May to mid-Jun
Messidor	'harvest'	mid-Jun to mid-Jul
Thermidor	'hot'	mid-Jul to mid-Aug
Fructidor	'fruitful'	mid-Aug to mid-Sep

According to some authorities, Thermidor was alternatively known as Fervidor (which meant roughly the same thing).

But in the new calendar there were only three 10-day weeks per month, at the end of each of which came the day of rest. Not only did this upset those who worked and were paid by the former 7-day week, it infuriated the Roman Catholic Church in abolishing Sundays as such. Most citizens of France therefore simply ignored the new calendar altogether, and the eventual official return to the Gregorian calendar as of 1 January 1806 was really no more than recognition of what had virtually always been the status quo. *See also* YEAR.

frequencies, musical *see* PITCH

frequency [physics] Rate of recurrence of a periodic vibration such as a wave motion, expressed as the number of cycles, oscillations, or vibrations in unit time. It is measured in units such as cycles per second or hertz. For a wave, it is inversely proportional to the wavelength, and for an electromagnetic wave (radio waves, microwaves, infra-red, visible light, ultraviolet light, X-rays, and gamma rays) it equals the speed of light divided by the wavelength.

frequency distribution [maths] For a set of data in statistics, the arrangement of the values in terms of the number of times they occur in the set, also called merely their distribution. Often distributions are arranged in tabular form (frequency table) or plotted on a graph (of value against frequency) to produce a curve of a characteristic shape. *See also* BELL CURVE, BELL-SHAPED CURVE.

frequency response [physics; telecommunications] For a piece of equipment handling audio or radio frequencies, the range of output powers for a given range of input frequencies, measured in watts, volts, or decibels below a particular level (such as the peak frequency response).

frequency table [maths] A table that orders a set of data in terms of the number of times a particular value occurs: *see* FREQUENCY DISTRIBUTION.

fresnel [physics] Unit of light frequency equal to 10^{12} HERTZ ($= 1$ terahertz). It was named after the French physicist Jean Augustin Fresnel (1788-1827).

fricative consonants [literary: phonetics] Consonants that rely on a compression of air between a deliberately narrowed opening between the lips, between the lips and teeth, or between the tongue and teeth. In English they correspond to:

> f [ph],   v,   s,   z

friction, coefficient of [physics] Friction is a force that makes a surface resist sliding or rolling on another surface. Up to a value known as the limiting friction, the force trying to cause movement is equal to the frictional force resisting it; any increase in the moving force beyond this value results in movement. The coefficient of friction is the limiting friction divided by the normal reaction between the sliding or rolling surfaces; it is a constant for any particular pair of materials.

frigid zones *see* ZONE

frigorie [physics] In continental Europe, a formerly little-used unit of refrigeration equal to a rate of heat extraction of 1 kilocalorie per hour.

It was named as the antithesis of calorie, a word that derives from Latin roots implying warmth.

frog units [medicine] In the mid- to late 1800s, a measure of the potency of digitalis doses (for heart irregularities) according to the number of live, captive frogs that were rendered comatose or killed.

Digitalis drugs are still very much in use today, and still prepared from plants of the foxglove family. The best-known are digoxin and digitoxin.

frost point [temperature] The temperature at which water vapour in air that is already at a temperature below freezing turns into frost. It depends on altitude, atmospheric pressure, and other factors such as windlessness.

Froude number [physics; hydrodynamics] In fluid dynamics, a number that describes the flow of a fluid in which there is a free surface, equal to the velocity of the fluid divided by the square root of the product of the length of flow and the acceleration of free fall. It was named after the British engineer and mathematician William Froude (1810-79).

frustum [maths] A solid figure formed by cutting through another solid with two parallel planes. The most common is the conical frustum (shaped like an upturned bucket).

f-stop *see* F-NUMBER

FTSE index, Financial Times Special index [comparative values] An INDEX relating to the average cost of living in Britain, based on both the values of shares on the stock exchange and on the actual cost of a set of specific household commodities in the shops. It is published as a single figure on week days in the business-economy newspaper the *Financial Times* (London), and is included in most full-length radio and television news broadcasts.

Fugio cent [comparative values] Earliest independent unit of currency in the United States: a copper coin issued in 1787. On the obverse was a depiction of the Sun and the Latin word *Fugio* 'I fly'; on the reverse was the motto 'We are one' surrounded by thirteen interlocking rings.

fulham [sporting term] In England – especially London – from Shakespearian times to the early 1900s, a dice weighted on one or two sides (a loaded or biased dice). A dice that was a 'high fulham' would inevitably land to show the 4, 5, or 6 uppermost; a 'low fulham' showed the 1, 2, or 3 uppermost. Either would be used as most convenient, provided detection was unlikely.

The term is thought to have derived from the London area of Fulham, once the site of scores of gaming houses of ill repute. Alternatively, the term may have been part of an expression in rhyming slang.

full house *see* POKER SCORES

full Moon *see* LUNAR PHASES

funal [physics] Unit of force in the little-used unit system in which the metre, tonne, and second are base units, now more properly called a STHÈNE: *see also* NEWTON.

> 1 sthène/funal   =   1,000 newtons

function [maths] A mathematical expression with one or more related variables.

functional [maths] A FUNCTION of which the value depends on all the values assumed by another function.

fundamental interval [physics] The range of temperatures between the two fixed points of a temperature scale. For example: in the Celsius scale (fixed points 0° and 100°) it is 100°; for the Fahrenheit scale (32° and 212°) it is 180°.

fundamental particle [physics] A subatomic particle that cannot be further divided (for example, gauge bosons, leptons, and quarks). *See also* ELEMENTARY PARTICLE.

fundamental tone [music] The basic note (tone) from which harmonics (overtones) derive. Also, the lowest-pitched note of a chord (alternatively – but far less commonly – called the fundamental bass).

fundamental units [physics] Units of length, mass, and time – the chief base units of any system of units – from which velocity, acceleration, force, work (energy), and power are directly derived.

furlong [linear measure] A measure of land that was once commonly used to express the length of the perimeter of a property, but that is now used almost exclusively to describe the various distances of horse races at racecourses.

$$
\begin{aligned}
1 \text{ furlong} \quad &= \quad \text{one-eighth of a MILE} \\
&= \quad 10 \text{ CHAINS, } 40 \text{ RODS (or poles), } 220 \text{ YARDS, } 660 \text{ feet} \\
&= \quad 201.168 \text{ metres, } 0.201168 \text{ kilometre} \\
&\qquad (200 \text{ metres} = 0.9942 \text{ furlong}) \\
8 \text{ furlongs} \quad &= \quad 1 \text{ mile}
\end{aligned}
$$

It may be useful also to note that

$$
\begin{aligned}
&1 \text{ square furlong} \\
&(220 \times 220 \text{ yards}) \quad = \quad 10 \text{ ACRES}
\end{aligned}
$$

– a measure that in fact gave the furlong its name, for the 10-acre square was the standard size of a medieval English field. Each furrow ploughed down the length of the field would then be 220 yards or 'one furrow long' – 1 furlong.

futures [comparative values] Commodities or stocks traded in dealings at a stock exchange that relate mostly to consumer perishables (such as wheat harvests) of future years and to the anticipated value of such goods at that time, taking into account intervening world events, inflation, etc.

futurity race [sporting term] In horse racing, a race (generally for two-year-olds) for which entries are registered at least two years in advance.

G

g, g *see* GRAM, GRAMME; GILBERT; GRAIN; GRAVITY, FORCE OF

Gaelic football measurements, units, and positions [sport] Gaelic football is rarely played outside Ireland. A fast-moving game in which two teams of fifteen players may kick the ball, fist it, or briefly catch and hold it, the object is to score as many points as possible by getting the ball between the opposing goalposts, either above the crossbar or below it.

The dimensions of the pitch:

maximum area: 160 x 110 yards (146.3 x 100.6 metres)

minimum area: 140 x 84 yards (128.0 x 76.8 metres)

the goal:

width: 21 feet (6.4 metres)

crossbar height: 8 feet (2.44 metres)

goalpost total height: 16 feet (4.88 metres)

Positions of players:

players are divided into (usually seven) forwards, (usually four) halfbacks, (usually three) backs, and a goalkeeper

Timing: (usually) 60 minutes, in two halves, separated by an interval of 10 minutes

top-class games in the finals of competitions may last 80 minutes

Points scoring:

ball kicked or fisted (beneath crossbar) into goal: 3 points

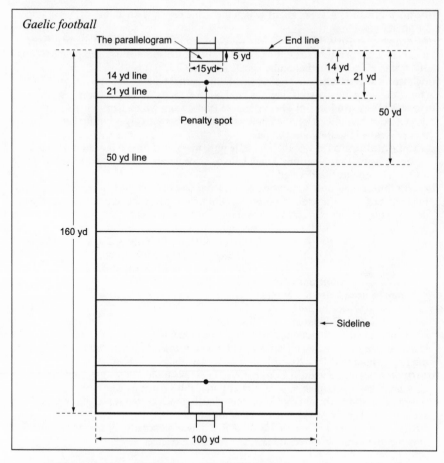

Gaelic football

ball kicked over crossbar between posts: 1 point
the team with the more points at the end of the game wins
Dimensions of the ball:
circumference: 27-29 inches (68.58-73.66 centimetres)
weight: 13-15 ounces (368.5-425.2 grams)

gaggle of geese [collectives] The term is apparently a variant of *cackle* (with the two initial consonants voiced), in reference to the sound a skein of geese flying overhead makes.

gait [sport] In horse racing and three-day eventing dressage, any of the set methods in which a horse must proceed either to stay within the rules or to earn (or not to lose) points. The main forms of gait in trotting races, for example, are the trot and the pace; in dressage, the walk, the canter, and the trot are of central importance (the rack is the form of gait in which a horse lifts its forelegs high off the ground); most other competitions and races rely on the gallop.

In dog shows and shows of farm animals, the trained way in which an animal walks with its owner in front of adjudicators, and on which its performance is judged.

The term is an abstract noun based on the verb 'to go' (which originally meant 'to walk').

gal, galileo [physics] An obsolete unit of acceleration, equal to 1 centimetre per second per second, once used for small differences in the ACCELERATION OF FREE FALL (which were generally expressed in milligals; 1 milligal = $\frac{1}{1000}$ gal). It was named after the Italian physicist and astronomer Galileo Galilei (1564-1642).

galactic coordinates [astronomy] A pair of spherical coordinates that locate a point on the GALACTIC PLANE. Galactic latitude is measured (positively) northwards from the galactic plane; galactic longitude is measured from the point with right ascension 17 hours 42.4 minutes and declination – 28 degrees 55 minutes (1950): *see* EPOCH.

galactic plane [astronomy] The plane represented by the huge disc of stars that is our Galaxy (the Milky Way), passing roughly through its centre at an inclination to the CELESTIAL EQUATOR of 62°.

galactometer [physics/chemistry] An instrument that measures and displays the fat content (the 'richness') in milk by determining the milk's specific gravity.

The first element of the term derives from ancient Greek *galact-* 'milk(y)' – which is why the Milky Way is the archetypal *galaxy* – etymologically the same word as *milk* (with m- for g-) and Latin *lact-* (which has lost the initial consonant altogether).

Galaxy, age of the *see* AGE

gale [meteorology] Wind of forces 8, 9, 10, and 11 on the BEAUFORT SCALE OF WINDSPEED, corresponding to a wind speed of 34 to 63 knots, 62 to 115 kilometres per hour, or 39 to 72 miles per hour.

gallon [volumetric measure] Everyday unit of volume in the English-speaking world, although to some extent as a measure of liquids overshadowed by the LITRE of the metric system in certain contexts in Europe. Less commonly, the gallon is additionally a measure of dry capacity. Unhappily, differing traditions have caused the unit to vary between British and North American usages (*see* WINE GALLON).

1 UK gallon	=	8 UK pints, 4 UK quarts
	=	32 UK gills or quarterns, 160 UK fluid ounces
	=	4.5459631 litres
		(9 litres = 1.9798 UK gallon)
	=	1.20095 US gallon
		(1 US gallon = 0.83267 UK gallon)
1 (dry) UK gallon	=	1 (liquid) UK gallon
	=	4,545.9631 cubic centimetres
	=	277.41 cubic inches, 0.1605 cubic foot
		(1 cubic foot = 6.2305 UK gallons)
2 (dry) UK gallons	=	1 (UK) PECK
8 (dry) UK gallons	=	1 (UK) BUSHEL
1 US gallon	=	8 US pints, 4 US quarts
	=	32 US gills, 128 US fluid ounces
	=	3.785306 litres
		(4 litres = 1.0567 US gallon)
	=	0.83267 UK gallon
		(1 UK gallon = 1.20095 US gallon)
1 (dry) US gallon	=	4.40476 dry litres (4,404.76 cubic centimetres)
	=	268.795 cubic inches, 0.1556 cubic foot
		(1 cubic foot = 6.4267 US dry gallons)
	=	0.9689 UK dry gallon
2 (dry) US gallons	=	1 (US) peck
8 (dry) US gallons	=	1 (US) bushel

The term derives in English from Old French/medieval Latin, and is thought by most authorities to refer to some unit borrowed from Celtic sources – that is, the 'Gallic/Gaelic' unit. In any event, it originally represented only a liquid measure.

gallons per minute [flow] Unit of flow in large quantities, as might be required for measurement in oil supply systems, water reservoirs, and suchlike. As with the standard gallon, however, the actual quantity is different in Britain and the United States.

1 UK gallon per minute	=	0.1333 UK pint per second
		(1 pint every 7.5 seconds)
	=	60 UK gallons per hour
	=	4.5459631 litres per minute
		(1 litre every 13.2 seconds)
	=	1.20095 US gallon per minute
1 US gallon per minute	=	0.1333 US pint per second
		(1 pint every 7.5 seconds)

$$= \quad 60 \text{ US gallons per hour}$$
$$= \quad 3.785306 \text{ liters per minute}$$
$$\quad (1 \text{ liter every } 15.85 \text{ seconds})$$
$$= \quad 0.83267 \text{ UK gallon per minute}$$

gallop *see* GAIT

Galois field [maths] In the study of special codes by which information can be mechanically or electronically transmitted without error, a mathematical field made up of a finite number of elements.

The field is named after the French mathematician Evariste Galois, (1811-32), who was killed as the result of a duel, aged 20.

galvanometer [physics] An instrument for measuring small electric currents. It was named after the Italian physiologist Luigi Galvani (1737-98), who performed some of the earliest experiments with electricity.

gam of whales [collectives] It is likely that, as a collective noun, the meaning of this term has been transferred to an assembly of whales from originally relating to an assembly of whalers. A gam is technically a potentially rowdy or boisterous conference of seamen from different ships (such as a fleet of whalers) out at sea. Akin to Scandinavian words denoting 'sport' (cf. English *game*), it is the stem from which *gamble* (and possibly even *gambol*) is a frequentive.

Whales also come in *pods* and *schools*.

gamma [photography; physics] In photography and television, a measure of the contrast in a reproduced image (made by comparing its luminance/brightness with that of the original).

In physics, either another name for a microgram (one-millionth of a gram), or a unit of magnetic flux density, equal to 10^{-9} tesla.

gamma radiation [physics] Penetrating, very short-wavelength electromagnetic radiation (shorter than X-rays) emitted by some radioactive elements: *see* ELECTRO-MAGNETIC SPECTRUM.

garnets [volumetric measure] Old measure of liquid capacity in Russia.

$$1 \text{ garnets} \quad = \quad 3.28 \text{ litres}$$
$$\quad (3 \text{ litres} = 0.9146 \text{ garnets})$$
$$= \quad 5.773 \text{ UK pints, } 6.931 \text{ US pints}$$
$$\quad (1 \text{ UK gallon} = 1.3858 \text{ garnets}$$
$$\quad 1 \text{ US gallon} = 1.1542 \text{ garnets})$$
$$3.75 \text{ garnets} \quad = \quad 1 \text{ VEDRO}$$
$$15 \text{ garnets} \quad = \quad 4 \text{ vedro, } 40 \text{ schtoffs, } 64 \text{ boutylki, } 400 \text{ charki}$$

gas constant [physics] The constant (R) in the gas equation, equal to 8.31434 joules per mole per KELVIN.

gas equation [physics] For *n* moles of an ideal gas, the product of the pressure and volume equals *n* times the product of the gas constant and the absolute temperature ($pV = nRT$). The equation is alternatively known as the ideal gas equation.

gas laws *see* BOYLE'S LAW; CHARLES'S LAW; GAS EQUATION; GAY-LUSSAC'S LAW (OF VOLUME)

gasoline/petrol units *see* BARREL; GALLON; LITRE, LITER; LITRES PER HUNDRED KILO-METRES; MILES PER GALLON

gauge [engineering; textiles] On railways, the gauge is the distance between the inside faces of the rails: *see* BROAD GAUGE; NARROW GAUGE; STANDARD GAUGE. Similarly, in model railways, the gauge is the distance between the inside faces of the rails: *see* N GAUGE; O GAUGE; OO GAUGE.

In motion picture film, the gauge is the width of the film (for example, 8-, 16-, 35- and 70-millimetre). Similarly, the width of magnetic (recording) tape is its gauge (for example, 8 millimetres, ¼ inch, ½ inch, ¾ inch, 1 inch, and 2 inches).

The term *gauge* is also used for the thickness of wire, rods, or metal plate: *see* WIRE GAUGES.

In the United States, the term is used to indicate the bore of a shotgun (in place of the word *bore* as used in British English): *see* CALIBRE, CALIBER.

In knitting (and especially in knitted or woven stockings), a measure of the fineness of the finished fabric, expressed as the number of loops for every 1½ inches (3.8 centimetres).

The word *gauge* referred originally (in medieval French) to a measuring device rather than to a dimensional measurement, but etymologists are divided over whether it was a pendant scale for measuring weight (and thus akin to the English word *gallows*) or borrowed directly from the Basque *galga* 'gauge'. *See also* BOURDON GAUGE; FEELER GAUGE; MICROMETER GAUGE.

gauge boson [physics] A subatomic particle that is involved in interactions between pairs of fundamental particles. There are four types:
  gluons (strong interactions)
  gravitons (gravitational)
  intermediate vector bosons (weak interactions)
  photons (electromagnetic)

gauge pressure [physics; engineering] The pressure of a fluid – gas or liquid – as indicated by a pressure gauge; the amount by which the pressure in the fluid exceeds that of the atmosphere. The true, or absolute, pressure is obtained by adding atmospheric pressure to the gauge pressure.

gauging rod [linear measure] A calibrated rod for measuring the capacity of an empty cask or barrel, or for measuring the depth of the contents inside one that is not empty but is not full either; a type of dipstick. Gauging rods are used by customs officers to determine liability to excise or duty.

gauss [physics] Unit of magnetic flux density in the centimetre-gram-second (CGS) system, equal to 1 maxwell per square centimetre.
$$1 \text{ gauss} = 10^{-4} \text{ tesla (SI units)}$$
It was named after the German physicist and mathematician Karl Gauss (1777-1855).

Gaussian distribution *see* BELL CURVE; NORMAL DISTRIBUTION

Gaussian units [physics] Obsolescent system of units in which electrical parameters were measured in electrostatic units (ESU) and magnetic ones were measured in electromagnetic units (EMU). They were named after the German physicist and mathematician Karl Gauss (1777-1855), and have now been superseded by SI units.

Gay-Lussac's law (of volume) [chemistry] When gases react chemically, the reacting volumes are in a simple ratio to each other and to the volume of the product(s) (at the same temperature and pressure). It was named after the French chemist and physicist Joseph Gay-Lussac (1778-1850).

gear ratio [engineering] The factor by which a pair of gears or a gearbox increases or decreases the speed of rotation of a shaft. For example: a gear ratio of 2:3 represents an increase in speed of one-and-a-half times (every two rotations of the drive shaft produce three rotations of the driven shaft); this is an example of gearing-up. A gear ratio of 5:1 (a fivefold decrease in the speed of rotation) is an example of gearing-down.

gee pound [physics] Another name for a SLUG.

Geiger counter [physics; radiology] An instrument (alternatively known as a Geiger-Müller counter) for detecting and measuring charged particles (such as alpha particles and beta particles), and thus radioactivity. It was named after the German physicist Hans Geiger (1882-1945).

gene [biology] A segment of DNA (deoxyribonucleic acid) on a chromosome (in the nucleus of a cell) which can be regarded as a unit of heredity. Each gene singly, or in combination, specifies a particular characteristic or trait that is passed on from parents to offspring.
    The number of times one (specific) gene is present within a group after a (specific) number of generations is known as the *gene frequency*. A representation of the arrangement of genes within a chromosome is known as a *gene map*.

general [military rank] Very senior commissioned rank in the army (and in the United States, the airforce).
    In the British army, a general ranks above a lieutenant general and (only) below a field marshal, and is the equivalent of an admiral in the Royal Navy and an air chief marshal in the Royal Air Force.
    In the United States Army, a general ranks above a lieutenant general and (only) below a general of the army (a five-star general); similarly, in the United States Airforce, a general ranks above a lieutenant general and (only) below a general of

the airforce. Both are equivalent to an admiral in the United States Navy.

In a number of European armies, the rank of general is the equivalent of a rank slightly subordinate to a general in the British or United States' armed forces, corresponding rather to the rank of lieutenant general.

A general is so called because, in former centuries, he (always 'he' in those days) was in charge of the forces on the battlefield and thus had general (total) command. His immediate assistants were – and in some armies still are – called the general staff.

See also GENERAL OF THE ARMY, OF THE AIRFORCE.

general formula [chemistry] A symbolic chemical formula for any member of a group of compounds, such as those in a HOMOLOGOUS SERIES. For example, C_nH_{2n+2} is the general formula for an alkane (paraffin).

general of the Army, of the Airforce [military rank] Most senior officer in the United States Army and Airforce, in each case ranking above a general, and entitled to wear five stars in his or her insignia. The rank is equivalent to that of fleet admiral in the United States Navy, to that of field marshal in the British army, and to that of marshal in many European armies.

general theory of relativity *see* RELATIVITY: THE MASS-ENERGY EQUATION

generation [time] The people in a family or group that were born, raised, and educated at roughly the same time.

As a period of time that corresponds to the years between birth and reproduction, the duration is often quoted as about 30 years – although the average age at which European, Australasian, and North American parents now tend to have their first children is around 24.

generation time [biology] The time it takes a a single-celled organism to divide (undergo fission) to form a pair of identical daughter cells.

genetically significant dose *see* X-RAYS

genetic code [biology] A GENE is made up of DNA, which has four chemical bases. There are sixty-four possible ways of arranging three bases out of the four, to form triplets that specify (or code for) each of the twenty amino acids that make up proteins. Other triplets have other coded functions. This overall genetic code applies universally throughout the plant and animal kingdoms.

genetic fingerprinting [biology; physics] Forensic method of identification by examination of a blood or other tissue sample containing whole cells from which the DNA can be extracted and analysed.

Only in identical twins are the basic constituents of DNA likely to be closely matched enough to cause potential confusion in a case of identification. In the ordinary way, the chances of one person's having a sequence of genetic material corresponding to a large number of chemical bases constituting individual genes in any way resembling another person's (unless a member of the same family) is several billion to 1 .

genius [biology] In psychology, a person whose IQ is measured to be 140 or more. The word corresponds to a Latin abstract noun meaning 'origination-power', 'creativity' which, in ancient Rome, was used as a carefully impassive term for a guardian spirit who inspired profound thoughts.

genus [biology: taxonomy] In the taxonomic classification of organisms, the major category between Family (or Tribe) and species.

For full list of taxonomic categories *see* CLASS, CLASS.

geocentric parallax *see* DIURNAL PARALLAX

geochronology [geology; time] Dating fossils and rocks to the era, period, and/or epoch of their origin by geological analysis, a process sometimes alternatively called geochronometry.

geocorona [meteorology] A thin stratum of ionized gases surrounding the Earth at a height from the surface of between 14,500 and 29,000 kilometres (9,000 and 18,000 miles).

geodesic [maths] Technically, describing a curve that has the form of part or all of the circumference of the planet Earth. But *see also* GEODESIC LINE.

geodesic line, geodetic line [maths] The shortest path between two points on a surface of any shape.

geographical mile [linear measure] A distance corresponding to 1 minute of
longitude at the equator:

    1 geographical mile = 1,853.99 metres, 1.854 kilometre
                       = 6,082.66 feet, 1.1520 mile

It is sometimes alternatively called the nautical mile and, as such, used in British
practice (but not American) to define the unit of velocity called the KNOT.

The US nautical mile – sometimes also called the geographical mile – was
officially established in 1954.

    1 (US) nautical mile = 6,076.11549 feet, 1.1508 mile
                       = 1,851.99 metres, 1.852 kilometre

More generally, a geographical mile can be regarded as 1 minute of longitude at
any latitude.

geological strata [geology; time] An imprecise term, corresponding in this sense to
the synonymous *beds*, that applies to any of the optically distinctive layers of rock
within a mass of rock of different origins.

Layers of rock that correspond to a geological period are known as a *system*.
Within a system, layers of rock that correspond to a geological epoch are known as
a *series*. And within a series, a layer of rock that corresponds to a subdivision of an
epoch (sometimes known as an age) is known as a *stage*. Divisions of a stage tend
to be classified according to the circumstances: a time of glaciation may be represented by a *stade*, a time of special palaeontological input is called a *zone*. *See also*
GEOLOGICAL TIME.

geological time [time] The science of geology divides the history of the Earth into
four ERAS, themselves made up of eleven PERIODS, some of which also incorporate
EPOCHS, and numerous STAGES that are specific to particular locations. The major
divisions, from youngest to oldest, are:

Era	Period	Epoch	millions of years ago
Precambrian (or Archaeozoic)			about 4,600 to about 590
Palaeozoic	Cambrian		about 590 to about 505
	Ordovician		about 505 to about 438
	Silurian		about 438 to about 408
	Devonian		about 408 to about 360
	Carboniferous		about 360 to about 286
	Permian		about 286 to about 248
Mesozoic	Triassic		about 248 to about 213
	Jurassic		about 213 to about 144
	Cretaceous		about 144 to about 65
Cenozoic	Tertiary	Palaeocene	about
		Eocene	65
		Oligocene	to
		Miocene	about
		Pliocene	2-3
	Quaternary	Pleistocene	from
		Holocene	2-3 on

geomagnetic axis [physics; geology] The axis of the Earth relating to the (north)
magnetic pole, presently at an angle of 12° relative to the geographical axis of the
planet.

geometric mean [maths] Of *n* positive numbers, the *n*th root of their overall
product. For example: the product of the three numbers 4, 6, and 9 is 216; their
geometric mean is therefore the cube root (the root to the power three) of
216, = 6. *See also* ARITHMETIC MEAN; ROOT MEAN SQUARE.

geometric progression/series [maths] A sequence of numbers such that the ratio between any two successive terms is always the same – that is, there is a common ratio. For example: the numbers 2, 6, 18, 54 are a geometric progression (with a common ratio of 3).

geometry [maths] The branch of mathematics that deals with points, lines, curves, surfaces, and solids. Ordinary, or flat, geometry involves no curved planes and is called Euclidean geometry. Coordinate, or analytic, geometry describes shapes in terms of algebraic expressions. Calculus is used (for example to determine areas and volumes) in differential geometry. The study of figures of which the properties do not change under projection is termed projective geometry.

geordie [comparative values] In Britain from the 1780s to the 1890s, a gold coin worth a guinea (one pound and one shilling, 21 shillings). It was so called because it had a representation of St George – patron saint of England – on one side.

Georgian age/era [time] Period from 1714 to 1830, during which four kings called George occupied the British throne.

George I	reigned from	1 August 1714	to 11 June 1727
George II		11 June 1727	25 October 1760
George II		25 October 1760	29 January 1820
George IV		29 January 1820	26 June 1830

The period is particularly noted for its style of architecture and of furniture.

Since that time, there have been two more kings named George on the throne of Britain (George V reigned from 6 May 1910 to 20 January 1936, George VI from 11 December 1936 to 6 February 1952), but neither period of rule is generally known as Georgian.

geostationary orbit [geology; space science] Path of an artificial Earth satellite that takes 24 hours to complete one orbit, during which time the Earth has turned once on its axis – which means that the satellite remains constantly over the same point on the surface of the Earth. Such an orbit – also called a synchronous orbit – is attained at a height of 35,900 kilometres (22,308.3 miles).

gerah [weight] The smallest unit of weight in ancient Israel.

1 gerah	=	0.42 GRAM
	=	0.015 ounce, 6.5625 GRAINS
5 gerahs	=	1 REBAH
10 gerahs, 2 rebahs	=	1 BEKAH
20 gerahs, 2 bekahs	=	1 (ordinary or 'holy') SHEKEL

GeV, Gev [physics] Abbreviation of giga-electron volt (= 10^9 electron volts): *see* ELECTRON VOLT.

GHOST [physics; meteorology] Global Horizontal Sounding Technique – a set of radiosonde balloons stationed at different heights in the atmosphere to radio back meteorological information.

giant slalom *see* SKIING DISCIPLINES AND EVENTS

gibbous phase *see* LUNAR PHASES

Gibbs free energy [physics; chemistry] The energy that is absorbed or liberated in a reversible process, equal to the heat content minus the product of temperature and entropy. It was named after the American physicist and chemist Josiah Gibbs (1839-1903).

giga- [quantitatives: prefix] Prefix which, when it precedes a unit, increases the unit to 1 UK thousand-million/1 US billion (10^9) times its standard size or quantity.

Examples:	1 gigahertz	=	1,000,000,000 hertz
	1 gigavolt	=	1,000,000,000 volts
	1 gigawatt	=	1,000,000,000 watts

The prefix derives from ancient Greek *gigas* 'giant', represented in Latin by *gigans* and possibly (via different sources) also by *Titan*.

gilbert [physics] Unit of magnetomotive force (MMF) in the centimetre-gram-second (CGS) system, equal to ¼ π abampere-turns.

$$1 \text{ gilbert} = \text{¹⁰}\!\!/_4 \, \pi \, (= 0.7958) \text{ ampere-turns (SI units)}$$

It was named after the English scientist William Gilbert (1544-1603), physician to Queen Elizabeth I.

gill [volumetric measure] Small unit of liquid capacity relating to the PINT and, like the pint, differing in actual quantity in Britain and the United States.

In Britain and some other English-speaking countries:

1 gill	=	0.25 (one-quarter) pint, 5 (UK) fluid ounces
	=	0.14206 litre, 14.206 centilitres
	=	1.20095 US gill, 0.30024 US pint
	=	4.815 US fluid ounces

In the United States:

1 gill	=	0.25 (one-quarter) pint, 4 (US) fluid ounces
	=	0.11829 liter, 11.829 centiliters
	=	0.83267 UK gill, 0.20819 UK pint
	=	4.169 UK fluid ounces

The gill is in Britain known alternatively – but not commonly – as a quartern.

The standard measure for tots of spirits in British bars is one-sixth of a (UK) gill.

In the United States, for cooking and for mixing drinks, the gill is regarded as equivalent to half a cup.

gillion [quantitatives] In Britain, a little-known term for what in the United States is a billion: one thousand-million – 1,000,000,000 or 10^9.

The term derives on analogy with million, the first element taken from GIGA-.

gin [weight] In Malaysia (then Malaya), a unit of weight corresponding to one used throughout the Far East during the colonial times of the mid-1800s, devised artificially as a measure for taxation purposes. Also known in Malaysia as the *kati*, the gin was called the *chin* or *catty* in China, the *kin* in Japan, and the *kon* or *catty* in Korea.

1 gin	=	16 TAHILS, 160 chee, 1,600 hoon		
	=	1.333 pound avdp.		
		(1 pound = 0.75 gin, three-quarters of a gin)		
	=	604.775 grams, 0.604775 kilogram		
3 gin	=	4 pounds (1.8143 kilogram)		
100 gin	=	1 PICUL	=	133.333 pounds (60.4775 kilograms)

Giorgi units [quantitatives] An alternative name for METRE-KILOGRAM-SECOND (MKS) UNITS. The system was named after the Italian physicist Giovanni Giorgi (1871-1950).

girsh *see* GUERSH, GUERCHE

glissando [music] Musical instruction: play or sing a run of notes merging each with the next, so that no note seems individual. On a piano this instruction is ordinarily taken to mean running one finger across a sequence of keys. The word is an Italianate variant of a French present participle that means 'gliding'.

global warming [physics; meteorology] Since the last ice age ended about 10,000 years ago, the overall temperature of the atmosphere at the Earth's surface has increased by 6-10°C (11-18°F), although the spread of agriculture – which, from about 8,000 years ago, greatly augmented the reflective brightness of the planet's surface – and the so-called Little Ice Age (in Europe peaking around the AD 1600s and 1700s) have both acted as a brake on this process.

During the twentieth century alone, however, global temperature has risen by approximately 0.5°C (1°F) because of the higher concentration of greenhouse gases in the upper atmosphere – gases such as carbon dioxide and methane which absorb radiation reflected off the surface – that retain the heat instead of allowing it to escape back into space. The cause of the higher concentration is partly the burning of fossil fuels, partly the reduction of the amount of oxygen being replaced in the atmosphere following the deliberate devastation of tropical forests, and partly other factors.

Some Arctic ice sheets have melted by 15 per cent since 1980. It is possible that mean temperatures may have risen by a further 1-3°C (2-6°F) by the year 2050, in which case the mean sea-level will have risen correspondingly by between 61 and 244 centimetres (2 and 8 feet). This is, it must be said, something of a worst-case scenario.

glug [physics] A proposed unit of mass in the centimetre-gram-second (CGS) system, equal to the mass accelerated by 1 centimetre per second per second by a force of 1 gram weight.

$$1 \text{ glug} = 980.65 \text{ grams (approx. 1 kilogram)}$$
$$(1 \text{ kilogram} = 1.0197 \text{ glug})$$

gluon [physics] A subatomic particle that is the agent for strong interactions between pairs of fundamental particles (for example, it holds quarks together); it is a type of GAUGE BOSON.

gm *see* GRAM, GRAMME

gnathic index [medicine; biology] In comparative anatomy, a measure of by how much the upper jaw projects over the lower, calculated as the length from the middle of the anterior border of the foramen magnum of the skull (the basion) to the middle point of the anterior surface of the upper jaw (the alveolar point) x 100, divided by the length between the basion and the midpoint of the area in which the nasal bones and frontal bone meet (the nasion).

The first element of the term derives from ancient Greek *gnath-* 'jaw'.

gnomon [time] The triangular pointer on a sundial, the shadow of which denotes the time of day. The term derives directly from the ancient Greek for 'information-provider' (akin to English *name*).

go [volumetric measure] Smallest unit of the liquid capacity measurement system in Japan.

$$1 \text{ go} = 10 \text{ shaku}$$
$$= 0.18039 \text{ litre}, 18.039 \text{ centilitres}$$
$$(1 \text{ litre} = 5.5435 \text{ go})$$
$$= 0.3175 \text{ UK pint}, 0.3806 \text{ US pint}$$
$$(1 \text{ UK pint} = 3.1496 \text{ go})$$
$$1 \text{ US pint} = 2.6274 \text{ go})$$
$$10 \text{ go} = 1 \text{ SHÔ}$$
$$100 \text{ go}, 10 \text{ shô} = 1 \text{ TO}$$
$$1,000 \text{ go}, 100 \text{ shô}, 10 \text{ to} = 1 \text{ KOKU}$$

The name of the unit may or may not be related to a Japanese word of identical pronunciation meaning 'figure', 'number', 'item', and thus 'unit'.

goal, goal-line *see* individual sports' and games' measurements and units

gold disc, golden disk [comparative values] In the (phonograph) recording industry, an award in the form of a metallic golden record to an artist or group of artists whose single or album has sold a specific high number of copies, or whose financial receipts for those sales total more than a specific high value.

In the United States and generally in Britain, a record has to sell one million copies for the artist to qualify for a gold disc, or the receipts have to total more than one million dollars.

Elsewhere in the world, the required figures are generally not so high. In Canada (which has just over one-tenth of the population of the United States), for example, a gold disc may be awarded following sales of only 50,000 copies.

golden mean *see* GOLDEN SECTION

golden number [astronomy] The position of a particular year in the nineteen-year METONIC CYCLE, found by adding 1 to the given year and dividing the result by 19; the remainder is the golden number (a remainder of zero counts as a golden number of 19).

For example: the year 2000 has a golden number of 6.

It can be used (with other calendrical information, such as the SUNDAY LETTER) to find the date of Easter in any year, and gets its name because, in early medieval Catholic calendars, the number was marked in gold. *See also* EASTER DAY, DATE OF.

golden section [maths] A highly esteemed proportion in (Renaissance) art and architecture, found by cutting a line in such a way that the ratio of the smaller length to the larger length is the same as the ratio of the larger length to the whole. The term is also applied (as it was by Euclid) to rectangles of which the sides are in this proportion.

Its alternative name – the golden mean – is, in effect, borrowed from an idea first outlined by the poet Horace (Quintus Horatius Flaccus, 65-8 BC) in a verse essay on contemporary ethics and manners. There, he propounded as an ideal the avoidance of all extremes, claiming that the most favourable and most civilized form of existence was the *aurea mediocritas*.

golden wedding anniversary *see* ANNIVERSARIES

gold standard [comparative values] The market price of gold used by a country's fiscal authority as the value against which the country's own currency value is fixed. There is virtually no nation in the world that now ordinarily uses coins in which there is any appreciable gold content. The gold standard is thus a unifying measure of values throughout the world, although there may occasionally arise international circumstances in which payment in the currency of one country might be so bulky or unwieldy – even in the form of bills or bonds of note – as still to require the transfer of gold bullion.

golf [sport] The object of the game in golf is to propel a small, hard ball into eighteen holes one at a time in sequence around a countryside course containing natural and artificial hazards in as few strokes as possible, using any of a number of predominantly metal 'clubs' or sticks.

There are two major forms of the game: in *stroke play*, the winner or winning side is the one who finishes having taken the fewest strokes to complete the entire course; in *match play*, each hole is played as a separate entity, won by one or other player or side, and the winner or winning side is the one who finishes having won the greatest number of holes. In match play there is also the Nassau system, by which the winner of the first nine holes gets an additional point, and the winner of the second nine holes another.

The dimensions of the (average) course:

18 holes over a total length of 6,500-7,000 yards (5.944-6.400 kilometres); each hole may be anything from 100 to 600 yards (91.44-548.6 metres) long, with intentional variety

Points scoring:

for convenience, a cumulative score is generally reckoned against the number of strokes officially set for each hole: *see* PAR

scoring may also take into account a player's individual handicap, as calculated on the player's own established standard and the difficulty of the course, and as acknowledged by opponents (maximum handicap for men: 24 strokes, or 4 strokes every 3 holes; maximum handicap for women: 36 strokes, or 2 strokes per hole)

if a player's ball is lost, or lands in an unplayable situation, there are various methods of restarting with and without penalty, depending on the circumstance

Dimensions of the equipment:

the ball:

maximum weight: 1.62 ounce (45.926 grams)

minimum diameter: US – 1.68 inch (4.267 centimeters) elsewhere – 1.62 inch (4.115 centimetres)

the clubs:

maximum number per player: 14

length and weight variable according to purpose

Golf is a game enjoyed by millions around the world. By tradition it was devised in Scotland, and the 'home' of the game is at St Andrews, in Fife (Region), on the coast between Edinburgh and Dundee.

The name of the game nonetheless derives from Middle English (not Scottish Gaelic) and Germanic sources, and seems to be a variant of the modern English word *club* (cf. German *Kolben*, Dutch *kolf*, 'club'); the vowel sound persists in the cognate (frequentive) English verb *clobber*.

gon *see* GRAD

Gondwanaland *see* SUPERCONTINENT

goniometer [physics] An instrument for measuring angles, such as those between the faces of crystals, and those between headwinds and navigation landmarks on an aircraft navigator's chart.

googly [sporting term] In cricket, a ball that is bowled by a right-handed bowler apparently with a leg-break action (to a right-handed batsman) but turns out to be an off-break. Similarly, for a left-handed bowler but with the ball breaking in the opposite directions to a right-handed batsman.

googol [quantitatives] The very large number 10^{100}, or 1 followed by 100 noughts. It was reputedly named by a child who was asked to name the largest number he could think of. The corresponding ordinal number is *googolth*.

See also GOOGOLPLEX.

googolplex [quantitatives] The extremely large number 10 to the power GOOGOL, so named by the inventor of the googol. The final element of the term is on analogy with *duplex*, *triplex*, and so forth.

gourde [comparative values] Unit of currency in Haiti.

$$1 \text{ gourde } = 100 \text{ centimes}$$

The term once applied to a coin formerly used in Haiti, Cuba, and Louisiana, contemporaneously equivalent to a US dollar, apparently so called because it was heavy or thick (French *gourde* 'heavy', 'numb'). *See also* COINS AND CURRENCIES OF THE WORLD.

gowpen [volumetric measure] As much of a liquid or substance as can be contained in both hands cupped together.

Now relegated to dialectal usage only in Scotland, the word is good Scandinavian in origin, and is probably akin to English *cup* (and thus *scoop*).

grad [arc measure] One-hundredth of a right-angle or quadrant, alternatively known as a *gon*.

$$1 \text{ grad } = 0.9 \text{ (nine-tenths of a) degree}$$
$$100 \text{ grads } = 90 \text{ degrees}$$
$$200 \text{ grads } = 180 \text{ degrees}$$
$$400 \text{ grads } = 360 \text{ degrees}$$

The unit is used particularly in the calculation of a course in maritime navigation, and for locating a vessel's position at sea.

The name of the unit relates to a grade (*see below*); as a gon, the term derives from the ancient Greek for 'angle' (as in polygon; akin to English *knee*).

grade [maths] Primarily another word for DEGREE and, in many European languages, the *only* word for 'degree'. [Etymologically, both grade and degree derive from Latin *(de)gradus* 'step (down)'.]

Despite the potential confusion, however, the term is additionally an alternative for GRAD or gon (nine-tenths of a degree).

In the United States, the word *grade* is also used as a synonym for GRADIENT.

gradient [maths; physics] In geometry, the gradient of a line is its slope (angle with the horizontal), and the gradient of a curve at a particular point is the slope of the tangent at that point. The gradient is, therefore, generally expressed in degrees.

In everyday usage, a gradient is more often expressed as a ratio or a percentage. For example: a road may have an uphill gradient of 1 in 8 (or 1:8) which means, strictly, that it rises 1 distance unit for every 8 units on the level. In practice, however, so that gradients can be measured on the ground, 1 in 8 means a rise of 1 unit for every 8 units along the road (that is, up the slope). In percentage terms, this gradient is 12½%.

A device used by surveyors for measuring gradients on site is, not unreasonably, called a gradienter and basically comprises a low-powered telescope, a spirit level, and a vertical arc calibrated in degrees, mounted on a tripod.

In physics, a gradient is the rise or fall in a variable property such as temperature or pressure. For example: temperature gradient may describe how temperature varies along the length of a bar heated at one end.

gradiometer [physics] A device that measures the gradient of the Earth's magnetic or gravitational field at a chosen location.

Graetz number [physics; engineering] A number characterizing heat transfer by streamline flow of a fluid in a pipe, equal to π times the product of the diameter of the pipe, REYNOLD'S NUMBER and PRANDTL'S NUMBER, divided by 4 times the length of the pipe. It was named after the American engineer L. P. Graetz (1856-1941).

Graham's law [physics; chemistry] The rate of diffusion of a gas is inversely proportional to the square root of its density. It was named after the Scottish chemist Thomas Graham (1805-69).

grain [weight] An extremely small mass or weight in several systems.

In the metric system, a grain is a mass equal to one-quarter of a metric carat:

$$1 \text{ grain } = 0.050 \text{ gram}$$

4 grains, 1 metric carat = 0.200 gram
20 grains, 5 metric carats = 1 gram
  = 0.03527396 ounce avdp.

In the avoirdupois, apothecaries', and troy weight systems, the smallest weight –
and the same weight in each system.

1 grain apoth. = $\frac{1}{20}$ scruple (= 0.0648 gram)
1 grain troy = $\frac{1}{24}$ pennyweight (= 0.0648 gram)
  = $\frac{1}{240}$ ounce apoth.
  = $\frac{1}{5,760}$ pound apoth. (or troy)
1 grain avdp. = $\frac{1}{7,000}$ pound avdp. (= 0.0648 gram)
  = 0.002285 ounce avdp.
      (437.5 grains = 1 ounce avdp.)

As the name of a unit of mass or weight, the term derives from the Latin *granum*
'seed', 'grain' (particularly a grain of wheat), apprehended as something of little or
no weight.

gram, gramme [weight] The fundamental unit of mass in the centimetre-gram-
second (CGS) system, now defined as one-thousandth of the mass of the international prototype kilogram. (It was originally defined as the mass of 1 cubic centimetre of water at its maximum density – that is, at 4°C.) The gram is fairly small for a
practical unit, and is often used in various multiples for masses that are common in
trade and industry.

1 gram = 15.432098 grains
  = 0.03527396 ounce avdp.
      (1 ounce avdp. = 28.349523125 grams)
100 grams = 3.5274 ounces avdp.
      (4 ounces, ¼ pound = 113.398 grams)
1,000 grams = 1 kilogram
  = 35.274 ounces, 2.2046 pounds avdp.
50,000 grams,
50 kilograms = 1 (metric) CENTNER
  = 110.23 pounds avdp.
1,000,000 grams,
1,000 kilograms,
20 centners = 1 (metric) tonne
  = 2,204.623 pounds avdp.
  = 0.9842116 ton avdp.

In scientific and medical usage, submultiples are often required:

$\frac{1}{10}$ (one-tenth) gram = 1 decigram
$\frac{1}{100}$ (one-hundredth) gram = 1 centigram
$\frac{1}{1,000}$ (one-thousandth)
gram = 1 milligram (10^{-3} gram)
$\frac{1}{1,000,000}$ (one-millionth)
gram = 1 microgram (10^{-6} gram)

The spelling 'gramme' represents the obsolescent English variant that is gradually vanishing as British usage conforms to the American spelling 'gram'. As the
name of a unit of mass, the term derives somewhat unexpectedly from the same
ancient Greek root as the English words *grammar* and *graven*: the original meaning
was '(a) legibly marked-off (division)', but the aspect of division (into units)
assumed more importance than the basic idea of legibly marking them off.

gram-atom [chemistry; physics] An amount of an element equal to its relative
atomic mass (atomic weight) in grams – that is, 1 MOLE of atoms.

gram-calorie [physics] Another name for a CALORIE.

gram-centimetre [physics] Unit of work equal to the work done in lifting a mass of
1 gram a vertical distance of 1 centimetre. *See also* WORK.

gram equivalent [chemistry] The equivalent weight of an element or compound in
grams.

gramme *see* GRAM, GRAMME

gram-metre [physics] Unit of work equal to the work done in lifting a mass of
1 gram a vertical distance of 1 metre. *See also* WORK.

gram-molecule [physics; chemistry] The relative molecular mass (molecular weight) of a substance in grams – that is, 1 MOLE.

gram-rad [physics: radiology] A former unit of absorbed radiation dose, equal to $\frac{1}{100}$ joule per kilogram (= 100 ergs per gram): *see* X-RAYS.

gram-röntgen [physics: radiology] A former unit of absorbed radiation energy, equal to the energy absorbed by 1 gram of air irradiated by 1 röntgen – approximately 3.8×10^{-7} joules (3.8 ergs): *see* X-RAYS.

grams per square metre (gsm) [paper] Unit of substance of paper or board, expressed as the weight in grams of 1 square metre of material. *See also* BASIS WEIGHT.

gram weight [physics] Unit of force equal to that given to a mass of 1 gram by the acceleration of free fall.

$$
\begin{aligned}
1 \text{ gram weight} \quad &= \quad 980.665 \text{ dynes (CGS units)} \\
&= \quad 9.80665 \times 10^{-3} \text{ newton (SI units)} \\
&= \quad 7.09315 \times 10^{-2} \text{ poundal (FPS units)}
\end{aligned}
$$

Grand Prix motor racing *see* MOTOR VEHICLE RACING

grand slam [sporting term] In contract bridge, winning all the tricks in a hand.
   By extension, any completely successful venture in any game or sport, especially over a series or sequence of events. In baseball, for example, a grand slam home run is a home run hit when the bases are loaded – so that every batter on the pitch scores.

graph [maths] A diagram that shows how two quantities vary in relation to each other. There may be a mathematical relationship between the two (for example, a graph of the equation $y = 2x + 3$ is a straight line relating all values of y to the corresponding values of x that satisfy the equation), or there may be no such mathematical link (for example, a graph of a patient's body temperature taken at hourly intervals). *See also* BAR CHART; HISTOGRAM; PIE CHART.

grapheme [literary] Smallest unit of a language in written or printed form – generally a single letter of the alphabet representing a vowel or a consonant, but occasionally a combination such as -th- , -ch, or ps- .

Grashof number [physics; engineering] In fluid mechanics, a number that describes free convection. Over a particular distance, it equals the product of the cube of the distance, the coefficient of cubical expansion, the temperature difference, and the acceleration of free fall (acceleration due to gravity), divided by the square of the kinematic viscosity. It was named after the German physicist and theoretical engineer F. Grashof (1826-93).

grass hockey *see* HOCKEY MEASUREMENTS, UNITS, AND POSITIONS

grav [physics] Unit of acceleration equal to the acceleration of free fall (acceleration due to gravity).

$$
\begin{aligned}
1 \text{ grav} \quad &= \quad 9.80665 \text{ metres per second per second} \\
&= \quad 32.17405 \text{ feet per second per second}
\end{aligned}
$$

gravimeter [physics; geology] An instrument, known also as a gravity meter, that measures small variations in the Earth's gravitational field from place to place, used primarily in prospecting for minerals.

gravimetric analysis [chemistry] Type of quantitative chemical analysis that relies on (instituting reactions and then) weighing samples. *See also* VOLUMETRIC ANALYSIS.

gravitation, law of universal *see* NEWTON'S LAW OF GRAVITATION

gravitational constant [physics] The proportionality constant in NEWTON'S LAW OF GRAVITATION, equal to 6.670×10^{-11} newton metre squared per square kilogram. The constant is alternatively known as the universal gravitational constant.

gravitational field [physics; geology] The region within which an object that has mass exerts a force of attraction on another object because of the force of GRAVITY.

gravity, force of [physics] The attractive force that exists between objects because of their masses; also called gravitational attraction, its strength is given by NEWTON'S LAW OF GRAVITATION.

gravity meter *see* GRAVIMETER

gray [physics; radiology] Unit of absorbed radiation dose in SI units, equal to supplying 1 joule of energy per kilogram.

$$
1 \text{ gray} \quad = \quad 100 \text{ rad (CGS units)}
$$

The unit is named after the X-ray and gamma-ray research physicist J. A. Gray. *See also* X-RAYS.

great circle [maths; geology] Path round a sphere created by a plane that cuts it and passes through the centre. The name is also given to an imaginary line on the surface of the Earth that cuts the globe in half. The shortest distance between any two points on the surface of a sphere (or the Earth) is an arc of the great circle that passes through them.

great gross [quantitatives] 1,728; twelve times 1 GROSS (12^3).

It is the number of cubic inches in a cubic foot.

great year [astronomy] Time taken for the equinoctal points (the two points at which the apparent path of the Sun intersects the celestial equator) to make one complete revolution in the heavens, equal to approximately 25,800 years. The great year is occasionally instead called a Platonic year.

green pound [comparative values] Artificial rate of exchange applied to the British pound sterling in relation to trading in consumer goods (especially products subject to the Common Agricultural Policy) within the European Community. The purpose of the green pound is to protect prices against fluctuations in the exchange rates of the real currencies of Community member states.

Other states of the European Community also have 'green' currencies in their national currency units (the green franc, the green mark, etc.): *see* COINS AND CURRENCIES OF THE WORLD.

Greenwich Mean Time (GMT) [astronomy] Time along the Greenwich meridian (longitude 0°), used as an international standard in scientific and navigational calculations.

Gregorian calendar [dates] The calendar currently in use in most countries, a revised version of the Julian calendar introduced by order of Pope Gregory XIII first in France in 1582. The revisions included making a 365-day year and the present system of LEAP YEARS. Sometimes called the New Style calendar (the Julian was the Old Style), it was introduced in Britain and its colonies only in 1752, and in Russia only in 1918. *See also* YEAR.

gries *see* GRY

groat [comparative values] Coin used in England from 1351 to 1662, and worth four pence (fourpence: one-third of a shilling). But the groat was well known as the name of various coins in Europe from the 1200s on, all originally based at a value of one-eighth of an ounce of silver. Early ones were rather thick and shapeless – and it was the thickness that gave it its name, as a 'gross' coin (cf. Dutch *groot* 'thick', 'big'; cognate with English *great*).

Two European countries, Austria and Poland, still use variants of the term for a unit of their currency: *see* COINS AND CURRENCIES OF THE WORLD; GROSCHEN; GROSZ, GROSZY.

groschen [comparative values] Unit of currency in Austria.

        100 groschen = 1 schilling

The name of the unit derives from its thickness as an ancient coin: *see* GROAT. There was once a coin of the same name used in Germany, where the term is now instead slang for a 10-pfennig coin. *See also* COINS AND CURRENCIES OF THE WORLD.

gross [quantitatives] 144; twelve times twelve, one dozen dozens, evidently apprehended (in medieval France, hence via Norman English) as a 'big' number (Latin *grossus* 'large'; cognate with English *great*). In view of this derivation, it is slightly ironic that the figure denoting 12 to the next power – that is, 12 x 12 x 12 – is known as a GREAT GROSS. *See also* BASE [maths].

gross domestic product (GDP) [comparative values] The total value of all goods and services produced within a country in a year. It equals the GROSS NATIONAL PRODUCT minus investment income from abroad.

gross national product (GNP) [comparative values] The total value of all goods and services produced by a country in a year (including investment income from abroad).

gross profit [comparative values] The profit calculated as the selling price per unit minus the unit cost of production. This calculation does not take into account other costs and expenses, such as the overheads – the costs incurred by way of wages, the everyday expenditure on premises, and so forth – which, when they are taken into account, contribute to a smaller NET PROFIT.

gross ton [weight; volumetric measure] As a unit of weight, a gross ton is merely another name for the standard (long) TON of 2,240 pounds (1.0160416 metric tonnes).

But the gross ton is also a unit of volume, as used in the shipping industry. In this sense,

$$
\begin{aligned}
1 \text{ gross ton} \;&=\; 100 \text{ cubic feet, } 3.7037 \text{ cubic yards} \\
&=\; 2.8317 \text{ cubic metres} \\
&\quad (3 \text{ cubic metres} = 1.0594 \text{ gross ton}) \\
&=\; 77.88 \text{ UK bushels, } 80.39 \text{ US bushels} \\
&=\; 2.857 \text{ displacement tons} \\
&\quad (1 \text{ displacement ton} = 0.35 \text{ gross ton})
\end{aligned}
$$

See also DISPLACEMENT TON.

gross tonnage [weight; volumetric measure] The weight of an empty merchant ship, reckoned in (long) TONS. The overall weight including the weight of crew, cargo, supplies for the voyage, spare parts, and passengers is the *deadweight tonnage*.

gross weight [weight] The total weight of an article, including any container and packaging; or the total weight of a vehicle, including fuel, lubricants, and engine coolant (also called kerb weight).

grosz, groszy [comparative values] Unit of currency in Poland; singular grosz, plural groszy.

$$100 \text{ groszy} \;=\; 1 \text{ zloty}$$

The name of the unit derives from its thickness as an ancient coin: *see* GROAT. *See also* COINS AND CURRENCIES OF THE WORLD.

ground log [speed] Means of measuring the speed of a ship or boat in shallow water. A weight on the end of a line is thrown into the water and allowed to sink to the bottom (ground). The speed can then be judged by the rate at which line is paid out over the side of the vessel.

ground speed [aeronautics] The speed of an aircraft or rocket relative to the ground (as opposed to AIR SPEED).

ground state [physics] The lowest ENERGY LEVEL of an atom or molecule (that is, not an excited state).

group [maths] In mathematics, a set and an operation that fulfil certain conditions: the operation is associative, the set is closed with respect to the operation (*see* CLOSED SET), the set has an identity element within it, and each element in the set has a unique inverse within the set.

group [chemistry] Column, or vertical row, of chemical elements in the PERIODIC TABLE OF ELEMENTS. The elements in a group have similar properties, and the group number (I to VIII and 0) indicates the number of electrons in the neutral atom's outermost shell.

group captain [military rank] Senior rank in the Royal Air Force and in several other predominantly Commonwealth airforces.

A group captain ranks above a wing commander and below an air commodore, and is equivalent in rank to a colonel or brigadier in the British army, and to a captain or commodore in the Royal Navy.

The equivalent in the US Airforce is a COLONEL.

groupement [military] A division numbering between 12,000 and 13,000 troops, as organized in the combined forces of the European Defence Community.

growth ring *see* DENDROCHRONOLOGY

gry [linear measure] Very small linear measure proposed for use in England in 1813 as part of a scheme to make all linear measurements decimal-based. None of the other suggested units was adopted, and this one only achieved longevity – albeit solely in technical dictionaries – presumably because it was smaller than any existing unit. Plural: gries.

$$
\begin{aligned}
1 \text{ gry} \;&=\; 0.1 \text{ (one-tenth of a) LINE} \\
&=\; 0.008333 \;(\tfrac{1}{120}) \text{ INCH, } 8.333 \text{ MILS} \\
&=\; 0.21666 \text{ MILLIMETRE}
\end{aligned}
$$

(1 millimetre = 4.6155 gries)
$$10 \text{ gries} = 1 \text{ line, 6 POINTS}$$
The name of the unit is taken from a Greek loan-word in the Latin writings of Plautus, its original meaning actually being 'speck of dirt under the fingernail', thus 'something inconsiderable'.

guarani [comparative values] Unit of currency in Paraguay.
$$1 \text{ guarani} = 100 \text{ centimos}$$
The name of the unit derives from that of the tribal natives – properly the Guaraní or Tupi-Guaraní – who lived mostly in eastern Paraguay until the Spanish conquest but who, since then, have grown to dominate much of the rural areas of the whole country. *See also* COINS AND CURRENCIES OF THE WORLD.

guersh, guerche [comparative values] Unit of currency in Saudi Arabia and in Ethiopia.
In Saudi Arabia,
$$20 \text{ guersh} = 1 \text{ rial, 100 hallalas}$$
In Ethiopia, the coin is of minimal value.
The unit is obviously a transliteration of the Arabic, and is sometimes alternatively spelled girsh. *See also* COINS AND CURRENCIES OF THE WORLD.

guilder, gulden [comparative values] Unit of currency in the Netherlands, and in some of the former Dutch colonies. The true Dutch form of the word is *gulden* in both singular and plural but, in English, they have been known as *guilders* (when they haven't been described as FLORINS) since the mid-1400s.
$$1 \text{ guilder/gulden} = 100 \text{ cents}$$
The name of the units in Dutch is a variant of the English word *golden*, describing what the coin was first made of. *See also* COINS AND CURRENCIES OF THE WORLD.

guinea [comparative values] Gold coin first struck in England in 1663 and, at that time, of a value of one pound, or 20 shillings. From 1717 to 1813, however, the coin had the value of 21 shillings (one pound one shilling) and, although the coin ceased to be legal tender at that date, the term has ever since been used to mean a sum or value of that quantity, as used, for example, at auctions of fine art, at sales of haute couture, and in horse racing for the evaluation of horses and the posting of prize money. In present-day terms,
$$1 \text{ guinea} = £1.05 \text{ (one pound and five pence)}$$
The coin was first minted 'in the name and for the use of the Company of Royal Adventurers of England trading with Africa', and was made of gold mined in the large area of West Africa then called by the native African name Guinea (now comprising the states of Guinea, Sierra Leone, Liberia, Ivory Coast, Ghana, Togo, and Benin). Its representation of a little elephant on one side made it a popular coin that became known very quickly as a 'guinea'.

guitar *see* STRINGED INSTRUMENTS' RANGE

gulden *see* GUILDER, GULDEN

gun bore, gun gauge *see* CALIBRE, CALIBER

Gunter's chain [surveying] A measuring chain of 100 links each 7.92 inches long, thus totalling 66 feet (20.116 metres) in overall length – that is, 1 chain long: *see* CHAIN.
It was named after its deviser, the English mathematician Edmund Gunter (1581-1626), who was particularly interested in surveying and navigation.

guz [linear measure] Former unit of length in the countries between Arabia and India. But it seems to have been part of a unitary system to some degree assimilated to exact numbers of standard (imperial) INCHES, which may explain why it varied between even relatively close localities. The guz in India, for example, varied between 27 inches (Bombay), 33 inches (Madras), and 37 inches (Bengal). The guz in Arabia was alternatively 25 inches long. In Iran the length of the guz was generally 41 inches (6,000 guz = 1 persakh) until, in 1925, the unit was instead assimilated there to the standard metre of the metric system.

G-value [chemistry] In radiochemistry, a constant that denotes the number of molecules that react after absorbing 100 electron volts of radiation energy.

gyrocompass [physics; navigation] Motor-driven form of compass that uses a GYROSCOPE instead of a magnetic needle, and, because it is unaffected by magnetic influences, points to geographic north rather than to magnetic north.

gyro horizon [physics; navigation] The instrument on an aircraft pilot's panel that shows the attitude of the aircraft in relation to the horizon.

gyromagnetic ratio [physics] For a spinning or orbiting magnet (including charged subatomic particles), the ratio of its magnetic moment to its angular momentum. For an orbiting electron, it is half the ELECTRON CHARGE/MASS RATIO.

gyroscope [physics; engineering] Mechanical device with a flywheel that rotates at high speed. It resists attempts to change its axis of rotation and, for this reason, is used in compasses, guidance systems, and stabilizers for ships.

H

h, H *see* HENRY; HOUR; PLANCK'S CONSTANT

ha *see* HECTARE

hadron [physics] An elementary particle that undergoes strong interaction with other particles (for example: a baryon or a meson).

haemacytometer, hemacytometer *see* HAEMOCYTOMETER, HEMOCYTOMETER

haematimeter, hematimeter [medicine] A microscope slide with a specific quantity of blood on it, used as a control against which other slides with an identical quantity of blood on them are compared, in order to measure the number of corpuscles contained in each sample.

The term derives primarily from ancient Greek *haimat-* 'blood'.

haemat(in)ometer, hemat(in)ometer [medicine] Any device that measures the h(a)emoglobin content in a sample of blood. Such measurement is important, for h(a)emoglobin is the red (iron-containing) component in the blood that conveys oxygen.

In some hospitals, the word is used also to mean a device that identifies and measures various constituents of a blood sample. Rarely, the word is used instead of a device to measure blood pressure (*see* SPHYGMOMANOMETER).

The term derives primarily from ancient Greek *haimat-* 'blood'.

haematocrit, hematocrit [medicine] A form of centrifuge used to spin a blood sample and so separate and measure the relative volume of blood cells and plasma within the sample.

The term derives primarily from ancient Greek *haimat-* 'blood'; the second element is Greek *krit-* 'judge'.

haematometer, hematometer *see* HAEMAT(IN)OMETER, HEMAT(IN)OMETER

haemocytometer, hemocytometer [medicine] Any device or apparatus that counts the number of blood cells in a blood sample.

haemogram, hemogram [medicine] Graph, tabulation, or other form of representation produced by a device or apparatus that counts the number of white blood cells (leucocytes) in a blood sample. Such a count is important, for the white blood cells are mainly responsible for body immunity against invasion by bacteria and other infective agents.

haiku *see* VERSE FORMS

haler [comparative values] Unit of currency in the Czech and Slovak Republics; plural: haleru.

$$100 \text{ haleru} = 1 \text{ koruna}$$

The name of the unit derives from a time when Austrian currency was used in much of the region, and 100 Austrian HELLER made 1 *krone*. *See also* COINS AND CURRENCIES OF THE WORLD.

half [quantitatives] ½. Although the fraction could technically be described as 'one-second', this would inevitably cause confusion with other meanings of the latter word.

$$½ = 0.5 = 50\%$$

The word was not originally any specific fraction: it derives, in fact, from a Germanic past participle of the verb corresponding to modern English *cleave* (cf. *halve*) – a half is something that has thus been *cleft* from the whole (cf. German *halfte* 'half'). Additional cognate English words include *clipped* and *scalped*; even closer in verbal form, icebergs are said to *calve* when they split off the end of glaciers (although the bovine sort of calf is not linguistically akin); and with *scalp* cf. *shelve*. *See also* the articles beginning HALF *below*; HEMI-; SEMI-.

half a crown *see* HALF-CROWN

half a dozen [quantitatives] 6: half of twelve – a quantitative unit still surprisingly commonly used despite the standard dominance of decimal counting. Eggs, buns, rolls, and other delicatessen products are frequently offered for sale in half-dozens. The expression can also be used casually to mean 'a handful', 'a small number'. *See also* DOZEN; BASE.

half after *see* HALF PAST, HALF AFTER

half a league *see* LEAGUE

half a mile [linear measure] Useful measure of distance in many English-speaking countries, especially when giving directions to drivers or walkers.

½ mile = 2,640 feet, 880 yards, 40 chains, 4 FURLONGS
= 804.672 metres, 0.804672 kilometre
(800 metres = 874 yards 2 feet 8.06 inches)

See also MILE.

half a pint [volumetric measure] Unit of volume common as a measure of soft or alcoholic drinks, or hot beverages, but differing slightly in quantity between Britain and the United States.

In Britain (and some other English-speaking countries),

½ pint = 2 GILLS or quarterns, 10 (UK) FLUID OUNCES
= 0.2841 litre, 28.41 centilitres
= 0.60047 US pint

In the United States,

½ pint = 2 gills, 8 (US) fluid ounces
= 0.2366 liter, 23.66 centiliters
= 0.4164 UK pint

See also PINT.

half a pound [weight] Smallish unit of weight mostly applied to measuring out quantities of consumer goods (meats, cheeses, sugar, flour, etc.) in shops and supermarkets.

½ pound = 8 ounces, 128 drams
= 226.796 grams, 0.226796 kilogram

See also POUND; OUNCE; AVOIRDUPOIS WEIGHT SYSTEM.

half a stone [weight] Unit used mostly in Britain (and comparatively rarely any-where else), generally in relation to a person's own (body) weight.

½ stone = 7 pounds (one-sixteenth of a HUNDREDWEIGHT)
= 3.1751 kilograms

See also STONE.

half a yard *see* EIGHTEEN INCHES

half-crown [comparative values] Former unit of currency in Britain, a coin worth two shillings and sixpence (30 pence, one-eighth of a pound, and thus half a 'crown' of five shillings). For many decades it was the coin of highest everyday denomination – the crown being issued as a commemorative coin and not really for ordinary use – but it was phased out when British currency was converted to decimal coinage and notes in 1971 and, instead of a 12½-pence coin, fiscal authorities chose to continue with the 10-pence coin that represented the former florin or two-shilling piece, and to introduce a 50-pence piece representing the former ten-shilling note. Later still, a pound coin replaced the once ubiquitous pound note. *See also* CROWN.

half-dime [comparative values] Another name for a NICKEL and, as half of a DIME, worth 5 cents (in the United States and Canada). Minted first in the United States in 1794, it was originally a silver coin and known only as a half-dime. The change of metal from silver (last issued 1873) to cupro-nickel in 1883 caused the change also of its popular name (although there had been an earlier 1 cent coin also called a 'nickel').

half-dollar [comparative values] Former unit of currency in the form of a silver coin, now one made of cupro-nickel in the United States, and a modern unit of currency in several other countries of the world today.

$$\tfrac{1}{2} \text{ dollar} = 50 \text{ cents}$$

In Britain, 'half-dollar' or 'half a dollar' was a slang expression for a HALF-CROWN, or two shillings and sixpence, until the conversion to decimal currency in 1971. *See also* DOLLAR.

half-eagle [comparative values] Former unit of currency in the United States, corresponding to a gold coin worth 5 dollars, thus half the value of an EAGLE.

half-guinea [comparative values] Old unit of currency in Britain valid between the reign of Charles II (ruled 1660-85) and the year 1813, and corresponding to a gold coin then worth half a GUINEA, or 10 shillings and sixpence.

half-hour *see* HALF PAST

half-life [physics] For a radioactive isotope, the time it takes for half of its nuclei to spontaneously decay, equal to $\log_e 2$ divided by the DECAY CONSTANT. It varies, depending on the isotope, from a few seconds to thousands of years, and can be used as a basis for geological dating (*see* RADIOCARBON DATING).

half-mile *see* HALF A MILE

half Moon *see* LUNAR PHASES

half-note *see* MINIM

half-past, half after [time] In telling the time, 30 minutes past the hour; half an hour after the previous hour. The English language (together with various other languages, such as French) is accustomed to referring to the previous hour rather than the next one, so that the colloquialism 'half three' is immediately understood as meaning 'half-past three', '3.30'. The practice is not universal. In other European languages, especially Germanic languages such as German and Dutch, 'half three' in fact means 'half an hour to three' – that is, 2.30 or half-past two.

halfpenny, ha'penny [comparative values] Former coin in British currency almost as old as the penny itself (and worth half of one), but phased out because of its eventual minimal value several years after the conversion to decimal currency in 1971. Sentimental regard for the pre-decimal coin meant that, although the earlier coin was customarily called a ha'penny (pronounced *hayp-ni*), the decimal coin was pronounced *half-penny* until its demise. *See also* PENNY.

half-pint *see* HALF A PINT

half-sovereign [comparative values] Former unit of currency in Britain, first issued in 1503 and worth 10 shillings, or half a SOVEREIGN (one pound).

half-step [music; speed] In music, a half-step in North America is what in Britain is called a SEMITONE: the interval that, on a piano, is represented by one note (key) and the note (key) immediately next to it up or down, black or white. Also known technically as a diminished second, the interval sounded or played as two notes together is a discord and cannot contribute to harmony.

In military marching in the United States, a half-step is a short pace of prescribed length amounting to exactly half the distance of the pace prescribed in both QUICK TIME and DOUBLE TIME. The effect is thus of a fast march that does not cover the ground particularly rapidly. In quick time a half-step is of 15 inches (38.10 centimetres), and in double time a half-step is of 18 inches (45.72 centimetres).

Hall effect [physics] The production of a potential difference (voltage) in a conductor or semiconductor carrying an electric current while it is in a strong transverse magnetic field. It was named after the American physicist Edwin Hall (1855-1938).

Halley's comet, orbital period *see* ORBITAL PERIOD

halocline [physics] The depth in sea water at which the salinity sharply decreases, generally at around 30 fathoms (180 feet; 54.9 metres). It may be marked as a contour on the charts of marine biologists.

The term derives on analogy with ISOCLINE, the first element instead representing the ancient Greek *halos* 'salt'.

Hamiltonian function [maths; physics] A function describing the positions and momenta of a group of particles, in terms of which their motion and energy may be expressed. It was named after the Irish mathematician William Rowan Hamilton (1805-65).

hammer, throwing the *see* ATHLETICS FIELD EVENTS AREA, HEIGHT, AND DISTANCE
MEASUREMENTS

hand [linear measure] Unit used by equine authorities to measure the height of
horses and ponies at an animal's shoulder.

$$1 \text{ hand} = 4 \text{ inches}$$
$$= 10.16 \text{ centimetres}$$
$$(10 \text{ centimetres} = 0.9842 \text{ hand})$$
$$3 \text{ hands} = 1 \text{ foot } (30.48 \text{ centimetres})$$

The term derives from the practice of using the width of one's own hand as a
measure, starting at ground level and consecutively placing the next hand on top of
the previous hand until the height is reached.

hand [sporting term] In card games, the set of cards dealt to one player for a round of
the game (and thus for a time kept in the hand); also, the round itself.

handball measurements, units, and positions [sport] The team game of
handball is fast and furious. Seven players on each side are on the court at any one
time, and their object is to get the ball in the opponents' goal as many times as
possible.

The dimensions of the court:
>maximum area: 44 x 22 metres (144 feet 4.32 inches x 72 feet 2.16 inches)
>minimum area: 38 x 18 metres (124 feet 8 inches x 59 feet 8 inches)
>radius of semicircular goal area: 6 metres (19 feet 16 inches)
>the goal:
>>width: 3 metres (9 feet 9 inches)
>>height: 2 metres (6 feet 6 inches)

Positions of players:
>all players apart from the goalkeeper are both attackers (offense) and defenders
>(defense), depending on which side is in possession of the ball

Timing: for men, 60 minutes in two halves separated by a 10-minute interval; for
>women, 50 minutes in two halves separated by a 10-minute interval
>in tournaments, men may play for a total of 30 minutes without interval,
>women for a total of 20 minutes without interval
>any player may be suspended from play for 2 minutes (first offence) or 5
>minutes (second offence) for unsportsmanlike conduct; a third offence entails
>disqualification
>in the event of a tie, two periods of extra time may be played after a 5-minute
>interval, each period for men of 5 minutes, for women of 3½ minutes; if the
>tie persists, another interval may be taken before two further periods of extra
>time

Points scoring:
>the team with the greater number of goals at the end wins

Dimensions of the ball:
>for men: weight – 425-475 grams (15-16.75 ounces); circumference – 58-60
>centimetres (22.8-23.6 inches)
>for women: weight – 325-400 grams (11.46-14.11 ounces); circumference –
>54-56 centimetres (21.26-22.05 inches)

Handball is the name also of a four-walled court game for two or four players in
which the ball is struck by gloved hands and the rules are much like those of
squash; a very similar game played with stubby wooden rackets is known as
paddleball. Both games are comparatively rare.

handbell ringing *see* CHANGE RINGING, CHANGES

handicap [sporting term] In some games and sports, a penalty imposed by the
authorities to even up the contestants in some way and make competition fairer. In
horse racing, for example, this may take the form of adding weight to the equipment
of a horse that has won many previous races. In club golf, the handicap is the
number of extra strokes that the more skilful opponents of a lesser player have to
take on their own score within a round to give everybody the same chance of
winning. In a handicap running race, runners known to be better may start at a set
distance behind the main field, each according to his or her previous record. A
variant on this is used in dragster racing, in which the two competitors may be

given independent countdowns (by way of coloured lights) so that the slower starts first.

The term apparently derives from a fourteenth-century practice in a wagering competition (called New Fair in *Piers Plowman*) in which a challenger offered an article of some value against an opponent's article of greater value, and quoted odds in doing so. Both challenger and opponent, and a chosen adjudicator, would all then deposit sums of money according to the odds quoted into a cap. The two competitors in turn would then put one hand each into the cap – thus, 'hand in cap' – and withdraw them; a full hand would indicate acceptance of the wager, an empty hand would signal that the bet was off.

The term was then transferred to horse racing, in races where again a challenger offered an article of one value against an opponent's article of another value, but this time where the disparity in values corresponded to some extent with the odds on the horses' winning or losing. And so the term was given its modern sense.

Han dynasty [time] In China, the period between 202 BC and AD 220. It was a time of cultural excellence, of religious renewal (the introduction of Buddhism), and of military conquest (of Mongolia).

hank [textiles] Unit of measurement in the yarn industry.

1 hank	=	7 SKEINS or LEAS
20 hanks	=	1 BUNDLE

The actual number of yards in each hank, however, differs according to the composition of the yarn.

In cotton yarn, for example,

1 hank	=	840 yards
1 bundle, 20 hanks	=	16,800 yards

In worsted yarn,

1 hank	=	560 yards
1 bundle, 20 hanks	=	10,200 yards

The term would seem to derive from Germanic words (especially Scandinavian) for a 'coil' or 'ring', especially of nautical rope.

haplo- [quantitatives: prefix] Prefix meaning 'single-', especially when unusually so.

The term derives from ancient Greek *haplous* 'single' and is etymologically identical to Latin *simplex*.

haploid [biology] Describing a cell in which the nucleus has half the normal number of chromosomes (for example, as in a gamete or sex cell: an ovum or a sperm) – that is, in which the chromosomes are single instead of being in pairs. Haploid cells are occasionally described alternatively as monoploid. *See also* DIPLOID.

hardness [physics; engineering] The resistance to abrasion, scratching, or indentation. Scratch-resistance is measured by the MOHS' HARDNESS SCALE. Other hardness scales generally measure resistance to indentation, such as the BRINELL SCALE, ROCKWELL SCALE, and VICKER'S SCALE.

hardness of water [chemistry] Hardness in water is caused mainly by the presence of calcium and magnesium salts. Temporary hardness, caused by bicarbonates (hydrogen carbonates), is removed by boiling the water; permanent hardness, caused by carbonates, remains even after boiling. Various measures have been used for expressing the hardness of water, which is now usually stated in parts of calcium carbonate per million parts of water.

Hare system of proportional representation [quantitatives] In an election, a system by which each voter numbers the candidates on a list in the order of his or her choice. Depending on the total number of electoral places available, all the first choices should be elected, and the remaining number of available places are then shared between the voters' second choices.

The system was devised by and named after the English lawyer Thomas Hare (1806-91).

harmonic [physics] The frequency of a wave, such as a sound wave, that is an exact multiple of the frequency of the fundamental wave. An alternative name for a harmonic is an overtone.

harmonicas' and accordions' range [music] These instruments rely on the passage of air past (generally metal) 'reeds' to produce musical notes.

In *harmonicas* or *mouth organs*, for the notes of the tonic or KEY chord the air is produced directly by the player's lungs and directed via the mouth through an arrangement of holes resembling a keyboard; when air is conversely drawn in through the same holes, reeds in the reverse position produce the remaining notes of the key SCALE.

Most harmonicas are tuned to specific keys. Common base keys are C (for amateurs), A, B♭, and E♭ (for professionals), although the more expensive soprano/treble varieties also have a chromatic key that allows a player to obtain semitones (half-steps) between the basic notes of the scale, and thus to reach all the notes available to the usual keyboard. It is less common for an alto or tenor harmonica to have the chromatic key; they are almost always tuned to a specific key, and may only span a range of two octaves in place of the more customary three.

The *concertina* is the most harmonica-like of the accordions. The air is pumped past the reeds by means of bellows which allow the device to be opened and shut as a whole telescopically. Most concertinas are also like harmonicas in that notes produced by air in one direction are those of the basic (key) scale, and notes produced by air in the other direction are those of the remainder of the notes of the same scale. The notes are selected by the player as buttons on one side or other of the concertina, some of which may also produce not notes but whole chords.

The oldest forms of accordion had lever-like keys instead of a proper keyboard. The type is still in use, mostly in reproduction, in parts of the United States where Cajun music is performed.

The modern *accordion* has a keyboard much like that of a piano (hence the more formal term 'piano accordion'). Beginners' models have a keyboard for the right hand that spans two octaves (usually from C to C), but may have one or more different registers (giving either a third octave or a change of tone in the two already available), and have twelve buttons ('basses') for the left hand, most often providing six bass notes and six related bass chords, positioned so that bass buttons are immediately outside their respective bass chord.

The professional accordion has a keyboard for the right hand of forty-one piano-style keys (from F at the bottom to A at the top), with at least three registers (giving at least one further octave and two changes of tone); for the right hand there are 120 buttons, arranged in six rows of twenty buttons, each row corresponding (from the outside in) to (1) the mediant note of the scale in the bass octave, (2) the bass note, (3) the major chord of the bass note, (4) the minor chord of the bass note, (5) the SEVENTH chord in the major, and (6) a diminished chord with the bass note as principal; the arrangement of the bass notes is such that the hand travels downwards through progressive dominants (C, G, D, A, E, etc., as if going through the sharp keys) and upwards through subdominants (C, F, B♭, E♭, A♭, etc., as if going through the flat keys), so that at all times the three basic chords of any key are together. There may also be separate registers for the left hand. The form of accordion known as a *melodeon* has buttons ('button-keys') instead of a piano keyboard, in an arrangement meant to make rapid finger movements even more accessible.

harmonic mean [maths] The reciprocal of the arithmetic mean (average) of the reciprocals of a series of given numbers. For example: the harmonic mean of 3, 4, and 5 is the reciprocal of the average of $\frac{1}{3}$, $\frac{1}{4}$, and $\frac{1}{5}$ ($\frac{20}{60}$, $\frac{15}{60}$, and $\frac{12}{60}$); now the average of $\frac{20}{60}$, $\frac{15}{60}$, and $\frac{12}{60}$ is $(47 \div 3)/60$; and the reciprocal of that (the harmonic mean) is $60 \times \frac{3}{47}$, or $\frac{180}{47}$ (= 3.8298).

harmonic minor *see* KEY, KEYNOTE

harmonic progression [maths] A sequence of numbers of which the reciprocals form an arithmetic progression – that is, they have a common difference. For example: $1 + \frac{1}{2} + \frac{1}{3} + \frac{1}{4} + \frac{1}{5} + \ldots$

harmonic series [maths; music] In mathematics, a series of numbers that forms a HARMONIC PROGRESSION. In music, a series of notes (tones) in which each is a harmonic of a particular fundamental – for example: the notes on a non-valve brass instrument, such as a bugle, or the series of harmonics obtained by touching the string of a stringed instrument at the nodes while playing it.

harmonium *see* KEYBOARD INSTRUMENTS' RANGE

harmony [music] In most forms of music in the English-speaking world, the use beneath a continuing melody line of other notes and chords that relate the melody line to a specific key and that add to musical effect and interest. Where the rhythm of the melody line corresponds precisely and continuously also to the rhythm of the other notes and chords played beneath, the result is said to be *block harmony*. Where the notes and chords beneath the melody line have their own independent rhythms and sequences, the result is said to be *polyphony*, *syncopation*, or, if the rhythms and sequences are so particularly different as to suggest a number of competing melody lines, *counterpoint*.

harp *see* STRINGED INSTRUMENTS' RANGE

harpsichord *see* KEYBOARD INSTRUMENTS' RANGE

Hartmann number [physics] A number that gives a measure of the opposition to viscous action in a conducting fluid flowing in a transverse magnetic field (in terms of the magnetic flux density, electrical conductivity, and viscosity).

hartree [physics] An obsolete unit of energy at the atomic scale, equal to $4\pi^2$ times the product of the mass of an electron and the fourth power of its charge, divided by the square of PLANCK'S CONSTANT.

$$1 \text{ hartree } = 110.5 \times 10^{-21} \text{ joule}$$

It was named after the physicist D. R. Hartree (1897-1958).

Harvard classification [astronomy] A classification of the spectra of stars in the Draper Catalogue, produced by Harvard Observatory in the United States.

hat trick [sporting term] In cricket, and by extension in other games in Britain, a score of three in a row by one player. The original hat trick in cricket was the achievement of a bowler who, in three consecutive balls, dismissed three batsmen. The expression is now used just as commonly in other field, rink, or pool team games for anyone who scores three goals during the course of a single match.

Despite intermittent but lengthy correspondence in the London *Times* newspaper during the 1970s-90s, the derivation of the term remains a mystery. (No hat or cap seems ever to have been awarded at an official level by cricketing authorities for a hat trick, although many dictionaries suggest the possibility.)

HCF *see* HIGHEST COMMON FACTOR (HCF)

hearing loss [medical] A measure of relative deafness, equal to the difference between the threshold of hearing (in decibels) and that of a healthy ear (at any particular frequency).

heartbeat, heart rate *see* PULSE; SINOATRIAL NODE, SINOAURICULAR NODE

hearts *see* SUITS OF CARDS

heat [physics] Heat is a form of energy, measured in such units as British thermal units, calories, joules, or therms.

heat [sporting term] A preliminary race or contest to eliminate competitors in preparation for a grand final event.

The term derives from the blacksmith's forge, in which metal had to be taken to the heat several times before the final forging and hammering.

A dead heat or tied race is so called because, if two contestants reached the finishing line at the same time (especially when only two competitors were involved), no one was eliminated: the heat was 'dead' and another would have to be undertaken.

heat capacity [physics] The amount of heat needed to raise the temperature of an object by 1 degree, expressed (in the SI system) in joules per kelvin. The heat capacity is known alternatively as the thermal capacity. *See also* SPECIFIC HEAT CAPACITY.

heat conductivity *see* THERMAL CONDUCTIVITY

heat content [physics] Another name for ENTHALPY.

heat equator [physics: geology] A thermal contour drawn at the centre of the belt around the Earth in which the daily temperature averages about 26°C (80°F) or more; it closely follows, but is not colinear with, the geographical equator.

heat of activation [chemistry] For a chemical reaction, the difference between the thermodynamic functions for the activated complex and the reactants (in their standard states). *See also* ACTIVATION ENERGY.

heat of atomization [physics; chemistry] The quantity of heat needed to convert 1 mole of an element into the gaseous state.

heat of combustion [physics; chemistry] The change in heat that takes place when 1 mole of a substance is burned completely. It can be measured using a BOMB CALORIMETER.

heat of formation [chemistry] The change in heat that takes place when 1 mole of a chemical compound is formed from its component elements (in their normal states). *See also* ENDOTHERMIC; EXOTHERMIC.

heat of neutralization [chemistry] The change in heat that occurs when 1 mole of hydrogen ions in dilute solution are neutralized by a chemical base.

heat of reaction [chemistry] The change in heat that occurs when molar quantities of substances react (as defined by the chemical equation for the reaction). *See also* ENDOTHERMIC; EXOTHERMIC.

heat of solution [chemistry] The change in heat that occurs when 1 mole of a substance dissolves completely in a large volume of water.

heat transfer factor [physics; engineering] For a fluid undergoing forced convection, the product of the STANTON NUMBER and the ⅔ power of the PRANDTL NUMBER.

Heaviside layer *see* E LAYER, E REGION

heavy meson *see* K-MESON

heavy mineral [physics: geology] Any mineral that has a specific gravity of more than 2.9 and occurs in small quantities within sedimentary strata. Ores of platinum, iron, and gold are heavy minerals.

heavyweight *see* BOXING WEIGHTS

hebdomad [quantitatives] The number 7; a group of seven elements – especially of days, thus one week. The adjective hebdomadal is almost always used simply to mean 'weekly'.
   The word derives from ancient Greek *hebdomas* 'seven in number'.

hect-, hecto- [quantitatives: prefix] Prefix which, when it precedes a unit, multiplies the unit by one hundred.
   Example: 1 hectostere   =   100 steres
   The prefix derives from a French corruption of the original ancient Greek *hekaton* 'one hundred'; very occasionally, it is seen also as *hekt-*. *See also* entries beginning HECT- *below*.

hectare [area] Unit of area in the metric system.

1 hectare	=	100 ARES
	=	100 x 100 metres, 10,000 square metres
	=	0.01 (one-hundredth) square kilometre
	=	2.471 acres
		(1 acre = 0.4047 hectare, 4,047 square metres
		5 acres = 2.0235 hectares)
	=	11,959.9 square yards, 0.00386 square mile
100 hectares	=	1 square kilometre, 1,000,000 square metres
	=	0.3861 square mile

The unit is most commonly used for the measurement of land in Continental Europe, and as such is much more convenient than the *are* on which it is technically based.

hectobar [physics; meteorology] Unit of pressure.

1 hectobar	=	100 BARS
	=	10^7 newtons per square metre (10^7 PASCALS)
	=	98.69 ATMOSPHERES

hectolitre, hectoliter [volumetric measure] Unit of dry or liquid capacity in the metric system.

1 hectolitre	=	100 LITRES
	=	22.00 UK gallons, 26.4179 US gallons
	=	2.7495 UK bushels, 2.8378 US bushels
	=	100,000 cubic centimetres
	=	0.1 (one tenth) cubic metre
	=	3.5304 cubic feet

Hegira *see* ERA, CALENDRICAL

Hehner number [chemistry: nutrition] The percentage of fatty acid and water in 1 gram of a fat.

Heian era [time] Period in Japan between the AD 800s and 1100s, in which a strongly artistic culture flourished, centred on the city now called Kyoto.

height [maths] In geometry, the perpendicular distance from the base to the apex of a triangle, cone, or pyramid, also called the altitude. The perpendicular distance between opposite parallel sides of a parallelogram is also called its height or altitude.

height, human body (standards and norms) [medicine; biology] In the English-speaking world, the average body height for men and for women has increased since 1900. For men, the rise has been from about 5 foot 10 inches (178 centimetres) to about 5 foot 11½ inches (181 centimetres); for women, the rise has been from about 5 foot 4 inches (163 centimetres) to about 5 foot 5½ inches (166 centimetres). In both cases, the increase has resulted through an overall improvement in the national diet and in health care and maintenance – and has been parallelled by a corresponding increase in average weight. *See also* WEIGHT, HUMAN BODY (STANDARDS AND NORMS).

Heisenberg uncertainty principle [physics] The exact position and momentum of an electron or other particle cannot be determined at the same moment in time; the product of the uncertainties in these two parameters must be greater than the DIRAC CONSTANT. It is an important principle in wave mechanics, and is known alternatively as the Heisenberg principle or just the uncertainty principle. It was named after the German physicist Walter Heisenberg (1901-76).

hekteus [volumetric measure] Little-known measure of dry capacity in ancient Greece, effectively existing only as a submultiple (one-sixth) of the medimnos (from which proportion it also got its name): *see* MEDIMNOS.

heliometer [astronomy] An instrument for measuring the angular distance between nearby celestial objects (and formerly used for measuring the Sun's diameter – hence its name).

helix [maths] A curve that cuts the surface of a cylinder or cone at a constant angle (like the shape of a screw thread).

heller [comparative values] Former coin in Germany and in Austria, now obsolete in both.

In Germany,

1 PFENNIG equalled 2 heller or häller

In Austria,

1 KRONE equalled 100 heller

It is thought that the coin got its name through first being minted at Hall (or Halle), a city in Swabia, southern Germany. *See also* HALER.

Helmert's formula [physics; geology] An empirical formula that gives the acceleration due to gravity (acceleration of free fall) at a particular location in terms of its latitude and altitude.

helmholtz [physics] An obsolete unit for the moment of an electrical double layer, equal to the product of the charge density (on each plane) and the distance between them. Its units were debye per square angstrom, equivalent to electrostatic units of charge per square metre. It was named after the German physicist Hermann von Helmholtz (1821-94).

Helmholtz free energy [physics] In thermodynamics, quantity equal to the difference between the internal energy of a system and the product of its temperature and entropy. It was named after the German physicist Hermann von Helmholtz (1821-94). *See also* GIBBS FREE ENERGY; ENTROPY.

hematimeter *see* HAEMATIMETER, HEMATIMETER

hemat(in)ometer *see* HAEMAT(IN)OMETER, HEMAT(IN)OMETER

hematocrit *see* HAEMATOCRIT, HEMATOCRIT

hematometer *see* HAEMAT(IN)OMETER, HEMAT(IN)OMETER

hemi- [quantitatives: prefix] Prefix denoting 'half'.

Example: hemisphere – half a sphere

The term derives from ancient Greek *hemisys*, *hemi-* 'half', which may in turn derive from the same source as Greek *hama-* 'together' if the original meaning of that word was 'doubled' – that is, both 'as a pair' and (doubled up) 'half the size'. The same correspondence is visible in Latin: *semi-* 'half' and *simi-*, *sym-* 'together'. *See also* SEMI-.

hemi-demi-semiquaver [music] In musical notation, a hemi-demi-semiquaver – known in the United States as a sixty-fourth note – is represented by the symbol ♬ and corresponds to

½ DEMI-SEMIQUAVER or thirty-second note

¼ SEMIQUAVER or sixteenth-note

⅛ QUAVER or eighth-note

1/16 CROTCHET or quarter-note

1/32 MINIM or half-note

The hemi-demi-semiquaver is rarely found in musical scores because there is seldom any need to make use of a note of such brief duration.

The term derives as a triple prefix in front of the word quaver, each element of which means 'half': *hemi-* Greek, *demi-* French, and *semi-* Latin. A hemi-demi-semiquaver is thus 'half of a half of a half' (that is, one-eighth) of a quaver.

hemina [volumetric measure] In ancient Rome, a measure of liquid capacity – especially in the preparation or preservation of wine and oil – and of dry capacity, approximating loosely to half a modern pint.

1 hemina	=	24 ligulae, 4 ACETABULA
	=	26.5625 centilitres, 0.265625 litre
	=	0.4675 UK pint, 0.5614 US pint
	=	0.4675 UK dry pint, 0.4820 US dry pint
	=	265.625 cubic centimetres
	=	16.2105 cubic inches
2 heminae	=	1 SEXTARIUS
12 heminae, 6 sextarii	=	1 CONGIUS
48 heminae, 24 sextarii, 4 congii	=	1 URNA
96 heminae, 48 sextarii, 8 congii, 2 urnae	=	1 AMPHORA

The name of the unit was borrowed from that of a Greek unit corresponding to half (Greek *hemi-*) another, in this case in reference to the measure's proportional size (one-half) of the basic Roman unit, the sextarius.

hemiplegia [medicine] Paralysis of one side of the body (sometimes including certain psychological functions of the opposite side of the brain). Paralysis of the four limbs and part of the trunk is called *paraplegia*, although the definition is frequently extended to paralysis from the neck down (thus including all of the trunk), which is better known by physicians as *quadriplegia*.

hemisphere [maths: geometry; geography] Half of a sphere, globe, or ball.

The volume of a hemisphere may be calculated as $\frac{2}{3}\pi r^3$ (where r is its radius).

The planet Earth is often considered a sphere for the purposes of cartography or other geographical reference. In particular, the EQUATOR is said to divide the Northern Hemisphere from the Southern Hemisphere. But there are additional ways in which the planet may be regarded as comprising two hemispheres; for example:

the Western Hemisphere and	the Eastern Hemisphere
(North and South America)	(all the other continents)
the Land Hemisphere and	the Water Hemisphere
(centred on London, England)	(centred on Auckland, New Zealand)

In an astronomical sense, the part of the celestial sphere that corresponds to the dome of the sky above the horizon (the 'vault of the heavens') is often also described as a hemisphere.

hemistich [literary] Half a line of verse, especially one that begins or ends with a caesura (pause); also a line of verse that is intentionally shorter than the metric scheme demands.

The term means 'half-line' in ancient Greek.

hemocytometer *see* HAEMOCYTOMETER, HEMOCYTOMETER

hemogram *see* HAEMOGRAM, HEMOGRAM

hendeca- [quantitatives: prefix] Prefix which implies an elevenfold unity.

Example: hendecasyllable – a line of verse containing eleven syllables

The prefix derives from ancient Greek, literally 'one (and) ten', thus 'eleven'.

hendecagon [maths: geometry] A POLYGON with eleven sides (and eleven angles).

henry [physics] Unit of inductance, equal to the inductance that produces an induced electromotive force (voltage) of 1 volt for a change in current flow of 1 ampere per second. It was named after the American physicist Joseph Henry (1797-1878).

Henry's law [chemistry; physics] When a gas dissolves in a liquid, the weight of dissolved gas is proportional to its pressure. It was named after the British chemist William Henry (1774-1836).

hept-, hepta- [quantitatives: prefix] Prefix which implies a sevenfold unity.
　　Examples: heptarchy – government by seven rulers
　　　　　　　heptastich – a poetic verse of seven lines
　　The prefix derives from ancient Greek *hepta* 'seven'.

heptad [quantitatives] The number 7; a group of seven.

heptagon [maths: geometry] A POLYGON with seven sides (and seven angles). The interior angles add to 900 degrees.

heptahedron [maths: geometry] A solid figure (POLYHEDRON) with seven plane faces.

heptameter [literary] In verse, a line of seven feet. In English a line of such length has traditionally been regarded as cumbersome.
　　But if the six lines of each verse in Lewis Carroll's *The Walrus and the Carpenter* were instead printed as three long lines, each would be a heptameter.

heptane *see* OCTANE NUMBER, OCTANE RATING

heptavalent, heptavalence [chemistry] Having a valence (valency) of 7; an alternative description is septivalent.

herd of cattle, of antelope, of whales, etc. [collectives] Strictly speaking, this collective noun should be used only of *horned* animals, for that is what the word originally referred to. But, over the centuries, the term was extended to cover collections of such hornless creatures as whales, horses, zebra and donkeys, camels, elephants, pigs, sheep, and even cranes. In the meantime, some of the animals themselves lost their horns as they were progressively domesticated – some breeds of cows, for example. *See also* COLLECTIVE NOUNS.

heredium [square measure] Unit of area in ancient Rome based upon the actus quadratus, and thus involving a measurement longer in one dimension than in the other.

$$
\begin{aligned}
\text{1 heredium} \quad &= \quad \text{4 acti quadrati, 2 JUGERA} \\
&= \quad \text{120 'feet' x 480 'feet', 57,600 square 'feet'} \\
&= \quad \text{116 feet 9.6 inches x 467 feet 1.7 inch} \\
&\qquad \text{(54,569 square feet, 6,063.2 square yards,} \\
&\qquad \text{just over 1¼ acre)} \\
&= \quad \text{35.604 metres x 142.416 metres} \\
&\qquad \text{(5,070.58 square metres, 0.5071 hectare)} \\
\text{100 heredia} \quad &= \quad \text{1 centuria} \\
\text{400 heredia, 4 centuriae} \quad &= \quad \text{1 saltus}
\end{aligned}
$$

　　The name of the measure reflects its association with the right granted by law to Roman citizens to own and bequeath land. On becoming a Roman citizen, every man was given 1 heredium of land to be his property and hereditary estate.

heroic verse [literary] Not a true VERSE FORM because the term describes different forms in different cultures, to each one nonetheless seemingly the essence of the heroic, the epic, the noble.
　　To the English-speaking literary world, heroic verse is in IAMBIC PENTAMETERS, in rhyming couplets or not. Iambic pentameters also constitute heroic verse in Italian and in German.
　　In France, however, heroic verse is traditionally in ALEXANDRINES: iambic hexameters.
　　In Latin and ancient Greek, the heroic form was also hexametric, but largely in dactyls or spondees rather than iambs (*see* METRE IN VERSE).

Hero's formula [maths] A formula for finding the area of any triangle. If the lengths of the sides are a, b, and c, the area equals the square root of $s(s-a)(s-b)(s-c)$, where $s = \frac{1}{2}(a+b+c)$. It was named after the Greek scientist Hero of Alexandra (*fl.* AD 62).

herschel [physics] A proposed – but seldom adopted – unit of radiant power (of a light source), equal to π times the number of watts per square metre per steradian. It was named after the British astronomer John Frederick Herschel (1792-1871), the son of the discoverer of the planet Uranus.

hertz [physics] Unit of frequency, equal to 1 cycle, oscillation, or vibration per second. It is often used in its multiples:

$$1 \text{ kilohertz (kHz)} = 1{,}000 \text{ hertz } (10^3 \text{ Hz})$$
$$1 \text{ megahertz (MHz)} = 1{,}000{,}000 \text{ hertz } (10^6 \text{ Hz})$$
$$1 \text{ gigahertz (GHz)} = 1{,}000{,}000{,}000 \text{ hertz } (10^9 \text{ Hz})$$

The unit was named after the German physicist Heinrich Hertz (1857-94).

Hertzian waves [physics] Another name for radio waves – that is, electromagnetic waves in the frequency range 10 kilohertz to 10 gigahertz (10^4-10^{10} hertz). They were named after the German physicist Heinrich Hertz (1857-94).

Hertzsprung-Russell (H-R) diagram [astronomy] A graph that plots stellar luminosity against spectral type for stars, which can provide information about the evolution and fate of various groups of stars. Above all, it makes visible how stars fall into fairly well-defined categories from which few exceptions emerge. The principal type of star that is neither a giant nor a dwarf – the ordinary sort of average star of which our Sun is an example – belongs in this way to a swathe of stars right across the diagram from top left to bottom right, known as the Main Sequence. Giants and supergiants are plotted in the top right corner; dwarfs are located at bottom left.

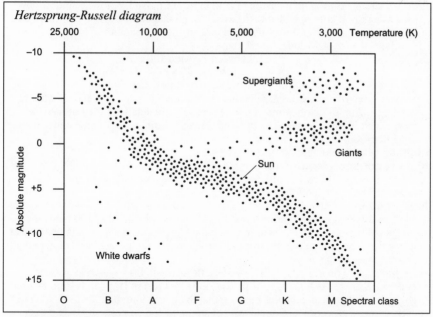

The diagram is named after the Danish astronomer Ejnar Hertzsprung (1873-1967) and the American astronomer Henry Russell (1877-1957), and is often called simply the H-R diagram.

Hess's law [chemistry] For a given chemical reaction, the total change in energy depends only on the initial and final states (and is independent of the route of the reaction). It was named after the Austrian-born American physicist Victor Hess (1883-1964).

hetero- [quantitatives: prefix] Prefix originally (in ancient Greek) meaning 'the other (of two)', but extended to mean also 'other' and thus 'different', 'unequal'.

Example: heteroploid – having a number of chromosomes different from an exact multiple of the number in the parent (haploid) germ cells

The word is etymologically identical with English *other* and *either*.

heteropolar [chemistry] Describing a chemical bond with an unequal distribution of electric charge (as, for example, in a covalent bond between unlike atoms).

hex *see* HEXADECIMAL

hex-, hexa- [quantitatives: prefix] Prefix which implies a sixfold unity.
> Examples: hexaemeron – the six days of Creation
>   hexagram – a six-pointed star
>   hexapod – with six feet (like a true insect)

The prefix derives from ancient Greek *hex* 'six'.

hexad [quantitatives] The number 6; a group of six. The adjective is *hexadic*.

hexadecimal [maths] Describing a number system with the base (radix) 16, commonly used in computers. Its digits are 0, 1, 2, 3, 4, 5, 6, 7, 8, 9, A, B, C, D, E, and F. In computer usage, hexadecimal notation is usually called simply hex.

hexagon [maths] A POLYGON with six sides (and six angles). The interior angles of a regular hexagon are each 120 degrees.

hexahedron [maths] A solid figure (POLYHEDRON) with six plane faces.

hexameter [literary] In verse, a line of six feet. If the meter is iambic, such a line is better known as an ALEXANDRINE. The corresponding adjective is *hexametric*.

hexane *see* OCTANE NUMBER, OCTANE RATING

hexavalent, hexavalence [chemistry] Having a valence (valency) of 6; an alternative description is sexavalent.

hexoctahedron [maths] A solid form (especially a crystal) with 48 equal triangular plane faces. The term derives from ancient Greek words that mean 'six (times) eight sides'.

hide [square measure] Old measure of land in rural England, not for centuries now used as a unit. It corresponded to the area of land that could support one free family and its dependants over the agricultural year, using one plough.

> 1 hide   =   4 VIRGATES
>   =   between 100 and 120 acres
>   (between 40 and 48.5 hectares)
> 100 hides   =   1 HUNDRED or WAPENTAKE

Extraordinarily, the word *hide* in this sense is etymologically identical with the word CITY. Anglo-Saxon *higid*- 'hide' is precisely cognate with the Latin *civitat*- from which English derives *city* (through French), both originally meaning something like 'settlement' or 'settled community'.

highball glass *see* DRINKING-GLASS MEASURES

highest common factor (HCF) [maths] The largest whole number that divides exactly into each of two or more other numbers. For example: the highest common factor of 18, 27, 36, and 45 is 9.

high frequency (HF) [physics; telecommunications] A radio frequency in the range 3,000 to 30,000 kilohertz (corresponding to wavelengths of 10 to 100 metres).

high gear [engineering] The gear in which the highest speed can be attained, and at which the gear ratio causes the driving wheels to turn the fastest in relation to the power output of the engine.

high jump *see* ATHLETICS FIELD EVENTS AREA, HEIGHT, AND DISTANCE MEASUREMENTS

Highland Games events [sport] In Scotland – not just in Highlands Region – the Highland Games is an athletics meeting that includes more than the usual athletics events, notably various additional tests of strength, and contests in playing the bagpipes and in Scottish dancing. It is usually presided over by the local clan chieftain, who is received with due deference and ceremony.

The usual (grass) track events are: 100 metres/yards sprint, 200 metres/220 yards sprint, 400 metres/440 yards race, and 800 metres/880 yards race, for men and for women. Hurdle races are contested over 110 metres, 200 metres, and (less commonly) 400 metres. There may also be races of greater distance around the track or around the perimeter of the Games area. Field events for men and women comprise the high and long/broad jumps, and the pole-vault; in the hammer-throw (for men only), the ordinary form of hammer is replaced by one with a wooden shaft attached to a solid iron ball; and in putting the shot (normally for men only), the shot used is often a round stone of approximately the usual weight.

Tests of strength (for men) include the subjectively judged event known as tossing the caber (*see* CABER-TOSSING), the tug-of-war (in which two teams each at the end of a rope strive to pull the other team across a dividing line), and any of various other fairly modern events featured in strongman contests on television (such as the picking up of progressively larger boulders to place them on a shoulder-high surface).

Bagpipe competition is usually between soloists (PIBROCH), and is also judged subjectively but to a much more rigorous and complex standard than is ordinarily realized. Occasionally, there are contests too for bagpipe bands (in which precision in marching while playing may also be judged).

Highland dancing contests are mostly confined to the reel, the (Highland) fling, the Seann Triubhas, and a sword dance. All of these dances were for men only until the 1900s; competitions for women have since then almost always required competitors to wear male dress – plaid, targe, kilt, and sporran.

Highland Games are also popular among Scottish communities (or communities with Scottish ancestry) all over the world, especially in Canada, New Zealand, and parts of southern England.

high pole [linear measure] In forestry and horticulture in the United States, a class of young tree specifically between 4 and 5 feet (1.22 and 1.52 metre) tall, and between 8 and 12 inches (20.3 and 30.5 centimetres) in diameter.

hin [volumetric measure] Unit of liquid capacity in ancient Israel.

1 hin	=	3 CAB or kab, 12 LOGS
	=	6.72 litres
	=	1.4784 UK gallon (11.827 UK pints)
	=	1.7753 US gallon (14.202 US pints)
2 hin	=	1 SEAH
6 hin, 3 seah	=	1 BATH
60 hin, 30 seah, 10 bath	=	1 KOR or homer

Hindu calendar *see* ERA, CALENDRICAL

histogram [maths] A type of bar chart on which quantities are proportional to the areas of the bars (rather than to their length).

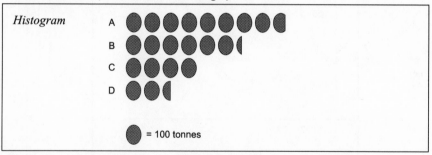

Histogram

A

B

C

D

● = 100 tonnes

hiyaka-me [weight] A unit of weight in the Japanese measuring system.

1 hiyaka-me	=	375.6584 kilograms
		(400 kilograms = 1.065 hiyaka-me)
	=	828.1766 pounds avdp., 59.155 stone
	=	7.394 (long) hundredweight
		(1,120 pounds, ½ ton = 1.352 hiyaka-me)

This is very close to being (and maybe meant to be) 100 KWAN.

100 kwan	=	625 KIN, 10,000 TAELS, 100,000 MOMME
	=	375.000 kilograms
	=	826.70 pounds avdp.

hiyak-kin [weight] A unit of weight in the Japanese measuring system.

1 hiyak-kin	=	60.1007 kilograms
		(60 kilograms = 0.9983 hiyak-kin)
	=	132 pounds 8 ounces avdp., 9.46 stone

This is evidently an approximation to the PICUL:

1 picul	=	16 KWAN, 100 KIN, 160 TAELS, 16,000 MOMME
	=	60 kilograms, 132 pounds 4.8 ounces avdp.

hockey measurements, units, and positions [sport] Hockey – known in some countries as field hockey to distinguish it from ice hockey – is very popular in a surprisingly diverse group of countries in Europe and Asia. Teams of ten outfield players try to get a fairly hard ball into the opponents' goal past a heavily padded and masked goalminder.

The dimensions of the pitch:
    length: 110 yards (91.44 metres)
    width: 60 yards (54.86 metres)
    radius of semicircular goal area ('the circle'): 16 yards (14.63 metres)
    the goal: width – 12 feet (3.66 metres)
              height – 7 feet (2.13 metres)
Positions of players: from the back forwards
    goalminder/goalkeeper
    full back
    halfbacks (usually 4)
    forwards (usually 5)
Timing: 70 minutes in two halves, separated by an interval of between 5 and 10 minutes (at the umpires' and timekeepers' discretion)

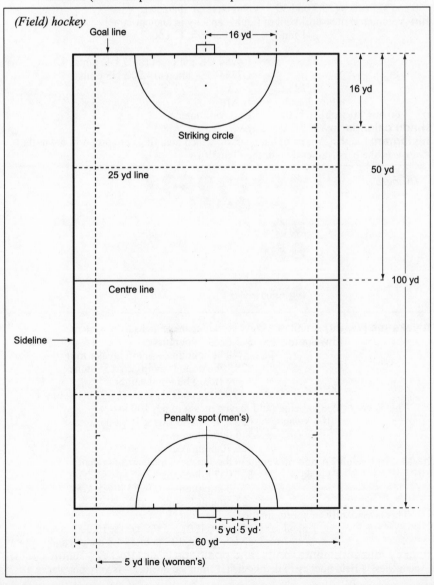

(Field) hockey

Goal line

16 yd

16 yd

Striking circle

25 yd line

50 yd

Centre line

100 yd

Sideline

Penalty spot (men's)

5 yd 5 yd

60 yd

5 yd line (women's)

Points scoring:
   the team that scores the greater number of goals wins
Dimensions of hockey stick and ball:
   the stick:
      width – must pass through a ring of diameter 2 inches (5.08 centimetres)
      maximum weight – for men, 28 ounces (793.8 grams) – for women, 23
         ounces (652.0 grams)
      the ball: weight – 5½-5¾ ounces (155.9-163.0 grams)
   The name of the game derives from the shape of the stick used to play it – it is
hooked (medieval French *hoqué* 'hooked', but from a Germanic original).

hodometer [linear measure] Device that measures the number of times a wheel
   revolves and, thus, in a wheeled vehicle, the same thing as an ODOMETER (in reality
   as well as by etymology). Hodometer is also the technical term, however, for the
   device used in surveying and cartography that utilizes a wheel to measure distance
   both on the ground and on a map. *See also* TACHOMETER.

hogline, hogscore [sporting term] In curling, a line across the rink at a distance of
   21 feet (6.40 metres) or at one-sixth of the rink's length from the tee, representing
   the minimum distance a stone must travel to remain in the game. A stone that does
   not cross the hogline is taken off the ice altogether.

hogshead [volumetric measure] A large cask or barrel of a size that differs according to the nature of its contents (as does the size of a BARREL). In the fifteenth
   century, the hogshead used for the transportation of wine and other alcoholic
   liquors, for example, had a capacity of 63 old wine gallons (63 present-day US
   gallons, 52.458 UK gallons; 238.474 litres). But, by the end of the same century,
   hogsheads of other liquids might contain as much as 140 imperial (UK) gallons
   (168 US gallons; 636.435 litres). In brewing, a hogshead contains 54 UK gallons
   (65 US gallons; 205 litres) of beer or cider.
   Why the cask was called a hogshead is not known.

Hohmann orbit, Hohmann ellipse [physics: space navigation] The elliptical path
   of a spacecraft from an orbit around one celestial body to an orbit around another
   that requires the least amount of energy (and preferably relying solely on gravitational forces).
   The term commemorates the German aerospace engineer Walter Hohmann
   (1880-1945).

Holocene epoch [time] The present geological EPOCH, specifically the post-glacial
   part of the QUATERNARY PERIOD following the Pleistocene epoch in which the Ice
   Ages occurred. This corresponds roughly to the last 10,000 years.
   The term *Holocene* derives from ancient Greek words that mean 'entirely new'.

homeopathic measures [medicine] In homeopathic medicine, great use is made
   of solutions, suspensions, and infusions containing truly minute quantities of herbs
   and other natural substances (generally aimed at reproducing symptoms similar or
   identical to a patient's current ill condition). The quantities involved are measured
   mainly in terms of weight or volume in parts per hundred, parts per thousand, or
   even parts per ten thousand of the solvent or suspension medium, and are specific to
   each substance or to each combination of substances.

homer *see* KOR, HOMER

home run, homer *see* BASEBALL MEASUREMENTS, UNITS, AND POSITIONS

homoeopathic measures *see* HOMEOPATHIC MEASURES

homogeneous function [maths] A function in algebra in which the terms have
   indices that add to the same number. For example: $2x^3 + x^2y - 3xy^2 + 3y^3$ (in which
   the indices of each term total 3).

homologation [engineering] Official process of measuring and checking the
   dimensions and proper functioning of a motor vehicle either before it is licensed for
   legal sale in a foreign country, or before it is authorized to compete in a race held in
   compliance with an international motor racing formula.

homologous series [chemistry] A family of organic chemicals that share the same
   general FORMULA.

homolographic projection [cartography] Any form of map projection in which
   equal areas of the ground surface are shown with equal areas on the map, no matter

how far north or south.

homopolar [chemistry] Describing a chemical bond with an equal distribution of electric charge (as, for example, in a covalent bond between identical atoms).

homoscedastic [maths: statistics] Describing two or more sets of statistics that show equal variability or the same standard deviation.

The term derives from ancient Greek words that mean 'of the same distribution'.

Hong Kong dollar [comparative values] Until 1997, the unit of currency in that British protectorate.

> 1 Hong Kong dollar = 100 cents

See also COINS AND CURRENCIES OF THE WORLD.

honours, honors [sporting term] In card games – especially bridge, especially if held or controlled by one hand, and especially of the suit that is currently trumps – the ten, knave (jack), queen, king, and ace, or, in a no-trump game, the four aces. In whist games the ten may not count as one of the honours.

honour score, honor score *see* CONTRACT BRIDGE CALLS (BIDS) AND SCORES

Hooke's law [physics] For an elastic material being stretched below its ELASTIC LIMIT, the extension (stress) is proportional to the force (strain) causing it. It was named after the British scientist Robert Hooke (1635-1703).

hook money [comparative values] Former units of currency in the islands off the coast of the Indian subcontinent and in Iran, comprising fishing hooks made of silver.

hoon [weight] Tiny unit of weight in Malaysia, dating from colonial times.

1 hoon	=	5.832 grains, 0.01333 ounce avdp.
		(1 ounce avdp. = 75 hoon)
	=	0.3779 gram
		(1 gram = 2.6462 hoon)
10 hoon	=	1 CHEE
100 hoon, 10 chee	=	1 TAHIL (1.333 ounce avdp.)
1,600 hoon, 160 chee, 16 tahils	=	1 GIN (1.333 pound avdp.)
160,000 hoon, 16,000 chee, 1,600 tahils, 100 gin	=	1 PICUL

hop, skip, and jump *see* ATHLETICS FIELD EVENTS AREA, HEIGHT, AND DISTANCE MEASUREMENTS

horizon [surveying; geology; astronomy] In general, the farthest one can see; the skyline, dividing the land from the sky. It is also known as the apparent horizon or visible horizon. But the term has various special meanings in science. In surveying, it is the plane at right-angles to the vertical (as defined by a plumb-line). In geology, it is the surface between two layers or strata of dissimilar rock, or a thin band within a layer.

In astronomy, the so-called sensible horizon is a circle formed by the intersection of the celestial sphere by a plane that is a tangent to the Earth's surface (at the point of observation). A plane parallel to this that passes through the centre of the Earth, and forms a great circle on the celestial sphere, is called the celestial horizon.

The term derives from the Greek words *horizon kuklos* 'limiting circle', the first word in turn a derivative of *horos* 'boundary'.

horizontal [maths] A line that is parallel to the Earth's surface; it is at right-angles to the vertical.

horizontal parallax [astronomy] For a member of the Solar System, its geocentric parallax (altitude correction to the Earth's centre) when it is on the observer's horizon.

horn *see* BRASS INSTRUMENTS' RANGE

horology [time] The study of the measurement of passing time, and the science of making instruments to effect such measurements.

The term derives from Latin words meaning 'telling the hour'.

horoscope [biology] Analysis of a person's character and disposition, and a forecast of the person's future, based upon the relative positions of Sun, Moon, major planets, and the zodiacal constellations at the precise time of birth. The zodiacal constellations are:

Aries	the Ram	birthday:	20 March-20 April (approx.)
Taurus	the Bull		20 April-20 May
Gemini	the Twins		20 May-20 June
Cancer	the Crab		20 June-20 July
Leo	the Lion		20 July-20 August
Virgo	the Virgin		20 August-20 September
Libra	the Scales		20 September-20 October
Scorpio	the Scorpion		20 October-20 November
Sagittarius	the Archer		20 November-20 December
Capricorn	the Goat		20 December-20 January
Aquarius	the Water-carrier		20 January-20 February
Pisces	the Fish		20 February-20 March

The positions of the Sun, Moon, and major planets as calculated for the time of birth are of great significance in that their relative positions are held to afford considerable influence on an individual's life and character.

Just how ancient this method of divination is can be evinced from the fact that, when the zodiacal constellations were first listed in the order that the Sun passed through them from the vernal equinox (the beginning of the ancient year), the Sun was in Aries at the equinox itself. After around four millennia, solar precession has taken the Sun at the vernal equinox right the way through Pisces and just into Aquarius.

The term derives from Latin words meaning 'viewing the hour'.

horse latitudes [geography] Two zones around the Earth, at around 30° north and south of the equator, in which there is generally little or no wind. Alternatively, two zones around the Earth, at around 25° north and 30° south of the equator, in which rainfall is markedly low.

The expression is thought to be a fifteenth- to sixteenth-century translation from the Spanish, but why the Spanish should have called the zones by this name is obscure.

horsepower [physics; engineering] Unit of power in the foot-pound-second (FPS) system, equal to the power needed to raise a weight of 33,000 pounds a distance of 1 foot in 1 minute (a rate of working of 33,000 foot-pounds per minute, or 550 foot-pounds per second).

$$1 \text{ horsepower } = 42.41 \text{ British thermal units per minute}$$
$$= 745.70 \text{ watts}$$

In the United States, the horsepower is often standardized as 746 watts.

Brake horsepower is a measure of the useful (or effective) horsepower of a motor or engine, as measured by a dynamometer or the resistance to a brake attached to its output shaft.

In the metric system used in Continental Europe, 1 *cheval-vapeur* (French) or *Pferdestärke* (German) is the power needed to raise a weight of 75 kilograms through a distance of 1 metre in 1 minute.

$$1 \text{ metric horsepower } = 0.986 \text{ horsepower}$$
$$(1 \text{ horsepower} = 1.014 \text{ metric hp})$$
$$= 735.5 \text{ watts}$$

The original horsepower unit was reputedly calculated in 1782 by the British engineer James Watt (1736-1819) as the power exerted by a dray horse pulling a capstan with a force of 180 pounds around a circle 24 feet across at a rate of 2½ times a minute (equal to 32,400 foot-pounds per minute).

horsepower-hour [physics; engineering]

Unit or work or energy in the foot-pound-second (FPS) system, equal to the work done by 1 HORSEPOWER in 1 hour.

$$1 \text{ horsepower-hour } = 2.686 \text{ x } 10^6 \text{ joules (SI units)}$$

hour [time; astronomy] In normal parlance,

$$1 \text{ hour } = 60 \text{ minutes } = 3,600 \text{ seconds}$$

and there are 24 hours in a day. In astronomy and in physics, however, an hour's precise value depends on the definition of 'day' or 'second' that is adopted: *see* DAY; SECOND.

The word *hour* in English derives ultimately from ancient Greek *hora* 'period' and thus both 'part of a day' and 'season of the year'. The Romans adopted it as pertaining generally to the daily period, the hour later accordingly applied particularly to the Roman Catholic Church offices (*see* CANONICAL HOURS). But cognate Germanic forms in fact hark back to the annual meaning and, in English, include the words *year* (Old English *gear*) and – with the meaning 'years past' – *yore*.

hour angle [astronomy] For a given celestial object, the angle that its HOUR CIRCLE makes with the meridian of the observer at the celestial pole, expressed in hours, minutes, and seconds measured westwards from the meridian.

hour circle [astronomy] The great circle that cuts the celestial equator at 90 degrees and passes through a celestial object and both celestial poles.

hourglass, sandglass, egg-timer [time] Device for measuring an exact period of time, involving two glass bulbs connected by a narrow waist through which sand (or other fine granular material) runs from the upper to the lower. In an hourglass, the sand takes precisely one hour for the sand to empty completely from the upper bulb into the lower. A sandglass may be timed to last for a longer or shorter duration, and an egg-timer generally has a maximum period of six minutes (and sometimes much less).

house [astronomy: astrology] Technical term for any one-twelfth segment (30°) of the starry night sky measured from a line running north-south to the horizon from the observer's position. It was in this way that the 'houses' 'belonging' to the zodiacal constellations were first allocated (*see* HOROSCOPE).

hp *see* HORSEPOWER

H-R diagram *see* HERTZSPRUNG-RUSSELL DIAGRAM

hu [volumetric measure] Unit of liquid capacity in China.

$$
\begin{aligned}
1 \text{ hu} \quad &= \quad 50 \text{ SHENG} \\
&= \quad 51.77 \text{ litres} \\
&\qquad (50 \text{ litres} = 0.9658 \text{ hu}) \\
&= \quad 11.388 \text{ UK gallons, } 13.677 \text{ US gallons} \\
&\qquad (12 \text{ UK gallons} = 1.0537 \text{ hu}) \\
&\qquad 12 \text{ US gallons} = 0.8774 \text{ hu})
\end{aligned}
$$

hubble [astronomy] Unit of distance in astronomy, equal to 10^9 (1 UK thousand million/1 US billion) light years. It was named after the American astronomer Edwin Hubble (1889-1953).

Hubble's constant [astronomy] For any galaxy, its distance (from Earth) divided by its rate of recession, expressed in kilometres per second per megaparsec. It is a measure of the rate at which the Universe is expanding, and is in the range 50 to 100 km/sec/Mps. It was named after the American astronomer Edwin Hubble (1889-1953).

Hubble's law [astronomy] The speed of recession (red shift) of a distant galaxy is proportional to its distance (from Earth). The proportionality constant is HUBBLE'S CONSTANT. The law was named after the American astronomer Edwin Hubble (1889-1953).

huffduff [physics: radio] A device for locating the direction from which high-frequency radio signals are emanating.

The term is an apparent attempt to create an acronym from High-Frequency Direction-Finder.

human morphology *see* MORPHOLOGY, HUMAN

humidity [physics; meteorology] A measure of the amount of water vapour in a gas, such as air; it is generally expressed as a percentage. At a given temperature, relative humidity is the ratio of the amount of water vapour in a gas to the maximum amount of vapour it would hold. Humidities are measured using a HYGROMETER, and may be regulated by means of a *humidistat*.

hundred [quantitatives] 100, 10 x 10, 10^2; five score.

$100\% \quad = \quad 1$ (the whole, the sum total)

One-hundred years is a century.

One-hundred runs or points is a century or ton.

One-hundred cents is one dollar, one-hundred pence is one pound

One-hundred centimetres is one metre

Ten hundred is one thousand
$$100 \times 100 \ (100^2) \ = \ 10,000$$
To multiply by 100 in the decimal system, take the decimal point two places to the right; to divide by 100 in the decimal system, take the decimal point two places to the left.

Examples:
$$1,234.5678 \times 100 \ = \ 123,456.78$$
$$1,234.5678 \div 100 \ = \ 12.345678$$

The prefix that most commonly denotes 'one-hundred times' a unit is *hect-* or *hecto-*, which derives through French from ancient Greek: *see* HECT-, HECTO-. In some compounds, however, the prefix CENTI- may be used instead.

The term derives as a combination of two Anglo-Saxon words, *hund* '100' and *-red* 'reckoning' – the first of which corresponds to and is cognate with Latin *cent-* '100' (an example of the Indo-European *centum* languages) and Sanskrit *satam* '100' (an example of the Indo-European *satem* languages).

The plural is also *hundred* if preceded by a number or an expression standing in for a less than general number, *hundreds* if not (and followed by the word *of*); thus:

six hundred years	many hundreds of years
several hundred years	several hundreds of years

See also HUNDREDTH.

hundred [square measure] In certain counties of England, the former division of a county known in other counties as a wapentake, and originally reckoned to contain about 100 families and to require its own court. It was basically an administrative division for the purposes of taxation and social discipline, and applied mainly in those areas of the country that were occupied by people of Anglo-Saxon (as opposed to Scandinavian) descent.

1 hundred	=	100 HIDES, 400 VIRGATES
	=	between about 10,000 and 12,000 acres (between about 4,000 and 4,850 hectares)
	=	between about 15.625 and 18.75 square miles (between about 40.47 and 48.56 square kilometres)

The system was in due course transported across to the New World, although it remains current only in the state of Delaware, where the chief officer (bailiff) and those liable to sit on juries were formerly known as 'hundreders'. *See also* WAPENTAKE; GORE.

hundredth [quantitatives] $\frac{1}{100}$, 10^{-2}; the element in a series between the ninety-ninth and the hundred-and-first, or the last in a series of one hundred.
$$\tfrac{1}{100} \ = \ 0.01 \ = \ 1\%$$
The prefix that most commonly denotes 'one-hundredth' of a unit is 'centi-', although this can on occasions alternatively mean 'one-hundred times' the unit (*see* CENTI-).

Old Hundredth is the name for the best-known tune sung to the hymn 'All people that on Earth do dwell'.

hundred thousand [quantitatives] 100,000, 10^5.
The square root of 100,000 is 316.22776 (to five decimal places).
The reciprocal is one hundred-thousandth, 0.00001 or 10^{-5}. *See also* LAKH.

hundredweight [weight] There are two forms of hundredweight in the avoirdupois system. The traditional hundredweight, used mainly in Britain and English-speaking countries other than in North America, relates to the original form of TON (sometimes called the long, or gross, ton in North America), and is also known as a quintal or centner. The more recent hundredweight, used mainly in North America, really does correspond to 100 lesser units, and accordingly relates to a ton consisting of a number of pounds that is much easier to remember (sometimes called the short ton), and is also known as a cental or quintal.

Thus:

1 (long, gross) hundredweight, quintal, or centner/sentner	= 112 pounds avdp., 8 stone, 4 quarters

$$(1 \text{ UK quintal} = 1.120 \text{ US quintals})$$

	= 50.80208 kilograms

(50 kilograms, also called a centner = 110.23 pounds)

20 (long, gross) hundredweight = 1 (long, gross) ton, 2,240 pounds

And:

1 short hundredweight, or cental, also known as a quintal = 100 pounds avdp.

(1 US quintal = 0.8929 UK quintal)

= 45.35924 kilograms

(45 kilograms = 99.208 pounds)

20 short hundredweight = 1 short ton, 2,000 pounds

The abbreviation for 'hundredweight' is *cwt*, in which the c- represents the Latin numeral denoting 100, and -wt is 'weight'.

For the metric, Mediterranean, and Latin American forms of the ancient Roman version of the hundredweight, *see* QUINTAL.

hurdles events [sport] In athletics, races predominantly over 100 metres for women, 110 metres for men, and 400 metres for men or women, in which competitors have also to leap over frail barriers (hurdles) at regularly spaced intervals.

In a 100-metres hurdles race for women, there are ten hurdles to be cleared, all of them 2 feet 9 inches (83.8 centimetres) high, the first hurdle 13 metres (42 feet 7.8 inches) from the starting line, each of the next nine hurdles 8.5 metres (27 feet 10½ inches) ahead of the last, leaving a sprint of 10.5 metres (34 feet 5.4 inches) to the finishing line.

In a 110-metres hurdles race for men, there are ten hurdles to be cleared, all of them 3 feet 3 inches (99.06 centimetres) high, the first hurdle 13.72 metres (45 feet) from the starting line, each of the next nine hurdles 9.144 metres (30 feet) ahead of the last, leaving a sprint of 14.02 metres (46 feet) to the finishing line. In the 400-metres hurdles race, there are ten hurdles to be cleared, those for men 3 feet (91.44 centimetres) high, those for women (usually) 2 feet 6 inches (76.20 centimetres) high, the first hurdle 45 metres (49.21 yards) from the starting line, each of the next nine hurdles 35 metres (38.28 yards) ahead of the last, leaving a sprint of 40 metres (43.74 yards) to the finishing line.

Further dimensions of hurdles:
width across track of top bar: 1.20 metres (3 feet 11½ inches)
vertical width of top bar: 7 centimetres (2¾ inches)
minimum pressure to upset: 3.46 kilograms (8 pounds)

In some indoor events, 60-metres hurdles races are staged but have fewer than the usual ten flights. Very rarely, and as a special event, 200-metres hurdles races are similarly staged (and do have the customary ten flights).

The word *hurdle* derives as a Middle English diminutive of a Germanic word meaning 'a woven panel', 'a movable plaited screen'.

hurling measurements, units, and positions [sport] Hurling is rarely played outside Ireland. A fast-moving game in which two teams of fifteen players may kick the ball, fist it, or briefly balance it on the stick (the 'hurley') before flipping it up and batting it on, the object is to score as many points as possible by getting the ball between the opposing goalposts, either above the crossbar or below it.

The dimensions of the pitch:
maximum area: 160 x 110 yards (146.3 x 100.6 metres)
minimum area: 140 x 84 yards (128.0 x 76.8 metres)
the goal:
width: 21 feet (6.4 metres)
crossbar height: 8 feet (2.44 metres)
goalpost total height: 16 feet (4.88 metres)
Positions of players:
players are divided into (usually 7) forwards, (usually 4) halfbacks, (usually 3) backs, and a goalkeeper

Timing: (usually) 60 minutes, in two halves, separated by an interval of 10
  minutes
Points scoring:
  ball hit or kicked (beneath crossbar) into goal: 3 points
  ball hit over crossbar between posts: 1 point
  the team with the more points at the end of the game wins
Dimensions of the hurley and the ball:
  the hurley:
    overall length: 3 feet (91.44 centimetres)
    maximum width: 4 inches (10.16 centimetres)
    weight: unrestricted
  the ball:
    circumference: 9-10 inches (22.86-25.40 centimetres)
    weight: 3½-4½ ounces (99.22-127.57 grams)
The name of the game is English (not Celtic) and is indeed the present particular
gerund of the verb 'to hurl' – but that verb originally meant not 'to throw' but 'to
beat on', 'to smack against' as applied particularly to winds in gusts – and as is
especially appropriate in this sense.

Huronian [time; geology] A geological stage within the Precambrian era (before the
Algonkian stage). Rocks found to the north of Lake Huron in Canada are typical of
this stage.

hurricane *see* BEAUFORT SCALE OF WINDSPEED

hydraulic gradient [physics; engineering] The difference in fluid pressure between
the ends of a pipe carrying a flowing liquid. The term derives from the Greek words
hydra- 'water' and *aulos* 'pipe'.

hydrogen ion [chemistry; physics] A positively charged hydrogen atom (H^+); a
proton. The concentration of hydrogen ions in a solution is a measure of its ACIDITY.

hydrometer [physics] An instrument for measuring the DENSITY of a liquid, consisting of a glass bulb (containing a weight) at the lower end of a glass stem. It floats in
a liquid with the stem upright, and the stem is graduated with a density scale. For
solutions of a single substance, density is generally proportional to CONCENTRATION
and, in such cases, the stem may be graduated directly in concentrations.

hydrophone [physics] Any device used to pick up and measure sound under or in
water, and to locate its source.

hyetograph [physics: meteorology] Graph or other diagrammatic representation of
the precipitation (rainfall, snow, frost, etc.) in an area.

hygrometer [physics] An instrument for measuring the HUMIDITY of a gas, such as
air. The type that makes a recording of humidity is called a *hygroscope*.

hyp-, hypo- *see* HYPO-, HYP-

hyper- [quantitatives: prefix] Prefix originally denoting 'above', 'more than', and
extended to mean 'more than usually', 'excessively', 'over-'.
    It is the ancient Greek equivalent of the Latin *super-*.

hyperbola [maths] A (pair of) curve(s) with an eccentricity greater than 1 – that is,
the locus of a point that moves so that the ratio of its distance from a fixed point
(focus) to its distance from a fixed line (directrix) is always more than 1. A hyperbola is also generated as a conic section when a plane cuts through the base of a
cone at an angle that is not parallel to the cone's slope (a parallel cut generates a
PARABOLA).
    The term derives from two ancient Greek words together meaning 'thrown too far'.

hyperbolic function [maths] One of six functions that are related to a HYPERBOLA in
a way analogous to that in which trigonometrical functions are related to a circle.
Thus, the trig functions sine, cosine, tangent, cosecant, secant, and cotangent
become the hyperbolic sinh, cosh, tanh, cosech, sech, and coth.

hyperbolic paraboloid [maths] In solid geometry, a type of paraboloid whose
sections in two planes at right-angles are parabolas but whose sections parallel to
the third coordinate plane are hyperbolas. The shape of the solid roughly resembles
that of a saddle.

hypergeometric distribution [maths] In statistics, a distribution showing the
probable frequency of a number of specified elements in a random sample selected

from a larger set of data containing a given number of such elements.

hyperon [physics] A short life-time (about 10^{-10} seconds) elementary particle of which the mass is greater than that of a neutron. Hyperons include any BARYON that is not a nucleon.

hypersonic [physics] Describing speeds in excess of Mach 5: *see* MACH NUMBER. *See also* ULTRASONIC.

hypertonic [physics; medical] Describing a solution in which the osmotic pressure is higher than a given standard (such as the osmotic pressure of blood).

hypo-, hyp- [quantitatives: prefix] Prefix originally denoting 'under', 'less than', and extended to mean 'less than usually', 'insufficiently', 'under-'.

It is the ancient Greek equivalent of the Latin *sub-*.

hypobenthos [biology] A term that collectively describes all organic life below a depth of 500 fathoms (3,000 feet, 914.4 metres) in the ocean.

In ancient Greek the term means literally 'below the (sea)bed'.

Hypophrygian mode [music] Medieval form of KEY, the SCALE of which is characterized on a modern piano by the white notes between one B and the next. In modern music this is a set of notes virtually impossible to use as a key, even its 'key chord' sounding as if requiring resolution.

Its name derives from one of the musical modes used in ancient Greece about which little or nothing is known (*hypo-phrygian-* actually means 'under-Phrygian': *see* PHRYGIAN MODE, although the two modes have virtually nothing in common). In any case, the ancient Greeks had completely different musical referents.

hypotenuse [maths] The longest side of a right-angled triangle (the side opposite the right-angle).

hypotonic [physics; medical] Describing a solution in which the OSMOTIC PRESSURE is lower than a given standard (such as the osmotic pressure of blood).

hypsometer [physics] An apparatus for measuring the boiling point of a liquid. Because boiling point varies with external pressure and therefore with altitude, it may be used for calculating altitude.

Hz *see* HERTZ

I

I (Roman numeral) [quantitatives] As a numeral in ancient Rome, the symbol I corresponded to *unus* 'one' – yet, in this sense, it did not derive from the ninth letter of the Latin alphabet, the I that was in turn derived from the same source as the (capital form of the) Greek letter *iota* (the lower-case form of which, to the Greeks as a numeral symbol, signified 10 or 10,000). As in many other numeral systems, ancient and modern, the symbol for 'one' was intended to be a simple, single stroke – the very essence of unity. The ancient Egyptians had used the same symbol for 'one' for at least 1,500 years before the Romans. The ancient Chinese used the same symbol, but wrote it horizontally rather than vertically. Perhaps the only other comparably simple representation of the number one is a dot (which in fact corresponds to the number one in the old Mayan script), and it may or may not be coincidental that, in our own modern numeral system, the alphabetical letter that is closest in shape to the numeral has a dot above it in its lower-case form.

In many of the ancient numeral systems, the symbols for 2 and 3 were II and III; most also followed this with a 4 in the form of IIII. *See also* ONE.

i [maths] The square root of –1 (meaning therefore that $i^2 = -1$). Itself an imaginary number, the product of *i* and any real number is also an imaginary number: *see* NUMBER.

iamb, iambic foot, iambus *see* FOOT; METRE IN VERSE

iambic pentameter *see* METRE IN VERSE; PENTAMETER; HEROIC VERSE

ice ages [time; geology] Throughout the history of the Earth there have been many times when ice has covered much of the planetary surface – so many in fact that they have not all been scientifically distinguished from one another. Certainly there were glacial episodes in the late Precambrian era that continued on into the Cambrian period; more occurred at the end of the Palaeozoic era during the

Permian period. The last, and best known, sequence of these ice ages began during the Pleistocene epoch, from only 2 or 3 million years ago or so, and continued well into human history. It is when the Ice Ages seem to have stopped that the modern geological time – the Holocene epoch – is said to have commenced.

During each ice age much of the sea was converted into ice, leaving considerable areas that had previously been seabed exposed. It was in this way that many of the animals of previous millennia managed to become so widespread, crossing land bridges that do not exist today.

ice hockey measurements, units, and positions [sport] When ice skating was a popular pastime enjoyed by families and friends, each rink might have its own ice hockey team that was well supported. Now ice hockey teams are comparatively scarce, and, although there still is a World Championship competition, standards have fallen so much that the same seven or eight international teams tend to reappear again and again in the series finals.

The object of the game is to score as many goals as possible. Teams comprise five players plus a goalminder/goalkeeper on the ice at any one time (unless one or more team-members are temporarily banished from the ice to the 'sin-bin' by way of penalty for unsporting conduct).

The dimensions of the rink:

maximum area: 200 x 100 feet (61 x 31.50 metres)

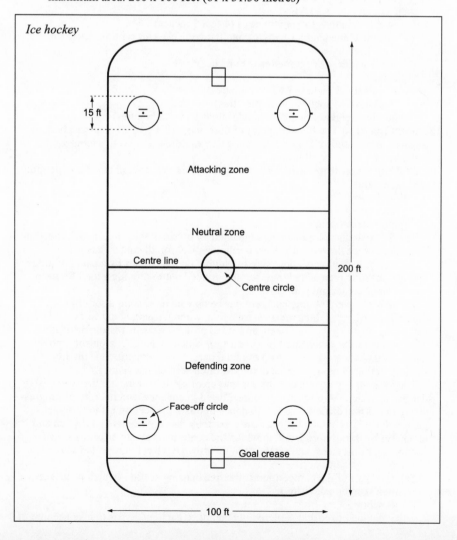

Ice hockey

15 ft

Attacking zone

Neutral zone

Centre line

Centre circle

200 ft

Defending zone

Face-off circle

Goal crease

100 ft

smaller areas in proportion
goal area: 6 x 4 feet (1.829 x 1.219 metres)
the goal:
width: 6 feet (1.829 metres)
height: 4 feet (1.219 metres)
Positions of players:
(usually) 3 forwards, (usually) 2 backs, goalminder
such positions may have to be abandoned during a power-play (when one
team-member has been temporarily dismissed from the ice); even the
goalminder may be replaced by an attacker towards the end of a game when
emergency measures are called for
Timing: 60 minutes of actual playing time (the clock stops when the puck is not
in play), in three 20-minute periods separated by two 10-minute intervals
Points scoring:
the team with the greater number of goals at the end wins
Dimensions of equipment:
the stick:
overall length – 135 centimetres (4 feet 5 inches)
blade length – 37 centimetres (1 foot 2½ inches)
blade width – 7.5 centimetres (3 inches)
the goalminder's stick:
overall length – 135 centimetres (4 feet 5 inches)
outside blade length (it angles into the vertical) – 98 centimetres (38.6
inches)
blade width – 9 centimetres (3½ inches)
the puck:
diameter – 3 inches (7.62 centimetres)
thickness – 1 inch (2.54 centimetres)
minimum height of boards around rink: 4 feet (1.22 metres)

ice point [physics] The freezing point of pure water (0°C, 32°F) at atmospheric
pressure, often used as a FIXED POINT on a temperature scale. At this temperature
water and ice are in equilibrium.

ice skating disciplines and events [sport] The three disciplines of competition
ice skating are:
solo skating
pairs skating
(pairs) ice dancing
The first two disciplines are accompanied by music of the contestants' selection;
the music for ice dancing is at least partly identical for all contestants.

All are marked on very subjective but highly complex lines by a panel of judges
whose marks are then added together and divided out to provide a total for each
contestant or contestant pair.

The *solo and pairs skating* competitions begin with three compulsory figures. The
figures are performed three times on each foot, without pause, along an axis indicated
before beginning by the contestant, and are of progressively increasing difficulty.

There then follows the short program(me), which includes compulsory movements with connecting steps. And the final part of the competition is the free
skating section (which is the section most often seen on television).

The *ice dancing* event has a similar line-up of sections. First are the compulsory
dances, of which there are usually three, listed in advance, and in order of progressively increasing difficulty; a fourth dance in this same section is the 'original set
pattern dance', for which the music is chosen by the contestants to a rhythm and
tempo set by the authority, and to which the contestants must perform a series of
connecting turns, steps, and rotations of their own device, but repeated several
times over.

The final part of the competition is the free dancing section (which is the section
most often seen on television).

Points scoring:
the method of scoring is incredibly complex; each judge has to take account of

every mark for the contestants already given in calculating the next mark (involving considerable mathematical workings); only in the free skating/ dancing sections is it customary for the judges to present their marks publicly (in two elements: one for technical merit, the other for artistic impression) out of 6, to the nearest single decimal

It is possible that ice skating may become a form of competition simultaneously open to both amateurs and professionals.

icosahedron [maths] A solid figure that has 20 faces; those of a regular icosahedron are equilaterial triangles.

ideal gas [physics] A hypothetical gas that exactly follows the gas laws; it is alternatively known as a perfect gas.

identity element [maths] In set theory, a member of a set that combines with any other member of the set and leaves it unchanged. For example: for the operation of addition, 0 is the identity element (because any number plus 0 equals the number); for multiplication, 1 is the identity element.

ides [time] Day, in the calendar of ancient Rome, that represented an approximate half-way mark through the month, which it thus divided in two (*idus* 'division'). In the original ten-month system, this meant that in the four 31-day months – March, May, July (Quinctilis), and October – the ides fell on the fifteenth day; and in the six 30-day months – April, June, August (Sextilis), September, November, and December – the ides fell on the thirteenth day.

In modern terms, then, the ides corresponded to: 15 March, 13 April, 15 May, 13 June, 15 July, 13 August, 13 September, 15 October, 13 November, 13 December.

Julius Caesar's much later addition of (the equivalent of) two more 30-day months at the end of the (Roman) year accordingly introduced two more ides: 13 January and 13 February.

It was with the date of the ides each month that the date of the nones also corresponded, and it was at the beginning of every month (on the CALENDS), on the NONES, and on the ides that the full Roman senate convened – which is why Julius Caesar was warned to beware of treachery on that date, when so many powerful enemies were all gathered together.

So early was the use of the word to mark the half-month that most authorities suggest that *idus* was adopted by the Romans as a borrowing from the Etruscans. (The English form of the word is in fact a French corruption of the original.)

Illinoian [time] Describing the third sequence of ice ages (glaciation) in North America, which occurred during the PLEISTOCENE EPOCH and lasted for about 55,000 years from about 230,000 years ago.

illuminance [physics] The amount of light or luminous flux that falls on a unit area of a surface, expressed in lux (1 lux = 1 lumen per square metre). It is inversely proportional to the square of the distance from the light source, and is sometimes instead known as the *illumination*.

imaginary number [maths] A number obtained by multiplying a real number by i (the square root of -1): *see* NUMBER.

immi [volumetric measure] Swiss unit of liquid capacity based on (or assimilated to) the metric system.

$$
\begin{aligned}
1 \text{ immi} \;=\; & 1.5 \text{ litres} \\
=\; & 2.6399 \text{ UK pints, } 3.1699 \text{ US pints} \\
& (3 \text{ UK pints} = 1.1364 \text{ immi} \\
& \;\; 3 \text{ US pints} = 0.9464 \text{ immi)} \\
2 \text{ immi} \;=\; & 3 \text{ litres} \\
& (1 \text{ UK gallon} = 3.0304 \text{ immi} \\
& \;\; 2 \text{ US gallons} = 5.0475 \text{ immi)}
\end{aligned}
$$

immunoassay [medicine] Chemical analysis of a person's body tissues and fluids to measure immunocompetence – a positive response by the immune system to infective organisms.

IMP *see* INTERNATIONAL MATCH POINT (IMP)

impedance [physics] The property of an electric component or circuit that makes it oppose the passage of a current. For direct current (DC), impedance equals RESIST-ANCE; for alternating current (AC), it equals REACTANCE; both are measured in ohms.

For a sound wave, the sound pressure in a particular medium divided by the rate of flow of the particles comprising the medium through a given area; the ratio of sound pressure to sound flux.

For a vibrating system, the mechanical force (in the direction of motion) divided by the velocity of the vibration that results – also called the *mechanical impedance*.

The term derives from a Latin verb literally meaning 'to set (someone's) feet in (shackles)', thus 'to hinder', 'to hold back'.

imperial [paper sizes] A size of paper. As a large sheet, imperial is

in Britain		22 x 30 inches (558.8 x 762.0 millimetres)
	or	22 x 32 inches (558.8 x 812.8 millimetres)
in the United States		23 x 31 inches (584.2 x 787.4 millimeters)

Folded for printing, imperial octavo is 7½ x 11 inches (190.5 x 279.4 millimetres); imperial quarto is 11 x 15 inches (279.4 x 381.0 millimetres).

imperial [comparative values] Former gold coin in Tsarist Russia, initially worth 10 r(o)ubles, and later worth 15.

The Roman Emperor Augustus (ruled 27 BC to AD 14) was the first to strike a coin of this name.

imperial system [quantitatives] A system of weights and measures based on the foot (linear), the pound (weight), the pint (volume), and their multiples and submultiples. The weights are those of the AVOIRDUPOIS SYSTEM. It was formerly the chief system for trade and industry in Britain and, with a few differences, still is in the United States, where it is known as standard, or customary, measures.

The system is so called because, from the sixteenth century onwards, it was introduced for use throughout the colonies of the British Empire (by statute from 1838).

impetus *see* MOMENTUM

implicit function [maths] The variable x is an implicit function of y when they are related by a non-explicit function (that is, one in which x is not expressed in terms of y): *see* EXPLICIT FUNCTION.

implosive consonants *see* PLOSIVE CONSONANTS

improper fraction [maths] A fraction that has an upper part (numerator) larger than the lower part (denominator); for example: ⅝, ¾,¹⁰⁄₇. It is, therefore, always larger than 1, and can be converted into a mixed number (for example, 2⅓, 2¼, 1³⁄₇).

impulse [physics] The change in MOMENTUM of a moving object resulting from a collision. It is the time integral of the force of reaction between the colliding objects.

A surge in electric current, or the 'on' part of a coded signal consisting of short periods when current/voltage is 'on' and 'off', is also sometimes called an impulse. The preferred term for either of these, however, is *pulse*.

inch [linear measure] Unit of length in the foot-pound-second (FPS) system, equal to one-twelfth of a FOOT.

1 inch	=	1,000 mils, 120 gries, 72 points, 12 lines
	=	25.40 millimetres, 2.54 CENTIMETRES
		(25 millimetres = 0.98425 inch)
12 inches	=	1 foot
36 inches, 3 feet	=	1 YARD

There are 63,360 inches in 1 mile.

The depth of rainfall or snow is commonly announced in inches; less commonly, atmospheric pressure may be measured in inches of mercury in a barometer. Normal pressure is 32 inches of mercury.

The inch derives directly from the Roman *uncia* – one-twelfth of a *pes* 'foot', and etymologically akin to the Latin *unus* 'one' – to which it is closely comparable in size: *see* UNCIA.

inch, cubic [volumetric measure] Cubic/volumetric unit within the imperial system.

1 cubic inch	=	16.387064 CUBIC CENTIMETRES/millilitres
		(1 cubic centimetre = 0.06102374 cubic inch
		1 cubic metre = 61,023.740 cubic inches)

                      1.733 cubic inch   =   1 UK fluid ounce   [for
                      1.806 cubic inch   =   1 US fluid ounce   most
                     28.92 cubic inches =   1 US pint   ordinary
                     34.67 cubic inches =   1 UK pint   practical
                     61.02 cubic inches =   1 litre   purposes]
                    1,728 cubic inches   =   1 cubic foot
                   46,656 cubic inches   =   1 cubic yard

inch, square [square measure] Unit of area within the imperial system.
              1 square inch   =   6.4516 SQUARE CENTIMETRES
                                  (1 square centimetre = 0.1550003 square inch
                                   1 square metre = 1,550.003 square inches)
            144 square inches =   1 square foot
          1,296 square inches =   1 square yard

inches per second [speed] Unit of speed or velocity.
            1 inch per second =   5 feet per minute, 100 yards per hour
                                   (1 foot every 12 seconds)
                              =   0.05682 mile per hour
                                   (1 mile every 17 hours 36 minutes)
                              =   2.54 centimetres per second
                              =   1.524 metre per minute
                                   (1 metre every 39.37 seconds)
                              =   0.09144 kilometre per hour

inches per year [speed] Unit of extremely slow speed or velocity occasionally
   used, for instance, in the measurement of the rate of annual tidal encroachment or of
   tectonic movement in the Earth's crust.
              1 inch per year =   1 line per month, 10 gries per month
                              =   10 inches per decade
                                   (1 yard every 36 years)
                              =   8 feet 4 inches per century
                              =   83 feet 4 inches per millennium
                                   (1 mile every 63,360 years)
                              =   2.54 centimetres per year
                                   (1 centimetre every 143 days 16.8 hours
                                    1 metre every 39 years 135 days)
                              =   2.54 metres per century
                              =   25.4 metres per millennium
                                   (1 kilometre every 39,370 years 285 days)

incidence, angle of *see* ANGLE OF INCIDENCE

inclination [maths; astronomy] In geometry, the relationship of two lines or two
   planes that are at an angle to each other; the angle itself.
       In astronomy, either the angle between the plane of a planet's orbit within the
   Solar System and the ecliptic (the apparent path of the Sun), or the angle between
   the plane of any satellite's orbit and the equatorial plane of the larger body around
   which it is orbiting.

inclinometer [physics; geology] An instrument for measuring MAGNETIC DIP, also
   known as a dip needle.

incommensurable numbers [maths] Numbers that have no common factor or
   multiple (usually because one or more of them is irrational or imaginary).

incongruent [maths: geometry] Describing plane or solid figures that are not the
   same although they may be similar; such figures are sometimes alternatively
   described as *incongruous*. A pair of CONGRUENT plane figures can be made to
   overlap exactly (one figure may require rotation or reflection first).

incubation period [biology; medicine] In zoology, the time that elapses between
   the laying of an egg (by a bird or reptile) and its hatching. It varies from a few days
   for a small bird to several weeks for a large tortoise.
       In medicine, the time that elapses between infection by a bacterium or virus and
   the onset of a disease (when the first symptoms show). The periods for some
   common diseases are:
           chickenpox   2-3 weeks   poliomyelitis   3-21 days

common cold	2-72 hours	rabies	2-6 weeks
diphtheria	2-5 days	rubella	14-21 days
gonorrhea	3-9 days	scarlet fever	1-3 days
influenza	1-3 days	syphilis	2-70 days
malaria	2 weeks	tetanus	4-21 days
measles	8-13 days	typhoid	1-3 weeks
mumps	12-26 days	typhus	7-14 days
paratyphoid	1-10 days	whooping cough	7 days

indefinite article *see* ARTICLE

independent clause [literary] An alternative name for the main clause in a sentence, described as 'independent' because it can stand by itself: *see* CLAUSE.

independent variable [maths] If the variable y is a function of another variable x, x is the independent variable of the function (y is the dependent variable).

indeterminacy principle [physics] A less common name for the HEISENBERG UNCERTAINTY PRINCIPLE.

index [maths] Either the number that indicates what root is required – so, for example, in the expression for a cube root, $\sqrt[3]{}$, the index is 3.

Or another name for an exponent, or power: *see* EXPONENT, POWER.

In both cases the term is thoroughly appropriate in its original Latin meaning, 'indicator', 'pointer (out)'.

index [comparative values] In economics, a value on a scale of values relating to a nation's financial matters and fiscal priorities – in particular, the cost of living, the existing value of the national currency in relation to foreign currencies, and corporate confidence in the activities of the stock exchange.

The index may be expressed either as an overall value, a value in comparison with other nations' values, or a percentage rise or fall from the value previously quoted or set. *See also* DOW-JONES INDEX; FTSE INDEX.

index fossil [time; geology] Any fossil found commonly enough over a sufficiently wide range to be used as a customary indicator of the date of the rock stratum within which it has been found.

index of refraction *see* REFRACTIVE INDEX

indicator [chemistry] A substance utilized to show when a chemical reaction is complete, usually by changing colour. For example: for the reaction between an acid and an alkali, the indicator litmus changes from red to blue as neutralization takes place. A substance that measures pH (acidity or alkalinity) by its colour is also called an indicator. One that operates over a very wide range of pH is called a universal indicator.

In medicine, other indicators are used for testing body fluids, for example to measure the glucose content of blood or urine in the monitoring of patients with diabetes.

individual medley *see* SWIMMING DISCIPLINES AND COMPETITIVE DISTANCES

inductance [physics] For an electric component or circuit, the property that makes it create a magnetic field and store magnetic energy. It is measured in henrys. Inductance is also a measure of electromagnetic induction (the production of a current in a conductor in a varying magnetic field).

inertia [physics] The resistance that an object offers to a change in its state of rest or to its motion; it results from, and is proportional to, the object's mass.

inertial mass [physics] An object's mass as determined by its MOMENTUM (as distinct from, but nevertheless equal to, its gravitational mass).

inferior [astronomy: astrology] Describing a body (such as either of the planets Mercury or Venus) that comes between the Sun and the Earth, especially in terms of astrological significance for a HOROSCOPE.

inferno [physics] A proposed, but not adopted, unit of star temperature, equal to 1,000 million kelvin (10^9 kelvin).

infinitesimal [maths] Describing a very small change (in the value of a variable quantity), which approaches a limiting value of 0.

infinity [maths] A quantity that is larger than any quantity that can be measured/counted. Mathematically, it can be considered to be the reciprocal of zero (that is, $\frac{1}{0} = \infty$).

inflation [comparative values] Consistent general rise in the cost of living, most often caused by a vicious circle in which rising costs are thought to demand rising salaries and wages to counteract them, but the increased salaries and wages themselves contribute to the increase in costs. The situation may be aggravated if banks and other credit organizations authorize monetary credits on too lax a basis, and if the national government issues too much paper currency.

During times of inflation, the value of the nation's currency goes down in relation to the currencies of other countries which are not experiencing inflation (or which are experiencing it at a lower rate), sometimes resulting in a official currency devaluation by the national bank.

inflection [maths] A point on a curve at which it changes from being convex to concave (or vice versa). An alternative spelling is *inflexion*.

infradian [time] Recurring in a cyclic pattern completed in well under 24 hours. The word is used especially in relation to body rhythms such as regular digestive processes and other neurologically organized functions of a periodic nature.

The term is as crassly derived as the related CIRCADIAN, in this case supposedly coming from Latin elements *infra* 'inside' and *dies* 'day' (the adjectival compound form of which is actually *diurnus*, *diurnal-*, from which come the English words *diurnal* and *journal*).

inframedian zone [oceanography] The area of the oceanic seabed that lies between 50 fathoms (300 feet, 91.44 metres) and 100 fathoms (600 feet, 182.88 metres) deep, supposedly the average depth of all the seabed.

infra-red radiation [physics] Electromagnetic radiation in the wavelength range 0.75 nanometre to 1 millimetre. It is heat radiation, occupying a region of the electromagnetic spectrum between visible light and microwaves.

infrasound [physics] Sound waves that have a frequency of 20 hertz or less, below the threshold of human hearing. *See also* ULTRASONIC.

inhour [nuclear engineering] For a nuclear reactor, a unit of reactivity equal to the reciprocal of the operating time in hours (so that, for example, a 3-hour period of operation gives ⅓ inhour reactivity).

ink-blot test *see* RORSCHACH TEST

inner [sporting term] In archery and rifle shooting, the innermost ring of the target surrounding the bull. It is the second-highest-scoring part of the target.

inner product *see* DOT PRODUCT

in-off *see* BILLIARDS MEASUREMENTS AND UNITS

input-output analysis [comparative values] In economics, a method of evaluating the financial efficiency of a nation's different industries by comparing the value of the goods manufactured or produced by each industry against the expense of the raw materials, labour, and overheads of the industry. The results of such an analysis may be presented as a table. The financial efficiency of one industry in relation to another industry can in this way be reduced to a single figure corresponding to how much more or less efficient one industry is, expressed as a positive or negative percentage. Such a figure is then called the *input-output coefficient*.

insolation [physics: meteorology] In the measurement of solar energy at the Earth's surface, particularly for the purpose of using solar radiation as a source of power, the measured radiation per unit surface area per unit time.

inst. [time] Abbreviation – meaning 'of this month' – of the word *instant*. From the mid-1500s to perhaps the 1970s a common method of referring to dates of the current month in formal correspondence, the expression is now rapidly becoming obsolete, and is generally regarded as stylistically unacceptable.

insulin unit *see* INTERNATIONAL UNIT

integer [maths] A whole number, positive or negative: *see* NUMBER.

integral [maths] Describing either an integer (a whole number), or the value that results from INTEGRATION.

integration [maths] A process in calculus of summing the series of infinitely small quantities (infinitesimals) that together make the difference between two values of a function; the inverse of DIFFERENTIATION. Among its many applications is determining the areas and volumes of shapes defined by equations (curves).

integrodifferential [maths] Describing a mathematical expression or process that involves both INTEGRALS and differentials (or derivative coefficients: *see* DIFFEREN-TIATION) of a function.

intelligence quotient (IQ) [medicine] A quantity that attempts to measure intelligence, defined originally as the ratio of a child's mental age to his or her chronological age, expressed as a percentage. 'Average' or 'normal' – when mental age is the same as chronological age – is therefore 100. IQ is now usually based on performances (scores) in tests, compared with norms for people of the same age (and thus applicable also to adults).

intelligence tests [medicine] For medical purposes, intelligence is usually tested in children for whom there is doubt about the extent of a mental handicap, mainly to arrive at a diagnosis on what the child should or may not be capable of doing through education, special or otherwise. Less commonly, a child's intelligence may be tested to find out whether he or she is unusually gifted – especially if the child is consistently bored with the normal playthings and occupations of his or her age, and even more if he or she is hyperactive.

The tests applied are generally those to determine an INTELLIGENCE QUOTIENT (*see above*).

intensity [physics] A general term to quantify the presence and effect of radiation (especially light or heat) or sound in relation to its surrounding volume (volumetric units) or area (square units). The depth or saturation of a colour is also sometimes called its intensity.

intercalary days and months *see* YEAR

interest [economics] Sum of money regularly repaid to somebody who invests money, or who loans it, generally defined as an annual percentage of the much larger amount originally invested or borrowed.

Simple interest is reckoned on the whole amount over the whole period. So if a sum s is borrowed for y years at a simple interest rate of r per cent per year, the total interest is $syr/100$ [and the sum that has to be repaid is $s + (syr/100)$].

Compound interest takes into account interest earned in the previous year or years. After y years, a sum s invested at r per cent per year compound interest totals $s[(100 + r)/100]y$. Sometimes compound interest rates are quoted for a shorter period, such as a month; so that these can be compared with the interest due over a year, they may also be stated as ANNUAL PERCENTAGE RATES (APR), or annualized percentage rates.

interferometer [physics] An instrument for measuring distances by making use of the phenomenon of *interference* (which occurs when two or more waves of the same frequency, from coherent sources, interact partly to reinforce or cancel each other).

interindustry analysis [comparative values] Another expression for INPUT-OUTPUT ANALYSIS.

interior angles [maths: geometry] In geometry, when two parallel lines are crossed by a single line, any of the four angles subtended within the parallel lines; interior angles on opposite sides of the crossing line are equal, interior angles on the same side of the crossing line are supplementary (add to 180°).

interlunation [astronomy] The 4- or 5-day period of each lunar month when the Moon is not visible from the Earth's surface: *see* LUNAR PHASES.

intermediate boson *see* INTERMEDIATE VECTOR BOSON

intermediate vector boson [physics] A subatomic particle, a type of GAUGE BOSON, that takes part in weak interactions. There are three types, designated W^+, W^-, and Z^0.

internal energy [physics] In thermodynamics, the sum of the kinetic and potential energies of the molecules in a system, expressed in joules. Its commonest manifestation is the temperature of the system. The heat absorbed by a system minus the external work done by it equals the change in internal energy (which is a statement of the first LAW OF THERMODYNAMICS).

internal friction [physics] For a fluid, another name for VISCOSITY.

international angstrom [physics] Unit of length, defined in such a way that, at a temperature of 15°C and a pressure of 760 millimetres of mercury, the red line in the cadmium spectrum has a wavelength of 6438.4696 international angstroms. It is very nearly the same as an ANGSTROM UNIT (10^{-10} metres).

international candle [physics] A former unit of luminous intensity (candle power), defined in terms of the light from the flame of a standard wax candle (weighing ⅙ of a pound, 2⅔ ounces, or 75.6 grams) burning at a rate of 120 grains (7.78 grams) per hour. It has been superseded by the CANDELA.

international date line [time; geography] The imaginary line running from North Pole to South Pole that represents the end of one day and the beginning of another. The line of longitude that is identical with the Greenwich meridian in London is held to be the line of 0°; the date line is for the most part accordingly the line of 180°. Thus: 12 noon on Sunday at Greenwich corresponds just to the west of the date line to the time of 12 midnight at the end of Sunday, almost the first thing Monday morning; and just to the east of the date line to the time of 12 midnight at the end of Saturday, almost the first thing Sunday morning.

international henry *see* HENRY

international match point (IMP) [sporting term] In contract bridge tournaments played between teams representing their countries over a series of matches in Europe, international match points are scored for winning games and beating opponents by specified margins, and, as each season progresses, are consistently listed in tables for comparison by enthusiasts.

The equivalent in the United States is master point.

international nautical mile *see* KNOT; MILE

international pitch [music] A fixed standard of 'concert' pitch based on the note A above middle C – the note to which an orchestra tunes – corresponding to 440 hertz (vibrations per second). It is known also as Stuttgart pitch (after the historic capital of Baden-Württemberg state in Germany).

international practical temperature scale [physics] A usable temperature scale made as close as possible to the (theoretical) thermodynamic temperature scale, generally using a platinum resistance thermometer.

International Standard Book Number [literary] Every single title published in book form, and every different edition of that title, is allocated an International Standard Book Number (ISBN) by which it can thereafter be identified (for instance, by libraries and by copyright authorities).

The number includes elements that denote the country of publication and the publisher. In most countries, publication of any book is illegal without its ISBN shown on the copyright page.

International Union of Biochemistry unit *see* ENZYME UNIT

international unit [medical] Unit for specifying quantities (often doses) of various medicinal substances, such as enzymes, hormones, vaccines, and vitamins. Its size varies, depending on the substance. For example, an insulin unit equals the activity of 0.0455 milligrams of a fixed standard type of insulin; for diphtheria antitoxin (vaccine) a unit is equivalent to 0.0628 milligrams of the substance. The sizes of units for some common vitamins are:

vitamin A	0.0006 milligrams
vitamin B complex	10 milligrams
vitamin C	0.05 milligrams
vitamin D	1 milligram

intersection [maths] In geometry, the point or surface at which two lines or planes cross each other. (*See also* BISECT.)

In set theory, a set of elements that are in both of two other sets; the area of overlap between two sets on a Venn diagram.

interval [music] Technical term describing the distance between two notes sounded simultaneously, measured as if counting the notes on the keyboard of a piano, including both (upper and lower) notes in the count, expressed as a reciprocal (fraction) and in relation to a key's harmony.

Thus, the interval between C and the G above it is (C-D-E-F-G) a fifth. And, because it can contribute to both major and minor harmonies in the key of C, it is a perfect fifth at that. Using the note C and notes immediately above it, the most commonly described intervals are:

a minor second	C and C#
a major second	C and D

a minor third	C and D#/E♭	
a major third	C and E	
a perfect fourth	C and F	
an augmented fourth	C and F#	on a piano, these two
a diminished fifth	C and G♭	are the same thing
a perfect fifth	C and G	
a minor sixth	C and G#/A♭	
a major sixth	C and A	
a minor seventh	C and A#/B♭	
a major seventh	C and B	
an octave	C and C above it	
a (major) ninth	C and D above the next C	
a (major) tenth	C and E above the next C	
a (perfect) eleventh	C and F above the next C	

and so on. In reality, the term *interval* in this sense describes not so much the distance of the notes from each other but the effect of their being sounded simultaneously.

The word *interval* derives in English through French from Latin *inter-vallum* '(the space) between the fortification-mounds or palings (and the garrison's tents)', thus any 'intervening space'. *See also* KEY, KEYNOTE; HARMONY; PITCH.

intonation pattern [literary] In speech, the rise and fall in the pitch of a person's voice according to the sense of what is being said. In English, for example, questions tend to end with a distinct rise in intonation whereas definitive statements tend to end with a distinct fall. Different languages have different intonation patterns.

The standard intonation patterns (or contours) of a language can be 'mapped' on a total-clause or total-sentence basis, and are sometimes programmed into the more advanced electronic speech decoders and voice reproduction systems.

intra-, intro- [quantatives: prefix] Prefix meaning 'inside' and thus sometimes 'less than normal', 'less than required', 'under-'.

inverse [maths] If, in set theory, a set has an IDENTITY ELEMENT under a particular operation, a member has an inverse if its product equals the identity element. For example: under the operation of addition the inverse of a positive whole number is the corresponding negative number (because they add to 0, and 0 is the identity element for addition). The inverse of a non-zero whole number under multiplication is its reciprocal (because their product equals 1, the identity element for multiplication).

inverse function [maths] A mathematical function that is opposite in nature or effect to another one, such as inverse hyperbolic functions, inverse trigonometrical functions.

inverse hyperbolic functions [maths] The inverse (reciprocals) of the hyperbolic functions – that is, $\sinh^{-1}$, $\cosh^{-1}$, $\tanh^{-1}$, cosech^{-1}, sech^{-1} and $\coth^{-1}$.

inverse square law [physics] A common type of relationship in physics, which quantifies the falling off of an effect with the square of the distance from the source. It applies, for example, to light and to electrostatic, gravitational, and magnetic fields (*see* for example ILLUMINANCE; NEWTON'S LAW OF GRAVITATION).

inverse trigonometrical functions [maths] The inverse (reciprocals) of the trigonometrical functions – that is, $\sin^{-1}$, $\cos^{-1}$, $\tan^{-1}$, cosec^{-1}, $\sec^{-1}$ and $\cot^{-1}$.

inversion layer [meteorology] A layer of air that is warmer than the layer of air below it, especially at an altitude at which this effect is unusual.

involute [maths] A curve that is the locus of the end of a tight string being unwound from another curve (like a piece of cotton being unwound from a fixed spool, which generates a spiral). *See also* EVOLUTE.

I/O analysis *see* INPUT-OUTPUT ANALYSIS

iodine number [chemistry] Measure of the amount of iodine that reacts with an organic compound such as a fat or oil, expressed as the number of grams of iodine absorbed by 100 grams of the fat. Alternatively known as the *iodine value*, it gives a measure of the number of unsaturated (double and triple) bonds present. The process by which such measurement is made is called *iodometry* or *iodimetry*.

ion [chemistry; physics] A positively or negatively charged atom or molecule, resulting from the loss or gain of one or more electrons. Many inorganic chemical compounds consist of ions, or dissociate into them when they dissolve in water. For example, sodium chloride (common salt) consists of sodium ions (Na^+) and chloride ions (Cl^-). During electrolysis and in gas discharge tubes, ions carry electric current.

Ions emitted by radioactive material may be detected on instruments such as a GEIGER COUNTER.

The term derives as a modern borrowing directly from ancient Greek *ion* '(thing) going'. *See also* NEGATIVE ION; POSITIVE ION.

Ionian mode [music] Medieval form of KEY, the SCALE of which is characterized on a modern piano by the white notes between one C and the next. Regarded as a *modus lascivus* [*sic*] by the medieval Roman Catholic Church because it was by far the most commonly used in secular music (as it still is today, now known better as the ordinary major scale), it was not officially sanctioned for ecclesiastical music in Europe until the 1500s. It is very occasionally known alternatively as the *Hypolydian mode*, a term relating it to a much earlier form of musical classification dating from the late (AD) 500s.

The name of the Ionian mode, however, derives from one of the musical modes used in ancient Greece, and held to be particularly soft and suited to the female voice or temperament. It is extremely unlikely, nonetheless, that the ancient Greeks ever had a notational sequence at all resembling today's major scale.

ionic concentration [chemistry; physics] The number of positive and negative ions (ion pairs) per unit volume of solution. Alternative names: ion concentration, ionic density.

ionic product [chemistry] The product of the activities (concentrations) of ions in a solution, expressed in moles per litre.

ionic radius [chemistry; physics] The radius of an ion in a crystalline substance, usually stated in angstrom units.

ionization chamber [physics] A closed gas-filled tube with a high voltage between a pair of electrodes arranged parallel to each other or coaxially. When ionizing radiation (such as radioactivity or cosmic rays) passes through the tube, the gas ionizes and a current flows between the electrodes. The value of the current is a measure of the rate of ion production in the tube. A very high-voltage version is used in a GEIGER COUNTER.

ionization constant [chemistry] The product of the activities (concentrations) of the ions produced by a substance divided by the activity (concentration) of its undissociated molecules.

ionization energy [chemistry; physics] For a gas, the energy need to produce an ION PAIR, expressed in electron volts.

ionization potential [physics] The energy needed to remove an electron from a neutral atom; the electron's bonding energy. It is expressed in electron volts.

ionosphere *see* ATMOSPHERE, COMPOSITION OF.

ion pair [physics] A positive and a negative ion, usually resulting from the simultaneous ionization of two uncharged particles (by the transfer of an electron from one to the other).

IQ *see* INTELLIGENCE QUOTIENT (IQ)

iron [printing] Unit of type measure, equal to about 25 EMS and used for defining a mixture of sorts (particular letters and symbols supplied to order) from a type founder or supplier.

Iron Age [time] Period of civilization following the Bronze Age, in which implements and weapons were cast or forged of the hard metal iron. The term refers not to any specific time but to the stage to which human culture had progressed. Some cultures even today observe elements of societal behaviour generally associated with the Iron Age.

irradiance *see* RADIANT FLUX DENSITY

irrational number [maths] A real number that cannot be stated in the form of a fraction (as the ratio of two whole numbers). Examples include e, π, and the square root of 2. *See also* NUMBER; SURD.

is-, iso- [quantatives: prefix] Prefix denoting 'equal', 'the same (amount)', and applied especially to dimensional units that can be charted on maps or diagrams.

Examples: isacoustic – of equal sound intensity

isothermal – of equal temperature

isallobar [meteorology; physics] In meteorology, a line drawn on a map to demarcate an area that experiences the same changes in atmospheric pressure (over a particular period of time). The term derives as a learned borrowing from ancient Greek elements meaning '(of) equally different pressure'.

ISBN *see* INTERNATIONAL STANDARD BOOK NUMBER

Islamic calendar *see* YEAR; ERA, CALENDRICAL

isobar [meteorology; physics] In meteorology, a line drawn on a map that demarcates and connects places that have the same atmospheric pressure (at a particular time).

In nuclear physics, one of a set of atomic nuclei that have the same total number of protons and neutrons – that is, the same mass number, or nucleon number. Their individual numbers of protons differ, and so they are different elements.

The term derives as a learned borrowing from ancient Greek elements meaning '(of) equal pressure'.

isobath [physics: cartography] In oceanography, a line drawn on a map to demarcate an area that is of the same depth beneath the surface of the sea or a lake.

The term derives as a learned borrowing from ancient Greek elements meaning '(of) equal depth'.

isobathytherm [physics: cartography] In oceanography, a line drawn on a vertical cross-sectional view of the sea or of a lake to demarcate an area that is of the same temperature.

The term derives as a learned borrowing from ancient Greek elements meaning '(of) equal depth (and) temperature'.

isochor, isochore [physics] A curve or graph that indicates the relationship between two different properties of a fluid at constant volume; the properties so quantified might, for example, be pressure and temperature.

The term derives as a learned borrowing from ancient Greek elements meaning '(of) equal position'.

isocline [geology; physics] A line drawn on a map that demarcates and connects places that have the same inclination (angle of dip of the Earth's magnetic field).

isodiapheres [physics] Two or more atomic nuclei that have the same neutron excess (difference between the numbers of neutrons and protons).

The term derives as a learned borrowing from ancient Greek elements meaning '(of) equal throughput'.

isodynamic [geology; physics] Describing lines drawn on a map that demarcate and connect places where the strength of the Earth's magnetic field is the same; an alternative term is *isomagnetic*.

The term *isodynamic* derives as a learned borrowing from ancient Greek elements meaning 'of equal power'.

isoelectric point [physics] The hydrogen ion concentration (pH of a solution) at which a system is electrically neutral; electrical conductivity and viscosity are also at a maximum.

isogeotherm [geology] A line drawn on a map to demarcate an area thought to have the same subterranean temperature.

The term derives as a learned borrowing from ancient Greek elements meaning '(of) equal soil temperature'.

isogloss [literary] A line drawn on a map to demarcate an area in which the inhabitants all speak the same language, use the same dialectal variants, or have some other linguistic characteristic of specified interest. Rarely, the word *isograph* is used alternatively.

The term derives as a learned borrowing from ancient Greek elements meaning '(of) equal tongue'.

isogon [maths] A plane figure (POLYGON) in which all the angles are equal.

isogonic line [geology] A line drawn on a map that demarcates and connects places that have the same magnetic declination (deviation between true and magnetic north).

The term *isogonic* derives as a learned borrowing from ancient Greek elements meaning 'of the same angle'.

isogram *see* ISOLINE

isograph *see* ISOGLOSS

isohel [geology] A line drawn on a map to demarcate an area on the Earth's surface that has the same numbers of hours of sunshine (over a particular time).

The term derives as a learned borrowing from ancient Greek elements meaning '(of) equal sun(light)'.

isohyet [meteorology; geology] A line drawn on a map to demarcate an area of the Earth's surface that experiences the same amount of rainfall (over a particular time).

The term derives as a learned borrowing from ancient Greek elements meaning '(of) equal rain'.

isoline [cartography] Any line on a map that demarcates an area in which a specified condition exists equally; less commonly, it may instead be called an *isogram*.

isomagnetic *see* ISODYNAMIC

isomer [chemistry; physics] In chemistry, one of two or more (different) compounds that have the same molecular composition, and thus the same chemical formula – that is, their molecules are made up of the same atoms, although they are joined in a different way in each isomer and therefore have different structures.

In physics, one of two or more atomic nuclei that have the same atomic numbers and mass numbers, but differ in energy states.

The term derives as a learned borrowing from ancient Greek elements meaning '(of) the same constituents'.

isometric [maths; physics; medicine] Having equal dimensions. In geometry (and drawing), an isometric projection has the horizontal lines of an object drawn at 30 degrees to the horizontal on the drawing, producing a three-dimensional representation with equal emphasis to all three planes.

In physics and crystallography, describing a crystal that has equal axes that are mutually perpendicular.

In physiology, describing a muscle contraction that does not shorten the muscle.

isophot [astronomy; physics] A line drawn on a map to demarcate an area of the Earth's surface that receives the same amount of light from a single source (often the Sun, but occasionally the Moon or even a star).

The term derives as a learned borrowing from ancient Greek elements meaning '(of) equal light'.

isopleth [maths] An alternative name for a NOMOGRAM or alignment chart.

The term derives as a learned borrowing from ancient Greek elements meaning '(of) equal total'.

isopycnic [physics] Describing fluids and solids of equal density.

The term derives as a learned borrowing from ancient Greek elements meaning '(of) equal compactness'.

isosceles [maths] In geometry, describing a triangle that has two sides equal in length (and two equal angles).

The term derives as a learned borrowing from ancient Greek elements meaning '(of) equal legs'.

isoseismal [physics] Describing an area on the Earth's surface subject to an equal degree of shock and tremor following the same seismic disturbance (earthquake).

ISO system *see* A SERIES; B SERIES; C SERIES; PAPER SIZES

isoteniscope [physics] A device used to measure the pressure at which a liquid and its vapour are at equilibrium at a specified temperature.

The first two elements of the term derive from a learned borrowing of ancient Greek elements meaning 'of equal tension'.

isotherm [physics; meteorology] In physics, a graph connecting points that have a particular temperature; in this sense it is known also as an *isothermal*.

In meteorology, a line drawn on a map to demarcate an area on the Earth's surface that has the same temperature.

The term derives as a learned borrowing from ancient Greek elements meaning '(of) equal warmth'.

isotonic [physics] Describing a solution of which the OSMOTIC PRESSURE is the same as a given standard (such as the osmotic pressure of blood, or plant sap).

The term derives as a learned borrowing from ancient Greek elements meaning '(of) equal straining'.

isotope [physics; chemistry] A form of an atom of an element that has the same atomic number but an atomic mass different from other forms of that atom; the isotopes of an element thus have the same number of protons but different numbers of neutrons. For example: the isotopes uranium-235 and uranium-238 both have 92 protons, but 143 and 146 neutrons, respectively.

The term derives as a learned borrowing from ancient Greek elements meaning '(of) the same place'.

isotopic number [physics; chemistry] The difference between the number of neutrons and protons in an isotope, alternatively known as the *neutron excess*. In uranium-235 for example it is $143 - 92 = 51$.

isotopic weight [physics; chemistry] An isotope's atomic weight (relative atomic mass), alternatively known as the *isotopic mass*.

Italian sonnet *see* VERSE FORMS

J

j, J *see* JOULE

jack *see* BOWLS MEASUREMENTS AND UNITS; KNAVE

Jacobean age/era [time] Period between 1603 and 1625 when James I occupied the English throne. James II occupied the throne for so short a period (1685-88) that his name is not associated with any particular style or fashion.

jacobus [comparative values] Gold coin issued in England from 1603 for a time, and worth initially 20 shillings, later 24 shillings. It was named after King James I of England: *see* JACOBEAN AGE/ERA.

jansky [physics; astronomy] In radioastronomy, a unit of received power (at a radio-telescope) from a cosmic source of radio waves, equal to 10^{-26} watt per square metre per hertz per steradian. It was named after the American radio engineer Karl Jansky (1905-50).

jar [physics] Obsolete unit of capacitance equal to 1,000 centimetres stat (the electro-static unit of capacitance), equivalent to $\frac{1}{9} \times 10^{-11}$ FARAD. It was named after the Leyden (Leiden) jar, an old type of capacitor, itself named after the Dutch city.

javelin *see* ATHLETICS FIELD EVENTS AREA, HEIGHT, AND DISTANCE MEASUREMENTS

jeroboam [volumetric measure] A champagne bottle that holds 100 UK fluid ounces, 104 US fluid ounces (3.0756 litres).

Alternatively, but amounting to very much the same thing, reckoning on a standard wine bottle containing 75 centilitres (0.75 litre),

1 magnum	=	the contents of 2 bottles
1 jeroboam	=	4 bottles
1 rehoboam	=	6 bottles
1 methuselah	=	8 bottles
1 salmanazar	=	12 bottles
1 balthazar	=	16 bottles
1 nebuchadnezzar	=	20 bottles

The jeroboam is named after Jeroboam, first King of Israel at the time when the kingdom of Solomon split into two, Israel (under Jeroboam) and Judah (under Rehoboam). His name derives from Hebrew *Yarob'am* 'Let the people become great'.

jigger, shot *see* SHOT, JIGGER

Jordan curve theorem [maths] A Jordan curve is a simple closed curve (that is, one that does not cross itself anywhere). The theorem states that such a curve divides a plane into two regions, and that a point on the outside of the curve cannot be joined to one on the inside without crossing the curve. The theorem may appear to be self-evident, but it has important consequences in mathematical analysis. It was named after the French mathematician Marie-Ennemond Jordan (1838-1922).

joule [physics] Unit of work and energy in the SI system, equal to the work done by a force of 1 newton moving though a distance of 1 metre in the direction of application of the force. It was named after the British physicist James Joule (1818-89).

1 joule	=	10^7 ergs (CGS units)
	=	6.242×10^{18} electron volts (particle energy)
	=	2.778×10^5 kilowatt-hours (electrical energy)
	=	0.2392 calorie (heat energy)
		(1 calorie = 4.1806 joules)
	=	9.478×10^{-4} british thermal unit (heat energy)
1,000 joules	=	1 kilojoule

Joule's law [chemistry; physics] There are three important relationships called Joule's law:

For a solid compound, the molar heat capacity is equal to the sum of the atomic heat capacities of the elements of which it is composed.

The internal energy of a given mass of gas depends only on its temperature, and is independent of its pressure and volume.

When an electric current flows through a resistance, the heat produced equals the square of the current times the resistance and the time for which the current flows.

joule-second [physics] The dimensions of Planck's constant (a fundamental physical quantity equal to the ratio of the energy of a quantum to its frequency). It has been proposed that it be renamed the *planck*, equivalent to 1 quantum – the basic unit – of 'action'.

judo/jujitsu measurements and units [sport] Originally a method of self-defence combining strength, balance, and the use of an opponent's momentum against himself or herself, and developed primarily in Japan, the sport proved so popular worldwide that it became an Olympic sport in 1964.

Contests are between two individuals on a square (usually) green mat around which is an extension of the mat (usually) marked in red (the danger area), outside which is a further extension (usually) in green (the safety area). If the contestants enter the danger area, the contest halts and both resume their initial placings.

The dimensions of the mat:

contest area: 8 x 8 metres (26 feet 3 inches x 26 feet 3 inches)

width of danger area: 1 metre (3 feet 3 inches)

width of safety area: 3 metres (9 feet 9 inches)

Starting position of competitors:

4 metres (13 feet) apart, on one axis of the contest area square

Timing: minimum 3 minutes, maximum 20 minutes (as previously arranged)

Scoring:

ippon (a throw, a lift, an unbreakable stranglehold): 1 point

waza-ari (partly successful throw, lift, stranglehold): ½ point (at the discretion of the two judges and the referee)

Methods of winning:

an ippon wins outright

all other decisions at the discretion of the judges and referee (in the light of any waza-ari scored, and any fouls awarded against contestants, including actual penalties)

As in boxing, there are several weight categories for competitors (applying mostly to male competitors).

		weight up to		
category	kilograms	pounds	stone	pounds
lightweight	63	138.89	9	12.89
light middleweight	70	154.32	11	0.32
middleweight	80	176.37	12	8.37
light heavyweight	93	205.03	14	9.03
heavyweight	more	more	more	
open		any weight		

The word *judo* is a combination of Japanese elements meaning 'way of softness', 'way of gentleness' (*-do* = *tao* 'way'); the word *jujitsu* is very much a variant on an identical theme: *ju* is again 'softness', 'gentleness', and *-jitsu* (properly *jutsu*) is 'the art (of)'.

jugerum [square measure] Unit of area in ancient Rome based upon the ACTUS quadratus, and thus involving a measurement longer in one dimension than in the other.

1 jugerum	=	288 scrupula, 2 acti quadrati
	=	120 'feet' x 240 'feet', 28,800 square 'feet'
	=	116 feet 9.6 inches x 233 feet 7.2 inches
		(27,284.5 square feet, 3,031.6 square yards, 0.6264 acre)
	=	35.604 metres x 71.208 metres
		(2,535.29 square metres, 0.2535 hectare)
2 jugera	=	1 HEREDIUM
	=	120 'feet' x 480 'feet', 57,600 square 'feet'
200 jugera, 100 heredia	=	1 centuria
800 jugera, 400 heredia, 4 centuriae	=	1 saltus

The *jugerum* was the most commonly used unit measure of land (real estate) in ancient Rome.

The name of the unit derives from its use as a measure of agricultural land: it is the Latin for 'yoke (of plough-oxen) area'. On becoming a Roman citizen, every man was given two jugera of land to be his property and hereditary estate.

Julian calendar *see* OLD STYLE; YEAR

Julian dates [astronomy] Dates reckoned in the number of days since 4713 BC (ignoring years or months), used for dating periodic occurrences in astronomy. The system was named after the French scholar Joseph Justus, or Julius, Scaliger (1540-1609), who introduced it in 1582.

jumbo [quantitatives] Extra-large, mammoth, bumper.

Rarely can an elephant have made such an impression on a national vocabulary as did Jumbo, the enormous tusker of the Barnum & Bailey Circus, later displayed in the Zoological Gardens in London, towards the end of the nineteenth century (1800s). An African bush elephant, his maximum height was measured at 10 feet 9 inches (3.277 metres). But the word had already existed in English since the 1820s, meaning 'heavy, dull, and clumsy' – probably as a fairly offensive reference to non-Caucasian foreigners of different languages and faiths (thus connected with mumbo-jumbo). Through the enormous popularity of Jumbo, however, the word came to express great size only. The classic form of jumbo jet, the 400-passenger Boeing 747, for example, is 71 metres/232 feet in length and has a maximum wingspan of 60 metres/196 feet. *See also* MAMMOTH.

jun [comparative values] Unit of currency in North Korea.

100 jun	=	1 WON

See also COINS AND CURRENCIES OF THE WORLD.

Jupiter: planetary statistics *see* PLANETS OF THE SOLAR SYSTEM

Jurassic period, Upper, Middle, and Lower [time] A geological PERIOD during the Mesozoic or Secondary era, after the Triassic period but before the Cretaceous, corresponding roughly to between 213 million years ago and 144 million years ago. The period is generally considered to comprise three divisions: first the Lower Jurassic (or Lias), and then the Middle (or Dogger) and Upper (or Malm) Jurassic both together also known as the Oolithic. The Lower Jurassic lasted from roughly 213 million years ago to about 183 million years ago; the Middle from roughly 183 million years ago to about 173 million years ago; and the Upper from roughly 173 million years ago to about 144 million years ago. In the Northern Hemisphere, the Jurassic was a period of mountain-building, coral growth in shallow seas, and an abundance of ammonites and other molluscs. On land the dinosaurs reached their maximum sizes, roaming among the cycads, conifers, and ferns.

The name of the period is derived from the Jura mountains, a large crescent-shaped range of limestone plateaus that line the borders of France, Switzerland, and Germany; the suffixial ending -ssic was added on analogy with the name of the Triassic period which preceded it.

For table of the Mesozoic era *see* TRIASSIC PERIOD.

K

K *see* KILO-; KELVIN SCALE OF TEMPERATURE

kab *see* CAB, KAB

kalends *see* CALENDS, KALENDS

kan [weight] In Hong Kong, a unit of weight corresponding to one used throughout the Far East during the colonial times of the mid-1800s, devised artificially as a measure for taxation purposes. In China it was called the *chin* (or *catty*), in Japan the *kin*, in Malaysia the *gin* (or *kati*), and in Korea the *kon* (or *catty*). Based on the AVOIRDUPOIS system, the unit corresponded to 1.333 POUNDS and was made up of 16 smaller divisions (each therefore of 1.333 OUNCES).

1 kan	=	16 TAELS
	=	1.333 pounds
		(1 pound = 0.75 kan: three-quarters of a kan)
	=	604.775 grams, 0.604775 kilogram
3 kan	=	4 pounds (1.8143 kilogram)
100 kan	=	1 PICUL = 133.333 pounds (60.4775 kilograms)

kantar [weight] In Arab countries, a weight corresponding roughly to the imperial unit the hundredweight (112 pounds, 50.80208 kilograms) and similarly based upon the old *centenarium* of ancient Rome, but differing slightly in different states and even between localities. In many parts of Egypt, for example, the kantar is reckoned to be (100 rotls) the equivalent of 99.0492 pounds (44.93 kilograms) but even then depends on exactly what goods are being weighed. *See also* CENTNER; QUINTAL.

karat [minerals] An alternative spelling of CARAT.

karate measurements and units [sport] Karate, unlike judo, is a genuine martial art, a fighting technique originally intended to disable opponents. As a sport, karate contests are necessarily closely supervised to ensure that all blows, kicks, and other forms of strike land within well-defined target areas of the body, and that they are 'pulled' so that full force is never delivered.

Competitions take place on a large square mat, under the supervision of four judges and a referee.

The dimensions of the mat: 8 x 8 metres (26 feet 3 inches x 26 feet 3 inches)
Starting position of competitors:
   3 metres (9 feet 9 inches) apart, on one axis of the mat
Timing: usually 2 minutes, although it may be extended to 3 or 5
Scoring:
   ippon (a blow or kick struck with vigour and alacrity, the right timing, and at the correct distance): 1 point
   waza-ari (partly successful blow or kick): ½ point (at the discretion of the judges and the referee)
Methods of winning:
   an ippon wins outright
   two waza-ari win outright
   all other decisions at the discretion of the judges and referee (in the light of any waza-ari scored, and any fouls awarded against contestants, including actual penalties)

The word *karate* is the Japanese for 'empty hand', 'naked fist', referring specifically to the lack of weapons or gloves.

kassaba(h) [linear measure] In Egypt, formerly a measure of a useful length but with no ordinary Western equivalent.

1 kassabah	=	3.8824 yards, 11.6472 feet (11 ft 7.77 in.)
		(1 yard = 0.2576 kassabah)
	=	3.55 metres
		(1 metre = 0.2817 kassabah)

kati *see* CATTY

kayak *see* CANOEING DISCIPLINES AND MEASUREMENTS

kayser [physics] In spectroscopy, the reciprocal of a spectral wavelength in centimetres – that is, the WAVE NUMBER in cm^{-1}.

1 kayser represents an energy of 123.977 x 10^{-6} ELECTRON VOLTS.

It was named after the German spectroscopist J. Kayser (1853-1940).

K-band [physics: telecommunications] A wide range of microwave frequencies between 12 and 40 gigahertz, used mainly in radar.

keddah [volumetric measure] In Egypt, a volumetric measure of liquids.

1 keddah	=	2.0625 litres
		(1 litre = 0.4848 keddah)
	=	0.4538 UK gallon, 3.630 UK pints
	=	0.5449 US gallon, 4.359 US pints
		(1 UK gallon = 2.2036 keddah
		1 US gallon = 1.8352 keddah)

keel [weight] An old unit of weight for the coal transported by barge or lighter, probably corresponding to one boat-load on the type of Dutch barge called a *kielboot* or 'keel-boat'.

1 keel	=	21.20 imperial (long) tons, 47,488 pounds
	=	21.54 (metric) tonnes, 21,540 kilograms

keg [volumetric measure] A small BARREL or cask, generally for containing beer, ale, or similar alcoholic beverage, of no specific capacity but almost always holding less than 10 gallons.

As a container for herring, however, it amounts to a unit of 60 fish.

keg [weight] In weighing out nails, a unit of 100 pounds.

1 keg	=	100 pounds, 1 short hundredweight or cental
		(1 hundredweight, 112 pounds = 1.12 keg)
	=	45.359 kilograms
		(50 kilograms, 1 centner = 1.1023 keg)

Kelvin ampere balance [physics] Instrument for measuring an electric current by balancing the electromagnetic force it produces between two coils against the gravitational force (weight) of a mass that can be moved along a beam. It was named after the British physicist Lord Kelvin (*see below*).

Kelvin scale of temperature [physics] Temperature scale devised by William Thomson, later Baron Kelvin of Largs (1824-1907), in 1848 and intended to bear no numerical relation to any specific substance (unlike other temperature scales which were based on the boiling point or freezing point of water). Using degrees identical in size to those of the CELSIUS SCALE, he made ABSOLUTE ZERO equal to 0 kelvin. This means that the freezing point of water on the Kelvin scale is 273.16 K, and the boiling point 373.16 K.

Note that, technically, the word 'degrees' (or the symbol for it) does not appear between the figures and K when quoting a temperature on the scale.

ken [linear measure] A linear measure in Japan and Thailand, but not the same in both countries.

In Japan, a linear measure approximating closely to a fathom.

1 Japanese ken	=	6 SHAKU, 600 BU
	=	5.965 feet (5 ft 11.58 in.), 1.988 yard
		(6 feet, 2 yards = 1.0059 Japanese ken)
	=	1.818 metre
		(1 metre = 0.55 Japanese ken
		2 metres = 1.10 Japanese ken)
60 Japanese ken	=	1 CHÔ (109.08 metres, 119.28 yards)

In Thailand, a much shorter measure approximating to the metre.

1 Thai ken	=	4 KUP, 48 NIN
	=	40 inches (3 ft 4 in.), 1.1111 yard
		(1 yard = 0.90009 Thai ken)
	=	1.016 metre
		(1 metre = 0.98425 Thai ken)
40 Thai ken	=	1 SEN (40.6 metres, 44.44 yards)

Kepler's laws [astronomy] Three important relationships to do with planetary motion:

The planets orbit in ellipses round the Sun, with the Sun at one of the focuses of the ellipse.

A line from the Sun to a planet (the radius vector) traces equal areas in equal intervals of time.

Planetary orbital times squared are proportional to average planet-Sun distances cubed.

The laws were named after the German astronomer Johannes Kepler (1571-1630).

kerat [linear measure] Former unit of length in Turkey.

1 kerat	=	28.575 millimetres, 2.8575 centimetres
	=	1.125 inch
16 kerat	=	1 short pik
25 kerat	=	1 long pik

The term evidently derives from the Arabic unit derived from the measurement of a 'bean' and also applied to a very small unit of weight which in English became the CARAT. *See also* PIK.

kerma [physics] In radiology, the total kinetic energy transferred to charged particles by ionizing radiation from a given mass of material divided by the mass of the material. The equivalent SI unit is the GRAY.

The kerma gets its name as an acronym of $\underline{k}$inetic $\underline{e}$nergy $\underline{r}$eleased (per unit) $\underline{ma}$ss.

KeV [physics] One thousand (K for KILO-) ELECTRON VOLTS.

key, keynote [music] The keynote is the basic note on which a musical SCALE is founded: the tonic. In the key of C, for example, the keynote (tonic) is C and the major scale represents all the white notes on a piano between one C and another. In the key of D, the keynote (tonic) is D and the major scale represents all the notes in the same sequence of tones and semitones as the key of C but transferred upward by one tone (and therefore involving two black notes, F# and C#). In the key of F, the keynote (tonic) is F and the major scale represents all the notes in the same sequence of tones and semitones as the key of C but transferred upward by two and a half tones (and therefore involving one black note, B♭).

major scale of C:

C		D		E		F		G		A		B		C
	tone		*tone*		*semitone*		*tone*		*tone*		*tone*		*semitone*	

major scale of D:

D		E		F#		G		A		B		C#		D
	tone		*tone*		*semitone*		*tone*		*tone*		*tone*		*semitone*	

major scale of F:

F		G		A		B♭		C		D		E		F
	tone		*tone*		*semitone*		*tone*		*tone*		*tone*		*semitone*	

Whereas the key of C thus has no sharps and no flats (other than introduced especially for effect), the key of D is therefore said to have two sharps or to *be* two sharps, and the key of F is said to have one flat or to *be* one flat.

Keys in sharps and flats (with the relevant accidentals):

	key	*accidentals in key*
no sharps or flats	C	
1 sharp	G	F#
2 sharps	D	F# + C#
3 sharps	A	F# + C# + G#
4 sharps	E	F# + C# + G# + D#
5 sharps	B	F# + C# + G# + D# + A#
6 sharps	F#	F# + C# + G# + D# + A# + F(E#)
1 flat	F	B♭
2 flats	B♭	B♭ + E♭
3 flats	E♭	B♭ + E♭ + A♭
4 flats	A♭	B♭ + E♭ + A♭ + D♭
5 flats	D♭	B♭ + E♭ + A♭ + D♭ + G♭
6 flats	G♭	B♭ + E♭ + A♭ + D♭ + G♭ + B(C♭)

The best-known mnemonic by which to remember the order of the sharp keys is: 'Go Down And Enter By Force'; to remember the flat keys in order, just reverse the mnemonic: 'Force By Enter And Down Go' (but remember, too, that the flat keys are all in flats after the first).

From the above table it is evident that the progression of sharp keys requires one more sharp to be added each time, and each time it is a leading note that is added (the note immediately beneath the tonic or keynote). Similarly, the progression of flat keys requires one more flat to be added each time and, in this case, it is the (flattened) subdominant (a perfect fourth above the tonic or keynote). It is evident in addition that, on the piano, the key of 6 sharps is identical with the key of 6 flats. (And, in fact, 7 sharps is identical with 5 flats, 8 sharps with 4 flats, and so on.)

A minor key differs from a major key in that the scale (most often) has a flattened mediant and submediant. In the so-called harmonic version of C minor, for example, instead of corresponding to all the white notes on a piano between one C and another (as in the key of C major), there is E♭ instead of E and A♭ instead of A. The result is a scale that somehow sounds 'sad' and lugubrious in contrast to the major. The harmonic version of D minor corresponds to the same sequence of tones and semitones as C minor but transferred upward by one tone.

The minor key that is most associated with the major key of C is not C minor but A minor. This is because the natural scale of A minor is also based on what on the piano are the white notes, and because, in the harmonization of even simple melodies in C major, chords of A minor may well be appropriate. A minor is, therefore, called the relative minor of C major.

harmonic minor scale of A:

A	B		C	D	E		F		G#		A
tone	*semitone*		*tone*	*tone*	*semitone*		*tone and a half*		*semitone*		

melodic minor scale of A:

A	B		C	D	E		F#		G#		A
tone	*semitone*		*tone*	*tone*	*tone*		*tone*		*semitone*		

natural minor scale of A:

A	B		C	D	E		F		G		A
tone	*semitone*		*tone*	*tone*	*semitone*		*tone*		*tone*		

The natural scale of a minor key is not used very much, except in melodies or chord sequences descending in pitch following the use of the melodic minor scale in the melody or previous chord sequence ascending in pitch. The melodic minor is probably the most-used form of the minor key, although the harmonic minor is popularly used to suggest Arabic or Oriental styles of melody.

See also RELATIVE MINOR KEYS.

keyboard instruments' range [music] Keyboard instruments have, in the late twentieth century, begun to dominate popular music thanks to electronic effects that can duplicate the sounds of virtually any other instrument (and virtually any kind of noise). Through the ages, however, keyboards have generally been constructed in all shapes and sizes, even within the definition of specific instruments. It is only possible therefore to give an outline of the general range of any instrument.

clavichord:	3 to 4 octaves	middle register only
spinet:	4 octaves	middle register only
virginal:	3½ octaves	middle register only (rarely, two manuals)
harpsichord:	5 octaves	full range (one or more manuals)
harmonium:	5 octaves	full range
church organ:	8 octaves	very full range (many tones)
concert organ:	8 octaves	very full range (many tones)
cinema organ:	7 octaves	full range (tones more important)
upright piano:	7½ octaves	very full range
grand piano:	7½ octaves	very full range (tone pedals)
celeste/		
glockenspiel:	3 to 4 octaves	upper register only

for piano accordion/melodeon *see* HARMONICAS' AND ACCORDIONS' RANGE

for xylophone/marimba/vibraphone *see* PERCUSSION INSTRUMENTS' RANGE

key signature [music] Sign at the beginning of a piece of written music that denotes the home key of the piece. It is located after the clef (which indicates the pitch-range of notes on the stave or staff) and before the time signature (which indicates the number and duration of beats per bar).

The key signature is made up of either one or more sharps or one or more flats.

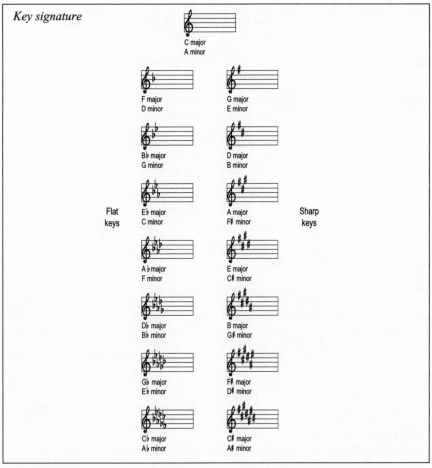

Key signature

C major
A minor

F major   G major
D minor   E minor

B♭ major   D major
G minor   B minor

Flat   E♭ major   A major   Sharp
keys   C minor   F♯ minor   keys

A♭ major   E major
F minor   C♯ minor

D♭ major   B major
B♭ minor   G♯ minor

G♭ major   F♯ major
E♭ minor   D♯ minor

C♭ major   C♯ major
A♭ minor   A♯ minor

Where there are no sharps or flats, the key is C. For progressive lists of keys in sharps and flats *see* KEY, KEYNOTE, where it is additionally noted that a major scale corresponds also in terms of its constituent accidentals to its relative minor. A key signature that suggests the home key is C major, therefore, may in fact refer instead to a home key of A minor.

kg *see* KILOGRAM/KILOGRAMME

kgm *see* KILOGRAM, METRE

kheme, cheme [volumetric measure] French-derived name for an ancient Greek measure of dry and liquid capacity that was a minor submultiple of the kyathys: *see* KYATHYS, CYATHYS.

khoinix, choenix [volumetric measure] Unit of dry capacity in ancient Greece, for most of the Classical period closely approximating to the modern (dry) quart and (dry) litre.

1 khoinix	=	96 mystra, 24 kyathoi, 16 oxybatha, 4 COTYLAI, 2 SEXTE (or xestes)
	=	1.8986 UK dry pint, 1.9576 US dry pint
	=	0.9493 UK dry quart, 0.9788 US dry quart
	=	1,078.78 cubic centimetres
	=	65.834 cubic inches
3 khoinikes	=	1 KHOUS
36 khoinikes, 12 khoes	=	1 (Greek) AMPHORA
48 khoinikes, 16 khoes	=	1 MEDIMNOS

One khoinix of grain was the standard daily allowance for a slave in ancient Greece. The term probably arose as the name of a rectangular wooden container on which the lid was fixed at both ends (for the word also means a plank fixed diametrically across a wheel through which the axle could be inserted – thus fixed to the wheel's circumference at both ends – and a short shackle or fetter to which both feet were attached, one at each end or side).

khoum [comparative values] Unit of currency in Mauretania.

$$5 \text{ khoum} = 1 \text{ ouguiya}$$

See also COINS AND CURRENCIES OF THE WORLD

khous, chous [volumetric measure] Unit of dry and liquid capacity in ancient Greece corresponding to the CONGIUS of ancient Rome.

1 khous	=	72 kyathoi, 48 OXYBATHA, 12 COTYLAI, 6 SEXTE
	=	3 KHOINIKES (dry measure only)
	=	3.2363 litres
	=	5.6958 UK pints, 6.8394 US pints
	=	5.6958 UK dry pints, 5.8728 US dry pints
	=	3,236.3 cubic centimetres
	=	197.502 cubic inches
12 khoes	=	1 (Greek) AMPHORA
16 khoes	=	1 MEDIMNOS (dry measure only)

The name of the unit is the ordinary Greek word for 'pitcher' or 'jug', and is related to the verb *kheō* 'I pour'.

kHz see KILOHERTZ

Kienböck unit, K unit [radiology] Unit of X-ray dosage, equal to one-tenth of the dose that produces slight reddening of the skin (the erythema dose). It was named after the Austrian radiologist Robert Kienböck (1871-1953).

kilderkin [volumetric measure] In the United States, a unit of liquid or dry capacity equal to half a barrel, named after the Dutch-style cask of that size and that name, representing in the Netherlands one-quarter of a tun. (The name itself derives from the medieval Dutch equivalent of 'quintal-kin': see QUINTAL.)

The actual capacity of the kilderkin thus varies with the contents, in the same way as the capacity of a barrel varies (see BARREL).

The term has no such connotations in British usage, being merely a word for a container that can hold between 16 and 18 UK gallons (19.22 and 21.62 US gallons, 72.74 and 81.83 litres).

kilerg, kiloerg [physics] Unit of energy in the centimetre-gram-second (CGS) system.

$$1 \text{ kiloerg} = 1,000 \text{ ERGS}$$
$$= 10^{-4} \text{ joule}$$

kilo see KILOGRAM, KILOGRAMME

kilo- [quantitatives: prefix] Prefix which, when it precedes a unit, multiplies the unit by 1,000 times its standard size or quantity. For examples see below.

The term derives from French *kilo-*, an adaptation of the ancient Greek *chilioi* 'thousand', which is thought to represent a rather odd dialectal variant in any case of a word that the Romans independently derived in the form *mille* 'thousand'.

kiloampere [physics] Unit of electric current.

$$1 \text{ kiloampere} = 1,000 \text{ AMPERES}$$

kilobar [physics: meteorology] Unit of pressure.

1 kilobar	=	1,000 BARS
	=	10^8 newtons per square metre (or pascals)
	=	approximately 1,000 atmospheres

kilobyte [electronics] Unit of information in computer technology.

$$1 \text{ kilobyte} = 1,000 \text{ BYTES}, 8,000 \text{ BITS}$$

kilocalorie, kiloCalorie see CALORIE

kilocycle [physics] Unit for measuring a repeating sequence of occurrences.

$$1 \text{ kilocycle} = 1,000 \text{ CYCLES}$$

It was once used for measuring frequencies in kilocycles per second (in common usage abbreviated to just 'kilocycles'), but has been largely replaced by the exactly equivalent KILOHERTZ.

kilodyne [physics] Unit of force in the centimetre-gram-second (CGS) system.

$$1 \text{ kilodyne} = 1,000 \text{ DYNES}$$

$=\ 10^{-2}$ newton

$=$ very approximately 1 gram weight

kiloerg *see* KILERG, KILOERG

kilogauss [physics] Unit of magnetic induction.

1 kilogauss $=$ 1,000 GAUSS

kilogram, kilogramme [weight] Unit of weight in the SI SYSTEM, and a measure in common usage over much of the world. The spelling 'kilogramme' represents the obsolescent English variant that is gradually vanishing as British usage conforms to the American spelling 'kilogram'.

1 kilogram $=$ 1,000 GRAMS

$=$ 2.204623 POUNDS (2 lb 3.274 oz avdp.)

(1 pound = 0.45359 kilogram, 453.59237 grams

14 pounds, 1 STONE = 6.350 kilograms)

50 kilograms $=$ 1 CENTNER (110.23115 pounds)

[100 lb, 1 short hundredweight = 45.359 kg

112 lb, 1 HUNDREDWEIGHT = 50.80208 kg

(in some countries a hundredweight is also

known as a centner)]

1,000 kilograms $=$ 1 (metric) TONNE

$=$ 2,204.623 pounds, 0.9842116 TON avdp.

(1 ton = 1,016.0416 kg, 1.0160416 tonne)

In scientific usage, the kilogram is the unit of mass and a fundamental unit in the SI system and the obsolescent metre-kilogram-second (MKS) system, in which it also doubles as a unit of force (*see below*). It corresponds to the mass of a platinum-iridium cylinder kept at the International Bureau of Weights and Measures at Sèvres in France – the international prototype kilogram – and is the only fundamental unit still defined in terms of a material 'object'.

kilogram force [physics] Unit of force in the metre-kilogram-second (MKS) system.

1 kilogram force $=$ 1,000 GRAMS FORCE

$=$ 980,665 dynes (CGS units)

$=$ 9.807 newtons (SI units)

(10 newtons = 1.0197 kilogram force)

$=$ 2.205 pounds force

(2 pounds force = 0.907 kilogram force)

$=$ 0.0685 poundals

(1 poundal = 14.598 kilograms force)

For approximate calculations, 1 kilogram force can be equated to 10 newtons.

kilogram-metre [physics] Unit of energy or work in the metre-kilogram-second (MKS) system.

1 kilogram-metre $=$ 1,000 GRAM-METRES

$=$ 1,000 ergs (CGS units)

$=$ 10^{-4} joule (SI units)

kilogram(me)s per square measure [physics] Units of force and pressure.

1 kg/cm^2 $=$ 14.2232 lb/sq in., 0.09877 lb/sq ft

(1 lb/sq in. = 0.07031 kg/cm^2, 70.31 g/cm^2

1 lb/sq ft = 0.00049 kg/cm^2, 0.49 g/cm^2)

1 kg/m^2 $=$ 0.2048 lb/sq ft, 0.02276 lb/sq yd

(1 lb/sq yd = 0.5425 kg/m^2, 542.50 g/m^2)

kilohertz [physics] Unit of frequency in the SI system.

1 kilohertz $=$ 1,000 HERTZ

$=$ 1,000 cycles per second

$=$ 1 kilocycle per second

kilohm [physics] Unit of electrical resistance.

1 kilohm $=$ 1,000 OHMS

kilojoule [physics] Unit of energy or work in the SI system.

1 kilojoule $=$ 1,000 JOULES

$=$ 10^{10} ergs (CGS units)

$=$ 2.778 x 10^8 kilowatt-hours

$=$ 0.9478 calorie

(1 calorie = 1.0551 kilojoule)

kilolitre, kiloliter [volumetric measure] Unit of volume in dry and liquid measure.

1 kilolitre	=	1,000 LITRES
	=	219.97 UK gallons, 264.17 US (liquid) gallons
	=	1 CUBIC METRE (1 m³), 1,000,000 cubic centimetres
	=	1.30795 cubic yards, 35.3147 cubic feet
		(1 cubic yard = 0.7645 kilolitres, 764.555 litres)
	=	27.495 UK bushels, 28.378 US bushels

kilometre, kilometer [linear measure] Standard unit of distance in the METRIC SYSTEM, and a measure in common usage over much of the world. The spelling 'kilometer' is the North American variant, whereas 'kilometre' derives directly from its usage in France and is the preferred spelling in Britain.

1 kilometre	=	1,000 METRES = 100,000 centimetres
	=	1,093.613 yards, 3,280.84 feet (3,280 ft 10.08 in.)
	=	0.62137 MILE
		(1 mile = 1.609344 kilometre)
1.5 kilometres, 1,500 metres	=	0.932 mile (1,640.32 yards)
3 kilometres, 3,000 metres	=	1.864 miles (1 mile 1,520.64 yards)
5 kilometres, 5,000 metres	=	3.107 miles (3 miles 188.32 yards)
10 kilometres, 10,000 metres	=	6.214 miles (6 miles 376.11 yards)

kilometre, cubic [cubic measure] Large and rather unwieldy cubic unit occasionally used in the measurement of water in oceans or lakes, or for astronomical calculations.

1 cubic kilometre (1 km³)	=	1,000,000,000 cubic metres or KILOLITRES
	=	1,307,951,000 cubic yards
	=	0.2399 CUBIC MILE
		(1 cubic mile = 4.16784 km³)

kilometre, square [square measure] Standard unit of territorial area in the METRIC SYSTEM, and a measure in common usage over much of the world.

1 square kilometre (1 km²)	=	1,000,000 square metres
	=	1,195,990 square yards
	=	100 HECTARES, 10,000 ares
		(1 hectare = 0.010 km²)
	=	247.1 ACRES, 61.775 roods
	=	0.3861022 SQUARE MILE
		(1 square mile = 2.589988 km²)

kilometres per hour [speed] Units of speed as regulated by motor traffic authorities over much of the world. The standard form of abbreviation is 'km/h'. The other major world unit of motor traffic speed is miles per hour ('mph').

km/h	mph	mph	km/h
1	0.6	1	1.6
5	3.1	5	8.0
10	6.2	10	16.1
15	9.3	15	24.1
20	12.4	20	32.2
25	15.5	25	40.2
30	18.6	30	48.3
35	21.7	35	56.3
40	24.9	40	64.4
50	31.1	50	80.4
60	37.3	60	96.5
70	43.5	70	112.6
80	49.7	80	128.7
90	55.9	90	144.8
100	62.1	100	160.9
110	68.4	110	177.0

120	74.6	120	193.1
130	80.8	130	209.2
140	87.0	140	225.3
150	93.2	150	241.4

1 km/h	=	16.666 metres per minute	= 27.77 centimetres per second
	=	54.659 feet per minute	= 10.936 inches per second
1,000 km/h	=	16.666 kilometres per minute	= 277.77 metres per second
	=	10.356 miles per minute	= 911.34 feet per second

See also KILOMETRES PER SECOND.

kilometres per litre [comparative values] Measure by a vehicle driver of how efficient the vehicle's engine is by reference to a relationship between the distance travelled and the volume of fuel consumed in travelling that distance.

This method of calculation uses the amount of fuel consumed as the constant: the distance travelled is the variable for comparison with other vehicles or other journeys. This is the system also preferred in those countries that purvey fuel in gallons and travel distances in miles: the measure in this case is in miles per gallon (mpg).

kilometres per litre	miles per gallon	
	UK	US
1	2.824	2.352
3	8.472	7.056

kilometres per litre	miles per gallon	
	UK	US
5	14.120	11.760
6	16.944	14.112
7	19.768	16.464
8	22.592	18.816
9	25.416	21.168
10	28.240	23.521
11	31.064	25.872
12	33.888	28.224
13	36.712	30.576
14	39.536	32.928
15	42.360	35.280
16	45.184	37.632
17	48.008	39.984
18	50.832	42.336
20	56.480	47.040
23	64.952	54.096
26	73.424	61.152
30	84.720	70.560
35	98.840	82.320
40	112.960	94.080

In many European countries in which the litre is the standard unit measure of liquid fuel, however, the method of calculation is instead in the form of LITRES PER HUNDRED KILOMETRES, a method that turns the kilometres per litre method on its head, for the distance travelled is then the constant, and the quantity of fuel to travel that distance is the variable used for comparison. See also MILES PER GALLON (MPG).

kilometres per second [speed] Unit of velocity used primarily in astronomical calculations.

1 kilometre per second		
(1 km/s)	=	1,000 metres per second
	=	60 kilometres per minute
	=	3,600 kilometres per hour
	=	3,280.84 feet per second
	=	37.282 miles per minute
	=	2,236.936 miles per hour
1 mile per second	=	1.609344 kilometre per second
	=	96.561 kilometres per minute
	=	5,793.638 kilometres per hour

kilometric waves [physics] Long radio waves that are in the wavelength range 1,000 to 10,000 metres (frequency range 30-300 kilohertz).

kilonewton [physics] Unit of force in the SI system.

$$1 \text{ kilonewton} = 1,000 \text{ NEWTONS}$$
$$= 10^8 \text{ dynes (CGS units)}$$
$$= 224.8 \text{ pounds force}$$
$$= 6.981 \text{ poundals}$$

kiloparsec [astronomy] Unit of astronomical distance.

$$1 \text{ kiloparsec} = 1,000 \text{ PARSECS}$$
$$= 3,262 \text{ LIGHT YEARS}$$

kilorad [physics] Unit of radiation dose, equal to 10 joules per kilogram of absorbing medium.

$$1 \text{ kilorad} = 1,000 \text{ RAD}$$
$$= 10 \text{ grays (SI units)}$$

kiloton [weight; physics] A unit of weight in the imperial (English or UK) measurement system.

$$1 \text{ kiloton} = 1,000 \text{ (imperial or long) TONS}$$
$$= 1,016.0416 \text{ (metric) TONNES}$$
$$(1,000 \text{ tonnes} = 984.2116 \text{ tons, } 0.9842 \text{ kiloton})$$

By extension, during the 1960s and 1970s, the unit came also to represent the energy released by 1,000 (imperial or long) tons of the high explosive TNT, as a unit of potential destruction by a nuclear weapon.

kilotonne [weight] A unit of weight in the metric system.

$$1 \text{ kilotonne} = 1,000 \text{ metric tonnes}$$
$$= 984.2116 \text{ (imperial or long) tons}$$
$$(1,000 \text{ tons} = 1,016.04 \text{ tonnes, } 1.016 \text{ kilotonne})$$

kilovolt [physics] Unit of electromotive force.

$$1 \text{ kilovolt} = 1,000 \text{ VOLTS}$$

kilovolt-ampere [physics] Unit of electrical power, used to express the rating of an electrical machine.

$$1 \text{ kilovolt-ampere} = 1,000 \text{ VOLT-AMPERES}$$
$$= 1 \text{ KILOWATT}$$

kilowatt [physics] Unit of electrical power in the SI system, equal to 1,000 joules per second.

$$1 \text{ kilowatt} = 1,000 \text{ WATTS}$$
$$= 1.341 \text{ horsepower}$$
$$(1 \text{ horsepower} = 0.7457 \text{ kilowatt, } 745.7 \text{ watts})$$

kilowatt-hour [physics] Unit of electrical energy, commonly called merely 'unit' when it refers to electricity consumption.

$$1 \text{ kilowatt-hour} = 1,000 \text{ WATT-HOURS}$$
$$= 3.6 \times 10^6 \text{ joules}$$

kin [weight] In Japan, a unit of weight corresponding to one used throughout the Far East during the colonial times of the mid-1800s, devised artificially as a measure for taxation purposes. In China it was called the *chin* (or *catty*), in Hong Kong the *kan*, in Malaysia the *gin* (or *kati*), and in Korea the *kon* (or *catty*). Based on the AVOIRDUPOIS system, the unit corresponded to 1.333 POUNDS and was made up of 16 smaller divisions (each therefore of 1.333 OUNCES).

In Japan, however, the unit was assimilated to one that was very close – but not the same.

$$1 \text{ kin} = 16 \text{ TAELS, } 160 \text{ momme, } 1,600 \text{ fun}$$
$$= 1.323 \text{ pound}$$
$$(1 \text{ pound} = 1.7559 \text{ kin})$$
$$= 600 \text{ grams, } 0.600 \text{ kilogram}$$
$$6\frac{1}{4} \text{ kin, } 100 \text{ taels} = 1 \text{ KWAN}$$
$$= 8.267 \text{ pounds, } 3.750 \text{ kilograms}$$
$$100 \text{ kin, } 16 \text{ kwan} = 1 \text{ PICUL}$$
$$= 132.3 \text{ pounds, } 60 \text{ kilograms}$$

kina [comparative values] Unit of currency in Papua New Guinea.

$$1 \text{ kina} = 100 \text{ toae}$$

See also COINS AND CURRENCIES OF THE WORLD.

kindle of kittens [collectives] Unusual collective noun, but one – like 'a litter of pups' – that refers to the animals' being given birth: Middle English *kindlen* 'give birth', related to German *Kind* 'child', 'one born', and Latin *gnatus* 'born'.

kine [physics] Unit of velocity, one proposed for use in the centimetre-gram-second (CGS) system, equal to a velocity of 1 centimetre per second.

The term derives from the ancient Greek verb *kinein* 'to (cause to) move', 'to set in motion'.

kinetic energy [physics] Kinetic energy is the energy of motion. For a given object moving in a straight line, the *translational kinetic energy* is equal to half its mass times the square of its velocity. For an object rotating about an axis, the *rotational kinetic energy* equals half its moment of inertia times the square of its angular velocity. The SI unit of kinetic energy is the JOULE, but ELECTRON VOLTS are used for high-speed particles, whose kinetic energy is given by the relativistic increase in mass times the square of the velocity of light (c). *See also* ENERGY.

kinematic viscosity [physics] For a fluid, the ratio of its coefficient of viscosity to its density, expressed in square metres per second (SI units), or stokes (= square centimetres per second, in CGS units).

king [sporting term] In cards, the court card of the highest value, above the queen. It is, nonetheless, in some games and at some times subordinate in value to the ACE or to other wild cards of the same suit. The king in some countries is known instead as 'the man' or 'the gentleman' (just as the queen is then called 'the lady', and the knave/jack 'the soldier').

In chess, the piece or man the irretrievable capture of which is the entire point of the game: *see* CHESS PIECES AND MOVES.

See also DRAUGHTS (CHECKERS) MOVES AND TERMS.

kingdom [biology] In the taxonomic classification of organisms, the first and widest category of all, corresponding to an organism's definition as a member of one of the five kingdoms. *See* full list of taxonomic categories *under* CLASS, CLASS.

king-size, king-sized [quantitatives] An expression meaning merely 'larger than standard' and initially applied mainly to cigarettes or cigars, although then extended to just about anything else (notably beds and headaches).

kip [engineering] Little-used unit of mass for measuring the load on a structure, equal to 1,000 pounds avdp. (half a short ton). It was named after the first letters of k̲ilo i̲mperial p̲ound.

kip [comparative values] Unit of currency in Laos. The term actually derives from Thai.

> 1 kip = 100 at(t) or cents

See also COINS AND CURRENCIES OF THE WORLD.

Kirchhoff's laws [physics] Two laws:

In a complex electrical circuit: at any junction, the sum of the currents is zero.

The sum of the EMFs (voltages) around any closed path equals the sum of the currents times the impedances. The laws are thus extensions of OHM'S LAW, and were named after the German physicist Gustav Kirchhoff (1824-87).

kite [weight] Basic unit of weight in ancient Egypt. Unfortunately, there is no known standard version of the measure, which seems to have varied from as little as 4.5 grams (just over one-sixth of an ounce, or about 2.6 drams avdp.) to a unit almost seven times that size, 29.9 grams (about 1.055 ounce, or just over 16 drams avdp.). But over the course of 3,500 years or more it is perhaps not so surprising that standards changed, as cultures, overlords, and capital cities did.

> 10 kite = 1 deben
> 100 kite, 10 deben = 1 sep

kJ *see* KILOJOULE

klafter [linear measure] Linear measure in Austria and Switzerland approximating to the fathom and used in marine measurement in addition to the measurement of timber. The unit differs between the two countries in actual length.

> 1 Austrian klafter = 2.074 yards = 6 feet 2.664 inches
> (1 fathom, 6 feet = 0.9643 Austrian klafter)
>   = 1.8965 metres = 189.65 centimetres
> (2 metres = 1.0546 Austrian klafter)

$$1 \text{ Swiss klafter} \quad = \quad 1.9685 \text{ yards} = 5 \text{ feet } 10.866 \text{ inches}$$
$$(1 \text{ fathom, 6 feet} = 1.0160 \text{ Swiss klafter})$$
$$= \quad 1.800 \text{ metre} = 180.00 \text{ centimetres}$$
$$(2 \text{ metres} = 1.1111 \text{ Swiss klafter})$$

K-meson [physics] A MESON with a rest mass nearly 1,000 times that of an electron; a type of boson, also called a *heavy meson*.

km/h *see* KILOMETRES PER HOUR

km/s, km/sec *see* KILOMETRES PER SECOND

knave [sporting term] In cards, the court card of value between the 10 and the queen, otherwise known as the jack, the court page, or the soldier.

The term 'knave' has nothing of the intrinsically villainous in its medieval derivation: Old English *cnafa* meant merely 'youngster', and, by medieval times, that connotation had changed to 'smart young man', in which the smartness applied both to clothes or uniform (court page, soldier, and, later in the 1700s, liveried manservant) and to streetwise guile (jack the lad). It was the Victorians who really put the boot in and made a knave the kind of scurvy villain who went around stealing tarts.

knot [speed] Properly, a knot is 1 nautical mile per hour – a speed and not a distance, and as such is commonly used in relation to the speed of aircraft and of winds. But it is quite frequently accepted as a simple synonym for the nautical mile itself, without reference to speed.

$$1 \text{ knot} \quad = \quad 1 \text{ nautical mile per hour}$$
$$= \quad \text{in Britain, 6,082.66 feet per hour}$$
$$\text{(sometimes rounded down to 6,080 feet: an Admiralty mile)}$$
$$= \quad 1.1520 \text{ (ordinary) miles per hour}$$
$$= \quad 1.854 \text{ kilometres per hour}$$
$$= \quad \text{in North America (since 1954), 6,076.11549 feet per hour}$$
$$= \quad 1.1508 \text{ (ordinary) miles per hour}$$
$$= \quad 1.852 \text{ kilometres per hour}$$

knots	UK mph	US mph	UK km/h	UK metres/sec	US ft/sec
1	1.152	1.151	1.854	0.515	1.688
5	5.760	5.755	9.270	2.575	8.441
10	11.520	11.510	18.540	5.150	16.881
15	17.280	17.265	27.810	7.725	25.322
20	23.040	23.020	37.080	10.300	33.763
25	28.800	28.775	46.350	12.875	42.203
30	34.560	34.530	55.620	15.450	50.644
35	40.320	40.285	64.890	18.025	59.084

The name of the unit derives from the knots physically tied at set intervals in the line attached to the nautical log that in earlier centuries was thrown overboard. The rate of knots disappearing over the side of the ship gave a measure of the ship's speed. *See also* NAUTICAL LOG.

knot [maths] A curve made in space by looping a piece of string, interlacing the ends, and joining them together, topologically equivalent to a circle.

Knudsen number [physics] Number describing the flow of gases at very low pressures, equal to the mean free path of the molecules of the gas divided by a number that is characteristic of the apparatus. If low-pressure gas flows through a hole or tube that is much smaller than the mean free path, resistance to motion is caused mainly by collisions of gas molecules with the walls of the apparatus (and ordinary viscosity constraints do not apply). It was named after the physicist M. Knudsen (1871-1949).

kobo [comparative values] Unit of currency in Nigeria. The term derives from the English description of lesser coinage: 'copper'.

$$100 \text{ kobos} \quad = \quad 1 \text{ naira}$$

See also COINS AND CURRENCIES OF THE WORLD.

köddi [volumetric measure] In Arab countries, a measure of liquid and dry goods.

    1 köddi = 7.58 litres
                  (10 litres = 1.3193 köddi)
             = 1.6676 UK GALLON, 2.0025 US gallons
                  (1 UK gallon = 0.5997 köddi
                  1 US gallon = 0.4994 köddi)
             = 0.8337 UK PECK, 0.8604 US peck
                  (1 UK peck = 1.1995 köddi
                  1 US peck = 1.1623 köddi)

koku [volumetric measure] In Japan, a large measure of liquid and dry goods.

    1 koku = 10 TO, 100 SHÔ, 1,000 GO
            = 180.391 litres
                  (200 litres = 1.1087 koku)
            = 39.686 UK gallons, 47.656 US gallons
                  (40 UK gallons = 1.0079 koku
                  50 US gallons = 1.0492 koku)
            = 4.9599 UK bushels, 5.1192 US bushels
                  (1 UK bushel = 0.2016 koku
                  1 US bushel = 0.1953 koku)

See also BARREL.

kon [weight] In Korea, a unit of weight corresponding to one used throughout the Far East during the colonial times of the mid-1800s, devised artificially as a measure for taxation purposes. Also called the *catty*, the kon was known in China as the *chin* (or *catty*), in Japan as the *kin*, in Malaysia as the *gin* (or *kati*), and in Hong Kong as the *kan*). Based on the AVOIRDUPOIS system, the unit corresponded to 1.333 POUNDS and was made up of 16 smaller divisions (each therefore of 1.333 OUNCES).

    1 kon = 16 TAELS
           = 1.333 pound
                  (1 pound = 0.75 kan: three-quarters of a kan)
           = 604.775 grams, 0.604775 kilogram
    3 kon = 4 pounds (1.8143 kilogram)
    100 kon = 1 PICUL = 133.333 pounds (60.4775 kilograms)

kopeck [comparative values] Unit of currency in Russia and some Russian-speaking or former Soviet Union states.

    100 kopecks = 1 r(o)uble

The term derives from a coin issued between 1535 and 1719 which showed a portrayal of Grand Duke Ivan IV (the Terrible) brandishing a lance (Russian *kop'je* 'spear', *kopeika* 'little spear'). *See also* COINS AND CURRENCIES OF THE WORLD.

kor, cor, homer [volumetric measure] In ancient Israel (and therefore the Bible), a unit of both dry and liquid capacity.

As a liquid measure,

    1 kor = 10 BATH, 30 SEAH, 60 HIN, 80 CAB or kab, 720 LOGS
           = 402.30 litres
                  (400 litres = 0.9943 kor)
           = 88.506 UK gallons, 106.279 US gallons
                  (90 UK gallons = 1.0169 kor
                  110 US gallons = 1.035 kor)

As a dry measure,

    1 kor = 2 lethech, 10 EPHAHS, 30 seah, 100 omers, 180 cab
           = 403,200 cubic centimetres, 0.4032 cubic metre
                  (1 cubic metre = 2.4802 kor)
           = 24,602.4 cubic inches, 14.238 cubic feet
                  [8 cubic feet (2 x 2 x 2 feet) = 0.5619 kor]
           = 11.159 UK bushels, 11.445 US bushels
                  (10 UK bushels = 0.8961 kor,
                  10 US bushels = 0.8737 kor)

In the ancient Greek version of the New Testament, the measure was transliterated in the form *koros*, a spelling that occurs in some Biblical dictionaries and commentaries.

korntonde [volumetric measure] In Norway a former unit of dry goods, especially grain (*korn*).

1 korntonde = 144 (Norwegian) pots, 8 skjeppe
       = 138.97 litres
       = 3.821 UK bushels, 3.944 US bushels
           (1 UK bushel = 0.2617 korntonde
           1 US bushel = 0.2535 korntonde)
       = 4.90625 cu. ft

The Danes also had an equivalent measure for dry goods, especially grain, also called the *tønde*.

1 Danish tønde = 144 (Danish) pots, 8 skaeppe, 4 fjerdings
       = 1.00094 Norwegian korntonde
       = 139.1 litres
       = 3.825 UK bushels, 3.947 US bushels

But as a liquid measure, the Danish tønde was instead the equivalent of 131.4 litres. The Norwegians themselves had a slightly larger unit measure of grain, the *korn-topmaal*.

1 korn-topmaal = 160 litres
       = 4.3992 UK bushels, 4.5406 US bushels
       = 5.64855 cu. ft

In relation to these sorts of quantities, *see also* BARREL.

koruna [comparative values] Unit of currency in the Czech and Slovak Republics. The term derives from the Latin *corona* 'crown', a word found in many guises relating to the currencies of a large number of European states.

1 korona = 100 haleru

See also COINS AND CURRENCIES OF THE WORLD.

kotyle *see* COTYLA, KOTYLE

koyan [weight] In Malaysia, a large unit of weight.

1 koyan = 4,000 GIN (or catty), 40 PICUL
       = 5,333.33 pounds, 2.38095 (imperial or long) tons
           (1 imperial or long ton = 0.420 koyan)
       = 2.41914 (metric) tonnes, 2,419.14 kilograms
           (1 metric tonne = 0.4134 koyan)

K particle *see* K-MESON

kran [comparative values] Unit of currency in Iran.

1 kran (or qran) = 1,000 dinars

See also COINS AND CURRENCIES OF THE WORLD.

krina [volumetric measure] In Bulgaria, a measure of volume in liquids.

1 krina = 2 VEDROS = 20 litres
       = 4.3994 UK gallons, 5.284 US gallons
           (5 UK gallons = 1.1365 krina
           5 US gallons = 0.9463 krina)

krona, króna, krone, kroon [comparative values] Units of currency in predominantly Germanic-speaking Europe present and past.

1 Swedish krona = 100 öre
1 Icelandic króna = 100 aurar
1 Danish krone = 100 øre
1 Norwegian krone = 100 øre
1 Estonian kroon = 100 senti
1 former German krone = 100 heller
1 former Austrian krone = 100 heller

In all cases, the term derives from Latin *corona* 'crown' (akin to ancient Greek *koronis* 'curved', 'hooked'). *See also* COINS AND CURRENCIES OF THE WORLD.

krugerrand [comparative values] A coin containing one full troy ounce of gold, issued from 1967 by the South African government not as legal tender but as a medium for international exchange and collection. Because the gold was in the form of a coin, it overcame the ban in many countries on the possession or collection of gold bullion. Its (high) value depends on the state of the gold market.

The coin's name derives from the surname of a former South African politician and statesman, Paul Kruger (1824-1904), together with the name for the South African coin, the *rand*.

külimet, kulmet [volumetric measure] A former volumetric measure of liquids in Estonia and Latvia.

1 Estonian külimet	=	11.48 litres
		(10 litres = 0.8711 külimet)
	=	2.5256 UK gallons, 3.0328 US gallons
		(1 UK gallon = 0.3959 külimet
		1 US gallon = 0.3297 külimet)
1 Latvian kulmet	=	10.93 litres
		(10 litres = 0.9149 kulmet)
	=	2.4046 UK gallons, 2.8875 US gallons
		(1 UK gallon = 0.4159 kulmet
		1 US gallon = 0.3463 kulmet)

K unit *see* KIENBÖCK UNIT, K UNIT

Kunitz unit [biology] Unit for measuring the activity or concentration of the enzyme ribonuclease (which acts on ribonucleic acid, RNA). RNA normally absorbs ultraviolet light (wavelength 300 nanometres). A Kunitz is the amount of enzyme that decreases the UV absorption of a 0.05 per cent solution of yeast RNA by 100 per cent per minute (at 25°C). It was named after the Russian-born American chemist M. Kunitz (b.1887).

kup [linear measure] In Thailand, part of the linear measuring system.

1 kup	=	12 NIN
	=	10 inches, 0.8333 foot
		(1 foot = 1.20 kup)
	=	25.40 centimetres, 0.254 metre
		(1 metre = 3.937 kup)
4 kup	=	1 (Thai) KEN
160 kup	=	40 (Thai) ken = 1 SEN

kurtosis [maths] In statistics, the extent to which a distribution peaks sharply at its centre, giving a measure of the concentration distribution around the mean value. For a normal distribution it equals 3, known as *mesokurtic* (less than 3 is *leptokurtic*, more than 3 is *platykurtic*).

The term derives from the ancient Greek *kurtos* 'rounded', 'humped' (and the prefixes *meso-*, *lepto-*, and *platy-* correspond to 'middle', 'slender', and 'flat' respectively).

kurus [comparative values] Unit of currency in Turkey, also known as a *piastre*.

100 kurus   =   1 lira (or Turkish pound)

See also COINS AND CURRENCIES OF THE WORLD.

kuwan *see* KWAN

kwacha [comparative values] Unit of currency in Malawi and Zambia. The term means 'dawn'.

1 Malawi kwacha   =   100 tambala
1 Zambian kwacha   =   100 ngwee

See also COINS AND CURRENCIES OF THE WORLD.

kwan [weight] In Japan, part of the weight measuring system that derived its terminology mostly from the colonial units propounded by European trading nations in the mid-1800s, but which the Japanese assimilated in slightly different actual quantities closer to their own standards.

1 kwan	=	6¼ KIN, 100 TAELS, 1,000 momme, 10,000 fun
	=	8.267 pounds (8 lb 4.272 oz avdp.)
		(1 pound = 0.12096 kwan)
	=	3.750 kilograms, 3,750 grams
		(1 kilogram = 0.26667 kwan)
16 kwan, 100 kin	=	1 PICUL
	=	132.3 pounds (9 stone 6.3 pounds)
	=	60 kilograms

The name of the unit is sometimes alternatively spelled *kuwan*.

kwanza [comparative values] Unit of currency in Angola after the colonial Portuguese pulled out in the early 1970s. The term is Swahili for 'to begin'.

$$1 \text{ kwanza} = 100 \text{ LWEIS}$$

See also COINS AND CURRENCIES OF THE WORLD.

kwh *see* KILOWATT-HOUR

kyat [comparative values] Unit of currency in Burma.

$$1 \text{ kyat} = 100 \text{ pyas}$$

See also COINS AND CURRENCIES OF THE WORLD.

kyathys, cyathys [volumetric measure] Unit of liquid and dry capacity in ancient Greece, approximating loosely to one-eleventh of a modern pint.

1 kyathys	=	12 kheme, 6 setier, 4 mystra
	=	44.949 millilitres, 0.044949 litre
	=	0.0791 UK pint, 0.09499 US pint
	=	1.5827 UK fluid ounces, 1.5185 US fluid ounces
	=	0.0791 UK dry pint, 0.0816 US dry pint
	=	44.949 cubic centimetres
	=	2.7431 cubic inches
12 kyathoi	=	1 SEXTE or xestes
24 kyathoi	=	1 KHOINIX (dry measure only)
72 kyathoi, 6 sexte	=	1 KHOUS or chous

None of the three submultiples of the kyathys is now called by its true Greek name; all comprise French adaptations of Latinate variants.

The term *kyathys* is the ordinary Greek word for 'ladle' or 'dipper', especially in relation to the deep ladle used for filling and refilling wine glasses from the large container called the *krater* in which wine and water were mixed to taste (cf. today's punch-bowl sets).

kymography [medicine] Technique of making a single X-ray that records the movements of a body organ (by taking into account movements caused by the pulse, respiration, etc.), using a *kymograph*, which makes a continuous recording of blood pressure changes, respiratory movements, and so on.

The first element of the term derives from ancient Greek *kuma* (*kyma*) 'surge', 'swell'.

L

L (Roman numeral) [quantitatives] As a numeral in ancient Rome, the symbol L corresponded to 50 – yet, in this sense, it did not derive from the twelfth letter of the Latin alphabet, the L that was in turn derived from the same source as the Greek letter *lambda* (which, to the Greeks, as a numerical symbol signified 30 or 30,000). Instead, it comprised the single vertical stroke that was the numeral '1', with a horizontal dash at its foot, thus corresponding in form to roughly half of the original version of the symbol c that stood for '100'.

£ *see* LIBRA

l, l. *see* LIRA, LIRE; LITRE, LITER

laari, laaree [comparative values] Unit of currency in the Maldive Islands.

$$100 \text{ laari} = 1 \text{ rufiyaa (rupee)}$$

See also COINS AND CURRENCIES OF THE WORLD.

Labanotation [literary] System or alphabet of symbols representing dance steps and other rhythmic movements, devised by the Hungarian-German dance teacher Rudolf von Laban and published in 1928. Further developed by the US Dance Notation Bureau, it was later used not only to teach dancing but to relate the movements required by some physiotherapeutic methods and to display stress-relief exercises for industrial workers.

labial consonants [literary: phonetics] Consonants that are produced mostly or entirely through the use of the lips (Latin *labia* 'lips'). In English these correspond to:

b,   f,   m,   p,   v,   w.

labour of moles [collectives] Moles do tend to seem preoccupied with toil, tunnelling patiently away with their outsize forepaws, ever watchful for a succulent worm – but whether that is why a group of moles is a 'labour' is obscure. It would surely be even more appropriate if the term was instead taken directly from Norman (and modern) French *labour* 'tillage', 'ploughing', 'excavation' (as opposed to French *labeur* 'labour'; note also that the word *mole* itself derives as the first element of an originally two-element Anglo-Saxon description meaning 'earth-turner').

lacrosse measurements, units, and positions [sport] Lacrosse is played in two versions, one for men and the other for women. The object in both versions is to score goals, in quest of which team members pass a ball between sticks (or rackets) fitted with nets, or kick the ball.

Men's lacrosse

The dimensions of the pitch: 100 x 70 yards (91.44 x 64 metres)
distance of goal from end line: 15 yards (13.72 metres)

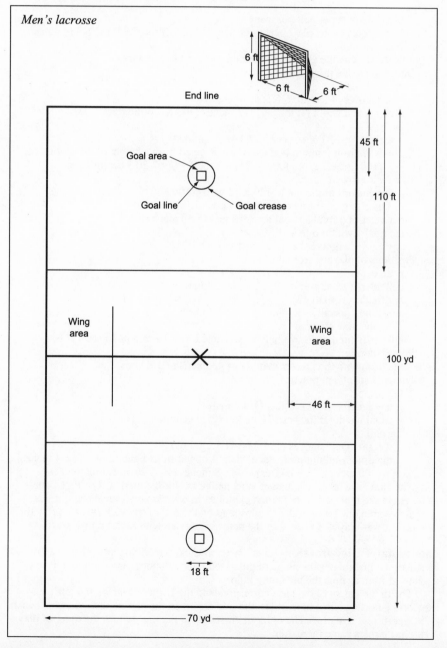

Men's lacrosse

diameter of (circular) goal area: 18 feet (5.48 metres)
the goal: width: 6 feet (1.83 metres)
    height: 6 feet (1.83 metres)
Positions of players: from the back
 goalkeeper
 defenders (usually 3)   total:
 midfielders (usually 3)   10
 attackmen (usually 3)
  up to 9 substitutes
Timing: 60 minutes in four 15-minute quarters; there are 2-minute intervals
between first and second quarter and third and fourth quarter, and a 10-minute half-
time interval
 two time-outs per half per team
 for personal fouls, players may be suspended (off the field) for 1-3 minutes
Points scoring:
 the team with the greater number of goals at the end wins
Dimensions of equipment:
 the stick:
  usual overall length: 6 feet (1.83 metres)
  maximum width (at the net): 12 inches (30.48 centimetres)
 the ball:
  weight: 4½-5 ounces (127.57-141.75 grams)
  maximum-minimum bounce when dropped from a height of 8 feet 4 inches
   (100 inches, 2.54 metres): 7¾ – 8 inches (19.7-20.3 centimetres)
Women's lacrosse
The dimensions of the pitch: 120 x 82 yards (109.7 x 75 metres)
 distance of goal from end line: 10 yards (9.144 metres)
 diameter of (circular) goal area: 18 feet (5.48 metres)
 the goal: width: 6 feet (1.83 metres)
     height: 6 feet (1.83 metres)
Positions of players: from the back
 goalkeeper
 defenders (usually 4)   total:
 midfielders (usually 3)   12
 attackmen (usually 4)
  1 substitute allowed
Timing: 50 minutes in two halves, separated by a 10-minute interval
Points scoring:
 the team with the greater number of goals at the end wins
Dimensions of equipment:
 the stick:
  any length: usual is 6 feet (1.83 metres)
  usual width (at the net): 10 inches (25.4 centimetres)
 the ball:
  weight: 4½-5 ounces (127.57-141.75 grams)
  maximum-minimum bounce when dropped from a height of 8 feet 4 inches
   (100 inches, 2.54 metres): 7¾ – 8 inches (19.7-20.3 centimetres)
The game is a relatively ancient one, native to North America. The first Cauc-
asians to see it played were French explorers in what is now Canada, and it was
they who renamed it after the rough scoop on the end of the stick (*la crosse*) with
which it was played. In this way the name is cognate with such English words as
crutch and *crook*: *see* CRICKET; CROQUET.

laevorotatory, levorotatory [chemistry; physics] Describing an optically active
substance that rotates the plane of polarized light anticlockwise – to the left – as
viewed from against the incoming light.
 The first element of the term corresponds to the Latin *laevus* 'on the left'.

lag line [sporting term] In marbles, the line drawn on the ground towards which each
player throws a marble. The order of closeness to the line defines the order of play
in the marbles game thereafter.

Lagrangian function [maths; physics] A function that describes the kinetic and potential energies of a set of particles in a system. It was named after the French mathematician Count Joseph Lagrange (1736-1813).

Lagrangian point [physics] With two celestial objects orbiting around their common centre of mass, a point on the plane of the orbit at which a third object (of negligible mass) can remain in equilibrium.

Alternatively, one of five points in such a system (such as the Earth and the Moon) where the combined gravitational fields of the two objects are zero.

lakh [quantitatives] In Hindi, a numeral unit of 100,000 (similar in many ways to the ancient Greek *myrias* '10,000', and just as likely to be used to mean 'a very great number').

The term may or may not be connected with the very great number of insects that leave their resinous excrescence on Indian trees, a substance known in English as *lac* and, when processed, *lacquer* or *shellac*.

lambda [volumetric measure] Tiny unit of volume equal to a microlitre (one-millionth of a litre).

lambda particle *see* HYPERON

lambda point [physics] Temperature below which liquid helium changes into a superfluid. It corresponds to 2.2 kelvin.

lambert [physics] Unit of luminance (surface brightness) of a reflector that emits 1 lumen per square centimetre. In most practical applications, the millilambert is used.

$$
\begin{aligned}
1 \text{ lambert} \quad &= \quad 1{,}000 \text{ millilamberts} \\
&= \quad 10^4/\pi \text{ candelas per square metre} \\
&= \quad 929.023 \text{ foot-candles} \\
&\qquad (1{,}000 \text{ foot-candles} = 1.0764 \text{ lambert})
\end{aligned}
$$

The unit was named after the German physicist Johann H. Lambert (1728-77).

Lambert conformal projection *see* CONFORMAL PROJECTION

Lambert's law [physics] The illuminance of a surface is inversely proportional to the square of the distance to the source (for light from a point source striking the surface at right angles). It was named after Johann Lambert (*see* LAMBERT *above*).

lance corporal [military rank] First step up the military rank ladder from private. It is a title that has different connotations in different armies, however. In the United States armed forces, a lance corporal ranks between a private first class and a full corporal. In the British army, a lance corporal remains technically a private (and is paid as such) although taking on the duties and responsibilities of a corporal, possibly only for a temporary period. (In the Royal Artillery, the rank is instead called lance bombardier.)

In a similar way, in the British army, a corporal who (is paid as such but who) takes on the duties and responsibilities of a sergeant, possibly only for a temporary period, is occasionally called a *lance sergeant*. *See also* CORPORAL.

langlauf *see* SKIING DISCIPLINES

langley [physics; astronomy] Unit of solar radiation equal to 1 calorie per square centimetre per minute.

$$1 \text{ langley} \quad = \quad 2.475 \times 10^6 \text{ watts per square metre}$$

The unit was named after the American astrophysicist S. Langley (1834-1906).

Langmuir probe [physics] An instrument used to measure the distribution of potential along an electric discharge in a plasma-filled tube, consisting of a pair of electrodes introduced into the plasma. It was named after the American chemist Irving Langmuir (1881-1957).

lap [sporting term] One full circuit of a circular, oval, or otherwise endless track, in athletics, cross-country, animal or vehicular races.

The term appears to derive from the word's meaning of 'fold' (or 'flap', as in 'overlap'), in the mathematical sense 'twofold', 'threefold', 'fourfold', and so forth.

lapse rate [geology; meteorology] Rate at which atmospheric temperature decreases as altitude increases, equal to about 0.5°C (0.9°F) per 100 metres (110 yards) for air saturated with moisture, and about 1°C (1.8°F) per 100 metres (110 yards) for dry air.

large intestine [medicine] The part of the digestive system that constitutes the alimentary canal between the small intestine and the anus, comprising the caecum (cecum) and appendix, the (ascending, transverse, and descending) colon, and the rectum.

The human large intestine is ordinarily about 1.5 metres (5 feet) long and from 4 to 7 centimetres (1.6 to 2.8 inches) wide. The small intestine is much thinner but much longer.

largo, larghetto [music] Musical instructions: *largo* – (play or sing) slowly, expansively, in a stately and dignified fashion, perhaps even as slowly as an ADAGIO; *larghetto* – (play or sing) rather slowly, but not quite as slowly as *largo*, and certainly more stately than an ANDANTE. Both terms derive directly from the Italian for 'expansive'.

The instruction for a rendition to become slower and louder is *allargando* (literally 'expanding').

larithmics *see* POPULATION STATISTICS, LARITHMICS

last [weight; volumetric measure] Old Germanic unit, primarily of weight but in certain countries relating to the weight of specific goods at a specified volume, and thus later also the volume involved. Not now used, the measure remains of interest in the United States, where it was mainly a unit of weight, and in Britain and the Netherlands, where it was mainly a unit of volume. Even in those countries, however, the measure was liable to varying definitions.

In the United States,

1 last	=	around 4,000 pounds, 40 centals, 2 short TONS
	=	around 1.8144 metric TONNES

except as a weight of wool in sacks, in which case

1 last	=	12 sacks at 364 pounds each	= 4,368 pounds
	=	1.95 (long) ton, 2.184 short tons	
	=	1.9813 metric tonne	

In Britain, in the measurement of grain and malt,

1 last	=	10 UK quarters (cartloads), 80 UK bushels
	=	82.568 US bushels
	=	102.72 cubic feet, 3.8044 cubic yards
	=	2,909.6 dry litres, 2.9096 cubic metres

but in the measurement of cod and herring,

1 last	=	12 UK barrels
	=	38.16 UK bushels, 39.396 US bushels
	=	48.996 cubic feet, 1.815 cubic yards
	=	1,387.2 dry litres, 1.3872 cubic metres

(equal in pilchards to between 10,000 and 13,200 fish).

In the Netherlands, in both dry and liquid capacity,

1 last	=	3,000 litres, 30 hectolitres, 3 KILOLITRES
	=	660 UK gallons, 792.5 US gallons
	=	3 CUBIC METRES
	=	82.485 UK bushels, 85.134 US bushels

Deriving from an old Germanic noun meaning 'load', the term is also the second syllable of the word *ballast* (literally 'bare load'). *See also* RIGA LAST.

last quarter *see* LUNAR PHASES

latent heat [physics] Latent heat is the energy needed to produce a change of state in a substance (that is, to melt a solid into a liquid, or to vaporize a liquid into a gas); temperature remains constant during the change. Latent heat is released when the substance changes back to its former state (that is, when a liquid freezes/solidifies, or when a gas condenses). These various heat changes are measured in CALORIES or JOULES. The specific latent heat of a substance is the difference in ENTHALPY of the two states.

The word *latent* derives from the Latin *latens* 'lying hidden' (which is the opposite of *patens* 'lying open', from which we get *patent*).

lateral velocity [astronomy] For a moving celestial object, the component of its velocity at right-angles to its line-of-sight velocity.

latitude [geography; surveying; astronomy] In geography, an imaginary line encircling the Earth parallel to the equator – in fact, an alternative name for (a line of) latitude is (a) parallel. There are 90° of latitude between the equator and the poles, each degree corresponding to approximately 110 kilometres (69 miles) on the ground. The latitude of any given place is the angle subtended at the centre of the Earth by the line (arc of longitude) joining that place to the equator. Together with LONGITUDE, latitude can be used to specify the position of any place on Earth.

In surveying, a positive latitude indicates the distance north (and a negative latitude a distance south) of a place from the point of origin. Together with *departure* (the distance east or west of the origin) it can be used to locate a point and relate it to true latitude or longitude.

In astronomy, the *celestial latitude* is one of a pair of coordinates (the other is the celestial longitude) that define position. It is the angular distance north or south of the ECLIPTIC.

lattice [crystallography; physics; maths] In crystallography, the regular arrangement in space of the points occupied by atoms in a crystal. This regular spacing makes the atoms act like a grating in the diffraction of X-rays.

In maths, a partly ordered set with paired elements that have a greatest lower bound (infimum) and least upper bound (supremum).

latus rectum [maths] In geometry, a line (chord), perpendicular to the major axis, that passes through the focus of a conic section (ellipse, parabola, or hyperbola). Ellipses and hyperbolas have two latera recta.

The term derives directly from Latin words meaning 'straight side'.

Laurasia *see* SUPERCONTINENT

law of constant proportions [chemistry] In any given (pure) chemical compound, the same elements are always combined in the same proportions (by mass). Alternative names for this law are: the law of constant composition, the law of definite proportions.

law of equivalent proportions [chemistry] If two elements each form a compound with another element, a compound of the first two elements contains them in the relative proportions they have in their compounds with the third element. This law is alternatively known as the law of reciprocal proportions.

law of mass action [chemistry] The speed of a (homogeneous) chemical reaction is proportional to the concentrations (activities, or active masses) of the reacting substances.

law of multiple proportions [chemistry] If two elements combine to form two different compounds, the different masses of the first element that combine with the same (fixed) mass of the second are in a simple ratio.

Lawson criterion [physics] In a thermonuclear (fusion) reaction, the minimum temperature, density, and confinement time for the plasma for fusion to generate power and become self-sustaining. The criterion was first proposed during the 1960s by the British physicist J. D. Lawson.

lb, lbs *see* POUND [weight]

L-band [physics: telecommunications] A band of microwave frequencies between 390 and 1,550 megahertz, used mainly for satellite communications and radar.

LCM, lcm *see* LOWEST COMMON MULTIPLE

LCD, lcd *see* LOWEST COMMON DENOMINATOR

LD 50 [medicine; biology] Abbreviation of lethal dose 50, the amount of a poison that causes death in 50 per cent of a large sample of test organisms.

lea [textiles] Measure of yarn, a unit length of different values depending on the type of yarn involved, and corresponding to a United States version of what in the UK is a SKEIN (although it must be said that a skein is a rolled or folded length of yarn whereas a lea represents just an ordinary linear unit).

In cotton and silk yarn,

$$1 \text{ lea} = 1 \text{ skein} = 120 \text{ yards, } 109.73 \text{ metres}$$

In worsted (wool) yarn,

$$1 \text{ lea} = 1 \text{ skein} = 80 \text{ yards, } 73.15 \text{ metres}$$

In linen yarn,

$$1 \text{ lea} = 300 \text{ yards, } 274.32 \text{ metres}$$

See also YARN

lead [linear measure] Weight on the end of a rope calibrated in units of length (feet), designed to be let over the side of a ship and down into the sea to gauge the depth of water under the keel. The man who took soundings in this way was known as the leadsman, and the actual weight was between 7 and 14 pounds (half a stone and 1 stone; 3.175 and 6.350 kilograms). The calibrations on the rope were known as 'marks', and, in dangerously shallow waters, the leadsman would continually be letting it over the side and drawing it up again, chanting as he did so 'By the mark – ' and the number of feet that the water was deeper than the DRAUGHT/DRAFT of the ship. [This is how American author Samuel Clemens derived his nom-de-plume, hearing the Mississippi boatmen calling 'By the mark twain (= two)'.]

The system has now long been obsolete thanks to the use of echo-sounders.

It is possible that the name of the metal lead derives from this usage – Germanic and Celtic roots suggest 'pendant' meanings (*see also* LOT) – and not vice versa, as might be expected.

leading note, leading tone *see* SCALE

league [linear measure] A measure of distance generally used only in poetic or romantic contexts, and accordingly varying in value in different countries. In English-speaking countries, however, the league is reckoned to equal 3 MILES – but even these miles are not the same in the United Kingdom and the United States, and in the United Kingdom they are not the same on land and at sea.

1 UK league on land	=	3 statute miles
		(half a league = 1½ miles)
	=	4.828 kilometres
	=	0.8691 US league
1 US league on land		
or at sea	=	3 geographical (US nautical) miles
	=	3.452 statute miles
	=	5.555 kilometers
	=	1.1507 UK league on land
	=	0.9989 UK league at sea
1 UK league at sea	=	3 UK nautical miles, 3,041 fathoms
	=	3.4557 statute miles
	=	5.561 kilometres
	=	1.0011 US league

Elsewhere in the world, where the UK terrestrial unit has not been adopted, the unit has been standardized within the metric system of measurement (as exemplified in France).

1 French league (*lieue*)	=	4 kilometres
	=	2.4855 (statute) miles
	=	0.8285 UK league on land
		(1 UK terrestrial league = 1.207 French league)
	=	0.7193 UK league at sea
		(1 UK marine league = 1.3902 French league)
	=	0.7201 US league
		(1 US league = 1.3887 French league)

The term as a unit derives from a borrowing by the Greeks in the first centuries AD of a Gaulish measure, in the original language apparently a 'division' of a yet larger unit.

league, square [square measure] A square measure generally used only in poetic or romantic contexts and, like the LEAGUE on which it is based, varying in value in different countries, and sometimes as a terrestrial or marine measure.

1 UK square league		
on land	=	9 square miles
	=	23.3099 square kilometres
	=	0.7553 US square league
	=	1.4569 metric square leagues
1 US square league on		
land or at sea	=	9 square US nautical miles
	=	11.9163 square (statute) miles
	=	30.8631 square kilometers
	=	1.3240 UK square league on land
	=	0.9986 UK square league at sea

	=	1.9289 metric square league
1 UK square league at sea	=	9 square UK nautical miles
	=	11.9419 square (statute) miles
	=	30.9293 square kilometres
	=	1.0021 US square leagues
	=	1.9331 metric square league
1 metric square league	=	16 square kilometres
	=	6.1776 square (statute) miles
	=	0.6864 UK square league on land
	=	0.5173 UK square league at sea
	=	0.5184 US square league

leang *see* LIANG, LEANG

leap year [time] A year of 366 days, the extra day being 29 February: *see* YEAR.

leash [quantitatives] In hunting or in judging animals at shows, a group of three animals (most often dogs, foxes, hares, rabbits, or game birds).

The term derives from Norman French: whereas a group of two animals is a BRACE, one for each arm (French *bras*), with three there must be (at least) one loose (Norman French *laisse*).

least common denominator *see* LOWEST COMMON DENOMINATOR

least common multiple *see* LOWEST COMMON MULTIPLE

Lecher wires [physics: telecomunications] A pair of insulated parallel wires (carrying microwave signals) with a sliding strip of copper that short-circuits them to form a tuned circuit, used for measuring short radio wavelengths (at high frequencies). As an apparatus, the wires are alternatively known as a parallel wire resonator.

The name of the wires should technically be pronounced with a Germanic (guttural) -ch- , as in *loch* or *ach*, for they take their name from that of the German physicist Ernst Lecher (1856-1926).

ledger line [musical] In musical notation, the staff (or stave) consists of five parallel horizontal lines, on or between which the notes are usually arranged to correspond to the melody. Notes to be played or sung that are above or below those on the staff have to have a brief portion of what would be the next staff line or even the following staff line shown through, above, or below them to represent their exact spatial relationship to the staff. These brief portions of line, parallel with the nearest staff line, are known as ledger lines.

The ledger line should probably be called no more than the *ledge(r)*, apparently deriving from Middle English *legge* 'a rod', 'a horizontal line' (and thus modern 'ledge' meaning 'a horizontal shelf'); the second word was presumably added once the meaning of the first had become obscure. By

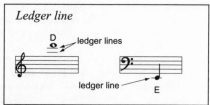

Ledger line

ledger lines
ledger line
D
E

derivation the term is only distantly akin to the type of ledger in which complicated records were kept by scribes.

legal cap [paper size] In the United States, a size of ruled paper 8 x 13½ inches (203 x 343 millimetres) with a fold at the top, so called because it was popular with lawyers. It is about half American foolscap (13 x 16 inches), and has been largely superseded by similar sizes in the A- and B-series of paper sizes.

leg-bye [sporting term] In cricket, a BYE resulting from the ball's bouncing off the batsman's leg while he was making a definite stroke at it (but missing), judged also to have been going to miss the stumps.

legion [quantitatives] A large, indefinite number. The term derives from the division of the Roman army, which usually comprised between 4,200 and 6,000 soldiers, at least the officers of whom were on horseback.

A legion in the Roman army was so called because it comprised men who enlisted themselves on a roll: their names were written down, they were contracted into the army for a set period at a set rate of financial reward and, if they were foreigners, with the promise of eventual citizenship of Rome. The word has not only to do with the Latin verb meaning 'to inscribe', however, but has overtones of an assembly of the chosen – that is, a collection of the (s)elect – overtones that were utterly deliberate on the part of Roman military authorities.

lei *see* LEU, LEI

lek [comparative values] Unit of currency in Albania.

1 lek = 100 qintar

See also COINS AND CURRENCIES OF THE WORLD.

lemniscate [maths] In coordinate geometry, a curve shaped like a figure-of-eight. The lemniscate of Bernoulli is a particular case with the equation $r^2 = a^2\cos2\theta$ (in polar coordinates).

lempira [comparative values] Unit of currency in Honduras.

1 lempira = 100 centavos

The term takes its name from that of a local native leader who gave his name also to a department (administrative area) of the country. *See also* COINS AND CURRENCIES OF THE WORLD.

lens formula [physics] For a thin lens, the reciprocal of the focal length equals the sum of the reciprocals of the object-lens distance and the image-lens distance. The size of the image divided by the size of the object is the magnification (magnifying power of the lens).

lento [music] Musical instruction: (play or sing) slowly – more slowly than any other similar instruction (such as LARGO or ADAGIO). The term derives directly from Italian, and is the equivalent of *lentamente*, also used as a musical instruction.

The instruction to begin slowing from a faster tempo is *lentando*.

Lenz's law [physics] The electric current induced in a wire moving in a magnetic field generates an additional magnetic field that opposes the movement. It was named after the German-born Russian physicist Heinrich Lenz (1804-65).

leone [comparative values] Unit of currency in Sierra Leone.

1 leone = 100 (Sierra Leone) cents

The term derives from the name of the country, Sierra Leone, the 'Lion Range (of mountains)'. *See also* COINS AND CURRENCIES OF THE WORLD.

lepton [comparative values] Unit of currency in Greece.

100 leptons = 1 drachma

The term derives from a word meaning 'thin (thing)', 'small (thing)'. *See also* COINS AND CURRENCIES OF THE WORLD.

lepton [physics] A fundamental (subatomic) particle that does not take part in strong interactions with other particles. Leptons include electrons, negative muons, tau-minus particles, and the respective neutrinos. Their antiparticles – positrons, positive muons, tau-plus particles, and antineutrinos – are known as *antileptons*. *See also* HADRON.

lepton number [physics] In a nuclear process, the number of LEPTONS minus the number of antileptons that take part in it.

leu, lei [comparative values] Unit of currency in Romania: singular leu, plural lei.

1 leu = 100 bani

The term is Romanian for 'lion' (cf. LEONE *above* and LEV *below*). *See also* COINS AND CURRENCIES OF THE WORLD.

lev, leva [comparative values] Unit of currency in Bulgaria: singular lev, plural leva.

1 lev = 100 stotinki

The term is a variant form of the Bulgarian for 'lion' (cf. LEONE and LEU *above*). *See also* COINS AND CURRENCIES OF THE WORLD.

Levant dollar [comparative values] A silver coin first issued in Austria and having on it a portrait of Empress Maria Theresa (1717-80), but now issued by any of a number of countries in the Near East as a medium of international trade.

levorotatory *see* LAEVOROTATORY, LEVOROTATORY

Lewis number [physics] For a diffusing gas or liquid, the diffusivity divided by the diffusion coefficient (diffusivity is the ratio of thermal conductivity to the product of specific heat and density). It was named after the American chemist Gilbert Lewis (1875-1946).

lexeme [literary: phonetics] A form of word corresponding to two or more MOR-PHEMES that together represent the minimum meaningful unit of information (the briefest way of presenting the information in an immediately comprehensible way). Most words correspond to lexemes: few words actually have redundant morphemes.

li [linear measure] In China, a measure of distance that varies in different parts of the country because the unit it is based on (the CH'IH) also varies.

$$1 \text{ li} = 360 \text{ PU, } 1{,}800 \text{ ch'ih}$$
$$250 \text{ li} = 1 \text{ tu}$$

Generally, however, it is reckoned that

$$1 \text{ li} = \text{about one-third of a mile, } 1{,}760 \text{ feet}$$
$$(1 \text{ mile} = \text{about } 3 \text{ li})$$
$$= \text{about } 536.5 \text{ metres}$$
$$(1 \text{ kilometre} = \text{about } 1.864 \text{ li})$$

liang, leang [weight] In China, part of the weight-measuring system that derived from that of the colonial powers of the nineteenth century, and based largely upon the AVOIRDUPOIS WEIGHT SYSTEM.

$$1 \text{ liang} = 1 \text{ TAEL (metal weight)}$$
$$= 1.3333 \text{ ounces, } 37.798 \text{ grams}$$
$$(1 \text{ ounce} = 0.75 \text{ liang}$$
$$50 \text{ grams} = 1.323 \text{ liang})$$
$$3 \text{ liang} = 4 \text{ ounces, one-quarter of a pound (113.4 grams)}$$
$$16 \text{ liang} = 1 \text{ CHIN or catty}$$
$$= 1.3333 \text{ pounds, } 604.775 \text{ grams}$$
$$1{,}600 \text{ liang, } 100 \text{ chin} = 1 \text{ tan}$$
$$= 133.333 \text{ pounds, } 60.4775 \text{ kilograms}$$

Lias [time] Alternative name for the geological period otherwise known as the Lower Jurassic period, during the earlier part of the MESOZOIC (or Secondary) era. The term is apparently the same word as *lees*, the dregs or grounds at the bottom of a bottle, in this case referring to the clayey bluish layers of sedimentary limestone that make up the rock strata characteristic of this period. *See* chart and information *under* TRIASSIC PERIOD.

libra [weight] A fundamental unit of weight first used as a measure in ancient Rome, where *libra* was primarily the balance or scales upon which things were weighed (cf. its diminutive *libellum* from which, through French, the English word *level* derives), and only later a specific weight.

In ancient Rome,

$$1 \text{ libra} = 12 \text{ unciae}$$
$$= 11.552 \text{ ounces avdp., } 0.722 \text{ POUND avdp.}$$
$$= 327.45 \text{ grams, } 0.32745 \text{ kilogram}$$

Linked to the libra but in the Roman currency system, 1 *as* was originally equal to the weight of 1 libra in the Roman currency metal, *aes*. But it was soon devalued, and so much so that the *as* at one stage was worth no more than half of one ounce (one-twenty-fourth) of its former weight in the metal. (For further information, *see* DENARIUS.)

Since then, the measure and its name have been used throughout the world in various contexts and with various values. It has often been incorporated into monetary systems – the English pound sterling is still abbreviated with a florid crossed L (£), short for *libra* – or into weight measuring systems that no longer contain its name, just its abbreviated form – the English pound avoirdupois is abbreviated as 'lb', yet again for *libra*.

As a weight, in present-day countries,

$$1 \text{ Argentinian libra} = 1.0128 \text{ pound avdp., } 16.20 \text{ ounces avdp.}$$
$$(1 \text{ pound avdp.} = 0.9874 \text{ Argentinian libra})$$
$$= 459.2623 \text{ grams}$$
$$(500 \text{ grams} = 1.0887 \text{ Argentinian libra})$$
$$100 \text{ Argentinian libras} = 1 \text{ QUINTAL}$$
$$1 \text{ Portuguese libra}$$
$$(\text{or arratel}) = 16 \text{ onças, } 2 \text{ marcos}$$
$$= 1.01194 \text{ pound avdp., } 16.19 \text{ ounces avdp.}$$
$$(1 \text{ pound avdp.} = 0.9882 \text{ Portuguese libra})$$
$$= 458.9788 \text{ grams}$$
$$(500 \text{ grams} = 1.0894 \text{ Portuguese libra})$$

128 Portuguese libras   =   1 quintal
1 Castilian (Spanish)
     or Mexican libra   =   1.0143 pound avdp., 16.23 ounces avdp.
                            (1 pound avdp. = 0.9859 Castilian libra)
                        =   460.0787 grams
                            (500 grams = 1.0868 Castilian libra)
100 Castilian libras   =   1 quintal

The *libbra* in Italy is, in fact, no more now than another name for the KILOGRAM although, in former centuries, it referred to a weight of 11.958 ounces avdp., 339 grams – not far from the original ancient Roman value. *See also* LIVRE.

libration point *see* LAGRANGIAN POINT

lieutenant [military rank] Commissioned rank in the armed forces. In most countries it is normal for an officer to be a 2nd lieutenant or sublieutenant before being promoted to full lieutenancy, when he or she may be called a lieutenant or a 1st lieutenant.

In the British army, a lieutenant ranks between a 2nd lieutenant and a captain; in the Royal Navy a lieutenant ranks between a sublieutenant and a lieutenant commander, but is equivalent to the army captain; in the Royal Air Force a flight lieutenant ranks between a flying officer and a squadron leader, and is also equivalent to an army captain.

In the United States Army and Airforce, a lieutenant ranks between a 2nd lieutenant and a captain; in the United States Navy, a lieutenant ranks between a lieutenant junior grade and a lieutenant commander, but is equivalent to the army and airforce captain.

The title derives from the French *lieu-tenant* 'holding the position', referring originally to an officer of a junior rank temporarily carrying out the duties of a senior officer.

lieutenant colonel [military rank] A senior commissioned rank in the armed forces.

In the British army, a lieutenant colonel ranks between a major and a full colonel, and is the equivalent of a commander in the Royal Navy and a wing commander in the Royal Air Force.

In the US Army and Airforce, a lieutenant colonel ranks between a major and a full colonel, and is the equivalent of a commander in the US Navy.

Many other armies in the world follow this system of ranking. The only significant alternative system is that in which a lieutenant colonel ranks between a 'commandant' and a colonel.

lieutenant commander [military rank] A senior commissioned rank in the navy.

In the Royal Navy, a lieutenant commander ranks between a lieutenant and a full commander, and is the equivalent of a major in the British army and a squadron leader in the Royal Air Force.

In the US Navy, a lieutenant commander ranks between a lieutenant and a full commander, and is the equivalent of a major in the US Army and Airforce.

lieutenant general [military rank] Very senior commissioned rank in the armed forces.

In the British army, a lieutenant general ranks between a major general and a general, and is the equivalent of a vice-admiral in the Royal Navy and an air marshal in the Royal Air Force.

In the US Army and Airforce, a lieutenant general ranks between a major general and a general, and is the equivalent of a vice-admiral in the US Navy.

lieutenant junior grade [military rank] Rank in the US Navy between ensign and full lieutenant. It is the equivalent of 1st lieutenant in the US Army and Airforce, and corresponds to sublieutenant in the Royal Navy, 'underlieutenant' or junior lieutenant in other navies and armies of the world.

life expectancy, life expectation [medicine] The length of time (in years) that an average person of a particular age is expected to live – his or her projected lifespan. Statistically, the older a person is the longer he or she is likely to live.

Life Master [sporting term] In the card game (contract) bridge, a player who gains 300 points or more in the United States national tournaments.

lift [physics: aerodynamics] For an aircraft in flight, the component of the aerodynamic forces that supports the aircraft, resulting from the flow of air over the lifting surfaces (wings or rotors). It acts upwards perpendicular to the undisturbed airflow. *See also* DRAG.

light [physics] Light is an electromagnetic radiation that we can see. It has various characteristics, such as wavelength (in the range 400 to 800 nanometres), frequency, and velocity (a constant, which it has in common with all other types of electromagnetic radiation). The amount of light emitted, reflected or absorbed by a substance/surface can be measured and is expressed in a whole range of units, including CANDELA, FOOT-CANDLE, FOOT-LAMBERT, LAMBERT, LUMEN, and LUX. *See also* ABSORBANCE; ABSORPTANCE; SPEED OF LIGHT.

light, speed of *see* SPEED OF LIGHT

light efficiency [physics] For an electric lamp, the total luminous flux produced divided by the electric power input, expressed in lumens per watt.

light flux [physics] The quantity of light that passes through a given area, measured in lumens (the light flux per unit area is the LUMINANCE).

light heavyweight *see* BOXING WEIGHTS

light meter *see* EXPOSURE METER

lightweight *see* BOXING WEIGHTS

light year [linear measure] The distance travelled by an object at the speed of light during 1 solar Earth year (365 days).

1 light year	=	63,271.47 ASTRONOMICAL UNITS
	=	5,880,534,262,081.1 miles
	=	9,467,660,160,950.6 kilometres
3.26 light years	=	1 PARSEC

It is important to note that a light year is a measure of distance, not of time.

ligne [linear measure] Unit of length used by watchmakers to measure the thickness of a watch movement. Small linear measure in France.

1 ligne	=	12 douziemes or twelfths
		(1 douzieme/twelfth = 0.188 millimetre)
	=	2.256 millimetres
		(1 centimetre = 4.433 lignes)
	=	0.08882 inch
		(1 inch = 11.259 lignes)

The name of the unit is French for 'line'. *See also* LINE.

ligula [volumetric measure] In ancient Rome, the smallest normal measure of liquid and dry capacity, approximating loosely to one-fiftieth of a modern pint.

1 ligula	=	11.0677 millilitres, 0.0110677 litre
	=	0.0195 UK pint, 0.0234 US pint
	=	0.3897 UK fluid ounce, 0.3739 US fluid ounce
	=	0.0195 UK dry pint, 0.0201 US dry pint
	=	11.0677 cubic centimetres
	=	0.6754 cubic inch
6 ligulae	=	1 ACETABULUM
24 ligulae, 4 acetabula	=	1 HEMINA
48 ligulae, 8 acetabula, 2 heminae	=	1 SEXTARIUS

The name of the unit describes exactly how small it is – just a *lick* [Latin *ligula* is a denasalized diminutive of *lingua* 'tongue', with which both the English words 'lick' and – perhaps surprisingly – *tongue* are cognate (in early Latin, *lingua* has an alternative form *dingua*, denasalized in verbal form as *dicere* 'to say', 'to point out')].

likuta [comparative values] Unit of currency in Zaïre: singular likuta, plural makuta.

100 makuta	=	1 zaire

See also COINS AND CURRENCIES OF THE WORLD.

lilangeni [comparative values] Unit of currency in Swaziland: singular lilangeni, plural emalangeni. Its value is linked to that of the South African RAND. *See also* COINS AND CURRENCIES OF THE WORLD.

limacon [maths: geometry] A heart-shaped curve that is the locus of a point on a line that remains a fixed distance from its intersection with a circle as it rolls round it. It was named by the mathematician Blaise Pascal (1623-62) after the French word for the non-edible snail (ultimately Latin *limax*, Greek *leimax* 'snail', appropriately akin to English *lime* and *slime*). *See also* CARDIOID; EPICYCLE.

limerick *see* VERSE FORMS

limit, limiting value [maths] The value to which a mathematical series or sequence tends as more and more terms are considered.

line [linear measure] Rarely used, small linear measure.

$$1 \text{ line} = 0.08333 \text{ (one-twelfth)} \text{ INCH}$$
$$= 2.11666 \text{ MILLIMETRES}$$
$$= 6 \text{ POINTS}$$
$$= 10 \text{ gries} (\textit{see} \text{ GRY})$$

Remarkably, this measure in fact derives from an old French unit, originally called a *ligne*: *see* LIGNE.

$$1 \text{ ligne} = 2.256 \text{ millimetres} (0.08882 \text{ inch})$$

And it was this measure that was also passed on to the Russians as their *liniya* (*see below*).

line [linear measure: buttons] Measurement of the thickness of buttons and of watch movements, also sometimes called a *ligne*. Slim, ladies' wristwatches are generally 3 to 8 lines thick; gentlemens' wristwatches are 10 to 13 lines; pocket watches are 17 to 19 lines.

$$1 \text{ line} = 0.025 \text{ (one-fortieth)} \text{ INCH}$$
$$= 0.635 \text{ millimetre}$$
$$(1 \text{ millimetre} = 1.575 \text{ line})$$

linear [maths] Involving only one dimension, or describing an (algebraic) equation of the first order (that is, in which the variable is raised only to the power 1 – there are no terms of higher powers).

linear absorption coefficient [physics] A measure of the ability of a medium to absorb radiation passing through it (ignoring any diffusion or scattering): *see* ABSORPTION.

linear attenuation coefficient [physics] A measure of the ability of a medium to diffuse and absorb radiation passing through it.

linear energy transfer [physics] When particles or electromagnetic radiation penetrate an absorbing medium, the linear rate of energy dissipation.

linear equation [maths] A first-order equation (*see* LINEAR). The equation of a straight line in coordinate geometry is a linear equation.

linear molecule [chemistry] A molecule with (three or more) atoms that are arranged in a straight line.

linear momentum [physics] For an object moving in a straight line, the product of its mass and velocity. *See also* ANGULAR MOMENTUM.

line-of-sight velocity [astronomy] The velocity at which a celestial object moves towards (or away from) an observer. To prevent variations arising from the Earth's motion, line-of-sight velocities are usually adjusted and stated as if the observer were at the Sun.

linguistic analysis [literary] The equivalent in stratificational grammar of PARSING in traditional grammar: analysing the words and word elements in a statement to distinguish the basic linguistic units, specifically in terms of phonemes, morphemes, combining forms, lexemes, and similar.

liniya [linear measure] Small unit of linear measure in Russia.

$$1 \text{ liniya} = 0.10 \text{ (one-tenth)} \text{ INCH}$$
$$(1 \text{ inch} = 10 \text{ liniya})$$
$$= 2.54 \text{ MILLIMETRES}$$
$$(1 \text{ centimetre} = 3.937 \text{ liniya})$$
$$17.5 \text{ liniya} = 1 \text{ VERCHOK}$$
$$= 1.75 \text{ inches, } 4.445 \text{ centimetres}$$
$$280 \text{ liniya, } 16 \text{ verchoki} = 1 \text{ ARSHIN}$$

The unit derives ultimately from a borrowing of the French LIGNE that was then assimilated to existing measures.

link [linear measure] A linear unit used mainly by engineers or surveyors. But the surveyors' (or Gunter's) link is not the same as the engineers'.

1 surveyors' link	=	7.92 inches (0.660 foot)
	=	20.117 centimetres (0.20117 metre)
100 surveyors' links	=	1 CHAIN, 22 yards, 66 feet
	=	20.117 metres
1,000 surveyors' links	=	1 FURLONG, 10 chains, 220 yards
8,000 surveyors' links	=	1 (statute) mile

The engineers' link is identical with the statute FOOT:

1 engineers' link	=	12 inches, 1 foot
	=	30.480 centimetres
5,280 engineers' links	=	1 (statute) mile

It is evident that the measure derives as a unit from the actual chains used by surveyors and engineers in physically gauging distances, although it would appear that these chains were made up of unusually extended links and must have been very bulky to carry around.

link, square [square measure] Unit of square measure used by surveyors, and thus based on the linear LINK (also called the Gunter's link) used in surveying, as opposed to the engineers' link.

1 square link	=	7.92 x 7.92 inches, 62.726 square inches
	=	0.4356 square foot, 0.0484 square yard
		(1 square foot = 2.2046 square links)
	=	20.117 x 20.117 centimetres, 404.694 cm^2
		(400 cm^2 = 0.9884 square link)

liquefaction temperature [physics] The temperature at which a gas changes into a liquid (equal to the BOILING POINT of the liquid).

lira, lire [comparative values] Unit of currency in Italy, Turkey, and formerly Israel, the name of which in every case derives from the ancient Roman *libra*.

1 Italian lira (plural: lire)	=	100 centesimi
1 Turkish lira		
(or Turkish pound)	=	100 kurus
1 former Israeli lira		
(now the shekel)	=	100 agorot (singular: agora)

See also COINS AND CURRENCIES OF THE WORLD.

lisente [comparative values] Unit of currency in Lesotho.

| 100 lisente | = | 1 loti |

See also COINS AND CURRENCIES OF THE WORLD.

list system [quantitatives] In political systems that employ proportional representation, the election of candidates listed in order of party prominence according to the proportion of votes cast for their political party.

liter *see* LITRE, LITER

litre, liter [volumetric measure] Standard measure of capacity in the METRIC SYSTEM, corresponding mainly to volumetric liquid units but including sometimes dry measure. It is technically defined as the volume of a kilogram of water at its maximum density (4°C). The spelling 'litre' is the usual in the United Kingdom and some other countries; 'liter' is the standard spelling in the United States.

1 litre	=	1,000 millilitres, 100 centilitres, 10 decilitres
	=	1.76 UK PINTS, 2.1134 US pints
		(2 UK pints, 1 UK QUART = 1.13636 litres
		2 US pints, 1 US quart = 0.94634 liter)
	=	0.2200 UK GALLON, 0.264179 US gallon
		(1 UK gallon = 4.5459631 litres
		1 US gallon = 3.785306 liters)
	=	35.20 UK fluid ounces, 33.76 US fluid ounces
	=	1,000 CUBIC CENTIMETRES (cc)
	=	0.8799 UK dry quart, 0.9073 US dry quart
		(1 UK dry quart = 1,136.5 cc
		1 US dry quart = 1,102.2 cc)
	=	61.02374 cubic inches
		(64 cubic inches = 1,048.77 cc)
10 litres	=	1 decalitre

$$100 \text{ litres} = 1 \text{ hectolitre}$$
$$1,000 \text{ litres} = 1 \text{ KILOLITRE}, 1 \text{ cubic metre}$$

In scientific usage, the term *litre/liter* is obsolescent, and in the process of being replaced by the cubic decimetre (dm^3).

The term derives from the ancient Greek *litra*, an equivalent to the ancient Roman *libra*, also meaning 'a pound weight', to which it may or may not in addition be etymologically akin.

litre-atmosphere [physics] An obsolete unit of work, equal to 101.325 joules.

litres per hundred kilometres [comparative values; engineering] Standard measure in non-English-speaking Europe of the efficiency of a vehicle's engine in relation to its fuel consumption: the larger the engine or the heavier its load, the more LITRES of fuel are required to cover a distance of one hundred KILOMETRES.

In the English-speaking world, however, the equivalent measurement is in the form of MILES PER GALLON (MPG) – which turns the above method of measurement on its head, in that the amount of fuel (not the distance covered) is the standard element: the larger the engine or the heavier the load, the shorter the distance that can be travelled using the standard amount of fuel. So:

litres per 100 km	miles per gallon		litres per 100 km	miles per gallon	
	UK	US		UK	US
3	97.0	80.8	8.5	34.2	28.5
4	72.7	60.5	9	32.3	26.9
5	58.2	48.5	10	29.1	24.2
5.5	52.9	44.1	11	26.4	22.0
6	48.5	40.4	12	24.2	20.2
6.5	44.8	37.3	13	22.4	18.7
7	41.6	34.6	14	20.8	17.3
7.5	38.8	32.3	15	19.4	16.2
8	36.4	30.3	20	14.5	12.1

See also KILOMETRES PER LITRE.

litter of pups, of piglets [collectives] The collective noun in this case reflects on the accouchement of the mother animal in giving birth to the creatures in question. Such a delivery generally takes place on some form of bed or litter (late Latin *lectaria*, Old French *litière*) of straw, rags or other soft material.

This derivation has also resulted in the other meaning of the word *litter* – rubbish, trash.

Little Ice Age [time] A minor reversal in the receding of the last true ice age (which lasted until about 10,000 BC). There was a period of slightly increased glaciation from about 3000 BC that continued right up until the AD 1600s and 1700s, when it appeared to peak. At that time, many rivers as far south as in England and the Netherlands froze solid every winter. Since then, the climate has, for the most part, gradually become warmer. *See also* ICE AGES.

livre [weight] Former measure of weight in France, now long superseded by the metric system, but not very different from the UK/US POUND weight, and translated 'pound' frequently in literature and drama from previous centuries.

$$1 \text{ livre} = 2 \text{ MARCS}, 16 \text{ ONCES}$$
$$= 489.50 \text{ grams}, 0.4895 \text{ kilogram}$$
$$= 1.07916 \text{ pounds } (17.2666 \text{ ounces})$$
$$(1 \text{ pound} = 0.9266 \text{ livre}$$
$$= 453.5924 \text{ grams}, 0.4535924 \text{ kilogram})$$

The term was in France additionally applied variously to a sum of money of specific value, and to a coin, but the worth of both of these varied over time and even according to where in the country it was being used.

The term is also used at the present time in Belgium, but there it is simply an alternative word for KILOGRAM.

By derivation it is yet another variant of the ancient Roman *libra*.

load [physics] In physics, either the (mechanical) force applied (e.g. to a lever), or the weight a structure supports or lifts.

In electrical engineering, either the demand for power made on a source of supply, or the impedance to which a source of supply is connected.

load [cubic measure] As a technical term in the road-building or landscaping industries,

$$1 \text{ load} = 1 \text{ cubic yard, 27 cubic feet (of earth or gravel)}$$
$$= 0.7646 \text{ cubic metre}$$
$$(1 \text{ cubic metre} = 1.308 \text{ loads, } 35.316 \text{ cu. ft})$$

load factor [physics; aeronautics; electrical engineering] In aeronautics, the external load divided by the weight of an aircraft (in terms of its structure). For example: rapid acceleration and manoeuvring may impose considerable loads on an airframe, usually measured in g (where g is the weight of the aircraft).

In electrical engineering, the average load over a given period divided by the peak load: *see* LOAD.

load line [shipping] Painted on the side of a ship at about half-way along its length, the load line marks where the water level is when the ship is fully laden. *See also* PLIMSOLL LINE.

locus [maths] The path traced by a point moving according to certain constraints. For example: the locus of a point that moves so that it is always a fixed distance from another fixed point is a circle.

The term corresponds to the Latin for 'place'. *See* CONIC SECTION.

log [volumetric measure] In ancient Israel, the smallest unit of dry and liquid capacity, approximating quite closely to the PINT.

In liquid measure,

$$1 \text{ log} = 0.56 \text{ LITRE}$$
$$= 0.9856 \text{ UK pint, 1.1834 US pint}$$
$$(1 \text{ UK pint} = 1.0146 \text{ log}$$
$$1 \text{ US pint} = 0.8450 \text{ log})$$
$$= 19.712 \text{ UK fluid ounces, 18.934 US fluid ounces}$$
$$4 \text{ logs} = 1 \text{ CAB or kab}$$
$$12 \text{ logs, 3 cab} = 1 \text{ HIN}$$
$$24 \text{ logs, 6 cab, 2 hin} = 1 \text{ SEAH}$$
$$72 \text{ logs, 18 cab, 6 hin,}$$
$$3 \text{ seah} = 1 \text{ BATH}$$

In dry measure,

$$1 \text{ log} = 0.56 \text{ litre, 560 cubic centimetres (cc)}$$
$$= 0.9856 \text{ UK dry pint, 1.0161 US dry pint}$$
$$(1 \text{ UK dry pint} = 1.0146 \text{ log}$$
$$1 \text{ US dry pint} = 0.9842 \text{ log})$$
$$= 34.17 \text{ cubic inches}$$
$$4 \text{ logs} = 1 \text{ cab or kab}$$
$$7.2 \text{ logs, 1.8 cab} = 1 \text{ OMER or issaron}$$
$$24 \text{ logs, 6 cab,}$$
$$3.3333 \text{ omers} = 1 \text{ seah}$$
$$72 \text{ logs, 18 cab,}$$
$$10 \text{ omers, 3 seah} = 1 \text{ EPHAH}$$

The log was the standard measure for containers of wine (which would normally then be drunk well watered).

logarithmic scale of pressure *see* DECILOG

logarithms [maths] Of a given number, the power to which the (logarithmic) base must be raised to equal the number. For example: the logarithm of 1,000 to the base 10 (written $\log_{10}3$) is 3, because $10^3 = 1,000$. Addition and subtraction of logarithms corresponds to the multiplication or division of the numbers they represent. In a logarithm such as 1.3010 (which is the logarithm of 20 to the base 10), 1 is called the *characteristic* and .3010 is the *mantissa*. Logarithms to the base 10 are called common, or Briggs', logarithms [after their inventor, the British mathematician Henry Briggs (1561-1631)]. Those that use the irrational number e (= 2.718. . .) as their base are called natural, or Napierian, logarithms after their inventor, the Scottish mathematician John Napier (1550-1617). It was, in fact, Napier who coined the word *logarithm* using ancient Greek elements *logos* 'meaningful form', 'ratio' and *arithmos* 'number'.

lone pair (of electrons) [chemistry] A pair of unshared electrons in an atom (in the same orbital but of opposite spin), which are available to provide both the electrons in a chemical bond – called a coordinate, or dative, bond.

long hundredweight *see* HUNDREDWEIGHT; QUINTAL

longitude [geography; astronomy] In geography, an imaginary line (alternatively known as a MERIDIAN) that joins the Earth's north and south poles. The line of longitude that passes through Greenwich, London – the Greenwich meridian – is designated 0 degrees. Other lines of longitude are specified in degrees east or west of the Greenwich meridian to a total of 180° in each direction; there are 360 degrees of longitude around the whole globe. Each degree is approximately 110 kilometres (69 miles) on the ground. Together with LATITUDE, longitude can be used to specify the position of any place on Earth.

In astronomy, the *celestial longitude* is one of a pair of coordinates (the other is the celestial latitude) that define position. It is the intersection on the latitude circle – around the ECLIPTIC – measured in degrees eastwards from the First Point of Aries.

long jump, broad jump *see* ATHLETICS FIELD EVENTS AREA, HEIGHT, AND DISTANCE MEASUREMENTS

long suit [sporting term] In card games, the suit of which one's hand contains the most cards, especially if it is predominant by some margin.

long ton *see* TON

long waves [physics: telecommunications] Radio waves of wavelengths greater than 1,000 metres (corresponding to low frequencies of less than 300 kilohertz).

loran [physics; navigation] Radar device by which two fixed stations retransmit back the radio signals beamed to them by a craft at some distance, the time taken for the craft to receive the signals (and any difference in the order in which they arrive) indicating the craft's exact position. The system works up to a distance of about 1,300 kilometres/800 miles during the hours of daylight, and up to about 2,600 kilometres/1,600 miles at night.

The term derives from the first syllables of <u>lo</u>ng <u>ra</u>nge <u>n</u>avigation. *See also* SHORAN.

Lorentz-Fitzgerald contraction [physics] For an object moving at near the speed of light, its contraction in length (in the direction of motion), relative to the frame of reference from which the observation is made. It was put forward independently by the Dutch physicist Hendrik Lorentz (1853-1928) and the Irish physicist George FitzGerald (1851-1901), and later shown to be a consequence of the theory of relativity. *See also* RELATIVITY.

Lorentz transformations [physics] A set of equations for relating coordinates of space and time in two inertial frames of reference (moving relative to each other with a constant velocity), a consequence of the special theory of RELATIVITY.

Lorentz unit [physics] In spectroscopy, a magnetic field splits spectral lines into two or three components – the Zeeman effect. The Lorentz unit is the difference in frequency (of the split lines) divided by the velocity of light (equivalent to the frequency difference expressed in wave numbers).

Loschmidt number [chemistry] For an ideal gas, the number of molecules in a unit volume (at standard temperature and pressure), equal to 2.687×10^{25} per cubic metre. The term is sometimes incorrectly used as a synonym for AVOGADRO'S NUMBER.

lot [weight] A small former unit of weight in German-speaking countries, approximating to half an OUNCE.

1 German lot	=	about 14 grams
		(25 grams = about 1.786 German lot)
	=	0.49383 ounce
		(½ ounce = 14.175 grams
		= about 1.1025 German lot)
1 Swiss lot	=	15.625 grams
		(25 grams = 1.600 Swiss lot)
	=	0.5512 ounce
		(½ ounce = 14.175 grams
		= 0.9072 Swiss lot)

The term appears to derive from the German for 'lead', the weight used to keep a free-hanging light line taut and vertical.

loti [comparative values] Unit of currency in Lesotho: singular loti, plural maloti. Its value is linked to that of the South African rand.

1 loti = 100 lisente

See also COINS AND CURRENCIES OF THE WORLD. .

louis, louis d'or [comparative values] Former unit of currency – a gold coin (*d'or* 'of gold') – in France, issued at two different times. The first and most famous was issued from 1640 to 1795 and was comparatively valuable. The second lasted for a shorter period and was not worth anything like so much.

The coin was named after King Louis XIII of France who was on the throne in 1640, having occupied it since 1610, but who then had only three more years to live.

Louis Quatorze [time] Style or fashion of 1643-1715 in France, the period during which King Louis XIV (*quatorze* 'fourteen'), the 'Sun King', occupied the French throne.

Louis Quinze [time] Style or fashion of 1715-74 in France, the period during which King Louis XV (*quinze* 'fifteen') occupied the French throne.

Louis Seize [time] Style or fashion of 1774-92 in France, the period during which King Louis XVI (*seize* 'sixteen') occupied the French throne.

love [sporting term] Score of zero in tennis and (contract) bridge. It is not known how this word came to have this specialized meaning, although it has been suggested that medieval games of tennis were often for money prizes staked by the players, and that games in which no prizes were forthcoming (especially in addition if the score was not kept) were 'for love' – that is, for love of the game, for love and not money. Not scoring might in this way ironically come to be scored as 'love'.

lowest common denominator (LCD) [maths] For a group of fractions, the lowest (or least) common denominator is the smallest number into which all their denominators will divide (that is, the denominators' lowest common multiple). For example: the lowest common denominator of $\frac{1}{2}$, $\frac{1}{3}$, and $\frac{1}{4}$ is 12. The LCD is used in adding fractions. In the above example, the three fractions can be rewritten as $\frac{6}{12}$, $\frac{4}{12}$, and $\frac{3}{12}$ and thereby added to give $\frac{13}{12}$, or $1\frac{1}{12}$.

lowest common multiple (LCM) [maths] For a group of whole numbers, the lowest (or least) common multiple is the smallest number that they will all divide into. For example, the LCM of 2, 3, and 4 is 12.

low frequency [physics: telecommunications] A radio frequency in the range 30 to 300 kilohertz (corresponding to long wavelengths of 1,000-10,000 metres).

low gear [engineering] The gear in which it is possible to start off and maintain a slow speed, and at which the gear ratio causes the driving wheels to turn the least rapidly (but the most powerfully) in relation to the power output of the engine.

low-test [physics] Having a comparatively high BOILING POINT.

loxodromic line [maths] Straight line on the surface of a sphere, bisecting all meridional lines at the same angle (like a ship's constant course at an angle to the equator).

lumberg [physics] Unit of luminous intensity in the centimetre-gram-second (CGS) system, equal to the light emitted (in lumens per watt) by 1 ERG of radiant energy.

lumen [physics] Unit of luminous flux in the SI system, equal to the amount of light emitted through a unit solid angle by a light source of intensity 1 CANDELA.

luminance [physics] The surface brightness of an (illuminated or light-emitting) object, expressed in terms of the luminous flux per unit solid angle per unit projected area (in such units as candela per square metre).

luminosity [physics: astronomy] For an illuminated object, its brightness as perceived by an observer, depending on the luminous intensity in a specific direction.

For a celestial object that emits light, the absolute amount of energy it radiates each second, generally expressed in absolute magnitude (rather than the formally correct SI units, watts).

luminous flux [physics] For a point source of light of 1 candela constant intensity, the flux emitted from unit solid angle of 1 steradian, expressed in LUMENS.

luminous intensity [physics] The intensity of light radiated from a source in a particular direction, expressed in CANDELA.

lunar day [astronomy] The lunar day, as far as observers on Earth are concerned, has a duration of 24 hours 50 minutes. From the point of view of the Sun, however, one lunar day lasts for 27 days 7 hours 40 minutes 48 seconds.

lunar month [astronomy] The lunar month, synodic month, or *lunation* is the period of all the lunar phases – that is, the time between two consecutive passages of the Moon through conjunction or opposition, equal to 29.53059 days. It is also called a *synodic month. See also* YEAR.

lunar phases [astronomy] The phases of the Moon are:

new Moon (Moon invisible)
crescent Moon (points to right)
first quarter (left-half-Moon)
gibbous Moon (incomplete on right)
full Moon
gibbous Moon (incomplete on left)
third quarter (right-half-Moon)
'crescent' Moon (points to left)
[new Moon (Moon invisible)]

lunar year *see* YEAR

lunation *see* LUNAR MONTH

lung capacity [biology; medicine] The actual capacity of the human lungs is difficult to ascertain because breathing changes so radically (is 'tidal') according to the prevailing circumstances. At rest, the amount of air breathed in and out (the 'tidal volume') may be as little as 0.5 litre (30.51 cubic inches); during strenuous exercise, however, the tidal volume may increase to near 4.5 litres (274.62 cubic inches), or nine times as much. Lung walls may stretch more to accommodate this volume during exercise than they would under static medical tests: for this reason, tests of lung capacity are almost always carried out with a respirometer while the patient walks on a treadmill. (The respirometer has the additional advantage of also monitoring the 'quality' of air breathed in and out, in terms of the proportions of gaseous constituents.)

lunge [physics] Alternative name for a degree on the BAUMÉ SCALE.

lustre, luster, lustrum [time] Period of five years. Adjective: *lustral*.

The term derives from an ancient Roman ritual of purification by washing (*Lustrum*, probably earlier *Laustrum*), undertaken at five-year intervals by the city censors immediately after the completion of a national census, and on behalf of the people, who were then deemed to be spiritually cleansed.

lux [physics] Unit of illumination in the SI system, equal to 1 lumen per square metre. The unit is alternatively called a *metre-candle*.

lwei [comparative values] Unit of currency in Angola.

100 lweis = 1 kwanza

See also COINS AND CURRENCIES OF THE WORLD.

Lydian mode [music] Medieval form of KEY, the SCALE of which is characterized on a modern piano by the white notes between one F and the next. In technical terms in ordinary Western music, its main effect is that of a major key with a virtually unusable diminished subdominant chord.

Its name derives from one of the musical modes used in ancient Greece, and held to be particularly soft and effeminate – but the ancient Greeks had completely different musical referents.

M

μ **(mu)** *see* MICRO-; MICRON

M (Roman numeral) [quantitatives] As a numeral in ancient Rome, the symbol M corresponded to 1,000 – yet, in this sense, it did not derive from the thirteenth letter of the Latin alphabet, a variant on the Greek *mu* (the twelfth letter of the Greek alphabet which, to the Greeks, signified either 40 or 40,000). Instead, it derived from the old Tuscan numeral much like a Greek capital *phi*, or a capital I within rounded parentheses: (I), a symbol that indeed denoted '1,000'. Not only was this assimilated very readily to the Latin capital M, but it was also conveniently the first letter of the Latin word *mille* 'thousand' – so conveniently that some authorities suggest that the M is, in fact, no more than the first letter of that word. *See also* D (ROMAN NUMERAL).

M *see* MEGA-

m, m. *see* METRE, METER; MILE; MILLI-; MILLION; MINUTE

ma, ma. *see* MILLIAMPERE, MILLIAMP

mA. *see* MILLIAMPERE, MILLIAMP; MILLIANGSTROM

Maass [volumetric measure] Obsolete, dialectal form of a measure of liquid capacity in Germany and Switzerland. The Swiss variant has evidently more recently been assimilated to units of the metric system anyway.

1 German Maass	=	1.837 litres
	=	3.233 UK pints, 3.882 US pints
		(3 UK pints = 0.928 German Maass
		4 US pints = 1.030 German Maass)
1 Swiss Maass	=	1.5 litres
	=	2.640 UK pints, 3.170 US pints
		(3 UK pints = 1.136 Swiss Maass
		3 US pints = 0.946 Swiss Maass)

The unit may or may not have anything to do with the standard German word *Mass* which can mean 'quart' (of beer), as well as 'measure'.

mace [weight] In China, an old measure in the weight of precious metals, notably silver.

1 mace	=	58.32 grains
	=	0.1333 ounce avdp., 3.778 grams
10 mace	=	1 TAEL
	=	1.333 ounce avdp. (583.2 grains)
160 mace, 16 taels	=	1 CATTY or chin
	=	1.333 pound avdp.

Mach angle [aeronautics] For an aircraft in supersonic flight, the angle between the line of flight and the supersonic shock wave: *see* MACH NUMBER.

Mach number [physics; aeronautics] Unit of speed, related to the speed of sound in a given medium; it is the cosecant of the mach angle.

For example: for dry air at sea-level,

Mach 1	=	1,229 kilometres per hour, 763.67 miles per hour
Mach 2	=	twice the speed of sound
	=	2,458 kilometres per hour, 1,527.3 miles per hour
Mach 3	=	three times the speed of sound
	=	3,687 kilometres per hour, 2,291 miles per hour and so on.

In moist air, or at altitude, the speed of sound is different and the Mach number differs accordingly.

Speeds with Mach numbers in excess of 1 are supersonic and cause a potentially destructive sonic boom: *see also* SOUND. The device on an aircraft that monitors and displays the aircraft's speed in relation to the speed of sound is called a *machmeter*.

It was named after the Austrian philosopher and physicist Ernst Mach (1838-1916).

Maclaurin's series [maths] A Taylor series expanded around zero, resulting in a series of powers that can also be obtained by applying Maclaurin's theorem (which involves infinite differentiation) to a function.

For example: the Maclaurin's series for sin x is given by
$$x - x^3/3! + x^5/5! - x^7/7! + x^9/9! - \ldots$$

It was named after the Scottish mathematician and physicist Colin Maclaurin (1698-1746).

MacMichael degree [physics] Unit of viscosity, as measured by a MacMichael viscometer. A cylinder is rotated in the liquid under test, and viscosity expressed in terms of the torque in the wire suspension of the cylinder.

macro- [quantitative: prefix] A prefix meaning 'large' or 'long' (in size or duration).

Example: macromodel – a large-scale model

The term derives from ancient Greek *makro-*, akin to Latin *macer* 'stretched', 'lean' (cf. English *emaciated*).

mag [comparative values] In England between the 1780s and late Victorian times, a slang word for a halfpenny, sometimes instead spelled *meg*. Why it was given this name is not known.

magic numbers [physics] A series of numbers, corresponding to the number of protons or neutrons that confer extra stability to the nucleus of an atom. The numbers are 2, 8, 20, 28, 50, 82, and 126.

magna cum laude [comparative values] Second highest grade in the marking of university papers, theses, and dissertations, as conferred by the authorities of many universities in continental Europe and some in North America and elsewhere.

The ordinary grade, the equivalent of a pass-mark, is given *cum laude* 'with praise'; the next grade is given *magna cum laude* 'with great praise'; and the top grade is given *summa cum laude* 'with the highest praise'.

magnetic constant [physics] Another name for the PERMEABILITY of free space.

magnetic declination [physics; geology] The difference between the direction of true north (or south) and that indicated by a compass, measured in degrees. The location of true north varies over a period of time, and so the magnetic declination – alternatively called the *magnetic deviation* – also varies.

magnetic dip [physics; geology] At any given location, the difference between the horizontal and the direction of the Earth's magnetic field (measured, in the vertical plane, in degrees, using an inclinometer). The magnetic dip is alternatively known as either the *angle of dip* or simply the *dip*.

magnetic equator [physics] The line around the surface of the Earth at which there is no MAGNETIC DIP, at which there is no difference between the horizontal and the direction of the Earth's magnetic field and so a magnetic compass needle remains balancing horizontally.

magnetic field [physics] The field of force that surrounds magnetic poles.

magnetic field strength [physics] *see* MAGNETIC INTENSITY

magnetic flux [physics] The total size of a magnetic field, now customarily measured in webers (joules per ampere). It equals the magnetic flux density multiplied by the area.

$$1 \text{ weber (SI units)} = 10^8 \text{ maxwells (CGS units)}$$

magnetic flux density [physics] For a material, the magnetic intensity multiplied by the permeability, now customarily measured in teslas (webers per square metre).

$$1 \text{ tesla (SI units)} = 10^4 \text{ gauss (CGS units)}$$

magnetic induction [physics] The creation of magnetization in a magnetic material, usually by placing it in a magnetic field (produced by another magnet or by an electromagnet). The term is also used as a (misleading) alternative for MAGNETIC FLUX DENSITY.

magnetic intensity [physics] The magnitude of a magnetic field, now customarily measured in amperes per metre, and known alternatively as the *magnetic field intensity* or the *magnetizing force*.

$$1 \text{ ampere per metre (SI units)} = 4\pi \times 10^{-3} \text{ oersted (CGS units)}$$

magnetic moment [physics] A measure of the magnetic strength of a moving charge, atom, molecule, magnet, or current-carrying coil. The product of the magnetic moment and the magnetic induction equals the turning force (torque) on the magnet. *See also* MAGNETON.

magnetic north [physics; geology] Location of the point to which a magnetic compass needle points, which does not coincide with true north and varies over a period of time: *see* MAGNETIC DECLINATION.

magnetic permeability [physics] An alternative term for PERMEABILITY.

magnetic pole [physics] One of two hypothetical points, north and south, from which and to which a magnetic field is directed.

magnetic potential *see* MAGNETOMOTIVE FORCE

magnetic saturation [physics] For a magnetizable material, the magnetic induction that produces complete magnetization.

magnetic susceptibility [physics] For a material, the relative permeability minus 1. It equals the ratio of the intensity of magnetization to the applied field.

magnetic temperature *see* CURIE TEMPERATURE SCALE

magnetic units [physics] *see* GAUSS; MAXWELL; MAGNETOMOTIVE FORCE; OERSTED; TESLA; WEBER

magnetization *see* MAGNETIC INDUCTION; MAGNETIC MOMENT; MAGNETIC SATURATION

magnetometer [physics] An instrument for measuring the strength of a magnetic field at a given location.

magnetomotive force (MMF, m.m.f.) [physics] For a closed magnetic path, the circular integral of the magnetic intensity round the circuit. It is analogous to electromotive force (EMF, e.m.f.) – voltage – for an electric circuit.

magneton [physics] For an electron, the unit for expressing its magnetic moment, equal to the product of the electron charge and Planck's constant divided by 4π times the product of the electron mass and the velocity of light. It is alternatively known as a Bohr-magneton, after the Danish physicist Niels Bohr (1885-1962).

$$\text{1 magneton} \quad = \quad 9.274 \times 10^{-25} \text{ joule per kelvin (SI units)}$$
$$= \quad 9.273 \times 10^{-19} \text{ erg per oersted (EMU)}$$

magnetosphere [astronomy] The region surrounding a planet that is occupied by its magnetic field. The Earth's magnetosphere extends out into space for about 40,000 miles (64,000 km), and is partly comprised of the VAN ALLEN RADIATION BELTS.

magnification [physics] For a lens, lens system, or optical instrument, the size of the image divided by the size of the object. For a simple lens, it is also equal to the image-lens distance divided by the object-lens distance. The magnification is sometimes alternatively called the *magnifying power*.

The term derives from Latin elements meaning 'making greater'.

magnitude [maths; astronomy] For a vector, its length when drawn on a diagram; its numerical value irrespective of its direction.

In astronomy, the brightness of a celestial object. The brightness as seen from Earth is the *apparent magnitude*, expressed on a logarithmic scale of from 1 (brightest) to 6 (least bright). The brightness as seen from a distance of 32.6 light years (10 parsecs) is the real magnitude (adjusted to take into account the distance to the object).

magnum [volumetric measure] Quantity of wine, especially of champagne. Reckoning on a standard bottle containing 75 centilitres (0.75 litre),

1 magnum	=	the contents of 2 bottles
1 jeroboam	=	4 bottles
1 rehoboam	=	6 bottles
1 methuselah	=	8 bottles
1 salmanazar	=	10 bottles
1 balthazar	=	16 bottles
1 nebuchadnezzar	=	20 bottles

The name of the magnum is derived simply from the Latin for a 'big (thing)'.

mahnd [weight] In Arabic-speaking countries, a unit of weight not very different from the kilogram of the metric system.

1 mahnd	=	2 (Arabian) RAT(T)ELS or rot(t)els
	=	2 pounds 0.64 ounces (2.04 pounds) avdp.
		(2 pounds = 0.9804 mahnd)
	=	925.3284 grams, 0.9253284 kilogram
		(1 kilogram = 1.0807 mahnd)

maiden over [sporting term] In cricket, an OVER (of six or eight balls bowled at a batsman) during which no run is scored (and no extras given away). Maiden overs are prized by bowlers, and included in their bowling figures (averages) for a match and for a season – although they do not in themselves represent points scored.

A maiden over in which one or more batsmen are out is known as a 'wicket maiden'.

main [sporting term] In some gambling dice games, a number called by the person who is to throw the dice that then becomes the subject for betting on the odds of his or her obtaining (or not obtaining) that number.

In other dice games, the stake wagered on a single throw.

In this sense the term probably derives from French *main* 'hand', referring to the number or the amount 'at hand' for betting.

main clause *see* CLAUSE; MATRIX SENTENCE

Main Sequence [astronomy] The Sun-like stars that form a diagonal band from upper left to lower right on the HERTZSPRUNG-RUSSELL DIAGRAM.

mains frequency [physics; engineering] The frequency of the alternating current in the mains electricity supply, equal to 50 hertz in Britain and 60 hertz in the United States and parts of continental Europe.

major [music] A major key corresponds to the ordinary, relatively cheerful, SCALE in which most music of the last three centuries has been written. The minor keys – of which there are actually more than major (*see* KEY, KEYNOTE) – sound more lugubrious, perhaps somehow 'sadder', than the major. The classic major key is the scale of C on the piano, represented by the white notes between one C and the next. (A classic minor key, that partly corresponds to C major in including the same notes, is the natural minor scale of A – thus represented on the piano by the white notes between one A and the next.)

major [military rank] Middle rank in the army and, in the United States, in the airforce.

In the British army, a major ranks above a captain but below a lieutenant colonel, and on a parallel with a lieutenant commander in the Royal Navy and a squadron leader in the Royal Air Force.

In the US Army and Airforce, a major ranks above a captain but below a lieutenant colonel, and on a parallel with a lieutenant commander in the US Navy.

The term itself is Latin for 'greater'.

major axis [maths] For a geometric figure with two axes of symmetry (such as an ellipse), the longer axis. (The other is the minor axis.)

major general [military rank] Senior rank in the army and, in the United States, in the airforce.

In the British army, a major general ranks above a colonel or a brigadier but below a lieutenant general, and on a parallel with a rear admiral in the Royal Navy and an air vice-marshal in the Royal Air Force.

In the US Army and Airforce, a major general ranks above a brigadier general but below a lieutenant general, and on a parallel with a rear admiral on the upper half of the navy list in the US Navy.

major penalty [sporting term] In ice hockey, banishment from the rink to the sin-bin for five minutes (as opposed to the more usual two-minute penalty) following a serious breach of discipline.

makuta *see* LIKUTA

maloti *see* LOTI

mammoth [quantitatives] Extra-large, jumbo-sized, colossal.

The word came into English in around 1700, but attained its metaphorical meaning a century later. Till then it applied only to the extinct woolly elephants of Europe and northern Asia (Russian *mammot*, adapted from the Ostyak *mamut*) and North America. The North American imperial mammoth was the largest – some had a 'shoulder' height of 4.267 metres/14 feet. Most Eurasian mammoths had a shoulder height of only 2.896 metres/9 feet 6 inches, and were thus smaller than present-day African bush elephants. *See also* JUMBO.

Manchu Dynasty [time] Period in China from 1644 until 1912, during which the country was dominated by the Tungusic-speaking Mongols of Manchuria (to the north-east).

man-day [time] The standard amount of work done by one worker (man or woman) in one full working day, constituting a unit for calculating costs and schedules in relation to output.

mandolin *see* STRINGED INSTRUMENTS' RANGE

man-hour [time] The standard amount of work done by one worker (man or woman) in one full working hour, constituting a unit for calculating costs and schedules in relation to output.

manifold [quantitatives; maths; engineering] Many times over; containing many elements. The term is the Germanic equivalent of the Latin-based *multiple(x)*.

In topology [maths], a space comprising elements of Euclidean space.

In engineering, a pipe that draws several inlets towards a single outlet or several outlets towards a single inlet.

manometer [physics] An apparatus for measuring fluid pressure (liquid or gas). At its simplest, it consists of a U-tube containing a liquid (for example, mercury or water). One arm of the U is connected to the pressure vessel, and pressure is indicated by the rise or fall of the liquid in the other arm (which may be open or sealed).

The first element of the term derives from the ancient Greek *manos* 'sparse', 'rarefied'. *See also* SPHYGMOMANOMETER.

manometric flame [physics] Device for making visible and/or measuring the sound vibrations in an organ pipe (or any other pipe intended to resonate). It comprises a diaphragm sited in a vent within the wall of the organ pipe; movement of the diaphragm by sound vibrations causes a gas flame linked to it to fluctuate. The amount of fluctuation corresponds to the amount of vibration.

ma non troppo [music] Second half of a musical instruction: Italian for ' . . . but not too much'.

manpower [physics] Apart from its general meaning of the workforce, staff, or crew involved in an operation, a unit of work or power equal to the power needed to raise a weight of 3,300 pounds a distance of 1 foot in 1 minute (a rate of working of 3,300 foot-pounds per minute, or 55 foot-pounds per second).

$$\begin{aligned}
1 \text{ manpower} \quad &= \quad 0.1 \text{ (one-tenth) HORSEPOWER} \\
&= \quad 4.241 \text{ British thermal units per minute} \\
&= \quad \text{in Britain } 74.57 \text{ watts; in the US } 74.60 \text{ watts}
\end{aligned}$$

But the unit is little used (or known) outside the United States.

man-shift [time] The standard amount of work done by one worker (man or woman) in one full working shift (term of duty in hours per day), constituting a unit for calculating costs and schedules in relation to output.

The more common units are MAN-DAYS or MAN-HOURS, but the man-shift allows for two or more shifts by one worker in a day and may thus also be more appropriate than dividing the work into man-hours.

mantissa [maths] The non-integral, fractional part of a logarithm (the other part is the characteristic).

For example: $\log_{10} 100$ = 2.3010 in which .3010 is the mantissa.

man-year [time] The standard amount of work done by one worker (man or woman) in one full working year, constituting a unit for calculating costs and schedules in relation to output.

map, mapping [maths] A transformation that causes an element in a set to correspond with another element in the same or another set. It thus shows relationships between elements, which may be one-one (that is, one-to-one or one:one), one-many, or many-one.

marathon [sporting term] A running race over a distance of 26 miles 385 yards (42.19499 kilometres), as standardized at the Olympic Games in Paris in 1924 corresponding to the distance used at the Olympic Games in London in 1908 (which was the distance between Windsor Castle and the finishing stadium in London).

At the reintroduction of the Olympic Games in 1896, the race was added to the programme at the behest of the French official Michel Bréal, its distance then 24 miles 1,500 yards (39.9959 kilometres). In Paris in 1900 and in St Louis (Missouri) in 1904, the Olympic marathon was set at 40 kilometres (24 miles 1,504 yards 16.2 inches); in Stockholm in 1912 the marathon distance was 24 miles 1,725 yards (40.2016 kilometres), and in Antwerp in 1920 the distance was 26 miles 990 yards (42.7482 kilometres).

The marathon is, in effect, a commemoration of the legendary feat of the Athenian courier Pheidippides who in 490 BC is said to have run from the site of the Battle of Marathon (between Greeks and Persians) some 36.7497 kilometres (22 miles 1,470 yards) back to Athens to proclaim the Greek victory. Having, in the previous few days, run no fewer than 160 kilometres (100 miles), however – and despite apparently having received verbal encouragement from the god Pan – he dropped dead on arrival at Athens after gasping out his message. The battle took its name from the Plain of Marathon, on which it was fought, and the plain was itself so called because it was overgrown with fennel (Greek *marathron*, Latin *marathrum*).

maravedi [comparative values] Medieval gold coin issued by the Murabiteen, the Moorish rulers in Cordoba, Spain, during their occupation of the territory in the eleventh and twelfth centuries AD. The name of the coin is a corruption of the dynastic name Murabit.

After the Moors had been expelled from Spain, the name was given to a smaller, copper coin that was then in common currency in Spain for centuries.

marc, mark, marco [weight] Obsolete unit of weight in much of continental western Europe, possibly originally defining the standard weight of a large coin, but apparently relating also to the old Roman *libra* or pound.

1 Swedish mark	=	210.571 grams
	=	7.428 ounces avdp.
1 Portuguese marco	=	8 onças, 0.5 LIBRA (or arratel)
	=	229.4894 grams
	=	8.095 ounces avdp.
		(8 oz avdp. = 0.988 Portuguese marco)
1 Spanish marco	=	230.036 grams
	=	8.114 ounces avdp.
1 French marc	=	8 onces, 0.5 LIVRE
	=	244.750 grams
	=	8.633 ounces avdp.
		(8 ounces avdp. = 0.9265 French marc)
1 Austrian mark	=	280.553 grams
	=	9.896 ounces avdp.

marco *see* MARC, MARK, MARCO

Maria Theresa dollar *see* LEVANT DOLLAR

marine league *see* LEAGUE

Mariotte's law [physics] The volume of a perfect gas, at a given temperature, varies inversely in relation to the pressure it is subjected to.

In most of the English-speaking world, this is known as Boyle's law, after the Irish physicist and philosopher Robert Boyle (1627-91). In France and some other countries of continental Europe, however, the law is named instead after the French physicist Edme Mariotte (*c.*1620-84) in whose writings – published 1717 – the law (which he evidently discovered independently) is stated.

mark [comparative values] Unit of currency in Finland (described in English as 'the Finnish mark' but more accurately the *markka*) and Germany (the Deutsche Mark).

In Finland,

1 markka = 100 penniä

[the partitive (equivalent to plural) is *markkaa*]

In Germany,

1 Mark = 100 pfennigs

During the late 1400s, the *mark* was also the name of a silver coin current in Scotland and worth 13 shillings and 4 pence (or two-thirds of a pound). At around the same time, although there was no such coin in England, the word could be used there to mean the same amount of money.

The 'mark' is a very ancient term for a coin that has been stamped out (Latin *marcatus*) for use at the market by merchants in commerce.

mark [weight] An obsolete unit in the measurement by weight of gold and silver.

1 mark = 8 ounces avdp.

This corresponds very well with other units of similar names but less specialized application, now equally obsolete, in northern and western Europe: *see* MARC, MARK, MARCO

markka *see* MARK [comparative values]

Markov chain, Markov process [statistics; physics] A succession of events that would be totally random but for the fact that each event happens only as a result of (or in a way that results from) the previous one.

It is named after the Russian mathematician Andrei Markov (1856-1922).

Mars: planetary statistics *see* PLANETS OF THE SOLAR SYSTEM

Marshal of the Royal Air Force [military rank] The most senior rank in the Royal Air Force, ranking higher than an air chief marshal, and equivalent to field marshal of the British army and admiral of the fleet in the Royal Navy.

The highest rank in the US Airforce is GENERAL OF THE AIRFORCE.

maser [physics] Device that generates or amplifies microwaves, mostly for applications in radio astronomy or long-distance radar.

The term is an acronym from microwave amplication (by) stimulated emission (of) radiation.

mass [quantitatives; physics] The amount of matter in an object, and a measure of its inertia (the extent to which it resists movement if acted on a by a force). Some units of mass are:

> KILOGRAM (SI units)
> GRAM (CGS units)
> POUND (FPS units)

and their multiples and submultiples.

mass action, law of *see* LAW OF MASS ACTION

mass defect [physics] The mass of the nucleus of an atom minus the combined masses of the particles of which it is composed, expressed in mass units. It is equivalent to the binding energy of the nucleus, and is alternatively known as the *mass decrement* or the *mass excess*.

mass-energy equation [physics] The equation $E = mc^2$ (where E = energy, m = mass, and c = the speed of light), which quantifies the equivalence of mass and energy, as stated in Einstein's special theory of relativity.

mass excess [physics] *see* MASS DEFECT

mass number [chemistry; physics] In the nucleus of an atom, the total number of protons and neutrons, alternatively known as the *nucleon number*.

mass ratio [engineering] In space technology, the ratio of the initial weight of a spacecraft before lift-off to the weight of the spacecraft once its fuel has been consumed and various engines that assisted it to leave the ground (if any) have been jettisoned.

mass spectrograph [physics; chemistry] An apparatus in which a stream of positive ions (charged atoms) passes through electric and magnetic fields in a vacuum such that they are separated in order of their charge-to-mass ratio. The fragmented ion stream can be recorded on photographic film, or be detected by an ionization counter (in a mass spectrometer). The technique provides an accurate method of measuring the relative atomic masses (atomic weights) of elements and their isotopes.

mass spectrum [physics] The range of atomic masses, or charge-to-mass ratios, recorded by a MASS SPECTROGRAPH.

master point [sporting term] In contract bridge national tournaments in the United States, master points are scored for winning games and beating opponents by specified margins, and, as each season progresses, are consistently listed in tables for comparison by enthusiasts.

The equivalent in Europe is INTERNATIONAL MATCH POINT.

master sergeant [military rank] Senior non-commissioned rank in the army and airforce, ranking just below sergeant-major.

In the United States forces, in some corps, a master sergeant may be promoted additionally to senior master sergeant and then to chief master sergeant.

match play [sporting term] In golf, a method of scoring according to the total number of holes won by each player or team – rather than according to the overall number of strokes taken to complete the course.

match point [sporting term] The position in a game – especially tennis – when the scoring of one more point by a player or team would be enough to win the match.

In tennis, when only one more point is required and there will be two or three opportunities to win that point before the opponent is in a position to equal the game score, there are said to be two or three match points to play for, respectively.

mathematics [maths] The science of quantification and numerical relationships.

The term derives from an ancient Greek adjective meaning 'of learning'.

matrix sentence [literary] The main clause in a sentence or statement of more than one clause.

For example: in the sentence

The girl who was just arrested is surely your daughter?

the matrix sentence (or main clause, also called the independent clause) is 'The girl is surely your daughter'; 'who was just arrested' is an adjectival subordinate clause.

maund [weight] Obsolescent unit of weight in India and a few localities in the Middle East. Because the word simply means 'measure', however, it has different values in different places.

In much of India,

$$
\begin{aligned}
1 \text{ maund} &= 40 \text{ seers} \\
&= 37.3241 \text{ kilograms} \\
&\quad (40 \text{ kilograms} = 1.0717 \text{ maund} \\
&\quad\ 50 \text{ kilograms, 1 centner} = 1.3396 \text{ maund}) \\
&= 82 \text{ pounds 4.576 ounces avdp.} \\
&\quad (84 \text{ pounds, 6 stone} = 1.0208 \text{ maund} \\
&\quad\ 100 \text{ pounds, 1 cental} = 1.2153 \text{ maund} \\
&\quad\ 112 \text{ pounds, 1 hundredweight} = 1.3611 \text{ maund})
\end{aligned}
$$

Elsewhere, even in other parts of India, the maund is instead between 23 and 28 pounds (10.4326 and 12.7005 kilograms).

maximum [maths] In statistics, the largest value of a set of values.

In coordinate geometry, a point on a curve such that all points immediately to the left and right of it have smaller values. At the maximum, the tangent to the curve has a slope of zero (the slope of the curve changes from positive to negative as it passes through the maximum): *see also* MINIMUM.

The term is the Latin for 'the greatest (thing)'.

maximum and minimum thermometer [physics; ecology] A U-shaped THER-MOMETER in which the capillary tubes contain small markers that indicate the highest and lowest temperatures that have occurred (since it was last reset).

maxwell [physics] Unit of magnetic flux in the centimetre-gram-second (CGS) system, equal to the flux through one square centimetre normal to a magnetic field of 1 gauss intensity.

$$1 \text{ maxwell} = 10^{-8} \text{ weber (SI units)}$$

It was named after the British physicist James Clerk Maxwell (1831-79).

Maxwell-Boltzmann distribution law [physics] A law that describes the distribution of energy among the molecules of a gas that is in thermal equilibrium. It was named after James Clerk Maxwell (*see above*) and the Austrian physicist Ludwig Boltzmann (1844-1906).

mayer [physics] An obsolete unit of heat capacity, equal to that of a substance 1 gram of which is raised through a temperature of 1°C by 1 joule of heat energy. It was named after the German physicist Julius von Mayer (1814-78).

mb, mb. *see* MILLIBAR

mc *see* MILLICURIE

Mcfd, MCFD [volumetric measure] An abbreviation for 'thousands of cubic feet per day', a unit of measurement used mainly in the supply and meterage of household water in the United States.

mean [maths] An average: *see* ARITHMETIC MEAN; GEOMETRIC MEAN; ROOT MEAN SQUARE.

mean daily motion [astronomy] The average angle a celestial object moves through during one day, equal to 360 degrees divided by its orbital period.

mean deviation [maths] In statistics, the sum of all the deviations from the arithmetic mean (average) by a group of numbers, divided by the quantity of numbers in the group.

For example: for the group of numbers 2, 3, 5, 7, 8, and 11, the arithmetic mean is $(2 + 3 + 5 + 7 + 8 + 11) \div 6 = 6$. The deviations from the mean are 4, 3, 1, 1, 2, 5, which sum to 16. The mean deviation is therefore $16 \div 6 = 2.667$.

mean distance [linear measure; astronomy] The average distance away of an object (such as another planet or the Sun) that regularly approaches and recedes, or in relation to which the observer regularly approaches and recedes.

The mean distance of the Earth from the Sun is 149.5 million kilometres (92.9 million miles).

mean free path [physics] The molecules of a gas are always colliding with one another. The mean free path is the average distance a molecule travels before two successive collisions. It depends on the viscosity and hence density of the gas.

mean free time [physics] The average time that elapses between successive collisions of gas molecules: *see* MEAN FREE PATH.

mean life [physics] The average time for which a radioactive atomic nucleus exists before it decays. *See also* DECAY TIME.

mean proportional [maths] In a comparison of two equal ratios, an element that appears in both. For example: if *a:b* is equal to *b:c*, *b* is the mean proportional.

mean sea level [surveying] A local reference datum for surveys (fixed in Britain as the mean level at Newlyn in Cornwall).

mean solar day [astronomy; time] The average value of a day during a complete solar year (the time between two consecutive transits of the Sun across the meridian).

mean solar time [astronomy] Time measured by the hour angle of mean Sun plus twelve hours, also called universal time. When it is referred to the Greenwich meridian (longitude 0 degrees), it is called Greenwich Mean Time (GMT).

measure [volumetric measure] In the Bible, according to some translations, a cubic volume of grain (specifically wheat). The Hebrew word used, however, is *seah*: *see* SEAH.

measurement ton *see* FREIGHT TON

mechanical advantage *see* FORCE RATIO, MECHANICAL ADVANTAGE

mechanical equivalent of heat [physics] The specific heat capacity of water (4.186 kilojoules per kilogram), originally thought of as a coefficient for converting mechanical work into heat energy (4.186 joules of work = 1 calorie of heat). The original concept was abandoned with the realization that work and heat are just two manifestations of energy.

mediaeval *see* MEDIEVAL

median [maths] In statistics, the middle number of a group of numbers arranged in ascending order (or the average of the middle two numbers if there is no middle one). *See also* MODE.

median lethal dose [toxicology] *see* LD 50

mediant *see* SCALE

medieval [time] Of the Middle Ages, generally reckoned to be the period from about AD 850 to about AD 1400-50. Before were the Dark Ages (or the Dark Age); after came the enlightenment that was the Renaissance (Renascence, or rebirth).

medimnos [volumetric measure] Large measure of dry capacity in ancient Greece.

1 medimnos	=	192 COTYLAI, 96 SEXTE, 48 KHOINIKES, 16 KHOES, 6 hekteis
	=	58.7814 dry litres
	=	11.3916 UK dry gallons, 11.7456 US dry gallons
	=	1.4240 UK bushel, 1.4682 US bushel
	=	58,781.4 cubic centimetres, 0.05878 cubic metre
	=	1.8284 cubic foot

This unit was the standard 'measure of corn' to the Greeks.

medium frequency [physics; telecommunications] A radio frequency in the range 300 to 3,000 kilohertz (corresponding to wavelengths of 1 to 10 metres).

medley relay [sporting term] In swimming competitions, a relay race between teams of four (or occasionally three) swimmers, each member of which swims the race distance using a different stroke according to a set sequence. As in all relay races, the second, third, and fourth team members begin their distances only at the end of the previous member's distance.

In some long-distance running and cross-country skiing relay races, a medley relay is a race in which team members in turn run or ski different distances around a circuit to complete a total overall required distance.

The word *medley* derives through an Old French corruption of the original Latin frequentive *misculare* 'to mix thoroughly' responsible also for such words in English as *meddle* and *melée*. The word *relay* is effectively by derivation the same word as *release*, in this case referring to the fact that, as each team member comes to the end of a stint, the next team member is set free to go (the opposite of *relay* is thus of course *delay*).

mega- [quantitatives: prefix] Prefix that, in a technical sense when it precedes a unit, multiplies the unit by 1 million times (x 10^6) its standard size or quantity; it is often abbreviated to 'M'.

Examples: 1 megabit   –   1,000,000 bits
1 megawatt   –   1,000,000 watts

By derivation, however, the prefix means merely 'large' (ancient Greek *megas*), and can also be used with this meaning in a non-technical sense in English (as for example in *megablast*). *See also* MACRO-.

megabar [physics] Unit of pressure.

1 megabar   =   1,000,000 bars, 1,000 kilobars
= 10^{11} newtons per square metre (10^{11} pascals)
=   986,900 atmospheres

See also BAR.

megabit [electronics] In computing, a unit of binary digits.

1 megabit   =   1,000,000 bits, 1,000 kilobits
=   125,000 bytes

See also BIT.

megabyte [electronics] In computing, a unit of binary digits.

1 megabyte   =   1,000,000 bytes, 1,000 kilobytes
=   8,000 kilobits

See also BYTE.

megacurie [physics] Former unit of radioactivity now superseded by the BECQUEREL.

1 megacurie   =   1,000,000 curies
= 3.7037×10^4 becquerel

See also CURIE.

megacycles per second [physics] Former unit of frequency now superseded by the MEGAHERTZ.

1 megacycle per second   =   1,000,000 cycles per second
=   1 megahertz (MHz), 1,000 kilohertz (kHz)

See also HERTZ.

megadyne [physics] Unit of force in the centimetre-gram-second (CGS) system.

1 megadyne   =   1,000,000 dynes, 1,000 kilodynes
=   72.33 poundals
=   10 NEWTONS (SI units)
=   1.0194 kilogram weight

See also DYNE.

megahertz [physics] Unit of frequency equal to a million hertz.

1 megahertz (MHz)   =   1,000,000 hertz, 1,000 kilohertz (kHz)

See also HERTZ.

megajoule [physics] Unit of work and energy in the SI system.

1 megajoule   =   1,000,000 joules, 1,000 kilojoules
= 10^{13} ergs (CGS units)
= 6.242×10^{24} electron volts (particle energy)
= 2.778×10^{11} kilowatt-hours (electrical energy)
=   239,200 calories (heat energy)
=   947.800 British thermal units (heat energy)

See also JOULE.

'megalithic yard' [linear measure] Unit proposed by the archaeological mathematician Professor Alexander Thom in the 1960s, based on his research among all the prehistoric stone circles in northern Europe, as a single and remarkably consistent linear measurement which almost all the circles seem to have been based on or scaled up from. It approximated fairly closely to the standard modern YARD (3 feet, 91.44 centimetres).

megaparsec [astronomy] Unit of astronomical distance.

1 megaparsec (Mpc)   =   1,000,000 parsecs, 1,000 kiloparsecs
=   3,262,000 light years
= 2.06265×10^{11} astronomical units

See also PARSEC.

megarad [physics] Unit of radiation dose.

$$1 \text{ megarad} = 1,000,000 \text{ rad, } 1,000 \text{ kilorad}$$
$$= 10,000 \text{ grays (SI units)}$$

See also RAD.

megaton [nuclear physics] Unit of the explosive energy of a nuclear weapon (bomb or warhead), equivalent to 1 million tons of TNT.

$$1 \text{ megaton} = \text{approximately } 5 \times 10^{19} \text{ joules}$$

See also KILOTON.

megavar [physics] Unit of electrical power in terms of the reactive voltage or current.

$$1 \text{ megavar} = 1,000,000 \text{ volt-amperes, } 1,000 \text{ kilovolt-amperes}$$

See also VAR, VAR.

megavolt [physics] Unit of electromotive force.

$$1 \text{ megavolt} = 1,000,000 \text{ volts, } 1,000 \text{ kilovolts}$$

See also VOLT.

megawatt [physics] Unit of electrical power in the SI system, equal to 1 million joules per second. It is abbreviated as 'MW'.

$$1 \text{ megawatt} = 1,000,000 \text{ watts, } 1,000 \text{ kilowatts}$$
$$= 1,341 \text{ horsepower}$$

See also WATT; KILOWATT.

megohm [physics] Unit of electrical resistance.

$$1 \text{ megohm} = 1,000,000 \text{ ohms, } 1,000 \text{ kilohms}$$

See also OHM.

megohmmeter [physics; engineering] An instrument for measuring high electrical resistances, frequently used for testing continuity and the resistance of the earth connections in a mains circuit.

Meiji era [time] In Japan, the years AD 1868 to 1912, during which the Emperor Mutsuhito was on the throne. It was a period that introduced considerable social and political change.

Meiji means 'enlightened rule', and was applied as a name for his reign by the Emperor himself.

mel [physics] An obsolete unit of subjective pitch of sound, equal to a thousandth of the pitch of a 1-kilohertz tone at a level 60 decibels above the listener's threshold of hearing.

The name is reputedly derived from the first syllable of *melody*.

melodeon *see* HARMONICAS' AND ACCORDIONS' RANGE

melting point [physics; chemistry] The temperature at which a solid begins to liquefy. It is significant in chemical analysis because it is a fixed (and generally known) value for a pure given substance and can be used to identify it.

Mendeleev's law [chemistry] If the chemical elements are listed in order of relative atomic masses (atomic weights), their properties are generally similar for every eighth element. This observation led to the development of the PERIODIC TABLE OF ELEMENTS. Named after the Russian chemist Dmitri Mendeleev (1834-1907), it is alternatively known as the *periodic law*.

meno mosso [music] Musical instruction: play or sing the music 'not so fast' – the meaning of the Italian words.

mensal [time] Of a month; lasting for a month; occurring once a month.

The term derives from an adjectival form of the Latin *mensis* 'month'.

menstrual period *see* PERIOD, MENSTRUAL

mental age [biology] For a child in an intelligence test, the average age of children who achieve the same score as the one being tested, expressed in years and months; *see* INTELLIGENCE QUOTIENT (IQ).

mental deficiency [biology] Legal and medical definitions of mental deficiency vary between countries, but it is possible to cite generally recognized degrees of mental deficiency in terms of low intelligence quotients (IQs). Mild mental deficiency corresponds to an IQ of between 85 and 70; more serious mental deficiency corresponds to an IQ between 70 and 50; profound mental deficiency corresponds to an IQ below 50. *See also* INTELLIGENCE QUOTIENT (IQ).

Mercalli scale [geology] A scale of intensity for earthquakes relating to the varying seismic effects at the ground surface. It may very briefly be summarized:

1	Perceived only by sensitive seismographic machines.
2	Marginally perceived by people lying on the ground.
3	Slight vibration; hanging objects swing.
4	Vibration; windows and crockery rattle.
5	Small objects displaced; doors swing.
6	Furniture moves; pictures fall off walls; masonry cracks.
7	Difficult to stand; masonry falls; waves on ponds.
8	Walls, chimneys, and towers partly collapse.
9	Many buildings collapse; conspicuous cracks in ground.
10	Most buildings destroyed with their foundations; landslides.
11	No buildings standing; great damage to underground structures.
12	Masses of rock displaced horizontally and vertically.

The scale was first devised by two physicists named M. D. de Rossi and F. A. Forel, and was then modified by Giuseppe Mercalli during the nineteenth century. In 1931 Mercalli's scale was, in turn, modified by H. O. Wood and Frank Neumann. The result is now perhaps better known as the Modified Mercalli Scale of Felt Intensity. *See also* RICHTER SCALE.

Mercator projection [physics: cartography] Method of cartographic presentation that is probably the most widely used in large-scale atlases today: *see* CYLINDRICAL PROJECTION; EQUAL-AREA PROJECTION.

It takes its name from its chief publicist and proponent, the Flemish mathematician and astronomer known as Gerardus Mercator – a Latinized form of his real name, Gerhard Kremer – (1512-94), although he did not devise the projection himself.

Mercury: planetary statistics *see* PLANETS OF THE SOLAR SYSTEM

meridian [geology; astronomy] A line of LONGITUDE; a great circle that passes through both poles (the north and south poles of the Earth, or the celestial poles), or a great semicircle that passes from one pole to the other. The term 'the meridian' or 'the prime meridian' usually refers to the Greenwich meridian, longitude 0 degrees.

The term derives from Latin elements meaning 'of midday'.

merimba *see* PERCUSSION INSTRUMENTS' RANGE

Merovingian era [time] Period in western continental Europe between AD 486 and 751, during which the Merovingian kings occupied the Frankish throne – traditionally regarded by the French as the first Kings of France.

The dynasty takes its name from one Merovech, the grandfather of Clovis I who was the real founder of an empire that at one stage controlled much of what is now France, Belgium, the Netherlands, Luxembourg, Germany, and Switzerland, and who was the victor over the Visigoth Alaric II in 507.

From 687, however, the Merovingian kings were effectively mere puppets in the hands of their Carolingian viziers. Eventually, the Carolingians took over altogether. *See also* CAROLINGIAN, CARLOVINGIAN.

mesh [square measure] In fishing nets, the square- or diamond-shaped grid outlined as the cord or line repeatedly crosses itself to form the net. In most English-speaking communities, the mesh is measured in relation to the (square) inch: in a half-inch mesh, for example, there are four squares or diamonds to the square inch.

The mesh is particularly significant in coastal and deep-sea fishing, subject to licence by legal authorities according to the type (and age) of fish being caught and the area being fished. Certain sizes of mesh are prohibited to allow younger fish to escape and so conserve fish stocks.

In screen printing, the grid formed by the bars of the screen, also generally related to the (linear) inch: a 60-mesh screen has 60 squares in 1 inch along its edge.

The term derives from Germanic sources, most of which imply the same meaning 'mesh', 'knotwork', 'network'; this makes it akin to such English words as *mass*, *mess*, *mix*, and similar. *See also* SIEVE MESH NUMBER.

meso- [quantitatives; prefix] Prefix meaning 'middle', 'medium (between two opposites)'.

For examples, *see* entries beginning MESO- *below*.

The term derives from ancient Greek *mesos* 'middle'.

mesocephalic [biology] In comparative anatomy, describing a skull that is neither exceptionally long (dolichocephalic), exceptionally broad (brachycephalic), nor abnormally small (microcephalic) – often said to correspond to a cranial capacity of between 1,350 and 1,450 cubic centimetres (82.38 and 88.48 cubic inches).

The term derives from the ancient Greek words *meso-* 'medium' and *kephalos* 'head'. *See also* CEPHALIC INDEX.

mesognathous [biology] In comparative anatomy, describing a lower jaw that neither juts out to an exceptional extent (is not prognathous) nor lies completely behind the upper jaw (is not opisthognathous), often said to correspond to a GNATHIC INDEX of between 98 and 103, and a FACIAL ANGLE of between 80 and 85 degrees.

The term derives from the ancient Greek words *meso-* 'medium' and *gnatho-* 'jaw' (the latter element akin to the *-gon* 'angle' of *polygon*, and to English *knee*).

Mesolithic [time] From prehistoric times, of the human cultural stage between the Palaeolithic and the Neolithic (Greek *meso-* 'middle', *lith-* 'stone'), also called the Middle Stone Age. At this stage, people lived in small nomadic groups and communities, existing primarily by hunting – especially fishing – and gathering, and using wooden or stone tools and weapons. *See also* NEOLITHIC; PALAEOLITHIC.

mesomorph [medicine] In HUMAN MORPHOLOGY, the human body shape that is neither thin and perhaps associated with a nervous disposition (ectomorphic) nor fattish and often associated with a jovial, outgoing character (endomorphic).

The term derives from the ancient Greek elements *meso-* 'medium' and *morph-* 'shape', 'form'.

meson [physics] An unstable subatomic particle which has a mass between that of an electron and that of a nucleon; a type of hadron. There are three kinds of mesons, called eta-meson, K-meson (kaon), and pi-meson (pion).

A meson – occasionally known alternatively as a mesotron – may be of positive, negative, or neutral electric charge, and exists for only a very short time (generally less than one-millionth of a second).

Beams of mesons – used to probe atomic nuclei – may be generated by a particle accelerator known as a *meson factory*.

mesosphere *see* ATMOSPHERE, COMPOSITION OF.

Mesozoic era [time] Geological era before the present (Cenozoic era) but following on from the Palaeozoic era. Divided into the Triassic, (Upper, Middle, and Lower), Jurassic, and the (Upper and Lower) Cretaceous periods – and occasionally known alternatively as the Secondary era – the Mesozoic era began around 248 million years ago and ended around 65 million years ago, at the dawn of the Tertiary period.

The era is characterized by considerable volcanic activity and mountain building (orogeny) on the planetary surface. Reptiles flourished and, from the start of the era, dominated the formerly ubiquitous arthropods, brachiopods, molluscs, and other marine and near-marine animals; by the end of the era, in fact, dominating all sizeable mobile organisms in the form of the dinosaurs. Nonetheless, the beginning of the era also marks the initial evolution of mammal species which, after the end of the era, were in turn to dominate.

The name of the era corresponds to the progression inherent in the names of the eras before and after: Palaeozoic 'ancient life', Cenozoic 'new (or modern) life', and thus Mesozoic 'middle life'. *See* chart of the Mesozoic era *under* TRIASSIC PERIOD.

message unit [comparative values] In the charging system of telephone companies in the United States, a unit based on the number of calls subscribers have made over a specified period but taking into account how many calls were long distance, how many were local, how many pre-paid, and other similar factors, to arrive at a reasonable (charge per) unit.

In a number of telephone company charging systems, the average cost of regular telephone service amounts to about 50 message units.

Messier number [astronomy] A combination of a letter and a number used as an identification of a star or other celestial object, as listed in the Messier Catalogue. It was named after the French astronomer Charles Messier (1730-1817), who published a catalogue of nebulae in 1771 and, in his lifetime, discovered no fewer than fifteen comets.

metacentre, metacenter [physics] In the cross-sectional representation of a laden ship rolling in high seas, the point at which a constantly vertical line intersects the centre of buoyancy of the ship. This point is significant, for the ship is stable if the metacentre is above the ship's centre of gravity – and unstable if the metacentre is below the ship's centre of gravity.

metameric [chemistry] Describing a pair of isomers in which the constituent molecules differ in the way in which different groups are attached to the same atom – for example: the isomers methyl propyl ether (methoxypropane), $CH_3OC_3H_7$, and dimethyl ether (ethoxyethane), $C_2H_5OC_2H_5$.

meter, metre *see* METRE, METER

meter in verse *see* METRE IN VERSE

meter-kilogram-second (MKS) system *see* METRE-KILOGRAM-SECOND (MKS) SYSTEM

methuselah [volumetric measure] A large bottle, often used to contain champagne, that holds 200 UK fluid ounces, 208 US fluid ounces (6.1512 litres).

Alternatively, but amounting to much the same thing, reckoning on a standard wine bottle containing 75 centilitres (0.75 litre),

1 magnum	=	the contents of 2 bottles
1 jeroboam	=	4 bottles
1 rehoboam	=	6 bottles
1 methuselah	=	8 bottles
1 salmanazar	=	10 bottles
1 balthazar	=	16 bottles
1 nebuchadnezzar	=	20 bottles

The methuselah is named after the Biblical patriarch (in Hebrew Metushelah or Methushael, in Greek Mathousala) who, enjoying a lifespan of 969 years, was the longest lived of all the long-lived patriarchal leaders before the Flood. His name means 'man of Shelah' – but whether 'Shelah' is the name of some tribal deity or a corrupt form of the ordinary Hebrew word for 'throwing-spear' is obscure.

metical [comparative values] Unit of currency (since 1980) in Mozambique.

1 metical = 100 centavos

The term derives through Portuguese – the language of the country's former colonial power – from Arabic *mithqāl* 'a measurement', 'a unit', a word used of various units of currency and of weight in Islamic states.

Metonic cycle [astronomy] A time period of nineteen years (approximately 235 synodic months), after which the phases of the Moon recur on the same days of the year. It was named after the ancient Greek astronomer Meton (*fl.* 433 BC), who was said – by the historian Diodorus Siculus around 400 years later – to have discovered it. *See also* GOLDEN NUMBER.

metre, meter [linear measure] Unit of length in the SI system (and the basis of the metric system), now defined by scientific preference as the distance light travels in vacuum in a time of $\frac{1}{299{,}792{,}459}$ of a second (0.0000003 second), but, during the 1960s-80s, generally expressed as 1,650,763.73 wavelengths of the orange-red light emitted by the radioactive isotope krypton-86.

1 metre/meter	=	1,000 millimetres, 100 centimetres
	=	0.001 (one-thousandth) kilometre
	=	1.093613 yard, 3 feet 3.3701 inches
		(1 yard = 0.9144 metre, 91.44 centimetres)
		1 foot = 0.3048 metre, 30.48 centimetres)
10 metres	=	1 decametre (10.96313 yards)
		(10 yards = 9.144 metres
		11 yards = 10.0584 metres)
100 metres	=	1 hectometre (109.6313 yards)
		(110 yards = 100.584 metres)
1,000 metres	=	1 kilometre (1,096.313 yards, 0.6213712 mile)
		(1 mile = 1.609344 kilometres)

The metre (*mètre*) was originally defined by scientists in the revolutionary France of 1799 as one ten-millionth of the Earth's quadrant (the distance on the surface between the North Pole and the equator), using as their guide calculations made on the linear distance/arc between Dunkirk (on the north coast of France) and Barcelona [on the east coast of Spain close to 1,000 kilometres (621 miles) due south]. Until 1960, a metal bar intended to represent the statutory metre's length was kept in special conditions of temperature and pressure in the International Bureau of

Weights and Measures' laboratory in Sèvres, near Paris. (The statutory weight known as the kilogram is retained there to this day.)

The term derives as a French corruption of the ancient Greek *metron*, Latin *metrum* 'measuring-rod', but is etymologically akin also to such English words as *measure*, *moon* (measurement of time, cf. *menstrual*), *meaning* (measurement of understanding, cf. *mind* and *meditate*) and *medical* (measurement of health, cf. *mend*).

metre, cubic [cubic measure] Unit of volume used mainly for dry goods but not uncommon also in liquid applications.

1 cubic metre	=	1 metre x 1 metre x 1 metre
	=	1,000,000 cubic centimetres
	=	1.307951 cubic yard, 35.31467 cubic feet
		(1 cubic yard = 0.764555 cubic metre
		=764,555 cubic centimetres)
	=	27.495 UK bushels, 28.378 US bushels
		(30 UK bushels = 1.091 cubic metre
		30 US bushels = 1.057 cubic meter)
	=	1 kilolitre, 1,000 litres
	=	219.97 UK gallons, 264.17 US gallons

See also METRE, METER.

metre, square [square measure] Unit of area in very common use all over the world.

1 square metre	=	1 metre x 1 metre
	=	10,000 square centimetres
	=	1.19599 square yards, 10.76391 square feet
		(1 square yard = 0.83613 square metre
		= 8,361.3 square centimetres)
100 square metres	=	(10 x 10 metres, =) 1 ARE
10,000 square metres,		
100 ares	=	(100 x 100 metres, =) 1 HECTARE
1,000,000 square metres,		
100 hectares	=	1 square kilometre

See also METRE, METER.

metre in verse [literary] The metre, or meter, in classic poetry corresponds to the number and type of the feet in each regular line of poetry. In verse described as iambic pentameter, for example, the metre comprises (lines of) five feet, all or most of which are iambs: *see* FOOT [literary].

Most classic English poetry is written in lines of four feet (tetrameter), five feet (pentameter), six feet (hexameter), or seven feet (heptameter); in other languages, more or fewer feet per line are not uncommon. The types of foot used in most English verse are

iambic (two-syllable stress pattern: unstressed-stressed),
trochaic (two-syllable, stressed-unstressed),
spondaic (two-syllable, stressed-stressed, or long-long),
anap(a)est (three-syllable, unstressed-unstressed-stressed, or short-short-long),
dactylic (three syllable, stressed-unstressed-unstressed, or long-short-short), and
amphibrach (three-syllable, unstressed-stressed-unstressed, or short-long-short).

For examples of these feet *see* FOOT [literary].

There is a number of other types of foot, each with a specific name, most of which effectively do no more than represent regular combinations of the above selection in four- or five-syllable patterns. The only other foot that occurs with any true frequency in English verse – hardly ever recurring to make up a line of poetry, however – is the cretic (three-syllable, stressed-unstressed-stressed, or long-short-long).

metre-kilogram-second (MKS) system [quantitatives] A system of units – the basis of, and superseded by, the SI system – in which the unit of length is the metre, of mass the kilogram, and of time the second.

metre-tonne-second system [quantitatives] A system of units in which the unit of length is the metre, of mass the tonne (= 1,000 kilograms), and of time the second.

metres per hour [speed] Unit for measuring the speed of fairly slow-moving objects or organisms.

1 metre per hour	=	1.6667 centimetres per minute
		(1 centimetre every 36 seconds
		1 millimetre every 3.6 seconds)
	=	0.001 kilometre per hour
		(1 kilometre every 41 days 16 hours)
	=	3 feet 3.3701 inches per hour
		(1 yard every 54 minutes 51.84 seconds)
	=	0.0006213712 mile per hour
		(1 mile every 67 days 1 hour 16 minutes 9.8 seconds)

metres per second [speed] Unit of speed in common use.

1 metre per second	=	3.6 kilometres per hour
		(1 kilometre every 16 minutes 40 seconds)
	=	86.4 kilometres per day, 604.8 per week
	=	3 feet 3.3701 inches per second
	=	196 feet 10.18 inches per minute
	=	2.2369 miles per hour
		(1 mile every 26 minutes 49.37 seconds)
	=	about 1.94 knots (exactly 1.94 in the UK)
		(1 knot = about 0.515 metres per second)
10 metres per second	=	36 kilometres per hour
	=	22.369 miles per hour
	=	about 19.4 knots

metric hundredweight [weight] An informal term, used mainly in North America, for 50 kilograms. *See also* CENTNER, SENTNER.

metric mile [linear measure] An informal term used mainly in athletics track racing or long-distance swimming contests.

1 'metric mile'	=	1,500 metres
	=	1,640.32 yards (4,920 feet 11.52 inches)
	=	0.932 mile
		(1 mile = 1,609.344 metres)

As is evident, the so-called 'metric mile' is a good way (just over 119 yards) short of the statute mile: 1,600 metres would in fact be closer (but still short). The unit approximates more closely, however, to the old Roman mile (*mille passus* '1,000 paces'): *see* MILE.

metric prefixes [units] A set of prefixes used to form submultiples and multiples of units in the METRIC SYSTEM and its derivatives (such as the CGS, MKS and SI systems).

prefix	symbol	multiple	prefix	symbol	multiple
atto-	a	x 10^{-18}	deca-	da	x 10
femto-	f	x 10^{-15}	hecto-	h	x 10^2
pico-	p	x 10^{-12}	kilo-	k	x 10^3
nano-	n	x 10^{-9}	mega-	M	x 10^6
micro-	μ	x 10^{-6}	giga-	G	x 10^9
milli-	m	x 10^{-3}	tera-	T	x 10^{12}
centi-	c	x 10^{-2}	peta-	P	x 10^{15}
deci-	d	x 10^{-1}	exa-	E	x 10^{18}

For derivations and examples *see* individual entries.

metric slug [physics] Unit of mass, equal to that given an acceleration of 1 metre per second per second by a force of 1 kilogram weight. It has also been called a *mug*. *See also* SLUG.

metric system [quantitatives] Unitary system initially applying to the measurement of length and mass only, of which the major characteristic is that it is based on the decimal system – that is, that units derivative from the major units are 10, 100, or 1,000 times more or less in quantity than the next units. The system took eight years to devise (largely because it was felt necessary to create basic units with some relevance to global dimensions) – an enterprise undertaken in France, at the time rent by revolution – and was officially adopted there on 10 December 1799. The unit of length was the *mètre* (or METRE, METER), intended to be 1 ten-millionth of a planetary quadrant (the distance from the North Pole to the equator); the unit of mass was the *kilogramme* (or KILOGRAM), defined as 1,000 times the mass of a cubic centimetre of water at its maximum density.

It was not until 1864, however, that use of the system was authorized in England (where it has only since the 1970s begun seriously – and officially – to supersede the older imperial measurement system). The United States government authorized its use in 1866.

From the 1970s, the derivative system of units known as the SI system (which includes many of the metric system's units but applies also to such fields as magnetism, force, work, and luminous intensity) has itself superseded the metric system in technical and authoritative senses throughout the scientific world. *See also* SI UNITS; METRIC PREFIXES.

metric ton [weight] Another name for a tonne (1,000 kilograms): *see* TONNE.

metronome [music] Clockwork device for providing a set tempo to which musicians can play or sing. The tempo can be adjusted for faster or slower speeds by moving a weight down or up an inverted pendulum swung to and fro by the clockwork mechanism.

The name of the device derives from Greek elements meaning 'regulation of the measure'.

MeV, mev [physics] Technically, an abbreviation of mega-electron volt (10^6 eV), but understood in the United States as an abbreviation of million electron volts (which is the same thing): *see* ELECTRON VOLT.

mezzo [music] An Italian word-element meaning 'middle', 'medium (between two opposites)'. Thus, the musical instruction *mezzo-forte* requires a passage to be sung or played at a volume mid-way between merely loud and extra loud; *mezzo-piano* at a volume between merely soft and extra soft. A mezzo-soprano has a voice range between those of a soprano and a contralto.

mezzo-forte, mezzo-piano, mezzo-soprano *see* MEZZO

mf *see* MILLIFARAD, MF

mg *see* MILLIGRAM, MG

mgd, m.g.d. [volumetric measure] Abbreviation for 'million gallons per day', used as a unit mainly in commercial water supply technology.

mh *see* MILLIHENRY

mho [physics] Unit of conductance in the centimetre-gram-second (CGS) system, equal to the reciprocal of an ohm, and therefore also known as a reciprocal ohm. The unit of conductance, of the same size, in the SI system is the siemens.

MHz *see* MEGAHERTZ

mic [physics] A suggested – but seldom adopted – unit of inductance.

$$1 \text{ mic } = 10^{-6} \text{ (one-millionth) henry}$$

michron [physics; time] The time of vibration of a wave of wavelength 1 micron (one-millionth of a metre).

The term derives from the first syllables of the ancient Greek words *mikros* and *chronos*, 'small' and 'time'.

micro- [quantitatives: prefixes] Prefix that in a technical sense, when it precedes a unit, reduces the unit to one-millionth (10^{-6}) of its standard size or quantity; it is sometimes abbreviated to μ.

Examples: 1 microfarad (μF) – 0.000001 farad
1 microgram (μg) – 0.000001 gram

By derivation, however, the prefix means merely 'small' (ancient Greek *mikros*), and can also be used with this meaning in a non-technical sense in English (as, for example, in *microscope*). *See also* MEGA-.

microampere, microamp [physics] Small unit of electric current equal to one-millionth of an ampere.

$$1 \text{ microampere } = 0.00001 \text{ ampere } (10^{-6} \text{ ampere}), 0.001 \text{ milliamp}$$

See also AMPERE; MILLIAMPERE, MILLIAMP.

microbalance [physics; chemistry] A balance that can weight extremely small amounts, down to 10^{-8} gram.

microbar [physics; meteorology] Unit of pressure equal to one-millionth of a bar.

$$1 \text{ microbar} = 0.00001 \text{ bar } (10^{-6} \text{ bar}), 0.001 \text{ millibar}$$
$$= 10^{-1} \text{ newtons per square metre (SI units)}$$
$$= 1 \text{ dyne per square centimetre (CGS units)}$$

See also BAR; MILLIBAR.

microbarn [chemistry; physics] Unit of area so small as to be virtually impossible to comprehend, equal to one-millionth of the effective cross-sectional area of the nucleus of an atom (that is, one-millionth of a barn).

$$1 \text{ microbarn} = 0.000001 \text{ barn } (10^{-6} \text{ barn}), 0.001 \text{ millibarn}$$
$$= 10^{-34} \text{ square metre}, 10^{-28} \text{ square millimetre}$$
$$= 1.55 \text{ x } 10^{-31} \text{ square inch}$$

See also BARN; MILLIBARN.

microcalorie [physics; medicine: nutrition] Minute unit of heat equal to one-millionth of a calorie.

$$1 \text{ microcalorie} = 0.000001 \text{ calorie } (10^{-6} \text{ calorie})$$

See also CALORIE.

microcephalic [biology] In comparative anatomy, describing a skull that is abnormally small – often said to correspond to a cranial capacity of less than 1,350 cubic centimetres (82.38 cubic inches).

The term derives from the ancient Greek words *mikro-* 'small' and *kephalos* 'head'. *See also* CEPHALIC INDEX; MESOCEPHALIC.

microcurie [physics] Small, obsolete unit of radioactivity equal to one-millionth of a curie.

$$1 \text{ microcurie} = 0.000001 \text{ curie } (10^{-6} \text{ curie}), 0.001 \text{ millicurie}$$
$$= 3.7037 \text{ x } 10^4 \text{ BECQUEREL}$$

See also CURIE; MILLICURIE.

microdensitometer [photography] An instrument by which the changes in brightness across a tiny area of a photographic negative can be measured.

microdyne [physics] Minute, obsolete unit of force equal to one-millionth of a dyne.

$$1 \text{ microdyne} = 0.000001 \text{ dyne } (10^{-6} \text{ dyne})$$
$$= 7.233 \text{ x } 10^{-11} \text{ poundal}$$

See also DYNE.

microfarad [physics] Commonly used unit of electrical capacitance equal to one-millionth of a farad. It is abbreviated as 'μF'.

$$1 \text{ microfarad} = 0.000001 \text{ farad } (10^{-6} \text{ farad}), 0.001 \text{ millifarad}$$
$$= 10^{-15} \text{ electromagnetic unit (EMU)}$$
$$= 9 \text{ x } 10^5 \text{ electrostatic units (ESU)}$$

See also FARAD; MILLIFARAD; PICOFARAD.

microgauss [physics] Small unit of magnetic flux equal to one-millionth of a gauss.

$$1 \text{ microgauss} = 0.000001 \text{ gauss } (10^{-6} \text{ gauss})$$
$$= 10^{-10} \text{ tesla (SI units)}$$

See also GAUSS.

microgram [weight] Tiny unit of mass equal to one-millionth of a gram, and known alternatively as a *gamma*.

$$1 \text{ microgram} = 0.000001 \text{ gram } (10^{-6} \text{ gram}), 0.001 \text{ milligram}$$
$$= 0.000015432098 \text{ grain}$$
$$= 0.00000003527396 \text{ ounce avdp.}$$

In Britain, the spelling 'microgramme' (still current in continental Europe) has never been particularly popular and, since the introduction there of metric measurements, has virtually disappeared altogether in conformity with the American spelling. *See also* GRAM, GRAMME; MILLIGRAM.

micrometer gauge [physics; engineering] An instrument for measuring small dimensions. It uses a graduated screw thread fitted with a VERNIER SCALE.

micrometre, micrometer, μm [linear measure] Unit of length equal to one-millionth of a metre, formerly called a *micron*.

$$1 \text{ micrometre} = 0.00001 \text{ metre } (10^{-6} \text{ metre}), 0.001 \text{ millimetre}$$
$$= 0.00003937 \text{ inch}, 0.03937 \text{ mil or thou}$$

See also MILLIMETRE, MILLIMETER; NANOMETRE, NANOMETER.

micromicro- [quantitatives: prefix] Former prefix meaning '1 million-millionth' (10^{-12}), now replaced by PICO-.

micromillimetre, micromillimeter [linear measure] Former tiny unit of length – also once known as a *millimicron* – equal to one-millionth of a millimetre, now replaced by the NANOMETRE.

  1 micromillimetre = 0.000000001 metre (10^{-9} metre)

  See also MICROMETRE, MICROMETER, μM.

micron, μ [linear measure; physics] Name of two units, both now superseded by others.

  Former unit of length equal to one-millionth of a metre (10^{-6} metre), now called a MICROMETRE.

  Obsolete unit of pressure equal to one-thousandth of a millimetre of mercury, approximately equivalent to 1 MILLITORR.

microsecond [time] Very small unit of time equal to one-millionth of a second.

  1 microsecond = 0.000001 second (10^{-6} second)

  See also MILLISECOND.

microtone [music] Any interval between two notes that is smaller than a SEMITONE or half-step.

microvolt [physics] Unit of potential difference or electromotive force equal to one-millionth of a volt.

  1 microvolt = 0.000001 volt (10^{-6} volt)

  See also VOLT; MILLIVOLT.

microwatt [physics] Unit of power equal to one-millionth of a watt.

  1 microwatt = 0.000001 watt (10^{-6} watt)

  = 1.332 x 10^{-9} horsepower

  See also WATT; MILLIWATT.

microwaves [physics; telecommunications] High-frequency radio waves in the wavelength range 1 to 30 millimetres (corresponding to frequencies of 100-3,000 megahertz). They lie between infra-red radiation and ordinary radio waves in the electromagnetic spectrum.

Middle Ages, the [time] Period in Europe and Asia Minor from about AD 850 to the beginning of the European Renaissance (about 1400 to 1450), corresponding to a period between ancient history and (relatively) modern history.

  The adjective meaning 'of the Middle Ages' is *medi(a)eval*.

middle-distance track events [sporting term] Running races over distances between 400 metres (or 440 yards) and 1,500 metres (or 1 mile): *see* ATHLETICS TRACK EVENTS RACE DISTANCES.

Middle Stone Age *see* MESOLITHIC

middleweight *see* BOXING WEIGHTS

miglio [linear measure] Obsolete unit of length in Italy, based – like the statute mile – on the *mille passus* ('1,000 paces') that constituted the old Roman unit distance.

  1 miglio = 1.4886 kilometre, 1,488.6432 metres

  = 0.925 statute mile, 1,628 yards

  (1 mile = 1.081 miglio)

  At 1,628 yards, the *miglio* was nonetheless slightly – about 10 yards – longer than the (generally accepted length of the) Roman mile. It was 1,000 of these units that constituted the intended distance of the famous motor race the *Mille Miglia* (the Thousand Miglia, although often translated as the Thousand Miles), the last of which was staged in 1957.

mil [quantitatives] A term commonly used to indicate one-thousandth of a unit or measure. Its uses include:

  one-thousandth of a quadrant (that is, $\frac{1}{1,000}$ of 90 degrees, or 0.09 degrees)

  one-thousandth of a radian (approximately 0.057 degrees)

  one-thousandth of a litre (= a millilitre)

  one-thousandth of a metre (= a millimetre)

  one-thousandth of an inch (= 0.0254 millimetres), also called a thou

mile [linear measure] Basic longer unit distance of the imperial (English) measurement system, deriving ultimately from the *mille passus* '1,000 (double-)paces' that constituted the old Roman unit and is generally accepted to have measured 1,618 yards.

  1 Roman *mille passus* = 5,000 'feet'

$$= \quad \text{1,618 yards, 1479.499 metres}$$

The plural of *mille passus* is *milia passuum*, and it was in these units that the Roman legionaries measured the marching distances along the characteristically straight roads across Roman territory.

But in the English-speaking world,

1 statute mile	=	320 rods, 80 chains, 8 furlongs
	=	5,280 feet, 1,760 yards
	=	1.609344 kilometre, 1,609.344 metres
		(1 kilometre = 0.6213712 mile)

This mile is considerably – 142 yards (8.8 per cent) – longer than the original Roman unit, but is nonetheless not as long as some other 'miles' also based on the Roman version. The old Scottish mile, for example, comprised 1,980 yards (1.8105 kilometres); the old Irish mile was 2,240 yards (2.0483 kilometres) long; the old Portuguese mile (*milha*) was of 2,282.75 yards (2.0873 kilometres). The word has also been borrowed to apply to some quite different unit distances. In Denmark, for instance, a 'mile' represents a distance of 7.5 kilometres (4.7 statute miles); in Sweden a 'mile' is 10 kilometres (6.2 statute miles).

Even in the English-speaking world, there are variants on the mile, depending on the purpose of the measurement or the surface over which the measurement is made.

1 UK nautical mile	=	6,082.66 feet (1.854 kilometres)
		(also called a 'geographical mile')
(1 Admiralty mile	=	6,080 feet, 1.853 kilometres)
1 US nautical mile	=	6,076.12 feet (1.852 kilometers)

(also called an 'international nautical mile' or a 'geographical mile'). *See also* GEOGRAPHICAL MILE; KNOT.

mile, cubic [cubic measure] Large volumetric unit used mainly in the measurement of the water in oceans.

1 cubic mile	=	5,451,476,000 cubic yards
	=	4.16784 cubic kilometres
		(4 cubic kilometres = 0.9597 cubic mile)

See also KILOMETRE, CUBIC.

mile, square [square measure] Unit of area commonly used in the description of the dimensions of countries, provinces, territories, and large estates.

1 square mile	=	64 square furlongs, 640 acres, 2,560 roods
	=	3,097,600 square yards
	=	2.589988 square kilometres
		(1 square kilometre = 0.3861 square mile)
	=	258.9988 hectares
		(250 hectares = 0.9652 square mile)

miles per gallon [linear measure; volumetric measure] Measure of how efficient a vehicle's engine is by reference to a relationship between the distance travelled and the volume of fuel consumed in travelling that distance. The standard form of abbreviation is 'mpg'. Worldwide comparison of engine efficiency in miles per gallon is complicated, however, by the quantitative difference between the British and United States gallons:

1 UK gallon	=	4.5459631 litres
		(1 UK gallon = 1.20095 US gallon)
1 US gallon	=	3.7853060 liters
		(1 US gallon = 0.83267 UK gallon)

miles per gallon		*miles per gallon*	
UK	US	US	UK
1	0.833	1	1.201
2	1.665	2	2.402
5	4.165	5	6.005
10	8.827	10	12.009
12	9.996	12	4.411
15	12.495	15	8.014
18	14.994	18	21.617

miles per gallon		miles per gallon	
UK	US	US	UK
20	16.654	20	24.019
22	18.319	22	26.421
24	19.985	24	28.823
26	21.650	26	31.225
28	23.316	28	33.627
30	24.981	30	36.028
32	26.646	32	38.430
34	28.312	34	40.832
37	30.810	37	44.435
40	33.308	40	48.038
44	36.639	44	52.842
48	39.970	48	57.646
54	44.966	54	64.851
60	49.962	60	72.057
70	58.289	70	84.067
80	66.616	80	96.076

This method of calculation uses the amount of fuel consumed as the constant: the distance travelled is the variable for comparison with other vehicles or other journeys. This is the system preferred in those countries that sell fuel in gallons and travel distances in miles (and the equivalent in the metric system is KILOMETRES PER LITRE: *see* comparison chart under that heading), but, in many European countries in which the litre is the standard unit measure of liquid fuel, the method of calculation is instead in the form of LITRES PER HUNDRED KILOMETRES – a method that turns the miles per gallon system on its head, for the distance travelled is then the constant, and the quantity of fuel to travel that distance is the variable used for comparison.

miles per hour [speed] Units of speed as regulated by motor traffic authorities over much of the world. The standard form of abbreviation is 'mph'. The other major world units of motor traffic speed are kilometres per hour ('km/h').

mph	km/h	km/h	mph
1	1.6	1	0.6
5	8.0	5	3.1
10	16.1	10	6.2
15	24.1	15	9.3
20	32.2	20	12.4
25	40.2	25	15.5
30	48.3	30	18.6
35	56.3	35	21.7
40	64.4	40	24.9
50	80.4	50	31.1
60	96.5	60	37.3
70	112.6	70	43.5
80	128.7	80	49.7
90	144.8	90	55.9
100	160.9	100	62.1
110	177.0	110	68.4
120	193.1	120	74.6
130	209.2	130	80.8
140	225.3	140	87.0
150	241.4	150	93.2

1 mile per hour	=	88 feet per minute
	=	17.599 inches per second
	=	26.8 metres per minute
	=	44.7 centimetres per second
	=	1.609344 kilometre per hour
1,000 mph	=	16.667 miles per minute
	=	1,466.66 feet per second
	=	1,609.344 kilometres per hour

$$= \quad 26.82 \text{ kilometres per minute}$$
$$= \quad 447 \text{ metres per second}$$

miles per second [speed] Unit of speed, used mainly in relation to satellite orbits or other astronomical calculations.

$$1 \text{ mile per second} \quad = \quad 3,600 \text{ miles per hour}$$
$$(100 \text{ miles every 1 minute 40 seconds})$$
$$= \quad 1.609344 \text{ kilometre per second}$$
$$(100 \text{ kilometres every 62.1 seconds})$$
$$= \quad 5,793.64 \text{ kilometres per hour}$$

military ranks [military rank] The warranted and commissioned ranks of the British (and some other countries') armed forces are:

British army	Royal Navy	Royal Air Force
	warrant officer	warrant officer
cadet	midshipman	cadet
2nd lieutenant		pilot officer
lieutenant	sublieutenant	flying officer
captain	lieutenant	flight lieutenant
major	lieutenant commander	squadron leader
lieutenant colonel	commander	wing commander
colonel or brigadier	captain or commodore	group captain
		air commodore
major general	rear admiral	air vice-marshal
lieutenant general	vice-admiral	air marshal
general	admiral	air chief marshal
field marshal	admiral of the fleet	marshal of the RAF

The ranks of the United States (and some other countries') armed forces are:

US Army and Air Force	US Navy
warrant officer	warrant officer
cadet	midshipman
2nd lieutenant	ensign
1st lieutenant	lieutenant junior grade
captain	lieutenant
major	lieutenant commander
lieutenant colonel	commander
colonel	captain
brigadier-general	rear admiral (lower half of list)
major general	rear admiral (upper half of list)
lieutenant general	vice-admiral
general	admiral
general of the army/air force	fleet admiral

See also entries under individual ranks.

mill [comparative values] In the United States, a notional unit of currency equal to one-tenth of a cent or one-thousandth of a dollar, used primarily in tax rate calculations to the nearest cent. A tax rate expressed in mills per dollar may then be termed the *millage*.

The term derives as an abbreviation of the Latin *millesimus* 'one-thousandth'.

mill [linear measure] In Turkey, a unit of length identical to the KILOMETRE.

$$1 \text{ mill} \quad = \quad 1,000 \text{ ARSHINS (metres)}$$
$$10 \text{ mills} \quad = \quad 1 \text{ pharoagh}$$

millenary [quantitatives; time] Of 1,000; comprising 1,000 elements or members.

Also, of 1,000 years, lasting for 1,000 years – especially (in Judaism and Christianity) of the messianic age to come that is to last for 1,000 years.

Also, a 1,000th anniversary.

The term derives initially from Latin *milleni* '1,000 each' but was very early confused with *mill-ennius* '1,000 years'.

millennium [time] A period or cycle of 1,000 years. The related adjective is *millennial* (but *see* MILLENARY).

In a religious sense, the Millennium is the (Judaeo-Christian) messianic age to come that is to last for 1,000 years.

milli- [quantitatives: prefix] Prefix that normally signifies 'one-thousandth' of the unit to which it is attached.

Example: millilitre – one-thousandth of a litre

But the prefix has in the past also been used to signify 'one thousand', and is occasionally still used in this sense today.

Examples: millennium – one thousand years

millipede – animal with '1,000' legs (actually 80-240)

The usual abbreviation for 'milli-' is 'm'.

milliampere, milliamp [physics] Small unit of electrical current equal to one-thousandth of an ampere. It is abbreviated as 'ma', 'ma.' or 'mA'.

1 milliampere = 0.001 ampere (10^{-3} ampere)

Current in milliamps may be measured by a *milliammeter*. *See also* AMPERE; MICROAMPERE.

milliangstrom [physics; linear measure] Very small unit of measurement of length equal to one-thousandth of an angstrom unit.

1 milliangstrom = 10^{-16} metre, 10^{-14} centimetre

= 10^{-10} micrometre

See also ANGSTROM, ANGSTROM UNIT.

milliard [quantatives] One thousand-million (1,000,000,000), known instead in North America – and elsewhere in some contexts – as one BILLION. (In ordinary senses in Britain, however, a 'billion' is one million-million, called a 'trillion' in the United States.)

The term – which in any event is now rarely, if ever, used – derives through French ultimately from Latin *milliarius* '1,000-fold'.

millibar [physics; meteorology] Unit of pressure equal to one-thousandth of a bar.

1 millibar = 0.001 bar (10^{-3} bar)

= 100 newtons per square metre (SI units)

= 1,000 dynes per square centimetre (CGS units)

34 millibars = about 1 inch of mercury

Standard atmospheric pressure measured at sea-level corresponds to 1,013.25 millibars (14.69 pounds per square inch).

See also BAR; MICROBAR.

millibarn [chemistry; physics] Tiny unit of area equal to one-thousandth of the effective cross-sectional area of the nucleus of an atom (that is, one-thousandth of a barn).

1 millibarn = 0.001 barn (10^{-3} barn)

= 10^{-31} square metre, 10^{-25} square millimetre

= 1.55×10^{-28} square inch

See also BARN; MICROBARN.

millicron [linear measure] Abbreviated form of *millimicron*: *see* MICROMETER.

millicurie [physics] Small, obsolete unit of radioactivity equal to one-thousandth of a curie.

1 millicurie = 0.001 curie (10^{-3} curie)

= 3.7037×10^7 becquerels

See also CURIE; MICROCURIE.

millième [comparative values] Unit of currency in Egypt and the Sudan (and formerly in Libya).

In Egypt,

1,000 millièmes = 100 PIASTRES = 1 (Egyptian) pound

In the Sudan,

1,000 millièmes = 100 piastres = 1 (Sudan) pound

The name of the unit derives directly from the French word for 'one-thousandth'.

See also COINS AND CURRENCIES OF THE WORLD.

millier [weight] An unusual alternative term for a TONNE or metric ton (1,000 kilograms). It derives through French from the Latin *milliarius* '1,000-fold'.

millifarad [physics] Unit of electrical capacitance equal to one-thousandth of a farad. It is abbreviated as 'mF'.

1 millifarad = 0.001 farad (10^{-3} farad)

= 10^{-12} electromagnetic unit (EMU)

$$= \quad 9 \times 10^8 \text{ electrostatic units (ESU)}$$

See also FARAD; MICROFARAD; PICOFARAD.

milligal [physics] Small, once commonly used but now obsolete unit of acceleration equal to one-thousandth of a gal or galileo (that is, one-thousandth of a centimetre per second).

$$1 \text{ milligal} \quad = \quad 0.001 \text{ gal} \ (10^{-3} \text{ gal})$$

See also GAL, GALILEO.

milligram [weight] Small unit of mass equal to one-thousandth of a gram. It is abbreviated as 'mg'.

$$
\begin{aligned}
1 \text{ milligram} \quad &= \quad 0.001 \text{ gram} \ (10^{-3} \text{ gram}) \\
&= \quad 0.020 \text{ metric grain} \\
&\qquad (1 \text{ metric grain} = 50 \text{ milligrams}) \\
&= \quad 0.015432098 \text{ grain avdp./troy/apoth.} \\
&\qquad (1 \text{ grain avdp./troy/apoth.} = 64.800 \text{ milligrams}) \\
&= \quad 0.00003527396 \text{ ounce avdp.} \\
&\qquad (1 \text{ ounce avdp.} = 28,349.523125 \text{ milligrams}) \\
200 \text{ milligrams} \quad &= \quad 1 \text{ (metric) carat}
\end{aligned}
$$

The spelling 'milligramme', still current in continental Europe, was formerly more common in British usage. The adoption of American spelling in many scientific disciplines from the 1950s on, and the greatly increased public familiarity with the unit engendered by the introduction in Britain of metric measurements in the early 1970s, have together caused the earlier spelling virtually to disappear. *See also* GRAM, GRAMME; MICROGRAM.

millilitre, milliliter [volumetric measure] Commonly used unit of volume equal to one-thousandth of a litre. It is abbreviated as 'ml' (and may be referred to as a 'mil').

$$
\begin{aligned}
1 \text{ millilitre} \quad &= \quad 0.001 \text{ litre} \ (10^{-3} \text{ litre}) \\
&= \quad 0.00176 \text{ UK pint, } 0.00211 \text{ US pint} \\
&= \quad 0.0352 \text{ UK fluid ounces, } 0.03376 \text{ US fluid oz} \\
&\qquad (1 \text{ UK fluid ounce} = 28.4 \text{ millilitres} \\
&\qquad 1 \text{ US fluid ounce} = 29.6 \text{ milliliters}) \\
&= \quad 1 \text{ cubic centimetre (for ordinary purposes)} \\
&= \quad 0.06102374 \text{ cubic inch} \\
&\qquad (1 \text{ cubic inch} = 16.387064 \text{ millilitres})
\end{aligned}
$$

See also LITRE, LITER; MICROLITRE, MICROLITER.

millimass unit, mu [physics] Very small unit of mass equal to one-thousandth of an atomic mass unit.

$$
\begin{aligned}
1 \text{ millimass unit} \quad &= \quad 0.001 \text{ atomic mass unit} \ (10^{-3} \text{ atomic mass unit}) \\
&= \quad 1,660 \times 10^{-30} \text{ gram}
\end{aligned}
$$

millime [comparative values] Unit of currency in Tunisia.

$$1,000 \text{ millimes} \quad = \quad 1 \text{ (Tunisian) dinar}$$

The name of the unit is an adaptation of the French *millième* 'one-thousandth'.
See also COINS AND CURRENCIES OF THE WORLD.

millimetre, millimeter [linear measure] Small unit of length equal to one-thousandth of a METRE. It is abbreviated as 'mm' (and is occasionally referred to as a 'mil').

$$
\begin{aligned}
1 \text{ millimetre} \quad &= \quad 0.001 \text{ metre} \ (10^{-3} \text{ metre}) \\
&= \quad 0.1 \text{ (one-tenth) CENTIMETRE} \\
&= \quad 0.03937 \text{ inch, } 39.37 \text{ MILS or thou} \\
&= \quad 4.6155 \text{ gries (}see \text{ GRY), } 0.46155 \text{ line} \\
&= \quad 2.83 \text{ points, } 2.65 \text{ Didot points}
\end{aligned}
$$

millimetre, cubic [cubic measure] Small measure of cubic capacity in the SI system.

$$
\begin{aligned}
1 \text{ cubic millimetre} \quad &= \quad 1 \text{ UK thousand-millionth/1 US billionth} \\
&\qquad \text{cubic metre} \\
&= \quad 10^{-6} \text{ (one-millionth) litre} \\
&= \quad 0.001 \text{ (one-thousandth) millilitre} \\
&= \quad 0.00000176 \text{ UK pint, } 0.00000211 \text{ US pint} \\
&= \quad 0.0000610 \text{ cubic inch} \\
&\qquad (1 \text{ cubic inch} = 16,387.064 \text{ cubic millimetres})
\end{aligned}
$$

1,000 cubic millimetres   =   1 cubic centimetre
See also CENTIMETRE, CUBIC.

millimetre, square [square measure] Small unit of area in the SI system.
1 square millimetre   = 10^{-6} (one-millionth) square metre
=   0.00155 square inch
(1 square inch = 645.16 mm²)
100 square millimetres   =   1 square centimetre
See also CENTIMETRE, SQUARE.

millimicron *see* MICROMETRE, MICROMETER, µM.

millimicrosecond [time] Former tiny unit of time equal to 1 UK thousand-millionth/ 1 US billionth of a second, now superseded by the NANOSECOND.
1 millimicrosecond   =   0.001 microsecond (10^{-3} microsecond)
=   0.000000001 second (10^{-9} second)
=   1 nanosecond
See also MICROSECOND.

milline [literary] Unit of advertising space equal to one line (generally printed in 5½-point type) across one column in a newspaper, magazine, or other periodical publication with guaranteed sale of at least 1 million copies.

million [quantatives] 1,000,000; one thousand-thousand; 10^6.
The prefix that ordinarily denotes 'one-million times' a unit is *mega-*, which derives from ancient Greek: *see* MEGA-.
The number derives its name from a curiously diminutive-looking medieval Italian variant of Latin *milia* 'thousands'. *See also* MILLIARD; BILLION; MILLIONTH.

millionaire, millionnaire [comparative values; quantitatives] A person who possesses more than 1 million major currency units (such as dollars or pounds sterling).

millionth [quantatives] ¹⁄₁,₀₀₀,₀₀₀; 10^{-6}; the last in a series of one-million elements.
¹⁄₁,₀₀₀,₀₀₀   =   0.000001
The prefix that ordinarily denotes 'one-millionth' of a unit is *micro-*, which derives from ancient Greek: *see* MICRO-.

milliradian [maths] Unit of angular measure equal to one-thousandth of a radian.
1 milliradian   =   0.001 radian (10^{-3} radian)
=   0.057296 degree
=   3.43776 minutes of arc
See also RADIAN.

millisecond [time] Small unit of time equal to one-thousandth of a second. It is abbreviated as 'ms'.
1 millisecond   =   0.001 second (10^{-3} second)
See also MICROSECOND.

millitorr [physics; meteorology] Unit of pressure equal to one-thousandth of a torr.
1 millitorr   =   0.001 torr (10^{-3} torr)
=   0.1333 newton per square metre
See also TORR.

millivolt [physics] Unit of potential difference or electromotive force equal to one-thousandth of a volt. It is abbreviated as 'mV'.
1 millivolt   =   0.001 volt (10^{-3} volt)
See also VOLT.

milliwatt [physics] Unit of power equal to one-thousandth of a watt. It is abbreviated as 'mW'.
1 milliwatt   =   0.001 watt (10^{-3} watt)
= 1.332×10^{-6} horsepower
See also WATT.

milometer [linear measure] Device in a motor vehicle that displays the number of miles travelled as a row of figures. A milometer that can be reset at zero to measure the distances of individual journeys (trips) is known as a *odometer, trip-milometer,* or *tripmeter.*

milreis [comparative values] Old Portuguese gold coin worth 1,000 (Portuguese *mil-*) reals (*-reis*). The term was later borrowed for a silver coin issued in Brazil, and similarly worth 1,000 reis; that coin is also no longer current.

min, min. *see* MINIM; MINOR; MINUTE.

mina [weight; metal] Unit of weight – especially in relation to precious metals and coins – in the ancient Babylonian, ancient Hebrew, and ancient Greek unitary systems.

In ancient Babylon, the weight differed at different periods of history. At one stage the value was about 978 grams (2.156 pounds); at another it was standardized at around 640 grams (1.411 pound). But it was because of the Babylonian use of base 60 – the sexagesimal system – that the Hebrew and Greek minas were one-sixtieth of a TALENT.

In ancient Israel,

1 mina	=	120 bekahs, 60 shekels (50 shekels of metal)
	=	about 504 grams, 1.111 pound
60 minas	=	1 talent or KIKKAR

In ancient Greece,

1 mina	=	600 oboloi, 300 diabolons, 100 drachma
	=	about 430 grams, 15.168 ounces (0.948 pound)
60 minas	=	1 talent

The unit was known in Babylon as the *manu*, in ancient Israel as the *minah* or *maneh*, and more commonly in Greek as the *mna*. The spelling *mina* is the Latin version of all three (and is the form best represented in English Bibles).

Mindel [time; geology] Describing the second of the ICE AGES that occurred in Europe during the PLEISTOCENE EPOCH.

Ming Dynasty [time] Period in China between AD 1368 and 1644, during which there was a golden age of art and ceramics, effectively representing the reflowering of the native culture after centuries of dominance by Mongol overlords. The name the Chinese rulers chose for their reign was *ming* 'bright'. After 276 years, however, the country then fell into the hands of the MANCHU DYNASTY.

minim [volumetric measure] Tiny unit of liquid capacity (approximating to half a drop) deriving from the apothecaries' fluid measure and different in UK and US usages.

In the United Kingdom,

1 minim	=	$\frac{1}{60}$ of a FLUID DRAM, $\frac{1}{480}$ of a fluid ounce
	=	$\frac{1}{9,600}$ of a (UK) pint, 0.000104 pint
	=	0.0592 MILLILITRE
		(1 millilitre = 16.892 minims)

In the United States,

1 minim	=	$\frac{1}{60}$ of a fluid dram, $\frac{1}{480}$ of a fluid ounce
	=	$\frac{1}{7,680}$ of a (US) pint, 0.000130 pint
	=	0.0616 milliliter
		(1 milliliter = 16.234 minims)

The name of the unit derives from Latin *minimus* 'the least'.

minim [music] In musical notation, a minim – known in the United States as a half-note – is represented by the symbol ♩ and corresponds to

½ SEMIBREVE or whole note

2 CROTCHETS or quarter-notes

4 QUAVERS or eighth-notes

The name of the note derives from the Latin *minimus* 'the least', which is ironic for, in most modern musical scoring, the minim corresponds to one of the longer elements of notation. But it seems probable that the note was initially no more than a simple vertical line – corresponding to the minim (the simplest stroke) in calligraphy: certainly, the present open, rounded head of the symbol is much later than the name.

minimax [maths] For a set of maximum values, the smallest of them.

minimum [maths] In statistics, the smallest value of a set of values.

In coordinate geometry, a point on a curve such that all points immediately to the left and right of it have larger values. At the minimum, the tangent to the curve has a slope of zero (the slope of the curve changes from negative to positive as it passes through the minimum): *see also* MAXIMUM.

The term is the Latin for 'the least (thing)'.

minimum lending rate [comparative values] The lowest rate of interest charged by a high-street bank in Britain (where the term is often abbreviated to 'MLR').

The corresponding term in the United States is the *prime rate*.

minor [music] A major key corresponds to the ordinary, relatively cheerful, SCALE in which most music of the last three centuries has been written. The minor keys – of which there are actually more than major: *see* KEY, KEYNOTE – sound more lugubrious, perhaps somehow 'sadder', than the major. The classic major key is the scale of C on the piano, represented by the white notes between one C and the next. A classic minor key, that partly corresponds to C major in including the same notes, is the natural minor scale of A – thus represented on the piano by the white notes between one A and the next.

minor axis [maths] For a geometric figure with two axes of symmetry (such as an ellipse), the shorter axis. (The other is the major axis.)

minor penalty [sporting term] In ice hockey, banishment from the rink to the sin-bin for a two-minute penalty (as opposed to the less common five-minute penalty) following a relatively less serious infraction of the rules.

minuend [maths] Technical term in ordinary arithmetical subtraction for the number from which a second number is to be subtracted. In $7 - 4 = 3$, for example, the minuend is 7.

The term derives from the Latin gerundive *minuendum* '(thing) to be diminished'.

minus [maths] Less, reduced by: the main feature of the arithmetical operation of subtraction, employing the symbol – (although it was not until the 1570s that the symbol became known as 'minus' or 'the minus sign'). Before a numeral (for example, -1), the symbol indicates a minus number – that is, a number less than (below) 0.

The term derives directly from the Latin for '(something) less', corresponding to the neuter form of the comparative of *minor* meaning 'little'.

minute [time; maths: angles] Basic unit of time equal to $\frac{1}{60}$ of an hour.

$$1 \text{ minute} = 60 \text{ seconds (60,000 milliseconds)}$$
$$60 \text{ minutes, 3,600 seconds} = 1 \text{ hour}$$

As a unit of (plane) angle, a minute is equal to $\frac{1}{60}$ of a degree, and is itself made up of 60 seconds. To distinguish it from the time unit of the same name, it is often referred to as a minute of arc.

$$1 \text{ minute (of arc)} = 60 \text{ seconds (of arc)}$$
$$= 0.2909 \text{ milliradian}$$
$$60 \text{ minutes (of arc),}$$
$$3,600 \text{ seconds (of arc)} = 1 \text{ degree}$$

The consistent recurrence of the units in 60s betrays the ancient Babylonian background of this form of measurement – the Babylonians used the base 60 (the sexagesimal system) in their calculations.

Nonetheless, the term derives directly from Latin *minuta* 'small (part)'.

Miocene epoch [time] A geological EPOCH during the Tertiary period of the Cenozoic era, following the Oligocene epoch but preceding the Pliocene, and corresponding roughly to between 25 million years ago and 5 million years ago. During the epoch, the sea-level continued to fall, opening up much of the land surface to the growth of grasses which, in turn, encouraged the spread of grazing mammals (and the carnivores that preyed upon them). But the average temperature also continued to fall.

The name of the epoch derives from ancient Greek elements meaning 'less recent'.

For table of the Cenozoic era *see* PALAEOZOIC EPOCH.

mired [physics] Unit of colour temperature (in photography), equal to 1 million divided by the temperature in kelvins (that is, $10^6/K$).

Mississippian [time; geology] The first of the two epochs that make up the CARBON-IFEROUS PERIOD, known outside North America as the Lower Carboniferous. (The other epoch, completing the Carboniferous, is in North America called the Pennsylvanian and elsewhere known as the Upper Carboniferous.) During this time, ferns and lichens formed forests that were later to become the coal deposits of today.

mixed number [maths] A combination of a whole number (integer) and a fraction, as for example: 1½, 5³⁄₇, 15²³⁄₂₄): *see also* IMPROPER FRACTIONS.

Mixolydian mode [music] Medieval form of KEY, the SCALE of which is characterized on a modern piano by the white notes between one G and the next. In technical
terms in ordinary Western music, it is the sort of scale heard mostly in bagpipe
music because it features a flattened leading note or seventh.

One of the earliest modes to be used in Christian church music, it takes its name
from a mode used in ancient Greece about which little or nothing is known (*mixolydian* actually means 'half-Lydian': *see* LYDIAN MODE, although the two modes have
only their closeness to the ordinary major scale in common). In any case, the
ancient Greeks had completely different musical referents.

mk *see* MARK [comparative values]

mkono [linear measure] Unit of length in Kenya, Uganda, Tanzania, and other East
African states.

$$1 \text{ mkono} = 18 \text{ inches, } 1\tfrac{1}{2} \text{ feet, half a yard}$$
$$= 45.72 \text{ centimetres, } 0.4572 \text{ metre}$$
$$(50 \text{ centimetres, half a metre} = 1.0936 \text{ mkono})$$

See also EIGHTEEN INCHES.

MKS system *see* METRE-KILOGRAM-SECOND (MKS) SYSTEM

ml *see* MILLILITRE, MILLILITER

MLR *see* MINIMUM LENDING RATE

mm *see* MILLIMETRE, MILLIMETER

MMF, m.m.f. *see* MAGNETOMOTIVE FORCE (MMF, M.M.F.)

mna [weight] Name in ancient Greek for the weight that the Romans called the *mina*:
see MINA.

But the term is occasionally used in modern Greek to refer instead to a weight of
1,500 grams, 1.5 kilograms (3 pounds 4.9 ounces) in the metric system. *See also*
OKA, OKE.

MNS blood groups *see* BLOOD GROUPING

modal music [music] Music written in any of the medieval modes – medieval forms
of KEY in which the SCALES are characterized by conforming to the white notes on a
modern piano between one note and its octave up or down. The Ionian mode
corresponds to the ordinary modern major scale, from C to C. The modes are:

IONIAN MODE	(the white notes between C and C)
	– effectively the same as the Hypolydian mode
DORIAN MODE	(D and D)
PHRYGIAN MODE	(E and E)
	– effectively the same as the Hypoaeolian mode
LYDIAN MODE	(F and F)
MIXOLYDIAN MODE	(G and G)
	– effectively the same as the Hypoionian mode
AEOLIAN MODE	(A and A)
	– effectively the same as the Hypodorian mode
HYPOPHRYGIAN MODE	(B and B)

mode [maths] In a group of numbers, the most frequently occurring number (there can
be more than one mode in a group). For example, in the series 1, 2, 4, 4, 4, 5, 7, 7, 8,
the mode is 4; in the series 1, 1, 2, 3, 3, 3, 5, 6, 7, 7, 7, 9, 10, the modes are 3 and 7.

moderato [music] Musical instruction: play or sing at a moderate tempo – generally
following a section or piece that is faster (for the Italian represents a past participle:
'moderated').

moderator [physics] In nuclear fission reactors, a substance (such as heavy water or
graphite) that is used to reduce the speed of neutrons, so making them more likely
to split atomic nuclei.

modern pentathlon [sport] The five (ancient Greek *pent-*) events of the modern
pentathlon are:

horseback riding, on a horse provided, over an 800-metre (880-yard) course that
includes fifteen fences or obstacles, within a time limit of 2 minutes; maximum
score: 1,100 points, from which all faults and penalties incurred are deducted;

fencing, with electronically monitored épées, in individual 3-minute bouts against
all other (or at least twenty other) competitors; victory in 70 per cent of the bouts

secures the maximum 1,000 points – further points are added or deducted in proportion to the number of bouts won or lost in relation to this percentage;

pistol shooting, calibre .22 (5.6 millimetres), maximum weight 1.26 kilograms (2 pounds 12.45 ounces), aiming twenty shots (in four series of five) at targets at a distance of 25 metres (82 feet); at a maximum possible 10 points per shot, maximum total is 200 target points – 194 target points are awarded 1,000 competition points, 22 competition points are lost or gained for each target point below or above 194 respectively;

swimming freestyle for 300 metres (330 yards), in heats according to the starting list and the number of lanes in the pool used; target time, 3 minutes 54 seconds, is awarded 1,000 competition points – every half-second slower or faster than this time loses or gains 4 competition points respectively;

cross-country run over 4 kilometres (2.437 miles) on a gently uphill overall gradient; target time, 14 minutes 15 seconds, is awarded 1,000 competition points – every second slower or faster than this loses or gains 3 competition points respectively.

Contestants compete both as individuals and as one of a three-member team, and scores are calculated for individual and for team events.

Modified Mercalli Scale of Felt Intensity *see* MERCALLI SCALE

modius [volumetric measure; cubic measure] In ancient Rome, a measure of dry capacity approximating closely to a modern PECK.

1 modius	=	16 sextarii, 32 heminae
	=	8.49999 (dry) litres
	=	1.87 UK (dry) gallon, 2.25 US (dry gallons)
	=	0.935 UK peck, 1.125 US peck
1½ modius, 4 CONGII	=	1 URNA
3 modii, 8 congii, 2 urnae	=	1 AMPHORA

The word in Latin meant 'measure-full'.

module [quantitatives] A basic and individual unit (in any discipline) that can be used as a standard by which to measure, or that has its own function while at the same time being part of a greater whole.

modulus [maths] Of a real number, its value irrespective of its sign (that is, irrespective of whether it is positive or negative). It is indicated by two vertical lines. For example: the moduli of 3 and −7 are written |3| and |7| . Also called the *absolute value*, it is often abbreviated as 'mod'.

Also, a number that divides into two other specified numbers leaving an identical remainder in each case; in this sense, the term is commonly also presented as 'modulo'. For example: the modulus of 4 and 7 is 3 because 3 divides into the others leaving a remainder of 1 both times. This may be written '4 ≡ 7 (modulus 3)' or '4 is congruent to 7 modulo 3'.

Also, in the operation to change a logarithm in one base to a logarithm in another base, the factor by which the first logarithm must be multiplied for it to become the second.

modulus of elasticity [physics] *see* ELASTIC CONSTANT

mohm [physics] Obsolete unit of mechanical mobility equal to the reciprocal of the mechanical ohm (just as the electrical mho is the reciprocal of the ohm). The term is said to have been derived from mobile ohm.

Moho, Mohorovičić Discontinuity [geology] The stratum beneath the surface of the planet Earth at which the Earth's crust is met by the planetary mantle (which in turn encloses the Earth's core). The depth of the stratum varies according to the composition of the crust, whether the crust is continental or oceanic, and whether tectonic faults or movements are present; it may therefore be anything from about 9.5 kilometres/6 miles deep to about 40 kilometres/24.5 miles deep. On average, however, the depth is about 35 kilometres/21.75 miles.

That the Discontinuity exists is known from its effect on seismic waves (which change dramatically on penetrating it). It is because these waves do not simply continue straight through to the mantle that the stratum is called the Discontinuity. It is named after the Serbian seismological historian Andrija Mohorovičić (1857-1936), whose studies of earthquake records revealed it in 1909.

Mohr cubic centimetre [physics] Unit of volume, equal to the volume of 1 gram of water at a temperature of 17.5°C. This unusual unit was once used in determining the concentrations of sugar solutions, and was named after the German pharmacist Karl F. Mohr (1806-79).

1 Mohr cubic centimetre = 1.000235 millilitre

(1 millilitre = 0.999765 Mohr cubic centimetre)

Mohs' scale (of hardness) [physics; mineralogy] Series of numbers that indicate the hardness of minerals and similar substances, ranging from 1 (the most soft) to 10 (the most hard). The numbers relate to actual minerals, each of which is hard enough to scratch any of those above it on the following list.

scale number	mineral
1	talc
2	gypsum
3	calcite
4	fluorite
5	apatite
6	fel(d)spar/orthoclase
7	quartz
8	topaz
9	corundum
10	diamond

The scale was named after the German geologist Friedrich Mohs (1773-1839) who devised it.

mohur [comparative values] Unit of currency in Nepal, and formerly in India.

In India during British colonial rule, the mohur was the major gold coin, worth 15 RUPEES. Its name, however, seems to have derived from its possible use as an official seal (Farsi *muhr* 'seal').

It remains a name for what is otherwise called the Nepalese rupee.

moidore [comparative values] An eighteenth-century Portuguese and Brazilian gold coin (Portuguese *moeda d'ouro* 'coin of gold', ultimately Latin *moneta aurea*) which was, for a time, accepted as legal tender (that is, valid currency) in England. So much so, that, even when the coin dropped out of circulation, its value – 27 shillings – retained the name.

mol *see* MOLE

molal [chemistry] Describing a solution that has 1 mole of solute in 1 kilogram (1,000 grams) of solvent.

molality [chemistry] The concentration of a solution, equal to the number of MOLES of solute in 1 kilogram (1,000 grams) of solvent.

molar [chemistry] Describing a quantity of a substance that is proportional to its molecular weight in grams (that is, 1 mole).

molar absorbance [physics] The absorbance of a molar solution (a solution containing 1 mole of solute in 1 litre of solution) in a cell 1 centimetre thick.

molar concentration *see* MOLALITY; MOLARITY

molar conductance [physics] The (calculated) conductance of a solution that contains 1 mole of solute between electrodes 1 centimetre apart. (It cannot be measured directly because it is impractical to make an electrolytic cell large enough to accommodate electrodes of the size that would be required.)

molar conductivity [physics; chemistry] The electrical conductivity of a solution (electrolyte) that has a concentration of 1 mole of solute per litre of solution, expressed in siemens per square centimetre per mole.

molar fraction [chemistry] For a mixture of substances, the number of moles of a given constituent divided by the total number of moles in the mixture.

molar heat capacity [physics] The amount of heat needed to raise the temperature of 1 mole of a substance by 1 kelvin, expressed in joules per kelvin per mole.

molarity [chemistry] Unit of concentration, equal to the number of MOLES of solute in 1 litre of solution.

molar volume [physics] Under specified conditions (such as at standard temperature and pressure), the volume occupied by 1 mole of a substance. For an ideal gas at standard temperature and pressure, the molar volume is 22.414 cubic decimetres (litres) per mole.

mole [chemistry] The amount of a substance that corresponds to its molecular weight in grams, or the amount that contains a number of particles (atoms, ions, or molecules) equal to Avogadro's constant (6.02253×10^{23}). It may be abbreviated to 'mol'. The term derives as a back formation from *molecule*, itself a seventeenth-century French diminutive based on Latin *moles* 'a mass'.

molecular concentration *see* MOLALITY; MOLARITY

molecular formula [chemistry] A short-hand way of describing the composition of a molecule of a compound; chemical symbols identify the atoms (*see* PERIODIC TABLE OF ELEMENTS) and numerical subscripts represent the numbers of atoms of each element in the molecule concerned. For example, H_2O and H_2SO_4 are the molecular formulae of water and sulphuric acid respectively. *See also* EMPIRICAL FORMULA.

molecular gas constant *see* BOLTZMANN'S CONSTANT

molecular mass *see* MOLECULAR WEIGHT

molecular orbital [chemistry] In a chemical compound, a covalent bond between two atoms holds two electrons, one from each combining atom. A molecular orbital is the region occupied by those electrons, which can be considered as resulting from the overlap of two ATOMIC ORBITALS. It can be represented in terms of the electrons' energy by a WAVE EQUATION.

molecular weight [chemistry] Another name for the RELATIVE MOLECULAR MASS.

molecule [chemistry; physics] An atom or, more often, two or more atoms combined in fixed proportions and held together by chemical bonds. It is the 'unit' of a chemical compound – the smallest amount of it that is capable of independent existence. The simplest molecules have only one atom (for example, helium and the other noble gases). Next in simplicity come diatomic molecules, such as those of oxygen (O_2) and carbon monoxide (CO). The most complicated, and heaviest, are MACROMOLECULES.

molto [music] Part of a musical instruction: the Italian for 'very'.

moment of a couple [physics] *see* COUPLE

moment of a force [physics] For a force acting about an axis, the perpendicular distance between the line of action of the force and the axis multiplied by the component of the force that acts in that direction. It exerts a turning effect called a torque. *See also* COUPLE.

moment of inertia [physics] For a rotating object, the sum of the products of the masses of each particle in it and the square of their perpendicular distances from the axis of rotation, equal to the mass of the object times the square of its radius of gyration. It represents the tendency of a rotating object to resist a change in its angular momentum.

momentum [physics] For a moving object, its mass multiplied by its velocity. It is a vector quantity, in the direction of motion. *See also* ANGULAR MOMENTUM.

momme [weight] Small unit of mass in Japan, formerly used particularly for measuring the weight of pearls.

1 momme	=	10 fun
	=	3.750 GRAMS
		(10 grams = 2.6667 momme)
	=	57.870 grains avdp., 0.1323 OUNCE avdp.
		(1 ounce = 7.5586 momme)
10 momme	=	1 TAEL
1,000 momme, 100 taels	=	1 KWAN

mon- *see* MONO-, MON-

monacid, monoacid [chemistry] A basic compound of which the molecules have a single hydroxyl (hydroxide) group that can be replaced by an acid radical, to form a salt. A monoacid reacts with a monobasic acid to form a salt and water.

monad [quantitatives; chemistry; philosophy] A unity; an entity comprising a single element. By extension, in chemistry, an atom or molecule that has a valence (valency) of one; and in philosophy, a unit of existence, of simple entity-ship (part of the theory – monadism – that the universe is composed of and organized by tiny units of this type).

monatomic [chemistry] Describing a molecule consisting of just one atom (as, for example, molecules of the noble gases).

money of account [comparative values] Notional form of currency (with no actual

physical form) used for official calculations, such as the MILL in the United States
and the GREEN POUND, green mark, and green franc of the European Community.

money supply, M-1 [comparative values] The total currency of a nation in actual
circulation (outside banks and other forms of deposit holdings).

mongo, mung [comparative values] Unit of currency in Mongolia.

     100 mongo or mung   =   1 togrog or tugrik

   See also COINS AND CURRENCIES OF THE WORLD.

mono-, mon- [quantitatives: prefix] A prefix denoting 'one' or 'single-'. For examples
see entries beginning MONO-, MON- *above and below*. The prefix derives from ancient
Greek *monos* 'alone', 'single' (also found in the form *mounos*), which may or may not
be a version of the word for 'one' (cf. Latin *unus*) with some very early strengthening
prefix of its own (cf. Greek *kath-hena* 'individually').

monoacid *see* MONACID, MONOACID

monobasic [chemistry] Describing an acid of which the molecules have a single
hydrogen atom that can be replaced by a metal (or ammonium ion) to form a salt.

monochromatic [physics] Describing light of one colour – that is, of a single
wavelength. (Art or photography that is in one colour only, perhaps in different tints
or shades, is instead described as *monochrome* or *monotint*.)

monoclonal [biology] Describing a pure, specific substance (such as an antibody)
that is produced by a single clone of cells grown in the laboratory. (Clones are
organisms or cells of exactly the same genetic constitution, propagated through
asexual reproduction from a common ancestor.)

monocyclic [biology; physics; chemistry] Comprising a single cycle; existing only
for one cycle. In organic chemistry, the molecules of a monocyclic compound (for
example, benzene) consist of a single 'ring' of atoms.

monomer [chemistry] A (small) single molecule that can combine with others of its
kind to form larger molecules, such as a dimer (two molecules) or polymer (many
molecules).

monoploid *see* HAPLOID

monotypical [biology] In the taxonomic classification of organisms, describing a
genus in which there is only one species. The species itself may then be described
as a *monotype*.

monovalent, monovalence [chemistry] Having a valence (valency) of one.

month [time; astronomy] Unit of time based originally on the period between two
successive new Moons equal to approximately 27.5 days; also equal to 4 weeks or
$\frac{1}{13}$ of a calendar YEAR.

   The Gregorian calendar observes the following months:

month	days	meaning of the name
January	31	'of the god Janus'
February	28/29	'of Februa, the feast of purification'
March	31	'of the god Mars'
April	30	'of (the earth's) opening' (?)
May	31	'of the goddess Maia'
June	30	'of the goddess Juno'/'of the Junian clan'
July	31	'of Julius Caesar'/'of the Julian clan'
August	31	'of Augustus Caesar'
September	30	'seventh (after the 1st month, March)'
October	31	'eighth (after the 1st month, March)'
November	30	'ninth (after the 1st month, March)'
December	31	'tenth (after the 1st month, March)'

  See also IDES.

   In scientific terms, just as there are various kinds of years, there are several kinds
of months.

anomalistic month	=	27.55455 days, the time between two successive passages of the Moon through perigee
sidereal month	=	27.323166 (sidereal) days, the time taken for the Moon to complete one orbit of the Earth
solar, or tropical, month	=	27.32158 days, the time taken for the Moon to return to the same longitude after making a

complete orbit around the Earth (that is, one
complete revolution of the Moon on its axis)

synodic, or lunar, month = 29.53059 days, the time between two successive
passages of the Moon through conjunction (or
opposition)

As a 'moon-th', the month (Greek *mene*, Latin *mensis*) is one of the earliest
divisions of time in conscious human experience, corresponding also to various
biological effects (notably *menstruation*). Etymologically, the word *moon* is akin to
such English words as *measure*, *mete*, and *meter* (and also, a little less closely, to
meditate, *medical*, and others: *see* METRE, METER). The question remains: which
came first in words and ideas – was the Moon the first and ultimate measurer, and
were all other measures 'mooned' (by *mensuration*) thereafter, or was the idea of
measurement first and the Moon accordingly named as the Great Measure?

Moon, moons of the Solar System *see* SATELLITES OF THE SOLAR SYSTEM

Moon's phases *see* LUNAR PHASES

morgen [square measure] Two different obsolete units of area stemming from
fifteenth- to eighteenth-century northern Europe.

In the Netherlands and Dutch colonies (including South Africa and early New
York),

1 morgen = 2.10 acres, 10,164 square yards
(= very close to 100 x 100 yards)
(2 acres = 0.9524 morgen)
= 8,498.7 square metres, 0.84987 hectare
(1 hectare = 1.1767 morgen)

The morgen in South Africa was later standardized at a slightly different value –
2.117 acres (10,246 square yards), 8,567.75 square metres (0.856775 hectare) – and
was revered as a unit of cultural history until comparatively recent times.

In Norway, Denmark, and Prussia,

1 morgen = 0.631 acre, 3,054 square yards
(1 acre = 1.5848 morgen)
= 2,553.66 square metres, 0.25537 hectare
(1 hectare = 3.9159 morgens)

In every case, in the local language the word *morgen* means 'morning', and it
would seem that the area involved represented the average area one yoke of oxen
could plough in a single morning's operation (cf. ACRE). The difference between the
two (major) units would in that case appear to correspond with the sort of terrain to
be ploughed in the various countries. The flat, fertile Dutch soil would certainly
allow a more efficient tillage over a morning's toil than, for example, the Prussian
or Norwegian land.

morph [literary: phonetics] In speech, a sound or combination of sounds that
corresponds to a minimum meaningful unit.

The term derives directly from the ancient Greek for 'form'.

morpheme [literary: linguistics] In a word, any element (a prefix, suffix, verb or
noun stem, or maybe just a case ending) that has or gives an independent meaning.
The study of morphemes and the manner in which they characterize a langauge is
known as linguistic *morphology*. *See also* PHONEME.

morphology, human [medicine] The distinction of three types of human body
shape – the ECTOMORPH, the MESOMORPH, and the ENDOMORPH – together with sup-
posed characteristic temperaments and attributes.

The term derives from ancient Greek elements meaning 'the science of form'.

Morse code [physics; literary] One of several codes used in telegraphy, each letter
or symbol made up of combinations of 'dots' and 'dashes' of current-on transmission separated by brief periods of current-off transmission. The duration of a 'dash'
was equivalent to the duration of three 'dots'. The duration of a space between
letters or symbols was that of three 'dots', and the duration of a space between
words was that of five or six 'dots'.

For all its initial application to telegraphy, the code was soon found highly
suitable for use in other methods of sequential transmission – notably by flashes of
light or bursts of sound.

Morse code

A	·—	M	——	Y	—·——
B	—···	N	—·	Z	——··
C	—·—·	O	———	1	·————
D	—··	P	·——·	2	··———
E	·	Q	——·—	3	···——
F	··—·	R	·—·	4	····—
G	——·	S	···	5	·····
H	····	T	—	6	—····
I	··	U	··—	7	——···
J	·———	V	···—	8	———··
K	—·—	W	·——	9	————·
L	·—··	X	—··—	10	—————

It was devised by the American artist and inventor Samuel F. B. Morse (1791-1872), but was slightly adapted in various regions of the world because of orthographic difficulties. There was, in addition, a number of other, similar but later, telegraphic codes. It was only in relation to the Morse code, however, that a sort of telex-style system of transmission by perforated tape punched by keys from a typewriter-style keyboard was in time invented by Charles Wheatstone (*see* WHEATSTONE BRIDGE) in 1858, utilizing first a clockwork motor and thereafter an electric motor. The Morse Code's final variant – the International Morse code – achieved its greatest popularity in the mid-1950s, after which its use declined rapidly as advances in telecommunications technology rendered it unnecessary.

mortality *see* DEATH RATE

Mossbauer effect, Mössbauer effect [physics] The recoil-less emission of a gamma-ray photon from the nucleus of an atom in a solid. The ray thus has the normal transition energy, which is not reduced because of the recoil of the atom concerned, as it is in gases.

The effect, named after the German physicist Rudolf L. Mössbauer (1929-), is used in solid-state physics and in its own specialized branch of spectroscopy.

motor vehicle racing [sport] The classes and categories for motorcar racing ('motor racing') are:

Racing and circuit cars:

Formula One up to 3,000 cc unsupercharged, 1,500 cc supercharged, max. 12 cylinders, some electronic and computer-operated assistance

Formula Three up to 2,000 cc, based on 4-cylinder production engine, performance limited by air restrictor

Formula Ford production 1,600 cc Cortina/Opel engines; standard road tyres

Formula Vee/SuperVee production 1,300 cc/1,600 cc Volkswagen engines; restriction on cost of tuning

Formula 5000 mass-production 5,000 cc engines, single camshaft

Indy Cars up to 2,999 cc supercharged, 4,490 cc unsupercharged; engines slanted to withstand sideforce of constant curve on oval tracks

Formula B/Formula Atlantic single-seater, production 1,100-1,600 cc engines, restrictions on tuning

Group One series-production 4-seater sports touring cars; virtually no modifications allowed

Group Two limited-production 4-seater sports touring cars; a fair degree of modification allowed

Group Three series-production 2-seater sports grand tourers; virtually no modifications allowed

Group Four limited-production 2-seater sports grand tourers; a fair degree of modification allowed

Group Five long-distance open 2-seater sports cars, all engine capacities (sometimes in individual classes)

Drag racing:

Dragsters thin elongated specials with light front wheels and vast rear wheels for racing over a strip 440 yards (400 metres) long in elimination rounds against individual opponents over time; several classes, including 'funny cars' which run on methanol

US Stock car racing:

Standard size, maximum engine capacity 7 litres (430 cubic inches), minimum wheelbase 119 inches (3.0226 metres); bodies as standard

Intermediate size, maximum engine capacity 7 litres (430 cubic inches), wheelbase 115-119 inches (2.921-3.0226 metres)

British Stock car racing:

Senior stocks, Formula Two stocks, Formula Two Superstocks, Hotrods, rebuilt road cars with glass removed and rollbars inserted; roof-colour denotes driver's experience/status; in some classes contact between cars is permitted

Rallying and cross-country racing

Motor rallying, cars of groups 1-4 above, with modifications for lightness, speed, and night lighting; driver and navigator follow a route card or sheet to cover the distance in a stated optimum time; points awarded for each stage completed, and penalty points subtracted therefrom

Hill-climb, classes according to the number and nature of entrants, as organized by the competition authority; course up a steepish hill for a distance of at least 1 mile (1.6 kilometres) on tarmac, against the clock

Hill-trial, specially built light cars in three classes (750, production, and specials) on a track usually within a quarry or other area with near-vertical slopes, some of which have to be climbed, others descended, others traversed, all between untouchable markers and against the clock

Rallycross, production sports and saloon (sedan) cars only, over a part tarmac, part grass track (often within the perimeter of a full-scale motor-racing circuit), for three laps, in heats four to six cars at a time, against the clock

Autocross, production sports and saloon (sedan) cars, buggies, and specials (as organized in classes – which may also be defined by engine capacity – by the competition authority), over a grass or rough-surface course laid out with untouchable markers, over a distance of 500-800 yards (457-731.5 metres), one at a time or in pairs or threes, against the clock

Autotest/Slalom, production cars and specials, in a course that tests the car's manoeuvrability and the driver's skill (between markers, and requiring reversing more than once) on tarmac, against the clock and with penalty points

Go-karts/Karting:

Junior class, for drivers aged 12 to 16: Class 100 National specification karts, one-cylinder two-stroke 100 cc, chain-driven, no gearbox

Class One, Class 100 National specification karts and Class 100 International specification karts, one-cylinder two-stroke 100 cc, chain-driven

Class Four, Class 125 International specification karts, Class 210 Villiers specification karts, and Class 250 International specification karts (the figures relating to engine capacity in cc), all one- or two-cylinder two-stroke engines, chain-driven

The classes and categories for motorcycle racing are:

Track-racing classes:

50cc, 125cc, 250cc, 350cc, 500cc, 750cc, engine capacity defines class, maximum number in race determined on the basis of one rider for every 100 centimetres (40 inches) width at the course's narrowest point. Rarely, races for scooters or mopeds are also held.

Sidecar combination (500 cc and 750 cc), engine capacity defines class, minimum diameter of motorcycle wheels 16 inches (40.64 centimetres) to outside of tyre, minimum diameter of sidecar wheel 10 inches (25.4 centimetres); maximum overall width 4 feet (1.22 metre), minimum overall width 2 feet 6 inches (76.2 centimetres); maximum overall height 4 feet (1.22 metre)

Speedway, four-member teams racing over a large number of heats incorporating a different selection of team members in each heat, on an ash or shale track of lap length (usually) 350-440 yards (320-365.8 metres), for cumulative points (3 for a win, 2 for second, and 1 for third)

Ice track-racing, a sort of speedway on ice (usually held on frozen lakes in northern countries), using spiked tyres; four-member teams on a track 300-400 metres (330-440 yards) long between high ice/snow banks

Drag and sprint racing: drag machines have enlarged rear wheels for racing over a strip 440 yards (400 metres) long in elimination rounds against individual opponents against the stopwatch; sprint classes are simply road races for production bikes over a short distance against the clock, classes defined by engine capacity – there are sometimes sidecar races

Off-track motorcycle racing:

Moto-cross/Scrambles, classes defined by engine capacity (usually 125 cc, 250 cc, and 500 cc, sometimes sidecar races), over a tortuous course around a rough-track circuit marked out with cables and flags; maximum number of solo riders determined on the basis of 3 feet 3 inches (1 metre) width for every starter on the starting line (half that number for sidecar entrants); races held over a set number of laps or over a set time

Trials riding, lightweight bikes ridden over a course that is part surfaced, part very rough indeed (often a disused quarry); penalties for rider's touching the ground, stopping, or missing the course; test is of skill, not of speed; occasionally also races for sidecars

Grasstrack racing, less formal style of racing with speedway or special machines, on a circular or sometimes tortuous course across a grass or rough field; sidecars in sidecar races correspond to a modified light framework that is not a real 'chair' at all

Other motor vehicle races:

Truck racing, track/circuit racing between the locomotive elements of articulated trucks/lorries; popular with spectators, but expensive for driver-operators

Caravan and trailer racing, usually tongue-in-cheek destructive races on tracks and circuits between cars and 4-tracks towing specially lightweight reconstructed trailers and caravans, most of which end up as matchwood by the end of the race

Destruction derbies, part of the hotrod scene, in which older stock cars are deliberately crashed into each other, and the last one moving is the winner; sometimes the race is a 'figure-of-eight race', in which the course has a crossover point at which spectacular collisions are common and expected

mou [square measure] Part of the square system of measurement adopted by European colonial authorities during the 1700s to late 1800s when trading with China. The *mou* used for Customs purposes was, however, different from the native Chinese measure.

1 Chinese mou	=	60 ching, 4 CHÜO
	=	806.666 square yards, 7,260 square feet
	=	674.476 square metres (almost 26 x 26 metres)
10 mou	=	1 CH'ING

The Customs variant of the *mou* was significantly larger:

1 Customs mou	=	920.417 square yards, 8,283.75 square feet (fractionally over 91 x 91 feet)
	=	769.959 square metres

mouth organ *see* HARMONICAS' AND ACCORDIONS' RANGE

mouton [comparative values] Gold coins struck by order of Kings Edward III and Henry V of England for use in their domains in France during the periods of their reigns, 1327-77 and 1399-1413. They were known with somewhat disrespectful irony as *moutons* 'sheep' because they had on one side a representation of the Lamb of God (the French for 'lamb' is *agneau*).

mpg *see* MILES PER GALLON

mph *see* MILES PER HOUR

mps *see* MILES PER SECOND

MRI scan *see* SCANNING SYSTEMS

m/s *see* METRES PER SECOND
MTS system *see* METRE-TONNE-SECOND SYSTEM
mu *see* MICRO-; MICRON; MILLIMASS UNIT, MU
multi- [quantitatives: prefix] A prefix denoting 'many', corresponding to a Latin
equivalent of the Greek *poly-*.
    Example: multiform (equivalent to polymorphic) – many shaped
    It is likely that the Latin *multus* 'much' is, in fact, etymologically akin to the
Greek *polys* 'much', and that the root meaning is thus 'fullness in number' as
opposed to 'greatness in quantity' (as in the English *much*).
multimeter [physics] A measuring instrument that can be used as an AMMETER,
GALVANOMETER, or VOLTMETER.
multimillionaire, multimillionnaire [comparative values; quantitatives] A person
who possesses many millions of major currency units (such as dollars or pounds
sterling).
multiple [quantitatives] Many times over; containing many elements. The term is the
exact Latin-based equivalent of the Germanic-based word *manifold*.
multiple [maths] A number is a multiple of another number if the second number can
be divided exactly into the first. For example: 12 is a multiple of 4 (because 4
divides exactly into 12).
multiple proportions *see* LAW OF MULTIPLE PROPORTIONS
multiplet [physics] In spectrometry, a group of two or more closely spaced lines in a
spectrum. In nuclear physics, a group of two or more elementary particles that are
identical in all properties except electric charge.
multiplication [maths] Multiplication, the inverse of DIVISION, has the effect of
repeated addition. For example: 3 x 4 is the same as $3 + 3 + 3 + 3$. Technically, it is
a mathematical operation in which a *multiplicand* is multiplied by a *multiplier* to
form a *product*, indicated by the multiplication sign x or by a dot. In the multiplication 9 x 3 = 27, 9 is the multiplicand, 3 is the multiplier, and 27 is the product.
    Writing two quantities next to each other without a space between them also
stands for multiplication (so that, for example, A x B, A.B, and AB all mean
'multiply A by B'). This remains true even if the second quantity is a binomial. For
example: 3(4 + 7) can be taken as either 3 x 11 or (3 x 4) + (3 x 7).
    Multiplication is a commutative operation: despite the technical terms above, the
result remains the same whichever quantity is regarded as the multiplicand. For
example: 6 x 8 = 8 x 6 . *See also* CROSS PRODUCT; DOT PRODUCT.
multiplication constant [physics] In a nuclear fission reaction, the total number of
neutrons produced in a certain time divided by the number that escape or are
absorbed (in the same time). The multiplication constant is alternatively known as
the *multiplication factor* or the *reproduction constant*.
multiplicative inverse [maths] The RECIPROCAL of a number, particularly in relation
to fractions. The multiplicative inverse of ³⁄₇, for example, is ⁷⁄₃; that of *x* is ¹⁄*x*.
mu-meson [physics] Former name for a MUON (which is not, as originally thought, a
type of meson).
mung *see* MONGO, MUNG
muon [physics] A negatively charged subatomic particle that occurs in weak
interactions, a type of lepton that decays rapidly (with a half-life of 2 microseconds)
to form an antineutrino, electron, and neutrino.
murder of crows [collectives] Although one species native to Britain is called the
carrion crow, crows in general worldwide are scavengers of all kinds of edible
scraps, particularly at waste tips and refuse burial sites. They are not normally
predatory, and not even as anxious to congregate around a dead animal as many
other carrion-eating birds. The 'murder' involved, then, may be of the blue variety –
crows in large numbers are extremely noisy.
murmuration of starlings [collectives] This is another collective that is meant to
be onomatopoeic – murmuration is a description of a constant murmur – but it is far
too polite for starlings. That is the fault of modern English, however: originally the
word was much more vivid – cf. Sanskrit *murmuras, marmaras* 'crackling',
'roaring', 'popping'.
musical notation [music] Modern musical notation derives from the medieval

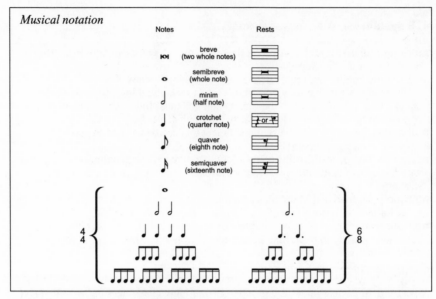

Musical notation

musical script used in monasteries to record the chants to which the Psalms and canticles were sung in the daily offices. The first requisite for writing down a range of notes is to have the equivalent of a 'keyboard' on which they may be located. From the very first, the principal method by which this was achieved was to use the stave or staff: four parallel lines in which each line and each space represent adjacent notes. Initially, the stave was not in any specific key and did not take account of the intervals between notes of the SCALE, merely representing the scale notes only (without relation to pitch). This is why, in later times, as the notion of temperament became established, there had to be recourse to the use of symbols for sharps and flats in addition, and to distinguish between actual KEYS which used sharps and flats.

The notes that correspond to the modern stave depend on which clef is being used. The most common clefs in use are the treble clef and the bass clef. In the treble clef, the lines of the stave from bottom to top represent the notes

> E   G   B   D   F

from a third above middle C to an octave-and-a-fourth above middle C, and the spaces between them accordingly represent the notes

> F   A   C   and   E.

In the bass clef the lines of the stave from bottom to top represent the notes

> G   B   D   F   A

from an octave-and-a-fourth below middle C to a third below middle C, and the spaces between them accordingly represent the notes

> A   C   E   and   G.

By around AD 1200, notes were written on staves (as they are today) but were *neumes*: square shaped, with or without descenders, and sometimes ligatured together to denote a combination of notes. Half a century later, time values (durations) began to be indicated: some of the neumes changed shape – the dia-mond-shaped *brevis* (which was to become the semibreve or wholenote) was introduced, and gradually, with florid ascenders, dots, and other scriptural embel-lishments, the notes to be found in modern musical scores came into general service, together with the symbols for the corresponding rests (or pauses).

The semibreve has for centuries now been considered the basic note, the note of fairly lengthy duration on which notes of shorter duration are based; in the United States, it is accordingly better known as a whole note. It is also the note that is the basis on which TEMPO is indicated in the two-figure tempo symbol at the beginning of a piece of music: the symbol 4/4, for example, denotes a piece of music in which each bar comprises 4 beats of notes that are one-quarter of a semibreve in duration (effectively thus four quarters), whereas the symbol 6/8 denotes a piece of music in which each bar comprises 6 beats of notes that are one-eighth of a semibreve in duration.

In order of duration (each one-half the preceding in duration), the shorter notes are:

minim, or half-note
crotchet, or quarter-note
quaver, or eighth-note
semiquaver, or sixteenth-note
demisemiquaver, or thirty-second note
hemi-demi-semiquaver, or sixty-fourth note

and the symbols for many of them (with those of the corresponding rests) are illustrated above.

The bars of 4/4 and 6/8 as illustrated are the simplest, most classic forms although there are other variants equally possible: *see* FOUR-FOUR TIME (4/4 TIME); SIX-EIGHT TIME (6/8 TIME).

muster of peafowl, mustering of storks [collectives] The word *muster* derives through French from late Latin *mostrare*, a variant of the earlier *monstrare* 'to show', 'to display' – a meaning entirely appropriate to the gaudy plumes of the peacock, though not quite so relevant to the peahen or the storks. Still, for them it may not be a matter of showing off so much as of showing up.

mutchkin [volumetric measure] An obsolete measure of liquid capacity in Scotland equal to one-quarter of an old Scottish pint.

1 mutchkin = about 0.426 litre, 42.6 centilitres
            = about 0.75 UK pint, 0.901 US pint
               (1 UK pint = 1.3333 mutchkin
               1 US pint = 1.1099 mutchkin)
            = about 16 UK fluid ounces, 14.41 US fluid ounces

As other liquid measures ending in -kin (firkin, kilderkin, nipperkin, etc.), the term derives from a medieval Dutch diminutive. In this case, it is a diminutive of the Dutch *mud* or *mutse*, which has since been assimilated to 1 hectolitre (100 litres, 22 UK gallons, 26.418 US gallons) in the metric system.

muton [biology] In genetics, the single gene that is, or small number of genes that as a group are together, responsible for a specific mutation in an individual.

muzzle velocity [speed] The speed at which a bullet or other projectile leaves the barrel of a gun, measured in metres per second or feet per second.

MV *see* MEGAVOLT

mV *see* MILLIVOLT

MW *see* MEGAWATT

mW *see* MILLIWATT

mya, m.y.a. Abbreviation for 'million years ago' in geological contexts.

myria-, myri- [quantitatives: prefix] A prefix denoting 10,000 – or just 'a great many'.

Example: myriametric – of 10,000 metres

The word derives from ancient Greek *myria-* '10,000', 'a host'.

myriad [quantatives] Technically 10,000, but, in everyday usage, just 'a great many', 'a host', 'countless': the word derives directly from the ancient Greek, which had the same diffuse meaning. *See also* LAKH; LEGION.

myriagram [weight] Uncommon unit of mass.

1 myriagram = 10,000 grams, 10 KILOGRAMS
             = 22.04623 pounds (22 pounds 0.74 ounces)

Measurement in kilograms renders the myriagram somewhat superfluous.

myriametric waves [physics; telecommunications] Radio waves with very long wavelengths (of the order of 10,000 metres), with correspondingly very low frequencies (about 30 kilohertz).

mystron [volumetric measure] French-derived name for an ancient Greek measure of dry and liquid capacity that was a minor submultiple of the kyathys: *see* KYATHYS, CYATHYS.

N

n [quantitatives] In algebra, an expression for an unknown or indefinite positive whole number. But, in ordinary parlance, an indefinite number approaching infinity – especially when taken 'to the *n*th degree'.

nadir [astronomy] Point on the celestial sphere directly below the observer's feet, and directly opposite the ZENITH.

nail [textiles] Old linear unit – dating from the 1460s and used in England for at least two centuries – in the measurement of cloth.

1 nail	=	one-sixteenth of a yard, 2¼ inches
	=	5.715 centimetres
4 nails	=	1 span or breadth, 9 inches
20 nails	=	1 (English) ELL, 45 inches

Like 'BOLT', the term presumably derives from the metal spike on which the cloth was wound.

In even earlier days, a nail was alternatively a square measure of land, or a unit of weight also called a clove (Latin *clavus*, medieval French *clouf* 'nail') and approximating to between 7 and 8 pounds avdp. (3.175 and 3.629 kilograms).

nails [quantitatives; minerals] In Britain and some ex-colonies, the standard sizes of domestic wood nails (wire nails or cut blacksmith's nails) are:

½-inch	(1.27 cm)
¾-inch	(1.91 cm)
1-inch	(2.54 cm)
1¼-inch	(3.18 cm)
1½-inch	(3.81 cm)
1¾-inch	(4.45 cm)
2-inch	(5.08 cm)
2¼-inch	(5.72 cm)
2½-inch	(6.35 cm)
3-inch	(7.62 cm)
3½-inch	(8.89 cm)
4-inch	(10.16 cm)
4½-inch	(11.43 cm)
5-inch	(12.70 cm)
6-inch	(15.24 cm)

In the United States, the standard lengths of common nails are expressed in penny sizes (originally referring to the price for 100 of each size of nail):

penny size	length		number of nails
	in	cm	per lb
2	1	2.54	875
3	1¼	3.18	550
4	1½	3.81	300
6	2	5.08	175
8	2½	6.35	100
10	3	7.62	65
16	3½	8.89	45
20	4	10.16	30

naira [comparative values] Unit of currency in Nigeria.

1 naira	=	100 kobos

The term is a deliberate contraction of the name of the country. *See also* COINS AND CURRENCIES OF THE WORLD.

nano- [quantitatives: prefix] Prefix which, when it precedes a unit, reduces the unit to 1 UK thousand-millionth/1 US billionth (10^{-9}) of its standard size or quantity.

Example: 1 nanosecond = 0.000000001 second

The prefix derives from the ancient Greek *nanos* 'dwarf'.

nanogram [weight] Tiny unit of mass equal to 1 UK thousand-millionth/1 US billionth of a gram.

1 nanogram	=	0.000000001 gram (10^{-9} gram)
	=	0.000000015432098 grain

nanometre, nanometer [linear measure; physics] Tiny unit of length in the SI
system equal to 1 UK thousand-millionth/1 US billionth of a metre, and used for
wavelengths of light and interatomic distances.

$$1 \text{ nanometre} = 0.000000001 \text{ metre } (10^{-9} \text{ metre})$$
$$= 0.000001 \text{ (one-millionth) millimetre}$$
$$= 10 \text{ angstroms}$$

nanosecond [time] Tiny unit of time equal to 1 UK thousand-millionth/1 US
billionth of a second.

$$1 \text{ nanosecond} = 0.000000001 \text{ second } (10^{-9} \text{ metre})$$

nanovolt [physics] Tiny unit of potential difference or electromotive force equal to
1 UK thousand-millionth/1 US billionth of a volt.

$$1 \text{ nanovolt} = 0.000000001 \text{ volt } (10^{-9} \text{ volt})$$

nanowatt [physics] Tiny unit of power equal to 1 UK thousand-millionth/1 US
billionth of a watt.

$$1 \text{ nanowatt} = 0.000000001 \text{ watt } (10^{-9} \text{ watt})$$
$$= 1.332 \times 10^{-12} \text{ horsepower}$$

Napierian logarithms *see* LOGARITHMS

napoleon [comparative values] Former coin issued and used in France. It was
Napoleon Bonaparte himself (1769-1821), styled Emperor Napoleon I, who first
struck the coin and had it named after him.

$$1 \text{ napoleon} = 20 \text{ francs}$$

The forename derives from the ancient Greek for 'of the new city', the new city
being any one of a number of that name, including, for example, Naples (Napoli),
Italy.

nappe [maths; geology] In mathematics, one of the two similar curved surfaces that
join at the vertex of a cone to form the cone itself. It is thus a sheet, which helps to
account for the term's derivation from the French word for 'tablecloth' (cf. the
English [diminutive] *napkin*, traditionally folded in the shape of a mathematical
nappe).

In geology, a folded body of rocks in which the limbs and the axes of the fold are
roughly horizontal – an overfold.

narrow gauge [engineering] In railway terminology, any distance between the rails
less than the standard 56½ inches (4 ft 8½ in., 1.4351 metre). Indonesia, Japan,
New Zealand, and South Africa all have railway networks based on a gauge of 3
foot 6 inches (1.0668 metre), and, in similarly mountainous areas of Europe –
notably Germany, Austria, and Switzerland – there are considerable distances of
even narrower gauges: 1 metre (39.37 inches), 2 ft 6 in. (0.762 metre) and 0.600
metre (1 ft 11.625 in.).

But most countries have – or have had – some short narrow-gauge railways,
generally built for industrial purposes. Today, quite a few such railways have been
restored and turned into popular tourist attractions (notably, for example, in Wales).

nasal consonants [literary: phonetics] Consonants of which the pronunciation
involves a hum through the nose. In English, these correspond to:

m,   n,   ng

nasal index [medicine] In comparative anatomy, the ratio of the greatest width of
the nasal opening of the skull in relation to its greatest height, expressed as a
percentage.

In comparative anthropology, the ratio of the greatest width of the nostrils at the
base of the (external) nose in relation to the nose's greatest length, expressed as a
percentage.

Nassau system *see* GOLF

natural [music] Generally, neither sharp (a semitone up) nor flat (a semitone down).
The symbol ♮ thus normally indicates that a note that has previously been specifically marked for playing or singing sharp or flat in relation to the home key should
now be played or sung as the note natural to the key. But, in describing a key (with
no further reference), natural is the equivalent of saying 'as of the white notes on a
piano' (on which the sharps and flats correspond to the black notes). So, for
example, the key of A natural is the key (with a keynote of A) of three sharps, not
the key of A flat (four flats) or of A sharp (two flats).

Also, of a brass instrument, having no valves, keys, finger-holes, or slide, and so producing only notes of the HARMONIC SERIES. *See also* NATURAL SCALE.

natural abundance [physics] The proportion of a particular isotope in a naturally occurring sample of an element, usually expressed as a percentage.

natural frequency [physics] For a vibrating system displaced from its neutral (rest) position, the frequency at which it then oscillates about that position. It is a characteristic of the particular system. *See also* RESONANCE.

natural numbers [maths] Any of the ordinary numbers 1, 2, 3, 4, 5, . . . etc.; 0 is also sometimes included in the definition. *See also* NUMBER.

natural scale [music] A scale of musical notes (tones) forming an octave in which the frequencies of the consecutive notes are proportional to the numbers 24, 27, 30, 32, 36, 40, 45, and 48. The notes of this scale are playable on non-fretted stringed instruments, but most other instruments and keyboard instruments play a compromise called the TEMPERED SCALE. *See also* SCALE; PITCH.

naught *see* NOUGHT, NAUGHT

Nautical Almanac [astronomy] An EPHEMERIS (containing tables of the positions of celestial objects on each day of the year) used by astronomers and navigators. It has been published annually since 1767, and the full edition has been renamed the *Astronomical Ephemeris*.

nautical league *see* LEAGUE

nautical log [shipping] A device for measuring the speed of a boat or ship. It consists of a small propeller which is towed through the water, so rotating the cable to which it is attached; the rate of rotation gives a measure of speed, usually measured in KNOTS.

The term derives from the fact that the original form of nautical log was indeed a log – a section of a tree-trunk – weighted so as to float upright when thrown overboard, and attached to a line that, in the very early days, had knots tied in it at regular intervals (the rate of knots disappearing over the ship's side then determined the ship's speed) or that later was attached to a spindle (the rate of rotation of which determined the ship's speed). The log would then be reeled back in on board the ship.

nautical mile *see* KNOT; MILE

neap tide [astronomy; geology] The high tide that occurs when the gravitational attraction of the Sun opposes that of the Moon (at the Moon's first and third quarters). It equals about 37 per cent of the height of the maximum, or spring, tide (*see* SPRING TIDE).

Nebraskan [time] First period of glaciation within the ICE AGES as experienced in North America, from about 1,200,000 years ago to about 1,136,000 years ago.

nebuchadnezzar [volumetric measure] Extremely large bottle (containing alcoholic drink) with a capacity of about 3.33 UK gallons, 4 US gallons (15.1412 litres). Reckoning on a standard wine bottle containing 75 centilitres (0.75 litre),

1 magnum	=	the contents of 2 bottles
1 jeroboam	=	4 bottles
1 rehoboam	=	6 bottles
1 methuselah	=	8 bottles
1 salmanazar	=	12 bottles
1 balthazar	=	16 bottles
1 nebuchadnezzar	=	20 bottles

Some of these terms relate more specifically to a net weight of the drink in fluid ounces, although the approximation to the overall volume by contents of bottles (as listed above) is remarkably accurate.

The nebuchadnezzar was named after the King of Babylon (ruled 605-592 BC), Nabu-kudurri-uzar [Akkadian: 'Nabu (god of wisdom, also spelled Nebo), protect (my) son'].

negative angle [maths; physics: navigation] Angle formed by a line rotating in a clockwise direction. In navigation at sea, compass directions are given by means of negative angles.

negative number [maths] A number that is less than 0, denoted by a minus sign (−). For example: −3, −0.56 and −½ are all negative numbers (numbers more than 0 are positive numbers). Zero is neither negative nor positive.

Neocene [time] In geological time, a collective name for the second part of the Tertiary period, including the Miocene and Pliocene EPOCHS, and lasting from about 25 million years ago to about 3 million years ago. According to some authorities, however, the term is synonymous with NEOGENE.

For chart of the Cenozoic era *see* PALAEOCENE EPOCH.

Neogene [time] In geological time, a collective name for the later part of the Tertiary period and all the time since (the so-called Quaternary period) – in other words, the last 25 million years.

Alternatively, according to some authorities, a collective name only for the second part of the Tertiary period, including the Miocene and Pliocene EPOCHS, and lasting from about 25 million years ago to about 3 million years ago. This definition makes it synonymous with NEOCENE (*see above*).

For geological information on the time (as defined in both respects) *see* HOLOCENE EPOCH; MIOCENE EPOCH; PLEISTOCENE EPOCH; PLIOCENE EPOCH.

Neolithic [time] From prehistoric times, of the later Stone Age or New Stone Age (Greek *neo-* 'new', *lith-* 'stone'). At this stage, settled farming was becoming the ordinary means of livelihood, whereas before this, hunting and gathering had been the more usual. The rudiments of agriculture were being learned, and the keeping of domesticated animals meant that, even in winter, there might be food and drink. Nonetheless, tools for the work – and weapons for defending the home farm – were still made of stone: metal-working had yet to be devised.

Neolithic cultures existed in various areas of the planet until relatively recently, notably in north-eastern South America, southern Africa, and New Guinea.

neper [physics] Unit of attenuation, the ratio of two impedances or powers expressed as a Napierian (natural) logarithm (to the base e).

$$1 \text{ neper} = 8.686 \text{ decibels}$$

It was named after the Scottish mathematician and inventor of natural logarithms John Napier (1550-1617).

nephelometry [chemistry] Method of quantitative analysis that relies on measuring the light absorption of a suspension to determine the concentration of the (pow-dered) suspended substance, and that is alternatively known as turbidimetric analysis.

The first element of the term derives from ancient Greek *nephele* 'clouds' (cognate with Latin *nebula*, German *Nebel*, etc.).

Nernst effect [physics] The production of an EMF (voltage) in a strip of metal in a magnetic field when heat flows through it. The heat flows in a direction across the magnetic lines of force, and the EMF is perpendicular to the lines of force and the direction of heat flow. It was named after the German physicist and chemist Walther Nernst (1864-1941).

Neptune: planetary statistics *see* PLANETS OF THE SOLAR SYSTEM

netball measurements, units, and positions [sport] A game played almost solely by girls and women, netball is a fast and active game rather like basketball but with more complex rules on positioning, shooting, and defending. The object of the game is for either of the two designated shooting members of the seven-member team to put the ball through a ring (underhung by a net) as many times as possible.

The dimensions of the court:
  area: 100 x 50 feet (30.48 x 15.24 metres)
  radius of shooting circle: 16 feet (4.88 metres)
  the ring:
    diameter: 1 feet 3 inches (38.1 centimetres)
    height above floor: 10 feet (3.048 metres)
    inset from edge of court: 6 inches (15 centimetres)
Positions of players:
  goalkeeper (allowed only in own third of court)
  goal defence (allowed in first and middle third)
  wing defence (allowed in first and middle thirds, but not own shooting circle)
  centre/center (allowed everywhere except shooting circles)
  wing attack (allowed in middle and opponents' thirds, but not in opponents' shooting circle)

goal attack (allowed in middle and opponents' thirds)

goal shooter (allowed only in opponents' third of court)

Timing: 60 minutes in four 15-minute quarters; the first and second quarters and the third and fourth quarters are separated by 3-minute intervals; there is a 10-minute interval at half-time

injury time may be added at the end of each quarter if appropriate

Points scoring:

the team with the greater number of goals at the end wins

Dimensions of the ball:

circumference: 28½-29 inches (72.4-73.7 centimetres)

weight: 14-16 ounces (396.9-453.6 grams)

internal pressure: 15 pounds per square inch (4.99 newtons per square metre)

net price [comparative values] Lowest price at which something is sold, and on which there can be no discount.

Earlier this century alternatively spelled *nett*, the term derives through French from Latin *nitidus* 'sparkling clean', 'transparently clear' and is thus the same word as the English *neat* in the senses both of 'clean and tidy' (cf. French *nettoyer* 'to clean up') and 'without dilution' (as of a strong drink).

net profit [comparative values] The profit of an enterprise calculated as the selling price minus the unit cost of production (up to this point, the gross profit), and less all other costs and expenses involved, such as wages, expenditure on commercial premises, and marketing and transportation costs.

For the etymology of *net* in this sense, *see* NET PRICE.

net tonnage [volumetric measure] Cargo-carrying capacity of a ship, calculated as the gross volume of enclosed deck space minus the space of crew and passenger accommodation, engines, and fuel storage, and expressed in 'tons' that actually correspond to unit measures of 100 cubic feet (2.8317 cubic metres).

Surprisingly, this definition means that net tonnage has nothing to do with a net ton, which is just another term for a short TON, a unit of weight (2,000 pounds, 907.18474 kilograms).

neum, neume *see* MUSICAL NOTATION

neutral [chemistry; physics] Describing a solution that is neither acidic nor alkaline (that is, with a PH of 7).

Describing something that has neither a positive nor a negative electric charge (for example: a neutron).

The term derives from the ancient Greek form of the modern English word *neither*.

neutretto [physics] The NEUTRINO associated with a MUON.

neutrino [physics] Uncharged (neutral) subatomic particle of zero mass that reacts only weakly with other particles. It is a type of LEPTON.

neutron [physics] An uncharged (neutral), comparatively massive subatomic particle found in the nucleus of every atom except that of hydrogen (which has a single proton in its nucleus). Its mass is 1.6748×10^{-27} kilogram. Free neutrons (outside a nucleus) are unstable, with an average life of about 12 minutes; they split up into a proton and an electron (beta particle).

neutron excess [physics] The difference between the number of neutrons and protons in the nucleus of an atom; it is alternatively called the *isotopic number*.

neutron flux, neutron flux density [physics] For a stream of neutrons (for example, as produced within a nuclear reactor), the number of neutrons that pass through unit area in unit time, expressed as the number per square metre per second. It is equal to the number of neutrons per unit volume times their average speed.

neutron number [physics] The number of neutrons in the nucleus of an atom. It is equal to the nucleon number (total number of protons and neutrons in an atom) of an element minus its atomic number.

new Moon *see* LUNAR PHASES

New Stone Age *see* NEOLITHIC

newton [physics] Unit of FORCE in the SI system, equal to the force that gives a mass of 1 kilogram an acceleration of 1 metre per second per second (in the direction of action of the force).

                    1 newton = 0.10197 kilogram force
                                     (1 kilogram force = 9.8068 newtons)
                               = 0.2249 pound force
                                     (1 pound force = 4.4464 newtons)
                               = 6.984 x 10^{-3} poundal

For practical purposes, 1 newton is approximately equal to a tenth of a kilogram.
The unit was named after the British mathematician and physicist Isaac Newton
(1642-1727).

Newton's law of cooling [physics] For a hot object losing heat by convection and
radiation, the rate of cooling is proportional to the difference in temperature
between the object and its surroundings.

Newton's law of gravitation [physics] For any two objects, the force of attraction
between them is proportional to the product of their masses divided by the square of
the distance between them. The proportionality constant is G, the gravitational
constant, equal to 6.6732 x 10^{-11} newton metre squared per square kilogram. It is a
basic relationship in physics (on the macro scale) and astronomy, one of several
inverse square laws, and is alternatively known as the *universal law of gravitation*.

Newton's laws of motion [physics] There are three Newton's laws of motion:

   An object at rest remains at rest, and an object moving in a straight line at
constant speed continues to do so, unless an external force acts on it.

   For any moving object, the rate of change of momentum is proportional to (and in
the same direction as) the force acting on it.

   When one object acts – that is, exerts a force – on another, there is an equal and
opposite reaction (force) by the second object on the first.

NGC number [astronomy] A combination of the letters NGC and a number used as
an identification of a nebula or galaxy, as listed in the *New General Catalogue* and
its supplement.

ngultrum [comparative values] Unit of currency in Bhutan.

   1 ngultrum (or tikehung) = 100 chetrum

See also COINS AND CURRENCIES OF THE WORLD.

ngwee [comparative values] Unit of currency in Zambia.

                    100 ngwee = 1 kwacha

The term describes the coin and means 'bright'.

See also COINS AND CURRENCIES OF THE WORLD.

nickel [comparative values] In the United States and Canada, half of a DIME, worth
5 cents. Minted first in the United States in 1794, it was originally a silver coin and
known only as a 'half-dime'. The change of metal from silver (last issued 1873) to
cupro-nickel in 1883 caused the change also of its popular name – although there
had been a 1 cent coin in the 1850s also called a nickel.

   The metal from which the coin gets its name is so called as an abbreviation of the
German *Kupfernickel* 'copper demon' – it looks like copper ore but copper cannot
be refined from it. With the second element, *nickel* 'dwarf', 'demon', 'troll', cf.
English '(Old) Nick', the devil, and *nixie*, a water-nymph.

nide of pheasants [collectives] This collective applies only to brooding pheasants
and pheasant families, for it derives through French from Latin *nidus* 'nest'. A
variant form is *nye*.

nil [sporting term] A score or rating of zero: 0.

   The term derives ultimately from Latin *nihil* 'nothing', but is found as a loan-
word in many languages in many forms.

nile [physics] In a nuclear reaction, a unit of reactivity.

                    1 nile = a reactivity change of 0.01

Because of the size of the nile, the more common practical unit is the millinile:

                    1 millinile = 10^{-3} nile
                                 = 10^{-5} reactivity change

nilpotent [maths] Describing a function or matrix that has some integral power equal
to zero.

nin [linear measure] In Thailand, the basic small linear measure.

                    1 nin = 0.8333 inch (ten-twelfths of an inch)
                                     (1 inch = 1.200 nin)

$$= \quad 2.1166 \text{ centimetres}$$
$$(2 \text{ centimetres} = 0.9449 \text{ nin})$$
$$12 \text{ nin} \quad = \quad 1 \text{ KUP } (10 \text{ inches, } 25.40 \text{ centimetres})$$
$$48 \text{ nin, } 4 \text{ kup} \quad = \quad 1 \text{ (Thai) KEN}$$

nine [quantitatives] 9.

$$^{9}\!/_{10} \quad = \quad 0.9 \quad = \quad 90\%$$
$$9 \text{ inches} \quad = \quad 1 \text{ span or breadth, } \tfrac{1}{4} \text{ yard}$$

A group of nine is a nonet or an ennead.

A POLYGON with nine sides (and nine angles) is a nonagon (or enneagon).

A solid figure with nine plane faces is an enneahedron.

As is evident from the above, the prefix meaning 'nine-' in English corresponds either to (Latin-based) 'non-', 'nona-' or to (Greek-based) 'ennea-'.

The name for the number derives from common Indo-European, attested over the entire range of languages from Sanskrit to Welsh. There would, however, seem to be some linguistic connection in these languages between *nine* and *new*, and it has been suggested that at an early time when fingers were used for counting – and not thumbs – nine would correspond then to the need for a new hand on which to count. *See also* NINTH.

ninth [quantitatives] ⅑ (one-ninth); the element in a series between the eighth and the tenth, or the last in a series of nine.

$$\tfrac{1}{9} \quad = \quad 0.1111 \ldots \quad = \quad 11.1111\%$$

ninth hour *see* NONES

nip [volumetric measure] Apparently a measure derived from the Dutch (medieval Dutch *nippe* 'sip'), and, if so, characteristic of those good people in that it varied according to whether applied to beer or to spirits. A nip of spirits today remains a small quantity, generally much less than a glass (and in the UK officially ⅙ of a GILL). But, even in nineteenth-century England, a nip of ale could be as much as half a (UK) pint, and thus served in a nipperkin.

nit [physics] Unit of luminance in the metre-kilogram-second (MKS) system, equal to 1 CANDELA per square metre.

The term derives from the Latin *nitor* 'brightness'.

NMR, nuclear magnetic resonance *see* SCANNING SYSTEMS

Nobel Prizes [comparative values] Series of large financial prizes that also have great worldwide prestige, instituted posthumously by the Swedish chemist and engineer Alfred Bernhard Nobel (1833-96), the inventor of dynamite. Initially there were five Nobel Prizes, to be presented annually unless no worthy recipient could be found. The five were for deeds and discoveries held to benefit the whole of humankind in the subjects of: physics; chemistry; physiology or medicine; literature; and peace.

Funded by Nobel's estate, four scientific institutions now regulate those awards: the Royal Swedish Academy of Sciences (physics and chemistry), the Royal Caroline (Swedish) Medico-Chirurgical Institute (physiology and medicine), the Swedish Academy (literature), and the Norwegian Nobel Committee, appointed by the Norwegian parliament in Oslo (peace). In council and with nominations from disinterested experts in the field, these institutions select the winners, and, at glittering public ceremonies in their own halls, present the prizes. The first prizes were presented in 1901.

In 1969 an additional Nobel Prize was established under the aegis of the Bank of Sweden, in the subject of: economic science.

This award is also in the gift of the Swedish Academy, and is regarded as equal in value and prestige to the others.

nobility [comparative values] The titles and ranks of British nobility have over the last fifty years become less of a matter of social status – they were never in any case a guarantee of what used to be called 'class' – and more a matter of hereditary consequence (often in the form of debts caused by a need to maintain stately homes now dilapidated but illegal to demolish). The activities of younger members of royalty and the posturing of senior politicians have similarly tarnished much of the former glory of the titles to which they both have acceded.

Nonetheless, the principal titles, from greater to lesser, are:

male title	form of address	female title	form of address
prince	your highness	princess	your highness
duke	your grace	duchess	your grace
earl/marquis	my lord	countess/marchioness	my lady
viscount	my lord	viscountess	my lady
baron	my lord	baroness	my lady
baronet/lord	my lord	lady	my lady
knight	Sir (forename)	lady	Lady (surname)

As is evident from the above, the Latin-based title 'count' is the equivalent of the Scandinavian-based title earl (formerly *jarl*). The title marquis is sometimes alternatively spelled 'marquess'. In medieval England and Scotland, the equivalent of the title viscount or baron (as applied especially to the jarl's son) was *thegn* or *thane*: Macbeth, for example, was created Thane of Cawdor. Members of the present British royal family ordinarily have at least three titles each, of descending consequence, generally reflecting inheritances in England, Scotland, Wales, and Northern Ireland. A lady born into her title may be addressed as Lady (forename). It is becoming common now to address all male title-holders up to the rank of duke as 'sir', and all female title-holders up to the rank of duchess as 'madam'.

noble [comparative values] Gold coin in use in England during the time of Edward III (ruled 1327-77) and for a few decades thereafter. Its name may have been intended to reinforce the prestige of the monarchy at a time when the power of the barons, earls, and other courtly princelings was threatening to get out of hand, for its value was set at 6 shillings and 8 pence – that is, at one-third of a SOVEREIGN (20 shillings, one pound).

node [physics; astronomy; maths] For a standing wave, such as a vibrating string, a stationary point on it (that is, a point with zero amplitude). Alternatively, it is the point of connection of two or more electrical conductors.

For an orbiting celestial object, a point where its orbit cuts the ecliptic or celestial equator. If the orbit is moving northwards it is the ascending node; moving southwards it is the descending node.

In mathematics, the point where two or more arcs in a network meet.

noggin [volumetric measure] In seventeenth-century England (and in a loose sense thereafter), both the container and the quantity of a small volume of alcoholic drink not always precisely defined. But most commonly,

$$1 \text{ noggin} = \frac{1}{4} \text{ pint} = 1 \text{ GILL or QUARTERN}$$
$$= 5 \text{ (UK) fluid ounces (4.815 US fluid ounces)}$$
$$= 142.06 \text{ millilitres (0.14206 litre)}$$

The chances are that the term derives from the specially strong ale brewed in East Anglia (primarily the English counties of Norfolk and Suffolk) at the time, called nog. Just as a nipperkin held a nip, so a nogkin (or noggin) might have been appropriate to nog.

nomogram, nomograph [maths] A diagram or chart consisting of two or more lines (straight or curved) marked with scales, against which measurements and calculations can be made. The straight-line version is also called an alignment chart.

The term derives from ancient Greek elements effectively meaning 'regulation graph'.

non-, nona- [quantitatives: prefix] Prefix which implies ninefold unity.

Example: nonagon – a POLYGON of nine sides (and nine angles)

The term derives ultimately from Latin *novem* 'nine' through its compound form *nonus* (for **nov-nus*) 'ninth'.

nonagenarian [time] Person who has passed the ninetieth birthday, but has not yet reached the hundredth.

nonagon [maths] A POLYGON with nine sides (and nine angles), alternatively known as an *enneagon*.

non-denominational number system [maths] A system in which the value of numbers does not depend on their position in a numeral (as they do with most number systems – in the numeral 223, the first 2 stands for 200 whereas the second 2 stands for 20). Roman numerals are an example of non-denominational numbers.

none [quantitatives; literary] Not one, no number; zero in quantity or number.

The term derives from an Anglo-Saxon equivalent of 'no(t) one' and is thus always followed by a SINGULAR verb.

nones [time] Old French version of two forms of the Latin *nonus* 'ninth', referring either to *nona (hora)* 'the ninth hour' or *nonae* 'the ninth day (before the *ides*)'.

The ninth hour in ancient Roman reckoning was the ninth hour after dawn – that is, around 3 p.m. (15.00). Through the inconsequential retiming of the early Catholic liturgical offices (among which nones is the fifth of seven canonical 'hours'), however, the service originally celebrated at this time was moved back three hours, to midday – which is why midday is alternatively known as *noon*.

The ninth day in the early days of ancient Rome was the equivalent either of a Sunday, intended for religious devotions, or of a Saturday, market day (Latin *nundinae*), although it actually occurred every eighth day because the last of one group of nine was the first of the next. To ensure that these special days fell equally within the ten months of the early Roman year, the nones of the four months containing 31 days [March, May, July (Quinctilis), and October] were celebrated on the seventh day before the ides, and the nones of the other six months which contained 30 days [April, June, August (Sextilis), September, November, December] on the ninth day before the ides.

In modern terms, this means that the nones corresponded to:

March	7th
April	5th
May	7th
June	5th
July	7th
August	5th
September	5th
October	7th
November	5th
December	5th

When Julius Caesar caused two further months to be added to the calendar much later, half-way through the first century BC, he reckoned that the two new months of January and February were of the less consequential 30-day variety, which (in modern terms) gave them nones corresponding to:

January	5th
February	5th

See also CANONICAL HOURS.

nonet [quantitatives; music] A group of nine, other than in musical terms alternatively called an *ennead*.

In music, also a group of nine, but generally of nine instrumentalists mostly playing different instruments of the same overall family (such as woodwind or brass).

Also, a composition for such a group.

non troppo *see* MA NON TROPPO

noon *see* NONES

no-par [comparative values] Of a stock, bond, banknote, or other form of realizable legal tender, having no face value; without a value specified on it.

normal [maths; physics] At right-angles to (a line or plane); perpendicular to (a base). For a curve, the normal at a point is at right-angles to the tangent at that point.

normal [chemistry] Describing a solution containing 1 gram equivalent weight of dissolved substance in 1 litre of solution; designated by the symbol N. Stronger solutions are designated as multiples of N (such as 2N – twice normal, 3N, and so on); weaker solutions as fractions of N (such as $N/10$ – decinormal, one-tenth normal, and so on). It has been largely superseded by MOLARITY (molar solutions).

normal curve, normal distribution *see* BELL CURVE, BELL-SHAPED CURVE

normal pressure [physics] A pressure of 101.325×10^3 newtons per square metre (or pascals), equivalent to 760 torr (or 760 millimetres of mercury). It is alternatively known as standard pressure: *see* STP; STP.

normal temperature [physics] A temperature of 0°C (or 273.16 K), alternatively known as standard temperature: *see* STP; STP.

normal temperature, medical *see* TEMPERATURE, MEDICAL
normal temperature and pressure *see* NTP
north *see* POINTS OF THE COMPASS
northing [astronomy; navigation; surveying] In astronomy, a (coordinate's) distance northwards from the celestial equator; a positive declination.

In navigation, the distance travelled northwards, usually expressed as a difference in LATITUDE.

In surveying, a northerly displacement from the observer or the origin (reference point); a north latitude.
north magnetic pole *see* MAGNETIC NORTH
note [music] In a very general sense, a musical note is a single example of what is heard when a musical instrument is played, or a voice sings; a melody line is thus made up of a string of such notes, and some chords contain specific notes that lend special effect. The word is, in addition, used for the written form of such a note: *see* MUSICAL NOTATION. And it is also used for a key (that is, a black or white lever) on a piano or any other keyboard instrument (hence black notes and white notes).

Finally, in the United States, a (whole) note is what is elsewhere generally known as a SEMIBREVE, a note of a specific duration, and the basis of the entire system of the notation of musical duration.
note frequencies *see* PITCH
nought, naught [quantitatives] The number zero: 0 .

The word derives from the Anglo-Saxon equivalent of 'no-whit' or 'no[t] aught'. Its adjective, *naughty*, originally meant 'worth nothing'.
noun [literary] A word for a person, place, or thing; any word that can have 'a'/'an' or 'the' in front of it.

Ordinary nouns, spelled without an initial capital letter, are known as *common nouns*. Names are *proper nouns*. In a slightly different method of distinction, *abstract nouns* are common nouns that do not refer to physical objects: examples are 'liveliness', 'despair', and 'intuition'.

Once mentioned in a sentence or series of sentences, a noun may be replaced by a PRONOUN. Thus, for example, in the sentence 'Egbert is my friend but he is a jerk', on the second occasion the proper noun 'Egbert' is replaced by the pronoun 'he'. For common nouns the replacing pronoun is ordinarily 'it'.

Nouns change according to whether they are singular or plural. The usual way to make a singular noun plural is to add -s , but there are exceptions: most nouns which, in the singular, already end in -s add a further -es to become plural. Thus

bonus (singular)	bonuses (plural)
bus (singular)	buses (plural)
class (singular)	classes (plural)

some (but not all) nouns ending in -o also take a plural in -es :

buffalo (singular)	buffaloes (plural)
potato (singular)	potatoes (plural)

some (but not all) nouns ending in -f takes a plural in -ves:

calf (singular)	calves (plural)
thief (singular)	thieves (plural)
wife (singular)	wives (plural)
wolf (singular)	wolves (plural)

and technical words may retain a classical or foreign plural form:

crisis (singular)	crises (plural)
criterion (singular	criteria (plural)
formula (singular)	formulae (plural)
goy (singular)	goyim (plural)
radius (singular)	radii (plural)

and there are quite a few downright irregular plurals, including:

man (singular)	men (plural)
woman (singular)	women (plural)
child (singular)	children (plural)
goose (singular)	geese (plural)
ox (singular)	oxen (plural)

mouse (singular)   mice (plural)

The word *noun* derives through French from the Latin *nomen* 'name' (literally 'what a thing/a person is known by/as').

nox [physics] Obsolete unit of very low levels of illumination.

$$1 \text{ nox} = 10^{-3} \text{ lux}$$
$$= 10^{-3} \text{ lumen per square metre}$$

The term derives directly from Latin *nox* 'night' (at a late stage deified as the Roman goddess of the hours of darkness).

noy [physics] Unit of noisiness (as perceived by a listener), equal to the noisiness of random noise in the band 910 to 1,090 hertz at a sound pressure level that is 40 decibels above 0.0002 microbars.

The term is (perhaps regrettably, perhaps faintly humorously) meant to correspond to a 'singular' form of the plural-sounding *noise*.

N-P-K system [chemistry: horticulture] In the composition of a fertilizer as listed on its container, a system intended to indicate in what type of plants the fertilizer is most suited to promoting growth. The three initials refer to the constituent proportions of nitrogen, phosphorus, and potassium (chemical symbols N, P, and K, respectively) in the fertilizer. According to the system, a 4-4-10 fertilizer therefore contains four parts of nitrogen and four parts of phosphorus to ten parts of potassium (potash) – and so is a 'high-potash' fertilizer best suited to promoting the growth of flowers and fruiting plants. High-nitrogen fertilizers are used to promote foliage, high-phosphor fertilizers to promote root growth and root crops.

NTP [chemistry; physics] Abbreviation of normal temperature and pressure, largely superseded by S.T.P. (STP): *see* NORMAL PRESSURE; NORMAL TEMPERATURE.

nuclear energy [physics] Strictly, the energy released on the disintegration of an atomic nucleus; the binding energy. More generally, the term is applied to the large amounts of energy released in (fission or fusion) nuclear reactions, also loosely – and incorrectly – called atomic energy.

The word *nuclear*, adjective of *nucleus*, derives from the Latin *nucleus* (earlier *nuculeus*), diminutive of *nux* 'nut', thus 'kernel' (cf. German *Kern* 'nucleus'), 'core'.

nuclear force [physics] The strong interaction force between certain subatomic particles (for example, between neutrons and protons in an atomic nucleus).

For the etymology of *nuclear/nucleus*, *see* NUCLEAR ENERGY.

nuclear magnetic resonance (NMR) *see* SCANNING SYSTEMS

nuclear magneton *see* MAGNETON

nucleon number [physics] The total number of neutrons and protons in the nucleus of an atom; the mass number.

number [maths] At its simplest, a number is merely a numeral – such as 1, 2, 3, 4, and so on, which are examples of whole numbers, or integers (as opposed to fractions), also called digits. In mathematical terms 0 (zero) is regarded as a number. The ordinary set of counting numbers are also called natural, or cardinal, numbers. More complicated mathematics and its various scientific and technological applications additionally require other kinds of numbers.

A *rational number* is any number that can be written as a fraction (as the ratio of two whole numbers).

An *irrational number* is a real number that cannot be written in the form of a fraction (as the ratio of two whole numbers). Examples include e, π, and $\sqrt{2}$. The square root of 2 is also an example of a *surd*, the name given to any number that involves such a root, which cannot be resolved (and thereby expressed as a rational number).

A *real number* is any positive or negative rational or irrational number (as opposed to an imaginary number).

An *imaginary number* is obtained by multiplying a real number by i, the square root of -1 (that is, $i^2 = -1$).

A *complex number* has the form $x + iy$, where x and y are real numbers and i is the square root of -1 (the term iy is an imaginary number).

numeral [quantitatives] The symbol for a number, especially a whole number, integer or digit.

The numerals in present everyday use are often referred to as 'Arabic numerals', and it was certainly through trading with Arab countries that the numerals passed into the culture of most European nations. But, in origin, the numerals most probably derive from the symbols first used in India in the third century BC. Numerals 1, 4, and 6 were current at that location in roughly their present form at that time; 2, 7, and 9 were to be found there a century later; and 3 and 5 were in use by the first century AD. It is possible that the 0 or zero was also used in India then, although evidence for it has not survived and the term *zero* derives immediately (through Italian and French) from Arabic.

The Indian (Hindi) numerals pased into Arabic usage during the AD 700s, shortly after which the Arab mathematician al-Khwarizmi (whose name is commemorated in the word *algorithm*) wrote a booklet on the subject. The earliest known European manuscript containing the numerals dates from 976, in Spain – some 150 years before al-Khwarizmi's booklet was translated into Latin by Abelard of Bath. *See also* ROMAN NUMERALS.

numerator [maths] In a fraction, the number above the line. (The number below the line is the denominator.)

Nummulitic *see* EOGENE

Nusselt number, biot number [physics] In fluid dynamics, a number that quantifies the transfer of heat between a moving fluid and a solid, equal to the product of the heat loss per unit area of solid and its length, divided by the product of the temperature difference between the solid and the fluid and the thermal conductivity of the solid. Its more common name is taken from that of the German engineer W. Nusselt.

nu value [physics] For a transparent medium, the inverse of its dispersive power, equal to its mean refractive index minus 1, divided by the difference between its refractive indices for red light (wavelength 656.3 nanometres) and blue light (486.1 nanometres). The nu value is alternatively known as the *constringence*.

nye of pheasants [collectives] This collective applies only to brooding pheasants and pheasant families, for it derives through medieval French *ni* from Latin *nidus* 'nest'. A less eroded variant is *nide*.

O

0 (zero, nought) [quantitatives; maths] The role of 0 in number systems is as a position indicator. For example: in the decimal system, the zeros in the number 1,020 indicate that the 1 is in the thousands (position) and the 2 is in the tens (position) – and that there are thus no hundreds or units in the number.

In maths, 0 is also the symbol for an empty set.

Although in certain contexts (especially in Britain in the recounting of telephone numbers), the symbol for zero may be verbally described as 'O', it does not derive from the alphabetical letter. At the same time, the form of the alphabetical letter actually corresponds to the position of the mouth in making the vowel sound it represents, while simultaneously being a fairly readily comprehensible ideogram for 'emptiness', the meaning of the mathematical symbol as borrowed from the medieval Arabs (*see* ZERO). *See also* LOVE; NIL; NOUGHT, NAUGHT.

oblate [maths] Describing something that is wider than it is long (cf. *oblong*). An oblate spheroid, for example, resembles a slightly flattened sphere – the shape of the Earth, which is slightly 'squashed' at the poles.

oblique *see* SOLIDUS, OBLIQUE, VIRGULE

oblique angle [maths] Any angle other than a right-angle (90°, 270°) or a straight line (180°, 360°). Both ACUTE ANGLES and OBTUSE ANGLES are oblique angles.

oblong *see* RECTANGLE, OBLONG

oboe *see* WOODWIND INSTRUMENTS' RANGE

obolus, obol [weight; comparative values] In ancient Greece, the name of a tiny unit of weight also applied to a coin of fairly low value (although the coin was initially struck in silver, later becoming bronze). In both respects,

$$1 \text{ obolus} = 8 \text{ khalkoi}$$

6 oboloi = 1 DRACHMA

In ancient Rome, the name only of a tiny unit of weight.

1 obolus = one-eighth of a solidus, one-twelfth of a siculus
48 oboli = 1 UNCIA

The name for the unit derives from the Greek for 'nail', presumably because the weight was that of one standard iron nail, but it is interesting that that meaning is itself a late extension of its original root meaning '(sharp object to be) thrown'. *See also* NAIL (last paragraph).

obtuse angle [maths] Angle of between 90° and 180° exclusively.

oceans [square measure; volumetric measure] The oceans cover more than 70 per cent of the entire surface of the Earth, accounting for some 140 million square miles/361 million km^2 area at the surface, and containing about 330 million cubic miles/1,322 million km^3 of water altogether – 97.5 per cent of all the water on the planet.

By European tradition, there are five oceans: the Pacific, the Atlantic, the Indian, the Arctic, and the Antarctic. But geographers normally count only three – the Arctic and Antarctic Oceans are held to be merely the most northerly and most southerly portions of the three other oceans.

	area in thousands		*average depth*		*greatest depth*		*annual expansion*	
	sq. mi.	km^2	ft	metres	ft	metres	in.	mm
Atlantic	41,147	106,570	±9,000	±2,743	30,200	9,200	0.6	15
Indian	28,617	74,118	12,598	3,840	26,400	8,047	0.5	12
Pacific	70,018	181,346	12,925	3,940	36,197	11,033	0.1	2.5

The average depth of the Atlantic in this table has to be expressed as an approximation because it includes the Arctic Ocean, which is much shallower (average depth 3,407 feet/1,038 metres) than the nine-times larger true Atlantic Ocean (average depth 11,730 feet/3,575 metres).

o' clock [time] Expression denoting the hour, especially the current hour, and especially at the stroke of the full hour. Short for 'of the clock', in its modern meaning it derives from a time no earlier than when clocks were actually in common use: the later part of the seventeenth century. Prior to that, although the expression did exist, it referred instead to the sounding of the bell at the hour – all the Germanic, Latinate, and Celtic cognates of English *clock* mean 'bell'. The shift of the meaning from 'bell' to 'clock' came about as a result of the invention of the chiming pocket watch.

oct-, octa-, octo- [quantitatives: prefix] Prefix that implies eightfold unity.
Examples: octagon – a POLYGON of eight sides (and eight angles)
octahedron – a POLYHEDRON with eight plane faces
octuple – eightfold, eight times, of eight elements

The term derives from Latin *octo* 'eight' but is cognate with virtually all Indo-European words for that number. It has also been suggested, however, that the ultimate root of the word contains elements meaning 'two times four', dating from a time when fingers – but not thumbs – might have been used for everyday arithmetic (*see also* NINE).

octad, ogdoad [quantitatives] Group or series of eight elements. The form 'ogdoad' is simply a Latinized Greek variant.

octagon [maths] A POLYGON of eight sides (and eight angles).

octahedron [maths] A solid figure (POLYHEDRON) with eight plane faces. A regular octahedron has eight equilateral triangles as its faces and takes the form of a pair of square-based pyramids joined base to base.

octal notation [maths] A number system in which the base (radix) is 8 – that is, in which the full set of digits is 0, 1, 2, 3, 4, 5, 6, 7.

octameter [literary: poetic measure] A line in verse that has eight feet or stresses.

octane number, octane rating [chemistry] Number that indicates the knock rating of a petrol fuel (for a spark-ignition internal combustion engine). 'Knock' or 'pinking' is the name given to pre-ignition, the premature ignition of the fuel before the spark actually occurs. The fuel under test is compared with a mixture of the alkane hydrocarbons *iso*-octane (2,2,4-trimethylpentane) and *n*-heptane. The percentage of *iso*-octane in the mixture that gives the same knock characteristics is the octane number.

Octane gets its name as the eighth (Latin *oct-* 'eight') member of the alkane (methane) homologous series of hydrocarbons, formerly known as paraffins. [The first ten members of the series are methane, ethane, propane, butane, pentane, hexane, heptane, octane, nonane, and decane. From pentane (Greek *pent-* 'five') onwards, they are named after the number of carbon atoms in their molecules.]

octant [maths] An eighth of a circle, or one of the eight parts of space created by the intersection of three planes at one point.

octant [astronomy] An instrument, with a 45° angular scale, used for measuring angles. The term is also used to describe a celestial object that is 45° from another object.

octavalent, octavalence [chemistry] Having a valence (valency) of eight.

octave [music; time; chemistry; quantitatives] In music, the eight notes that make up the chromatic (major) SCALE, between a KEYNOTE (or tonic) and the next corresponding keynote (tonic) up or down, represented on a piano, for example, by the white notes between one C and the next C. Also the interval between the first and last of these notes (that is, between one C and the next, one D and the next, and so on); technically, the frequency (pitch) of the lower note is exactly half that of the upper. Also the sounding of such an interval (two Cs, two Ds, two Es, and so on).

As a measure of time, the octave is the eighth day of a festival or fast as observed by the Church primarily in former centuries. Many feasts and fasts in the Church calendar are still observed by the most devout not just for the day as specified but for the whole week thereafter, a practice that is known as 'keeping the octave'.

In chemistry, a group of eight consecutive elements. It became significant in Newlands' law of octaves: in a list of elements in order of their atomic masses (weights), any one element resembles those eight places behind it and eight places in front of it in the list. It was first stated by the British chemist John Newlands (1837-98) who, in 1864, anticipated the Russian chemist Dmitri Mendeleev in observing periodicity in the properties of the elements: *see* PERIODIC TABLE OF ELEMENTS.

The term can also describe any group of eight items (especially lines in verse, or parries in fencing), but derives in fact from the Latin adjective *octavus* 'eighth'.

octavillo [volumetric measure] In Spain, a volumetric measure of dry capacity.

1 octavillo	=	289 cubic centimetres
	=	0.5086 UK dry pint, 0.5244 US dry pint
	=	17.636 cubic inches
4 octavillos	=	1 CUARTILLO
16 octavillos, 4 cuartillos	=	1 ALMUDE
192 octavillos, 48 cuartillos, 12 almudes	=	1 FANEGA

The term derives from the Spanish for 'eighth', although the unit is in fact one-sixteenth of an almude and one-fourth of a cuartillo (which itself means 'quarter').

octavo, eightvo., 8vo [paper size] A paper size that results from folding a designated sheet three times (to form eight pages).

The term derives from the Latin *in octavo* 'in one-eighth (of a sheet)'. *See also* BOOK SIZES; PAPER SIZES.

octennial [time] Once every eight years; lasting for eight years.

octet [quantitatives; music] A group of eight, especially in music a group of eight players of different instruments of the same family (such as woodwind or brass); also a composition for such a group.

octo- *see* OCT-, OCTA-, OCTO-

octodecimo [paper size] An alternative term for eighteenmo (18mo), an eighteenth of a sheet (of paper). The term derives directly from Latin, meaning 'as an eighteenth'.

octogenarian [time] Person who has passed the eightieth birthday, but has not yet reached the ninetieth.

octuple [quantitatives] Eightfold, eight times; comprised of eight elements.

ocular tests *see* OPHTHALMIC TESTING

odds, calculation of [maths; sporting term] The odds that a particular event will happen is mathematically the same as the probability – the difference is only the way in which they are expressed.

Two examples may help to clarify the difference. When a single coin is tossed, the probability that it will land heads (or tails) is 0.5 (= ½), which is the same as saying that, in the long run, a head (or tail) will occur once in every two throws, or that the odds are evens that a head (or tail) will occur in any one throw (sometimes written as '1:1'). When a single dice is thrown, the probability that any nominated number will turn up is 0.1666... (= ⅙), equivalent to a chance of 1 in 6, or odds of 5 to 1 against ('5:1'). *See also* PROBABILITY.

odds-on [sporting term] Term describing an event that is so likely to occur that the odds, if calculated by a professional bookmaker, would turn out to be less than 1:1 . When applied to a wager on a favourite to win in a horse race, the resultant reward if the favourite won would be (the premium originally bet returned plus) less than the premium.

Rather than quote such odds as (for example) 'one-half to one', the convention is to state them as 'two to one on', written as '1-2'.

ode *see* VERSE FORMS

odometer [linear measure] Device in a car or other motor vehicle that records and displays the number of kilometres or miles travelled, usually to the nearest tenth. One that records and displays the mileage is alternatively known as a milometer. On most dashboards the odometer appears within or beside the speedometer, and (in a mechanical odometer) is linked to the same shaft that revolves the speedometer magnet.

An additional odometer that can be manually reset to zero, so that the distances of individual journeys (trips) can be measured, is more often called a tripmeter or 'the trip'.

The term derives as an originally American borrowing of a French word, ultimately from ancient Greek *hodo-* 'road', 'way', *metron* 'measure'. The French were responsible for the loss of the initial *h-* . *See also* HODOMETER; TACHOGRAPH

oersted [physics] Unit of electromagnetic field strength in the centimetre-gram-second (CGS) system. A circular coil of 1 centimetre radius carrying a current of 1 abampere (= 10 amps) has a magnetic field of 2π oersteds at its centre. The equivalent SI unit is the ampere per metre.

$$1 \text{ ampere per metre} = 4\pi \times 10^{-3} \text{ oersted}$$

The unit was named after the Danish physicist Hans Christian Ørsted (1777-1851), the 'discoverer' of electromagnetism, and is now spelled in German fashion.

officers in the armed forces *see* MILITARY RANKS; and individual armed forces ranks

O gauge [linear measure] In model railway systems, a gauge in which the rails are 1¼ inches (31.75 millimetres) apart – 0.022 of the 56½-inch (1.435-metre) standard gauge for railroads in most of Europe and all of Northern America – and the locomotives and rolling stock are built to a scale of 8 millimetres to the foot. *See also* OO GAUGE.

ogdoad *see* OCTAD, OGDOAD

ogive [maths] In statistics, a graph – of which the shape resembles the top half of an onion – representing the distribution of cumulative frequencies (of a random variable).

The term derives through French but is otherwise of unknown etymology.

ohm [physics] Unit of electrical resistance, equal to the resistance of a conductor which, with a potential difference of 1 volt across it, passes a current of 1 ampere. It was named after the German physicist Georg Simon Ohm (1787-1854) and adopted two years after his death – making it the oldest electrical unit.

ohm-centimetre [physics] Unit of resistivity in the centimetre-gram-second (CGS) system.

$$1 \text{ ohm-centimetre} = 0.01 \text{ (one-hundredth) ohm-metre (SI units)}$$

ohm-metre [physics] Unit of resistivity in SI units, equal (for a given material) to the product of its resistance and cross-sectional area divided by its length.

$$1 \text{ ohm-metre} = 100 \text{ ohm-centimetres (CGS units)}$$

Ohm's law [physics] A basic relationship in electricity which states that the voltage across a conductor equals the current flowing through it times its resistance. In other words,

current	=	voltage divided by resistance, and
resistance	=	voltage divided by current.

oitavo [volumetric measure] In Portugal, an old measure of liquid capacity.

1 oitavo	=	1.730 litres
		(1 litre = 0.578 oitavo)
	=	3.0447 UK pints, 3.6560 US pints
		(3 UK pints = 0.9853 oitavo
		3 US pints = 0.8206 oitavo)
2 oitavos	=	1 Portuguese QUARTO
32 oitavos	=	1 Portuguese FANEGA

The fact that 2 oitavos ('eighths') make a quarto ('quarter') strongly indicates the existence at one time of a unit comprising 8 oitavos (13.84 litres/3.0447 UK gallons, 3.656 US gallons) – although none is listed in reference sources.

oka, oke [weight; volumetric measure] Obsolescent unit of weight in many of the countries of the eastern Mediterranean and Near East; in some of them also a unit of capacity. Deriving through Arabic ultimately from the old Roman UNCIA, the unit has, over the centuries, come to differ slightly between the countries that use it – most notably Greece and Cyprus, Turkey, Bulgaria, and Egypt.

As a measure of weight:

1 oka in Egypt	=	1.248 KILOGRAMS, 2.751 pounds avdp.
		(1 kilogram = 0.801 Egyptian oka
		3 pounds avdp. = 1.090 Egyptian oka)
1 oka in Greece/Cyprus	=	400 Greek DRACHMAS
	=	1.280 kilogram, 2.822 pounds avdp.
		(1 kilogram = 0.781 Greek/Cypriot oka
		3 pounds avdp. = 1.063 Greek/Cypriot oka)
1 oka in Bulgaria	=	1.282 kilograms, 2.826 pounds avdp.
		(1 kilogram = 0.780 Bulgarian oka
		3 pounds avdp. = 1.0615 Bulgarian oka)
100 Bulgarian oka	=	1 TOVAR
1 oka in Turkey	=	400 Turkish drachmas
	=	1.284 kilogram, 2.830 pounds avdp.
		(1 kilogram = 0.779 Turkish oka
		3 pounds avdp. = 1.060 Turkish oka)

As a volumetric measure:

1 oka in Turkey	=	0.6057 LITRE
		(1 litre = 1.6510 Turkish oka)
	=	1.066 UK PINT, 1.28 US pint
		(1 UK pint = 0.9381 Turkish oka
		1 US pint = 0.7813 Turkish oka)
1 oka in Bulgaria	=	1.28 litre
		(1 litre = 0.7813 Bulgarian oka)
	=	2.2527 UK pints, 2.7050 US pints
		(2 UK pints = 0.8878 Bulgarian oka
		2 US pints = 0.7394 Bulgarian oka)

Okun's law [maths: economics] As the gross national product (GNP) of a country declines (because of recession, debt, or inefficiency), there is a proportional rise in unemployment (measured in thousands per billion dollars' decline), statistically calculable according to specific fiscal indices. The law is named after its deviser, the American economist Arthur M. Okun (1928-80).

old-fashioned glass *see* DRINKING-GLASS MEASURES

Old Stone Age *see* PALAEOLITHIC

Old Style [time] Usage of the Julian calendar (*see* YEAR).

In most countries of Europe, usage of the Julian calendar came to an end in 1582 when the Gregorian calendar was instituted to replace it. It was not until 1752, however, that the Gregorian calendar was adopted in Britain (requiring a shifting of the date then by eleven days). And the Old Style was still current in Russia until 1918.

oleometer [physics] A type of HYDROMETER for measuring the density of an oil to determine its purity.
   The first element of the term derives from Latin *oleum* 'oil'.

olfactometer [medicine] An instrument for testing the (power of the) sense of smell.
   The first element of the term derives from the Latin verb *olefacere* 'to cause to smell' (literally 'to make an odour').

olig-, oligo- [quantitatives: prefix] Prefix meaning 'a few', and sometimes 'too few', 'too little'.
   Examples: oligarchy – rule by a (select) few
             oligotrophy – too little by way of nutrition
   The term derives from ancient Greek *oligos* 'a few', 'a little'.

Oligocene epoch [time] A geological EPOCH during the Tertiary period of the Cenozoic era, corresponding roughly to between 38 million years ago and 25 million years ago. During the Eocene epoch prior to this, there had been a general warming of the Earth's surface and atmosphere so that, by the end, tropical and subtropical vegetation abounded in many forms still visible today. But the Oligocene reversed many of these trends. The overall temperature decreased markedly over the millennia, and the seas retracted by a significant margin. Mammals became the dominant terrestrial life form, and the first primates appeared.
   The name of the epoch derives from ancient Greek *oligo-* 'few', 'short', together with the suffixial element -cene (from Greek *kainos* 'new') common to all epochs. But the Oligocene is by no means the shortest epoch of the Cenozoic era (the Palaeocene is shorter, and the Pliocene the shortest), so the name presumably refers to the reduction in the overall number of organic life forms that occurred during it.
   For table of the Cenozoic *see* PALAEOCENE EPOCH.

omega-minus particle [physics] An elementary particle that carries a negative charge – energy 1,672 million electron volts (1,672 MeV) – and the most massive HYPERON.

omer [volumetric measure] In ancient Israel, and thus in the Bible, a unit of dry capacity.

1 omer or issaron	=	1.8 CAB or kab, 7.2 LOGS
	=	4.032 litres
		(4 litres = 0.9921 omer)
	=	7.096 UK pints, 7.316 US pints
		(1 UK GALLON = 1.1274 omer
		1 US gallon = 1.0935 omer)
3.333 omers, 6 cab, 24 logs	=	1 SEAH
10 omers, 3 seah, 18 cab, 72 logs	=	1 EPHAH
50 omers, 5 ephahs	=	1 lethech
100 omers, 10 ephahs, 2 lethechs	=	1 HOMER or KOR

onça [weight] In Portugal, a unit of weight.

1 onça	=	28.688 grams
		(25 grams = 0.8714 onça)
	=	1.01194 ounce avdp., 442.724 grains
		(1 oz avdp. = 0.9882 onça)
8 onças	=	1 Portuguese MARCO
16 onças, 2 marcos	=	1 Portuguese LIBRA (or arratel or pound)

The unit is obviously based yet again on the old Roman UNCIA, although – as with the English ounce – there are here 16 to the 'pound' as opposed to the 12 that constituted the Latin 'pound'. *See also* OUNCE.

once [weight] Former unit of weight in France.

1 once	=	30.590 grams
		(25 grams = 1.2236 once)
	=	1.0792 ounce avdp., 472.133 grains
		(1 oz avdp. = 0.9266 once)
8 onces	=	1 MARC

16 onces, 2 marcs = 1 LIVRE

The unit is obviously based yet again on the old Roman UNCIA, although – as with the English ounce – there are here 16 to the 'pound' as opposed to the 12 that constituted the Latin 'pound'. *See also* OUNCE.

ondograph, ondometer, ondoscope [physics] Various instruments for recording, measuring, or displaying high-frequency alternating currents (microwaves).

The first element of the term in each case derives through French *onde* from Latin *unda* 'wave'.

one [quantitatives] 1.

Prefixes implying singleness, singularity, individuality are 'mono-', 'haplo-' (from ancient Greek) and 'un-/uni-' (from Latin); prefixes implying union, multiplex unity are 'homo-' (from Greek), and 'con-/com-/col-', 'syn-/sym-/syl-' (from Latin).

Prefixes implying the corresponding ordinal number, 'first', are 'proto-', 'ant-/ ante-' (from Greek), 'pre-/prem-/prim-/prin-' (from Latin), and 'fore-' (from Germanic).

The name of the number is identical with the indefinite article *a, an*, the second form of which, in fact, represents the earlier version. In other Germanic tongues – both modern German and Dutch, for example – a single form is retained both for the number and the indefinite article. Cognates appear in virtually all Indo-European languages. In English a further cognate is the word *any*.

It is probable that the expression 'Number One', referring to oneself, is not so much the generic use of 'one' (which is comparatively late Norman-French) as a broadly semi-literate attempt to be jocular about using the word 'I'.

ons [weight] This Dutch variant on the old Roman UNCIA has now been assimilated to the metric system.

1 ons = 100 grams
= 3.5274 ounces avdp.
(4 ounces = 1.1340 ons)
10 ons = 1 kilogram (2.204623 pounds avdp.)

OO gauge [linear measure] In model railway systems, a gauge in which the rails are ⅝ inch (15.875 millimetres) apart and the models are built to a scale of 4 millimetres to the foot. *See also* O GAUGE.

opacity [physics] The optical density of a medium, equal to the luminous flux incident on the medium divided by the transmitted luminous flux (that is, the reciprocal of the transmission ratio).

The term derives from Latin *opacitas*, abstract noun formed from the adjective *opacus* 'shady'.

open set [maths] In set theory, a set that is the complement of a closed set.

open window unit (o.w.u.) [physics] Obsolete term for a unit of sound absorption, equal to the (complete) absorption of low-frequency sound waves that takes place in an open window of area 1 square foot (0.0929 square metre). It is now known as the SABIN.

operand [maths] The quantity on which a mathematical OPERATOR acts.

operations analysis *see* SYSTEMS ANALYSIS, OPERATIONS ANALYSIS

operator [maths] A term or symbol (such as + or x) that stands for a mathematical operation – which is carried out on an OPERAND.

ophthalmic testing [medicine] There are many tests that can be performed on the eyes to check various aspects of vision. The most common, known as optometry or, simply, an 'eye test', involves looking at a chart of letters of various sizes to enable an optician (US optometrist) to diagnose the correct lenses for spectacles (glasses) or contact lenses. An ophthalmic surgeon (US ophthalmologist, or oculist) has access to many kinds of instruments to examine a patient's eyes, from an ophthalmoscope (for examining the retina at the back of the eyeball) to a tonometer (for measuring the fluid pressure within the eyeball).

opisometer [physics] A device for measuring curved lines (such as roads on a road map).

The first element of the term derives from ancient Greek *opiso* 'backwards'.

opposite (angle, side) [maths] Of the four angles created when two straight lines intersect, opposite angles are just that – they lie opposite each other (and there are two pairs of them).

In a right-angled triangle, the longest side (opposite the right-angle) is known as the hypotenuse. If one of the other angles of the triangle is taken as the subject, the side of the triangle not involved in defining that angle is known as the opposite side (and the third side is the adjacent side).

opposition [astronomy] The time when any two celestial objects differ in longitude by 180 degrees (usually referring to the position of the Moon or a planet and the Sun).

optic [volumetric measure] In Britain and certain former colonies, a mechanical measure of spirits or of flavoured cordial, attached in bars to an upturned bottle. The spirit or cordial accumulates by gravity within the device's glass or transparent plastic container: a valve at the bottom of the container on pressure releases the measured volume into a drinking-glass.

Of spiritous liquor, the standard measure of volume in an optic is most commonly one-sixth of a GILL. (The unit is not the same in the United States as it is in Britain.)

one-sixth of a UK gill = one twenty-fourth of a UK pint
= five-sixths of 1 UK fluid ounce
= 23.675 millilitres

The device is so called because, thanks to the container's transparent sides, the measure remains totally visible (ancient Greek *optikos*) as it is being poured, and there can be no question of short measure.

optical density [physics] An alternative, and imprecise, term for TRANSMITTANCE.

optical square [surveying] A hand-held instrument for setting out right-angles, fitted with a pair of mirrors (like a sextant) to give lines of sight at 90 degrees to each other.

optic axis [physics] In physical optics, a line normal (at right-angles) to the surface of a lens and passing through its centre of curvature. Rays of light on the optic axis are not reflected or refracted, but pass straight through the lens. The optic axis is sometimes alternatively called the principal axis.

In crystallography, a line in a doubly refracting crystal along which no double refraction occurs (both ordinary and extraordinary rays move at the same velocity). Uniaxial crystals have one optic axis; biaxial crystals have two.

optometry *see* OPHTHALMIC TESTING

opus number [music] In a catalogue of musical compositions by one composer, a number that is intended to relate a specific work to the works that were composed chronologically before it and after it. Opus number one is generally the first (extant) work of any composer.

The term derives from Latin *opus* 'work'. In English, especially American English, the plural is *opuses* rather than *opera*.

orbit [astronomy; physics; chemistry] In astronomy and space science, the path of a heavenly body or artificial satellite around another heavenly body resulting from gravitational attraction between the two. The orbiting object has to maintain a certain minimum velocity – the orbital velocity – to remain in orbit (if it moves faster it will gradually orbit farther and farther away; if it moves more slowly it will gradually spiral closer to the body it is orbiting). Stable orbits are typically ellipses.

In a simple model of an atom, the path of an electron circling the nucleus is termed its orbit: *see* ORBITAL.

The term derives from Latin *orbita* 'wheel-track', 'course', itself a derivative of *orbis* 'circle', 'wheel', 'globe'.

orbital [physics] The region surrounding the nucleus of an atom in which there is a high probability of locating an electron. This probability is expressed by the wave function of the electron (defined by a set of four QUANTUM NUMBERS).

orbital index [medicine] In comparative anthropology and medicine, the ratio of the horizontal width of the orbit (the skull socket of the eye) to its vertical length.

orbital period [astronomy] The time it takes for a celestial body within a gravitational field (such as a planet or comet) to complete one orbit around its gravitational focus (such as the Sun). Of the planets and comets in the Solar System, the orbital period depends upon such factors as orbital velocity, orbital eccentricity, and the aphelion and perihelion (maximum and minimum distances from the Sun respectively), expressed in astronomical units or in millions of kilometres or miles.

For the orbital period of the planet Earth: *see* YEAR.

As an example of the orbital period of a comet, that of Halley's comet is seventy-six years and about thirty-seven days (its orbital eccentricity is 0.97; aphelion 35 astronomical units – farther than Neptune; perihelion 0.6 astronomical units – inside the orbit of Venus). Its next appearance is due in early 2063.

Order, order [biology: taxonomy] In the taxonomic classification of life-forms, the category between Class and Family.

For full list of taxonomic categories *see* CLASS, CLASS.

ordinal numbers [maths; quantitatives] Numbers that indicate rank or order, such as 1st, 2nd, 3rd, 4th, and so on. Ordinary counting numbers (1, 2, 3, 4, and so on) are known as cardinal numbers.

ordinate [maths] The y coordinate of a point in coordinate geometry (that is, its distance from the x-axis). The other coordinate is the abscissa.

Ordovician period [time] A geological PERIOD during the Palaeozoic or Primary era, after the Cambrian period but before the Silurian period, corresponding roughly to between 505 million years ago and 438 million years ago. During this comparatively early division of the Earth's history, the planetary surface was subject to considerable volcanic activity and mountain-building. The invertebrate animals known as graptolites flourished; limpet-, clam-, or lampshell-like brachiopods diversified mightily; gastropods and cephalopods abounded; and the trilobites developed to much more advanced forms. Vertebrate life-forms appeared for the first time in the seas of what is now North America.

The period is named after an ancient Celtic (British) tribe of central and northern Wales – an area particularly associated with rocks and fossils of this period – known as the Ordovices, who were overrun by the Romans in the AD 70s. The tribe's main base was probably at Dinorwic (which is thought itself to mean 'Stronghold of the Ordovices'), between present-day Caernarfon and Bangor, and between Snowdon and the Menai Strait.

For table of the Palaeozoic era *see* CAMBRIAN PERIOD.

öre [comparative values] Unit of currency in Denmark, Norway, and Sweden.

$$
\begin{array}{ll}
\text{1 Danish or} & \\
\text{Norwegian øre} \quad = & \text{1 krone} \\
\text{1 Swedish öre} \quad = & \text{1 krona}
\end{array}
$$

The term derives through Germanic borrowing from ancient Celtic words for 'shiny metal' (as, for example, Old English *ar* 'copper', 'brass') that gave rise to such modern English words as *ore* and, through the adjective (Old English *aeren* 'brazen'), *iron*. See also AUR, AURAR; COINS AND CURRENCIES OF THE WORLD.

organ range *see* KEYBOARD INSTRUMENTS' RANGE

origin [maths] In coordinate geometry, the intersection of the x- and y-axes, from which the coordinates of all other points are measured; the origin has the coordinates (0, 0).

orrery [physics: astronomy; maths] Complex instrument incorporating a number of clockwork motors that operate rotating metal circles on which globes representing the planets revolve around a globe corresponding to the Sun. The device was intended to illustrate and explain the movement of the known planets across the sky, and required many extremely complex mathematical computations to get the apparent planetary motions correctly aligned with each other.

First designed by the inventor George Graham, the instrument was named after his patron Charles Boyle, Fourth Earl of Orrery (1674-1731), First Baron Boyle of Marston, Somerset.

orthocentre [maths] For a triangle, the point at which its three altitudes intersect (an altitude is a line at right-angles to a side that meets the opposite vertex).

orthogonal [maths] In geometry, describing a figure that has right-angles, or a line that is perpendicular (to another line).

Also, describing a set of functions that have a product of zero.

Also, describing a pair of vectors whose scalar product (DOT PRODUCT) is zero.

The term derives from the ancient Greek adjective *ortho-gonos* 'right-angled' (an 'orthogon' is an unwieldy alternative word for 'rectangle').

orthorhombic [maths; crystallography] Describing a solid figure or crystal of which the three axes – all of different lengths – are at right-angles to each other.

oscillation [physics] A vibration, the repeated and regular displacement of a point from one limit to another (that is, a periodic motion). Half the total displacement is the motion's *amplitude*. There are many examples in physics, from the motion of a simple pendulum to the rapidly varying voltage of an alternating current. High AC frequencies for radio broadcasting are produced by a circuit called an oscillator.

The term derives from the Latin *oscillatus*, past participle of *oscillare*, a derivative of the noun *oscillum* 'a little face', thus 'a mask': masks of the god Bacchus were traditionally hung on trees and bushes in vineyards, where they swung to and fro in the wind – they oscillated. *See also* SIMPLE HARMONIC MOTION.

oscilloscope [physics] An instrument for displaying and measuring waveforms. The waveform is converted to a varying voltage (if it is not already in that form) and, using a regular time-base, made to control the movement of an electron beam in the display on the screen of a cathode-ray tube.

For the etymology of the first element of the term *see* OSCILLATION.

osmol, osmole [physics] Unit of osmolality and osmolarity, used to express the concentrations of solutions that exert osmotic pressure. The osmolality (or osmolarity) of a solution is the molality (or molarity) of a non-dissociating solution that exerts the same osmotic pressure. *See also* OSMOLALITY, OSMOLARITY.

osmolality, osmolarity [physics] In osmosis – the movement of a solvent, through a semi-permeable membrane, from a dilute solution to a more concentrated solution – the osmolality is the degree of the pressure involved in the process, and the osmolarity is the tendency for the process to take place and the extent to which it does. *See also* OSMOTIC PRESSURE; OSMOL, OSMOLE.

osmotic pressure [physics] Osmosis is the movement of a solvent, through a semi-permeable membrane, from a dilute solution to a more concentrated solution. The 'vigour' of this movement is expressed as the osmotic pressure, which is the pressure that would have to be applied to prevent osmosis between the solution and pure water. It may be monitored and measured on an *osmometer*.

The term *osmosis* derives from the ancient Greek *osmos* 'impulse', 'thrust'. *See also* OSMOL, OSMOLE.

ostentation of peacocks [collectives] An entirely appropriate – if somewhat chauvinistic – collective for the male peafowl, and, in its way, simply a translation of the other collective for peafowl, a *muster*.

The word *ostentation* derives through French from a Latin verb meaning 'to extend in front of', thus 'to exhibit'.

ostmark [comparative values] Former unit of currency in East Germany, when that was a sovereign (if Soviet-dominated) state, now replaced by the DEUTSCHE MARK. The prefix 'ost-' corresponds to the German for 'east'. *See also* COINS AND CURRENCIES OF THE WORLD.

ottava rima *see* VERSE FORMS

ouguiya [comparative values] Unit of currency in Mauritania after it achieved independence from France in 1973. The term is nonetheless a French corruption of an Arabic word.

> 1 ouguiya = 5 khoum

See also COINS AND CURRENCIES OF THE WORLD.

ounce [weight] Different units of weight in the AVOIRDUPOIS WEIGHT SYSTEM and the APOTHECARIES' WEIGHT SYSTEM, the latter of which corresponds also to a unit in the TROY WEIGHT SYSTEM.

In the avoirdupois weight system:

1 ounce	=	16 DRAMS, 437.5 grains
	=	28.349523125 GRAMS
		(25 grams = 0.8818 ounce avdp.)
	=	0.9115 apothecaries'/troy ounce
16 ounces	=	1 POUND
		(1 pound avdp. = 453.59236 grams)
35.274 ounces (2.2046 pounds)	=	1 KILOGRAM

In the apothecaries'/troy weight system:

1 ounce	=	20 pennyweight, 24 SCRUPLES, 480 grains
	=	31.103475 grams
		(25 grams = 0.8038 ounce apoth.)
	=	1.09709 ounce avoirdupois
12 ounces	=	1 pound
		(1 pound apoth. = 373.2417 grams)
32.151 ounces		
(2.6792 pounds)	=	1 kilogram

The term derives from the old Roman UNCIA which, as a measure of weight, was one-twelfth of a larger unit (the LIBRA) just as the apothecaries'/troy ounce is. But the Roman *uncia* was, in fact, closer in value to the avoirdupois ounce.

1 ounce avdp.	=	1.0388 uncia
		(1 uncia = 0.9625 ounce avdp.
		= 27.2875 grams)
1 ounce apoth.	=	1.1398 uncia
		(1 uncia = 0.8773 ounce apoth.)

Other similarly close descendants of the Roman *uncia* are the Portuguese ONÇA and the French ONCE – although only the Portuguese is still extant as a measure.

1 ounce avdp.	=	0.9882 Portuguese onça
		(1 onça = 1.01194 ounce avdp.
		= 28.688 grams)
1 ounce avdp.	=	0.9266 French once
		(1 once = 1.0792 ounce avdp.
		= 30.590 grams)

See also OKA, OKE; ONS.

ounce, fluid [volumetric measure: weight] Standard very small liquid measure, different in UK and US usages.

In the United Kingdom:

1 fluid ounce	=	one-twentieth of a PINT
	=	one-fifth of a GILL
	=	8 FLUID DRAMS
	=	0.0284 litre, 28.4 millilitres
		(100 ml = 3.52 fl.oz)
	=	0.959459 US fluid ounce
5 fluid ounces	=	1 gill or quartern, ¼ pint
20 fluid ounces	=	1 pint (0.5682 litre)

In the United States:

1 fluid ounce	=	one-sixteenth of a pint
	=	one-quarter of a gill
	=	8 fluid drams
	=	0.0296 litre, 29.6 milliliters
		(100 ml = 3.38 fl.oz)
	=	1.042254 UK fluid ounce
4 fluid ounces	=	1 gill, ¼ pint
16 fluid ounces	=	1 pint (0.4732 liter)

outer [sporting term] In archery and rifle shooting, the part of the target outside the third concentric ring: the non-scoring (or least-scoring) edge of the target.

over [sporting term] In cricket, a sequence of six balls (or eight, in many Australian matches) bowled by one bowler at one end of the wicket. At the end of the over most of the fielders change position to the inverse of where they were previously (they change *over*: hence the term), and another bowler bowls the next over from the opposite end at the opposite wicket. For any *no ball* called by the umpire during an over, an extra delivery by the bowler is required to complete the over – some overs can thus last for several balls more than the nominal six or eight, if the bowler persists in bowling badly.

Accordingly, it is for the umpire to call 'Over!' at the end of each over. To enable them to be sure of keeping proper count, most umpires keep six or eight smooth pebbles or coins in one hand at the beginning of each over, releasing one into a pocket every time a legal delivery is bowled.

The innings of both sides in most one-day professional cricket matches is regulated to a set number of overs (often 60 or 55) to prevent an inevitable draw through lack of time to get both full teams out. In matches lasting more than one day, the over rate – the number of overs bowled per hour – may be monitored by the umpire in case gamesmanship becomes apparent in unduly time-consuming bowling and lethargic change-over between overs.

overcall [sporting term] In (contract) bridge, to outbid a previous bidder (usually by a small value, especially when a partner has passed without bidding).

overproof measurement [chemistry] Describing an alcoholic drink that is more than 100 proof (in the United States, more than 50 per cent alcohol by volume): *see* ALCOHOL CONTENT.

overthrow [sporting term] In cricket, a ball returned towards the wicket by a fielder that evades the wicketkeeper (or other fielder at the wicket) so that the batsmen can safely make another run. If the ball actually crosses the boundary after being thrown in this way, four runs are scored automatically (in addition to any runs scored before the ball was overthrown). Runs following overthrows are not counted as Extras, but constitute part of the score of the batsman who made the initial hit.

oxidation number [chemistry] The number of electrons that must be added to a positive ion (or removed from a negative ion) to produce a neutral atom, equal to the charge on the ion. Free elements are given an oxidation number of 0. An oxidized element has a positive oxidation number; a reduced element has a negative oxidation number. Thus, the oxidation number of ferric iron, Fe^{3+}, is +3; that of chloride, Cl^-, is –1. Oxidation numbers incorporated into the names of inorganic compounds are stated in Roman numerals; for example: sulphuric(VI) acid, in which sulphur has an oxidation number of 6 .

The term *oxidation* derives as a back formation from *oxide* in eighteenth-century French, a deliberate blending of *oxygène* and *acide*. But the English word *oxygen* itself derives from ancient Greek elements meaning 'acid-producer', chosen by its discoverer Joseph Priestley in 1774 in the mistaken belief that all acids contained oxygen.

oxidation potential [chemistry] Measure of an element's tendency to oxidize – that is, to lose electrons – expressed in volts. Hydrogen is assigned the (arbitrary) oxidation potential of zero. *See also* ELECTRODE POTENTIAL; STANDARD ELECTRODE POTENTIAL.

oxidation-reduction potential, redox potential [chemistry] Every chemical reaction that involves oxidation must also involve a simultaneous reduction (and vice versa). The reaction is – at least to some extent – reversible and involves the transfer of electrons. The associated voltage is the oxidation-reduction (redox) potential. *See also* OXIDATION POTENTIAL; ELECTRODE POTENTIAL.

oxybathon [volumetric measure] Small unit of liquid and dry capacity in ancient Greece, approximating roughly to one-eighth of a modern pint.

1 oxybathon	=	18 kheme, 9 setier, 6 mystra, 1½ KYATHYS
	=	67.4238 millilitres, 0.0674238 litre
	=	0.1187 UK pint, 0.1425 US pint
	=	2.374 UK fluid ounces, 2.278 US fluid ounces
	=	0.1187 UK dry pint, 0.1223 US dry pint
	=	67.4238 cubic centimetres
	=	4.1159 cubic inches
4 oxybatha	=	1 kotyle or COTYLA
8 oxybatha, 2 cotylai	=	1 SEXTE or xestes
16 oxybatha, 4 cotylai, 2 sexte	=	1 KHOINIX (dry measure)
48 oxybatha, 12 cotylai, 6 sexte	=	1 KHOUS

The name of this unit derives from the fact that it represents a capacity of around the same as the standard Greek household (saucer-like) container (*bathon*) for the sour (*oxys*) condiment – in other words, the bowl in which the vinegar was kept. Although of a similar capacity, it was apparently much wider and shallower than the equivalent Roman vessel: *see* ACETABULUM.

oz *see* OUNCE

ozone layer, ozonosphere [geology; meteorology] Ozone is an allotrope (form)
of oxygen that has three atoms in its molecules (as opposed to the usual two in
ordinary oxygen). The gas occurs naturally in the upper atmosphere and is found in
greatest concentration in the ozone layer (ozonosphere) at an altitude of 15 to 30
kilometres (9.3-18.6 miles). Ozone filters out much of the Sun's ultraviolet radiation which, if it were not attenuated in this way, could endanger life on Earth. The
concentration of ozone can be measured using an instrument (called an ozonometer)
carried aloft by a balloon (an 'ozonesonde'). Such measurements, and satellite
photography, have indicated depletion of the ozone layer particularly over the South
Pole, possibly caused by atmospheric pollution with CFCs (chlorofluorocarbons)
used as refrigerants and propellant gases in aerosol sprays.

The word *ozone* derives through German from the ancient Greek *ozon* 'emitting a
smell' (thus acknowledging the characteristic pungent odour of the gas).

P

p, p. *see* PENNY, PENCE; PESETA; PESO; PICO-; PIANO, PIANISSIMO

p.a. [time] An abbreviation for *per annum* 'per year', 'annually'.

pa'anga [comparative values] Unit of currency in Tonga.

1 pa'anga = 100 seniti

The term was apparently the native word for the seed of a certain local plant,
which is shaped like a coin. When coins were introduced, they were naturally called
by this name. *See also* COINS AND CURRENCIES OF THE WORLD.

pace [linear measure] The distance of one stride. The ancient Roman pace (*passus*),
however, was in fact the length of two strides or steps (*gradus*), and it was one-
thousand of these double-paces (*milia passuum*) that formed the basis of the linear
measure later to be known as the MILE.

1 passus	=	60 'inches' (*unciae*), 5 'feet' (*pedes*), 2 steps
	=	147.85 centimetres, 1.4785 metre
	=	4 feet 10.2 inches, 1.6167 yard
2 passus, 10 'feet'	=	1 decempeda
125 passus, 625 'feet'	=	1 STADIUM
1,000 passus, 100 decempeda	=	1 'mile'

In horse racing and three-day eventing dressage, the term *pace* is used either as a
synonym for GAIT (any of the set methods by which a horse must proceed to stay
within the rules and earn points), or for a specific gait in which the horse uses both
feet on one side at a time.

With no reference to walking, *pace* can mean simply 'speed', especially when the
speed is being regulated or monitored.

packen [weight] Obsolete unit of mass in Russia.

1 packen	=	1,200 funte, 30 pudi
	=	491.413 kilograms
		(500 kilograms = 1.0175 packen)
	=	1,083 pounds 6.08 ounces (9.673 hundredweight)
		(1,000 pounds = 0.9230 packen)

packing fraction [physics] For an isotope of an element, its MASS DEFECT divided by
its MASS NUMBER. It is an indication of how the actual mass of the isotope differs
from the combined masses of all the subatomic particles of which it is composed.

pack of dogs, cards, thieves, lies [collectives] A *pack* is by etymology simply a
bundle, especially a bundle put into a container – the word is probably cognate with
bag. But just as a collection of birds shot for sport can constitute a bag, so dogs
(especially wolves and hyenas), thieves, and lies can presumably assemble in packs.
Compare semantically the word *load* as applied similarly to a collection of elements.

A pack of cards (in North America, a DECK), however, is more likely to be a *packet* or carton of cards, just as in 'a pack of cigarettes'.

pagination [literary] In publishing or editing, the operation of numbering the pages in a book or typescript, or the arrangement and typesetting of the page numbers (folios) themselves. A right-hand (recto) page has an odd-numbered folio; a left-hand (verso) page has an even-numbered folio.

The term derives from the Latin *pagina* 'trellis', thus 'latticework of script', thus 'page (of writing)'.

pahlavi [comparative values] Former gold coin in Iran during the time that Reza Khan Pahlavi was the Shah of Persia (1925-41).

1 pahlavi = 20 rials

See also COINS AND CURRENCIES OF THE WORLD.

pair [quantitatives] Two; a couple, a brace, a duo.

The word derives through French from the Latin *paria* 'equal (things)' and is thus akin to such words as *parity*, *par*, and *peer*.

pairs (one, two) see POKER SCORES

paisa [comparative values] Unit of currency in Bangladesh, India, Nepal, and Pakistan.

In Bangladesh,

100 paisa = 1 taka

In India, Nepal and Pakistan,

100 paisa = 1 rupee

The term is the Hindi for 'copper coin' but may be related to Sanskrit *pada* meaning both 'quarter' and 'foot'. See also PICE; COINS AND CURRENCIES OF THE WORLD.

palaeo-, paleo- [time: prefix; geology] Prefix denoting 'ancient', in fact usually referring not merely to prehistoric times but to geological times.

The prefix derives through Latin *palaeus* from ancient Greek *palaios* 'ancient', but close cognates appear only in Celtic languages of the Indo-European family (cf. Welsh *pell* 'distant', 'afar').

Palaeocene epoch [time] A geological EPOCH during the Tertiary period of the Cenozoic era, corresponding roughly to between 65 million years ago and 55 million years ago. It began at the apparently cataclysmic end of the Upper CRETACEOUS PERIOD of the Mesozoic era (when the dinosaurs and other groups seem rapidly to have become extinct), and ended when the warmer EOCENE EPOCH started. During the Palaeocene – also spelled Paleocene – there was a general draining of the shallow inland seas. Animal evolution took most of the epoch to recover from the loss of life sustained in the cataclysm but it was, nonetheless, during the Palaeocene that the first primates evolved. See opposite for table.

The stages shown in this chart are meant as examples only: they represent predominantly north-west European terms for which there are sometimes corresponding east European and North American equivalents.

The name of the epoch derives through French from ancient Greek *palaios* 'ancient' *kainos* 'new', presumably reflecting that the epoch is both the first and the oldest part of the Tertiary period (which can be said to be continuing today).

palaeochronology [archaeology; time] Deriving an age for fossilized animals and plants by identifying the stage of evolution reached in each case.

Palaeogene see EOGENE

Palaeolithic [time] From prehistoric times, of the very first Stone Age (Greek *palaeo-* 'ancient', *lith-* 'stone') form of human culture – the stage at which the early *Homo sapiens* and other hominid species began to learn how to make and use tools that they could fashion for themselves not only out of wood and bone but also out of the harder varieties of rock around them (particularly flint). During this period, even the use of fire may not have been well known. See also MESOLITHIC; NEOLITHIC.

palaeomagnetism [physics] The north and south magnetic poles of the Earth have not always been where they are today, and the north-south polarity has itself reversed several times during the history of the planet. This is particularly evident in the igneous rocks of the sea-bed in certain areas of the oceans in which tectonic activity has caused two-dimensional expansion of the sea-bed as new rock welled

Era	Period	Epoch	Stages
C I O N O Z O N E C (CENOZOIC)	Quaternary (the last 2 to 3 million years)	Holocene	
		Pleistocene	
	Tertiary (from 65 million years ago to 3 million years ago)	Pliocene *(Neogene)*	Astian
			Plaisancian
		Miocene	Pontian
			Tortonian
			Helvetian
			Burdigalian
			Aquitanian
		Oligocene *(Palaeogene / Eogene)*	Chattian
			Stampian
		Eocene	Bartonian
			Lutetian
			Ypresian
		Palaeocene	Thanetian
			Montian
			Danian

up between tectonic plates. In this way, the prevailing direction of the Earth's magnetic field became 'frozen' in the rocks. The historical variations in magnetic alignment and polarity are now so well recorded that it is possible to date rocks (and some geological events) by analysis of the residual magnetism.

Palaeozoic era [time] Geological era that followed the Precambrian, Archaeozoic, or Proterozoic era and was in turn succeeded by the Mesozoic or Secondary era. Divided into the Cambrian, Ordovician, Silurian, Devonian, (the Lower and Upper, or Mississippian and Pennsylvanian) Carboniferous, and Permian periods – and occasionally known as the Primary era – the Palaeozoic began around 590 million years ago and ended around 248 million years ago.

The era is characterized by those very early forms of organic life now found as fossils within rock strata – ferns and fern-like trees, among which roamed early insects, primitive reptiles, and amphibians still emerging from seas inhabited by primitive fishes. During the era, the seas rose until much of the land surface was covered and the most numerous species were fishes. And, by the end of the era, it was still the marine and near-marine animals – the molluscs, arthropods, and brachiopods – that were most numerous, although the waters had by then receded once more.

The name of the era corresponds to the progression inherent in both of the succeeding eras: Palaeozoic 'ancient life', Mesozoic 'middle life', and Cenozoic 'new (or modern) life'.

See chart of the Palaeozoic era under CAMBRIAN PERIOD.

palatal consonants [literary: phonetics] Consonants that are produced mostly or entirely by raising the front or middle of the tongue up to (or toward) the hard palate. In English these correspond to:

          j (dzh)   l   n   y

paleo- see palaeo-
Paleocene epoch *see* PALAEOCENE EPOCH
Paleogene *see* EOGENE
Paleolithic *see* PALAEOLITHIC
Paleozoic era *see* PALAEOZOIC ERA
palm, palame, palmo [linear measure] From the earliest times, a measure of length

between the DIGIT or FINGER and the FOOT – although not now current in the linear units of English-speaking countries.

The palm was a unit first in ancient Egypt, ancient Israel, ancient Greece, and then ancient Rome.

In ancient Egypt,

1 palm (or *shep*)	=	4 digits
	=	2.95 inches, 7.49 centimetres
7 palms, 28 digits	=	1 royal cubit

In ancient Israel,

1 palm (or *tefah*)	=	4 digits or fingers
	=	2.95 inches, 7.49 centimetres
3 palms, 12 digits	=	1 span (or *zeret*)
6 palms, 2 spans	=	1 ordinary cubit (or *ammah*)
7 palms, 28 digits	=	1 royal cubit

In ancient Greece,

1 palm	=	4 digits or fingers
	=	3.04 inches, 7.70 centimetres
3 palms, 12 digits	=	1 span
4 palms, 16 digits	=	1 'foot'
6 palms, 2 spans	=	1 cubit

In ancient Rome,

1 palm	=	4 digits, 3 'inches' (*unciae*)
	=	2.92 inches, 7.40 centimetres
4 palms, 12 'inches', 16 digits	=	1 'foot'
5 palms, 16 'inches', 20 digits	=	1 'palm-foot'
6 palms, 18 'inches', 24 digits	=	1 cubit
10 palms, 30 'inches', 40 digits	=	1 step (*gradus*)
20 palms, 5 'feet', 2 steps	=	1 'pace' (*passus*)

There are a few modern descendants of these ancient measures, but mostly in name only, having been assimilated to other modern units of measurement. In today's Greece, for example, the *palame* is now equal to 1 inch (2.54 centimetres); in Holland, what is called the *palm* in fact represents 10 centimetres (3.94 inches). Even farther away from the ancient measures are the *palmo* of Spain and Portugal.

In Spain,

1 palmo	=	9 pulgada
	=	20.9 centimetres, 8.23 inches
		(1 metre = 4.785 Spanish palmos
		1 foot = 1.458 Spanish palmo)
4 palmos	=	1 vara, 3 pie, 36 pulgada
8 palmos	=	1 braza

In Portugal,

1 palmo	=	22.0 centimetres, 8.66 inches
		(1 metre = 4.545 Portuguese palmos
		1 foot = 1.386 Portuguese palmo)
3 palmos	=	1 covado (cf. 'cubit')
5 palmos	=	1 vara
10 palmos	=	1 braça

Whereas in Spain the palmo has been successfully integrated with a system of 'feet' (*pie*) and 'inches' (*pulgada*, by derivation literally 'thumbs'), it has not happened in Portugal, where the tradition of 'feet' (*pé*) and 'inches' (*polegada*) is equally and independently obsolescent.

These two variants of the 'palm' coincide well with the use of the word in North America for the length of the whole hand (from wrist to middle-fingertip) as a rough unit of measure. In Britain and other English-speaking areas of the world, it is more common to consider the palm as the width of the hand: *see* HAND.

The palm is probably so called because it is the flat part (cf. Greek *platys*, Latin *planus* 'flat', 'level') of the hand, although some authorities derive the word instead from the root of Latin *palpare* 'to sense by touching' (akin to English *feel*).

pan-, panto- [quantitatives: prefix] Prefix meaning 'all' or 'every'.

Examples: panchromatic – (sensitive to light of) every colour
pandemic – (illness affecting) all of a people

The term derives directly from ancient Greek (but may be related to Latin *quant-* 'how much?', 'however much').

Pangaea, Pangea *see* SUPERCONTINENT

pantograph [engineering] An instrument for copying drawings, either same-size, enlarged, or reduced in scale. It consists of four rods pivoted in the form of a parallelogram, the rods overlapping at two of the four corners, with an anchoring arrangement at one overlapping end, a pointer at another, and a pen at an intermediate corner. The principle is incorporated into more complex apparatus for 'copying' three-dimensional shapes – for example, for making mouldings for casting them.

By analogy, the term also describes the diamond-shaped frame used as a sliding current collector on the roof of an electric tram or train.

The term derives from ancient Greek elements that proclaim the device's ability to copy 'all drawings' at any size.

paper sizes [paper] Most countries now use the metric ISO paper sizes, in which the basic size (designated A0, B0, and so on) is 1 square metre in area. For example, A0 is 841 x 1,189 millimetres; the dimensions are in the proportion 1 to the square root of 2. For other ISO sizes *see* A SERIES; B SERIES.

Formerly – although still used in most of North America – paper sizes were given in inches and had their own names (some from the watermark they originally bore). Examples include (in inches):

size	standard	quarto (4to)	octavo (8vo)
atlas	26 x 34	17 x 13	13 x 8½
imperial	22 x 30	15 x 11	11 x 7½
elephant	20 x 27	13½ x 10	10 x 6¾
royal	20 x 25	12½ x 10	10 x 6¼
small royal	19 x 25	12½ x 9½	9½ x 6¼
medium	18 x 23	11½ x 9	9 x 5¾
demy	17½ x 22½	11¼ x 8¾	8¾ x 5⅝
crown	15 x 20	10 x 7½	7½ x 5
foolscap	17 x 13½	8½ x 6¾	6¾ x 4¼
pot	12½ x 15½	7¼ x 6¼	6¼ x 3⅞

Before the advent of the ISO sizes, the corresponding metric paper sizes were (in millimetres):

size	quad sheet	quarto (4to)	octavo (8vo)
metric royal	960 x 1,272	318 x 240	240 x 159
metric demy	888 x 1,128	282 x 222	222 x 141
metric large crown	816 x 1,056	264 x 204	204 x 132
metric crown	768 x 1,008	252 x 192	192 x 126

par [sporting term] In golf, the number of strokes set as standard for a hole (commonly between 3 and 6), differing between holes according to the distance between tee and green, and the degree of difficulty involved. Also, the set number of strokes for a complete round of golf (at many golf courses totalling 72). Both types of par are based upon the proficiency of a good to very good amateur player.

One below par for a hole – one fewer strokes than the set number – is called a *birdie*; two below is an *eagle*; three an *albatross*. One over par is called a *bogey*; two over is *double bogey*.

The term derives directly from Latin *par* 'equal', 'matching'.

par [comparative values] The face value of a banknote, a bond, stock-share or any other form of security.

Also, the value of the major currency unit of one country in terms of the currency of another country, especially if that value has remained fairly constant over a time. *See also* PARITY.

para [comparative values] Former unit of currency in Turkey and Yugoslavia, in both cases finally becoming worthless through inflation.

In Turkey,

40 paras = 1 piastre (now kurus)

In Yugoslavia,

100 para = 1 (Yugoslav) dinar

The term is actually the word for 'money', 'coin(age)' in Turkish (and was borrowed from Turkish in Yugoslavia, a large part of which was once under Turkish rule).

parabola [maths] A plane curve with an ECCENTRICITY of 1. It is the locus of a point that remains equidistant from a fixed point (the focus) and a fixed line (the directrix). A parabola is also generated as a conic section when a plane cuts through the base of a cone parallel to the cone's slope. Most projectiles – bullets and missiles – follow a parabolic trajectory.

paraboloid [maths] A solid figure obtained by rotating a parabola about its axis. It is the preferred shape for a headlamp reflector (a light source at the focus of the paraboloid is projected as a parallel beam) and for a 'dish' aerial used for radars, microwave communications, and radio-astronomy.

parallax [physics; astronomy] The apparent change in an object's position when it is observed from two different viewpoints. In astronomy, the change in the apparent position of a celestial object is usually caused because the Earth moves between consecutive observations. Such parallax is defined as the angle at a celestial object between straight lines that join it to the two different viewpoints. It may be diurnal parallax (because of the Earth's movement, due to rotation on its axis, during a day), annual parallax (because of the Earth's movement, due to orbiting the Sun, during a year), or secular parallax (because of the Earth's movement in space, along with the rest of the Solar System). Annual parallax is proportional to distance, and so in astronomy the term 'parallax' may be used (rather imprecisely) to mean distance.

The related adjective is *parallactic*.

The term derives from ancient Greek elements meaning 'leaving behind'.

parallel [maths] Describing two or more lines that are everywhere the same distance apart. Parallel lines never meet or cross.

The term derives from ancient Greek elements meaning 'beside one another'.

parallel [geology] An alternative name for (a line of) LATITUDE.

parallelepiped [maths] A solid figure that has six parallelograms as its faces, with opposite faces that are identical and parallel.

The term derives from ancient Greek elements effectively meaning 'with parallel planes'.

parallelogram [maths] A four-sided plane figure with opposite sides that are equal and parallel. Its area equals the length multiplied by the vertical height (altitude). A parallelogram with internal angles that are right-angles is called a *rectangle* or *oblong*.

The term derives from ancient Greek elements effectively meaning 'with parallel lines'.

parallelogram of forces [physics] A method of resolving two or more forces (which are vectors) using the method called the PARALLELOGRAM OF VECTORS.

parallelogram of vectors [maths] Vectors have magnitude and direction, and can be represented graphically by straight lines. Two different vectors from a common point can be drawn as a pair of lines forming an angle with each other. The resultant (a single vector having the sum of the other two) can be found by adding two more lines to make a parallelogram, and drawing the diagonal from the original point. The length and direction of the diagonal is the vector sum. If there are more than two vectors, they can be added in pairs (as above) until only two are left, and the final two resolved.

paramagnetic [physics] Describing a material that becomes magnetized when placed in a magnetic field (due to alignment of its molecular magnets), but whose magnetization disappears when the magnetic field is removed. *See also* DIAMAGNETIC; FERROMAGNETIC.

parameter [maths] A variable quantity that can be used to express other unrelated variables, then treated as if they were dependent on the parameter. The term has also come to mean any quantification that assists in defining an object.

paraplegia [medicine] Technically, paralysis of the four limbs and part of the trunk of the body, although the definition is often extended to paralysis from the neck down (thus including all of the trunk), which is better known by physicians as *quadriplegia*. Paralysis of one side of the body (sometimes including certain psychological functions of the opposite side of the brain) is called *hemiplegia*.

parasanges, parasang, farsang [linear measure] Ancient Babylonian (Persian) unit of distance originally called a *farsang*, but better known as an ancient Greek unit based upon it.

In ancient Babylon, the farsang is said by most authorities to have been equal to 'about 4 miles' (about 6.44 kilometres).

The ancient Greek unit of distance was assimilated to a unit constituting 30 stadia.

1 parasang(es)	=	30 stadia, 300 'cables', 18,000 'feet'
	=	5.5602 kilometres
	=	3.455 miles (3 miles 2,402 feet 4.8 inches)

By the time the unit was in due course borrowed by Latin authors, the original distance had been forgotten, as was the Greek variant. The Latin measure was accordingly anything between one and two times the Greek unit.

The unit was still in use by the Persians (Iranians) centuries later; known as a *persakh*, it was very close in value – if not identical – to the original farsang.

1 persakh = 6.244 kilometres, 3.88 miles

The term by derivation probably means no more than 'Persian (measure)'.

pari-mutuel, parimutuel [maths; comparative values] Mechanical betting system – known in Britain and elsewhere alternatively as the totalizator or 'tote' – by which those who correctly bet on the winner, second, and third in a race (at odds set according to the numbers of those betting and the amounts staked) share all the money bet (including the stakes of all who bet on other runners), less a small proportion for the management's costs and overheads.

The term derives from modern French elements meaning 'mutually equal'.

parity [maths: computing] The property of a number that makes it odd or even; for example: 4 and 82 have even parity, 3 and 95 have odd parity.

parity [comparative values] A consistent ratio between the value of the major currency unit of one country and the proportional value of the currency of another country, especially if that ratio has remained fairly static over a time.

Also, an established relationship between the financial value of a farmer's produce and the lesser value of the farmer's costs, calculated to ensure that the basic level of overall earnings (and their purchasing power or index-linked value) remains at a specified level after subsidy. If these values are expressed not in current prices but in relation to the prices of a previous year regarded as a base for calculation (according to a *parity index*), the use of a *parity ratio* is required. *See also* PAR.

parity [physics] In quantum mechanics, the behaviour of a wave function when it is reflected in space. If the sign of the function remains unchanged, parity is even; if the sign changes, parity is odd.

parity bit [electronics: computing] An extra binary digit that is added to a binary number (*see* BINARY SYSTEM) to make its parity odd (or even) – that is, to make the total number of 1s or 0s in the number odd (or even). The number can then be confirmed in a parity check; numbers with the wrong parity are assumed to be errors.

parlay [sporting term; comparative values] In North America, one or a series of gambles or bets in which the stake is all the money won on an initial and successful gamble or bet immediately beforehand; in Britain it is known as an *accumulator*.

parse *see* PARSING

parsec [astronomy] Unit of astronomical distance, equal to the distance at which a length of 1 ASTRONOMICAL UNIT subtends an angle of 1 second of arc.

1 parsec	=	206,265 astronomical units
	=	3.262 light-years
	=	3.09×10^{13} kilometres

$$= 1.92 \times 10^{13} \text{ miles}$$

It is frequently used in its multiples kiloparsec and megaparsec:

$$1 \text{ kiloparsec} = 1,000 \text{ parsecs}$$
$$1 \text{ megaparsec} = 1,000,000 \text{ parsecs}$$

Its name derives from parallax second (*see* PARALLAX).

parsing [literary] The analysis of a sentence or piece of prose in terms of defining the parts of speech (noun, verb, adjective, adverb, etc.), grammatical functions (subject, object, etc.), and semantic elements (clause, phrase, etc.) that it contains.

partial derivative [maths] The derivative of a function with respect to one of its variables (found by differentiation), any other variables being taken as constant. The partial derivative is also known as the *partial differential*.

partial fraction [maths] One of two or more component fractions that another fraction can be separated into; the sum of the partial fractions equals the original fraction. For example: ⅓ and ⅕ are partial fractions of ⁸⁄₁₅.

partial pressure [physics] The pressure exerted by a given gas in a mixture of gases: *see* DALTON'S LAW.

partial tone [music] Another term for a HARMONIC or overtone, especially one that is clearly audible above the fundamental tone.

particle size [geology] The average size of the individual grains that make up a type of sediment or rock. According to the Wentworth-Udden scale, the sizes are:

clay	less than ¹⁄₂₅₆ millimetre in diameter
silt	¹⁄₂₅₆ to ¹⁄₁₆ millimetre
sand	¹⁄₁₆ to 2 millimetres
gravel	2 to 4 millimetres
pebbles	4 to 64 millimetres
cobbles	64 to 256 millimetres
boulders	more than 256 millimetres

partition coefficient [chemistry] The ratio of the concentrations of a dissolved substance (solute) in two different immiscible solvents (at equilibrium). Alternative name: distribution coefficient.

parts of speech [literary] The grammatical functions of words as classified according to their use in speech or writing. The categories are:

a NOUN	a word or name for a person, place, or thing
a VERB	a word that implies activity or being
an ADJECTIVE	a word that describes/qualifies a noun
an ADVERB	a word that describes/qualifies a verb
a PREPOSITION	a word that describes position or possession
an ARTICLE	the words *the* and *a* or *an*
a CONJUNCTION	a word that connects two ideas
an interjection or exclamation	

Many individual words can function in two or more of these categories, depending on the chosen context. For example, the word *well* can be a noun, a verb, an adjective, an adverb, and an interjection.

parts per million (ppm) [physics; volumetric measure] Unit of concentration, used for trace quantities of a substance. For example, in the fluoridation of drinking water, the amount of fluorine added is about 1 part per million.

$$1 \text{ part per million} = 0.0001 \text{ per cent}$$

pascal [physics] Unit of pressure in the SI system of units, equal to 1 newton per square metre.

$$1 \text{ pascal} = 10^{-5} \text{ bar}$$
$$= 0.01 \text{ millibar}$$

The unit was named after the French mathematician and physicist Blaise Pascal (1623-62).

Pascal's law [physics] In hydrostatics, the pressure in a fluid is the same in all directions, and, when pressure is applied to a fluid in a confined space, it is transmitted equally in all directions (unless gravity also takes an effect).

This law was first proposed by the French mathematician and physicist Blaise Pascal (1623-62).

Pascal's triangle [maths] A pattern of numbers in which each number, from the third row downwards, is the sum of the two numbers above it. The first eight rows are:

$$
\begin{array}{ccccccccccccccc}
 & & & & & & & 1 & & & & & & & \\
 & & & & & & 1 & & 1 & & & & & & \\
 & & & & & 1 & & 2 & & 1 & & & & & \\
 & & & & 1 & & 3 & & 3 & & 1 & & & & \\
 & & & 1 & & 4 & & 6 & & 4 & & 1 & & & \\
 & & 1 & & 5 & & 10 & & 10 & & 5 & & 1 & & \\
 & 1 & & 6 & & 15 & & 20 & & 15 & & 6 & & 1 & \\
1 & & 7 & & 21 & & 35 & & 35 & & 21 & & 7 & & 1 \\
\end{array}
$$

The numbers in each row of the triangle are the coefficients of the expansion of the binomial expression $(x + y)^n$.

pataca [comparative values] Unit of currency in Macao.

$$1 \text{ pataca } = 100 \text{ avos}$$

The term derives from the name of a Portuguese coin worth 5.5 escudos. *See also* COINS AND CURRENCIES OF THE WORLD.

Pauli exclusion principle [physics] Every electron moving around an atom's nucleus has a characteristically different set of four quantum numbers. It is named after the Austrian physicist Wolfgang Pauli (1900-58).

payload [aeronautics] The weight of revenue-earning cargo or passengers carried by a commercial aircraft, or the weight carried by a rocket (ignoring the weight of the rocket itself and its fuel).

pé [linear measure] In Portugal, a derivative of the old Roman *pes* 'foot'.

$$
\begin{aligned}
1 \text{ pé } &= 12 \text{ polegada} \\
&= 33.324 \text{ centimetres, } 13.12 \text{ inches (1 ft 1.12 inch)}
\end{aligned}
$$

According to reference sources, however, this system of 'feet' and 'inches' (by derivation literally 'thumbs') is close to, but not quite in sequence with, the larger Portuguese measurements the *covado* (66 centimetres, not 66.648 centimetres = 2 pé) and the *vara* (1.10 metre, between 3 and 4 pé). *See also* PALM, PALAME, PALMO.

peak value [physics] The maximum value (positive or negative) of a quantity that alternates: *see* AMPLITUDE.

peal [quantitatives] A set of 4, 6, 8, 10, or 12 bells: *see* CHANGE RINGING, CHANGES.

peck [cubic measure; volumetric measure] Unit of dry capacity; the US peck is not the same as the UK peck.

$$1 \text{ peck } = 2 \text{ (dry) gallons, 16 (dry) pints}$$

1 UK peck	=	.9.0925 litres	1 US peck	=	8.8095 liters
	=	9,092.5 cc		=	8,809.5 cc
	=	0.321 cu. ft		=	0.311 cu. ft
	=	554.825 cu. in		=	537.550 cu. in
	=	1.0321 US pecks		=	0.9689 UK peck

$$4 \text{ pecks } = 1 \text{ bushel}$$

The term is thought to derive from a Latin variant of the ancient Greek word *bikos* 'wine-vase', and thus to be the same word as English *beaker* and *pitcher*.

pedometer [linear measure] Device fitted to a person's leg that registers the number of steps taken and, set to the average length of pace, thus measures the distance walked.

pelorus [physics: navigation] Form of gyroscopic compass used aboard a ship to take bearings.

The term derives from the personal name of the Carthaginian Pelorus, Hannibal's navigator on the journey back to Carthage from Italy.

pelvimetry [biology; medicine] Measurement of a person's pelvic girdle – the ilium, ischium, and pubis bones – generally of a pregnant woman's to foresee any possible problems during childbirth.

pence *see* PENNY, PENCE

pendulum [physics] The time of swing of a simple pendulum (its period) equals 2π times the square root of the ratio of its length to the acceleration of free fall (acceleration due to gravity). It is independent of the weight of the pendulum bob.

penetrometer [physics] An instrument that measures the density and homogeneity of the sedimentary layers that make up the sea-bed.

pengö [comparative values] Former unit of currency (until 1946) in Hungary, since replaced by the FORINT. The term means 'clinking', 'ringing' in Hungarian.

Penicillin units [medical] *see* INTERNATIONAL UNIT.

penni, penniä [comparative values] Unit of currency in Finland. The singular is *penni*.

100 penniä = 1 markka

The name of the unit appears to have been borrowed directly from English: *see* PENNY, PENCE. *See also* COINS AND CURRENCIES OF THE WORLD.

Pennsylvanian [time; geology] The second of the two epochs that make up the CARBONIFEROUS PERIOD, known outside North America as the Upper Carboniferous. (The other epoch, the first of the Carboniferous, is in North America called the Mississippian and elsewhere known as the Lower Carboniferous.) During this time, club-mosses and horsetails formed swampy forests that were later to become coal and oil deposits of today.

penny, pence [comparative values] Unit of currency in the United Kingdom.

100 pennies

(formerly 'pence') = 1 pound (sterling)

The name of the unit is also informally used in the United States, Canada, Australia, and New Zealand for a cent.

The term derives as the name of a coin that was initially either described as 'pan-shaped' or cast in moulds called 'pans', for it is a variant of *panning* – just as the German cognate the *pfennig* is closely associated with *pfanne* 'pan'. *See also* COINS AND CURRENCIES OF THE WORLD; DENARIUS.

penny sizes *see* NAILS

pennyweight [weight] A very small unit of weight in the troy weight system.

1 pennyweight = 24 grains
= 1.552 gram
(1 gram = 0.644 pennyweight)
= 0.05484 ounce avdp.
(1 ounce = 18.235 pennyweight)
20 pennyweight = 1 troy ounce, 24 scruples

penta-, pent- [quantatives: prefix] Prefix which implies a fivefold unity.

Examples: pentameter – having five metrical feet
pentatonic – comprising five tones or notes

The prefix derives from ancient Greek *pente* 'five'.

pentad [quantitatives; astronomy] A group of five; a sequence of five.

A period of five years.

A period of five days, equal to $\frac{1}{73}$ of a normal year. *See also* QUINTET.

pentagon [maths] A plane figure (POLYGON) with five sides (and five angles). The interior angles add to 540 degrees; each interior angle of a regular pentagon equals 108 degrees.

pentahedron [maths] A solid figure (POLYHEDRON) with five plane faces (such as a square-based pyramid).

pentameter [literary] In verse, a line of poetry consisting of five metrical feet: *see* FOOT. Pentameters composed of iambic feet (iambic pentameters) are said to correspond to HEROIC VERSE in English, Italian, or German.

pentane *see* OCTANE NUMBER, OCTANE RATING

pentathlon [sport] The events of the pentathlon are:

first day:
100-metre hurdles
shot-put
high jump
first or second day:
long (broad) jump
200-metre sprint
Points scoring:
according to tables issued by the International Amateur Athletic Federation (IAAF), relating to ideal times, heights, and distances, in thousands of points

The pentathlon is an event for women only; the equivalent for men is the ten-event DECATHLON. *See also* MODERN PENTATHLON.

pentatonic scale [music] A musical SCALE that, instead of having seven notes between one keynote and its octave, has five. (In technical terms, the subdominant and leading note are both omitted; in the key of C, the notes F and B are omitted and the scale comprises C, D, E, G, and A.) The most famous tune written in the pentatonic scale is *Auld Lang Syne*.

pentavalent, pentavalence [chemistry] Having a valence (valency) of 5; an alternative description is quinquevalent.

per annum [time] Yearly, per year.
    The expression derives directly from Latin.

percentage, per cent [maths] A fraction expressed in hundredths; for example: 30 per cent (30%) equals 30 hundredths ($^{30}/_{100}$). Any fraction can be expressed as a percentage by converting it into hundredths.
    For example,

¼	=	25/100
	=	25 per cent
⅓	=	33.333/100
	=	33.333 per cent
⅘	=	80/100
	=	80 per cent

percentage composition [chemistry] A way of giving the composition of a chemical compound in terms of the percentages (by mass) of each of the elements in it. It is determined by expressing the ratio of the relative atomic mass (atomic weight) of each element times the number of atoms of that element to the relative molecular mass (molecular weight) as a percentage. For example: the substance ethane has the chemical formula C_2H_6 and a relative atomic mass of 30. The percentage of carbon is 2 x 12 (its relative atomic mass) divided by 30 = 80 per cent; of hydrogen it is (6 x 1)/30 = 20 per cent. The percentage composition of ethane is thus 80 % carbon and 20 % hydrogen.

percentile [maths] In statistics, a hundredth part of a range of values of equal frequency. For example: the sixtieth percentile is the value below which 60 per cent of all the values fall. The twenty-fifth and seventy-fifth percentiles are the lower and upper QUARTILES; the fiftieth percentile is the MEDIAN.

perch [linear measure] In North America, a unit of length that is the same as a pole or rod. Elsewhere, this measure is also known as a rod or pole, and a perch is a square measure (*see below*).

1 (linear) perch	=	16½ feet, 5½ yards
	=	5.0292 metres
		(5 metres = 0.9942 perch)
4 perches	=	1 chain (22 yards)
40 perches	=	1 furlong (220 yards)
320 perches	=	1 mile

The term derives through French from the Latin *pertica* 'measuring-rod'.

perch [square measure] In Britain, a unit of area known also as a square rod and equal to 1 square US (linear) perch.

1 (square) perch	=	16½ x 16½ feet, 30¼ square yards
	=	25.2929 square metres
		(25 square metres = 0.9884 perch)
16 perches	=	1 square chain
40 perches	=	1 rood (1,210 square yards)
160 perches	=	1 acre (4,840 square yards)

The term derives through French from the Latin *pertica* 'measuring-rod'.

perch [volumetric measure] Unit of volume for masonry, equal to a piece of masonry 16½ feet [that is, 1 US linear perch (*see above*)] x 1½ feet x 1 foot.

1 (volumetric) perch	=	42,768 cubic inches
	=	24.75 cubic feet
		(1 cubic yard = 1.0909 perch)
	=	0.700425 cubic metres
		(1 cubic metre = 1.4277 perch)

The term derives through French from the Latin *pertica* 'measuring-rod'.

perche [linear measure] Obsolete unit of length in Francophone Canada.

$$1 \text{ perche} = 18 \text{ pieds}$$
$$= 19 \text{ feet } 2.22 \text{ inches}$$
$$= 5.8476 \text{ metres}$$

The term derives from the Latin *pertica* 'measuring-rod'.

perche [square measure] Obsolete unit of area in France.

$$1 \text{ perche} = 22 \text{ square pieds-de-roi}$$
$$= 7.1456 \text{ square metres}$$
$$= 8.5461 \text{ square yards}$$

The term derives from the Latin *pertica* 'measuring-rod'.

percussion instruments' range [music] Percussion instruments rely on the striking of one surface on another to produce a sound. Some produce genuinely musical sounds, and can be tuned to do so; others are for strictly rhythmic purposes, and, although the sounds they produce may be modified one way or another, no musical 'note' is actually involved.

Those that can be tuned and played as an instrument include:

marimba:	(usually) 3 octaves	middle register
xylophone:	(usually) 3-4 octaves	middle and upper registers
vibraphone:	(usually) 3 octaves	middle register
tubular bells/(US) chimes:	(usually) 1½ octaves	(middle register)
metallophone/glockenspiel:	1-1½ octaves	upper register

Caribbean steel drums: set of six, ranging from one of a simply rhythmic function, to a harmonic set of bass, baritone ('guitar'), cello, second melody ('second pan'), and melody ('ping pong') drums tuned accordingly

timpani/kettle-drums: (usually) set of three or more, tuned specifically for fundamental note and harmonics

Those that are not so much tuned as set up for a repeated rhythmic sound include:

bass drum

snare drum, tambourine

cymbals, triangle, and hi-hat

gong/tam-tam

tom-tom

bongo(s)

block

The percussion player is expected also to put his or her hand to various other chiefly rhythmical instruments (such as the castanets and the whip), and one or two items that provide special effects (such as the rattle and the Swannee whistle).

per diem [time] Per day, each day.

The expression derives directly from Latin.

perennial [time] Throughout the year, lasting at least one year.

The term is an adjectival derivative of PER ANNUM.

perfecta [sporting term] Another expression for EXACTA.

perfect fourth, perfect fifth [music] Technical term for the musical intervals of a fourth and a fifth, so called because they can contribute to either major or minor harmony (although not a harmony themselves).

perfect number [maths] A number that is the sum of all its factors (excepting the number itself). The first three perfect numbers are:

$$6 = 1 + 2 + 3$$
$$28 = 1 + 2 + 4 + 7 + 14$$
$$496 = 1 + 2 + 4 + 8 + 16 + 31 + 62 + 124 + 248$$

perfect ream [paper] In printing paper, 516 sheets (amounting to 21½ quires at 24 sheets per quire): *see* REAM.

perfect year *see* ABUNDANT YEAR

periastron [astronomy] For a pair of double stars orbiting each other, the point on the orbit of one star at which it is nearest the other star.

perigee [astronomy] For a moon or artificial satellite orbiting a planet, the point on its orbit at which it is nearest the planet. *See also* APOGEE.

perihelion [astronomy] For an orbiting planet or comet, the point on its orbit at which it is nearest the Sun. *See also* APHELION.

perimeter [maths] The distance around (the boundary of) a geometric figure; for a closed curve it is also called the *circumference*.

period [physics; astronomy; chemistry; geology] The time taken to complete one cycle of a regular movement; for example, the time for one complete swing of a pendulum, or the time taken for a moon to make one complete orbit of its parent planet. For an alternating or oscillating quantity, the period is the reciprocal of the frequency.

In chemistry, one of the horizontal rows of the PERIODIC TABLE OF ELEMENTS.

In geology, a subdivision of an ERA; periods are themselves divided into EPOCHS.

period, menstrual [medicine] The flow of blood and other tissues from the vagina of an adult non-pregnant woman every four weeks or so, representing the shedding of the lining of the womb as the final process of one menstrual cycle. During a period there are other physical and psychological effects that may cause pain or distress. Average blood loss is reckoned at about 60 millilitres (2.113 UK fluid ounces, 2.027 US fluid ounces). The average duration of a menstrual period is about five days, although some women experience periods of just a single day, and others have periods of up to eight days.

The onset of menstrual periods (generally between the ages of eleven and sixteen) heralds puberty; the gradual and final cessation of menstrual periods (generally between the ages of forty-five and fifty-six) defines the menopause.

Ovulation – the time at which a woman is fertile, corresponding to the release of an ovum (egg) from one ovary – occurs half-way between one menstrual period and the next, and lasts for about four days. *See also* RHYTHM METHOD.

periodic function [maths] Any function that regularly returns to the same value; for example, any of the trigonometrical functions, such as sine or cosine.

periodicity [chemistry] For the chemical elements, the regular variations in the physical values of elements with similar chemical properties: *see* MENDELEEV'S LAW.

Periodic Table of elements, with formulae [chemistry] A table that sets out the chemical elements in order of increasing atomic number. Elements with similar chemical properties (and outer electronic configurations) occupy the same *groups* (vertical columns). The horizontal rows of elements are called *periods*. The following table lists the elements, with their symbols, atomic numbers, and relative atomic masses (atomic weights):

element	symbol	at. no.	r.a.m.
actinium	Ac	89	(227)
aluminium	Al	13	26.9815
americium	Am	95	(243)
antimony	Sb	51	21.75
argon	Ar	18	39.948
arsenic	As	33	74.9216
astatine	At	85	(210)
barium	Ba	56	137.34
berkelium	Bk	97	(247)
beryllium	Be	4	9.0122
bismuth	Bi	83	208.9806
boron	B	5	10.81
bromine	Br	35	79.904
cadmium	Cd	48	112.40
caesium	Cs	55	132.9055
calcium	Ca	20	40.08
californium	Cf	98	(251)
carbon	C	6	12.001
cerium	Ce	58	140.12
chlorine	Cl	17	35.453
chromium	Cr	24	51.996
cobalt	Co	27	58.9332
copper	Cu	29	63.546
curium	Cm	96	(247)
dysprosium	Dy	66	162.50
einsteinium	Es	99	(254)

erbium	Er	68	167.26
europium	Eu	63	151.96
fermium	Fm	100	(257)
fluorine	F	9	18.9984
francium	Fr	87	(223)
gadolinium	Gd	64	157.25
gallium	Ga	31	69.72
germanium	Ge	32	72.59
gold	Au	79	196.9665
hafnium	Hf	72	178.49
hahnium	Ha	105	-
helium	He	2	4.0026
holmium	Ho	67	164.9303
hydrogen	H	1	1.0080
indium	In	49	114.82
iodine	I	53	126.904
iridium	Ir	77	192.22
iron	Fe	26	55.847
krypton	Kr	36	83.80
lanthanum	La	57	138.9055
lawrencium	Lr	103	(257)
lead	Pb	82	207.19
lithium	Li	3	6.941
lutetium	Lu	71	174.97
magnesium	Mg	12	24.305
manganese	Mn	25	54.9380
mendelevium	Md	101	(258)
mercury	Hg	80	200.59
molybdenum	Mo	42	95.94
neodymium	Nd	60	144.24
neon	Ne	10	20.179
neptunium	Np	93	(237)
nickel	Ni	28	58.71
niobium	Nb	41	92.9064
nitrogen	N	7	14.0067
nobelium	No	102	(255)
osmium	Os	76	190.2
oxygen	O	8	15.9994
palladium	Pd	46	106.4
phosphorus	P	15	30.9738
platinum	Pt	78	195.09
plutonium	Pu	94	(244)
polonium	Po	84	(209)
potassium	K	19	39.102
praeseodymium	Pr	59	140.9077
promethium	Pm	61	(145)
protactinium	Pa	91	231.0359
radium	Ra	88	226.0254
radon	Rn	86	(222)
rhenium	Re	75	186.20
rhodium	Rh	45	102.9055
rubidium	Rb	37	85.4678
ruthenium	Ru	44	101.07
rutherfordium	Rf	104	-
samarium	Sm	62	150.35
scandium	Sc	21	44.9559
selenium	Se	34	78.96
silicon	Si	14	28.086
silver	Ag	47	107.868

Periodic Table

IA	IIA	IIIB	IVB	VB	VIB	VIIB		VIII		IB	IIB	IIIA	IVA	VA	VIA	VIIA	0
1 H																	2 He
3 Li	4 Be											5 B	6 C	7 N	8 O	9 F	10 Ne
11 Na	12 Mg		←		transition elements						→	13 Al	14 Si	15 P	16 S	17 Cl	18 Ar
19 K	20 Ca	21 Sc	22 Ti	23 V	24 Cr	25 Mn	26 Fe	27 Co	28 Ni	29 Cu	30 Zn	31 Ga	32 Ge	33 As	34 Se	35 Br	36 Kr
37 Rb	38 Sr	39 Y	40 Zr	41 Nb	42 Mo	43 Tc	44 Ru	45 Rh	46 Pd	47 Ag	48 Cd	49 In	50 Sn	51 Sb	52 Te	53 I	54 Xe
55 Cs	56 Ba	57* La	72 Hf	73 Ta	74 W	75 Re	76 Os	77 Ir	78 Pt	79 Au	80 Hg	81 Tl	82 Pb	83 Bi	84 Po	85 At	86 Rn
87 Fr	88 Ra	89† Ac															

*lanthanides		57 La	58 Ce	59 Pr	60 Nd	61 Pm	62 Sm	63 Eu	64 Gd	65 Tb	66 Dy	67 Ho	68 Er	69 Tm	70 Yb	71 Lu
†actinides		89 Ac	90 Th	91 Pa	92 U	93 Np	94 Pu	95 Am	96 Cm	97 Bk	98 Cf	99 Es	100 Fm	101 Md	102 No	103 Lr

sodium	Na	11	22.9898
strontium	Sr	38	87.62
sulphur	S	16	32.06
tantalum	Ta	73	180.9479
technetium	Tc	43	(99)
tellurium	Te	52	127.60
terbium	Tb	65	158.9254
thallium	Tl	81	204.39
thorium	Th	90	232.0381
thulium	Tm	69	168.9342
tin	Sn	50	118.69
titanium	Ti	22	47.90
tungsten	W	74	183.85
uranium	U	92	238.029
vanadium	V	23	50.9414
ytterbium	Yb	70	173.04
yttrium	Y	39	88.9059
zinc	Zn	30	65.38
zirconium	Zr	40	91.22

Relative atomic masses in brackets are those of the most stable isotope.

The compilation of the Periodic Table (*see opposite*) was proposed by the Russian chemist Dmitri Mendeleev (1834-1907) in 1869: *see* MENDELEEV'S LAW.

permanent set [physics] The permanent elongation of an elastic material after it has been stretched beyond its elastic limit.

permeability [physics] The rate at which a gas or liquid under pressure diffuses through a porous material, expressed as the rate per unit area or unit thickness.

Also, the magnetization developed in material that is placed in a magnetic field, equal to the ratio of the magnetic flux density to the magnetic intensity. Known alternatively as the *absolute permeability*, it is measured in henry per metre: *see also* RELATIVE PERMEABILITY.

permeability of free space [physics] A magnetic constant equal to $4\pi \times 10^{-7}$ henry per metre.

Permian period [time; geology] A geological PERIOD during the Palaeozoic or Primary ERA. The last period of the era, immediately following the Upper Carboniferous or Pennsylvanian period, it was followed by the Triassic period of the Mesozoic era, and corresponded roughly to between 286 million years ago and 248 million years ago. The period is characterized by severe glaciation of the landmasses of the Southern Hemisphere. The cold may well also have contributed to the extinction of the previously numerous trilobites. As it was, the animals that left the water and went on to the land survived best, especially the reptiles. Similarly, the swamp-loving ferns and clubmosses of the Carboniferous were quickly replaced by forests of dry-land conifers.

In Europe, because of the difficulty in distinguishing the effects, it is common to group the Permian and Triassic periods together in what is called the Permo-Triassic System.

The period was named in 1841 after the province of Russia called after its major city, Perm, on the western flanks of the central Urals.

For table of the Palaeozoic era *see* CAMBRIAN PERIOD.

per mil [quantitatives] Per thousand, for each thousand; per one-thousandth of an inch. Also, per millilitre; per millimetre.

permittivity [physics] For a material in an electric field, the electric displacement developed in it, equal to the ratio of the electric flux density to the electric field strength. It is measured in farad per metre, and is alternatively called the *absolute permittivity*. *See also* RELATIVE PERMITTIVITY.

permittivity of free space [physics] An electric constant equal to 8.854×10^{-12} farad per metre.

Permo-Triassic System *see* PERMIAN PERIOD

permutation [maths] The number of ways a set of numbers or items can be ordered and arranged. For n different numbers, there are $n!$ ways of arranging them n at a time ($n!$ = factorial n: *see* FACTORIAL) and $n!(n-r)!$ ways of arranging them r at a time. For example: if there are 6 items, the number of ways of arranging them 6 at a time is

$$6! = 6 \times 5 \times 4 \times 3 \times 2 \times 1$$
$$= 720$$

The number of ways of arranging them, say, 3 at a time is

$$6! \times 3! = 6 \times 5 \times 4 \times 3 \times 2 \times 1 \times 3 \times 2 \times 1$$
$$= 4{,}320$$

Note that in permutations, a different order (of the same numbers) is a different permutation. If order does not matter, the arrangement is called a COMBINATION.

perpendicular [maths] Describing a line that is at right-angles to another line or plane. It is not necessarily upright (as is a vertical).

persakh *see* PARASANGES, PARASANG, FARSANG

per second per second ($/s^2$, $/sec^2$) [quantitatives] Rate of increase or decrease in a process measured by the second, for every further second that the process continues. These units are most commonly applied to ACCELERATION or deceleration. An acceleration of 10 metres per second per second, for example, means that an object travels 10 metres in the first second, 20 metres in the second second, 30 metres in the third, and so on.

person [literary] In most languages other than English, the 'person' corresponds to the specific form of the verb that matches the noun or pronoun that is its subject. In English, however, the form of the verb generally does not change much but, for the sake of comparison, there are said to be three persons in singular and in plural, corresponding to pronouns. For example:

	singular	*plural*
1st person	I (do, make, was)	we (do, make, were)
2nd person	you (do, make, were)	you (do, make, were)
	[formerly: thou (dost, makest, wert)]	
3rd person	he/she/it (does, makes, was)	they (do, make, were)
	[formerly: he/she/it (doth, maketh, was)]	

personal identification number (PIN) [computing] A unique number allocated to a person to give him or her access to a computer system (such as those that work cash-dispensing machines). The PIN is often (invisibly) encoded on a plastic card.

person-day [time] The standard amount of work done by one worker in one full working day, constituting a unit for calculating costs and schedules in relation to output.

perspective [physics: optics; maths] Method of representing three-dimensional objects in a two-dimensional drawing, based on establishing a horizon (on the viewer's eye level) and one or more vanishing points to which parallel lines apparently converge.

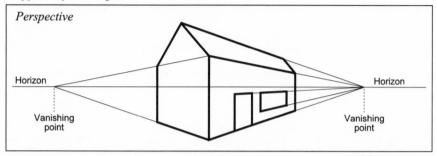

Perspective

pes [linear measure] The basic linear unit of ancient Rome: Classical plural, *pedes*. It is the basis for many present-day European measures, including the statute foot (which is, of course, its original meaning).

1 pes	=	16 digits, 12 UNCIAE ('inches'), 4 palmae
	=	11.68 (statute) inches, 29.67 centimetres
1½ pes	=	1 cubit
2½ pes	=	1 'step' (*gradus*)
5 pes	=	1 (double-)pace (*passus*)
10 pes	=	1 decempeda
625 pes, 125 paces	=	1 stadium
5,000 pes	=	1 'mile' (*mille passus*)

    See also FOOT.

peseta [comparative values] Unit of currency in Spain (and consequently also in Andorra).

                    1 peseta = 100 centimos

During Spanish colonial times, the name of the unit was used for a coin worth 2 reales, one-quarter of a PESO or DOLLAR: it was valid also in North America, where it was sometimes alternatively called a *pistareen*.

The term derives as a diminutive from Spanish *pesa* 'weight', the equivalent of *peso* – a coin that was of a specific weight. *See also* COINS AND CURRENCIES OF THE WORLD.

pesewa [comparative values] Unit of currency in Ghana.

                  100 pesewas = 1 cedi

    See also COINS AND CURRENCIES OF THE WORLD.

peso [comparative values] Unit of currency in Argentina, Bolivia, Chile, Colombia, Cuba, the Dominican Republic, Guinea-Bissau, Mexico, the Philippines, and Uruguay.

In Argentina, Bolivia, Colombia, Cuba, the Dominican Republic, Guinea-Bissau, Mexico, and the Philippines,

                  1 peso = 100 centavos

In Chile,

                  1 peso = 1,000 escudos

In Uruguay,

                  1 peso = 100 centesimos

The term is borrowed from the name of a sixteenth- to nineteenth-century Spanish coin of specific weight (Spanish *pesa* 'weight'), commonly known as 'a piece of eight' (because it was worth 8 *reales*), and valid all over the colonial world including North America, where it was alternatively called a 'dollar'.

$$1 \text{ old Spanish peso} = 8 \text{ reales}$$
$$4 \text{ pesos} = 1 \text{ pistole}$$
$$8 \text{ pesos} = 1 \text{ doubloon}$$

See also DOLLAR; DOUBLOON.

peta- [quantitatives: prefix] Prefix which, when it precedes a unit, increases the unit by 1 UK thousand-billion (one-thousand million million)/1 US quadrillion (10^{15}) times its standard size or quantity.

Example: petahertz – 1,000,000,000,000,000 hertz

The term derives from ancient Greek *peta-* 'widely extended'.

petrol/gasoline units *see* BARREL; GALLON; LITRE, LITER; LITRES PER HUNDRED KILO-METRES; MILES PER GALLON

PET scan *see* SCANNING SYSTEMS

pfennig [comparative values] Unit of currency in Germany.

$$100 \text{ pfennigs} = 1 \text{ Deutsche mark}$$

The term derives as the name of a coin that was either 'pan-shaped' or cast in moulds called 'pans', for it is a derivative of *pfanne* 'pan' – just as the English cognate *penn* is a variant of the word *panning*. *See also* COINS AND CURRENCIES OF THE WORLD.

pfund [weight] Obsolete unit of mass in Germany, Austria, and Switzerland, based – like the standard POUND – on the ancient Roman variant, the LIBRA, but assimilated since the 1870s (under Customs regulations) to the metric system.

Before the 1870s, in Germany,

$$1 \text{ pfund} = 16 \text{ unzen, } 32 \text{ lot}$$
$$= \text{ between 1 pound and 1 pound 4 ounces}$$
$$= \text{ between 453 and 566 grams}$$

Before the 1870s, in Austria,

$$1 \text{ pfund} = 16 \text{ unzen}$$
$$= 1 \text{ pound 3.75 ounces}$$
$$= 560 \text{ grams}$$

After the 1870s, in Germany, Austria, and Switzerland,

$$1 \text{ pfund} = 500 \text{ grams}$$
$$= 1 \text{ pound 1.64 ounces}$$
$$(1 \text{ pound} = 0.9072 \text{ pfund})$$

The term, like *pound*, derives from the Latin description of the measure as *pondo* 'by weight'.

pH [chemistry] A measurement of acidity, for a solution equal to the negative logarithm of the hydrogen ion concentration: *see* ACIDITY.

phagocytic index [medicine; biology] The average number of bacteria destroyed or absorbed by each bacteria-attacking body cell (that is, each phagocyte – such as a white blood cell) in a culture of phagocytes and bacteria in serum.

Phanerozoic [time; geology] Describing the entire duration of time in the Earth's history incorporated by the Palaeozoic, Mesozoic, and Cenozoic eras (sometimes called the Phanerozoic aeon or eon), from about 590 million years ago to the present day.

It is so called because only for these eras is there evidence (ancient Greek *phaneros* 'evident') of major geological change and organic evolution. Before these eras was the true prehistory of the Proterozoic or Archaeozoic PRECAMBRIAN era.

pharoagh *see* ARSHIN; MILL

phase [astronomy; chemistry; physics] In astronomy, the changes in shape of the visible part of a moon or planet lit by the Sun. The phases are caused by changes in the relative positions of the moon (or planet), the Sun, and the observer (that is, the Earth). The Earth's Moon undergoes a complete cycle of phases as it changes from New to Full and back to New again every month: *see* LUNAR PHASES.

In chemistry, a physically distinct, homogeneous component of a chemical system separated from other components by definite boundaries (interfaces). For example: in a mixture of ice and water, the ice and water are different phases.

In physics, for a periodically varying quantity (such as a cyclic wave motion), the part of the cycle that has been completed at a particular time. It is expressed as an angle, there being 360 degrees or 2π radians in one full cycle. When two varying quantities are 'in phase', the *phase angle* is 0 (or 360) degrees. When they are out of phase, one is said to lead or lag the other (by the phase angle). This is equally true in astronomy, in which the phase angle is also defined as the angle formed by the Sun and the Earth as seen by an observer on another planet: when the Sun and the Earth are 'in phase', the two are visually aligned.

phase rule [chemistry] For a heterogeneous chemical system at equilibrium, the sum of the number of phases and degrees of freedom equals 2 more than the number of components: *see* DEGREE OF FREEDOM; PHASE.

phi meson [physics] A short-lived, massive, subatomic particle with a zero charge.

pH meter [chemistry] An instrument for measuring the pH (acidity or alkalinity) of a solution by measuring the potential difference (voltage) between reference electrodes immersed in the solution. The potential difference is indicated on a millivoltmeter calibrated directly in pH.

phon [physics] Unit of objective sound loudness, measured in decibels above a reference of known frequency and intensity (usually 1 kilohertz at 0.002 microbar).

phone [literary: phonetics] From the 1910s to the 1950s, before the classifications of Noam Chomsky (and PHONEMES), a subdivision of speech sounds akin to the syllable, apprehended as a minimum speech unit.

phoneme [literary: phonetics] Unit of speech, a distinctive sound that forms one element in the pronunciation of a word, often – but not always – corresponding to a consonant or a vowel, as written.

The word derives through French from ancient Greek *phonema* 'a sound'.

phonon [physics; crystallography] For a crystal, a quantum of vibrational (thermal) energy in the lattice. The energy is equal to the product of the frequency of vibration and Planck's constant. It should not be confused with a PHOTON.

phot [physics] Unit of illumination in the centimetre-gram-second (CGS) system, equal to 1 lumen per square centimetre.

$$1 \text{ phot} = 10,000 \text{ lumens per square metre (SI units)}$$

photoelectric constant [physics] For a photoelectron (an electron emitted from the surface of certain materials when illuminated by light), Planck's constant (6.6262×10^{-34} joule-second) divided by the charge on an electron (1.602×10^{-19} coulomb). It provides an experimental way of determining Planck's constant.

photometer [physics] An instrument for measuring the intensity of light, usually by comparing light of unknown intensity with a source of known intensity. The light sources are adjusted until they provide equal illumination on similar surfaces, and the unknown intensity calculated from the knowledge that the ratio of the two intensities is equal to the ratio of the squares of their distances from the surfaces. *See also* INVERSE SQUARE LAW.

photon [physics] A quantum ('packet') of electromagnetic radiation, such as light or X-rays. Its energy equals the product of its frequency (in hertz) and Planck's constant (6.6262×10^{-34} joule-second).

photosphere *see* ATMOSPHERE OF THE SUN, OF A STAR

photosynthetically active radiation (PAR) [biology] Light in the wavelength range 400 to 700 nanometres, which brings about photosynthesis in green plants.

photovoltaic [physics; astronomy] Describing a substance that produces a voltage or electric current when exposed to light or other similar electromagnetic radiation; such substances (including for example silicon) are the basis of solar cells.

phrase [literary; music] In grammar, a short combination of words that does not include an actively used verb and that effectually represents a single idea (mostly describing a noun in an adjectival way). For example: in

The house on the hill is for sale.

the words 'on the hill' correspond to an adjectival phrase qualifying 'The house'.

In music, a phrase is a short passage within a piece, usually corresponding to a specific fraction or proportion of a melody or harmonic sequence. In jazz, such a musical phrase may be called a *riff*, especially if it is part of the main theme that is repeated several times and is the central focus of interest in the piece.

phrase markers [literary] In transformational grammar, abstract units of meaning that represent a stage of sentence analysis beyond the finding of phonemes, morphemes, and similar detailed elements in analysing a sentence, and that may combine them. Discerning such phrase markers is part of what is sometimes called *phrase-structure grammar*.

phrenology [medicine] 'Science' popular in the nineteenth century by which a person's character and disposition was claimed to be evident from the irregularities ('bumps' and 'dips') on his or her skull. At a time when social authorities were hoping to find some definitive bodily manifestation of criminality, this was one of many lines of investigation eagerly pursued until discredited.

But character reading through phrenology has not totally disappeared.

The first element of the term derives from ancient Greek *phren* 'the mind', especially 'the heart', and thus also 'the diaphragm' – which may or may not be akin to English *brain* [with the possible root meaning 'the enclosing/enclosed (*thing*)'].

Phrygian mode [music] Medieval form of KEY, the SCALE of which is characterized on a modern piano by the white notes between one E and the next. In technical terms in ordinary Western music, its main effect is that of a minor key with a virtually unusable dominant chord (because of the flattened supertonic, although the flattened leading-note could be tolerated).

Its name derives from one of the musical modes used in ancient Greece, about which little or nothing is known – but the ancient Greeks had completely different musical referents anyway.

pH value *see* PH

Phylum [biology] In the taxonomic classification of life-forms, the category of animal life between Kingdom and Class – one of the widest forms of classification – equivalent in the plant world to the category Division. *See* full list of categories under CLASS, CLASS.

phyton [biology] The smallest part of the leaf, stem or root of a plant that can be successfully grown on independently: a 'plant unit'.

π, pi [maths] The ratio of the circumference of a circle to its diameter, equal to the irrational number 3.1415926536 . . . (very approximately equal to $^{22}/_7$). It is also equal to half the number of radians in a complete circle (that is, in 360 degrees).

The term corresponds to the symbol and name of the Greek equivalent of the alphabetical letter 'p'.

piano, pianissimo [music] Musical instruction: play or sing softly (Italian *piano*; symbol p) or as softly as possible (*pianissimo*; symbol ppp). The opposite is FORTE, FORTISSIMO.

The term derives ultimately from Latin *planus* 'plain', thus 'smooth', thus 'easy', 'effortless', thus 'soft'.

piano accordion *see* HARMONICAS' AND ACCORDIONS' RANGE

piano range, pianoforte range *see* KEYBOARD INSTRUMENTS' RANGE

piastre, piaster [comparative values] Unit of currency in Egypt, Lebanon, the Sudan, and Syria (and formerly in Libya, Turkey, and South Vietnam).

In Egypt, Lebanon, and the Sudan
$$100 \text{ piastres} = 1 \text{ pound}$$

In Syria,
$$1 \text{ piastre} = 100 \text{ centimes}$$

In the years of the Spanish colonies in America, it was also another name for the PESO or DOLLAR.

The name of the unit derives in a roundabout way from Latin *emplastrum* 'something moulded by hand' (thus English *plaster* – in the senses both of a filling material to cover walls and a covering material to protect wounds, akin to *plastic*) which, in medieval Italian, became *piastro* '(a sheet of) malleable metal', and thus, through French, 'coin'.

pibroch [music] Bagpipe playing, especially the music of the bagpipe when played in solo competition.

The term derives from the Scottish Gaelic *piobaireachd* 'the piper's (music)'.

pica [printing measure] Unit used to measure type in printing, more specifically to measure the width of book or newspaper columns, and to give some idea of the average number of characters per line.

$$1 \text{ pica} = \text{ ⅙ inch, 12 points}$$
$$= 4.233 \text{ millimetres}$$
$$= 0.9380 \text{ CICERO (metric type units)}$$
$$(1 \text{ cicero} = 1.066 \text{ pica, } 4.512 \text{ millimetres})$$
$$6 \text{ picas} = 1 \text{ inch, 72 points}$$

The name of the unit derives, it is thought, from the type first used in printing the book of daily services for the medieval Church (a book known in Church Latin as a *Pica*). The book probably only got that name because the type used gave it such clarity in black and white, and 'black-and-white' in Latin is *picus, pica* (hence English *pie, pied* as in 'piebald', 'pied piper'). But the meaning 'black-and-white' in Latin derives in turn from the predominantly black-and-white bird, the woodpecker (Latin *picus*) and the magpie (Latin *pica*), which have notoriously hard, prominent bills – Latin *pic-* is the same as the English *beak, peck, pike,* and *pick*.
See also DIDOT POINT SYSTEM; EM.

piccolo *see* WOODWIND INSTRUMENTS' RANGE

pice [comparative values] Former unit of currency in India and Pakistan. Singular and plural: *pice*.

$$4 \text{ pice} = 1 \text{ anna}$$

The unit was replaced in India in 1957 and in Pakistan in 1961 by the cognate PAISA.

pickoff [physics: electronics] In automated engineering and control systems, an electronic device that monitors a geometrical pattern and responds with a signal when the pattern changes.

pico- [quantitatives; prefix] Prefix which, when it precedes a unit, reduces the unit to 1 UK million-millionth (1 billionth)/1 US trillionth (10^{-12}) of its standard size or quantity. Prior to its introduction as the metric prefix, the prefix that stood instead for this fraction was *micromicro-*.

For examples *see* entries *below* beginning 'pico-'.

The prefix derives from Spanish *pico-* '(a) small number', the root meaning of which is 'tapering to a minimum', 'coming to a point'; it is thus etymologically cognate with PICA and English *peak*.

picocurie [physics] Former unit of radioactivity equal to 1 UK billionth/1 US trillionth of a curie.

$$1 \text{ picocurie} = 0.000000000001 \text{ curie } (10^{-12} \text{ curie}).$$
$$= 3.7037 \times 10^{-2} \text{ becquerel}$$

See also CURIE.

picofarad [physics] Unit of capacitance equal to 1 UK billionth/1 US trillionth of a farad.

$$1 \text{ picofarad (pF)} = 0.000000000001 \text{ farad } (10^{-12} \text{ farad})$$
$$= 9 \times 10^{-1} \text{ electrostatic unit (ESU)}$$

See also FARAD.

picogram [weight] Unit of mass equal to 1 UK billionth/1 US trillionth of a gram.

$$1 \text{ picogram} = 0.000000000001 \text{ gram } (10^{-12} \text{ gram})$$

See also GRAM; NANOGRAM.

picosecond [time] Unit of time equal to 1 UK billionth/1 US trillionth of a second.

$$1 \text{ picosecond} = 0.000000000001 \text{ second } (10^{-12} \text{ second})$$

See also SECOND; NANOSECOND.

pictogram [maths] In statistics, a way of representing a set of data graphically using rows of symbols in which each symbol represents a particular number of items of data. It is a pictorial type of bar chart.

picul [weight] Unit of weight corresponding to one used throughout the Far East during the colonial times of the mid-1800s, mainly devised artificially as a measure for customs and taxation purposes but, in some areas, assimilated since to a unit within a different system.

In Hong Kong,

$$1 \text{ picul} = 100 \text{ kan}$$
$$= 133.33 \text{ pounds avdp. (9 stone 7.4 pounds)}$$
$$= 60.4788 \text{ kilograms}$$

In Malay(si)a,

$$1 \text{ picul} = 100 \text{ gin or kati}$$

$$= \quad 133.33 \text{ pounds avdp. (9 stone 7.4 pounds)}$$
$$= \quad 60.4788 \text{ kilograms}$$

In Japan,

$$1 \text{ picul} \quad = \quad 1,600 \text{ taels, } 100 \text{ kin, } 16 \text{ k(u)wan}$$
$$= \quad 132.3 \text{ pounds avdp. (9 stone 6.3 pounds)}$$
$$= \quad 60 \text{ kilograms}$$

The term derives from a Malay or Javanese expression for 'the load that can be carried by 1 man'.

pie [linear measure] In Spain and Italy, an obsolete derivative of the old Roman *pes* 'foot'.

In Spain,

$$1 \text{ pie} \quad = \quad 12 \text{ pulgada, } 1.333 \text{ palmo}$$
$$= \quad 27.864 \text{ centimetres, } 10.97 \text{ inches}$$
$$3 \text{ pie, } 4 \text{ palmos} \quad = \quad 1 \text{ vara}$$
$$6 \text{ pie, } 8 \text{ palmos} \quad = \quad 1 \text{ braça}$$

In Italy,

$$1 \text{ pie} \quad = \quad 29.794 \text{ centimetres, } 11.73 \text{ inches}$$

The Italian *pie* does not seem to be related to the Italian *braccio d'ara* (70 centimetres, 27.56 inches) and is probably connected instead with the old Roman form of CUBIT. *See also* PALM, PALAME, PALMO.

pie [comparative values] Former unit of currency in India, in the form of a bronze coin.

$$3 \text{ pies} \quad = \quad 1 \text{ pice}$$
$$12 \text{ pies, } 4 \text{ pice} \quad = \quad 1 \text{ anna}$$

It would seem that the terms *pie* and *pice* are cognate, and that they in turn are etymologically identical with *paisa*: *see* PAISA.

pieces of eight *see* PESO

pie chart [maths] In statistics, a way of representing a set of data graphically using a circular diagram divided into sectors. The size of a sector (given by its angle at the centre of the 'pie') represents a particular amount of data, as a percentage of the whole.

The pie chart is sometimes alternatively called a *circle graph*.

Pie chart

pied [linear measure] Obsolete unit of length in Francophone Belgium and Canada, based on the old Roman *pes* 'foot'.

In Belgium,

$$1 \text{ pied} \quad = \quad 10 \text{ pouces ('thumbs')}$$
$$= \quad 30 \text{ centimetres}$$
$$= \quad 11.81 \text{ inches}$$

In Canada,

$$1 \text{ pied} \quad = \quad 32.48 \text{ centimetres}$$
$$= \quad 12.79 \text{ inches}$$
$$18 \text{ pieds} \quad = \quad 1 \text{ perche}$$

The Belgium *pied* has evidently been assimilated to a metric measure of similar value (despite the subdivision into 'pouces'). The Canadian foot is identical with the intentionally larger form of French foot, the PIED-DE-ROI or 'royal foot'.

The term is the French for 'foot'.

pied-de-roi [linear measure] Obsolete unit of length in France, based on the old Roman *pes* 'foot' but intentionally larger, as the name 'royal foot' indicates.

$$1 \text{ pied-de-roi} \quad = \quad 12 \text{ pouces ('thumbs')}$$
$$= \quad 32.48 \text{ centimetres}$$
$$= \quad 12.79 \text{ inches}$$
$$6 \text{ pieds-de-roi} \quad = \quad 1 \text{ toise}$$

The first system of measurement to contain a 'royal foot' was that of the ancient

Egyptians, a parallel of the ordinary 'foot' of the ancient Babylonians (*see* FOOT).

pièze [physics] Unit of pressure in the metre-tonne-second system, equal to 1 sthène per square metre.

$$1 \text{ pièze} = 1,000 \text{ newtons (1 kilonewton) per square metre}$$

The term is a French adaptation of the ancient Greek word meaning 'pressure'.

pik [linear measure] Former unit of linear measure in Turkey, Egypt, Greece, and Cyprus, originally (in Greece and Turkey) of a minimum and a maximum value – a long pik and a short pik.

In Turkey,

$$
\begin{aligned}
1 \text{ short pik} &= 16 \text{ kerat} \\
&= 18 \text{ inches (1 foot 6 inches; } 1\tfrac{1}{2} \text{ foot)} \\
&= 45.72 \text{ centimetres (0.4572 metre)} \\
1 \text{ long pik} &= 25 \text{ kerat} \\
&= 28.1 \text{ inches (2 feet 4.1 inches)} \\
&= 71.37 \text{ centimetres (0.7137 metre)}
\end{aligned}
$$

Under the alternative name of *diraa*, however, the Turkish pik was standardized so that

$$
\begin{aligned}
1 \text{ pik or diraa} &= 27.90 \text{ inches (2 feet 3.9 inches)} \\
&= 70.87 \text{ centimetres, 0.7087 metre}
\end{aligned}
$$

As the *diraa*, the Turkish pik was standardized in Egypt at a value nearly halfway between the two short and long Turkish extremes, so that

$$
\begin{aligned}
1 \text{ Egyptian pik or diraa} &= 22.835 \text{ inches (1 foot 10.8 inches)} \\
&= 58.0 \text{ centimetres, 0.580 metre}
\end{aligned}
$$

The Greek pik, also known as a *pic* or *picki*, had a minimum and a maximum value like the Turkish pik, but not of such great difference.

In Greece,

$$
\begin{aligned}
1 \text{ short pik} &= 25.20 \text{ inches (2 feet 1.2 inches)} \\
&= 64 \text{ centimetres (0.64 metre)} \\
1 \text{ long pik} &= 26.38 \text{ inches (2 feet 2.38 inches)} \\
&= 67 \text{ centimetres (0.67 metre)}
\end{aligned}
$$

In Cyprus, no doubt because of the influence of British measures, the unit was assimilated to the imperial measurement system.

$$
\begin{aligned}
1 \text{ Cypriot pik} &= 24 \text{ inches (2 feet)} \\
&= 60.96 \text{ centimetres (0.6096 metre)}
\end{aligned}
$$

The name of the unit derives from ancient Greek *pekhys* 'the ulna', 'the forearm', and is thus by semantic derivation identical with the Latin-based CUBIT and ELL – which also explains why the measure is so variable as a unit.

pi-meson, pion [physics] A massive subatomic particle (of about 270 times the mass of an electron) that can have positive, negative, or zero charge.

PIN *see* PERSONAL IDENTIFICATION NUMBER (PIN)

pin [volumetric measure] Old unit of liquid capacity originally equal to half of 1 UK FIRKIN, one-quarter of a kilderkin, but borrowed as a measure in the United States as an equivalent number of gallons, and thus of an overall value different from that of the UK unit.

In Britain,

$$
\begin{aligned}
1 \text{ pin} &= 4.5 \text{ UK gallons, 36 UK pints} \\
&= 5.4 \text{ US gallons, 43.2 US pints} \\
&= 20.097 \text{ litres} \\
&= 0.833 \text{ US pin}
\end{aligned}
$$

In the United States,

$$
\begin{aligned}
1 \text{ pin} &= 4.5 \text{ US gallons, 36 US pints} \\
&= 3.75 \text{ UK gallons, 30 UK pints} \\
&= 17.034 \text{ liters} \\
&= 1.201 \text{ UK pin}
\end{aligned}
$$

The term derives from the shape of the cask or keg that contained the pin of liquid: one-eighth the size of a big, round barrel (the size of the present-day international beer barrel), it was in comparison a short, comparatively thin and stubby cylinder – like a pin, peg, or nail.

pine-tree shilling [comparative values] Unit of currency in the Massachusetts Bay

Colony of the 1600s, a coin intended to have the same value as the English shilling (12 pence, one-twentieth of a pound) although containing a slightly smaller quantity of silver.

It was so called because it had a depiction of a pine tree on one side.

pint [volumetric measure] Standard unit for many household measures, particularly in containers of milk and other liquid foods, although in Europe now much overshadowed by the LITRE and half-litre. The pint itself differs in quantity between European and North American usages.

As a liquid measure:

		1 pint	=	4 GILLS, one-eighth of a GALLON	
1 UK pint	=	0.5682 litre		1 US pint	=	0.4732 liter
	=	568.2 millilitres			=	473.2 milliliters
		(0.5 litre = 0.880 UK pint)				(0.5 liter = 1.057 US pint)
	=	20 UK FLUID OUNCES			=	16 US fluid ounces
	=	160 UK FLUID DRAMS			=	128 US fluid drams
	=	1.20095 US pint			=	0.83267 UK pint
					=	2 cups

2 pints	=	1 quart
8 pints	=	1 gallon

As a dry measure:

		1 pint	=	one-sixteenth of a PECK	
1 UK pint	=	568.2 CC		1 US pint	=	551.1 cc
	=	34.675 CU. IN.			=	33.625 cu. in.
	=	1.03112 US dry pint			=	0.96982 UK dry pint
		(500 cc = 0.880 UK dry pint)				(500 cc = 0.907 US dry pint)

16 dry pints	=	1 peck
64 dry pints, 4 pecks	=	1 BUSHEL

In the United Kingdom, the volume of a pint thus remains the same for both liquid and dry goods; in the United States there is a 16% difference between liquid and dry measures with the same name.

It is possible that the name of the unit derives as a French-based variant of the originally Germanic measure responsible also for the unit known as the PIN.

pion *see* PI-MESON, PION

pionium [physics] A short-lived, neutral, subatomic particle consisting of a combined PI-MESON and MU-MESON.

pipa [volumetric measure] Old Portuguese measure of liquid capacity, now assimilated to a metric unit.

In former times,

1 pipa	=	7.75 fanega, 25.7 almudes, 124 quartos
	=	429 litres
	=	94.38 UK gallons, 113.33 US gallons

Today, however,

1 pipa	=	500 litres
	=	110.00 UK gallons, 132.09 US gallons

The term is evidently borrowed from the old shipping measure known in English as the PIPE. In present-day terms, then,

1 pipa	=	1.048 pipe	1 pipe	=	0.954 pipa

pipe [volumetric measure] Unit of wet and dry capacity mostly used in marine freight transportation. Technically, it originated as the amount equivalent to 4 BARRELS, 3 TIERCES, or 2 HOGSHEADS – and thus varied, as does the barrel still, according to the nature of the contents.

But as a standard unit,

1 pipe	=	126 old wine (or modern US) gallons
	=	104.92 imperial (UK or English) gallons
	=	476.95 litres
		(500 litres = 1.048 pipe)
2 pipes	=	1 TUN

The name of the unit derives from the cylindrical nature of the container – that is, a tube, like any other kind of pipe. But it was the musical type of pipe (or fife, cf. German *Pfeife* 'pipe', 'whistle') that came first, the sort of light instrument that could imitate the birds' *peeping* (Latin *pippa* 'chirp', *pipiare* 'to tweet'). *See also* PIPA.

pipette [chemistry] A piece of apparatus used in volumetric analysis for measuring out a precise volume of liquid. It consists of a graduated glass tube usually with a swelling at the centre (to increase its volume); liquid is drawn into the tube up to the graduation.

pistareen *see* PESETA

pistol calibres *see* CALIBRE, CALIBER

pistol shooting *see* SHOOTING MEASUREMENTS AND UNITS

pistole [comparative values] Unit of currency of colonial Spain between the sixteenth and nineteenth centuries, in the form of a gold coin.

$$1 \text{ pistole } = \quad 4 \text{ pesos, } 32 \text{ reales}$$
$$2 \text{ pistoles } = \quad 1 \text{ doubloon}$$

All these coins were valid also in North America, where some were known under alternative names (the PESO – or 'piece of eight' – as the original DOLLAR, for example).

Various other countries also issued gold coins of this name (or as the cognate *pistolet*) at the time, but none so valuable as the Spanish. Why the coin should be given the name, however, is not known.

pitch [music] Pitch is defined technically as the frequency at which sound waves oscillate through the air. People who are said to have 'perfect pitch' or 'absolute pitch' can immediately and accurately identify musical notes on hearing them, without recourse to a musical instrument; for most people who can do this, it is a matter more of having a specialized aural memory, especially following a long-standing acquaintance with the notes of a particular instrument at home, than of having an intuitive overall grasp of pitch. But a few people really do have fully developed absolute pitch.

To tune instruments to a standard pitch for playing in musical performances, what is known as CONCERT PITCH (or standard pitch) recognizes 440 cycles per second (hertz) as the official frequency of the note A above middle C . To some extent this represents a compromise between the value of the note A in northern Europe from about 1500 to about 1670, which was 466 hertz (a semitone or half-step above concert pitch A), and the value of the note A in northern Europe from about 1670 to about 1770, which was 415 hertz (a semitone or half-step below concert pitch A).

It is evident, then, that about 50 hertz represents an interval of a tone or whole step at the middle register. But it is not the number of cycles per second that matters so much as the ratio of frequency between one note and another: the frequency ratio of one note to another that is one tone above it in pitch is 8:9 . So that if A is 440 hertz, B is (440 x $\frac{9}{8}$ =) 495 hertz. The frequency ratio of one note to another that is only a semitone or half-step above it in pitch is 15:16 – so if B is 495 hertz, C above middle C is (495 x $\frac{16}{15}$ =) 528 hertz. And the frequency ratio of one note to another that is an octave below it in pitch is 2:1, so that if C above middle C is 528 hertz, middle C itself is (528 x $\frac{1}{2}$ =) 264 hertz.

Frequency ratios of musical intervals

interval	ratio	example
semitone or half-step	15:16	middle C up to C# (264 to 282 hertz)
tone or whole step	8:9	middle C up to D (264 to 297 hertz)
minor third	5:6	middle C up to E♭ (264 to 317 hertz)
major third	4:5	middle C up to E (264 to 330 hertz)
perfect fourth	3:4	middle C up to F (264 to 352 hertz)
augmented fourth/		
diminished fifth	45:64	middle C up to F# (264 to 375 hertz)
perfect fifth	2:3	middle C up to G (264 to 396 hertz)
minor sixth	5:8	middle C up to A♭ (264 to 422 hertz)
major sixth	3:5	middle C up to A (264 to 440 hertz)
minor seventh	9:16	middle C up to B♭ (264 to 469 hertz)
major seventh	8:15	middle C up to B (264 to 495 hertz)
octave	1:2	middle C up to C (264 to 528 hertz)

From this it also follows that, if middle C is 264 hertz, the next C up is 528 hertz, the next C is 1056 hertz, and the next 2112, and so on. In a similar fashion, the frequencies of all the notes on a piano or any other instrument can be calculated from the table above.

For those people who are not blessed with perfect or absolute pitch, there are various means of obtaining a true note to which to tune an instrument, the most common of which are the tuning-fork (which is struck to vibrate at a set note, usually A) and the pitch pipe (which is blown for a reed to produce a set note, usually A but sometimes D or C).

pitch [aeronautics; shipping] The forward distance travelled by a propeller during one revolution.

Pitot tube [physics; aeronautics] Device for measuring the speed of flow of a moving fluid (gas or liquid), as in an aircraft's air-speed indicator or a ship's surface-speed indicator. It consists of an open-ended circular tube pointing into the flow, with a hole in the side of the tube at the other end. The difference between the dynamic pressure at the open end and the static pressure at the side hole is a measure of the speed of flow. It was named after the French physicist Henri Pitot (1695-1771).

piu [music] Part of a musical instruction: 'more . . . '. The term derives directly from the Italian.

pK value [chemistry] The negative logarithm of the equilibrium constant for the dissociation of an electrolyte into ions (in solution). Weak electrolytes have low equilibrium constants and high pK values.

place [sporting term] In betting on horse or greyhound racing, the final position of the runner-up or the third, as posted officially.

place value [maths] In arithmetic, the value of an individual figure as indicated by its position within the overall sequence of digits that make up a number. For example: in the number 44.4, the first 4 has a place value of 4 x 10; the second 4 has a place value of 4 x 1; and the third 4 has a place value of 4 x 0.1.

plack [comparative values] Former unit of currency in Holland and Scotland, in the form of a coin of very little value (and renowned for being so).

The coin's name is akin to the better-attested form *plaque* 'something stuck on' (cf. PIASTRE, PIASTER), thus 'a patch', 'a piece'.

planck *see* JOULE-SECOND

Planck's constant [physics] Fundamental physical constant, equal to 6.6262 x 10^{-34} joule-second. It is important in quantum theory, in which it is equal to the energy of a quantum divided by its frequency. The symbol for Planck's constant is h . It was named after the German physicist Max Planck (1858-1947).

plane [maths] A flat surface.

planets of the Solar System [astronomy] The planets of the Solar System are, in order of the nearest to the Sun to the farthest from it:

	deity/attribution	root meaning
Mercury	Roman/messenger, commerce	'reciprocality'
Venus	Roman/evening star, beauty	'desire'
the Earth		'tillable soil'
Mars	Roman/agriculture, war	'stamper'
Jupiter	Roman/ruler of gods and mortals	'god-father'
Saturn	Roman/food and plenty	'seed/satisfaction'
Uranus	Roman/the heavens	? 'sun-kissed sea'
Neptune	Roman/ruler of the oceans	? 'water'
Pluto	Roman/ruler of the underworld	'wealth(-giver)'

Although the planets' names (apart from the Earth) all derive from Roman deities, the deities themselves were by no means all of Roman origin. 'Mars', for instance, was almost definitely borrowed and adapted from a Middle Eastern deity of similar name; 'Neptune' was borrowed from an Etruscan god of inland streams; 'Pluto' was taken from the Greek superstitious nickname for Hades, the 'Unseen', the god of the underworld (understood by the Romans instead to be the kingdom over which Pluto ruled).

Between Mars and Jupiter lie the belt of planetary fragments known as the asteroids.

Planetary statistics

The elements listed below as main atmospheric constituents are given in alphabetical order, not in order of proportional constituency.

MERCURY

sidereal orbital period ('year')		87.95 days
mean synodic period		115.88 days
mean distance from the Sun	(astronomical units)	0.387
	(million kilometres)	57.9
	(million miles)	34.7
orbital velocity	(kilometres per second)	48.0
	(miles per second)	29.0
orbital eccentricity		0.2056
orbital inclination		7° 00'
rotation period		59 days
equatorial diameter	(kilometres)	4,880
	(miles)	2,928
mass (Earth = 1)		0.0553
mean density		5.4
mean daytime surface temperature	(°C)	400
	(°F)	752
main atmospheric constituents		helium, hydrogen
number of known moons		0

VENUS

sidereal orbital period ('year')		224.70 days
mean synodic period		583.92 days
mean distance from the Sun	(astronomical units)	0.723
	(million kilometres)	108.2
	(million miles)	64.9
orbital velocity	(kilometres per second)	35.0
	(miles per second)	21.0
orbital eccentricity		0.0068
orbital inclination		3° 24'
rotation period		retrograde, 243 days
equatorial diameter	(kilometres)	12,104
	(miles)	7,262
mass (Earth = 1)		0.815
mean density		5.2
mean daytime surface temperature	(°C)	470
	(°F)	878
main atmospheric constituents		carbon monoxide, carbon dioxide, fluorine, hydrochloric acid, nitrogen, oxygen, water
number of known moons		0

EARTH

sidereal orbital period ('year')		365.26 days
mean synodic period		—
mean distance from the Sun	(astronomical units)	1.0
	(million kilometres)	149.6
	(million miles)	89.8
orbital velocity	(kilometres per second)	30.0
	(miles per second)	18.0
orbital eccentricity		0.0167
orbital inclination		0
rotation period		23 hours 56 minutes

equatorial diameter	(kilometres)	12,756.3
	(miles)	7,653.8
mass (Earth = 1)		1.0
mean density		5.52
mean daytime surface temperature	(°C)	20
	(°F)	68
main atmospheric constituents		argon, carbon monoxide, carbon dioxide, hydrogen, nitrogen, oxygen, water
number of known moons		1

MARS

sidereal orbital period ('year')		686.98 days
mean synodic period		779.94 days
mean distance from the Sun	(astronomical units)	1.524
	(million kilometres)	227.99
	(million miles)	136.79
orbital velocity	(kilometres per second)	24.0
	(miles per second)	14.0
orbital eccentricity		0.0934
orbital inclination		1° 51'
rotation period		24 hours 37 minutes
equatorial diameter	(kilometres)	6,787
	(miles)	4,072
mass (Earth = 1)		0.1074
mean density		3.97
mean daytime surface temperature	(°C)	0
	(°F)	32
main atmospheric constituents		argon, carbon monoxide, carbon dioxide, hydrogen, water
number of known moons		2

JUPITER

sidereal orbital period ('year')		11.86 years
mean synodic period		398.88 days
mean distance from the Sun	(astronomical units)	5.203
	(million kilometres)	778.37
	(million miles)	467.02
orbital velocity	(kilometres per second)	13.0
	(miles per second)	8.0
orbital eccentricity		0.0485
orbital inclination		1° 18'
rotation period		9 hours 55 minutes
equatorial diameter	(kilometres)	142,200
	(miles)	85,320
mass (Earth = 1)		317.892
mean density		1.33
mean daytime surface temperature	(°C)	−103
	(°F)	−153
main atmospheric constituents		ammonia, helium, hydrogen, methane, nitric acid, nitrous oxide, water
number of known moons		16

SATURN

sidereal orbital period ('year')		29.46 years
mean synodic period		378.09 days
mean distance from the Sun	(astronomical units)	9.539
	(million kilometres)	1,427.0
	(million miles)	856.2
orbital velocity	(kilometres per second)	10.0
	(miles per second)	6.2
orbital eccentricity		0.0556
orbital inclination		2° 29'
rotation period		10 hours 24 minutes
equatorial diameter	(kilometres)	119,300
	(miles)	71,580
mass (Earth = 1)		95.168
mean density		0.69
mean daytime surface temperature	(°C)	−138
	(°F)	−216
main atmospheric constituents		ammonia, helium, hydrogen, methane
number of known moons		17

URANUS

sidereal orbital period ('year')		84.01 years
mean synodic period		369.66 days
mean distance from the Sun	(astronomical units)	19.18
	(million kilometres)	2,869.5
	(million miles)	1,721.7
orbital velocity	(kilometres per second)	7.0
	(miles per second)	4.0
orbital eccentricity		0.0472
orbital inclination		0° 46'
rotation period		17 hours 46 minutes
equatorial diameter	(kilometres)	51,800
	(miles)	31,080
mass (Earth = 1)		14.559
mean density		1.27
mean daytime surface temperature	(°C)	−218
	(°F)	−360
main atmospheric constituents		helium, hydrogen, methane
number of known moons		15

NEPTUNE

sidereal orbital period ('year')		164.79 years
mean synodic period		367.49 days
mean distance from the Sun	(astronomical units)	30.07
	(million kilometres)	4,497
	(million miles)	2,698
orbital velocity	(kilometres per second)	5.0
	(miles per second)	3.1
orbital eccentricity		0.0086
orbital inclination		1° 46'
rotation period		16 hours 7 minutes
equatorial diameter	(kilometres)	49,500
	(miles)	29,700
mass (Earth = 1)		17.617
mean density		1.64
mean daytime surface temperature	(°C)	−224
	(°F)	−371

main atmospheric constituents		helium, hydrogen, methane
number of known moons		8

PLUTO

sidereal orbital period ('year')		247.7 years
mean synodic period		366.74 days
mean distance from the Sun	(astronomical units)	39.44
	(million kilometres)	5,900
	(million miles)	3,540
orbital velocity	(kilometres per second)	5.0
	(miles per second)	3.1
orbital eccentricity		0.25
orbital inclination		17° 10'
rotation period		153 hours 18 minutes
equatorial diameter	(kilometres)	2,200
	(miles)	1,320
mass (Earth = 1)		perhaps 0.17
mean density		not accurately known
mean daytime surface temperature	(°C)	−200
	(°F)	−328
main atmospheric constituents		not accurately known
number of known moons		1

planimeter [engineering] An instrument for measuring areas on plans, maps, and diagrams. It is a mechanical device with a pointer that is traced around the boundary of the area to be measured; a rotating wheel indicates the area.

planisphere [astronomy] A chart of the night sky showing the stars and constellations in a rotatable frame calibrated with the dates of the year and times of the night, so that the specific night sky of any day and time can be presented (in relation to an established latitude).

Plantagenet [time] Describing the period in English history between AD 1154 and 1485, when members of the Plantagenet family were on the throne. The kings involved were:

Henry II	ruled	1154-89	(35 years)
Richard I		1189-99	(10)
John	ruled	1199-1216	(17)
Henry III		1216-72	(56)
Edward I		1272-1307	(35)
Edward II		1307-27	(20)
Edward III		1327-77	(50)
Richard II		1377-99	(22)
Henry IV		1399-1413	(14)
Henry V		1413-22	(9)
Henry VI		1422-61	(39)
Edward IV		1461-83	(22)
Edward V		1483	(75 days)
Richard III		1483-85	(26 months)

The first Plantagenet king, Henry II, who succeeded to the throne on the death of his cousin Stephen of Blois, was the son of Geoffrey, Count of Anjou, surnamed Plantagenet (Latin *planta genista* 'the broom plant'), and grandson of Henry I (son of William the Conqueror). Both the houses of Lancaster and York belonged to this wide-ranging family, and it was the defeat of the house of York by Henry VII, son of Edmund Tudor, that brought the dynasty to an end.

plasmasphere [astronomy; physics] An envelope of highly ionized gas surrounding a planet: *plasma* is the term for a 'fluid' made up of nearly identical quantities of positive ions and free electrons.

plastochron [time; biology] An interval between stages of progressive organic development, apprehended as a unit of developmental progress itself.

Platonic year *see* GREAT YEAR

platoon [military] In the army, a group under a leader (often a lieutenant) selected
for a special duty or task.

In the United States army, a platoon comprises two or more squads; four platoons
commonly make up one company.

The term derives from French *peloton* 'bunch', originally a double-diminutive of
the French equivalent of *ball*.

playing-card suits *see* SUITS OF CARDS

Pleistocene epoch [time; geology] The CENOZOIC ERA – the latest era in geological
time – is divided into the TERTIARY and QUATERNARY PERIODS; the Quaternary period
is then subdivided into the Pleistocene and Holocene epochs. But many of these
divisions are somewhat false. What 'separates' the end of the Tertiary period's final
epoch, the Pliocene, from the Quaternary's first, the Pleistocene, is no more than
the appearance, some 2-3 million years ago, of hominids on the surface of the Earth
(although mammals had been present for many millions of years already). And
what 'distinguishes' the present Holocene epoch from the previous Pleistocene is
no more than the absence of the ice ages at least four of which characterized the
earlier epoch – although because the Holocene is reckoned only to have begun
about 10,000 years ago, perhaps it merely represents one of the ordinary periods of
remission between ice ages within the still on-going Pleistocene.

Flatter areas of northern Europe, Asia, and Canada still show the marks of the
severe glaciation of the Pleistocene ice ages, notably in a multiplicity of lakes and
low ridges.

The term *Pleistocene* derives from ancient Greek elements meaning 'most
recent'. *See also* HOLOCENE EPOCH; PLIOCENE EPOCH.

plethron [linear measure; square measure] Unit of length and of area in ancient
Greece.

As a measure of length,

1 plethron	=	100 'feet'
	=	30.8 metres, 101 feet 4 inches
6 plethra	=	1 stadion
24 plethra, 4 stadia	=	1 ride

As a measure of area,

1 plethron	=	100 x 100 'feet', 10,000 square 'feet'
	=	948.64 square metres, 9.4864 ares
	=	10,268.38 square feet, 1,140.9 square yards

The term was used in ancient Greece also for a (running) racetrack – presumably
a track of the outer dimensions of the square measure and not of the linear measure
[for 31 metres (32 yards) seems an absurdly short distance for any athlete to race].

plimsoll line, load line [volumetric measure; weight] A mark on the side of a ship
that shows the highest waterline (to which the ship may be loaded) compatible with
stability in heavy seas. It is mandatory on all ships registered in Britain and is
required also by the marine authorities of many other countries (but not all).

The line – also called the plimsoll mark – is named after the British Member of
Parliament and reformer of maritime law Samuel Plimsoll (1824-98) who, after
years of political struggle with ship-owners in the British House of Commons,
managed in 1876 to get the Merchant Shipping Bill enacted to fix the maximum
loading line by law.

Pliocene epoch [time; geology] The last geological EPOCH of the TERTIARY PERIOD
of the CENOZOIC ERA, corresponding roughly to between 5 million years ago and 2-
3 million years ago. During this epoch the average temperature at the planetary
surface began the long decline towards the freezing cold of the ice ages that were
to characterize the succeeding PLEISTOCENE. The result was the lowering of sea-
levels around the world, and the migration of animals to warmer climates even if
it meant actually wandering from one continent to another. Mammals excelled at
such migration – among them, the anthropoid apes that were the ancestors of
humankind.

The Miocene and Pliocene epochs are sometimes together known as the Neocene;
the two epochs together with the Pleistocene and Holocene epochs of the Quater-

nary PERIOD are sometimes known as the Neogene.

The name of the epoch derives through French from ancient Greek *pleion* 'more', *kainos* 'recent' – in relation to the succeeding Pleistocene, the 'most recent'.

For table of the Cenozoic era *see* PALAEOCENE EPOCH.

plosive consonants [literary: phonetics] In phonetics, the consonants voiced or unvoiced that rely on compression of the lips or of the tongue against the teeth to make a slightly explosive sound. In English they correspond to:

b,   c(k),   d,   k,   p,   t

plural [literary] In grammar, the form of a noun, pronoun, or verb that indicates that more than one (person, place, or thing) is concerned. In English there is only one form of plural; in some other languages there are forms of nouns, pronouns, or verbs that indicate that specific numbers (of persons, places, or things) are concerned. *See also* PERSON.

plurality [quantitatives] Being plural – more than singular (one).

Also, in an election involving more than two candidates, the number of votes by which the winner wins if it is not an overall majority in relation to the total number of votes cast.

plus [maths] Added together with, in addition to: the main feature of the arithmetical operation of addition, employing the symbol + (although it seems that it was not until the 1660s that the symbol became known as 'plus' or 'the plus sign' – some ninety years after the minus sign was so called). Before a numeral (for example, +1), the symbol indicates a positive integer – that is, a number above 0.

The term derives directly from the Latin for '(something) more', corresponding to the irregular comparative of *multus* 'much'.

Pluto: planetary statistics *see* PLANETS OF THE SOLAR SYSTEM

pluviometer [meteorology] Another name for a RAIN GAUGE.

poco [music] Part of a musical instruction: 'a little (more, less, faster, slower, louder, softer, etc.)'. The term derives directly from Italian.

poetic metre *see* METRE IN VERSE; VERSE FORMS

point [maths] In geometry and coordinate geometry, a point is regarded as having no dimensions, and its position in space is represented by its coordinates.

point [printing] In typesetting in the English-speaking world, the basic unit measure of type.

1 point	=	$\frac{1}{72}$ inch, 1.6284 gry, 0.16284 line
	=	0.013889 inch, 13.889 mils or thou
	=	0.3528 millimetre
		(1 millimetre = 2.8345 points)
	=	0.9375 Didot point (continental European unit)
		(1 Didot point = 1.0666 points)
12 points	=	1 pica, 12.8 Didot points

The size of a typeface is expressed generally as the number of points from the base line of the type to the X-HEIGHT: 6-point type (as used in many telephone directories) thus has an x-height of $\frac{1}{12}$ inch. *See also* PICA.

point [weight] Unit of weight used for gems, equal to 0.01 (one-tenth of a) carat: *see* CARAT.

point of inflection/inflexion [maths] A point on a curve at which its slope changes from convex to concave (or vice versa). At the point of inflection the slope of the curve is zero. *See also* MAXIMUM; MINIMUM.

points of the compass On a normal compass card giving bearings over the complete range (or 'compass') of 360°, there are 32 points or individual directions, each therefore 11° 15' apart. They are:

North, south, east, and west are known as the cardinal points, but their present orientations are not as universal in history as might be imagined. The maritime navigation charts of the Vikings, for example, reckoned north as what we would now describe as north-west, despite the fact that they used virtually the modern term *north*. Moreover, that same word *north* probably first meant 'to the east', 'from the left'. The words *east* and *west* by derivation mean 'sunrise', 'dawn' and 'sunset', 'evening' respectively, but *south* is also 'sun-th', 'the direction of sunnier weather' (as far as the northern Germanic peoples were concerned).

Points of the compass

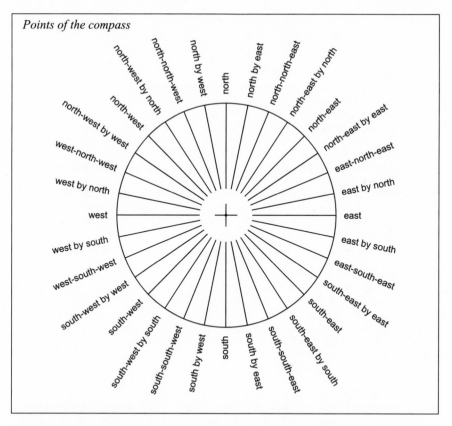

poise [physics] Unit of viscosity in the centimetre-gram-second (CGS) system, equal to the viscosity when a tangential force of 1 dyne per unit area maintains unit velocity gradient between parallel planes in the fluid. It is a large unit, and is usually employed as its submultiple the centipoise. It was named after the French physicist Jean Poiseuille (1797-1869).

$$1 \text{ poise} = 100 \text{ centipoise}$$
$$= 1 \text{ dyne per square centimetre second}$$
$$= 10^{-1} \text{ newton per square metre second (SI units)}$$

Poiseuille's formula [physics] A relationship that gives the rate of flow of a fluid through a narrow (capillary) tube as π times the product of the pressure and the fourth power of the tube's radius divided by 8 times the product of the fluid's viscosity and the length of the tube. It was named after the French physicist Jean Poiseuille (1797-1869).

Poisson distribution [maths: statistics] A distribution that represents a number of events that occur randomly at a given average rate (in a fixed time).

Poisson's ratio [physics] For a material being stretched (which gets narrower as it gets longer), an elastic constant equal to the contraction per unit thickness divided by the extension per unit length. It was named after the French mathematical physicist Siméon Denis Poisson (1781-1840).

poker scores [sport] In the game of poker, the scoring combinations are: one pairp; two pairs; three of a kind; straight (four/five cards in numerical sequence); flush (four/five cards of the same suit); full house (three of a kind plus a pair); straight flush (numerical sequence in the same suit); fours, four of a kind; royal flush (numerical sequence, same suit, up to ace).

There are many regional and local variations in whether this order of scoring is maintained, whether four or five cards should count towards a straight and a flush, and whether a joker or wild card can be used in the more important scoring combinations (in which case fives – five of a kind – becomes a possibility). In this betting game, it is as well for players to be absolutely clear on the regional or local rules in force before they start playing . . .

The name of the game is probably akin to a medieval Dutch verb meaning 'to boast', thus 'to bluff'.

polar axis [astronomy] A sphere's diameter that passes through its poles.

polar bond [chemistry] A type of covalent chemical bond in which the two bonding electrons are shared unequally between the two atoms forming the bond. It generally results in a polar molecule, which has a DIPOLE MOMENT.

polar coordinates [maths] The coordinates in a system in which the position of a point on a plane is given in terms of its distance from a fixed point (the origin) and its angle from a fixed line. If the distance is r and the angle is θ, the polar coordinates of the point are (r, θ). Three-dimensional polar coordinates are called *spherical coordinates*.

polarimetry [chemistry] The measurement of a substance's optical activity (ability to rotate the plane of polarized light), using a polarimeter (or polariscope).

polarity [physics] The extent to which the electric charge on something is positive or negative.

polarizing angle [physics: optics] The ANGLE OF INCIDENCE at which the greatest possible polarization of incident light occurs.

polar molecule *see* POLAR BOND

polarography [chemistry] A method of chemical analysis for a dilute solution, using a polarograph. The solution is electrolysed (has an electric current passed through it) using mercury electrodes, and the current continuously measured as the voltage between the electrodes is gradually increased. Characteristic changes in the graph of current against voltage indicate electrode potentials that identify elements in the solution.

pole [linear measure; square measure] *see* ROD

pole [geology; physics] The north end or south end of the Earth's axis: *see* GEO-GRAPHIC NORTH; MAGNETIC NORTH; MAGNETIC POLE.

polegada, pulgada [linear measure] Former smallish unit of length in Portugal and Spain, approximating farily closely to the INCH and similarly a subdivision of a unit based on the old Roman *pes* 'foot'.

In Portugal,

1 polegada	=	one-twelfth of a pé
	=	27.77 millimetres, 2.777 centimetres
	=	1.0933 inch
		(1 inch = 0.9147 polegada)
12 polegada	=	1 pé ('foot')

In Spain,

1 pulgada	=	one-twelfth of a pie, one-ninth of a palmo
	=	23.2156 millimetres, 2.32156 centimetres
	=	0.914 inch
		(1 inch = 1.0941 pulgada)
9 pulgada	=	1 palmo
12 pulgada	=	1 pie
36 pulgada, 4 palmos, 3 pie	=	1 vara
72 pulgada, 8 palmos, 6 pie, 2 vara	=	1 braça

The name of the unit in both cases derives from Latin *pollex* 'thumb', and is thus cognate with the French equivalent *pouce*.

pole strength [physics] The strength of a magnet, defined as the force exerted by a magnetic pole on a unit pole located unit distance away.

poll [quantitatives] A referendum, plebiscite, or numeric gauge of public opinion.

By derivation, it is a 'head-count' (Middle English *pol* or *polle* 'the head').

pollen count [biology; medical] The amount of pollen in the air, most commonly estimated by trapping it on a filter through which a known volume of air is drawn. In Britain and Europe the pollen count is expressed (in weather forecasts) simply as 'high', 'medium', or 'low', as appropriate. In North America it may alternatively be

expressed as the average number of pollen grains in 1 cubic yard of air. The count
is highly significant to sufferers of allergies such as hay fever.

polo measurements and units [sport] Polo is popular among certain groups of
people, especially those who ride and keep horses and ponies. The game is played
by two teams of four riders, and the object is for the riders to strike a ball with long-
handled sticks between the opponents' goal posts as many times as possible.

The dimensions of the ground:
  the playing area:
    length: 300 yards (274.3 metres)
    width: 160 yards (146.3 metres)
  safety zone: 40 yards (36.58 metres) behind each goal, 20 yards (18.29 metres)
      outside the (optional) polo boards down each side of the playing area
  the goals:
    width: 24 feet (7.315 metres)
    goal post height: 10 feet (3.048 metres), although notionally infinite
  the polo boards:
    overall height: 11 inches (28 centimetres)
  Timing: maximum 56 minutes divided into eight 7-minute chukkas, but may be
      reduced by organizers; 3-minute interval between chukkas except for 5-
      minute interval at half-time
    in the event of a tie at full time, extra time is played (after a 5-minute
      interval) and may go on for two or more further chukkas
  Points scoring:
    the team with the greater number of goals at the end of the game wins

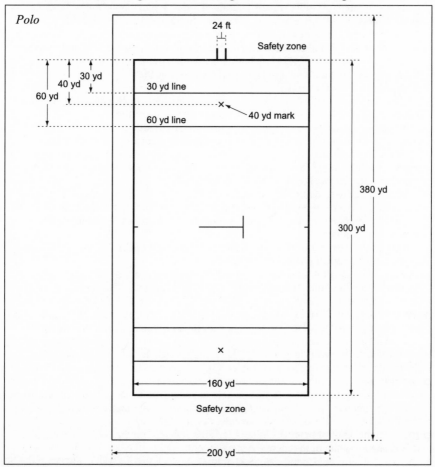

Polo

24 ft

Safety zone

30 yd
40 yd
60 yd

30 yd line

60 yd line   × 40 yd mark

380 yd

300 yd

×

160 yd

Safety zone

200 yd

Dimensions of equipment:
the stick: variable to suit individual player
the ball: weight – 4¼-4¾ ounces (120.5-134.7 grams), made of ash or
bamboo

The name of the game appears to derive from a word in the Tibeto-Burman
language called Balti spoken mainly in Kashmir (and may well be the only Balti-
derived word in the English language). Nonetheless, its meaning is no more than
'ball', and its closeness in linguistic form to such Indo-European words as *ball*
suggests either that Balti is another strand of the Nostratic superfamily of languages
or that even in Balti the word was borrowed from local Indo-European tongues
(such as Kashmiri).

poly- [quantitatives; prefix] A prefix denoting 'many', corresponding to an ancient
Greek equivalent of the Latin *multi-*.
Examples: polycentric – having many centres
polychromatic – having many colours
The Greek *polys* 'much' – which may well be etymologically akin to the Latin
multus 'much' – has as its root meaning the idea of 'fullness in number' rather than
'greatness in quantity' (as in the English *much*).

polybasic [chemistry] Describing an acid with two or more replaceable hydrogen
atoms per molecule (an acid with two such hydrogens may be called *dibasic*).

polycyclic [quantitatives; chemistry] Having many cycles.
In chemistry, describing a molecule with more than one ring of atoms linked by
chemical bonds.

polygon [maths] A plane figure with many sides (and angles). The common polygons are: triangle – 3 sides; quadrilateral – 4 sides; pentagon – 5 sides; hexagon – 6
sides; heptagon – 7 sides; octagon – 8 sides; decagon – 10 sides; d(u)odecagon – 12
sides; icosagon – 20 sides.
The formula for finding the total number of degrees in the internal angles of a
regular polygon is $(2n - 4)90°$, where n is the total number of angles. For example:
the angles of a regular pentagon add to $[(2 \times 5) - 4] \times 90° = 6 \times 90° = 540°$, and
each angle is therefore 108°.
The second element of the term derives from ancient Greek *gonia* 'angle', akin to
English *knee* (the angle of the leg) and *gnaw* (the angle of the jaw).

polyhedron [maths] A solid figure with many faces or sides. The common polyhedra are: tetrahedron (triangular-based pyramid) – 4 faces; pentahedron (e.g.
square-based pyramid) – 5 faces; hexahedron (e.g. cube, cuboid, parallelepiped) – 6
faces; octahedron – 8 faces; decahedron – 10 faces; dodecahedron – 12 faces;
icosahedron – 20 faces.
There are only five regular polyhedra (that is, with identical faces): tetrahedron
with faces that are – 4 equilateral triangles; cube – 6 squares; octahedron – 8
equilateral triangles; dodecahedron – 12 regular pentagons; icosahedron – 20
equilateral triangles
The second element of the term derives from ancient Greek *hedron* 'side', 'base',
cognate with English *seat*.

polymer [chemistry] A substance in which the molecules are made up of a large
number of smaller units (called monomers) joined together. They are types of
MACROMOLECULES.

polynomial [maths] An expression in algebra that comprises two or more individual
terms. For example: $2x^2 + 3xy - 4y$ is a quadratic polynomial (comprising three
individual but dependent terms).

polyvalent, polyvalence [chemistry] Having a valence (valency) of more than 1,
or having more than one valence (valency).

poncelet [physics] An obsolete unit of power in the metric system, equal to the work
done in 1 second by a force (which is able to accelerate a mass of 100 kilograms by
1 metre per second per second) acting over a distance of 1 metre. It was named after
the French engineer Jean-Victor Poncelet (1788-1867).

pond [weight] Unit of mass in Holland, originally based on the POUND but since
assimilated to units of the metric system.
1 pond = 500 grams

$$= \quad 17.637 \text{ ounces avdp., 1 pound } 1.637 \text{ ounces}$$
$$(1 \text{ pound} = 0.9072 \text{ pond})$$
$$2 \text{ ponden} \quad = \quad 1 \text{ kilogram}$$

Like all the other variants of the pound, the pond gets its name from the old Roman measurement *pondo* 'by weight'.

pontoon scores [sport] The game of pontoon is known also as *blackjack*, *vingt-et-un*, and *twenty-one* (the latter being a translation from the French version before it). A betting game, the object is to collect a hand of cards with a total numerical value of as close to 21 as possible (all court cards counting as 10, ace high or low – that is, worth 11 or 1) without going over ('bust'), one at a time after being dealt the first two, having placed a bet on sight of just the first card dealt to the hand by the dealer ('banker').

Rules vary around the world, especially in relation to the value of the combinations of a court card plus an ace (10 + 11, that is 21, in this form mostly called 'pontoon') for which the banker has to refund twice or three times (or whatever the local variant) the original bet, and hand over the dealership.

In most versions of the game, however:

a score of 14 in the first two cards merits an optional new hand
lowest total at which the hand can 'stick': 16
five-card trick (21 or less over five cards): twice bet back (even if banker also has a hand of 21, unless banker also a five-card trick)
two cards of identical numerical value as the first two of the hand can be split into two individual hands

In the casino version (blackjack), the cards are dealt out of a black box, or shoe, face down, whereas, in household card games, the cards apart from the first two are dealt face up (hence the instruction for a further card is 'twist') unless the original bet is added to.

The word *pontoon* as used for the name of this game is a regrettable corruption of the French for '21', *vingt-et-un*. The name for the game as played in casinos, *blackjack*, may or may not have something to do with the black box from which cards are dealt.

pood *see* PUD, PUDI

pool measurements and units [sport] The object of the game of pool is to score by cueing the white ball to knock designated numbered balls into designated pockets around the edge of the table.

The dimensions of the table:

maximum area: 10 x 5 feet (3.048 x 1.524 metres)
minimum area: 7 x 3½ feet (2.134 x 1.067 metres)
(width = one-half length)

Points scoring:

designated ball in designated pocket: 1 point
any other object ball pocketed in the same stroke: 1 point
points against:
failure to pocket ball, or to drive cue ball and two object balls to a cushion: lose 2 points
cue ball pocketed ('scratched'): lose 1 point, unless third scratch in a row, when lose 15 points
cue ball jumps off table (another form of 'scratch'): lose 1 point unless third scratch in a row, when lose 15 points
the game is made up of an agreed number of 'blocks' of points:
the first player to the agreed block points total (usually 125 or 150) wins the block

The name of the game is thought to derive from the rack of object balls perceived as the means of winning points, the collective stakes for which the players play, the assets which are then *pooled* between the contestants. *See over* for illustration.

population statistics, larithmics [quantitatives; biology] Figures listing the total number of people resident in a country, province, state, or territory, together with other details relevant to specific authorities and interests (such as males/females, adults/children, natives/immigrants, birthrate/mortality rate).

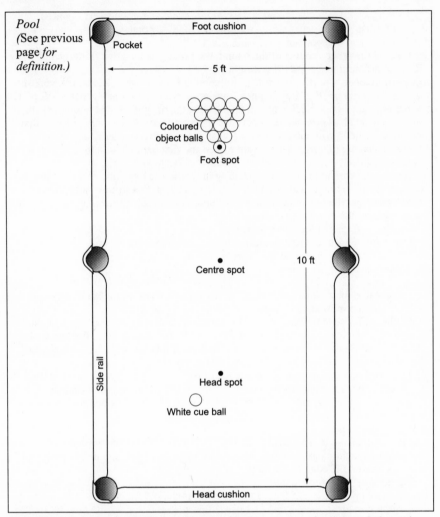

Pool (See previous page for definition.)

Foot cushion

Pocket

5 ft

Coloured object balls

Foot spot

Centre spot

10 ft

Side rail

Head spot

White cue ball

Head cushion

porosity [geology; physics] In a substance that has pores or tiny internal channels, the ratio of the total volume of pores to its total volume.

positional number system [maths] Method of showing numbers in which the value of a digit depends on its place (position) in the number. For example: in the number 232, the first 2 (because of its position) represents 200, the 3 represents 30 and the second 2 represents 2 units: the number is $200 + 30 + 2 = 232$. *See also* NON-DENOMINATIONAL NUMBER SYSTEM.

position circle [astronomy] For somebody making an observation, the circle that has its centre at the observer and a radius that makes the circumference of the circle pass through the object or point being observed.

positive number [maths] A number that is greater than 0 . For example: 0.02, 1½, and 29 are all positive numbers (numbers less than 0 are negative numbers). Zero is neither positive nor negative.

positron [physics] A fundamental particle equal in mass to that of an electron but carrying a positive charge; it is an electron's ANTIPARTICLE. The term comes from positive electron.

posology [medicine] The study of the quantification of medically therapeutic doses – how much of a drug should be administered to have an optimal effect on an individual patient.

The first element of the term derives from ancient Greek *posos* 'how much?'.

post- [time: prefix] Prefix denoting 'after'.

Example: post-meridian – after midday, of the afternoon

post-actinide [chemistry] Any element of which the atomic number is greater than that of lawrencium (element 103): *see* PERIODIC TABLE OF ELEMENTS.

postcode, postal code [geography] In Britain and Canada, the six- or seven-character coding involving numerals and letters that plots where in each country any specific address is located.

The equivalent in the United States is the ZIP CODE.

postposition [literary] In grammar, a word that qualifies another word or expression and that normally would come before the other word or expression but is, in this case, used after the other word or expression. For example: in 'my whole life through', *through* is a postposition (whereas, in 'through my whole life', it is a PREPOSITION); but other parts of speech can also be postpositional, as for example the adjective *general* in Secretary-general.

pot [volumetric measure] An obsolete measure of liquid and dry capacity in Norway, Denmark, Switzerland, and Belgium.

In Norway,

1 pot	=	96.50 centilitres, 0.965 litre
		(1 litre = 1.0363 Norwegian pot)
	=	1.698 UK pint, 2.039 US pints
	=	0.849 UK dry quart, 0.875 US dry quart
18 pots	=	1 skjeppe
144 pots, 8 skjeppe	=	1 korntonde

In Denmark,

1 pot	=	96.60 centilitres, 0.966 litre
		(1 litre = 1.0352 Danish pot)
	=	1.700 UK pint, 2.042 US pints
	=	0.850 UK dry quart, 0.876 US dry quart
18 pots	=	1 skaeppe
36 pots, 2 skaeppe	=	1 fjerding
144 pots, 8 skaeppe, 4 fjerdings	=	1 tønde

In Switzerland and Belgium, the unit has been assimilated to units of the metric system.

In Switzerland,

1 pot	=	1.50 litre, 1,500 cubic centimetres
	=	2.64 UK pints, 3.17 US pints
		(3 UK pints = 1.1364 Swiss pots
		3 US pints = 0.9464 Swiss pot)
	=	1.3198 UK dry quart, 1.3609 US dry quart
	=	91.5356 cubic inches
2 pots	=	3 litres

The Belgian unit as a measure of dry capacity is equal in value to the Swiss, but as a measure of liquid capacity is only one-third the value.

In Belgium,

1 liquid pot	=	50 centilitres, 0.50 litre
	=	0.880 UK pint, 1.0567 US pint
		(1 UK pint = 1.1364 Belgian liquid pot
		1 US pint = 0.9464 Belgian liquid pot)
2 liquid pots	=	1 litre

potassium-argon dating [geology] A method of determining the age of rocks based on measuring the amount of radioactive potassium (potassium-40) in it, which decays to argon-40. Potassium-40 has the extremely long HALF-LIFE of 1.30×10^9 years.

potential difference [physics] The difference in electric potential between two points in a circuit carrying current, measured in volts. It is equivalent to the work done (or received) in moving a unit positive charge between two points in an electric field, and is alternatively known as voltage: *see* VOLT.

potential energy [physics] The energy that an object has because of its position or because it is in tension (like a compressed or stretched spring), measured in joules.

For example: because of its position in a gravitational field, any object above the surface of the Earth has a potential energy equal to the product of its mass, its height, and the acceleration of free fall (acceleration due to gravity). *See also* KINETIC ENERGY.

potentiometer [physics] An instrument for measuring potential difference (voltage), often taking the form of a voltage divider.

pottle [volumetric measure] Obsolete term for the liquid capacity of 2 QUARTS (or 4 pints), or a vessel that could contain that amount.

It derives as a diminutive form of the Old French variant of the word *pot*.

pouce [linear measure] Unit of length in fourteenth- to eighteenth-century France, approximating closely to the modern INCH and similarly one-twelfth of (the equivalent of) a foot.

1 pouce	=	27.07 millimetres, 2.707 centimetres
	=	1.0657 inch
		(1 inch = 0.9384 pouce)
12 pouces	=	1 pied-de-roi
72 pouces, 6 pieds-de-roi	=	1 toise

The term derives from Latin *pollex* 'thumb'.

pound [weight] Unit of mass in the foot-pound-second (FPS) system, now defined as a mass equal to 0.45359237 kilogram (formerly the mass of a cylinder of platinum called the Imperial Standard Pound). It is the basic unit of the avoirdupois (avdp.), troy, and apothecaries' weight measurement systems.

In avoirdupois,

1 pound	=	16 ounces, 256 drams, 7,000 grains
	=	453.5924 grams, 0.4535924 kilogram
		[500 grams = 1.1023 pound (1 lb 1.637 oz) avdp.
		1 kilogram = 2.2046 pounds (2 lb 3.274 oz) avdp.]
	=	1.215278 pound (1 lb 2.583 oz) troy/apothecaries'
14 pounds	=	1 stone (6.35 kilograms)
28 pounds	=	1 quarter
100 pounds	=	1 short hundredweight or cental
112 pounds	=	1 hundredweight, quintal, or centner
2,000 pounds	=	1 short ton
2,240 pounds	=	1 (long) ton

In troy weights,

1 pound	=	12 ounces, 240 pennyweight, 5,760 grains
	=	373.2420 grams, 0.373242 kilogram
		[500 grams = 1.3396 pound (1 lb 4.075 oz) troy
		1 kilogram = 2.6792 pounds (2 lb 8.15 oz) troy]
	=	13.1657 ounces avdp.

In apothecaries' weights,

1 pound	=	12 ounces, 96 drachms, 288 scruples, 5,760 grains
	=	373.2420 grams, 0.373242 kilogram
		[500 grams = 1.3396 pound (1 lb 4.075 oz) apoth.
		1 kilogram = 2.6792 pounds (2 lb 8.15 oz) apoth.]
	=	13.1657 ounces avdp.

As a term for a weight, the *pound* derives directly from the old Roman *libra pondo* '1 pound by weight'. Many cultures adopted variants, those of northern Europe mostly taking the name of their units from the description 'by weight': *see* PFUND; POND; PUND. Some of these have since been assimilated to units in the metric system. Cultures of southern and western Europe instead tended to retain *libra* for the name of their unit of weight: *see* LIBRA; LIVRE.

pound [comparative values] Unit of currency in the United Kingdom and certain of its protectorates (where it is, by tradition and technically, known as the pound sterling), and in Cyprus, Egypt, the Lebanon, and the Sudan (in some of which the unit also has alternative local names).

In the United Kingdom,

1 pound sterling	=	100 pence (until 1971: 240 pence)

In Cyprus,

                                   1 pound   =   100 cents
                   In Egypt, the Lebanon, and the Sudan,
                                   1 pound   =   100 piastres
The major units of currency in Israel and Syria are also sometimes called the
pound. The unit of currency in the Israel of New Testament times called the *minah*
or *maneh* (as opposed to the unit of weight of the same name) is in some Bibles
translated as 'pound'.

   The name of the unit derives from its originally being part of a measurement 'by
weight', Latin *pondo*.

poundage [quantitatives; comparative values] A total or rate expressed in pounds
(weight or units of currency).

poundal [physics] Unit of force in the foot-pound-second system, equal to the
force that produces an acceleration of 1 foot per second per second on a mass of
1 pound.

                      1 poundal   =   0.138255 newton (SI units)
                                   =   13,825.5 dynes (CGS units)
                   32.2 poundals   =   1 pound force (1 pound weight)

pounds per month *see* REPAYMENT RATES

pounds per square foot [physics] Unit of pressure in the foot-pound-second
system.

         1 pound per square foot   =   0.00694 pound per square inch
                                   =   478.843 dynes per square centimetre (CGS units)
                                   =   4.88267 kilograms force per square metre (MKS
                                        units)
                                   =   47.8843 newtons per square metre (SI units)
                                   =   47.8843 pascals

pounds per square inch [physics] Unit of pressure in the foot-pound-second
system.

         1 pound per square inch   =   3.3253 dynes per square centimetre (CGS units)
                                   =   0.33253 newton per square metre (SI units)
                                   =   0.33253 pascal

              14.72 pounds per
                   square inch   =   1 atmosphere

pounds per week *see* REPAYMENT RATES

power [physics] The rate of doing work, measured in joules per second, equivalent to
watts (SI units). *See also* HORSEPOWER; WATT.

   Also, the extent to which a lens magnifies, measured in dioptres: *see* DIOPTRE;
MAGNIFICATION.

power [maths] *see* EXPONENT, POWER

power factor [physics; engineering] In an electric circuit, the total power (in watts)
dissipated divided by the total energy supplied (in volt-amperes). It is equal to the
cosine of the phase angle (*see* PHASE).

power loading [engineering] In the monitoring of freight transportation by propel-
ler-driven aircraft, the overall weight of the laden aircraft divided by the horsepower of its engines.

ppm *see* PARTS PER MILLION (PPM)

Prandtl number [physics] A number concerned with heat exchange in fluid
mechanics, equal to the product of a liquid's kinematic viscosity and specific heat
capacity divided by its thermal conductivity. It was named after the German
physicist Ludwig Prandtl (1875-1953).

Precambrian era [time] In the history of the planet Earth, the extended period
during which only primitive life forms existed. The Solar System came into
existence some 4,600 million years ago; the earliest fossils with hard parts giving
evidence of the flourishing of life date from about 590 million years ago, at the
beginning of the CAMBRIAN PERIOD of the PALAEOZOIC or PRIMARY ERA. Between those
'dates' stretched the Precambrian (or Archaeozoic or Proterozoic) era.

   Evidence for the geological events of those millenniums is visible at sites in
northern Canada, Scotland, western England, Finland, and Sweden, and in many
different areas of Africa.

precession [physics; astronomy] For a spinning object, such as a top, the gradual circular movement of its axis (which increases with decreasing speed and makes the top wobble).

For a revolving celestial object, such as a planet, the gradual change of the direction in which its axis points. The Earth's axis precesses every 25,800 years to describe a circle nearly 47 degrees across, causing an apparent change in the positions of the stars (as seen from Earth) over the centuries. Precession of the Earth also causes the equinoctial points to move westwards along the ecliptic by about 50 seconds of arc per year, a phenomenon known as the precession of the equinoxes.

precious stones [comparative values] Stones that for reasons of crystalline form, colour, lustre, transparency, scarcity, and therefore commercial value, are technically classified not merely as gemstones but also as precious stones.

They include: diamond, ruby, opal, some forms of emerald, and some forms of sapphire. Pearls are also regarded as precious stones.

pre-Columbian [time] In American history, before the arrival of Christopher Columbus in 1492.

predynastic Egypt [time] In the history of ancient Egypt, before 3100 BC – when dynastic rule became the norm.

pre-exilic, pre-exilian [time] In the history of the ancient Israelites, before 586 BC – when the Israelites were deported en masse to (the Exile in) Babylon.

preference shares, preferred stock [comparative values] In share/stock trading, shares/stock on which dividends must be paid, at a preset rate, before dividends are paid on common shares/stock (at possibly a different rate).

preferential voting system [quantitatives] System in which electors mark on their ballot papers their first, second, third, and possibly further choices of candidates. If no candidate has an overall majority, the second and third choices may then be taken into account.

preferred stock *see* PREFERENCE SHARES, PREFERRED STOCK

pregnancy testing [medicine] Most modern forms of testing for pregnancy involve the detection and measurement of specific hormones in a woman's blood or urine.

preposition [literary] In grammar, a part of speech indicating relative position or possession (but not an adverb or a pronoun). For example:

under	over	beside	through	beyond	by
on	with	for	by	off	of

prescriptions [medicine] Doctors in writing prescriptions employ a short-hand (pseudo-medieval) form of Latin or Greek. Some of these terms stand for quantities, times or frequencies (apart from actual mass and volume units, which are often in apothecaries' measure). For example:

abbreviation	Latin in full	meaning
alt. dieb.	alternis diebus	every other day
alt. hor.	alternis horis	every other hour
alt. noc.	alternis nocte	every other night
b.i.d.	bis in die	twice daily
b.i.n.	bis in noctu	twice a night
bis	bis	twice
bis in 7d.	bis in septem diebus	twice a week
c.m.	cras mane	tomorrow morning
c.m.s.	cras mane sumendus	to be taken tomorrow morning
c.n.	cras nocte	tomorrow night
d	dies	day
d.d. in d.	de die in diem	from day to day
dieb. alt.	diebus alternis	every other day
dieb. tert.	diebis tertiis	every third day
dim.	dimidius	one-half
hor 1 spat.	horae unius spatio	one hour's time
hor. som, h.s.	hora somni	at bedtime
in d.	in die	daily
m. et n.	mane et nocte	morning and night
noct.	nocte	at night

noct. maneq.	nocte maneque	night and morning
omn. bid.	omnibus bidendis	every two days
omn. bih.	omni bihoris	every second hour
omn. hor.	omni hora	every hour
omn. noct.	omni nocte	every night
om. ¼ h.	omni quadrante horae	every 15 minutes
omn. mane vel noc.	omni mane vel nocte	every morning or night
part. aeq.	partes aequales	equal parts
pulv.	pulvere	in powdered form
Q.h.	quaque hora	every hour
Q. 2h.		every two hours
Q. 3h.		every three hours
q.i.d.	quater in die	four times a day
t.i.d.	ter in die	three times daily
t.i.n.	ter in nocte	three times a night

pressure [physics] Force applied to a surface, measured as the force per unit area. The various measurement systems have their own units, for example:

FPS units	pounds per square foot
CGS units	dynes per square centimetre
MKS units	kilograms per square metre
SI units	pascal (newton per square metre)

Under the surface of a liquid, the (hydrostatic) pressure is equal to the product of the depth, the liquid's density, and the acceleration of free fall (acceleration due to gravity).

In a bubble in a liquid, the excess pressure (that is, the pressure in excess of the surrounding hydrostatic pressure) equals twice the surface tension of the liquid divided by the radius of the bubble.

Gas pressures (including atmospheric pressure) may be measured in bars, millibars, or torr (millimetres of mercury).

pressure gauge [physics] An instrument for measuring pressure, such as a BARO-METER or MANOMETER.

pressure gradient [physics; engineering] The difference in pressure between two locations, usually expressed as the change per unit distance.

pressure head [physics; engineering] The pressure at a point in a liquid, expressed as the height of its surface above that point.

presto, prestissimo [music] Musical instructions: play or sing 'fast' and 'as fast as possible' respectively – in both cases at a speed greater than ALLEGRO. The term derives directly from Italian (but ultimately from Latin *praesto* 'readily').

price-earnings ratio [comparative values] In commercial dealings on the stock market, a ratio between the market price of a stock or share, and the earnings made on that stock or share (in the same currency units). On Wall Street, for example, a market price of 108 dollars with share earnings of 9 dollars (that is, 9 dollars per share) has a price-earnings ratio of 12 to 1 (12:1).

price index *see* CONSUMER PRICE INDEX; INDEX

pride of lions [collectives] One of the earliest collectives, dating in English from the 1450s when few lions had been seen (and when the magnificence of the 'king of beasts' was consequently perhaps somewhat overrated). But the expression gained in currency again during the nineteenth century, when big-game hunting became popular and it was in the hunter's interest to suggest that the animal was especially grand and noble.

The word *pride* derives as a back-formation from earlier versions of the adjective *proud*, itself the equivalent of a past participle of a verb meaning 'to be foremost', 'to stand out' (which is why a ship is likewise *prowed*), of which the related abstract noun was/is *prowess*.

primary colours [physics; medicine] Three colours that can be mixed to make all other colours. For coloured light, the primaries are red, green, and blue (of wavelengths 640, 537, and 464 nanometres, respectively); the correct mixture of all three – an additive process – gives white. For coloured pigments (for example, paints), the primaries are cyan (a light greenish blue), magenta (a purplish red), and yellow;

a mixture of all three – a subtractive process – gives black. Additive primary colours are used in transparency colour film and colour television; subtractive primary colours are used in colour printing and colour print film.

As used in psychological tests and tests for colour-blindness, however, there are four 'primary' colours: red, yellow, green, and blue.

prime [quantitatives] Any equal part corresponding to a subdivision of a unit; the mark (') indicating such an equal part, used most commonly of 1 minute of one degree or of one inch.

prime meridian [geology; astronomy] The line of longitude between the north pole the south pole that passes through Greenwich, London, England: longitude 0 degrees. *See also* MERIDIAN.

prime number [maths] A whole number (integer) – other than 1 – that can be divided only by 1 and by itself (that is, it has only those two factors). There is an infinite number of primes, the first few being 2, 3, 5, 7, 11, 13, 17, 19, 23, 29, 31, 37, 41, . . . (all except 2 are odd numbers).

primeval [time] Describing the earliest times (particularly the earliest ages in the history of the planet Earth).

The term derives from Latin elements meaning '(of) the first age'.

primordial [time] Describing the earliest state or condition (particularly the first stage in the geological history of the planet Earth).

The term derives from Latin elements meaning '(of) the first order'.

prism [maths] A solid figure that has similar polygons as its ends and rectangles or parallelograms as its sides (with the same cross-section along its whole length). The simplest example is a triangular prism, with rectangular sides.

The volume of a prism equals the product of the area of one end and its length.

The term derives from an ancient Greek past participle meaning '(thing) sawn off'.

private first class [military rank] In the United States Army, a rank above a private but beneath a corporal.

probability [maths] The likelihood that a given event will occur. A certainty is ascribed a probability of 1 and an impossibility has a probability of 0; all other possibilities have a probability between 1 and 0. For example: the probability of throwing a 3 (or any other specified number) on a dice is $1:6 = 0.16666 . . .$ *See also* ODDS.

probability curve [maths] Another name for the BELL CURVE, BELL-SHAPED CURVE.

product [maths] The result of MULTIPLICATION.

profit *see* GROSS PROFIT; NET PROFIT

progression [maths] A sequence or series of mathematical terms: *see* ARITHMETIC PROGRESSION/SERIES; GEOMETRIC PROGRESSION/SERIES.

projection *see* AZIMUTHAL PROJECTION; CONICAL PROJECTION; CYLINDRICAL PROJECTION

pronoun [literary] A part of speech corresponding to a word that replaces a noun ('pro-noun'), usually so that the noun does not have to be repeated.

The personal pronouns are:

nominative	*accusative*	*emphatic*
I	me	myself
you	you	yourself
he	him	himself
she	her	herself
it	it	itself
we	us	ourselves
(you)	(you)	yourselves
they	them	themselves

The possessive pronouns are:

mine
yours
his
hers
its
ours
theirs

(Note that none of these pronouns has an apostrophe, and that *it's* is short for 'it is' or 'it has'.)

Interrogative pronouns are:

nominative	*emphatic*
what (as in 'what are you doing?')	whatever
which ('which of them is tomorrow's?')	whichever
who ('who can tell?')	whoever
whose ('whose lunch have I eaten?')	whoever's

(Note that *who's* is short for 'who is' or 'who has'.)

accusative

whom (as in 'whom can I blame?')   whomever

Relative pronouns are:

that (as in 'the lunch that I ate')
which (' – a detail which I'd neglected')
who ('a girl who was there')
accusative: whom (' – a person whom I know well')
    whose ('someone whose love was true')

Demonstrative pronouns are:

this (as in 'this is not right')
that ('that is not right')

Distributive pronouns are:

each (as in 'to each his own')
either ('either is incorrect')
neither ('neither is right')
both ('both were bald')

And, according to many authorities, there is a final category of pronoun known as indefinite pronouns:

any (as in 'I don't have any')
some ('give him some')
other(s) ('there was one other/were two others')

proof spirit *see* ALCOHOL CONTENT

proper fraction [maths] A fraction in which the denominator (the number below the line) is greater than its numerator (the number above the line); for example: ½, ¾, ²⁷⁄₃₆. *See also* IMPROPER FRACTION.

proper motion [astronomy] For a star moving in space, the component of its motion that is at right-angles to the line of sight (from Earth) – that is, any detectable difference in the position of a star after a period of exactly one year. The proper motion of most stars is not discernible by the naked eye even over hundreds of years. The greatest known proper motion (10.3 arcsecs per year) is that of the comparatively nearby Barnard's Star, which even then takes about 180 years to complete a visual distance across the sky equivalent to the width of the full Moon.

proportional representation [quantitatives] System of voting in an election in which candidates are elected to representative office in precise proportion to the total number of votes cast for their respective parties.

proportions *see* RATIO

pro rata [quantitatives] An expression meaning 'according to the rate (established)', 'in those proportions', generally referring to scaling up or scaling down one unit measure per another unit measure (as, for example, relating an hourly rate of pay to a weekly rate). In the United States, this calculation may be described as *prorating*. The expression derives directly from Latin.

Proterozoic era [time; geology] Another name for the PRECAMBRIAN ERA.

Some authorities, however, differentiate between the Precambrian (or Archaeozoic) – which is thought of as the earliest time in the history of the planet Earth – and the Proterozoic, during which the most primitive form of organized organic life (such as sponges and worms) appeared. All authorities agree that the Proterozoic era ended as the Cambrian period of the PALAEOZOIC ERA began.

Protolithic [time] Another word for EOLITHIC, describing the earliest part of the Stone Age.

proton [physics] A fundamental particle with a positive charge (equal and opposite to the charge on an electron) and a mass of 1.67252 x 10^{-27} kilogram (the lightest of the BARYONS). It forms the nucleus of a hydrogen atom, and the nuclei of all other atoms contain protons; the number of protons in the nucleus (the atomic number) gives an element its identity.

The term corresponds to the ancient Greek for 'the first (thing)'. *See also* ISOTOPE.

proton number [physics] Another name for ATOMIC NUMBER.

protractor [maths] A device for measuring angles, consisting of a semicircle (of transparent plastic) calibrated in degrees.

provost marshal [military rank] Officer in the army or navy responsible for the maintenance of order and discipline. The post is not of any specific rank, however, and the rank of the officer given the duty depends upon the size and importance of the military establishment at which he or she is based.

In the army, the provost marshal is head of the military police; in the navy the duty is more of responsibility for confining detainees until their courts martial.

psi, p.s.i. *see* POUNDS PER SQUARE INCH

psi-meson, psion, psi-particle [physics] A type of long-lived MESON with no charge, alternatively known as a J-particle.

psychrometer [physics; meteorology] A device for monitoring the relative humidity of the air. It primarily comprises just two thermometers – one a wet-bulb thermometer, the other a dry-bulb thermometer. The relative humidity is indicated by the difference between the two temperature readings at any given time.

The first element of the term derives from ancient Greek *psychros* 'cold (air)', the root meaning of which is 'air cold enough to vaporize the breath'.

pta *see* PESETA

pu [linear] Linear unit in China adopted by European colonial authorities during the 1700s to late 1800s when trading with China. The unit varied slightly in different Chinese ports. However, in general,

1 pu	=	5 CH'IH
	=	70.5 inches, 5 feet 10½ inches
	=	179.07 centimetres, 0.1791 metre
2 pu	=	1 CHANG
360 pu, 180 chang	=	1 LI
	=	about one-third of a mile, 536.5 metres
90,000 pu, 250 li	=	1 tu
	=	about 83.3 miles, 134.1 kilometres

pud, pudi [weight] Obsolete unit of mass in Russia, also transliterated as *pood*.

1 pud	=	40 funte
	=	16.381 kilograms
	=	36 pounds 1.81 ounce (2 stone 8.113 pounds)
30 pudi	=	1 PACKEN

The term is a borrowing in Russian from the Scandinavian forms of *pound* (*see* PUND) that ultimately relate to the old Roman form of measurement *pondo*'by weight'.

pul [comparative values] Unit of currency in Afghanistan.

100 puls	=	1 afghani

See also COINS AND CURRENCIES OF THE WORLD.

pula [comparative values] Unit of currency in Botswana.

1 pula	=	100 thebe

The name of the unit originates as a ceremonial greeting in Setswana (the language of the Tswana people), meaning 'rain' – something comparatively rare and valuable in parts of the country. *See also* COINS AND CURRENCIES OF THE WORLD.

pulgada *see* POLEGADA, PULGADA

pulley ratios [physics; engineering] The force ratio or mechanical advantage of a system of pulleys – that is, the number of times less effort by which the system permits a load to be moved – is equal to the number of pulleys in the system, if the pulleys are all the same size: *see* FORCE RATIO, MECHANICAL ADVANTAGE. For every additional pulley added to a system, the distance by which the load is moved by the effort moving a given distance is halved.

On a wheel-and-axle pulley in which the effort is on a rope attached to the wheel and the load is on a rope attached to the much smaller axle at its centre, the mechanical advantage is equal to the number of times that the wheel's circumference is larger than the axle's circumference. (This is effectively the same as a GEAR RATIO, and is representative also of a system that uses pulleys of different sizes.)

pulmometry *see* LUNG CAPACITY.

pulsatance [physics] An alternative name for angular frequency.

pulse [medicine] The rate of heartbeat. It can be felt directly with the hand over the area of the heart (left centre of the chest) as the contractions of the heart muscle, driving blood all around the body, or indirectly with the fingers over the wrist, temple, or beneath the earlobe as the wave of distension caused by the heartbeat reaching the arteries near the skin in those areas.

For accurate assessment, the pulse rate must be measured over at least 20 seconds, and, if not actually measured over 1 full minute, must be mathematically calculated as a rate per minute. Medical reference books generally avoid giving an average human pulse rate, mostly because it is no real guide to blood pressure, which is of much greater significance to overall health. (Devices such as the *pulsimeter* are designed to measure both the rate and strength of the pulse.) At rest, however, the normal human heart beats between 60 and 80 times per minute, increasing during exercise to a maximum safe rate (in excellent health) of perhaps 200 times per minute.

The term derives from the Latin verb *pulsare*, a frequentive of the verb *pellere* 'to beat', 'drive'.

puncheon [volumetric measure] A word that originally applied to a large cask or barrel, which was – like the present-day BARREL – of different actual volume depending on the nature of the contents (which could be liquid or dry). In the early 1500s, therefore, the puncheon was thus the equivalent of a unit measure in certain marketable wares, but all that can be said now is that it constituted a volume of anything between 70 to 120 old wine-gallons (= modern US gallons; 58 to 100 UK gallons).

The term would seem to be akin in etymological derivation to the much smaller volumetric measure, the PIN.

pund [weight] Obsolete unit of weight in Norway, Denmark, and Sweden, evidently deriving from the old Roman measurement 'by weight' (Latin *pondo*) as does the POUND. In Norway and Denmark, however, the old unit has been assimilated to units of the metric system (although it had a well-established independent value until the late nineteenth century).

In Norway and Denmark,

1 pund	=	500 grams
	=	17.637 ounces avdp., 1 pound 1.637 ounce
		(1 pound = 0.9072 pund)
2 pund	=	1 kilogram

and in this respect, the Norwegian/Danish pund is identical in every way with the Dutch POND.

In Sweden, where the unit remains obsolete,

1 pund	=	425.1 grams
	=	14.995 ounces avdp.
		(1 pound = 1.0670 pund)

See also LIBRA; PUD, PUDi.

punt [sporting term] In some card games that involve both a banker and a betting system, to bet against the banker. It is following this usage that the word *punter* has come to mean a person who stakes a bet in any form of gambling.

punt [comparative values] Unit of currency in the Republic of Ireland, until 1979 officially identical with the (British) pound sterling, together with which it went decimal in 1971, although sometimes then spelled *pount*. Today, however,

$$1 \text{ punt } = 100 \text{ pence}$$

Like many other variants, the term derives from the old Roman measurement *pondo* 'by weight'. *See also* COINS AND CURRENCIES OF THE WORLD.

pursuit cycling *see* CYCLING DISCIPLINES AND DISTANCES

putting the shot *see* ATHLETICS FIELD EVENTS

pya [comparative values] Unit of currency in Burma.

$$100 \text{ pyas} = 1 \text{ kyat}$$

See also COINS AND CURRENCIES OF THE WORLD.

pycnometer [physics] Another name for a density bottle: *see* DENSITY.

pyramid [maths] A solid figure that has a polygon as its base and three or more triangular sides that meet at a point (the apex). A triangular-based pyramid is called a *tetrahedron*.

The volume of any pyramid is ⅓ the area of its base times its perpendicular height.

The term derives in English through ancient Greek probably ultimately from ancient Egyptian, in which the first element (*pi-*) seems to correspond to 'house' and the other two elements may or may not mean 'of the son of Ra'.

pyrgeometer, pyrheliometer *see* PYROMETER

pyrometer [physics] An instrument for measuring high temperatures, generally by making use of optical or radiation effects. One that makes use of optical effects may alternatively be called a *pyrophotometer*. Other variant pyrometers include the *pyrgeometer*, which measures heat radiated into space from the Earth's surface, and the *pyrheliometer*, which measures heat received at the Earth's surface from the Sun.

The word element *pyr* in ancient Greek meant 'fire' (an English word with which the Greek is cognate) but, in technical terms in scientific English, is borrowed to mean 'high temperature'.

Pythagoras's theorem [maths] In a right-angled triangle, the square of the length of the hypotenuse (the longest side) is equal to the sum of the squares of the lengths of the other two sides.

The smallest numbers of whole units in the sides of right-angled triangles are (hypotenuse last):

$3, 4, 5 \quad (3^2 + 4^2 = 5^2)$
$5, 12, 13 \quad (5^2 + 12^2 = 13^2)$
$7, 24, 25 \quad (7^2 + 24^2 = 25^2)$
$8, 15, 17 \quad (8^2 + 15^2 = 17^2)$
$9, 20, 21 \quad (9^2 + 20^2 = 21^2)$

It was named after the Greek religious philosopher Pythagoras (*fl. c.*530 BC), who sought esoteric knowledge through numbers but who appears to have written nothing himself, leaving the several geometrical discoveries now associated with his name to be attributed to him by later followers (such as Euclid).

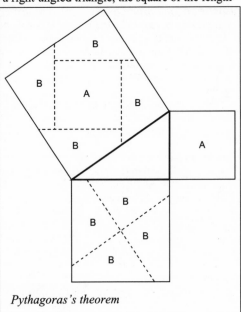

Pythagoras's theorem

Q

Q [engineering; physics] In engineering, a symbol used for various kinds of through-put or flow, generally expressed in cubic metres per second.

In physics, the symbol for electric charge: *see* COULOMB.

Q, quad, Q unit [physics; geology] Obsolete unit with which to express extremely large amounts of heat energy, particularly with reference to the world's fuel reserves.

$$1 \text{ quad} = 1 \text{ US quadrillion/1 UK thousand billion } (10^{15})$$
British thermal units

Q-band [physics; telecommunications] A band of radio frequencies in the range 36 to 46 gigahertz, coincident with the high end of the K-band and used by radar equipment.

Q-factor, quality factor [physics] Symbol for the selectivity, or 'quality', of a tuned circuit or resonant system, equal to 2π times the product of the frequency and the inductance, divided by the resistance. It may be measured using a *Q-meter*.

qintar [comparative values] Unit of currency in Albania.

$$100 \text{ qintar} = 1 \text{ lek}$$

The term derives from the Latin *centenarius* 'hundredth' (*See also* KANTAR, QUINTAL). *See also* COINS AND CURRENCIES OF THE WORLD.

Q scale [geology: seismology] Measure of how long earthquake vibrations take to die down at the surface. The term derives from (an abbreviation) of the German *Querwellen* 'transverse waves'.

qt *see* QUART

qto *see* QUARTO

quad *see* Q, QUAD, Q UNIT

quadr-, quadra-, quadri-, quadro- [quantitatives: prefix] Prefix which, when it precedes a unit, implies a fourfold unity.

Examples: quadraphonic – with sound from four sources, channels
quadrennial – once every four years, lasting four years
quadrilateral – having four sides

quadrant [maths] In geometry, a quarter of a circle, or one of the four parts into which a plane is divided by (Cartesian) coordinates at right-angles.

quadrant [astronomy; navigation; surveying] An instrument resembling a sextant for measuring angles (for example, the elevations of stars), covering angles up to 90 degrees.

quadrant [physics] A former name for a unit of inductance, equal in magnitude to its successor the henry. The term was also once used for a distance of 10 million metres (approximately equal to a quadrant, or quarter, of the Earth's circumference).

quadrat [biology; ecology] Random area of ground (usually 1 square metre), marked out with string or tape and examined for the various species it encompasses.

quadrate [physics; astronomy] In astronomy, describing the relative positions of two heavenly bodies at 90 degrees to each other. Two such bodies are said to exhibit *quadrature*.

quadratic, quadric [maths: algebra] In algebraic equations, involving one or more variables that are squared, but involving no higher power.

quadratrix [maths] A curve that is used in geometry when constructing a square that has the same area as a given curved figure.

quadrennial [time] Once every four years; lasting for four years.

quadric *see* QUADRATIC, QUADRIC

quadrilateral [maths] A four-sided plane figure. Special cases include the square, rectangle (oblong), parallelogram, diamond, rhombus, and trapezium.

quadrillion [quantitatives] In the UK a quadrillion is a million million million million (10^{24}, or 1 followed by 24 noughts), whereas in North America and parts of continental Europe, notably France, a quadrillion is a thousand thousand thousand million (10^{15}, or 1 followed by 15 noughts).

quadriplegia [medicine] Paralysis of all four limbs and the trunk of the body from the neck down. The condition is often instead known as *paraplegia*, although that is a technical term only for paralysis of the four limbs and part of the trunk (not all). Paralysis of one side of the body (sometimes including certain psychological functions of the opposite side of the brain) is termed *hemiplegia*.

quadrivalent, quadrivalence [chemistry] Having a valence (valency) of four. An alternative, and more common, description is tetravalent/tetravalence (the first element from ancient Greek rather than Latin).

quadruple [quantitatives] Fourfold, four times; of four parts.

quadruplet [music] In a melody, four notes played or sung in the time of three.

quadruple time [music] An expression for the time signature 4/4, in which there are four beats to the bar, each a crotchet (quarter-note) in duration, the first and third of which are accentuated.

quality factor *see* Q FACTOR, QUALITY FACTOR

quanta *see* QUANTUM

quantic [maths] An algebraic expression of two or more variables in which all terms have coefficients that are rational whole numbers (integers).

quantity [physics] A term with various applications in different fields of physics. A quantity of electricity is the current flow multiplied by the time for which it flows (for example, in ampere-hours). A quantity of light is the luminous flux multiplied by the time for which it occurs (for example, in lumen-hours). A quantity of radiation is the intensity multiplied by time for which it is maintained.

quantum [physics] A single indivisible 'packet' of energy, central to quantum theory. For example: a quantum of electromagnetic energy (such as light or X-rays) is the photon; its energy (expressed in joules) is equal to the product of the frequency and Planck's constant (6.6262×10^{-34} joule-second).

quantum numbers [chemistry; physics] According to quantum theory, certain physical quantities that apply to atoms and subatomic particles (such as energy levels and spin of nuclei) can have only specific (quantized) values, which are denoted in terms of integral or half-integral quantum numbers.

quark [physics; chemistry] One of twelve hypothetical subatomic particles that combine in various ways to form all other particles. For example, different combinations of three quarks form PROTONS and NEUTRONS.

    The term is borrowed (as of 1964, first by the physicist Murray Gell-Mann) from a word used by James Joyce in his novel *Finnegan's Wake* (published in 1939).

quart [volumetric measure] Unit of volume in liquid and dry goods, but varying in actual quantity between usage in the United Kingdom, Canada, and some other ex-colonial countries, and usage in the United States.

    As a liquid measure:

	1 quart	=	2 PINTS, one-quarter of a GALLON		
1 UK quart	=	1.1365 LITRE	1 US quart	=	0.9463 liter
	=	40 UK FLUID OUNCES		=	32 US fluid ounces
	=	1.20095 US quart		=	0.83267 UK quart
(1 litre = 0.8799 UK quart)			(1 litre = 1.0567 US quart)		

As a dry measure:

	1 quart	=	one-eighth of a PECK		
1 UK quart	=	1,136.5 CC	1 US quart	=	1,102.2 cc
	=	69.35 CU. IN.		=	67.25 cu. in.
	=	1.03112 US quart		=	0.96982 UK quart
(1,000 cc = 0.8799 UK quart)			(1,000 cc = 0.9073 US quart)		

In the United Kingdom, the volume of a quart thus remains the same for both liquid and dry goods; in the United States there is a 16% difference between liquid and dry measures with the same name.

quartan [time; medicine] Recurring every fourth day, inclusively – that is, with two days' interval between recurrences, or every 72 hours. Fevers symptomatic of some tropical diseases (notably malaria) recur like this.

quarter [quantitatives] ¼ (one-FOURTH). *See also* articles beginning QUARTER *below.*

quarter [weight; volumetric measure] An imperial measure of weight originating in the AVOIRDUPOIS WEIGHT SYSTEM, but traditionally adapted in the United States to a slightly diminished though related size.

1 UK quarter		
(of a HUNDREDWEIGHT)	=	2 stones, 28 POUNDS
	=	12.7006 KILOGRAMS
	=	1.1200 US quarter

1 US quarter (of a SHORT HUNDREDWEIGHT, or cental)	=	25 pounds
	=	11.3398 kilograms
	=	0.89286 UK quarter

In former centuries in Britain, the quarter was also a volumetric measure corresponding to a cartload of grain.

    In this sense:

1 quarter	=	8 UK BUSHELS (8.257 US bushels)
	=	290.96 litres, 0.2909 cubic metre

                                    =   10.272 cubic feet

quarter [comparative values] Unit of currency in the United States and Canada.
    1 quarter (of a dollar)   =   25 cents
    See also COINS AND CURRENCIES OF THE WORLD.

quarter [time] Most commonly, a quarter of an hour, 15 minutes. But sometimes (as
    in calculating leasehold rental payments) a quarter of a year, 3 calendar months (*see*
    QUARTER DAYS).

quarter-back *see* AMERICAN FOOTBALL MEASUREMENTS, UNITS, AND POSITIONS

quarter days [time] Four days in the year that divide it into four roughly equal
    quarters, for the purpose of calculating salaries, taxes, leasehold rents, or other
    financial arrangements. In England, the traditional quarter days are: Lady Day – 25
    March; Midsummer (St John's) Day – 24 June; Michaelmas – 29 September;
    Christmas Day – 25 December; which were evidently based on the equinoxes and
    solstices of the astronomical (solar) year.

quarterly [time] Every three months.

quartern [weight; volume] An old form of the word *quarter*, used in a deliberately
    archaic sense in Britain to mean one-fourth of several types of measure: in weight,
    one-quarter of a STONE or of a POUND; in volume, one-quarter of a PECK or of a PINT.

¼ stone	=	3½ pounds (3 lb 8 oz avdp.)
	=	1.5876 kilogram
¼ pound	=	4 OUNCES avdp., 64 DRAMS
	=	113.3981 grams
¼ (UK) peck	=	2 (UK) QUARTS, 138.70 cubic inches
	=	2,273.0 cubic centimetres
	=	2.06224 US quarts
¼ (UK) pint	=	1 (UK) GILL, 5 (UK) FLUID OUNCES
	=	0.30024 US pint
	=	0.14206 litre, 142.06 millilitres
	=	1.20095 US gill, 4.815 US fl. oz
		(1 US gill = 0.83267 UK gill)

quarter-note *see* CROTCHET

quarter of a mile [linear measure] The standard distance of the old athletics tracks,
    mostly replaced now by tracks of 400 metres in length (a distance quite unsuited to
    race lengths in multiples of 500 metres) because of the similarity in scale.

¼ mile	=	440 YARDS = 1,320 FEET = 15,840 inches
	=	20 CHAINS = 2 FURLONGS
	=	402.336 metres, 0.402336 kilometre
		(400 metres = 437.445 yards, 1,312 ft 4.03 in.)

    See also MILE.

quarter of an acre [square measure] Unit of area in the imperial scale.

¼ acre	=	1 ROOD, 40 SQUARE RODS or UK PERCHES
	=	1,210 SQUARE YARDS, 10,890 square feet
	=	0.025 square furlong
	=	1,011.75 SQUARE METRES, 0.101175 hectare

    See also ACRE.

quarter of an inch [linear measure] Imperial linear measure that is commonly
    considered the smallest unit in ordinary household usage.

¼ inch	=	0.635 centimetre, 6.350 MILLIMETRES
	=	18 POINTS (in typography), 1½ pica EM
	=	250 MILS

    See also EIGHTH OF AN INCH; INCH.

quarter of a pound [weight] Imperial measure formerly generally regarded as the
    least quantity available for sale at shopping counters of many types of perishable
    (consumer) goods.

¼ pound	=	4 OUNCES, 64 drams avdp.
	=	113.39809 GRAMS, 0.11339809 kilogram
		(100 grams = 3.5274 ounces)

    See also AVOIRDUPOIS WEIGHT SYSTEM; POUND.

quarter section [square measure] In the United States and Canada, a unit of land area that is generally square. It is so called because it corresponds to one-quarter of a SQUARE MILE or SECTION.

1 quarter section	=	¼ mile x ¼ mile (804.67 x 804.67 metres)
	=	160 ACRES, 16 square furlongs
	=	64.752 HECTARES

quartet [quantitatives; music] Group of four, all independently contributing to a unity.

In music, the most common quartets are those comprising four singers or four stringed-instrument players. The singers are generally one each of the voices soprano, alto, tenor, and bass (SATB), but sometimes instead are all female or all male, with vocal ranges accordingly. A (male) barbershop quartet, for example, ordinarily comprises a counter-tenor, a tenor, a baritone, and a bass. The usual stringed-instrument quartet is made up of two violins, a viola, and a cello, although occasionally the line-up is a violin, a viola, a cello, and a double-bass (reflecting the much earlier grouping of a 'family' of viols, of which the double-bass is the last surviving common member).

quartic [maths] An algebraic expression that has powers of the (single) variable up to the power 4.

quartile [maths] For a set of data in statistics, the first, or lower, quartile is the value below which a quarter of the set lies, equivalent to the twenty-fifth percentile. The second quartile is the median (fiftieth percentile), and the third, or upper, quartile is the seventy-fifth percentile.

quart major [sporting term] A technical term in card games, applying to a straight (or run) of four cards in a hand or played as a sequence corresponding to the ace, the king, the queen, and the jack or knave, in that order or in the reverse.

quarto [volumetric measure] A unit of volume very different in the two countries that use the term: Italy and Portugal. Nonetheless, both derive as 'one-fourth' of a larger measure.

1 Italian quarto	=	73.6 litres
	=	16.192 UK gallons, 19.4436 US gallons
		(20 UK gallons = 1.235 Italian quarto
		20 US gallons = 1.029 Italian quarto)
Portuguese quarto	=	2 oitavos ('eighths')
	=	3.46 litres
	=	6.09 UK pints, 0.7612 UK gallon
	=	7.313 US pints, 0.9141 US gallons
		(1 UK gallon = 1.3137 Portuguese quarto
		1 US gallon = 1.0940 Portuguese quarto)
16 Portuguese quartos	=	1 Portuguese FANEGA
	=	55.364 litres
	=	12.180 UK gallons, 14.626 US gallons
124 Portuguese quartos	=	1 old Portuguese PIPA
	=	429 litres
	=	94.38 UK gallons, 113.33 US gallons

quarto, 4to [paper size] A quarter of a sheet, or a sheet folded in four (to make four leaves or eight pages).

quasi particle [physics] Name sometimes used for a hypothetical particle (quantum of energy) needed to apply quantum theory to, for example, atoms and electrons in crystal lattices: *see* PHONON.

The term *qua-si* derives immediately from Latin, meaning 'as if'.

quasquicentennial, quasquicentenary [quantitatives; time] Of 125 years.

The term is American, and derives on an analogy with the regular English word *sesquicentenary/sesquicentennial*, 'of 150 years', which is based on Latin *sesqui-* '(one-)and-a-half' + 'century'. 'Quasqui-' is thus an attempt at constructing an element to mean 'one-and-a-quarter'.

quaternary [quantitatives] As a group of four; numbering four in total; the fourth.

Quaternary period [time; geology] The present geological period, which began between 3 and 2 million years ago, and comprises the glacial Pleistocene epoch (the ice ages) followed by the post-glacial Holocene epoch. In fact, however, the only major difference between the Pleistocene epoch and the preceding Pliocene epoch (that is, between the end of the TERTIARY PERIOD and the whole of the Quaternary period) is not a geological one at all. The Quaternary period represents no more really than the time humanoids have lived on the planet – and, for this reason, the period is constantly being redefined as evidence accumulates for earlier human ancestry.

The period is called the Quaternary ('fourth') simply because it is held to follow the Tertiary ('third').

quaternion [maths] A four-component complex number consisting of one real and three imaginary terms.

quatorzain [literary] A poem or verse of fourteen lines that is like a sonnet in all but strict verse form. The term derives from French *quatorzaine* '14 in number' (compare *douzaine* '12 in number', or dozen, and QUATRAIN below). *See also* VERSE FORMS.

quatrain [literary] A poem or verse of four lines (of any, but constant, length and rhythm). The rhyming scheme is most commonly *a b a b* , but not obligatorily.

quattrocento *see* -CENTO

quaver [music] In musical notation, the quaver – known in the United States as an eighth-note – is represented by the symbol ♪ or ♪ and corresponds to
> ½ crotchet or quarter-note
> ¼ minim or half-note
> ⅛ semibreve or whole note

The term derives from the same stem as the verbs *quiver* and *quake*, the root meaning in this case being that the note is of such brief duration as to be no more than a shake on to it and off again.

queen [sporting term] In cards, the court card of value between the KING and the jack or KNAVE, known in some countries alternatively as the 'lady' or 'dame'. In a very few games the Queen of Spades has a central role as the highest trump or as a wild card.

In chess, the piece with the greatest freedom of movement, a player's most valuable piece on the board. The capture of an opponent's queen does not, however, indicate the end of the game, which occurs as a result of the irretrievable capture of the opponent's KING. *See* CHESS PIECES AND MOVES.

Queen Anne [time] Style of a period between 1702 and 1714, during which Queen Anne was on the English throne.

quetzal [comparative values] Unit of currency in Guatemala.
> 1 quetzal   =   100 centavos

The original Nahuatl word *quetzalli* actually means 'tail-feather' because the feathers stick out: Nahuatl *quetza* 'to protrude', 'stand out'. Quetzalcoatl, the plumed serpent deity after whom the coin is named, is in fact *quetzalli-cohuatl* 'tail-feather harmful-snake', and thus 'plumed serpent'. *See also* COINS AND CURRENCIES OF THE WORLD.

quick time, quick march [speed] The standard marching pace for infantry formations. The actual speed varies between regiments and between countries, however, largely as a result of tradition but also because of the differing average strides of distinct racial groups. In the United States, for example, quick time is measured as 2 steps per second (120 steps per minute), each of the regulated distance of 30 inches (2 ft 6 in.; 76.2 centimetres, 0.762 metre; a speed of 3.4091 mph, 5.4864 km/h). By contrast, in most units of the British army quick time is geared to an overall speed of 4 mph (6.4374 km/h), which requires 128 steps per minute, each of a regulated 33 inches (2 ft 9 in.; 83.82 centimetres, 0.8382 metre).

'Quick march' is the instruction to march in quick time, or a piece of music played while a group marches in quick time. The dance known as the quickstep was also originally based upon music intended for a military formation to march to in quick time. *See also* HALF-STEP.

quincunx [maths] An arrangement of five objects in the shape of the 'five' on a dice: four at each corner of a square or rectangle, and one in the middle.

quincunx [physics; astronomy] An aspect of two planets at which they appear to be 150 degrees distant from each other.

The term derives from the Latin *quinqu-* 'five', *unci-* 'twelfth': 150 degrees is five-twelfths of 360 degrees.

quindecagon [maths] A POLYGON that has fifteen sides (and fifteen angles).

quiniela [sporting term] In betting on horses or greyhounds in North America, a system by which the better must correctly nominate the first two animals home but without any reference to which comes first or which second. There is also the exacta system (known in Britain as a forecast) by which the better nominates the first and second animals home in the correct order, to win the bet.

quinqu- [quantitatives: prefix] Prefix which, when it precedes a unit, implies a fivefold unity.

> Examples: quinquangular – having five angles
> quinquennial – once every five years, lasting five years
> quinquepartite – having five parts, in five parties

quinquagenarian [biology] Someone aged between 50 and 59, in his or her fifties.

quinquennial [time] Once every five years; lasting for five years.

quinquevalent, quinquevalence [chemistry] Having a valence (valency) of five. An alternative, and more common, description is pentavalent/pentavalence (the first element from ancient Greek rather than Latin).

quint [sporting term] A sequence of five, especially as a term in card games.

quint [music] The interval of a perfect FIFTH. On an organ, the organ stop of this name causes the playing of the note one perfect fifth above that of the key depressed on the keyboard (manual), or occasionally the note of the key depressed in addition to the note one perfect fifth above.

quintal [weight] A term used for specific weights in several countries but effectively ultimately deriving from one single weight: 100 pounds in ancient Rome (Latin *centenarius* '(a) hundred(weight)', thus Arabic *qintar*, thus medieval Mediterranean *quintale*). Accordingly, in English-speaking countries, the term has been assimilated to the standard HUNDREDWEIGHT.

In Britain and some other English-speaking countries,

1 quintal	=	1 hundredweight, or centner
	=	112 POUNDS, 8 STONE
	=	50.80208 KILOGRAMS
		(50 kg, or 1 centner = 110.231 lb)
	=	1.120 US quintal
	=	0.5080 metric quintal
		(1 metric quintal = 1.968 UK quintal)
20 quintals	=	1 (long) ton, 2,240 pounds
	=	1,016.0416 kilograms, 1.0160 (metric) tonne

In the United States,

1 quintal	=	1 (short) hundredweight, or cental
	=	100 pounds
	=	45.35924 kilograms
		(50 kg, or 1 centner = 110.231 lb)
	=	0.8929 UK quintal
	=	0.4536 metric quintal
		(1 metric quintal = 2.205 US quintals)
20 quintals	=	1 (short) ton, 2,000 pounds
	=	907.18474 kilograms, 0.9072 (metric) tonne

In the metric system,

1 quintal	=	100 kilograms
	=	220.4623 pounds, 15.75 stone
	=	1.968 UK quintal, 2.205 US quintals
		(224 lb, 2 UK quintals = 1.0160 metric quintal
		200 lb, 2 US quintals = 0.9072 metric quintal)
10 quintals	=	1 (metric) tonne
	=	2,204.623 pounds, 0.9842116 (long) ton
		(1 ton = 10.1604 metric quintals)

In Spain and Mexico,

1 quintal	=	100 Castilian pounds (libras)
	=	101.4 pounds
	=	45.9943 kilograms
		(50 kg, or 1 centner = 110.231 lb)

	=	0.9054 UK quintal, 1.014 US quintal
		(1 UK quintal = 1.1045 Spanish quintal
		1 US quintal = 0.9862 Spanish quintal)
	=	0.4599 metric quintal
		(1 metric quintal = 2.1744 Spanish quintals)
20 quintals	=	1 TONELADA, 2,000 Castilian libras
	=	919.886 kilograms
	=	2,028 pounds

In Argentina,

	=	
1 quintal	=	100 Argentinian pounds (libras)
	=	101.28 pounds
	=	45.9398 kilograms
		(50 kg, or 1 centner = 110.231 lb)
	=	0.9043 UK quintal, 1.0128 US quintal
		(1 UK quintal = 1.1058 Argentinian quintal
		1 US quintal = 0.9874 Argentinian quintal)
	=	0.4594 metric quintal
		(1 metric quintal = 2.1768 Argentinian quintals)

In Portugal,

	=	
1 quintal	=	128 Portuguese pounds (libras or arratels)
	=	129.526 pounds
	=	58.752 kilograms
	=	1.1565 UK quintal, 1.2953 US quintal
		(1 UK quintal = 0.8647 Portuguese quintal
		1 US quintal = 0.7720 Portuguese quintal)
	=	0.5875 metric quintal
		(1 metric quintal = 1.7021 Portuguese quintal)
13.5 quintals	=	1 tonelada, 1,728 Portuguese pounds
	=	793.15 kilograms
	=	1,748.597 pounds

quintan [time] Recurring every fifth day, inclusively – that is, with three days'
   interval between recurrences, or every 96 hours. Fevers symptomatic of some
   tropical diseases recur like this.

quintet [quantitatives; music] Group of five, all independently contributing to a unity.
   In music, the most common quintets are those comprising five instruments from
   the same division of the orchestra – five stringed instruments, for example, or five
   woodwind instruments – in a range of pitches, allowing for a variety of tone with
   complete audibility of the individual instruments.

quintic [maths] An algebraic expression that has powers of the (single) variable up to
   the power 5.

quintile [astronomy] Describing an angle of 72 degrees between two heavenly
   bodies: 72 degrees is one-fifth of a complete circle of 360 degrees.

quintillion [quantitatives] In the UK a quintillion is a million million million million
   million (10^{30}, or 1 followed by 30 noughts), whereas in North America and parts of
   continental Europe, notably France, a quintillion is a thousand thousand thousand
   thousand million (10^{18}, or 1 followed by 18 noughts).

quintuple [quantitatives] Fivefold, five times; of five parts.

quipu [maths] In the cultures of the central Andes (including that of the Incas, but
   continuing even into the 1960s), a method of recording mathematical statistics. It
   comprised one long, thick, horizontal, central cord, from which other strings of
   various colours were suspended vertically. The colours apparently denoted the
   subjects of the statistics: yellow for gold (by weight or bars), black for time (in
   nights elapsed), red for military action involving bloodletting (possibly in casualties
   or in prisoners flayed), and so on. The actual numbers of the statistics were displayed by knots or loops in the suspended strings, the specific distance down each
   string away from the central cord and the presence of single or multiple knots or
   loops corresponding to units, tens, hundreds, thousands, and so on.
   The quipu was solely a device for recording statistics; it could not be used for
   calculation. Nonetheless, it seems to have been possible to record censuses, laws

relating to permitted numbers of items, and even genealogies, in perfectly readable form.

The term derives from Quechua *quipu* 'knot'.

quire [paper] One-twentieth of a ream of paper – effectively equal to 25 sheets, but technically only 24 sheets: *see* REAM.

Q unit *see* Q, QUAD, Q UNIT

quorum [quantitatives] The number of members of a group, society, or political organization that must be present for decisions made to count as official.

In certain religious denominations, a quorum is by extension a subdivision of one chapter of the priesthood.

quotidian [time] Daily; liable to occur or recur every day. In medical practice, this term describes a fever that may recur irregularly but frequently.

quotient [maths] The result of a simple division sum. For example: 6 divided by 3 gives a quotient of 2.

The term derives from Latin *quotiens* 'how many times?'.

qurush [comparative values] Unit of currency in Saudi Arabia, also spelled *guersh*, *guerche*, or *girsh*.

$$20 \text{ qurush} \quad = \quad 1 \text{ (Saudi) rial, 100 hallalas}$$

See also COINS AND CURRENCIES OF THE WORLD.

R

R, R. *see* RANKINE SCALE OF TEMPERATURE; RÉAUMUR SCALE OF TEMPERATURE; RÖNTGEN

r, r. *see* CORRELATION COEFFICIENT; RESISTANCE; RÖNTGEN

race [biology] In the taxonomic classification of life-forms, a category below (of greater refinement than) that of species, better known as a subspecies. It is inaccurate to use the word in such a sense in relation to humans, however, in whom racial differences do not amount to such a genetic distinction (and more often centre on cultural and linguistic dissimilarities rather than on ancestral heredity). *See* full list of categories *under* CLASS, CLASS.

racing car *see* MOTOR VEHICLE RACING

racing odds [sport] The relative probabilities of the horses or dogs in a race either of winning or of coming second or third, as quoted by the official betting authorities. For how those odds are calculated, *see* ODDS, CALCULATION OF; PROBABILITY.

There is often a difference between the odds quoted days before a race ('quoted odds', 'listed odds') and the odds at the actual start of the race, thereafter regarded as fixed ('the starting price', 'fixed odds'). Such differences occur particularly at horse-racing meetings, as the potential punters and the betting authorities have an opportunity to gauge the condition of both the horses and the racecourse.

rack [sporting term] In the dressage section of the three-day event (horse-riding competition), a slowish GAIT in which a horse lifts its forelegs high off the ground at each step, or a faster gait in which the horse allows only one foot to be placed on the ground at a time (in both cases perhaps as if going over a series of bars on the ground).

The term derives apparently from a medieval Germanic word *rec* 'framework', 'bars' that is akin to English *reach* and *stretch*.

rad [maths] *see* RADIAN, RAD

rad [physics] For ionizing radiation (such as X-rays), a unit of absorbed dose, equal to 100 ergs per gram of absorbing material. It is reputed to have been chosen because it is the X-ray dose that will kill a mouse.

$$
\begin{aligned}
1 \text{ rad} \quad &= \quad 1.1 \text{ röntgen (approx.)} \\
&\quad [1 \text{ röntgen} = 0.9091 \text{ rad (approx.)}] \\
&= \quad 0.01 \text{ (one-hundredth) joule per kilogram} \\
&= \quad 0.01 \text{ (one-hundredth) gray (SI units)}
\end{aligned}
$$

The term derives as an abbreviation of radiation.

radial velocity [astronomy] The velocity at which a celestial object moves towards (or away from) an observer, alternatively known as the *line-of-sight velocity*.

radian, rad [maths] Unit of plane angle, equal to the angle subtended at the centre of a circle by an arc that is the same length as the radius.

1 radian = 57.29578 degrees
(1 degree = 0.017454 radian
360 degrees = 2π radians)

radiance [physics] Of a luminous surface, the RADIANT FLUX emitted (radiated) per unit area.

radian frequency [physics] Another name for ANGULAR FREQUENCY.

radiant [astronomy] A point on the celestial sphere from which meteors in a shower (or any other parallel tracks) appear to come.

radiant flux [physics] The flow of radiant electromagnetic energy (such as light) per unit time from or to an object. *See also* LUMINOUS FLUX.

radiant flux density [physics] For a surface, the radiant power that flows per unit area, alternatively known as the *irradiance*. *See also* LUMINOUS FLUX DENSITY.

radiant intensity [physics] For a radiating object, the radiant energy emitted per second per unit solid angle. *See also* LUMINOUS INTENSITY.

radiation belt(s) *see* VAN ALLEN (RADIATION) BELTS

radiation counter [physics] A device for detecting and measuring charged atomic and subatomic particles. *See also* GEIGER COUNTER.

radiation flux density [physics] Another name for RADIATION INTENSITY.

radiation intensity [physics] For an irradiated surface, the rate of flow of energy through unit surface area at right-angles to the irradiating beam.

radiation pressure [physics] The minute pressure exerted by photons ('particles') of electromagnetic radiation when they strike a surface. For example: sunlight striking the Earth's surface has a radiation pressure of 10^{-5} newton per square metre.

radiation pyrometer [physics] An instrument for measuring high temperatures of a distant heat source (such as a fire or furnace). Radiated heat is focused on to a thermocouple, and the voltage generated is a measure of temperature (often displayed on a millivoltmeter calibrated directly in degrees).

radiation unit [physics] Unit of radioactivity equal to the number of disintegrations (of a radioisotope) per second. Formerly measured in CURIES, the SI unit is the BECQUEREL.

radical [chemistry] In the molecule of a chemical compound, a group of atoms that maintain their identity during chemical changes that affect the rest of the molecule. Most radicals cannot have an independent existence (but *see* FREE RADICAL).

radical, radical sign [maths] Technical term for the sign ($\sqrt{}$) denoting the root of a number (for example: the square root, the cube root). The number that is to be reduced to its root is known as the *radicand*.

The word *radical* derives from Latin *radix, radic-* 'root' (with which the English word is cognate).

radical axis [maths] In geometry, the straight line that joins the two points at which two circles intersect. Also, the straight line between two circles that do not intersect, perpendicular to a line that that joins the centres of the two circles.

radicand *see* RADICAL, RADICAL SIGN

radioactive decay [physics] The rate at which a radioisotope changes into another element (by emitting alpha particles, beta particles, and/or gamma rays) is expressed as its HALF-LIFE.

radioactive standard [physics] A radioisotope of known half-life (rate of radioactive decay) used to calibrate instruments that measure radioactivity.

radioactivity [physics] Radioactivity, and other forms of ionizing radiation, are (and have been) measured in a large variety of units, some of which appear above. For others *see* BECQUEREL; CURIE; GRAY; RÖNTGEN; RUTHERFORD; SIEVERT. *See also* X-RAYS.

radio alphabet *see* RADIO SIGNALLING ALPHABET

radio altimeter [physics; aeronautics] An instrument on board an aircraft for measuring its height above ground by detecting the delay before the return of a reflected radio signal (an echo), as in radar, or the change in frequency of the echo caused by the DOPPLER EFFECT.

radio beacon, radio beam [physics; navigation] A continuous radio signal emitted as a beacon or guide to navigation for shipping and aircraft. The direction of the beam as received on board ship or aircraft may be determined by use of a *radio compass* (*radiogoniometer*, or direction-finder).

radiocarbon dating [geology; archaeology] A method of measuring the age of carbon-containing objects that are up to 50,000 years old by detecting the amount of the radioisotope carbon-14 they contain. All living plants absorb some carbon-14 that occurs naturally in minute quantities in atmospheric carbon dioxide; the absorption stops when the plants die, and their age – or that of products from animals that have eaten the plants – can be calculated from the ratio of normal carbon-12 to carbon-14 in their remains. The method is also known simply as *carbon dating*. *See also* POTASSIUM-ARGON DATING; RUBIDIUM-STRONTIUM DATING.

radio compass *see* RADIO BEACON, RADIO BEAM

radio frequency (RF) [physics; telecommunications] The frequency of electromagnetic radiation that can be used for radio transmissions, in the range 10 kilohertz to 300,000 megahertz (corresponding to wavelengths of 30,000 metres to 1 millimetre). The range of radio frequencies is known alternatively as the *radio spectrum*.

radiogoniometer *see* RADIO BEACON, RADIO BEAM

radioimmunoassay [medicine] Method of determining the presence or quantity of a substance in the human body by harmlessly 'labelling' it with a radioactive element, combining it with an antibody, and reintroducing it to the blood circulation, to monitor the (strength of the) resultant reaction of the body's immune system.

radio interferometry [astronomy] The use of two or more radio telescopes to produce interference effects – to measure, for example, the distances to remote radio sources in the Universe.

radio signalling alphabet [literary] An alphabet devised to ensure clarity when relaying individual letters verbally over a radio network, in which each letter is given a name different in vocalic sound from any other. The modern form of the alphabet – used particularly by the police, the military, and air traffic controllers – is:

A	alpha	N	november
B	bravo	O	oscar
C	charlie	P	papa
D	delta	Q	quebec
E	echo	R	romeo
F	foxtrot	S	sierra
G	golf	T	tango
H	hotel	U	uniform
I	india	V	victor
J	juliet	W	whisky
K	kilo	X	x-ray
L	lima	Y	yankee
M	mike	Z	zulu

radio sonde, radiosonde [meteorology] An instrument, carried in a balloon, that measures humidity, pressure, and temperature at various heights in the atmosphere and transmits back the findings by radio.

radio spectrum *see* RADIO FREQUENCY (RF)

radio wave [physics] Electromagnetic wave (of wavelength in excess of 1 millimetre) used for radio transmissions: *see* RADIO FREQUENCY (RF).

radius [maths] The distance from the centre of a circle to its circumference; half the diameter.

radius of curvature [maths] For a point on a curve, the radius of the circle that touches (and has the same curvature as) the curve at that point. *See also* CURVATURE.

radius of gyration [physics] For a rotating object, the distance from its axis to the point at which its total mass could be concentrated without changing the object's moment of inertia: *see* MOMENT OF INERTIA.

radius vector [maths; astronomy] In polar coordinates, the distance from a point to the origin. If the point has coordinates (r, θ), it is r.

For an orbiting celestial object, the distance from the object to the focus of its orbit. For example: for a planet it is its distance from the Sun.

radix [maths] The base of a number system: *see* BASE.

The word is the Latin for 'root'.

railway/railroad gauges of the world *see* BROAD GAUGE; NARROW GAUGE; STANDARD GAUGE

rainfall [meteorology] Rainfall is usually stated as the amount of rain (in inches or centimetres) that falls in a given period (such as a day or a year). It is measured using a RAIN GAUGE.

rain gauge [meteorology] An instrument for measuring rainfall, also known as a *pluviometer*. The simplest type consists of a funnel that collects rain into a graduated cylinder. The area of the funnel and width of the cylinder are taken into account in calibrating the cylinder, which may be graduated in inches or centimetres.

rallentando [music] Musical instruction: gradually slow the tempo. Italian: 'slowing', 'braking'. *See also* RITARDANDO.

rally [sporting term] In motor sport, a competition over many stages and several days, in which the time taken to complete each stage (to the nearest hundredth of a second) is crucial. The first rally competitions actually began at a number of different starting points, from which entrants had then to come together – to rally (the root meaning of the word).

Raman effect, Raman lines [physics] The scattering of monochromatic (single-wavelength) light by the molecules of a transparent medium into a range of characteristic wavelengths. It was named after the Indian physicist Chandrasekhara Raman (1888-1970).

rand [comparative values] Unit of currency in South Africa (generally abbreviated as 'R').

$$1 \text{ rand } = 100 \text{ cents}$$

The name of the unit is taken from that of the goldfield (Witwatersrand, known locally as 'the Rand') on which much of the country's economy is based, and from which gold is mined that goes into the prestige coin called the KRUGERRAND.

In Afrikaans the term means '(field) border', but derives from a root originally meaning 'visible boundary', thus also 'shoreline' (English *strand*). *See also* COINS AND CURRENCIES OF THE WORLD.

range [maths; engineering; telecommunications] In statistics, the difference between the largest and smallest values in a set of data.

In engineering, the farthest distance a bullet, shell, missile, or other projectile can be fired by a firing mechanism.

In radio broadcasting, the maximum distance from the transmitter at which reception is acceptable. It varies, depending on the terrain and atmospheric conditions. Similarly, in laser emission, the maximum distance from the laser device at which the laser beam is detectable.

range [surveying] In the United States, a row of grid-planned townships established between two meridians 6 miles (9.656 kilometres) apart, each township occupying an area 6 miles x 6 miles (36 square miles; 9.656 x 9.656 kilometres, 93.2396 square kilometres).

In Canada, a (rectilinear) subdivision of a township.

rank [maths] In statistics, the position of a given number in a set of numbers when they are arranged in ascending order. If the same number occurs twice, its rank is ½ more than the position of its first occurrence. For example, the set of numbers 8, 5, 10, 0, 3, 7, 1, 13, 8 can be rearranged as 0, 1, 3, 5, 7, 8, 8, 10, 13 (in ascending order). Their ranks are:

number	rank
0	1
1	2
3	3
5	4
7	5
8	6½
8	6½
10	8
13	9

A procedure carried out on ranks instead of actual values is called a rank test.

rank, armed forces *see* MILITARY RANKS

Rankine scale of temperature [physics] A (thermodynamic) temperature scale on which absolute temperatures are expressed in degrees Fahrenheit. Thus, absolute zero (0 K) is 0°R, the freezing point of water (0°C, 32°F) is 491.67°R and the boiling point of water (100°C, 212°F) is 671.67°R. To convert Fahrenheit temperatures to Rankine temperatures, add 459.67. (To convert Celsius to Fahrenheit – and thus to Rankine – *see* CELSIUS.) The scale was named after the British physicist and engineer William Rankine (1820-70).

Raoult's law [chemistry] Dissolving a substance in a solvent lowers the vapour pressure of the solvent. Raoult's law states that, for an ideal solution at a given temperature, the relative lowering of vapour pressure is proportional to the mole fraction of the solute (dissolved substance). In other words, the vapour pressure of a solution equals sums of the products of the vapour pressures of each component and their mole fractions. It was named after the French scientist François Raoult (1830-1901).

rap [comparative values] In the early 1800s in England, a term for the coin of the lowest denomination (which then passed into the language in the expression 'I don't care a rap').

rappen *see* CENTIME

rate [quantitatives] A relation between (one quantity in) one form of units and (another quantity in) another form of units; for example: metres per second, miles per gallon, dollars per kilogram, pounds per square inch. In scientific usage, the term is reserved for quantities that change with time (so that units of time constitute the second form of units); for example: litres per hour, feet per minute.

rate constant [chemistry] For a chemical reaction with reactants whose activities (concentrations) are unity, the speed of the reaction expressed in moles of change per cubic metre per second, alternatively known as the *velocity constant*.

rate-determining step [chemistry] The slowest stage in a chemical reaction, which determines the overall rate of reaction. Thus, the order of reaction, for example, is that of the rate-determining step.

ratel *see* RAT(T)EL, RATTLE, ARRATEL, ROTL, ROT(T)OL

rate law [chemistry] The speed of a chemical reaction equals the RATE CONSTANT times the product of the activities (concentrations) raised to a power corresponding to their order of reaction.

ratio [maths] A way of comparing two quantities – stating their proportion – by considering their quotient (that is, by dividing them). Thus, the ratio 3:4 (expressed as '3 to 4') is equivalent to the quotient ¾ (a fraction). In this example, 3 is the *antecedent* and 4 is the *consequent*. *See also* DIVISION; FRACTION.

rationalized units [physics] System of electrical units in which the values of the permeability and permittivity of free space are modified (by a factor 4π) to avoid the factor's appearing in common formulae and equations.
The problem is avoided in SI units.

rational number [maths] Any number that can be expressed as a ratio of two whole numbers (integers) – that is, as a fraction. *See also* IRRATIONAL NUMBER; NUMBER.

ratio of specific heat capacities [physics] An important parameter in physics, denoted by the symbol γ, equal to the specific heat capacity of a gas at constant pressure divided by its specific heat capacity at constant volume. For monatomic (single-atom) gases its value is 1.67; for diatomic gases it is 1.4.

rat(t)el, rattle, arratel, rotl, rot(t)ol [weight; volumetric measure] Unit of weight deriving from the trading centres of the Arab world of North Africa and the western and eastern Mediterranean, and approximating closely to the old Roman *libra pondo* 'pound by weight'.

1 Portuguese arratel		
(or libra)	=	16 onças, 2 MARCOS
	=	1 pound 0.19 ounce (1.01194 pound)
		(1 pound = 0.9882 arratel)
	=	459.2623 grams, 0.4593 kilogram
		(500 grams = 1.0894 arratel)
1 Arabian rat(t)el		
or rot(t)ol	=	1 pound 0.32 ounce avdp. (1.02 pound)
		(1 pound = 0.9804 rat(t)el)
	=	462.6642 grams, 0.4627 kilogram

		(500 grams = 1.0807 rat(t)el)
2 Arabian rat(t)els	=	1 MAHND
1 Iranian ratel	=	1 pound 0.224 ounce avdp. (1.014 pound)
		(1 pound = 0.9862 ratel)
	=	460 grams, 0.460 kilogram
		(500 grams = 1.0870 ratel)
1 Egyptian rotl or ratel	=	15.848 ounces avdp. (0.9905 pound)
		(1 pound = 1.0096 rotl)
	=	449.2833 grams, 0.4493 kilogram
		(500 grams = 1.1129 rotl)
100 Egyptian rotls	=	1 kantar

In Turkey, the unit was borrowed as a (now obsolete) volumetric measure:

1 Turkish rottol	=	1.6 litre, 1,600 cubic centimetres
	=	2.816 UK pints, 3.381 US pints
	=	0.352 UK gallon, 0.4227 US gallon
	=	97.6380 cubic inches

The term derives as an Arabic corruption of the ancient Greek *litra*, itself from the same root as the Latin variant *libra* 'pound'.

rawin, rawinsonde [meteorology] The measurement of the speed and direction of the wind by tracking a balloon (with or without a radio sonde device aboard) with radar or a radio compass.

The term *rawin* is made up from the first letters of ra̲d̲ar and wi̲n̲d̲.

rayl [physics] Unit of specific impedance for sound, equal to the effective sound pressure at a surface divided by the effective particle velocity, measured in newton-seconds per cubic metre (MKS units) or dyne-seconds per cubic centimetre (CGS units). It is equivalent to the density of a gas times the velocity of sound in it. It was named after the British chemist and physicist John William Strutt, third Lord Rayleigh (1842-1919).

rayleigh [physics] An extremely small unit of luminous intensity equal to the intensity produced by 1,000,000 photons per square centimetre of illuminated area. It was devised for measuring the brightness of aurorae and the night sky (for which it is about 250 rayleigh), and was named after the British physicist Robert John Strutt, fourth Lord Rayleigh (1875-1947).

Rayleigh disc [physics] A small circular disc suspended by a monofilament thread (glass or quartz) at an angle to a progressive sound wave. Sound pressure deflects the disc, which rotates on its suspension through an angle that is a measure of the sound intensity (dependent, in turn, on the square of the sound's velocity). It was named after the British chemist and physicist John William Strutt, third Lord Rayleigh (1842-1919).

Rayleigh limit [physics] For a lens system, the maximum extent to which the optical path from a point on the object differs from that to the same point on the image, equal to a quarter of a wavelength for perfect definition. It was named after the third Lord Rayleigh (*see above*).

reactance [physics] The property of a capacitor or inductor in an AC circuit that makes it oppose the passage of current, and thereby store energy; it is the imaginary part of the IMPEDANCE, measured in ohms. For a capacitor it equals $\frac{1}{2}\pi$ times the product of the AC frequency and the capacitance (in farads). For an inductor, it equals 2π times the product of the frequency and the inductance (in henrys).

reactance voltage [physics] The voltage produced by current flowing in a reactance, equal to the current (in amperes) times the reactance (in ohms): *see* REACTANCE.

reaction rate [physics] In a nuclear reactor, the rate at which fission takes place. *See also* RATE OF REACTION.

reaction time [biology; medicine] For a person or animal presented with a (visual or audible) signal, the time it takes to respond. Normal reaction times in humans are slowed (increased) by fatigue, by mental stress or boredom, by any medical condition causing raised temperature, and by the effects of alcohol and various other drugs. Certain drugs may, conversely, temporarily reduce reaction times, but there is rarely any medical need for it.

reactive power [physics] For an electric circuit, the product of the effective current in amperes and the effective voltage in volts (that is, the volt-amperes reactivity) multiplied by the sine of the angular phase difference between them. Its unit is the var (or VAr).

reactivity [physics] The amount by which a nuclear reactor's multiplication factor differs from unity, expressed as a percentage or measured in such units as CENT, DOLLAR, NILE, or INHOUR.

real, reales [comparative values] Unit of currency of the colonial Spain of the sixteenth to nineteenth centuries, widely used in both South and North America, in the form of a small silver coin.

$$\begin{array}{rcl} 8 \text{ reales} & = & 1 \text{ PESO ('piece of eight')} \\ 32 \text{ reales, 4 pesos} & = & 1 \text{ pistole} \\ 64 \text{ reales, 8 pesos,} & & \\ 2 \text{ pistoles} & = & 1 \text{ doubloon} \end{array}$$

Whereas, in the American colonies, the peso became known as a 'dollar', the real was the first coin in North America to be given the name the *bit*.

The real was so called because it was issued by the royal (Spanish *real*) mint.

real axis *see* ARGAND DIAGRAM

real number [maths] Any positive or negative number from among all irrational and rational numbers (including 0); only imaginary numbers are excluded: *see* NUMBER.

real part [maths] The non-imaginary part of an imaginary number: *see* ARGAND DIAGRAM; NUMBER.

real-time [electronics] In electronics, describing a data-processing system that processes continuously varying data (such as that relating to air traffic control) so fast that it can be employed to control the source of the data. In lay terms, data that are available in real time are available immediately.

ream [paper] Quantity of paper, properly comprising 20 quires at 24 sheets per quire, thus 480 sheets altogether, but most often equal to 500 sheets (to allow for spoilage).

$$1 \text{ ream} = 20 \text{ quires}$$

Accordingly, a ream of 500 sheets may be (inaccurately) considered to comprise 20 quires at 25 sheets per quire.

By tradition, in the printing and publishing trades, a *perfect ream* (also known as a *printers' ream*) allows for even more spoilage and consists of 516 sheets, corresponding to 21½ quires.

The term derives through French or Spanish ultimately from the Arabic *rizmah* 'bundle (of cloth, rags or paper)'.

rear admiral [military rank] Senior commissioned rank in the navy, ranking above a captain or commodore but below a vice-admiral.

In the United States Navy, there is a difference in status within the rank between rear admirals on the lower half of the list (immediately senior in rank to a captain, and equal in rank to a brigadier-general in the US Army and Airforce) and rear admirals on the upper half of the list (immediately junior to a vice-admiral, and equal in rank to a major general in the US Army and Airforce).

In the (British) Royal Navy – and in some other navies of the world – there is no such distinction: a rear admiral is equal in rank to a major general in the army and an air vice-marshal in the airforce.

Réaumur scale of temperature [physics] An obsolete scale of temperature on which the fixed points are the freezing point of pure water (0°R) and the boiling point of water (80°R). It was named after the French physicist and biologist René Réaumur (1683-1757), who devised it in about 1730. (That the boiling point was set at 80° came about quite coincidentally, through a combination of the relatively arbitrary calibration Réaumur used on his thermometric equipment and the impurities in the alcohol he utilized in his thermometer.)

rebah [weight] Small unit of weight in ancient Israel.

$$\begin{array}{rcl} 1 \text{ rebah} & = & 5 \text{ GERAHS} \\ & = & 2.1 \text{ grams} \\ & = & 0.074 \text{ ounce (1.185 dram avdp., 32.8125 grains)} \\ & & (1 \text{ ounce} = 13.5135 \text{ rebahs}) \\ 2 \text{ rebahs} & = & 1 \text{ BEKAH} \\ 4 \text{ rebahs, 2 bekahs} & = & 1 \text{ SHEKEL} \end{array}$$

240 rebahs, 120 bekahs,
60 shekels = 1 MINAH

The term is the Hebrew for 'quarter', referring to its proportional value in relation to the shekel.

reciprocal [maths] A number obtained by dividing another number into 1. For example: the reciprocals of 2 and 3 are ½ (0.5) and ⅓ (0.333 . . .). And technically, the reciprocals of ½ and ⅓ are conversely 2 and 3. Zero has no true reciprocal, although its value is sometimes taken as infinity.

The root meaning of *reciprocal* is 'complementarily opposed' – literally 'both backward and forward', the re- representing 'backward' and the -pro- 'forward' – extended in this mathematical context to mean 'inversely related'.

reciprocal ohm *see* MHO

reciprocal proportions *see* LAW OF EQUIVALENT PROPORTIONS

recon [biology] The smallest molecular unit of genetic material, out of which the larger units – the muton and cistron – are built. The term derives as an abbreviation of *recombinant. See also* CISTRON; MUTON.

recorder *see* WOODWIND INSTRUMENTS' RANGE

rectangle, oblong [maths] A quadrilateral (four-sided plane figure) with opposite sides equal and parallel, and right-angles between each pair of adjacent sides. Its diagonals are equal in length, and its area equals the product of the lengths of two adjacent sides.

The word *rectangle* reflects on the figure's dependence on right-angles; the word *oblong* suggests a figure longer than it is wide (*see* OBLATE).

rectangular axes [maths] Axes at right-angles to each other, as are normal Cartesian coordinate axes: *see* CARTESIAN COORDINATES.

rectangular coordinates [maths] Another name for CARTESIAN COORDINATES.

rectangular prism [maths] Another name for a CUBOID.

rectangular hyperbola [maths] A HYPERBOLA of which the asymptotes are at right-angles (typically the x- and y-axes, in CARTESIAN COORDINATES).

rectilinear [maths] Describing something that is characterized by, bounded by, or moves in straight lines. For example: the principle of the rectilinear propagation of light means simply that light travels in straight lines.

recurring decimal [maths] A decimal fraction that has a number, or group of numbers, that repeat indefinitely. For example: fractions that are thirds, sixths, ninths, or elevenths give recurring decimals:

⅓	=	0.333333 . . .	1/9	=	0.111111 . . .
⅔	=	0.666666 . . .	7/9	=	0.777777 . . .
⅙	=	0.166666 . . .	1/11	=	0.090909 . . .
⅚	=	0.833333 . . .	7/11	=	0.636363 . . .

The recurring numbers can be indicated by a dot over them, so that these same examples become $0.\dot{3}$, $0.\dot{6}$, $0.1\dot{6}$, $0.8\dot{3}$. $0.\dot{1}$, $0.\dot{7}$, $0.\dot{0}\dot{9}$, and $0.\dot{6}\dot{3}$, respectively.

recursive function [maths] A function that allows values or meaning to be determined in at least two different ways, so tending towards assured accuracy in the result. Technically, a recursive function may be described as a mathematical function produced by addition, multiplication, the selection of one element from an ordered set, and the resolving of whether one quantity is greater than another (when both quantities are values that tend towards accuracy in the result).

Such functions are significant in computer data-handling.

red alert [quantitatives] State of maximum preparedness to repel attack or to employ emergency defensive measures.

Some national armed forces and security organizations recognize a series of such states, generally along the lines of *yellow* or *amber alert* (a high degree of readiness), *orange alert* (extreme readiness), and red alert (maximum readiness).

red cent [comparative values] Former unit of currency in the United States in the form of a 1-cent coin made of copper. *See also* CENT.

red dwarf, red giant *see* HERTZSPRUNG-RUSSELL DIAGRAM

red heat [physics] Loose term for the temperature of a hot metal or ceramic, in the range 500 to 1,000°C.

redox potential *see* OXIDATION-REDUCTION POTENTIAL, REDOX POTENTIAL

red shift [astronomy] For galaxies and quasars that are moving rapidly away from the Earth, a displacement of their spectral lines towards the red end of the spectrum, representing a wavelength change caused by the DOPPLER EFFECT. Its existence is cited as proof that the Universe is expanding. *See also* HUBBLE'S LAW.

reduce [maths] To simplify a fraction, an algebraic expression or an equation to its least complex form. The fraction $^{50}/_{100}$ can be reduced thus:

$$^{50}/_{100} \; (= \, ^{25}/_{50} = \, ^{5}/_{10} \;) = \, ^{1}/_{2}$$

reduced frequency *see* STROUHAL NUMBER

red-wine glass *see* DRINKING-GLASS MEASURES

reed [linear measure] According to some versions of the Bible, a unit of length in ancient Israel.

1 reed	=	3 'paces', 6 cubits, 12 'spans'
	=	2.6976 metres, 8 feet 10.2 inches (2.95 yards)
		(3 metres = 1.1121 reed
		3 yards, 9 feet = 1.0169 reed)
10 reeds	=	1 plethron
60 reeds	=	1 stadion
500 reeds	=	1 'mile'

re-entrant, re-entering [maths] Describing an angle of more than 180 degrees. Also, describing a polygon that includes an internal angle of more than 180 degrees (viewed from inside the figure).

reference electrode [chemistry; physics] A standard electrode against which a varying EMF (voltage) can be measured: *see* ELECTRODE POTENTIAL.

reflectance [physics] For a mirror or other reflecting surface, the intensity of the reflected radiation (for example, the luminous flux of light) divided by the intensity of the incident radiation. The reflectance is known alternatively as the *coefficient of reflection* or *reflection factor. See also* REFLECTIVITY.

reflection, angle of *see* ANGLE OF REFLECTION

reflection, laws of [physics] Two laws:

For a ray of light (or other radiation) reflected at a surface, the angle of incidence equals the angle of reflection (the angles are measured between the rays and the normal – the line at right-angles to the surface).

The incident ray, reflected ray, and normal all lie in the same plane.

reflectivity [physics] For radiation reflected at a surface, the proportion of its incident energy that is returned, measured by a *reflectometer*.

reflex [maths] Describing an angle that is more than 180 degrees but less than 360 degrees.

reflexes, testing of [medicine; biology] Many functions within the human body are carried out by reflex – an automatic response to a stimulus sensed and reacted to by the nervous system without reference to the conscious. Breathing rates, sweating and shivering, sneezing and hiccuping, and most of the digestive and excretory functions are mostly controlled by such reflexes. Testing of reflexes thus measures the functioning of much of the body's metabolism.

Reflex testing normally centres on monitoring the knee-jerk reflex, the plantar reflex (the curling of the toes as the sole of the foot is stroked), and the pupillary reflex (the constriction of the pupil of the eye in response to increased light); a full neurological examination may include the testing of other reflexes. New-born babies have several additional reflexes that disappear within a few weeks of birth.

refraction, angle of *see* ANGLE OF REFRACTION

refraction, laws of [physics] Two laws:

For a ray of light (or other radiation) passing from one transparent medium to another, the sine of the angle of incidence divided by the sine of the angle of refraction is a constant, called the *refractive index* (the angles are measured between the rays and the normal – the line at right-angles to the surface). This statement is also called Snell's law, after the Dutch mathematician and physicist Willebrord Snell van Roigen (1591-1626), who is better known in continental Europe under the Latinized form of his name, Snellius.

The incident ray, refracted ray, and normal are all in the same plane.

Refraction may be measured using an instrument called a *refractometer*.

refractive index [physics] For a transparent medium, the speed of electromagnetic radiation in vacuum divided by its speed in the medium. For a refracted light ray, it equals the sine of the angle of incidence divided by the sine of the angle of refraction. The refractive index is alternatively known as the *index of refraction* or *refractive constant*, and may be measured using either of two instruments called a *refraction circle* and a *refractometer*. *See also* ANGLE OF REFRACTION.

Regency period [time] Period in Great Britain from 1811 to 1820, during which George, Prince of Wales (later King George IV), acted as regent for his father King George III, who was unable to reign due to mental imbalance (now thought to be) caused by the disease porphyria. George IV acceded to the throne on 29 January 1820 and died (of alcoholic cirrhosis) ten years later. *See also* GEORGIAN AGE/ERA.

Regency style [architecture; engineering; time] Of a style relating either to the REGENCY PERIOD in Britain, or to the period from 1715 to 1723 in France, during which Philippe, Duke of Orléans (nephew of the 'Sun King' Louis XIV), acted as regent for the child Louis XV until he came of age.

regiment [military] Unit or force of an army, comprising several battalions, under the unified command of a colonel or officer of similar rank. Two or more regiments may in turn form a BRIGADE.

The term derives from a Latin abstract noun meaning 'command'.

region [geography] In addition to meaning an area on the Earth's surface of particular significance (in terms of administration, biology, or ecology), the word *region* is also a technical term for an underwater stratum in the oceans, and for a stratum in the atmosphere.

register ton [cubic measure] Unit measure of the overall internal capacity (not weight) of a ship, as detailed in its national registration documents.

1 register ton	=	20 BARRELS BULK, 2½ FREIGHT TONS
	=	about 4.65 x 4.65 x 4.65 feet
	=	100 cubic feet
	=	3.7037 CUBIC YARDS
		(3 cubic yards = 0.8100 register ton)
	=	2.8317 CUBIC METRES
		(3 cubic metres = 1.0594 register ton)
	=	77.88 UK bushels, 80.39 US bushels
	=	2.8571 displacement tons
		(3 displ. tons = 1.0500 register ton)

See also DISPLACEMENT TON.

Regnault's hygrometer [physics] A hygrometer consisting of a silvered flask containing ether (ethoxyethane) and a thermometer. Air is bubbled through the ether and its temperature noted when moisture condenses on the flask. The instrument was named after the French chemist and physicist Henri Regnault (1810-78).

regular polygon [maths] A many sided plane figure in which all the sides (and all the internal angles) are equal: *see* POLYGON.

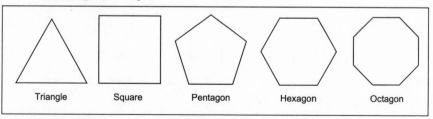

| Triangle | Square | Pentagon | Hexagon | Octagon |

regular polyhedron [maths] A many faced solid figure of which the faces are all regular polygons: *see* POLYHEDRON.

regular year [time] In the Jewish calendar a year of 354 days, representing the average form of the three common types of year: the other two types are the ABUNDANT YEAR and the DEFECTIVE YEAR.

rehoboam [volumetric measure] A champagne bottle that holds 150 UK fluid ounces, 156 US fluid ounces (4.26 litres).

Alternatively, but amounting to very much the same thing, reckoning on a standard wine bottle containing 75 centilitres (0.75 litre),

1 magnum	=	the contents of	2 bottles
1 jeroboam	=		4 bottles
1 rehoboam	=		6 bottles
1 methuselah	=		8 bottles
1 salmanazar	=		12 bottles
1 balthazar	=		16 bottles
1 nebuchadnezzar	=		20 bottles

The name of the bottle derives from that of Rehoboam, first King of Judah at the time when the kingdom of Solomon split into two: Israel (under Jeroboam) and Judah (under Rehoboam). His name derives from Hebrew *Rehab'am* 'The nation (has taken up an) enlarged space' or – less likely – 'The uncle (has given) me enlarged space'.

reichsmark, reichspfennig [comparative values] Units of currency in Germany from 1924 to 1948. *See also* RIXDOLLAR, RIJKSDAALDER, RIKSDALER, RIGSDALER; COINS AND CURRENCIES OF THE WORLD.

reis [comparative values] Former unit of currency in Portugal and Brazil.
The term represents an irregular plural of *real*: *see* REAL, REALES.

relations and relatives [biology] The list of relations/relatives within a family is:

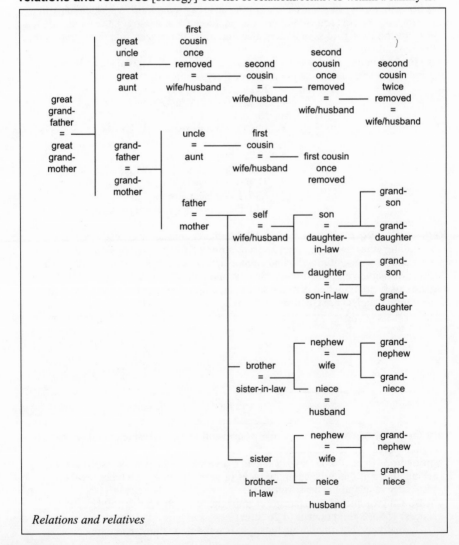

Relations and relatives

relative atomic mass (r.a.m.) [chemistry] The preferred term for *atomic weight*. It equals the mass of an atom relative to the mass of the carbon-12 isotope (which is assumed to be exactly 12). For a list of relative atomic masses *see* PERIODIC TABLE OF ELEMENTS.

relative density [physics] For a given substance, its density divided by the density of a reference substance (usually water at 4°C, for liquids), equivalent to the mass of a given volume of the substance divided by the mass of an equal volume of water. Relative density was once better known as *specific gravity*.

relative frequency [maths] In statistics, the ratio between the number of times that an event actually happens and the total possible number of times that it could happen.

relative humidity [physics] At a particular temperature, the vapour pressure of moist air divided by the vapour pressure of air saturated with water (the saturation vapour pressure), expressed as a percentage.

relative minor keys [music] A minor key that uses virtually the same notes in its SCALE as a different major key is said to be the major key's relative minor. It is sometimes useful when transposing a piece of music from one key to another to know what the relative minor is in the new key.

major		relative minor		
C	(no sharps or flats)	A minor	(with G# as leading-note)	
D♭	(five flats)	B♭ minor	(	A)
D	(two sharps)	B minor	(	A#)
E♭	(three flats)	C minor	(	B)
E	(four sharps)	C# minor	(	C)
F	(one flat)	D minor	(	D♭)
F#	(six sharps)	D# minor	(	D)
G♭	(six flats)	E♭ minor	(	D)
G	one sharp)	E minor	(	D#)
A♭	(four flats)	F minor	(	E)
A	(three sharps)	F# minor	(	F)
B♭	(two flats)	G minor	(	G♭)
B	(five sharps)	G# minor	(	G)

This table can be used conversely to find the major key to which a minor key is the relative minor.

See also KEY, KEYNOTE.

relative molecular mass [chemistry] The preferred term for *molecular weight*. It equals the sum of the relative atomic masses (atomic weights) of the component elements in one molecule of a chemical compound.

relative permeability [physics] For a material in a magnetic field, its permeability divided by the permeability of free space: *see* PERMEABILITY.

relative permittivity [physics] For a material in an electric field, the flux density produced in it divided by the permittivity in free space produced by the same field strength: *see* PERMITTIVITY.

relatives *see* RELATIONS AND RELATIVES

relativistic mass [physics] The increased mass of a particle travelling at near the speed of light, equal to its rest (stationary) mass divided by the square root of 1 minus the ratio of the square of its velocity to the square of the velocity of light.

relativity: the mass-energy equation [physics] A conclusion of the special theory of relativity is the equivalence of mass and energy, embodied in the statement that energy equals the product of mass and the square of the velocity of light ($E = mc^2$).

relay race [sporting term] A race in which two or more members of a team take part one after another.

The word *relay* is effectively by derivation the same word as *release*, in this case referring to the fact that, as each team member comes to the end of a stint, the next team member is set free to go (the opposite of *relay* is thus of course *delay*). *See also* MEDLEY RELAY.

reluctance [physics] For a magnetic circuit, the total magnetic flux when a magnetomotive force is applied (analogous to resistance in an electric circuit). *See also* RELUCTIVITY.

reluctivity [physics] The reciprocal of magnetic permeability, alternatively known as the *specific reluctance*.

rem [physics] Abbreviation of RÖNTGEN EQUIVALENT MAN.

remainder [maths] The quantity left over in a division sum when the divisor (dividing number) does not go exactly into the dividend (number being divided): *see* DIVISION.

remanence [physics] For a magnetizable material, the magnetization that remains after the magnetizing field is removed. *See also* RESIDUAL FLUX DENSITY; RETENTIVITY.

remanié [geology; time] Of a fossil or a rock stratum, dating from a time earlier than the bed of rock in which it was found: somehow misplaced in time.

The word is a French past participle meaning 're-treated'.

Renaissance [time] Describing the style and culture in Europe during the 1400s and 1500s – a period in which there was a great rebirth (French *renaissance*) of interest in the sciences and a flowering of art and architecture.

rep [physics] Abbreviation for RÖNTGEN EQUIVALENT PHYSICAL.

repeating decimal [maths] Another name for RECURRING DECIMAL.

replacement set [maths] In an equation involving the variables x and y, the set of all numbers that can be assigned as values of x, also known as a *domain*. It is called a replacement set because the members of such a set may take the place of the variable in a given relation.

representative fraction [literary; geography] The scale of a map expressed not as a ratio but as a fraction – for example, not as 1:250,000 but as 1/250,000.

réseau [astronomy] A grid of specific size on a photographic plate used in photography of parts of the night sky. Pictures developed from the plate then automatically have the grid superimposed to give some idea of scale and some means of comparative measurement.

The term derives directly from the French for 'mesh', 'grid'.

reserve buoyancy [aeronautics] The buoyancy of an amphibian aircraft that is in excess of that needed for it to float (equal to the downward force that would be required to sink it).

residual flux density [physics] For a magnetizable material, the magnetic induction that remains after the magnetizing field is removed. *See also* REMANENCE; RETENTIVITY.

residual resistance [physics] In theory a material's electrical resistance falls to zero at a temperature of absolute zero. In practice there is a residual resistance because of impurities and crystal defects in the material.

residual volume [medical] The volume of air left in the lungs after a forceful exhalation (equal to about 1,500 cubic centimetres, 91.54 cubic inches for an adult man).

resistance [physics] The property of a material that makes it oppose the passage of an electric current. When a potential difference (voltage) of 1 volt causes a current of 1 ampere to flow through a conductor, its resistance is 1 ohm. It is the reciprocal of CONDUCTANCE. *See also* IMPEDANCE; OHM'S LAW.

resistance thermometer [physics] Type of thermometer that registers changes in temperature by detecting consequent variations in the electrical conductivity of a metal such as platinum. A *thyristor* is a resistance thermometer that utilizes conductivity changes in a semiconductor.

resistivity [physics] A quantity that expresses the electrical resistance of a conductor in terms of its dimensions. For a given conductor, it equals the product of its resistance and its cross-sectional area divided by its length, expressed in ohm-metres.

resolution of vectors [maths] A method of dividing vectors into components that act in specified directions (usually at right-angles). *See also* RESULTANT.

resolving power [physics] For a lens or camera, the number of paired black and white lines per millimetre that it can resolve. It is a measure of its ability to distinguish between closely spaced objects.

For an astronomical telescope, the smallest angular separation (of two apparently nearby celestial objects) that it can distinguish.

resonance [chemistry; physics] In chemistry, the state of a molecule that exists as a mixture of two or more structures between which it alternates; resonance of this type is also called *mesomerism*.

In a vibrating system that produces sound, a state of minimum acoustical impedance in which a periodic driving force sets up forced vibrations of maximum
amplitude (at the natural frequency of the system).

In an electrical circuit, the state of maximum (or minimum) impedance for a
capacitor and inductor in parallel (or series), which occurs at a particular frequency
(the resonant frequency).

In a nuclear reaction, a temporary state in which extremely short-lived (a half-life
of about 10^{-23} seconds) and unstable particles are created as intermediates in the
reaction. Such particles are undetectable, and their existence can only be inferred.

resonance potential [physics] The potential needed to raise an electron (in an
atom) from one orbital in one energy level to an orbital at a higher energy level; it is
known alternatively as the excitation potential.

respiratory quotient (RQ) [biology] For an organism, the amount of carbon
dioxide it produces divided by the amount of oxygen it consumes. An RQ of about
1 indicates that carbohydrate is being oxidized; a lower RQ indicates that protein or
even fat is being metabolized; and a high RQ indicates that anaerobic respiration is
probably taking place.

respirometer [medicine] A machine that measures the quantity and contents of the
breath as both inhaled and exhaled, particularly in a patient undertaking exercise on
a treadmill.

rest [music] A pause of a defined duration between notes played or sung. *See also*
MUSICAL NOTATION.

resting potential [biology] For a nerve not conducting an impulse, the potential
difference (voltage) between its inner and outer membranes (usually of the order of
−60 to −80 millivolts). When a nerve impulse passes, the potential rapidly increases
to a positive value (the action potential).

restitution *see* COEFFICIENT OF RESTITUTION

rest mass [physics] The mass of a particle when it is at rest or moving only slowly:
see RELATIVISTIC MASS.

Restoration period [time] The period in Britain between 1660 and 1688. The end of
the Commonwealth (republic) under the Lord Protector, Oliver Cromwell, and his son
occurred on 29 May 1660 when Charles II – elder son of Charles I, executed on 30
January 1649 – was restored to the throne on his birthday at the age of thirty. He then
ruled for twenty-five years before succumbing to uraemia and mercury poisoning. He
had no legitimate issue, so his brother James II succeeded to the throne. Three years
later, in 1688, James was held to have abdicated by fleeing to France as plans were put
into effect by the statesmen and notable generals of the day to present the throne to
William (of Orange, James's nephew) and Mary (James's elder daughter).

resultant [maths] A single vector (such as a force) that has the same effect as two or
more other vectors: *see* PARALLELOGRAM OF VECTORS.

retentivity [physics] For a magnetizable material, the magnetic induction remaining
after a saturating magnetic field is removed. *See also* REMANENCE.

reverberation time [physics] The time it takes for the level of a sound of average
intensity to die down by 60 decibels (that is, to one-millionth of its original amplitude), measured in seconds.

reversing layer *see* ATMOSPHERE OF THE SUN, OF A STAR

Revolutionary calendar *see* FRENCH REVOLUTIONARY (REPUBLICAN) CALENDAR

revolutions per minute (rpm) [engineering] Units used (for example) in measuring the constant speed of a record-player turntable or the variable speed of rotation
of an internal combustion engine ('revs').

    1,000 revolutions per minute = 16.666 revolutions per second

reyn [physics] An obsolete unit of dynamic viscosity in the foot-pound-second (FPS)
system, once used for lubricating oils. It was named after the British physicist
Osborne Reynolds (1842-1912).

Reynolds number [physics; engineering] A dimensionless number concerned with
fluid flow: the internal force divided by the viscous force in a liquid flowing
through a cylindrical tube. It equals the product of the diameter of the tube, the
liquid's linear velocity, and its density, divided by its viscosity. As fluid velocity
decreases, the critical Reynolds number is reached, when turbulent flow changes to

laminar flow. It was named after Osborne Reynolds (*see above*).

rH [chemistry] A measure of the strength of an oxidizing agent. For a particular oxidation-reduction (redox) system, the negative logarithm of the hydrogen gas pressure that would produce the same electrode potential in a solution of the same pH. The larger the rH value, the greater is the oxidizing power of the system. *See also* PH.

rhe [physics] Unit of fluidity, equal to the reciprocal of the dynamic viscosity of a fluid. *See also* POISE; STOKES.

rhesus blood groups *see* BLOOD GROUPING

rhm [physics] Unit of effective strength of gamma-rays, equal to the strength of a source that produces a dose rate of 1 röntgen per hour at a distance of 1 metre, approximately equivalent to 1 curie. The term is short for röntgen per hour at 1 metre.

Rh-negative, Rh-positive *see* BLOOD GROUPING

rhomboid [maths] A parallelogram with opposite sides equal. A rectangle or oblong is a special type of rhomboid.

The term derives as an adjective meaning 'rhombus-like'.

rhombus [maths] A parallelogram with four equal sides; its diagonals intersect at right-angles. It is known also as a *rhomb* or *rhombohedron*; the related adjective is *rhombic*.

The word derives directly from an ancient Greek word meaning 'spinner', thus 'top'.

Rh-positive, Rh-negative *see* BLOOD GROUPING

rH scale, rH value *see* RH

rhyming scheme, rhyme scheme *see* VERSE FORMS

rhythm *see* BEAT; TEMPO

rhythm method [medicine; biology] Unreliable method of contraception relying on abstinence from sexual intercourse while a woman is fertile. Ovulation – the release of an egg from an ovary – occurs in every adult woman (until the time of the menopause) half-way between the final flow of blood in one menstrual period and the onset of the flow in the next menstrual period. The egg may then take two or three days to move from the ovary to the wall of the womb (uterus), where it remains viable for another two or three more days. A woman is thus fertile for between four and seven days during each menstrual cycle – but menstrual cycles may be of variable duration, and to monitor the exact time of fertility is notoriously difficult (even with the knowledge that a woman's body temperature temporarily decreases when ovulation takes place).

ri [linear measure] Unit of length in Japan.

$$
\begin{aligned}
1 \text{ ri} \ &= \ 36 \text{ CHÔ, 2,160 ken} \\
&= \ 4,294.4 \text{ yards, 2.440 miles} \\
&\quad (2\tfrac{1}{2} \text{ miles} = 1.0246 \text{ ri} \\
&\quad \ 5 \text{ miles} = 2.0492 \text{ ri}) \\
&= \ 3,926.88 \text{ metres, 3.9269 kilometres} \\
&\quad (4 \text{ kilometres} = 1.0186 \text{ ri})
\end{aligned}
$$

The term is sometimes translated in English as a 'Japanese league', and approximates quite closely to the metric league.

$$
\begin{aligned}
1 \text{ ri} \ &= \ 0.9817 \text{ French metric league} \\
&\quad (1 \text{ French metric league} = 4 \text{ kilometres} \\
&\quad\quad\quad\quad\quad\quad\quad\quad\quad\quad\quad = 1.0094 \text{ ri})
\end{aligned}
$$

See also LEAGUE.

rial [comparative values] Unit of currency in Iran, the Sultanate of Oman, and the Yemen.

In Iran,
$$1 \text{ rial} \ = \ 100 \text{ dinar}$$

In Oman,
$$1 \text{ rial saidi} \ = \ 1,000 \text{ baiza}$$

In the Yemen,
$$1 \text{ rial or riyal} \ = \ 100 \text{ fils}$$

The unit takes its name from the Spanish/Portuguese coins of the sixteenth to nineteenth centuries: *see* REAL, REALES; REIS. *See also* COINS AND CURRENCIES OF THE WORLD.

richesse of pine martens [collectives] This collective would seem to have been introduced into English by the Norman French, to whom the word meant 'wealth' (it was later apprehended as a plural, becoming *riches*). In fact, the term *marten*, too, is a Norman French word (although deriving ultimately from Old Germanic sources), and it is quite possible the animals were deliberately brought into England at the time of the Norman Conquest (AD 1066) and shortly after, when they were especially valued for their *rich* fur which lent an aspect of opulence to a wearer.

Richter scale [geology] A logarithmic scale for expressing the strength of shocks from an earthquake, corresponding to the varying seismic effects at the ground surface. In part it closely approximates to the MERCALLI SCALE, and may briefly be summarized:

2.0-2.9	perceived only by sensitive seismographic machines
3.0-3.9	slight vibration; hanging objects swing
4.0-4.9	vibration; crockery rattles; small objects displaced
5.0-5.9	furniture moves; masonry cracks and falls; waves on ponds
6.0-6.9	difficulty standing; walls and chimneys partly collapse
7.0-7.9	buildings collapse; cracks in ground; landslides
8.0-8.9	damage to underground structures; masses of rock displaced

The scale was named after the American seismologist Charles Richter (1900-85).

riding [square measure] An area or administrative region of a county, province, or state, especially one divided into three parts. The word was classically used of the three divisions of the county of Yorkshire in north-eastern England until the 1970s, but is used also of similar divisions of territory in Canada and New Zealand.

The fact that there were three parts of Yorkshire has great bearing on the word itself, for it derives as a 'third-ing' (just as a 'fourth-ing' was another form of measurement: *see* FJERDING; FARTHING; FIRKIN). The initial th- was lost by assimilation to the final consonant -t and -th in the preceding elements 'East', 'West', and 'North'.

riel [comparative values] Unit of currency in Cambodia.

$$1 \text{ riel} = 100 \text{ sen}$$

The term may or may not represent a borrowing of the name for a coin elsewhere spelled *rial, riyal, real,* and so forth. *See also* COINS AND CURRENCIES OF THE WORLD.

riff *see* PHRASE

rifle shooting *see* SHOOTING MEASUREMENTS AND UNITS

Riga last [cubic measure] Cubic unit in the measurement of timber.

1 Riga last	=	80 cubic feet of square-sawn timber
	=	62.304 UK bushels, 64.312 US bushels
	=	2.264 cubic metres
or	=	65 cubic feet of round timber
	=	50.622 UK bushels, 52.254 US bushels
	=	1.8395 cubic metre

The first element of the term is the name of the seaport capital of Latvia, Riga, formerly well known for the export of timber (notably deal and oak) and wood products. The second element is borrowed from the Germanic unit of weight and volume the *last*, but is slightly smaller in value (although the *last* varied according to the actual material it was the weight or volume of): *see* LAST.

right-angle [maths] An angle of 90 degrees, one-quarter of a complete revolution.

right-angle	=	90 degrees
	=	$\frac{1}{2}\pi$ radian, 1.570788 radian

Two right-angles correspond to a straight line; a line leading at right-angles off a straight line thus leaves it in the most straight or direct manner: the adjective *right* is etymologically akin to both *straight* and *direct*. (Compare also the adjectives *straight, right,* and *normal,* all of which can mean both 'ordinarily honest' and 'perpendicular'.)

right ascension [astronomy] For a celestial object, the angular distance eastwards from the vernal equinox to the hour circle that passes through the object, measured in hours, minutes, and seconds (1 hour = 15 degrees). It is the equivalent of longitude, and, together with declination (the equivalent of latitude), it specifies the position of an object on the CELESTIAL SPHERE.

right-hand rule [physics] Either of two rules:

If the thumb of a person's right hand points in the direction of flow of an electric current along a wire, the fingers curl round in the direction of the induced magnetic field (which is concentric with the wire). This is also called the *right-hand grip rule*.

If a person's right hand is held with the thumb, first finger, and second finger at right-angles, the thumb points in the direction of movement, the first finger in the direction of the magnetic field, and the second finger in the direction of the induced current, along a wire. The rule thus gives the relationship between the three parameters involved in electromagnetic induction. This rule is known alternatively as *Fleming's right-hand rule*, after the British physicist John Ambrose Fleming (1849-1945).

rigidity modulus [physics] For an elastic material, a measure of its resistance to shearing, equal to the ratio of the shearing force per unit area to the angular deformation. It is alternatively known as the *shear modulus*. *See also* ELASTICITY; POISSON'S RATIO.

rigsdaler, rijksdaalder, riksdaler *see* RIXDOLLAR, RIJKSDAALDER, RIKSDALER, RIGSDALER

rin [comparative values] Notional unit of currency in Japan, used only for taxation and similar purposes when decimal fractions of the actual coins of the realm are calculated in multiplication.

$$1 \text{ rin} = 0.1 \text{ (one-tenth of a) sen}$$

The term is sometimes translated into English as 'farthing', but originally applied to a real coin made of phosphor copper (or bronze, Japanese *rin*).

ring [chemistry] A closed chain of atoms linked by chemical bonds. It may be represented graphically in the form of a circular diagram.

ring [maths] A non-empty set (of elements) endowed with the operations of addition and multiplication, which is commutative under addition and associative under multiplication, with multiplication distributive over addition.

ring [quantitatives] In the numerical measurement of wooden boards or staves from which to make casks,

$$1 \text{ ring} = 4 \text{ SHOCKS} = 240 \text{ boards}$$

The term presumably derives from the metal ring put around the boards to keep them together until required for use.

ring, tree ring *see* DENDROCHRONOLOGY

Ringelmann smoke chart [physics] A method of estimating the density of smoke by comparing it with one of six cards shaded from white, through four tones of grey, to black (numbered 0 to 6).

ringgit [comparative values] Unit of currency in Malaysia.

$$1 \text{ ringgit (or}$$
$$\text{Malaysian dollar)} = 100 \text{ SEN}$$

See also COINS AND CURRENCIES OF THE WORLD.

ring of bells *see* PEAL

riometer [physics] A machine that monitors and records the intensity of residual radio noise and how much it is absorbed by the IONOSPHERE.

The first element of the term comprises the initial letters of relative ionospheric opacity.

Riss [time; geology] During the ice ages of the PLEISTOCENE EPOCH, the third episode of glaciation in Europe.

The name is that of an Alpine river in south-western Germany on which the glaciation appears to have centred.

ritardando [music] Musical instruction: gradually slow the tempo. Italian for 'holding back', 'delaying'. The instruction has virtually the same meaning as RALLENTANDO, but the reduction in speed should perhaps be more gradual.

ritenuto [music] Musical instruction: sing or play at a much reduced tempo. Italian for 'retained', 'held back'. The instruction calls for an immediately slower speed, whereas RITARDANDO and RALLENTANDO both require a gradual and consistent slowing.

rixdollar, rijksdaalder, riksdaler, rigsdaler [comparative values] The *rixdollar* is a North American variant spelling for the former unit of currency in northern

Germanic countries known there as a 'state's dollar' (Dutch *rijks-daalder*, Danish *rigs-daler*, Swedish and Icelandic *riks-daler* – the equivalent of a non-existent German *reichs-thaler*). In each case, the unit took the form of a silver coin, and, in the case of the Scandinavian currencies, it constituted multiples of the SKILLING. *See also* DOLLAR; REICHSMARK, REICHSPFENNIG.

riyal [comparative values] Unit of currency in Saudi Arabia and the Yemen.

In Saudi Arabia,

$$1 \text{ riyal} = 20 \text{ guersh or qurush} = 100 \text{ hallalas}$$

In the Yemen,

$$1 \text{ riyal or rial} = 100 \text{ fils}$$

The unit takes its name from the Spanish/Portuguese coins of the sixteenth to nineteenth centuries: *see* REAL, REALES; REIS. *See also* COINS AND CURRENCIES OF THE WORLD.

R meter *see* RÖNTGENOMETER, R METER

r.m.s. power [physics] Root-mean-square power, the mean power level of an AC supply of electricity: *see* ROOT MEAN SQUARE.

r.m.s. value [physics] For an alternating (cyclical) wave form, the root-mean-square value of the varying quantity (such as current or voltage), for a sine wave equal to the peak value divided by the square root of 2. *See also* ROOT MEAN SQUARE.

roaring forties [geography; meteorology] A zone across the southern Atlantic, the Pacific, and the Indian Oceans between latitudes 40° and 50° South, in which boisterous westerly winds are characteristic, causing a high ('roaring') sea swell.

In modern times, the expression has also been used of a zone in the Atlantic Ocean between similar latitudes north of the equator, although the winds and the high surge are far less common there.

Despite the attempts of some commentators to suggest that the verb *roar* is purely echoic (and thus that the roaring forties are characterized by continuously howling winds), other Germanic cognates are all to do with stirring and surging (German *ruhren*, Dutch *roeren* 'to stir'; cf. English *rear*).

Roche limit [astronomy] The lowest altitude at which a moon can orbit its parent body (below which destructive internal tidal forces cannot be resisted). For a moon of the same density as its parent planet, the limit is equal to a distance from the planet's centre of 2.44 times the planet's radius. Saturn's rings may well represent the fragments of a moon that penetrated within the Roche limit.

The limit was named after the French mathematician Édouard Roche, who calculated it in 1850.

Rockwell number [metallurgy; engineering] A scale of numbers representing the hardness of a metal. Under standard conditions of loading, a diamond cone is pressed into the metal being tested, and the Rockwell number is related to the depth of penetration. A letter following the number corresponds to the size of cone used. *See also* BRINELL SCALE (OF HARDNESS).

rod [linear measure] Unit of length that is the same as a pole (and, in the United States, a perch).

$$1 \text{ rod} = 16\tfrac{1}{2} \text{ feet, } 5\tfrac{1}{2} \text{ yards}$$
$$= 5.0292 \text{ metres}$$
$$(5 \text{ metres} = 0.9942 \text{ rod or pole})$$
$$4 \text{ rods} = 1 \text{ chain (22 yards)}$$
$$40 \text{ rods} = 1 \text{ furlong (220 yards)}$$
$$320 \text{ rods} = 1 \text{ mile}$$

The term derives from the use of a rod as a measuring tool.

rod, square [square measure] Unit of area known also as a square pole and equal to 1 square US (linear) perch.

$$1 \text{ square rod} = 16\tfrac{1}{2} \times 16\tfrac{1}{2} \text{ feet, } 30\tfrac{1}{4} \text{ square yards}$$
$$= 25.2929 \text{ square metres}$$
$$(25 \text{ square metres} = 0.9884 \text{ square rod})$$
$$16 \text{ square rods} = 1 \text{ square chain}$$
$$40 \text{ square rods} = 1 \text{ rood (1,210 square yards)}$$
$$160 \text{ square rods} = 1 \text{ acre (4,840 square yards)}$$

The term derives from the use of a rod as a measuring tool.

roede [linear measure] Obsolete unit of length in Holland now assimilated to a unit of the metric system.

$$1 \text{ roede} = 10 \text{ METRES, } 1 \text{ decametre}$$
$$= 10.96313 \text{ yards, } 32 \text{ feet } 10.67 \text{ inches}$$
$$100 \text{ roeden} = 1 \text{ kilometre}$$

This unit was originally probably akin to the now obsolete linear measure in England called the rood: *see* ROOD.

roentgen *see* RÖNTGEN

roentgenometer *see* RÖNTGENOMETER, R METER

rolling period [time: shipping] The time a ship takes to roll (from an upright position, to one side, then over to the other side, and back upright).

Roman calendar *see* IDES; NONES

Roman numerals The numerals used in ancient Rome. They were:

Roman	arabic	Roman	arabic	Roman	arabic	Roman	arabic
I	1	XIV	14	LX	60	CC	200
II	2	XV	15	LXX	70	CCC	300
III	3	XVI	16	LXXX	80	CD	400
IV	4	XVII	17	LXXXIX	89	CDXCVIII	498
V	5	XVIII	18	XC	90	ID	499
VI	6	XIX	19	IC	99	D	500
VII	7	XX	20	C	100	DIC	599
VIII	8	XXX	30	CX	110	DC	600
IX	9	XXXIX	39	CIL	149	CM	900
X	10	XL	40	CL	150	CMXCVIII	998
XI	11	IL	49	CXC	190	IM	999
XII	12	L	50	CXCVIII	198	M	1,000
XIII	13	LI	51	CIC	199	MM	2,000

The system is pleasantly logical and easy to grasp, but it has some severe drawbacks. It is a non-denominational number system – that is, the position of a digit within a number does not denote its numerical rank (in thousands, hundreds, tens or units) – lengthy numbers may be followed by brief numbers that are higher in value. And there is no zero or its equivalent. Both these factors make arithmetical calculation in Roman numerals extremely difficult.

For etymologies of the numerals *see* I; V, V; X; L; C; D; M.

rondeau, rondel, rondelet *see* VERSE FORMS

röntgen [physics] Unit of radiation, equal to the amount of ionizing radiation (for example, X-rays, gamma-rays) that will produce 2.58×10^{-4} coulomb of electric charge in 1 kilogram of dry air. Alternatively spelled *roentgen*, (the preferred form in some countries), it was named after the German physicist Wilhelm Röntgen (1845-1923).

röntgen equivalent man (rem) [physics] Obsolescent unit of radiation dose equal to the dose that will produce in a human being the same effect (on tissues) as 1 röntgen of X-rays or gamma-rays. It has been superseded by the effective dose equivalent (measured in sieverts): *see* RÖNTGEN; X-RAYS.

röntgen equivalent physical (rep) [physics] Obsolescent unit of radiation dose equal to the amount of ionizing radiation that transfers 93 ergs of energy to 1 gram of living tissue.

röntgenometer, R meter [physics] An instrument that monitors and measures the intensity of X-rays or gamma-rays.

röntgen rays [physics] Another name for X-rays.

rood [square measure] Obsolescent measure of area equal to one-quarter of an ACRE.

$$1 \text{ rood} = 40 \text{ square rods (or UK perches), } 2\frac{1}{2} \text{ square chains}$$
$$= 1,210 \text{ square yards, } 10,890 \text{ square feet}$$
$$= 1,011.75 \text{ square metres, } 0.101175 \text{ hectare}$$
$$(1 \text{ hectare} = 9.8839 \text{ roods})$$
$$4 \text{ roods} = 1 \text{ acre, } 10 \text{ square chains, } 4,840 \text{ square yards}$$
$$40 \text{ roods} = 1 \text{ square furlong}$$
$$2,560 \text{ roods} = 1 \text{ square mile}$$

In the United States, the term is occasionally used instead as a synonym for the square rod: *see* ROD, SQUARE. This usage is not surprising, for the etymological derivation of both *rood* and *rod* is the same – a pole (for measuring), a stick (and thus also a crucifix). And, in former centuries, the rood was additionally a linear measure (like the rod), but of variable length, between 6 and 8 yards (18 and 24 feet, 5.5 and 7.3 metres), slightly longer than a rod or pole.

root [maths] A number (specified in terms of another number) that, multiplied by itself a given number of times, equals the other number. For example, the square root of a number multiplied by itself equals the number (for example: the square root of 9 is 3; 3 x 3 = 9). *See also* CUBE ROOT; SQUARE ROOT.

root [literary; music] In grammar, another word for *stem* – a basic form of a word on to which prefixes and suffixes may be attached.

In etymology, an ancient form of a word (that may or may not ever have existed independently but) which is thought to have given rise to a set of linguistically related modern words.

In music, the note on which a chord is based – also known as the FUNDAMENTAL TONE.

By etymological derivation, a root represents a gro<u>w</u>th that <u>ra</u>diates <u>ra</u>dically.

root mean square (r.m.s.) [maths] For a set of numbers or quantities, the sum of the squares of individual values divided by the number of values in the set; it is a type of average. *See also* R.M.S. POWER; R.M.S. VALUE.

Rorschach test [medicine] In the psychological testing of a person's imagination, background knowledge, and general disposition, a test in which a subject is confronted in turn by ten 'inkblots' – standardized designs that are fairly amorphous but symmetrical left and right – and required to respond verbally to each one, suggesting possible interpretations and revealing associations. Popular as the test is in many countries, the accuracy and validity of the resultant analysis remain highly debatable.

The test is named after the Swiss psychiatrist Hermann Rorschach (1884-1922), who devised it.

rosary [quantitatives] In the Roman Catholic Church, a string of 165 beads divided into fifteen individual units of ten small and one large bead, used to assist a worshipper's memory (and to gauge progress) in the recitation of fifteen sets of ten Hail Marys (*Aves*), each set separated by a Lord's Prayer (*Paternoster*) and a doxology (*Gloria*).

There is a smaller rosary of 55 beads (the lesser rosary) used similarly in the recitation of five sets.

rose noble [comparative values] Gold coin in use in England in the time of Edward IV (ruled 1461-83) and for a few decades thereafter, also called a RYAL. Unlike the noble issued by Edward III in the previous century, it was worth 10 shillings (or half of one pound). It was called the rose noble because there was a depiction of a rose on one side of the coin: Edward IV was the first Yorkist King of England – it was his York rose on the coin, and he came to the throne by conquest during the Wars of the Roses.

rotation, speed of [physics] For a rotating object, the number of rotations in unit time, expressed in such units as revolutions per minute or radians per second. It is measured using a TACHOMETER or rev(olution) counter.

rotation speed [aeronautics] For an aircraft taking off, the speed at which the nosewheel leaves the ground (before the aircraft is 'rotated' into its initial climb).

rotl *see* RAT(T)EL, RATTLE, ARRATEL, ROTL, ROT(T)OL

roton [physics] According to quantum theory, rotational energy can be considered to be quantized (to exist in discrete levels of energy), and the quantum of such energy is called a roton. *See also* PHONON.

rottol *see* RAT(T)EL, RATTLE, ARRATEL, ROTL, ROT(T)OL

rouble, ruble [comparative values] Unit of currency in Russia and Byelorussia.

$$1 \text{ r(o)uble} = 100 \text{ kopecks}$$

The name of the unit seems to relate to an ancient root meaning '(metal made into) coinage', and thus to be akin to RUPEE. *See also* COINS AND CURRENCIES OF THE WORLD.

rounders measurements, units, and positions [sport] Rounders is an original form of baseball mostly played by nine-member teams of schoolchildren (and especially schoolgirls) in England. As in baseball, a batter attempts to hit a ball into the field but away from fielders, and then to run to and past as many bases (in this game called 'posts') as possible, up to the fourth post.

The dimensions of the pitch:

distance from batting square to 1st post, 1st post to 2nd post, and 2nd post to 3rd post: 12 metres (39 feet 3 inches)

distance from 3rd post to 4th post: 8.5 metres (27 feet 8 inches)

distance from front of bowling square to front of batting square: 7.5 metres (24 feet 5 inches)

area of bowling square: 2.5 x 2.5 metres (8 feet 2 inch square)

area of batting square: 2 x 2 metres (6 feet 6 inch square)

Positions of fielding players:

bowler

back stop (behind batter)

one fielder at each post (base)

the remaining three in deep field positions

Timing: the game lasts as long as it takes for both teams to play two innings (unless the game is won before the second team's second innings is completed)

Points scoring:

having hit the ball, a completed run around all four posts: 1 rounder

not having hit the ball, a completed run around all four posts: ½ rounder

a penalty ½ rounder is scored for various bowling or fielding faults

the team with the greater number of rounders at the end of the game wins

Dimensions of the equipment:

the bat or stick:

maximum length: 46 centimetres (1 foot 6 inches)

maximum thickness: 17 centimetres (7 inches)

maximum weight: 370 grams (13 ounces)

the ball:

circumference: about 19 centimetres (7½ inches)

weight: 70-85 grams (2½-3 ounces)

The game is named after the fact that to complete a run entails making a full circuit around the posts.

rounding off, rounding up, rounding down [maths] A method of approximating numbers, usually by reducing the number of non-zero digits. A number or quantity may be rounded off to, say, the nearest whole number, or the nearest metre, etc. For example: rounding off 45.78 to the nearest whole number gives 46; 23.2 metres rounds off to 23 metres (to the nearest metre). Or a number may be rounded off to, say, four decimal places; for example the value of π, 3.1415926535 . . ., rounded off to four places is 3.1416 . Numbers may also be rounded up (to the nearest higher significant number) or rounded down (to the nearest lower significant number).

round number [quantitatives] In general parlance, a number ending in one or more zeros. *See also* ROUNDING OFF, ROUNDING UP, ROUNDING DOWN.

rowing disciplines and measurements [sport] There are two principal methods of racing in rowing: one in knock-out heats and finals that together comprise a regatta, and the other in the form of a time trial, where boats begin serially and their times over the course are crucial. In either case, the usual distances for races are:

for men: 2,000 metres (2,187.23 yards)

for young men and boys: 1,500 metres (1,640.42 yards)

for women: 1,000 metres (1,093.61 yards)

and the events are for:

single sculls, one rower, two oars/sculls, typical boat length around 27 feet (8.2 metres)

double sculls, two rowers, two oars/sculls each, typical boat length around 34 feet (10.4 metres)

coxless pairs, two rowers, one oar each, typical boat length around 34 feet (10.4 metres)

coxed pairs, two rowers, one oar each, one 'passenger' who may also steer and who encourages vociferously, typical boat length around 35 feet (10.7 metres) *coxless fours*, four rowers, one oar each, typical boat length around 44 feet (13.4 metres)

coxed fours, four rowers, one oar each, one 'passenger' who may also steer and who encourages vociferously, typical boat length around 45 feet (13.7 metres)

eights, eight rowers, one oar each, one 'passenger' who may also steer and who encourages vociferously, typical boat length around 62 feet (18.9 metres)

Within each class of boat, competitors are categorized by experience, weight, or age; previous form (by way of competition results) and overall expertise are also taken into account. Losers in some regatta heats may yet go through to the finals on a repechage. *See also* CANOEING DISCIPLINES AND MEASUREMENTS.

rowland [physics] Obsolete unit of wavelength approximately equal to 0.1 nanometres, superseded first by the angstrom unit and then by the nanometre. It was named after the American physicist Henry Rowland (1848-1901).

royal *see* PAPER SIZES

royal cubit *see* CUBIT

royal flush *see* POKER SCORES

royal foot *see* PIED-DE-ROI

rpm, r.p.m. *see* REVOLUTIONS PER MINUTE (RPM)

ruba'i, rubaiyat [quantitatives; literary] The word *ruba'i* is Arabic for 'foursome', 'quartet' and therefore 'a verse of four lines', 'quatrain'; *rubaiyat* is a plural form, applying thus to a collection of quatrains.

rubato [music] Musical instruction: sing or play with such expression as to ignore the time values of the notation, shortening or lengthening the notes at will. The term derives from the Italian expression *tempo rubato* 'appropriated time'.

rubber [sporting term] In contract bridge and certain other card games, in backgammon, and in bowls, a set of three or five games. Formerly, the term was used in this sense only of two games out of three, or three games out of five, that had been won by one side. Earlier still, the term was used for the final game of three or five when both sides had won an equal number of games up to that point – the decider was the 'rubber' as the two sides 'rubbed' against each other to win.

rubidium-strontium dating [geology] A method of determining the age of rocks based on measuring the amount of radioactive rubidium in them (rubidium-87), which decays to strontium-87, of which the rate of decay (half-life) is known. *See also* POTASSIUM-ARGON DATING.

ruble, rouble *see* ROUBLE, RUBLE

rugby league football measurements, units, and positions [sport] Rugby league is an exciting team game for professional players, and is played mainly in England, Wales, Australia, and New Zealand. There are thirteen members of each team on the field at any one time, and the object of the game is to amass as many points as possible by scoring tries (and kicking subsequent conversions) and by kicking goals.

The dimensions of the field:

    field length (tryline to tryline): 110 yards (100 metres)

    field width: 75 yards (68.62 metres)

    goal area (behind goal posts, in which tries may be touched down): 12 yards (10.98 metres), to dead ball line

    the posts:

        width: 18 feet 6 inches (5.64 metres)

        crossbar height: 10 feet (3.048 metres)

Positions of players: from the back

    fullback

    right wing/left wing

    right centre/left centre

    stand-off

    scrum-half

    tight flanker/loose flanker

    number eight

    props (2)

    hooker

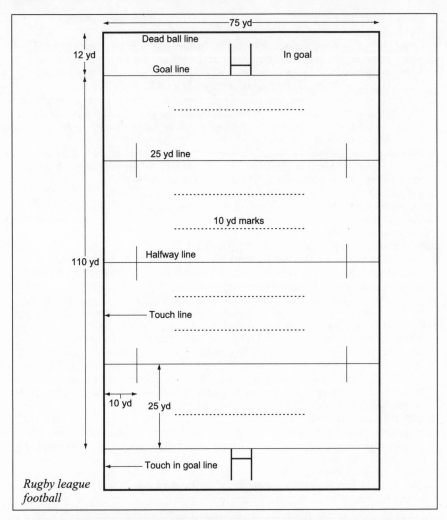

Rugby league football

Timing: 80 minutes in two 40-minute halves, separated by an interval of up to 5
   minutes (or, if the game is televised, up to 15 minutes)
Points scoring:
   a try: 4 points
   a penalty goal: 2 points
   a drop-goal during play or a conversion after a try: 1 point
      the team with the greater number of points when the game ends wins
Dimensions of the ball:
   overall length: 10¾ inches (27.3 centimetres)
   weight: 13½-14½ ounces (382.72-411.07 grams)

rugby union football measurements, units, and positions [sport] Rugby
union is an exciting team game for amateur players, and is played in many countries
of Europe, Oceania, southern Africa, and South America, in Canada, and in Japan.
There are fifteen members of each team on the field at any one time, and the object
of the game is to amass as many points as possible by scoring tries (and kicking
subsequent conversions) and by kicking goals.
   The dimensions of the field:
      field length (tryline to tryline): 110 yards (100 metres)
      field width: 75 yards (68.62 metres)
      goal area (behind goal posts, in which tries may be touched down): 25 yards
         (22.87 metres), to dead ball line

the posts:
    width: 18 feet 6 inches (5.64 metres)
    crossbar height: 10 feet (3.048 metres)
Positions of players: from the back
    fullback
    right wing/left wing
    right centre/left centre
    stand-off
    scrum-half
    tight flanker/loose flanker
    loose forwards (2)
    number eight
    props (2)
    hooker

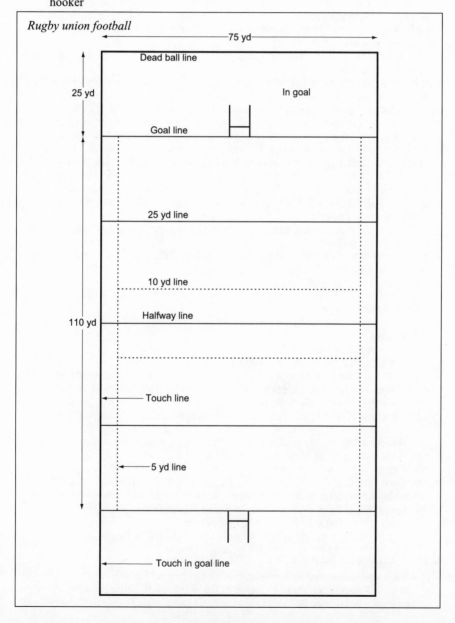

Rugby union football

Timing: 80 minutes in two 40-minute halves, separated by an interval of up to 5 minutes

Points scoring:

a try: 5 points

a penalty goal: 3 points

a drop-goal during play: 3 points

a conversion after a try: 2 points

the team with the greater number of points when the game ends wins

Dimensions of the ball:

overall length: 11 inches (28 centimetres)

weight: 13½-15½ ounces (382.72-439.42 grams)

This version of the game prides itself on being less evolved (and thus closer to the original game) than rugby league which, as a movement, definitively split from rugby union early in the 1920s. By tradition, however, the game of rugby started as a result of an incident during a game of soccer at the English public school in Rugby, Staffordshire, during the 1860s, when a player actually picked the ball up and ran with it.

rule of Fajans *see* FAJANS' RULES

ruling gradient, ruling grade [engineering] For a road or railway, the maximum permissible gradient.

rummer [volumetric measure] Another name for a large, tall goblet, or for a highball glass: *see* DRINKING-GLASS MEASURES.

run *see* CRICKET MEASUREMENTS, UNITS, AND POSITIONS

rundlet, runlet [volumetric measure] Obsolete form of cask or tun that varied in size according to the nature of the contents (as the present-day barrel still does), and even then varied according to the location. As a measure of alcoholic drink, for example, a rundlet could contain as little as 12 UK gallons, 14.4 US gallons (54.552 litres) or as much as 18.5 UK gallons, 22.2 US gallons (84.100 litres). As a measure of gunpowder, a rundlet was most often equal to 15 UK gallons, 18 US gallons (68.2445 litres, 2.4075 cubic feet, 1.875 UK bushel, 2.250 US bushels). Miniature rundlets had a capacity of only a pint.

The term derives as a diminutive of the Old French *rondelle*, itself a diminutive of *ronde* '(round) barrel'.

rupee [comparative values] Unit of currency in India, Pakistan, Sri Lanka, Bhutan, Nepal, and Mauritius.

In India, Pakistan, and Nepal,

1 rupee = 100 paisa

In Sri Lanka and Mauritius,

1 rupee = 100 cents

In Bhutan,

1 rupee is equal to

1 ngultrum/tikehung = 100 chetrum

The name of the unit derives from a very ancient root (the Sanskrit variant is *rupya*) meaning '(metal made into) coinage', and thus Hindi *rupiah* 'silver'. *See also* COINS AND CURRENCIES OF THE WORLD.

rupiah [comparative values] Unit of currency in Indonesia.

1 rupiah = 100 sen

The name of the unit is evidently borrowed from that of the RUPEE current in various countries to the north-west (*see above*). *See also* COINS AND CURRENCIES OF THE WORLD.

rutherford [physics] Obsolete unit of radioactivity equal to the mass of radioactive substance that undergoes a million disintegrations per second. It was named after the New Zealand-born British physicist Ernest Rutherford (1871-1937).

1 rutherford = 0.027 millicurie

(1 millicurie = 37.037 rutherfords)

= 10^6 becquerel (SI units)

ryal [comparative values] Another name for a ROSE NOBLE.

Also, a coin once used in Scotland, made of either gold or silver.

The coin's name represents a version of the word *royal*, either because it was issued by the royal mint or as a borrowing from the contemporaneous Spanish and

Portuguese REAL.

rydberg [physics] Obsolete unit of wave number equal to the number of wavelengths per centimetre, superseded by the equally obsolete kayser (having the same value). In SI units, the wave number is the number of wavelengths per metre (that is, the reciprocal of the wavelength expressed in metres). It was named after the Swedish physicist Johannes Rydberg (1854-1919).

Rydberg constant [physics] A universal physical constant, originally related to the wave number of lines in an atomic spectrum, equal to 1.09677×10^7 per metre. It was named after the Swedish physicist Johannes Rydberg (1854-1919).

S

s., sec *see* SECOND

/s², /sec² *see* PER SECOND PER SECOND

sabin [physics] Obsolete unit of sound absorption (by architectural materials), equal to the absorption by an open window of 1 square foot area (0.0929 square metre) for low-frequency sound waves. (The absorption of an open window was assumed to be total, or perfect.)

The unit was named after the American physicist Wallace Sabine (1868-1919).

Sabine reverberation formula [physics] For an enclosure, the reverberation time of a sound equals 0.16 times the volume (in cubic metres) divided by the total absorption in the enclosure (that is, the product of the absorption coefficient and the surface area in square metres).

The formula was devised by the American physicist Wallace Sabine (1868-1919).

saccharimetry, saccharometry [chemistry] The determination of the sugar concentration of a solution (usually as a percentage) by measuring its optical activity or density, using a polarimeter or HYDROMETER.

sadzhen, sagène [linear measure] In Russia, an old unit of distance not normally now used.

1 sadzhen	=	3 ARSHIN, 48 verchoki
	=	2.1336 metres
		(2 metres = 0.9374 sadzhen)
	=	84 inches (exactly), 7 feet (exactly)
500 sadzhen'	=	1 VERST or versta

SAE numbers [physics] Method of classifying lubricating oils based on their viscosity, named after the American Society of Automotive Engineers. Light oil is SAE 10; heavy oil is SAE 40.

safety *see* AMERICAN FOOTBALL MEASUREMENTS, UNITS, AND POSITIONS

sagène *see* SADZHEN, SAGÈNE

salient angle [maths] For a closed figure (polygon), an angle of less than 180 degrees, which points outwards. *See also* RE-ENTRANT, RE-ENTERING.

salinometry [physics] The measurement of the concentration of salt (sodium chloride) in a solution – such as sea water – using a salinometer (a type of HYDROMETER, also occasionally called a salimeter or salometer).

salmanazar [volumetric measure] Large bottle (containing alcoholic drink, especially wine, especially champagne) with a capacity of about 300 UK fluid ounces (15 UK pints), 312 US fluid ounces (18 US pints; 8.52 litres).

Alternatively, but amounting to very much the same thing, reckoning on a standard wine bottle containing 75 centilitres (0.75 litre),

1 magnum	=	the contents of 2 bottles
1 jeroboam	=	4 bottles
1 rehoboam	=	6 bottles
1 methuselah	=	8 bottles
1 salmanazar	=	12 bottles
1 balthazar	=	16 bottles
1 nebuchadnezzar	=	20 bottles

Some of these terms relate more specifically to a net weight of the drink in fluid ounces, although the approximation to the overall volume by contents of bottles (as listed above) is remarkably accurate.

The salmanazar is named after the Biblical King of Babylon Shalmaneser, properly Shulman-Ashered V (ruled 727-22 BC), son of the Assyrian ruler Tiglath-Pileser III, and captor of the nineteenth and last King of Israel, Hoshea. His name appears to mean (in Akkadian) '(the god) Shulman (is) dominant'.

saltus [square measure] Largest unit measure of area in the ancient Roman system of measurement based fundamentally on the ACTUS quadratus.

$$
\begin{aligned}
1 \text{ saltus} \quad &= \quad 800 \text{ jugera, } 400 \text{ heredia, } 4 \text{ centuriae} \\
&= \quad 23,040,000 \text{ square 'feet' } (pedes) \\
&= \quad 202.84 \text{ hectares, } 2.0284 \text{ square kilometres} \\
&= \quad 501.0909 \text{ acres, } 0.7829 \text{ square mile}
\end{aligned}
$$

As a unit, it was used to define the area of large tracts of wild woodlands – and this accounts for its name, which is the Latin for 'forest-land'. *See also* HEREDIUM; JUGERUM.

sandglass *see* HOURGLASS, SANDGLASS, EGG-TIMER

Sangamon [time; geology] Describing the period of time regarded as the third interglacial stage of the PLEISTOCENE EPOCH in North America.

The term derives from the name of a river in Illinois.

sans-culottides *see* FRENCH REVOLUTIONARY (REPUBLICAN) CALENDAR

saponification number [chemistry] Saponification is the process of treating a fat or oil (chemically an ester) with an alkali to convert it into glycerol and a soap. The saponification number is the amount of potassium hydroxide (in milligrams) needed to saponify 1 gram of fat or oil.

saros [astronomy] A period of 223 synodic months, equal to a cycle of 18 years 11 days and 8 hours, or 19 eclipse years, used since ancient times for predicting eclipses. After each saros the relative positions of the Sun, Moon, and NODE are the same.

The term derives in English directly from an ancient Greek transliteration of a word in Assyro-Babylonian for the mystic number 3,600 (corresponding to the Babylonian base 60 times itself). Although the saros cycle has certainly been known to astronomers since the age of the Babylonians, however, that name has been applied to it only in modern times.

satang [comparative values] Unit of currency in Thailand, in the form of a bronze coin.

$$100 \text{ satang} \quad = \quad 1 \text{ baht}$$

The term is a variant on the Thai word for 'hundredfold'. *See also* COINS AND CURRENCIES OF THE WORLD.

satellites of the Solar System [astronomy] The natural satellites of the planets of the Solar System (now often called moons to distinguish them from the artificial satellites sent into planetary and lunar orbits since 1957) are:

name	disco-vered	mean sidereal period – days	mean distance from planet – thousand km	miles	dimensions/ diameter km	miles	density (water = 1)
EARTH							
Moon		27.322	384.40	230.60	3,476	2,085	3.33
MARS							
Phobos	1877	0.319	9.35	5.61	27 x 21 x 19	16 x 12 x 11	2.00
Deimos	1877	1.262	23.49	14.09	15 x 12 x 11	9 x 7 x 6.5	2.00
JUPITER							
Metis	XVI 1979	0.294	127.6	76.56	40	24	?
Adrastea	XV 1979	0.297	128.4	77.04	40	24	?
Amalthea	V 1892	0.498	181.0	108.60	270 x 170 x 150	162 x 102 x 90	?
Thebea	XIV 1979	0.678	222.4	133.44	80	48	?

name		disco-vered	mean sidereal period – days	mean distance from planet – thousand km	mean distance from planet – thousand miles	dimensions/ diameter km	dimensions/ diameter miles	density (water = 1)
Io	I	1610	1.769	421.6	252.96	3,632	2,179	3.53
Europa	II	1610	3.551	670.9	402.54	3,126	1,876	3.03
Ganymede	III	1610	7.155	1,070.0	642.00	5,276	3,166	1.93
Callisto	IV	1610	16.69	1,880	1,128	4,820	2,892	1.79
Leda	XIII	1974	239	11,094	6,656	10	6	?
Himalia	VI	1904	251	11,480	6,888	180	108	?
Lysithea	X	1938	259	11,720	7,032	20	12	?
Elara	VII	1905	260	11,737	7,042	80	48	?
Ananke	XII	1951	631*	21,200	12,720	20	12	?
Carme	XI	1938	692*	22,600	13,560	30	18	?
Pasiphaë	VIII	1908	735*	23,500	14,100	40	24	?
Sinope	IX	1914	758*	23,725	14,235	10	6	?
SATURN								
Atlas	XV	1980	0.602	137.7	82.62	20 x 40	12 x 24	?
Prometheus	XVI	1980	0.613	139.4	83.64	140 x 100 x 80	84 x 60 x 48	?
Pandorus	XVII	1980	0.628	141.7	85.02	110 x 90 x 70	66 x 54 x 42	?
Epimetheus	XI	1980	0.694	151.4	90.84	140 x 120 x 100	84 x 72 x 60	?
Janus	X	1966	0.695	151.5	90.90	220 x 200 x 160	168 x 120 x 96	?
Mimas	I	1789	0.942	188.2	112.92	390	234	1.20
Enceladus	II	1789	1.370	240.2	144.12	510	306	1.20
Calypso	XIV	1980	1.888	294.6	176.76	34 x 22 x 22	20 x 13 x 13	?
Tethys	III	1684	1.888	294.7	176.82	1,060	636	1.20
Telesto	XIII	1980	1.888	294.7	176.82	34 x 28 x 26	20 x 17 x 16	?
Dione	IV	1684	2.737	377.4	226.44	1,120	672	1.40
Helena	XII	1980	2.737	377.4	226.44	36 x 32 x 30	22 x 19 x 18	?
Rhea	V	1672	4.517	527.1	316.26	1,530	918	1.20
Titan	VI	1655	15.95	1,221	733.14	5,150	3,090	1.90
Hyperion	VII	1848	21.28	1,481	888.60	410 x 260 x 220	246 x 156 x 132	?
Iapetus	VIII	1671	79.33	3,561	2,136	1,460	876	1.20
Phoebe	IX	1898	550*	12,954	7,772	220	132	?
URANUS								
Cordelia	VI	1986	0.33	49.7	29.82	40	24	?
Ophelia	VII	1986	0.37	53.8	32.28	50	30	?
Bianca	VIII	1986	0.43	59.2	35.52	50	30	?
Cressida	IX	1986	0.46	61.8	37.08	60	36	?
Desdemona	X	1986	0.47	62.7	37.62	60	36	?
Juliet	XI	1986	0.49	64.6	38.76	80	48	?
Portia	XII	1986	0.51	66.1	39.66	80	48	?
Rosalind	XIII	1986	0.56	69.9	41.94	60	36	?
Belinda	XIV	1986	0.62	75.3	45.18	60	36	?
Puck	XV	1986	0.76	86.0	51.60	170	102	?
Miranda	V	1948	1.41*	129.9	77.94	480	288	1.30

name		disco-vered	mean sidereal period – days	mean distance from planet – thousand		dimensions/ diameter		density (water = 1)
				km	miles	km	miles	
Ariel	I	1851	2.52*	190.9	114.54	1,160	696	1.60
Umbriel	II	1851	4.14*	266.0	159.60	1,190	714	1.40
Titania	III	1787	8.71*	436.3	261.78	1,580	948	1.60
Oberon	IV	1787	13.46*	583.4	350.04	1,526	916	1.50
NEPTUNE								
N6 (Naiad)		1989	0.29	48.0	28.8	50	30	?
N5 (Thalassa)		1989	0.31	50.0	30.0	80	48	?
N3 (Despina)		1989	0.34	52.5	31.5	180	106	?
N4 (Galatea)		1989	0.43	62.0	37.2	150	90	?
N2 (Larissa)		1989	0.56	73.6	44.2	190	114	?
N1 (Protea)		1989	1.12	117.6	70.6	400	240	?
Triton	I	1846	5.88*	354.8	212.9	2,705	1,623	2.05
Nereid	II	1949	365.20	5,513.4	3,308.0	340	204	?
PLUTO								
Charon	I	1978	6.3867	19.7	11.82	1,200	720	?

* retrograde motion

saton *see* SEAH

Saturn: planetary statistics *see* PLANETS OF THE SOLAR SYSTEM

sau, xu [comparative values] Unit of currency in Vietnam.

100 sau or xu = 10 hao = 1 dong

The term is a borrowing from the French *sou* (the French were formerly the colonial power of the area). *See also* COINS AND CURRENCIES OF THE WORLD.

savart [music] A mathematically derived unit of frequency formerly used to quantify musical scales. An octave was divided into 301 savart (often simplified to 300 per octave). The unit was named after the French physicist F. Savart (1791-1814).

saxophone *see* WOODWIND INSTRUMENTS' RANGE

Say's law [comparative values] The theory that, in a system of free commercial enterprise, supply increases or decreases in direct proportion to demand, even as demand increases or decreases in direct proportion to supply, with the result that total demand and total supply are equal.

The 'law' is named after the French economist Jean-Baptiste Say (1767-1832), who devised it.

S-band [physics; telecommunications] In ultra-high-frequency radio, a band of frequencies between 1,550 and 5,200 megahertz (1.55×10^9 hertz and 5.2×10^9 hertz, corresponding to wavelengths between about 20 centimetres and 6 centimetres).

scalage [cubic measure; quantitatives] An overall estimate of the cubic measure of usable wood in a tree or stand of trees, expressed in BOARD FEET, and sometimes alternatively called a *scale*.

In more general terms, also an amount (in weight purveyed or price charged) deducted for normal wastage in any consumer goods.

scalar [maths] A quantity that has magnitude only – that is, an ordinary number (unlike a vector, which has direction as well as magnitude).

scalar product [maths] *see* DOT PRODUCT

scale [maths] A set of graduations and figures that calibrate a measuring device such as a ruler or thermometer.

On a map or technical drawing, a graduated bar-line or set of figures that indicates the proportion between the map/drawing and the actual size of the area being represented. It is expressed, for example, as 1 inch to the mile or 1 centimetre to the kilometre: *see* SCALE FACTOR.

An exact model of something much larger in real life may be described as a *scaled-down* version (for example, a model railway); a product made in the likeness of a much smaller three-dimensional model, on which testing can be carried out, may be described as a *scaled-up* version (like a prototype aircraft).

The term derives in all these senses from the Latin *scalae* 'steps', in reference to the graduations and calibrations involved.

scale [music] The notes that make up an octave from keynote to keynote in a key:
see KEY, KEYNOTE. In the key of C, for example, the keynote (tonic) is C and the
major scale represents all the white notes on a piano between one C and another. In
the key of D, the keynote (tonic) is D and the major scale represents all the notes in
the same sequence of tones and semitones as the key of C but transferred upward by
one tone (and therefore involving two black notes, F# and C#).

There are at least two useful ways of describing a scale in any key, however. One
is to use the technical musical terms for each note in a scale, and another is to use
the 'names' of the notes in what is called the *tonic sol-fa*.

Using the scale of C for reference only (any key would do),

	technical terms	*tonic sol-fa*
C	tonic (keynote)	doh, do, ut
D	supertonic	re
E	mediant	me, mi
F	subdominant	fa
G	dominant	soh, so
A	submediant	lah, la
B	leading-note	te
C	tonic (keynote)	doh, do, ut

In the technical terms, the tonic is so called because it is the tone (or keynote) of
the key; the supertonic is the note above the tonic; the mediant is in the middle of
the three-note key chord, between the tonic and dominant; the subdominant is the
note below the dominant; the dominant is the next most significant note of the key
after the tonic; the submediant is in the middle of the three-note chord formed with
the subdominant and tonic; and the leading-note leads up to the higher tonic.

The names of the notes in the tonic sol-fa derive from the first syllables of the
lines in a medieval hymn in Church Latin (written by Guido d'Arezzo and addressed to John the Baptist): Ut queant laxis/resonare fibris,/Mira gestorum/famuli
tuorum,/Solve polluti/labii reatum,/Sancte/ Johannes.

It was apparently Guido d'Arezzo himself (died 1050) who used these 'Aretinian
syllables' as the names for the notes. As is evident, what is now known as 'te' was
once 'sa', and it was from 'jo' that we now have the form 'doh' or 'do'. Use of the
tonic sol-fa is also known as *solmization* or *solfeggio*.

The scale in minor keys is rather more complex. A minor key differs from a
major key in that the scale (most often) has a flattened mediant and submediant. In
the so-called harmonic scale of C minor, for example, instead of corresponding to
all the white notes on a piano between one C and another (as in the key of C major),
there is E♭ instead of E and A♭ instead of A. The result is a scale that somehow
sounds 'sad' and lugubrious in contrast to the major.

The minor key that is most associated with the major key of C is not C minor but
A minor. This is because the natural scale of A minor is also based on what on the
piano are the white notes, and because, in the harmonization of even simple
melodies in C major, chords of A minor may well be appropriate. A minor is
therefore called the RELATIVE MINOR of C major.

The word scale derives from the Latin *scalae* 'steps'.

scale factor [maths] The ratio between the length of a distance on a map or plan and
the actual distance it represents: the reduction (or enlargement) of the real area that
is mapped: *see also* RATIO; SCALE.

scalene [maths] Of a triangle, having sides all of different lengths.

Of a solid figure (such as a cone), having an axis that is (not perpendicular to but)
at an angle to the base.

The term derives through Latin from the ancient Greek *skalenos* 'uneven'
(literally, 'limping').

scalogram [medicine: psychology] A series of problems or questions of gradually
increasing complexity, designed to measure the time of response and the consist-
ency of answers, as part of a psychological examination.

scan, scansion [literary] For a verse of poetry to scan, each line must consist of a
specified number of regular feet: *see* FOOT. A syllable too few, or too many, and the
line does not scan.

To *scan* poetry is also to mark off the feet within each line, and the stressed and unstressed syllable or syllables within each foot; the result is the poem's *scansion*.

scanning frequency [telecommunications] The number of times an area (such as a television or radar picture) is scanned per second.

scanning systems [medicine] Various forms of energy or radiation can be used to scan body tissues as an aid to diagnosis or to monitor treatment. In computerized axial tomography (CAT) or computerized tomography (CT), a narrow beam of X-rays is aimed at a part of the body from various angles. The strengths of rays reaching detectors are processed by a computer to produce an image (on a VDU screen) of a thin 'slice' through the body. A nuclear magnetic resonance (NMR) or magnetic resonance (MR) scanner detects changes in the magnetic properties of hydrogen atoms in the body when they are exposed to high-frequency electromagnetic (radio) energy. Again a computer builds up a picture of the tissues so scanned. Positron emission tomography (PET) relies on the detection of photons emitted by positively charged electrons (positrons) within substances that have been 'labelled' with them and introduced to the body, where they concentrate in areas of tissue that are more metabolically active. In ultrasonic scanning, commonly used for non-invasive examination of a foetus, a transducer detects echoes of ultrasonic sound waves and a computer displays the resulting sound 'picture'. Similar echograms are used to examine the heart and major blood vessels.

scansion *see* SCAN, SCANSION

scantling [linear measure; volumetric measure] In the building and ship-building industries, a general term for the overall dimensions of certain regular components used in large numbers for construction.

The term derives with fine irony from Old French *escantillon* 'splinter'.

scheffel [volumetric measure] Former measure of dry capacity in Germany, in modern times applied to a unit in the metric system in both liquid and dry measure.

1 scheffel	=	50 litres
	=	11 UK gallons (exactly), 13.209 US gallons
	=	1.3748 UK bushel, 1.4189 US bushel
		(1 UK bushel = 0.7274 scheffel
		1 US bushel = 0.7048 scheffel)
	=	1.765 cubic foot

The usual translation of *scheffel* in English is 'bushel' – even to the extent of the Biblical expression suggesting that one should not hide one's light under it. The same is true of the Dutch *schepel*, the Danish SKAEPPE and the Norwegian skjeppe – all three of which must be etymologically identical – and none of which is anything like as close in value to a UK/US bushel as is the German *scheffel* (which is not very).

schepel [volumetric measure] Former measure of dry capacity in Holland, in modern times applied to a unit in the metric system in both liquid and dry measure.

1 schepel	=	10 litres, 1 decalitre
	=	2.2 UK gallons, 2.64179 US gallons
		(2.5 UK gallons = 1.13636 schepel
		2.5 US gallons = 0.94633 schepel)
	=	1.0999 UK peck, 1.135 US peck
		(1 UK peck = 0.9092 schepel
		1 US peck = 0.8811 schepel)
	=	0.27495 UK bushel, 0.28378 US bushel

See also SCHEFFEL.

schilling [comparative values] Unit of currency in Austria, in the form of a cupro-nickel coin.

1 schilling	=	100 groschen

There was formerly a unit of currency of this name in Germany and in Britain (the latter with a slightly different spelling), each worth 12 lesser units. The name of the unit probably derives from the shield that was depicted on very early examples (as a 'shield-ing'), although some etymologists suggest an echoic derivation (cf. German *Schall* 'resonance'; semantically cf. also PENGÖ). *See also* COINS AND CURRENCIES OF THE WORLD.

Schmidt number [physics]A dimensionless number that describes flow in a fluid, equal to the ratio of the coefficient of viscosity to the product of the fluid's density and diffusion coefficient.

school figure [sporting term] In ice-skating competitions, any of the set figure exercises to be skated as perfectly as possible.

school of fish, of marine mammals [collectives] Both *school* and *shoal* derive in this sense from a Germanic verb meaning 'to distinguish', 'to tell apart', thus implying a corporate number of individual animals. The ability 'to distinguish' in this way is represented by the cognate English word *skill*.

school tuna [biology; weight] A tuna (fish) weighing between 20 and 100 pounds (9.07 and 45.36 kilograms) when caught.

schooner [volumetric measure] In Britain, a short-stemmed concave-sided wine glass in which sherry is served: *see* DRINKING-GLASS MEASURES.

In North America and Australasia, a large cylindrical tumbler (a sort of highball glass) containing about 13½ (US) fluid ounces, just over three-quarters of 1 (US) pint, in which beer is customarily served.

Why this name is used, for two different types of glass, is not known.

schoppen [volumetric measure] Former unit of liquid capacity in Germany and Switzerland. The German measure has more recently been assimilated to metric units.

In Switzerland,

1 schoppen	=	37.5 centilitres, 0.375 litre
		(1 litre = 2.6667 schoppen)
	=	half a standard wine bottle
	=	0.660 UK pint, 0.7925 US pint
		(1 UK pint = 1.5152 schoppen
		1 US pint = 1.2618 schoppen)
4 schoppen	=	1 immi, maass or (Swiss) pot
40 schoppen	=	1 (Swiss) viertel

In Germany,

1 schoppen	=	50 centilitres, 0.500 litre
	=	0.880 UK pint, 1.0567 US pints
		(1 UK pint = 1.13636 schoppen
		1 US pint = 0.94634 schoppen)
2 schoppen	=	1 litre

In many German-English dictionaries, the term is translated as 'half a pint' – which is badly inaccurate in both cases, worse in the German.

Schrödinger (wave) equation [physics] An equation that describes an electron (in an atom) as a three-dimensional stationary wave, giving the location of the electron in terms of the probability of its being in a particular location (probability is the square of the amplitude of the wave function). It also shows that the electron can occupy certain discrete energy levels – that is, that energy is quantized (*see* QUANTUM). It was named after the Austrian physicist Erwin Schrödinger (1887-1961). *See also* WAVE EQUATION.

schtoff [linear measure] Former volumetric unit in Russia, approximating very loosely to the LITRE.

1 schtoff	=	10 charki
	=	1.230 litre
		(1 litre = 0.8130 schtoff)
	=	2.16 UK pints, 2.59 US pints
		(2 UK pints = 0.9259 schtoff
		2 US pints = 0.7722 schtoff)
10 schtoffs	=	1 (Russian) VEDRO

The schtoff coincides in value with an old Dutch unit, the *mingel*, but whether that is a matter of pure chance is not known.

Schuler pendulum [physics] A hypothetical ideal pendulum that is the radius of the Earth long (and is therefore unaffected by the Earth's rotation). It would swing extremely slowly, with a period (time of swing) of 84 minutes.

Schwartzchild radius [physics; astronomy] As a dying star collapses in upon itself, its gravitational forces increase to the point that, at a specific radius, the star permits the escape of no radiation, not even light – it has become a black hole. That critical radius is the Schwartzschild radius. (For our Sun, the radius is calculated to be about 3 kilometres, 1.86 miles.)

It is named after the American astrophysicist Martin Schwartzchild (1912-).

scientific notation [quantitatives] The use in scientific applications of the formula x x 10^n units , in which x corresponds to a number between 0.01 and 9.99, and n to the power to which 10 is raised or lowered accordingly. In this way, for example, the measurement 1,550 megahertz (1,550,000,000 hertz) may be expressed in scientific notation as 1.55 x 10^9 hertz, and the measurement 0.0012345 millivolt (0.0000012345 volt) may be expressed in scientific notation as 1.2345 x 10^{-6} volt.

In some areas of the world, scientific notation is better known as *standard form*.

scintillation counter [physics] An instrument that detects and measures intensities of high-energy (particularly ionizing) radiation. Incoming particles strike a phosphor layer, and the flashes of light so produced are detected by a photomultiplier whose output current pulses are counted electronically.

sclerometry [physics; engineering] The measurement of the hardness of a material: *see* BRINELL SCALE; MOHS' SCALE; ROCKWELL NUMBER.

score [quantitatives; sporting term] 20; twenty.

The term derives from the practice of cutting notches ('scoring') on a tally-stick, especially in cutting one notch to represent a set of 20 elements or items (particularly sheep) at a time. It was in this way that points in competitions were first 'scored'.

But it was not always in twenties that items were notched up. In weighing pigs and oxen during the later 1400s in England, for example, a score represented the weight of 1½ stones, 21 pounds (9.525 kilograms), although this was at times assimilated to a weight of 20 pounds.

score [music] The printed (or handwritten) form of a piece of music for more than one voice or player, especially a form presenting a stave for each voice or player.

The term in this sense was probably first used because the notes as written vertically on horizontal lines resembled the notches cut across the tally-stick in the original form of 'scoring'. Alternatively, the word may have been used with equal relevance to its basic meaning of 'cutting' in reference to the music's separate presentation of the parts for each voice or player.

Scoville scale [biology] The Scoville scale is a little-known gastronomic measure of the 'heat' of peppers, chillies, curries, and other forms of highly spiced food. An ordinary chilli pepper is about 5,000° on the Scoville scale; a really hot one about 8,000°.

scratch [sporting term] In golf and some other sports, describing a player without a handicap (or with a handicap of zero).

screen printing mesh *see* MESH

screw threads [engineering] There have been – and still are – many standards for screw threads, which are specified in terms of diameter, number of threads per unit length (or its reciprocal, the pitch), and form or contour (as profile angle, in degrees). There has been an attempt to standardize in Britain on ISO metric threads and ISO unified inch threads, although others are still in use.

British Association (BA): Nos. 0-25, 6.0-0.25 mm diameter, 47½° profile

British Standard brass (BSB): Whitworth profile, 26 threads/inch (on outside of brass tubes)

British Standard cycle (BSC): ⅛ – ¾ inch diameter

British Standard fine (BSF): Whitworth profile but finer pitch, 3/16 inch upwards

British Standard pipe (BSP): Whitworth profile but designated by bore of pipe (for gas fittings), ⅛ to 4 inches

British Standard Whitworth (BSW): ⅛ inch upwards, profile 55°

Metric: diameter and pitch in millimetres, 60° profile

Sellers (US Standard Screw, USS): diameter and pitch in inches, profile 60°

Unified: combines BSW and USS, coarse and fine versions, diameter (¼ inch upwards) and pitch in inches, profile 60°

The SI system is virtually the same as metric, but some European countries persist with their own standards (France CNM, Germany DIN, and Switzerland VSM). The Society of Automotive Engineers (SAE), American Standard coarse (ASC), and American Standard fine (ASF) are still found in the United States. British Association (BA) – dating from 1884 – remains popular for screws of very small diameter in scientific instruments and fine mechanisms.

scrip [comparative values] Paper money (bills, notes, or bonds) that can be used like ordinary currency but which, for one reason or another, is not the same. The term is used especially of paper money issued by wartime governments or occupying powers, or of paper money issued by banks and counting for denominations for which there are no national coins or notes (as in the case of hyperinflation).

The term derives from the fact that such notes were originally either handwritten or signed by hand (script).

scruple [weight] Unit of weight in the apothecaries' weight system.

1 scruple	=	20 GRAINS
	=	1.295986 GRAM, 0.045714 ounce avoirdupois
		(1 gram = 0.7716 scruple)
	=	0.04166 troy ounce
3 scruples	=	1 drachm (US dram)
24 scruples	=	1 troy ounce, 1 ounce apoth.
	=	1.09709 ounce avdp.
		(1 ounce avdp. = 21.875 scruples)

The term derives from the name of a unit of weight in ancient Rome that was very close in value to that of the 'modern' weight. It was also the name of a Roman gold coin (the value of which varied in accordance with the value of gold).

But the name of the Roman unit was also the diminutive form of *scrupus* 'a sharp pebble', of the type that might creep into your sandal and cause you in future to be *scrupulous* about where you put your feet.

scruple, fluid [volumetric measure] Rare, small unit of liquid capacity.

1 fluid scruple	=	1.18298 MILLILITRE, 0.00118298 litre
		(1 millilitre = 0.8453 fluid scruple)
	=	0.333 (one-third of a) UK FLUID DRAM
	=	0.320 US fluid dram
	=	0.002 (⅟₅₀₀) UK pint
	=	0.0025 (¼₀₀) US pint

See also SCRUPLE.

scudo [comparative values] Former unit of currency in Italy, once in the form of a gold coin, and later in the form of a silver coin. It was so called because the original version had a shield (Italian *scudo*, Latin *scutum*; cf. English *escutcheon*) depicted on one side. *See also* ESCUDO.

se [square measure] Unit of area in Japan, closely approximating to the (relatively uncommon) metric unit the ARE.

1 se	=	30 BU or tsubo, 3,000 shaku
	=	0.9918 are, 99.180 square metres
		[1 are (10 x 10 metres; 100 square metres)
		= 1.00827 se]
	=	118.615 square yards
		[121 square yards (11 x 11 yards; 4 square rods)
		= 1.02011 se]
100 se	=	1 chô

seah [volumetric measure] In ancient Israel (and therefore the Bible) a measure of both liquid and dry capacity.

As a liquid measure,

1 seah	=	3 HIN, 6 CAB or kab, 24 LOGS
	=	13.44 litres
		(10 litres = 0.7440 seah)
	=	2.9568 UK gallons, 3.5506 US gallons
		(3 UK gallons = 1.0146 seah
		3 US gallons = 0.8449 seah)

$$3 \text{ seah} = 1 \text{ BATH}$$
$$30 \text{ seah, } 10 \text{ bath} = 1 \text{ KOR or homer}$$

As a dry measure,

$$1 \text{ seah} = 6 \text{ cab or kab, } 24 \text{ logs}$$
$$= 13{,}440 \text{ cubic centimetres}$$
$$= 820.08 \text{ cubic inches}$$
$$= 1.4781 \text{ UK peck, } 1.5256 \text{ US peck}$$
$$3 \text{ seah} = 1 \text{ EPHAH}$$
$$30 \text{ seah, } 10 \text{ ephahs} = 1 \text{ kor or homer}$$

The Hebrew *seah* was transliterated by the ancient Greeks in the adapted form *saton*, a spelling that occurs in some Biblical dictionaries and commentaries.

sea level [geography; surveying] The level of the surface of the sea. The datum for measuring heights (altitudes) is usually taken as mean sea level (midway between high and low tides), and most maritime countries base their surveys on a nominated standard. For example, in the British Ordnance Survey the datum for mean sea level was fixed at Newlyn, in Cornwall. The Survey's BENCH MARKS relate to this datum.

seat-mile [quantitatives; comparative values] A unit in costing out air, bus, or coach transportation, equivalent to (the cost/profit/loss of) 1 passenger seat per unit mile, calculated as a proportion of (the total incoming revenue and the total outgoing expenditure in relation to) the total number of passenger seats filled and the total number of miles travelled.

sec, sec. *see* SECANT; SECOND

secant [maths] For an angle in a right-angled triangle, the length of the hypotenuse divided by the length of the side adjacent to the angle; it is the reciprocal of the cosine.

Also, a straight line that intersects a curve at two (or more) points. It is this meaning that gives a clue to the word's etymology: Latin *secant-* 'cutting', 'intersecting'.

sech *see* HYPERBOLIC FUNCTION

secohm [physics] Obsolete unit of inductance equal to 1 'legal' ohm times 1 second – about the same size as a henry. [A legal ohm, as established in 1883, was the resistance of a 106-millimetre (4.17-inch) thread of mercury 1 square millimetre across at 0°C, or 0.09972 absolute ohm.]

second [quantatives: ordinal] In English, both cardinal and ordinal numbers are virtually all based on Germanic roots: the ordinal *second* is the exception. Instead of 'two-th' [cf. Dutch *tweede*, German *Zweit(er)*], or the equally non-numeric but still Germanic '(an)other' (cf. Swedish *andre*, Icelandic *annar*), English has borrowed a Latin-based equivalent of *sequent* literally meaning 'the following' or 'next'.

But the numerical meaning was present even in Latin (especially in the use of *Secundus* as a forename). And it was in its numerical meaning that the word gained its other applications – to denote the second subdivision of a primary unit. The second that is a unit of time, for example, was the second subdivision because the minute was the first subdivision (of the hour). The second that is a unit of arc was likewise the second subdivision because the minute was the first subdivision (of the degree, and also ⅓,₆₀₀).

second [physics] An overworked unit with several meanings.

In time measurement, it is the fundamental unit of time in the SI, CGS, MKS, and FPS systems, now defined as the time it takes for 9,192,631,770 cycles of resonance vibration of the caesium-133 atom. A former definition was ⅟₃₁,₅₅₆,₉₂₅.₉₇₄₇ of the tropical year for the EPOCH 0 January 1900 at 12 hours EPHEMERIS TIME.

$$1 \text{ second} = 10^9 \text{ nanoseconds, } 10^6 \text{ microseconds, } 1{,}000 \text{ milliseconds}$$
$$= \tfrac{1}{60} \text{ minute, } \tfrac{1}{3{,}600} \text{ hour}$$
$$= \tfrac{1}{86{,}400} (0.0000115) \text{ mean solar day}$$
$$= \tfrac{1}{31{,}536{,}000} (0.0000000316) \text{ mean solar year}$$
$$60 \text{ seconds} = 1 \text{ minute}$$

The time measurement is used in many scientific descriptions of rates of progress (such as velocity or flow): *see* PER SECOND PER SECOND; FEET PER SECOND; FOOT-SECOND; METRES PER SECOND; MILES PER SECOND; RATE; and *see below*.

In the measurement of angle, a second is 1/60 of a minute of arc, which is in turn 1/60 of a degree (*see* ANGLE).

$$1 \text{ second} = \frac{1}{60} \text{ minute (of arc)}$$
$$= \frac{1}{3,600} \text{ degree}$$
$$60 \text{ seconds} = 1 \text{ minute (of arc)}$$

Both the above types of seconds (time and angle) can be represented by the symbol " .

In the measurement of length using feet and inches, a second is 1/12 of an inch (also called a LINE), and can be represented by the symbol ''' .

$$1 \text{ second (in length)} = \frac{1}{12} (0.08333) \text{ inch}$$
$$= 2.11666 \text{ millimetres}$$
$$= 10 \text{ gries, 6 points}$$

In the measurement of viscosity, a second is used to measure flow times. There are various systems for such measurement, named after their inventors. For example, the Engler second (used in Continental Europe) is the viscosity of a fluid that takes 1 second for 200 cubic centimetres to flow through an Engler viscometer. The Redwood second is used in Britain, and the Saybolt Universal second in the United States. *See also* POISE.

second [music] Musical interval that, in the form of two notes one semitone (half-step) apart, is a discord but, in the form of two notes one tone (step) apart, may be either a discord or part of a harmony (but which in both cases requires resolution).

Also an instrumental or vocal part which, although written for the same instrument or voice as the first part, is at a lower pitch (and generally of a less melodic nature). In an orchestra, the musicians who play the second violins have instruments identical to those of the first violins, but play notes that contribute to harmonies beneath the melodic line of the first violins (and may be of equal or greater complexity).

secondary colour [physics] A colour produced by mixing two primary colours. For example: the primary pigments blue and yellow mix (subtractive process) to give the secondary colour green: *see* PRIMARY COLOURS; TERTIARY COLOURS.

second cousin *see* RELATIONS AND RELATIVES

second-degree burn *see* BURNS

second lieutenant *see* LIEUTENANT

second person singular/plural *see* PERSON

secpar Another name for a PARSEC.

section [square measure] In the United States and Canada, a unit of land area that is generally square, and corresponds in total to 1 square mile.

$$1 \text{ section} = 4 \text{ quarter sections}$$
$$= 1 \text{ SQUARE MILE, 640 acres (64 square furlongs)}$$
$$= 259.005 \text{ hectares, 2.59 SQUARE KILOMETRES}$$

To qualify as a township, an urban area has usually to occupy 36 or more sections of civic territory.

section [maths] *see* CROSS-SECTION

sector [maths] A plane figure formed by two radii of a circle and the arc of circumference between them (resembling a slice of pie): *see* CIRCLE.

Also, a geometrical device with which to draw angles, made up of two calibrated rulers joined at one end.

sedimentation rate [physics; medicine] The rate at which the particles suspended in a liquid settle, either under the action of gravity or in a centrifuge, from which the particle size can be calculated. Sedimentation rates for erythrocytes (red blood cells) can also be used in the diagnosis of disease. *See also* SVEDBERG.

see [square measure] In Church denominations that have bishops, a large administrative area including many individual parishes, over which a bishop is responsible for the 'cure of souls'. It is known also as a *diocese* or *bishopric*.

The term derives as a Norman French term for the official residence or 'seat' (medieval French *sie*, Latin *sedes*) of the bishop.

seed [sporting term] In sporting competitions, especially tennis tournaments, to seed players is to list the best players in order of recorded skill, and to separate them widely over the heats or first rounds of the competition so that the final rounds include as many of them as possible. It is a method of avoiding the elimination of some of the best players at an early stage, in heats or rounds that might well be of a much higher standard than the eventual final – all of which would alienate both players and (paying) spectators.

The process is called *seeding* because of the initial scattering or broadcasting of the favoured players over the heats or first rounds: the players have been seeded (sown as seeds) for later flowering.

seer, ser [weight] Former unit of weight in India, (probably) originally the weight of grain within a specified container. But, because the size of the container varied from locality to locality, so did the weight. The result was that the seer could be valued at anything between 9½ ounces avdp. and 2 pounds 3 ounces avdp. (270 grams and 1 kilogram). There was, however, an official value set by central government:

$$\begin{aligned}
\text{1 (government) seer} \quad &= \quad 80 \text{ tolas} \\
&= \quad 2 \text{ pounds } 0.9 \text{ ounce} \\
&\qquad (2 \text{ pounds} = 0.9723 \text{ seer}) \\
&= \quad 933.04 \text{ grams, } 0.99304 \text{ KILOGRAM} \\
&\qquad (1 \text{ kilogram} = 1.0070 \text{ seer}) \\
\text{40 (government) seers} \quad &= \quad 1 \text{ MAUND}
\end{aligned}$$

Also, in northern India and in Sri Lanka, a former unit of dry capacity.

$$\begin{aligned}
\text{1 Bengali/Sri Lankan seer} \quad &= \quad 1.06 \text{ litre} \\
&= \quad 0.9327 \text{ UK dry quart} \\
&= \quad 0.9617 \text{ US dry quart} \\
&= \quad 64.685 \text{ cubic inches}
\end{aligned}$$

The term probably reflects the agricultural nature of the produce that was weighed/contained, seemingly akin to Sanskrit *ksetra* 'field'.

segment [maths] A plane figure corresponding to the part of a circle cut off by a chord, or a solid figure corresponding to the part of a sphere cut off by a plane. A chord cuts a circle (or a plane cuts a sphere) into two parts; the smaller part is the minor segment and the larger part is the major segment. If the chord (or plane) goes through the centre, the segment is a semicircle (or hemisphere): *see* CIRCLE.

seicento *see* -CENTO.

seichometer [linear measure; geography] An instrument that measures the rise and fall in the surface level of a lake. Not only do lakes tend to rise and fall according to seasonal conditions (rising with increased rainfall, and ebbing with increased evaporation), but there are occasional local fluctuations in water level apparently resulting from variations in atmospheric pressure. These fluctuations (known as *seiches*) may represent the slight tilting of the entire body of lake water from one side to the other and back in a rhythmic oscillation.

seidel [volumetric measure] Former unit of liquid capacity in Austria.

$$\begin{aligned}
\text{1 seidel} \quad &= \quad 35.4 \text{ centilitres, } 0.354 \text{ litre} \\
&\qquad (50 \text{ centilitres, } 0.5 \text{ litre} = 1.4124 \text{ seidel}) \\
&= \quad 0.623 \text{ UK pint, } 0.748 \text{ US pint} \\
&\qquad (1 \text{ UK pint} = 1.605 \text{ seidel} \\
&\qquad 1 \text{ US pint} = 1.337 \text{ seidel})
\end{aligned}$$

In many German dictionaries the word *seidel* is translated predominantly as 'mug', but secondary meanings are listed as 'pint' and 'half-litre' – neither of which corresponds well to the (former) Austrian measure (or to each other). By derivation, the word is borrowed from Latin *situla* 'bucket'.

seismic disturbance [geology] *see* MERCALLI SCALE; RICHTER SCALE

self-inductance [physics] An electromotive force (voltage) is induced in a circuit when the magnetic flux linked with it changes, resulting from a change in the current flowing. The induced EMF – a back EMF – flows in the direction that tends to oppose the change (this is a statement of Lenz's law). This resistance to a change in current in a circuit is self-inductance, measured in henrys.

self-rating scale [medicine] In psychiatry, a scale on which a patient is invited to rate himself or herself in terms of personality, general disposition, ability, intelligence, and overall health, with a view to discovering the patient's ambitions, fears, and mental strengths and weaknesses.

selsyn [engineering] A device for synchronizing two or more operations, at least one of which is electric or electronic.

The term is derived from a trademark for such a machine, representing the first elements of the words self-synchronous.

semanteme [literary] In linguistics, a morpheme (a word element that has independent meaning although it may or may not correspond to a whole word) that corresponds to a whole word.

semaphore [literary] Signalling system utilizing the positioning of two arms (or the equivalent) in combinations of any of eight positions.

There is – from an English point of view – a disappointing lack of continuity among the codings for the letters of the alphabet: the letters J, V, and Y are evidently out of sequence with the rest. (This is sometimes attributed to the fact that the code was originally devised by a Frenchman, Claude Chappe, in the 1790s. But French uses all three of those letters, and commonly.)

The term derives from two ancient Greek elements meaning 'conveying a signal'.

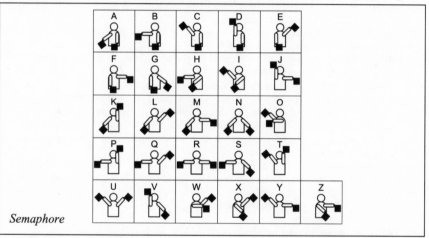

Semaphore

sememe [literary] The meaning of a morpheme that is not a semanteme – that is, the meaning of a word element that is a prefix, or a suffix, or some grammatical element that does not correspond to a complete word. The suffixes -ess and -euse , for example, are both part of the sememe that implies 'female counterpart of a male'.

semester [time] Period or term of work at school, college, or university. Different countries observe educational terms of differing duration, and even differing numbers of such terms (mostly two, but in some countries three or even four) per year.

In Germany, however, a semester represents a half-year of a school or university course, including any holidays (vacations) within the half-year. This usage corresponds to the root meaning of the word, derived from late Latin *se-mestris* 'of six months'.

The related adjective is *semestral* or *semestrial*.

semi- [quantitatives: prefix] Prefix denoting 'half'.

Examples: semicircle – half a circle
semitone – half a tone
semiconductor – material with a conductivity that is higher than that of an insulator but lower than that of a conductor

The term derives through Latin, and corresponds to the Greek *hemi-* which appears to have the root meaning 'doubled' – that is, both 'as a pair', 'alike', and (doubled up) 'half the size'. The first of these two meanings is, for example, evident in the akin English words *same* and *seem*. *See also* HEMI-.

semiannual [time] Once every six months (half-year); lasting for six months.

semibreve [music] In musical notation, a semibreve – known in the United States as a whole note – is represented by the symbol o and corresponds to

    2 MINIMS or half-notes
    4 CROTCHETS or quarter-notes
    8 QUAVERS or eighth-notes
    16 semiquavers or sixteenth-notes
The name of the unit also corresponds to its duration, as half of one BREVE.

semicentenary, semicentennial [time] A fiftieth anniversary (and its celebration).

semicircle [maths] A plane figure formed when a chord passes through the centre of a circle (the chord is a diameter); half a circle (with or without a diameter, although a semicircle without a diameter may instead be called a *semicircumference*).

semiconsonant, semivowel [literary] A semiconsonant is a vowel that may have virtually consonantal pronunciation: for example, the y in *yet*; a semivowel is a consonant that may have virtually vocalic pronunciation: for example, the w in *wet*.

semicylindrical [maths] Describing a solid figure that has a semicircular cross-section (*see* SEMICIRCLE).

semidiurnal [time] Once every twelve hours (half a day); lasting for twelve hours.

semigroup [maths] A SET that has a binary operation that is ASSOCIATIVE – usually addition – under which it is closed.

semilogarithmic [maths] Describing a graph that has a logarithmic scale along one axis and a linear scale along the other axis.

semimajor axis [astronomy] The mean distance of a planet from its parent star (corresponding to half of the mean diameter of its orbit).

semiprecious stones [comparative values] Stones that, for reasons of crystalline form, colour, lustre, transparency, scarcity, and therefore commercial value, are technically classified not as precious stones but are still regarded as gemstones.

    They include garnet, tourmaline, olivine, topaz, various forms of quartz (including amethyst, smoky quartz, rose quartz, aventurine, and citrine), and carnelian (also called cornelian).

semiquaver [music] In musical notation, the semiquaver – known in the United States as a sixteenth-note – is represented by the symbol ♪ or ♪ and corresponds to
    ½ QUAVER or eighth-note
    ¼ CROTCHET or quarter-note
    ⅛ minim or half-note
    1/16 semibreve or whole note
The name of the unit corresponds to its duration, as half of one quaver.

semitone [music] A musical INTERVAL that is half a tone, corresponding on a modern piano to the interval between a black note and an immediately adjacent white note, and vice versa, or between two white notes that do not have an intervening black note – also called a minor second, and better known in the United States as a *half-step*.

    In more technical terms, on an even-tempered musical scale, a semitone is the difference in pitch between two tones whose frequency ratio is the twelfth root of 2.

semivowel *see* SEMICONSONANT, SEMIVOWEL

sen [comparative values] Unit of currency in Cambodia and Indonesia; formerly a unit of currency in Japan, and still used there as a notional currency (a currency of account) in calculating multiples of fractions of the present currency units for taxation purposes.

    In Cambodia,
        100 sen = 1 riel
    In Indonesia,
        100 sen = 1 rupiah
    In Japan, nominally,
        100 sen = 1 yen

    The name of the Cambodian and Indonesian currency units is borrowed directly from the Japanese. The Japanese term may or may not be etymologically associated with the Japanese word *sen* that means '1,000'. *See also* COINS AND CURRENCIES OF THE WORLD.

sen [linear measure] A unit of length in Thailand.
        1 sen = 40 (Thai) ken, 160 kup, 1,920 nin
             = 1,600 inches, 133 feet 4 inches

$$= \quad 44.444 \text{ yards, } 40.64 \text{ metres}$$
$$(50 \text{ yards} = 1.125 \text{ sen}$$
$$400 \text{ yards} = 9 \text{ sen}$$
$$50 \text{ metres} = 1.2303 \text{ sen}$$
$$450 \text{ metres} = 11.073 \text{ sen})$$

senary [quantitatives] Of the number 6; having 6 as the base; comprising 6 elements.
The term derives from the Latin *seni* (for **sex-ni* – cf. the careless but common pronunciation of 'sixth' as 'sicth', which is half-way to losing the -x- similarly) 'six each'.

sengi, senghi [comparative values] Unit of currency in Zaïre.

$$100 \text{ sengi} \quad = \quad 1 \text{ likuta}$$
$$10,000 \text{ sengi, } 100 \text{ makuta} \quad = \quad 1 \text{ zaire}$$

See also COINS AND CURRENCIES OF THE WORLD.

seniti [comparative values] Unit of currency in Tonga.

$$100 \text{ seniti} \quad = \quad 1 \text{ pa'anga}$$

The term is borrowed from the *cent* of other currency systems. *See also* COINS AND CURRENCIES OF THE WORLD.

sennight [time] One week. The term is a contraction from the Old English for 'seven nights'. Much better known (especially in Britain) is the corresponding contraction of the Old English for 'fourteen nights': fortnight. Both these expressions are so ancient as to refer to a time when it was easier to talk about nights than to talk about days, which had no set hours.

sensation unit [physics] Unit of loudness based on the false belief that (subjective) loudness increases logarithmically, and therefore defined as twenty times the logarithm of the ratio between the sound pressure being measured to the pressure at the threshold of hearing. The error was corrected with the introduction of the DECIBEL.

sensitometer [physics] A device that measures the sensitivity to light of the human eye, or of photographic plates or film.

sentence [literary; maths] In grammar, a meaningful combination of words, technically including at least a VERB and the subject of that verb. There may be additional words and clauses, and there is often an object; the end of a sentence is marked by a full point or period.

As a complete statement of an idea, the sentence is found also in mathematics, usually in relation to equations. A closed sentence is an equation in which there is no requirement for further elucidation: for example, $3 + 6 = 9$. An open sentence includes a further requirement for elucidation: for example, $5x + 7 = 22$.

sep [weight] Unit of weight in ancient Egypt. Unfortunately, there is no known standard version of the measure, a multiple of two other units, but over the course of 3,500 years or more it is perhaps not so surprising that values changed, as cultures, overlords, and capital cities did.

$$1 \text{ sep} \quad = \quad 10 \text{ deben, } 100 \text{ kite}$$

separation energy [physics] The energy needed completely to remove a nucleon (proton, neutron, etc.) from an atom's nucleus, usually measured in electron volts.

sept-, septa-, septi- [quantitatives: prefix] Prefix that implies sevenfold unity.
Example: septennial – occurring or lasting seven years
The term derives from Latin *sept-* 'seven', but is cognate with virtually all Indo-European words for that number. *See also* HEPT-, HEPTA-.

septavalent, septivalent [chemistry] Having a valence (valency) of seven, alternatively described as being *heptavalent*.

septenary [quantitatives; time] Of the number 7; having 7 as the base; comprising 7 elements. Also, as an equivalent of *septennial* and *septennate*, occurring or lasting for seven years. Also, as an equivalent of *septennial*, a seventh anniversary.
The term derives from the Latin *septeni* 'seven each'.

septennial, septennate [time] Once every seven years, lasting for seven years; (as a noun) a septennate (or *septennium*) is also a period of seven years, and a septennial (or *septenary*) is a seventh anniversary (and its celebration).
The terms derive from Latin elements meaning 'seven years'.

septennium [time] A period of seven years: a synonym for *septennate*.

The term derives from Latin elements meaning 'seven years'.

septentrional [geography] To the north, northern: *see* POINTS OF THE COMPASS.

This now somewhat archaic term derives from Latin *septem-triones*, the 'seven plough-oxen' – that is, the seven stars that form The Plough, or Ursa Major (the Great Bear, the Big Dipper), two of which are The Pointers that 'point' to Polaris, the Pole Star – north.

septet [quantatives; music] A group of seven (elements), especially a group of seven musicians or singers; a piece of music for seven players or singers.

septimal [quantitatives; maths] Of the number 7; having 7 as the base. *See also* SEPTENARY.

septivalent *see* SEPTAVALENT, SEPTIVALENT

septuagenarian [time] A person who has passed the seventieth birthday, but has not yet reached the eightieth.

septuple [quantitatives] Sevenfold, seven times; of seven elements.

sequence [sporting term] In card games, particularly poker, another term for a *straight* or three or more cards of consecutive denomination (whether or not they are also in the same suit).

sequin [comparative values] Ancient Venetian unit of currency, in the form of a gold coin, the brightness of which gave rise to the present-day meaning of 'spangle'.

The name of the coin derives through a French adaptation of an Italian diminutive (*zecchino*) of what was originally an Arabic word for 'the mint'.

ser *see* SEER, SER

sergeant [military rank] In the military, a non-commissioned officer that ranks higher than a corporal (or in some airforces, an airman first class), but (in many armed forces) that ranks below a staff sergeant, a sergeant first class, and a master sergeant (or equivalently progressively senior sergeant).

In many police forces, the term is also used of an officer ranking above a constable (or trooper or patrolman) but below an inspector (in Britain and elsewhere) or a lieutenant or captain (in North America).

The term is an excruciating French corruption of the same word that has in modern English become (public/civil) *servant*.

sergeant major [military rank] In many armed forces, the highest rank of non-commissioned officer, with particular responsibilities for discipline and drill, especially at an army or airforce base. *See also* SERGEANT.

series [chemistry] A sequence of chemical elements, as they appear in the PERIODIC TABLE, that have common properties that vary (in an increasing or decreasing manner) along the sequence. For example: the lanthanides (or lanthanoids) – elements of atomic numbers from 58 (cerium) to 71 (lutetium), inclusive – constitute such a series, as do the actinides (actinoids) – 90 (thorium) to 103 (lawrencium).

series [geology] In rock strata made up of beds of related rocks, the number of layers (in various stages) that together represent a geological EPOCH or a major subdivision of a geological epoch. *See also* STAGE.

series [maths] A sequence of terms that can each be represented as an algebraic function of its position (such as an ARITHMETIC PROGRESSION/SERIES and a GEOMETRIC PROGRESSION/SERIES). A series with a fixed number of terms is called a finite series; one with an infinite number of terms – such as the exponential series (in which the terms have the general expression x^n/n) – is an infinite series.

serotype [biology; medicine] A 'standard' micro-organism, defined by the types and combinations of antigens present in its cells.

service ceiling [engineering; physics] The greatest altitude at which an aircraft can still climb at a rate faster than 100 feet per minute (30 metres per minute).

sesqui- [quantitatives: prefix] Prefix meaning 'one-and-a-half times'.

Examples: sesquipedalian – 1½ (metrical) feet in length

sesquiplane – a fixed-wing aircraft (biplane) with one pair of wings much longer than the other pair

The term derives from the Latin *sesqui-* 'one and a half', which derives as an easing of **semis-que* 'and a half', thus 'half again'.

sesquicentennial, sesquicentenary [quantitatives; time] Of 150 years, (celebrating) a 150th anniversary.

The term derives from Latin elements meaning 'one-and-a-half centuries'. *See also* QUASQUICENTENNIAL, QUASQUICENTENARY.

sesquipedalian [literary] In poetic scansion, describing (a word or expression that takes up) one-and-a-half metrical feet.

By extension, in everyday terms, describing a person who is overly given to using long words (especially when verbal construction utilizing less amplification might represent a more naturally efficacious phraseology!).

The term largely derives from Latin elements meaning 'one-and-a-half feet' (the final -ian is an English addition).

sesterce [comparative values] Unit of currency in ancient Rome, in the form first of a silver coin, but later of a bronze coin. Originally, and for much of the duration of the Roman Empire,

1 sesterce (*sestertius*)	=	2½ asses
4 sesterces, 10 asses	=	1 denarius
1,000 sesterces,		
250 denarii	=	1 sestertium

– but, whereas the value of the as and the sesterce remained fairly constant, the value of the denarius varied according to the overall wealth and prosperity of the nation: in times of war and periods of bad inflation, the denarius occasionally collapsed to a value of 16 asses – a value that could not be exactly measured in sesterces.

The name of the coin, *sestertius*, represents an easing of *semis-tertius 'half third', meaning 'one-half less than three', thus 'two-and-a-half'. The Romans were accustomed, with certain numbers, to counting back from the next higher major figure: in this way, after all, the number IX represented I less than X .

sestet [literary; music] In verse, a stanza of six lines or a distinctively final six lines of a longer verse or entire poem.

In music, like SEXTET, a group of six instrumentalists or vocalists.

sestina *see* VERSE FORMS

set [maths] In mathematics, a group of items (called elements) that have at least one thing in common; there may be a finite (that is, countable) number or an infinite number of elements in a set. Examples of sets, which are written between { and } brackets, are:

> {odd numbers}
> {metals}
> {boys whose names begin with J}

A set is usually denoted by a capital letter:

> O = {odd numbers}
> M = {metals}
> B = {boys whose names begin with J}

The symbol ϵ denotes that an element is a member of a given set:

> 5ϵO   thus '5 is an element of the set of odd numbers'
> steelϵM   thus 'steel is an element of the set of metals'
> JohnϵB   thus 'John is an element of the set of boys whose names begin with J'

Thus also trumpetϵ{brass instruments} means that a trumpet is an element of the set of brass instruments. *See also* UNIVERSAL SET; VENN DIAGRAM.

set [sporting term] *see* TENNIS MEASUREMENTS AND UNITS

setier [volumetric measure] French-derived name for an ancient Greek measure of dry and liquid capacity that was a minor submultiple of the kyathys: *see* KYATHYS, CYATHYS.

set square [maths] A flat plastic or metal right-angled triangle used in technical drawing and in constructing geometrical figures. Its other angles are either 60° and 30°, or both 45°.

sett [square measure] Any of the identifiable squares in the pattern of squares that all together make up a tartan, so called because each square in a tartan plaid has to be individually set up on the loom before the tartan is woven.

settecento *see* -CENTO.

seven [quantitatives] 7.

$\frac{7}{10}$ = 0.7 = 70%

7 days = 1 week

A group of seven is a septet or heptad.

A polygon with seven sides (and seven angles) is a heptagon.

A solid figure with seven plane faces is a heptahedron.

As is evident from the above, prefixes meaning 'seven-' in English correspond to (Greek-based) 'hept-', 'hepta-' and (Latin-based) 'sept-', 'septa-', 'septi-'.

The name for the number derives from common Indo-European, attested over the entire range of languages from Sanskrit to Welsh, and found in akin forms also in some of the non-Indo-European Nostratic languages.

In continental Europe, the figure seven is crossed (that is, it has a short bar horizontally across it half-way up). This may well be a result of the traditional form of teaching of non-ranging numerals in schools from the last half of the 1800s. In this system, the number 7 extends below the base line (other figures sit upon the line). The bar may thus represent where the figure used to cross the base line. Alternatively, the bar may simply be a means of distinguishing a 7 from the heavily hooked Continental figure 1 . Either way, so important has this bar become in Europe that an open 4 (a 4 written carelessly so as not to come to a point) may be mistaken for a 7 simply because of its central cross.

seven-a-side rugby, rugby sevens [sporting term] The game of rugby, played by two teams of only seven players instead of the full complement of fifteen (rugby union) or thirteen (rugby league).

seven-league boots [linear measure] Magic boots that allow the wearer to travel seven leagues at each step. Unfortunately, the league is not a constant measure, and seven of them amount to anything between 28 kilometres (17.4 miles) and 24.2 miles (38.9 kilometres), depending on whether the wearer calculates by the metric league (the shortest league), the US league, the UK league on land, or the UK league at sea (the longest league): *see* LEAGUE.

A wearer must remember to take the boots off before going indoors.

sevens *see* SEVEN-A-SIDE RUGBY, RUGBY SEVENS

seventh [quantatives] $\frac{1}{7}$ (one-seventh); the element in a series between the sixth and the eighth, or the last in a series of seven.

$\frac{1}{7}$ = 0.142857 = 14.286%
$\frac{2}{7}$ = 0.285714 = 28.571%
$\frac{3}{7}$ = 0.428571 = 42.857%
$\frac{4}{7}$ = 0.571428 = 57.143%
$\frac{5}{7}$ = 0.714285 = 71.429%
$\frac{6}{7}$ = 0.857142 = 85.714%

(The decimal fractions contain the same sequence of constituent numbers, each beginning at a different number.)

seventh [music] Musical interval that in its major form corresponds to the interval between a keynote and the leading-note above it (that is, the note in the scale immediately below the keynote above).

However, it is the minor seventh that is the more famous, and which is meant if just the word *seventh* is used in a technical sense. This involves a flattened leading-note, the classic 'blue note' that, to purists, requires resolution by a subdominant chord after it, and the note that almost always sounds a little awkward in bagpipe renditions of well-known melodies. Thus, the chord commonly known as C7 ('C seventh') is a chord of C (C E G C) that also includes the flattened seventh B♭.

sex-, sexa- [quantatives: prefix] Prefix that implies sixfold unity.

Examples: sexennial – occurring or lasting for six years

sextet – a group of six (musicians)

The word derives from Latin *sex* 'six', but is cognate with virtually all Indo-European words for that number. *See also* HEX-, HEXA-.

sexagenarian [time] A person who has passed the sixtieth birthday, but has not yet reached the seventieth.

sexagesimal system [maths] A number system with the base (radix) 60. It was introduced by the Persian rulers of ancient Babylon, and its remnants can still be seen in the units of angle (60 seconds of arc = 1 minute; 60 minutes of arc = 1 degree; 360 degrees = 1 complete revolution) and time (60 seconds = 1 minute; 60 minutes = 1 hour). It has the advantage of having many factors – 2, 3, 4, 5, 6, 10, 12, 15, 20, and 30 are all factors of 60.

sexcentenary [time] Of 600 (elements or constituents); once in 600 years, lasting for 600 years, (celebrating) a 600th anniversary.

sexennial [time] Once every six years, lasting for six years; (as a noun, also a *sexennium*) a period of six years, a sixth anniversary.

The terms derive from Latin elements meaning 'six years'.

sex ratio [biology] The ratio of males to females born in a specific population (of animals or humans), normally expressed as the number of males per 100 females (to the nearest whole number).

sextan [time] Recurring every sixth day inclusively – that is, with four days between recurrences, or every 120 hours. Fevers symptomatic of various tropical diseases (notably some forms of malaria) recur like this.

sextant [navigation] A navigational instrument that is used to measure altitudes (elevation angles) of celestial objects, such as the Sun and stars, to calculate latitude – from a knowledge of the angle and the exact time, historically supplied by a CHRONOMETER. *See also* POSITION CIRCLE.

sextarius [volumetric measure] In ancient Rome, a basic measure of liquid capacity – especially in the preparation or preservation of wine and oil – and of dry capacity, very closely approximating to the modern pint.

1 sextarius	=	48 ligulae, 8 acetabula, 2 heminae
	=	53.125 centilitres, 0.53125 litre
	=	0.935 UK pint, 1.1227 US pint
	=	0.935 UK dry pint, 0.9639 US dry pint
	=	531.25 cubic centimetres
	=	32.421 cubic inches
6 sextarii	=	1 CONGIUS
24 sextarii, 4 congii	=	1 URNA
48 sextarii, 8 congii,		
2 urnae	=	1 AMPHORA

Despite the fact that this measure corresponded to a basic unit, its name derives from its proportional size (one-sixth) in relation to the congius. *See also* SEXTE.

sexte [volumetric measure] Basic unit of liquid and dry capacity in ancient Greece, very closely approximating to the modern pint.

1 sexte	=	144 kheme, 72 setier, 48 mystra, 12 KYATHOI,
		8 OXYBATHA, 2 COTYLAI
	=	53.939 centilitres, 0.53939 litre
	=	0.9493 UK pint, 1.1399 US pint
	=	0.9493 UK dry pint, 0.9788 US dry pint
	=	539.39 cubic centimetres
	=	32.917 cubic inches
6 sexte	=	1 KHOUS
72 sexte, 12 khoes	=	1 (Greek) AMPHORA

In dry measure only, there were further multiples of the sexte:

2 sexte	=	1 KHOINIX
96 sexte, 16 khoes	=	1 medimnos

Like some of the names of its submultiples, the (present) name of the unit is not true Greek but based on a French rendition of a Latin variant originally meaning 'sixth', and so called – despite the fact that the measure represents a basic unit – because of its proportional size in relation to the khous (or chous). The original Greek name was the *xestes* (a bad attempt at *sextenarius*).

sextet [quantatives; music] A group of six (elements), especially a group of six musicians or singers; a piece of music for six players or singers.

sextile [astrology] Describing the aspect of two celestial objects that have an ANGULAR DISTANCE of 60 degrees, and in astrology the effects of such positioning on the life of someone born contemporaneously.

The term derives from the Latin word for 'sixth', referring to the angle that is one-sixth of the 360-degree zodiac.

sextodecimo, 16mo [paper sizes] A paper size that results from folding a designated sheet four times (to form sixteen pages).

sextuple [quantitatives] Sixfold, six times; of six parts.

shaku [linear measure] Unit of length in Japan, the name of which is used also for a unit of area (*see below*) and a volumetric measure (*see below*). The linear measure approximates very closely to the standard imperial (English) foot.

1 shaku	=	100 bu
	=	30.30 centimetres, 0.3030 metre
		(1 metre = 1.3003 shaku)
	=	11.93 inches, 0.9942 foot
		(1 foot = 1.0058 shaku)
6 shaku	=	1 ken
360 shaku, 60 ken	=	1 chô

The term in Japanese means simply 'measure', 'scale', 'rule', and in this linear context is commonly and appropriately translated into English as 'the Japanese foot'.

shaku [square measure] In Japan, apart from the shaku that is the linear measure (*see above*) and the shaku that is the volumetric measure (*see below*),

1 shaku	=	330.6 square centimetres, 0.03306 square metre
		(1 square metre = 30.25 shaku)
	=	51.2431 square inches, 0.3559 square foot
		(1 square foot = 2.8098 shaku)
100 shaku	=	1 bu or tsubo
3,000 shaku, 30 bu	=	1 se
300,000 shaku, 3,000 bu, 100 se	=	1 chô

The term in Japanese means simply 'measure', 'scale', 'rule'.

shaku [volumetric measure] In Japan, apart from the shaku that is the linear measure (*see above*) and the shaku that is the square measure (*see above*), a unit of dry and liquid capacity.

1 shaku	=	18.039 millilitres, 0.018039 litre
		(1 litre = 55.4354 shaku)
	=	0.03175 UK pint, 0.03806 US pint
		(1 UK pint = 31.4961 shaku
		1 US pint = 26.2743 shaku)
	=	18.039 cubic centimetres
	=	1.1008 cubic inch
		(1 cubic inch = 0.9084 shaku)
10 shaku	=	1 GO
100 shaku	=	1 shô
1,000 shaku, 10 shô	=	1 to
10,000 shaku, 100 shô, 10 to	=	1 koku

The term in Japanese means simply 'measure', 'scale', 'rule'.

sharp [music] A note that is technically a semitone (or half-step) higher in pitch than a given note: the note C# is thus one semitone (half-step) higher than the note C .

On a piano keyboard, such notes are not always represented by black keys: the note E#, for example, corresponds to the note F (because the note F is only a semitone or half-step higher than E anyway).

Many keys incorporate sharps as part of their tonic scales: for a list of the sharp keys *see* KEY, KEYNOTE.

But the term as an adjective or adverb in less technical musical terminology means merely at a slightly higher pitch. So to sing sharp is to sing out of tune by reaching a note fractionally above the desired one. Unaccompanied choirs in cold halls, especially if nervous, quite often sing progressively sharper and sharper. *See also* DOUBLE FLAT, DOUBLE SHARP.

shear modulus *see* RIGIDITY MODULUS

shed [physics] An extremely small unit of area (nuclear cross-section), equal to 10^{-52} square metre.

1 shed = 10^{-24} barn

The unit was given the name as a 'smaller version' of the barn.

shekel [weight] Unit of weight that originated in the Persian markets of ancient Babylon and was adopted by the ancient Israelites when they were deported there from about 598 BC.

1 'ordinary' shekel	=	20 GERAHS, 4 REBAHS, 2 BEKAHS
	=	8.4 grams
		(100 grams = 11.9 shekels)
	=	0.296 ounce (4.741 drams avdp.)
		(1 ounce = 3.378 shekels)
60 shekels	=	1 MINAH
3,600 shekels, 60 minahs	=	1 TALENT

The ordinary or 'holy' shekel was used for everyday purposes in relation to consumer goods, but there were slightly different equivalences in relation to measurements in gold and silver.

In gold and silver,

50 'profane' shekels	=	1 minah
3,000 shekels, 60 minahs	=	1 talent

The first coin of this weight was the gold *daric* issued by Darius I around 515 BC, which soon became known instead (at least in Palestine) as the 'shekel of gold'. The silver coin issued at the same time, to be a submultiple of the daric, then became known as the 'silver shekel' although it was, in fact, only about two-thirds of that weight. In this way, the name of the weight was transferred to the coinage: *see below*.

The name of the unit is either Hebrew for 'weight' or has been assimilated from a Babylonian word to that Hebrew word.

shekel [comparative value] Unit of currency in Israel.

1 shekel = 100 agorot

The unit was formerly equal to 10 Israeli pounds. It is named after the silver coin of the ancient Hebrews, current for many centuries until AD 70. That, however, in turn got its name from a unit of weight used by the ancient Israelites and adopted by them during the Exile in Babylon (about 598 to 538 BC): *see above*.

shell, shell number [physics] Of an atom, the orbit or orbital in which electrons are located may be called a shell, and the number of such shells is the atom's shell number.

sheng [volumetric measure] Unit of liquid capacity in China.

1 sheng	=	1.035 litre
		(1 litre = 0.9662 sheng)
	=	1.8216 UK pint, 2.1874 US pints
		(2 UK pints = 1.0980 sheng
		2 US pints = 0.9143 sheng)
50 sheng	=	1 hu

sherry glass *see* DRINKING-GLASS MEASURES

shift [physics; astronomy] For the lines in a spectrum (for example, the spectrum of a star), a change in wavelength, often due to the Doppler effect because the light source (the star) is moving towards or, more usually, away from the observer. *See also* RED SHIFT.

shih [weight] Unit of weight in China.

1 shih	=	71.6177 kilograms
		(100 kilograms = 1.3963 shih)
	=	157.89 pounds
		(154 pounds, 11 stone = 0.9754 shih)

The unit appears to be unrelated to other Chinese measures in this book.

shilling [comparative values] Unit of currency in Kenya, Somalia, Tanzania, and Uganda, and formerly a unit of currency in the British Isles and some other British ex-colonies (such as the thirteen American colonies).

In Kenya, Somalia, Tanzania, and Uganda,

1 shilling = 100 cents

Formerly in Britain,

1 shilling (1/-) = 12 pence
20 shillings = 1 pound sterling

In Ireland during the 1810s-1830s,

1 shilling = 13 copper pennies

During the 1600s there was a coin of the same name current in Scotland: it was, however, worth only 1 English penny at the time.

The size and shape of the shilling coin were retained in Britain when the currency was decimalized in 1971 but, because it was no longer such an individual unit submultiple with its own submultiples, it became known almost overnight simply as the '5-pence piece'. In 1992, the final link with the past was lost when the 5-pence piece became an altogether much smaller coin.

The name of the unit probably derives from the shield that was depicted on very early examples (as a 'shield-ing'), although some etymologists suggest an echoic derivation (cf. German *Schall* 'resonance'; semantically cf. also PENGÖ). *See also* SCHILLING; COINS AND CURRENCIES OF THE WORLD.

shipping stowage *see* BARREL BULK; DISPLACEMENT TON; FREIGHT TON; REGISTER TON

shipping ton *see* FREIGHT TON

ship's bells *see* BELLS; WATCH

shirt and blouse sizes [quantitatives; linear measure] There are three major world systems for classifying the size of shirts and blouses: the systems of Britain (the UK), of the United States, and of continental Europe.

In women's blouses,			*In men's shirts,*	
UK size	*US size*	*European size*	*UK/US size*	*European size*
32	30	38	14	36
34	32	40	14½	37
36	34	42	15	38
38	36	44	15½	39
UK size	*US size*	*European size*	16	40
40	38	46	16½	41
42	40	48	17	42
44	42	50	17½	43
46	44	52		

The UK/US shirt sizes originally corresponded to the circumference of the neck aperture with the top button done up, in inches. Although, at the smaller measures, the European numbers correspond fairly accurately with the equivalent value in centimetres, at the larger measures, the figures do not correspond.

shô [volumetric measure] Unit of liquid capacity in Japan.

1 shô = 100 SHAKU, 10 GO
= 1.8039 litre
(2 litres = 1.1087 shô)
= 3.175 UK pints, 3.806 US pints
[3 UK pints = 0.9449 shô
4 US pints (half a US gallon) = 1.0510 shô]
10 shô = 1 to
100 shô, 10 to = 1 KOKU

shoal of fish, of troubles [collectives] Both *shoal* and *school* derive in this sense from a Germanic verb meaning 'to distinguish', 'to tell apart', thus implying a corporate number of individual animals or elements. The ability 'to distinguish' in this way is inherent in the cognate English word *skill*.

shock [quantitatives] A word like *dozen* or *gross* referring to a specific number of units.

1 shock = 60 (units)

and in relation to the boards or staves that make up a cask,

4 shocks = 1 RING

(In the United States, however, in both these senses, the variant *shook* is at least as common as *shock*.)

The term was apparently first used in relation to specific items of merchandise imported from abroad, but was then applied to a heap or bundle of objects all identical (such as stalks of corn, or hairs on the head).

shoe sizes [quantitatives; linear measure] There are three major world systems for classifying the size of shoes: the systems of Britain (the UK), of the United States, and of continental Europe. The US and European systems in addition define men's and women's shoes differently.

In women's shoes,			In men's shoes,		
UK	*US*	*European*	*UK*	*US*	*European*
2	4	32-35	6½	7	38-39
3	5	35-36	7	7½	40
4	6	36-38	7½	8	41
5	7	38-39	8	8½	42
6	8	40	8½	9	43
7	9	41-42	9	9½	43-44
8	10	42-44	9½	10	44
			10	10½	44-45
			10½	11	45
			11	11½	45-46
			11½	12	47

shook *see* SHOCK

shooting measurements and units [sport] The most common forms of shooting contest centre on either rifle shooting or pistol shooting, although there are other less common events, such as shotgun shooting and crossbow shooting (the latter of which is in any case more often included in archery events).

Rifle shooting

The three basic classes of rifle competition are:

air rifle
small-bore
large-bore (in the United States, bigbore)

There are three small-bore competitions: the small-bore standard event, the small-bore free event, and the small-bore free three-position event. Similarly, large-bore competitions comprise the large-bore standard event and the large-bore free event. In small- and large-bore competitions, events may include firing at targets in any of three shooting positions: prone, kneeling, or standing.

air rifle events – usual range: 10 metres (30.48 feet)
  – target diameter: 46 millimetres (1.811 inches)
  – calibre/caliber: 4.5 millimetres (0.177 inch)
  – 10 practice (sighting) shots
  – 40 shots standing, over maximum 1 hour 30 minutes
small-bore events– usual range: 50 metres (55 yards)
  – target diameter: 162.4 millimetres (6.394 inches)
  – calibre/caliber: 5.6 millimetres (0.22 inch)
small-bore standard event
  – maximum rifle weight: 5 kilograms (11 pounds)
  – 6 practice (sighting) shots
  – 20 shots in each position (total: 60 shots), over maximum 2 hours 30 minutes; 4-member team competition maximum duration 10 target hours
small-bore free event (English match)
  – maximum rifle weight: 8 kilograms (17 pounds 10.2 ounces)
  – 15-20 practice (sighting) shots
  – 60 shots in the prone position, over maximum 2 hours
small-bore free three-position event
  – maximum rifle weight: 8 kilograms (17 pounds 10.2 ounces)
  – 10 practice (sighting) shots in each position
  – 40 shots in each position (total: 120 shots), maximum duration prone 1 hour 30 minutes, standing 2 hours, and kneeling 1 hour 45 minutes (total: 5 hours 15 minutes); 4-member team competition maximum duration 21 target hours
large-bore events – usual range: 300 metres (330 yards)
  – target diameter: 1 metre (30.48 inches)
  – maximum calibre/caliber: 8 millimetres (0.315 inch)

large-bore standard event
– 6 practice (sighting) shots in each position
– 20 shots in each position (total: 60 shots), over maximum 2 hours 30 minutes; 4-member team competition maximum duration 10 target hours
large-bore free event
– 10 practice (sighting) shots in each position
– 40 shots in each position (total: 120 shots), maximum duration prone 1 hour 30 minutes, standing 2 hours, and kneeling 1 hour 45 minutes (total: 5 hours 15 minutes)

Pistol shooting

There are two main forms of pistol shooting competition. In rapid-fire (or 'silhouette') shooting, competitors fire at targets that present themselves for a specified time (of 4, 6, or 8 seconds) from a distance of 25 metres (82 feet). In free pistol shooting, competitors fire at a fixed target over a range of 50 metres (55 yards).

rapid-fire/silhouette shooting
– 5 targets per competitor, 75 centimetres (29½ inches) apart
– target scoring: 1 (outside) to 10 (centre)
– calibre/caliber: 5.6 millimetres (0.22 inch)
– maximum pistol weight: 1.26 kilogram (2 pounds 12.4 ounces)
– 5 practice (sighting) shots, 1 at each target
– 60 shots in two courses of 30, in sets of 5
free pistol shooting
– target scoring: 1 (outside) to 10 (centre)
– calibre/caliber: 5.6 millimetre (0.22 inch)
– maximum pistol weight: 1.26 kilogram (2 pounds 12.4 ounces)
– 15 practice (sighting) shots
– 60 shots, in 6 series (new target per series), over maximum 2 hours 30 minutes

Shotgun shooting

The three major disciplines of shotgun shooting are:
Olympic trench (clay-pigeon) shooting
skeet (clay-pigeon) shooting
down-the-line (clay-pigeon) shooting

All three involve the use of the same equipment in both weapons (any shotgun, 12-gauge bore or smaller) and targets (ceramic discs of diameter 11 centimetres/ 4.33 inches, ejected forcefully upwards into the air by a strongly sprung 'trap').

In Olympic trench shooting and down-the-line shooting, targets are ejected from directly ahead in any of three directions away from the competitor; in skeet shooting, targets are fired from left or right across the visual field of the competitor. *See also* BIATHLON; CALIBRE, CALIBER.

shoran [physics; navigation] Radar device by which two fixed stations retransmit back the radio signals beamed to them by an approaching craft, the time taken for the craft to receive the signals (and any difference in the order in which they arrive) indicating the craft's exact position.

The term derives from the first syllables of <u>sho</u>rt <u>ra</u>nge <u>n</u>avigation. *See also* LORAN.

short hundredweight *see* HUNDREDWEIGHT; QUINTAL

short-swing [time] In commercial transactions (especially in the United States), describing a period for the completion of a transaction of six months or less.

short ton *see* TON

short waves [physics; telecommunications; medical] Radio waves in the wavelength range 10 to 200 metres (corresponding to frequencies of 1.5 to 30 megahertz). They are used for long-distance radio transmissions and (in the range 6 to 30 metres) for the heat treatment of rheumatic and other muscular disorders (by short-wave diathermy).

shot, jigger [volumetric measure] Small unit measure of alcoholic drink, served in a glass that may or may not also be called a jigger (and which may have the measure marked upon it).

$$1 \text{ shot/jigger} = 44.5 \text{ millilitres } (0.0445 \text{ litre})$$

=   just over 1½ UK fluid ounce (0.0783 UK pint)
=   1½ US fluid ounce (0.0940 US pint)

The term *shot* in this sense is used much as the same word is used to mean '(hypo-dermic) injection' (cf. *slug*). The jigger was probably the glass before it was the measure within the glass – but represented the measuring vessel, and was therefore always going up and down (jigging) between the bottle and the customer's glass.

shot-put(t) *see* ATHLETICS FIELD EVENTS

shower unit [physics] For cosmic rays as they pass through matter, the average path length that results in a 50 per cent reduction in their energy. For example, for air it is 230 metres (250 yards) and for human tissue it is about 30 centimetres (about 1 foot). *See also* CASCADE UNIT; COSMIC RAYS.

show jumping measurements and units [sport] The major dimensional elements of a show jumping course are:
maximum length in metres: number of obstacles (jumps) x 60
maximum height of any obstacle (except in puissance events): 1.7 metres (5feet 6.93 inches)
normal maximum spread (apart from water jumps): 2 metres (6 feet 6 inches)
maximum spread of water jump: 4-5 metres (13 feet 1½ inch to 16 feet 4.85 inches)
the horse and rider who complete the course with the fewest number of faults (in fences jumped improperly or jumps refused, or in time over a restricted duration: *see* FAULT) or, rarely, with the highest number of points, win
in the event of a tie, there may be a jump-off over a shorter but more difficult course
Show jumping also forms the third (day's) discipline of the three-day event: *see* THREE-DAY EVENTING MEASUREMENTS AND UNITS.

shrinkage [biology; quantitatives] In the slaughter of animals for meat, a technical term for the proportion by weight of an animal that is lost in initial transportation to the abattoir, in the actual slaughter, in the removal of unwanted and unusable parts, and finally in the curing of the meat.

sice [quantitatives] The number 6 on a dice, and especially the number 6 on one dice in a throw of two dice when the other dice is not a 6.
It should be (but generally is not) pronounced 'cease', which is then rather confusing when talking of the sice on a dice, but which reflects the word's immediate derivation from the French version of *six*.

sideband [physics; telecommunications] At slightly higher and slightly lower frequencies than the frequency of the carrier wave of a radio broadcast are sidebands, which also carry the transmitted information signal (for example, sound, video). In FM broadcasting, the main frequency is generally suppressed and the signal carried by one of the sidebands. *See also* FREQUENCY MODULATION.

sidereal day [astronomy] The time that elapses between consecutive passages of the vernal equinox (across the same meridian), equal to 23 hours, 56 minutes, and 4.091 seconds of mean solar time, counted from one noon to the next. *See also* SIDEREAL TIME.

sidereal month [astronomy] The time it takes for the Moon to make one complete orbit of the Earth (relative to distant stars), equal to 27.3217 sidereal days (27 days 5 hours 55 minutes 49.44 seconds). *See also* SIDEREAL TIME.

sidereal period [astronomy] The time it takes for an orbiting celestial object to return to the same position in its orbit (relative to distant stars). *See also* SIDEREAL TIME.

sidereal time [astronomy; time] Time counted in terms of the rotation of the Earth relative to distant stars (rather than with respect to the Sun, as in solar time).
The word *sidereal* derives through French from Latin *sider-* 'stars', a word that might have been borrowed from ancient Egyptian or a Semitic language (and that might represent a variant form of the word *star*).

sidereal year [astronomy] The time that elapses between two successive passages of the Earth through the same point in its orbit (relative to distant stars), equal to 365.25636 days. *See also* TROPICAL YEAR; SIDEREAL TIME.

siderostat [astronomy] An instrument that makes the light from a moving celestial object (such as the Sun) shine in a constant direction for a significant period of time. It consists of a mirror rotated (usually by clockwork) about two axes simultaneously.

Siegbahn unit [physics] Unit of X-ray wavelength, equal to 100.2 x 10^{-15} metre (approx. 10^{-4} nanometre). Named after the Swedish physicist Karl Siegbahn (1886-1978), it is known alternatively as an *X unit*.

siemens [physics] Unit of conductance in the SI system, equal to the conductance of a material of resistance 1 ohm when there is a potential difference (voltage) of 1 volt between its ends. It superseded the mho (reciprocal ohm), which has exactly the same value. It was named after the German-born British electrical engineer Karl Wilhelm (later William) Siemens (1822-83).

sieve mesh number [physics] Method of specifying the size of particles according to the size of sieve (that is, the aperture in a metal mesh) that they will pass through. Britain has adopted numbers that correspond to the number of meshes to the inch (with the size of the wire forming the mesh specified); in the United States, sieve number indicates the size of the aperture. The two are similar; for example: a mesh number of 100 equals an aperture of 152 micrometres (0.006 inch), whereas a sieve number of 100 equals an aperture of 149 micrometres.

sievert [physics] Unit of radiation dose equivalent in the SI system, equal to the dose in 1 hour from 1 milligram of radium (encased in 0.5 millimetres of platinum) at a distance of 1 centimetre.

$$1 \text{ sievert} = 8.4 \text{ röntgens (approx.)}$$
$$= 1 \text{ joule per kilogram}$$
$$= 21.6 \text{ curies per kilogram}$$

It was named after the Swedish physicist Rolf Sievert (1896-1966). *See also* X-RAYS.

sigma particle [physics] A relatively massive, short-lived elementary particle, a type of HYPERON with a rest mass equivalent to 1,190 mega-electron volts (1.19 x 10^9 electron volts).

sign [maths] A mathematical symbol, such as + (plus sign), – (minus sign), x (multiplication sign) and ÷ (division sign). Other examples include the factorial sign (!), square root sign ($\sqrt{}$) and integral sign ($\int$).

Also, a symbol that indicates whether a number is positive or negative (for example: +31 is a positive number, –13 is a negative number). By extension, the same signs are used in physics and chemistry to indicate positive and negative charge (for example: Na$^+$ is a positive ion, $-e$ is a negative charge).

Also, an old angular measure equal to 30 degrees (like the signs of the zodiac, after which it is named, occupying one-twelfth of a circle).

signal-to-noise ratio [physics] For an audio or video signal, the difference between the required signal and unwanted background 'noise', expressed in decibels.

signs and symptoms *see* SYMPTOMS AND SIGNS

signs of the zodiac *see* HOROSCOPE

Sikes [physics] An obsolete unit of specific gravity, devised in 1794 for use with a hydrometer for measuring aqueous solutions of alcohol. Gravities were expressed in degrees Sikes.

Silurian Period [time; geology] A geological PERIOD during the Palaeozoic or Primary era. The third period of the era, immediately following the Ordovician period, it corresponded roughly to between 438 million years ago and 408 million years ago, and was then followed by the Devonian period. Life forms of the Silurian period were almost entirely marine – trilobites, primitive slugs and snails, and brachiopods were common, and the first jawed fishes evolved. But this was also the period during which plants first appeared on land, and the mountains of Scotland and Scandinavia were formed.

The period is named after the ancient Celtic tribe known to the Romans as Silures, who lived mainly in the area that is now represented by the Welsh counties of Gwent and Glamorgan. Rocks – especially limestones – characteristic of the period are, however, more common in the area that surrounds the southern half of the Wales-England border.

For table of the Palaeozoic era *see* CAMBRIAN PERIOD.

silver anniversary, silver jubilee *see* ANNIVERSARIES

silver dollar [comparative values] A dollar coin made of, or principally containing, silver: *see* DOLLAR.

similar figures [maths] In geometry, two or more plane figures (polygons) whose corresponding angles are the same and whose corresponding sides are equal or in the same proportion. Solid figures may also be similar.

simple equation Another term for a LINEAR EQUATION.

simple harmonic motion (s.h.m., SHM) [physics] A type of periodic motion in which a particle's acceleration is always proportional to its distance from a fixed point (and always directed towards it). The time for one such oscillation is the *period*, and the maximum distance from the fixed point is the *amplitude*. There are many practical examples of s.h.m., such as a mass bouncing at the end of a light coiled spring, a swinging pendulum, and a vibrating guitar string. For each of these, a graph of displacement plotted against time (period) is a sine curve – another characteristic of s.h.m.

simple interest [maths] *see* INTEREST

simplex [maths] In (a Euclidean space of) a given number of dimensions, the figure or unit that has the minimum number of boundary points. Thus, in two dimensions (that is, in two-dimensional space), the simplex is a triangle because it is the figure with the fewest boundary points (3) that can be achieved in two dimensions. In three dimensions, the simplex corresponds to a tetrahedron.

simplify [maths] To shorten an algebraic expression by gathering like terms, or to reduce the numbers in a fraction to their lowest possible. For example: $4x^2 + 3x - x^2 + 2x$ simplifies to $3x^2 + 5x$, and the fraction $9/12$ simplifies to $3/4$.

simultaneous equations [maths] In algebra, a set of two or more (usually linear) equations incorporating two or more unknowns. Because the number of equations is equal to the number of unknowns, the equations can be solved by elimination and the values of the unknowns determined.

For example:

$$
\begin{aligned}
(1) \qquad y &= 4x + 3 \\
(2) \qquad 2y &= x - 1
\end{aligned}
$$

Substituting the value of y from (1) in (2),

$$
\begin{aligned}
2(4x + 3) &= x - 1 \\
\text{therefore} \qquad 8x + 6 &= x - 1 \\
\text{therefore} \qquad 7x &= -7 \\
x &= -1
\end{aligned}
$$

Substituting this value of x in (1),

$$
\begin{aligned}
y &= -4 + 3 \\
\text{therefore} \qquad y &= -1
\end{aligned}
$$

In terms of coordinate geometry, a pair of simultaneous linear equations in x and y (say) represents two straight lines, and the values of x and y that satisfy both equations (the solutions) represent the coordinates of the point at which the two lines intersect. So, for example, in respect of the values of x and y that satisfy the the simultaneous equations above ($x = -1$, $y = -1$) if the equations represent straight lines, they intersect at the point with coordinates $(-1, -1)$.

sine (sin) [maths] For an angle in a right-angled triangle, the length of the side opposite the angle divided by the length of the hypotenuse (longest side).

The term *sine* derives as a translation into medieval Latin of the Arabic *jaib* 'breast', 'bosom' (cf. modern French *sein*), referring to an arc or curve (originally in a shirt or blouse): *see* SINE CURVE.

sine curve [maths] The graphic representation of a SINE WAVE.

sine wave [physics] Periodic oscillations of constant frequency (such as simple harmonic motion) have a waveform that is a sine wave – the plot of amplitude against wavelength is the same shape as a plot of sine against angle.

single, singleton [quantitatives; sporting term] One, only (one), sole – and especially in various sports and games, one point, one trick, one card, one run, one base, etc.

Most etymologists derive the word *single* as a modification in Latin (*singulus*) of *simul* and *simil-* meaning 'one and the same' (in which case *single* is closely akin to *simple*; certainly the term *singleton* was devised – no earlier than the 1880s – on analogy with *simpleton*). At least possible, however, is its alternative derivation as a Latin diminutive of *signum* 'a distinguishing feature', 'a unique quality', thus 'a sign', 'something that makes (a person, a place, a thing) signal' . . .

single bed [square measure] The standard size for a single bed is 74 x 36 inches (6 ft 2 in. x 3 ft; 187.96 x 91.44 centimetres), although smaller and larger sizes are, of course, available.

single entry [comparative values] In old-fashioned ledger bookkeeping, a system by which each entry – whether a credit or a debit – occurs just once, and the latest entry always represents an overall cumulative balance.

singular *see* PERSON

sinh *see* HYPERBOLIC FUNCTION

sinking speed [speed] The rate of vertical descent of an aircraft or bird, without reference to its forward motion.

sinoatrial node, sinoauricular node [medicine; biology] The natural pacemaker of the heart, together with the limiting operation of the vagus nerve responsible for the PULSE.

sinusoidal [maths; physics] Describing a SINE WAVE or SINE CURVE.

siriometer [astronomy] An obsolete unit of extreme length, equal to 1 million astronomical units.

$$
\begin{array}{rcl}
1 \text{ siriometer} & = & 1{,}000{,}000 \text{ a.u.} \\
& = & 1.496 \times 10^{14} \text{ kilometres} \\
& = & 9.2957 \times 10^{13} \text{ miles} \\
& = & 16 \text{ LIGHT YEARS} \\
& = & 5 \text{ PARSECS}
\end{array}
$$

It was proposed for measuring the distances to stars (hence the allusion in its name to the Dog Star, Sirius), and in German-speaking countries was known as the *Siriusweit*.

Siriusweit [astronomy] The German name for the SIRIOMETER.

sistroid [maths] The area formed by the intersection of the convex sides of two curves.

sitar *see* STRINGED INSTRUMENTS' RANGE

SI units [units] Système International d'Unités, the modern system of units used in science and technology, and increasingly in trade and commerce. Its seven basic units are the metre (length), kilogram (mass), second (time), kelvin (temperature), ampere (electric current), mole (quantity of substance), and candela (luminous intensity). The radian (plane angle) and steradian (solid angle) are supplementary units, and there are eighteen derived units: becquerel, coulomb, farad, gray, henry, hertz, joule, lumen, lux, newton, ohm, pascal, siemens, sievert, tesla, volt, watt, and weber. All have separate articles in this book. *See also* METRIC PREFIXES, for the prefixes that form multiples and submultiples of SI units.

six [quantitatives] 6.

$$ ^6\!/_{10} \;=\; 0.6 \;=\; 60\% $$

A group of six is a sextet or hexad.

A POLYGON with six sides (and six angles) is a hexagon.

A solid figure (POLYHEDRON) with six plane faces is a hexahedron.

As is evident from the above, prefixes meaning 'six-' in English correspond to (Greek-based) 'hex-', 'hexa-', and (Latin-based) 'sex-', 'sexa-'.

The name for the number derives from common Indo-European, attested over the entire range of languages from Sanskrit to Welsh, and found in akin forms also in some of the non-Indo-European Nostratic languages. *See also* HALF A DOZEN.

six-eight time (6/8 time) [music] In music, a style that stresses the first of every six beats (and that therefore has six beats to the bar, each of a duration of a quaver or eighth-note); there is generally a secondary stress on the fourth beat of each six. A classic bar in 6/8 time thus comprises

♪♪♪ ♪♪♪

although of course it may be broken up into different units:

♩ ♪ ♩.   ♩♪♪ ♪♩

The most common result is like a jig, a fast waltz, or a waltz in double time. *See also* TIME SIGNATURES; THREE-FOUR TIME (3/4 TIME).

six feet, six-foot [linear measure] A useful measure in non-metric countries, applied to timber lengths, as a general measure in relation to the average height for men, and equal to 1 FATHOM.

$$6 \text{ feet} = 72 \text{ inches, 2 yards, 1 fathom}$$
$$= 182.88 \text{ centimetres, 1.8288 metres}$$
$$(2 \text{ metres} = 6 \text{ feet } 6.74 \text{ inches})$$

See also HEIGHT, HUMAN BODY; WEIGHT, HUMAN BODY.

six inches [linear unit] A useful short measure of length in non-metric countries.

$$6 \text{ inches} = 72 \text{ lines, one-half of 1 FOOT, one-sixth of 1 YARD}$$
$$= \tfrac{2}{3} \text{ of 1 span, } 1\tfrac{1}{2} \text{ hand}$$
$$= 15.24 \text{ centimetres, 0.1524 metre}$$

See also INCH.

sixpence [comparative values] Former unit of currency in Britain, abolished on the introduction of decimal coinage in 1971, but until then equal to (6 pence) half of 1 shilling, $\tfrac{1}{40}$ of 1 pound sterling.

The related adjective, *sixpenny*, is ironically still in use in North America, in relation to a specific size of household nail: a sixpenny nail is 2 inches (5.08 centimetres) long: *see* NAILS.

sixteen [quantitatives] 16.

$$16\% = 0.16$$
$$16 \text{ ounces avdp.} = 1 \text{ pound}$$
$$16°C = 61°F$$

The number system that has a base (radix) of 16, commonly used in computers, is called the hexadecimal system: notation in the system is ordinarily described as 'hex'.

The name of the number derives through Germanic sources and is literally '6, 10'.

sixteen-millimetre [photography] A gauge of cine film, used mainly by amateurs and by news-gathering teams (although now largely superseded in this role by video cameras). It was also the origin of the amateur 8-millimetre size, which was made originally by slitting 16-millimetre film down the centre.

sixteenmo *see* SEXTODECIMO, 16MO

sixteenth-note *see* SEMIQUAVER

sixteenth of an inch [linear measure] Smallest measure of length on an ordinary foot rule.

$\tfrac{1}{16}$ inch			=	1.5875 millimetres
$\tfrac{2}{16}$ inch	=	$\tfrac{1}{8}$ inch	=	3.1750 millimetres
$\tfrac{3}{16}$ inch			=	4.7625 millimetres
$\tfrac{4}{16}$ inch	=	$\tfrac{1}{4}$ inch	=	6.3500 millimetres
$\tfrac{5}{16}$ inch			=	7.9375 millimetres
$\tfrac{6}{16}$ inch	=	$\tfrac{3}{8}$ inch	=	9.5250 millimetres
$\tfrac{7}{16}$ inch			=	1.11125 centimetres
$\tfrac{8}{16}$ inch	=	$\tfrac{1}{2}$ inch	=	1.27000 centimetres
$\tfrac{9}{16}$ inch			=	1.42875 centimetres
$\tfrac{10}{16}$ inch	=	$\tfrac{5}{8}$ inch	=	1.58750 centimetres
$\tfrac{11}{16}$ inch			=	1.74625 centimetres
$\tfrac{12}{16}$ inch	=	$\tfrac{3}{4}$ inch	=	1.90500 centimetres
$\tfrac{13}{16}$ inch			=	2.06375 centimetres
$\tfrac{14}{16}$ inch	=	$\tfrac{7}{8}$ inch	=	2.22250 centimetres
$\tfrac{15}{16}$ inch			=	2.38125 centimetres

In typography,

$$\tfrac{1}{16} \text{ inch} = 4\tfrac{1}{2} \text{ points}$$
$$= 4.222 \text{ Didot points}$$

sixth [quantitatives] $\tfrac{1}{6}$ (one-sixth); the element in a series between the fifth and the seventh, or the last in a series of six.

$\tfrac{1}{6}$			=	0.1666	=	16.6666%
$\tfrac{2}{6}$	=	$\tfrac{1}{3}$	=	0.3333	=	33.3333%
$\tfrac{3}{6}$	=	$\tfrac{1}{2}$	=	0.5	=	50%
$\tfrac{4}{6}$	=	$\tfrac{2}{3}$	=	0.6666	=	66.6666%
$\tfrac{5}{6}$			=	0.8333	=	83.3333%

See also ANNIVERSARIES; SIX.

sixth [music] In music, the interval between a keynote (tonic) up to the submediant of the key (*see* SCALE), whether in the major or in the minor. For example: in the key of C major, the interval between C and the A above it (and the sound of those notes when played) is a major sixth; in the key of A minor, the interval between A and the F above it (and the sound of those notes when played) is a minor sixth.

Like a THIRD (to which it is effectively the obverse), the sixth is thus a harmony that defines whether a chord to which it contributes is major or minor.

sixth of a gill [volumetric measure] In England and Wales, and in some other countries (but not the United States), the standard measure for a tot of spiritous liquor, as sold in bars.

$$
\begin{aligned}
\text{\frac{1}{6} (UK) gill} \ &= \ \text{\frac{1}{24} UK pint, \frac{5}{6} UK fluid ounce} \\
&= \ \text{0.0417 UK pint, 0.8333 UK fluid ounce} \\
&= \ \text{2.3677 centilitres, 0.023677 litre} \\
&= \ \text{just over \frac{1}{20} US pint} \\
&= \ \text{0.8025 US fluid ounce}
\end{aligned}
$$

    See also GILL.

sixty [quantitatives] 60.

$$
\begin{aligned}
60\% \ &= \ 0.6 \ = \ \tfrac{3}{5} \\
60 \text{ seconds} \ &= \ 1 \text{ minute} \\
60 \text{ minutes} \ &= \ 1 \text{ hour}
\end{aligned}
$$

The number 60 was the base (radix) of the Persian number system of ancient Babylon: *see* BASE; SAROS.

The name of the number derives from the Germanic variants of the common Indo-European term combining elements that mean '6 (times) 10'. *See also* SIX.

sixty-fourth note *see* HEMI-DEMI-SEMIQUAVER

skaal-pund [weight] Former unit of weight in Sweden, approximating fairly closely to the avoirdupois pound, and better known simply as the PUND.

skaeppe, skjeppe [volumetric measure] Former measure of dry capacity in Denmark and Norway, but, unlike the equivalent measures in Germany and Holland, not since assimilated to units in the metric system.

    In Denmark,

$$
\begin{aligned}
1 \text{ skaeppe} \ &= \ 18 \text{ (Danish) pots} \\
&= \ 17.407 \text{ (dry) litres, } 17,407 \text{ cubic centimetres} \\
&= \ 3.8291 \text{ (dry) UK gallons, } 3.9482 \text{ (dry) US gallons} \\
&= \ 0.4786 \text{ UK bushel, } 0.4935 \text{ US bushel} \\
&= \ 0.6145 \text{ cubic foot} \\
2 \text{ skaeppe} \ &= \ 1 \text{ fjerding} \\
8 \text{ skaeppe, 4 fjerdings} \ &= \ 1 \text{ (dry) tønde}
\end{aligned}
$$

    In Norway

$$
\begin{aligned}
1 \text{ skjeppe} \ &= \ 18 \text{ (Norwegian) pots} \\
&= \ 17.370 \text{ (dry) litres, } 17,370 \text{ cubic centimetres} \\
&= \ 3.8214 \text{ (dry) UK gallons, } 3.9398 \text{ (dry) US gallons} \\
&= \ 0.4777 \text{ UK bushel, } 0.4925 \text{ US bushel} \\
&= \ 0.6134 \text{ cubic foot} \\
8 \text{ skjeppe} \ &= \ 1 \text{ korntonde}
\end{aligned}
$$

The usual translation of both terms in English is 'bushel' – even to the extent of the Biblical expression that one should not hide one's light under it. The same is true of the German SCHEFFEL and the Dutch SCHEPEL – all four of which must be etymologically identical, although the English/American bushel is by far the largest.

In medieval English there was a container called a *skep*, which was the English equivalent of these measures. Its only use today is as a term for a type of beehive of the old cubic measure, but its original meaning was as a container that had been fashioned or *shaped*. Of a different (but no specific) size, the mining *skip* or container (and hence the general container for refuse and rubble) is also cognate. (Perhaps compare also *scoop*, and thus *cup* and *gowpen*.)

skating *see* ICE-SKATING DISCIPLINES AND EVENTS

skein [textiles, collectives] The yarn industry measures its wares in yards (or less commonly metres) of yarn in skeins (or, in the United States, leas), seven of which together form a HANK, twenty of which in turn make up a BUNDLE. The actual number of yards (or metres) in the skeins differs according to the composition of the yarn.

In cotton yarn, for example,

1 skein	=	120 yards of yarn
7 skeins	=	1 hank, 840 yards of yarn
140 skeins, 20 hanks	=	1 bundle, 16,800 yards of yarn

Whereas in worsted yarn,

1 skein	=	80 yards of yarn
7 skeins	=	1 hank, 560 yards of yarn
140 skeins, 20 hanks	=	1 bundle, 10,200 yards of yarn

The only difference between the skein and the lea is that the skein by definition is a rolled or folded length of yarn whereas the lea is a simple linear unit. The fold appears to be important to the term for, as a collective noun for birds in flight, it specifically refers to the V-formation that small flocks of geese or ducks tend to adopt. But, although the word comes to English via medieval French, its earlier etymology is unknown.

skep *see* SKAEPPE, SKJEPPE

skew distribution [maths] In statistics, a set of data that is asymmetric about its central value – that is, the distribution curve is skewed. *See also* BELL CURVE.

skiing disciplines and events [sport] The principal skiing contests are:

Alpine skiing
  the downhill race
  the slalom
  the giant slalom
  the supergiant slalom
  the combined (downhill, slalom, and giant slalom)
ski jumping
cross-country skiing (skilaufen)
  the Nordic cross-country race
  the cross-country relay race
  the combined (ski jumping and cross-country)
  the BIATHLON
freestyle skiing
  the moguls
  the aerials
  ski ballet

All events except ski jumping, the aerials, and ski ballet are (at least partly) scored on timing: the skier who completes the course in the shortest time, to the satisfaction of the officials, wins.

The word *skiing* is backformed from the noun *ski*, which is adopted in English from a Norwegian word deriving from Old Norse *skith* 'a split log', found in present-day English also in the form *skid*, a runner on which a sled glides or an aircraft lands on ice; with the root meaning 'split', 'separate', it is ultimately akin to (Greek-based) English words beginning with the elements *schis-* and *schizo-*, and thus additionally to such words as *scatter*, *shatter*, and *shed*.

skilling [comparative values] Former unit of currency in Sweden (till 1858), Denmark (till 1875), Iceland (till 1876), and Norway (till 1877), in all cases in the form of a copper coin.

In Sweden,

48 skilling banco	=	1 riksdaler

In Denmark,

96 rigsbank skilling	=	1 rigsdaler

In Iceland,

96 skilling	=	1 riksdaler

In Norway,

120 skilling	=	1 speciedaler

The countries at the above respective dates changed their currency to the *öre/ krone* (or equivalent) system still in operation today.

The name of the unit is evidently the same as that of the SHILLING that is part of the currency of Kenya, Somalia, Tanzania, and Uganda at the present time, and which was formerly part of the currency system of the United Kingdom (till 1971).

It probably derives from the shield that was depicted on very early examples (as a 'shield-ing'), although some etymologists suggest an echoic derivation (cf. German *Schall* 'resonance'). *See also* RIXDOLLAR, RIJKSDAALDER, RIKSDALER, RIGSDALER.

skin depth [physics; telecommunications] For a high-frequency radio wave (which travels mostly in a thin layer of metal at the surface of a conductor), the depth at which the strength (amplitude) of the signal has fallen to $1/e$, or 0.3679, of its surface strength.

skip, skep *see* SKAEPPE, SKJEPPE

skjeppe *see* SKAEPPE, SKJEPPE

skot [physics] Obsolete unit of luminance in poor lighting levels, equal to 3.1931 x 10^{-4} candela per square metre. It was used in Germany during World War II to measure lighting during the blackout. The term derives from the German form of the English word *scotopic*, meaning 'to do with night vision' (the first element is ancient Greek *skotos* 'darkness').

$$1 \text{ skot} = 10^{-3} \text{ apostilb}$$
$$= 10^{-3} \text{ lumen per square metre}$$

slalom skiing (and giant slalom/supergiant slalom) *see* SKIING DISCIPLINES AND EVENTS

slam [sporting term] In card games in which tricks are taken, the taking of all or almost all of the available tricks in a hand. In contract bridge, for example, the taking of twelve tricks of the thirteen possible is a small (or little) slam; the taking of all thirteen is a grand slam.

The word is a Germanic intensive (*s-*) on the verb *lam*.

slant height [maths] The distance from the base of a cone or pyramid to its apex, measured along the sloping side: *see* CONE.

slate sizes [square measure] Traditional roofing slates come in various sizes, described – as befits the highest part of a building – predominantly by the titles of nobility.

name	inches			millimetres		
countess	20	x	10	508	x	254
double	13	x	6	330	x	152
empress	26	x	16	660	x	406
imperial	33	x	24	838	x	610
lady	16	x	8	406	x	203
marchioness	22	x	12	559	x	305

slew, slue [quantitatives] Informal term in the United States meaning a large quantity or number. It may or may not be etymologically akin to Irish *slua* (often less correctly spelled *sluagh*) 'crowd', 'host', 'multitude'.

slide rule [maths] An instrument for doing calculations, particularly multiplication and division. It consists of two adjacent logarithmic scales that slide past each other. Multiplication is achieved by adding the scales (at the point they correspond to the multiplier and the multiplicand), and division by subtracting the scales (when they represent the dividend and the divisor). Other scales are calibrated in trigono-metrical ratios, square roots, and so on.

slue, slew *see* SLEW, SLUE

slow neutron [physics] In the products of nuclear fission, a low-energy neutron which travels sufficiently slowly to be readily absorbed by a nucleus and thus sustain a chain reaction. In a nuclear reactor, fast neutrons are slowed by a mod-erator to achieve this effect.

slug [physics] Unit of mass in the foot-pound-second (FPS) system, being the mass that, acted on by 1 pound-force, is given an acceleration of 1 foot per second per second. It is therefore also numerically equal to the value of the acceleration of free fall (acceleration due to gravity) in the FPS system: hence its alternative name of *gee pound*.

$$1 \text{ slug} = 32.1740 \text{ pounds}$$
$$= 14.594 \text{ kilograms}$$

The term in this sense derives from an old Germanic verb meaning 'to hit' (cf. modern German *schlagen* 'to smite', English *slog*).

slugging average [sporting term] In baseball, the total number of bases reached in a season by a batter divided by the total number of times at bat, expressed as a percentage. This is a measure of how effective the batter is at making extra-base hits.

small-bore [linear measure] An imprecise term applied to small-calibre competition rifles (as opposed to large-calibre weapons, which are termed full-bore). For example: the popular calibre of .22 inches is regarded as small-bore, whereas .300 or .303 inches is classified as full-bore. *See also* CALIBRE, CALIBER.

small circle [maths] A circle formed by sectioning a sphere with a plane that does not go through its centre. *See also* GREAT CIRCLE.

small intestine [medicine] The part of the digestive system that constitutes the alimentary canal immediately following the stomach, comprising the duodenum, the jejunum, and the ileum. Material being digested passes from the small intestine into the large intestine on its way to the rectum.

The human small intestine, which occupies much of the abdominal cavity and is coiled many times, is ordinarily about 6.4 metres (21 feet) long and 3.5 centimetres (1.38 inch) in diameter. The large intestine is much wider but much shorter.

small pica [literary] In typography, a term denoting 11-point type (as opposed to PICA, which is 12-point type). *See also* POINT.

snatch *see* WEIGHTLIFTING BODYWEIGHTS AND DISCIPLINES

snellen [physics; medicine] Unit of the human eye's visual power. It was named after the Dutch ophthalmologist Herman Snellen (1834-1908).

Snell's law [physics] For a ray of light passing from one transparent medium to another, the product of the refractive index of the first medium and the sine of the angle of incidence equals the product of the refractive index of the second medium and the angle of refraction. It was named after the Dutch mathematician and physicist Willebrord Snell van Roigen (1591-1626), also known as Snellius. *See also* REFRACTION, LAWS OF.

snooker measurements and units [sport] The object of the game of snooker is to score by cueing the white ball to knock designated balls into pockets around the edge of the table. A red ball must first be pocketed, after which any ball not coloured red or white may be pocketed, scoring accordingly; another red ball should then be pocketed before another colour, and so on until the player fails to pocket a ball. The opponent then takes his or her turn likewise, and the game continues until only the white cue ball remains on the table. While there are red balls still on the table, each coloured ball pocketed is replaced on its specific spot. When only coloured balls are left, the balls must be pocketed in order of successively increasing value.

The dimensions of the table:
   full-size table (table-top measurements, inside cushion of green baize): 12 feet x 6 feet 1½ inches (3.66 x 1.86 metres); smaller tables in proportion
   string line: across the width 2 feet 5 inches (73.7 centimetres) from the top cushion
   semicircle ('D') on string line: radius – 6 inches (15.2 centimetres)
   pocket widths:
      corner pockets – 3½ inches (8.89 centimetres)
      middle pockets – 4 inches (10.16 centimetres)
Timing: as long as it takes for an agreed number of frames (or rounds) to be completed
Points scoring:
   red ball pocketed:1 point
   (two reds pocketed in one shot: 2 points)
   coloured balls' values: yellow – 2 points
                           green – 3 points
                           brown – 4 points
                           blue – 5 points
                           pink – 6 points
                           black – 7 points
   points against (and end of break):
      cue ball misses red, hits nothing: lose 4 points
      cue ball misses red, hits/pockets colour: lose 4 points or more valuable

colour's points

cue ball misses colour, hits nothing: lose 4 points or more valuable colour's
  points

cue ball misses colour, hits/pockets red: lose 4 points

cue ball pocketed: lose 4 points

ball forced off the table: lose 4 points or more valuable colour's points

cue ball aimed at second red without colour between: lose 7 points

coloured ball hit after being incorrectly replaced on table: lose 4 points or
  more valuable colour's points

red or coloured ball used as cue ball by mistake: lose 4 points or more
  valuable colour's points

the game is made up of an agreed number of frames; the first player to more
  than half of the total number of frames wins

The name of the game (originally called *snooker pool*) derives from the practice,
when it is impossible to pocket a ball, to at least make it extremely difficult for the
opponent to pocket one, by contriving that the cue ball is left behind a coloured ball
that obscures all the red balls. This is 'snookering' the opponent.

snow line [geography; biology] The contour on a high mountain above which the
snow never melts.

Also, the varying latitude around the North and South Poles within which there is
always snow or ice.

soccer measurements, units, and positions [sport] Soccer is probably the
most popular field team game in the world, and its technical terms are probably
equally international. The object of the game is to score goals by kicking or heading
the ball into the opponents' net; hands may not be used. Each team comprises
eleven players on the field, including a goalkeeper.

The dimensions of the pitch:

length limits: 100-130 yards (91.44-118.87 metres)

width limits: 50-100 yards (45.72-91.44 metres)

  this allows considerable diversity in playing areas

centre circle diameter: 20 yards (18.29 metres)

penalty area: 44 x 18 yards (40.23 x 16.46 metres)

goal area: 20 x 6 yards (18.29 x 5.486 metres)

the goal:

  width: 24 feet (7.315 metres)

  height: 8 feet (2.438 metres)

radius of corner arc by corner-flags: 3 feet (91.44 centimetres)

Positions of players: from the back

goalkeeper

sweeper ('libero'; not always a set position)

right back, centre-back, left back

right-half, centre-half, left-half

right wing, centre-forward (sometimes 2), left wing

Timing: 90 minutes in two halves, separated by a 15-minute interval

in some specific competitions, in the event of a tie, two 15-minute periods of
  extra time (in the United States, overtime) are played, separated by an
  interval of 3 minutes

Points scoring:

the team that has a greater number of goals at the end of the game wins; in
  most matches, a tie or draw may represent a result

in some specific competitions, if the tie still persists, a penalty competition is
  then held, 5 penalties being taken against respective opponent goalkeepers by
  five different members of each team – and if the tie still persists, the penalty
  competition thereafter becomes a sudden-death contest in pairs of penalties –
  the first team to fail to score when the other team has already scored or then
  goes on to score, loses

Dimensions of the ball:

circumference: 27-28 inches (68.58-71.12 centimetres)

weight: 14-16 ounces (397-453.6 grams)

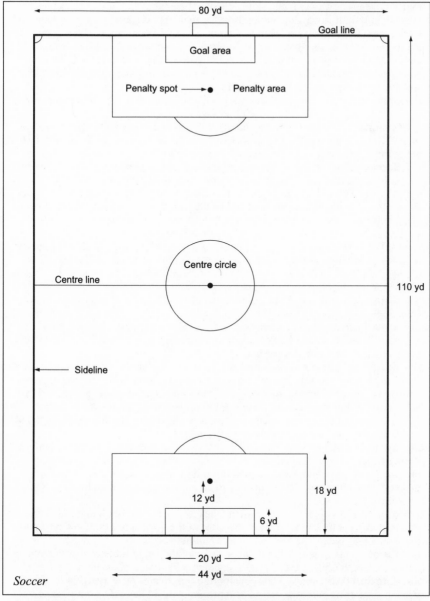

Soccer

internal pressure: 15 pounds per square inch (718.26 newtons per square metre)

The name of the game is technically 'Association football' (that is, the football for which the English authority called the Football Association was responsible), and *soccer* is really a sort of pet nickname derived from the first of those words. In many regions of the world, the game is known simply as 'football' anyway.

soda [sporting term] In (the card-game) faro, the top card in the dealing box as play begins, displayed face-up.

The derivation of the term used in this sense is unknown.

soft radiation [physics] Low-energy radiation, such as X-rays, with little penetrating power (as opposed to high-energy hard radiation). The energy of X-rays can be varied by adjusting the high voltage delivered to the X-ray tube.

soil horizons, soil layers [geology; chemistry] Geologists distinguish three layers of soil above true bedrock, known as the A horizon, the B horizon, and the C horizon. The topmost is the A horizon, which comprises soil through which moisture seeps downward, dissolving chemical elements like sodium, magnesium, and calcium. In this way, the minerals in the moisture eventually enter streams and rivers, from where they are taken to the sea. The middle layer, the B horizon, comprises soil containing primarily insoluble substances such as iron oxides and clay, deposited over time through the A horizon. At the bottom is the C horizon, which comprises a combination of decomposed rock and shaly materials in which soil as such has not as yet begun to form.

sol [comparative values] Unit of currency in Peru, in the form of a silver coin.

$$1 \text{ sol } = 100 \text{ centavos}$$

It is probable that the Peruvian unit takes its name directly from Spanish *sol* 'sun', with reference to the importance of the solar deity to the Quechua civilization at the time of the arrival of the Spanish *conquistadores*.

In former centuries, the sol was additionally a unit of currency in France, worth ¹⁄₂₀ of the somewhat unstable LIVRE, and later to turn into the SOU.

Both these French units derive their name ultimately from that of the Latin coin the *solidus* 'the big (solid coin)', through Old French *sold* 'cash payment', especially 'military pay' – which is how a man who received military pay became a *soldier*. *See also* COINS AND CURRENCIES OF THE WORLD.

sol [time] One Martian day – the length of the day on Mars, specifically 24 hours 37 minutes 22 seconds, corresponding to one rotation of the planet on its axis. There are 669.774 Martian days to 1 Martian year (the equivalent of 686.98 Earth days in time). *See also* PLANETS OF THE SOLAR SYSTEM.

solar antapex [astronomy] The point on the celestial sphere directly opposite the solar apex.

solar apex [astronomy] The point on the celestial sphere, in the constellation Hercules, towards which the Solar System is moving (at a speed of 20 kilometres per second, or 44,739 miles per hour).

solar constant [physics; astronomy] A measure of the Sun's energy that reaches the outer edge of the Earth's atmosphere – on a clear day very nearly the same as the amount of energy reaching the Earth's surface – per square metre, per second. Not actually constant, it is equal to around 1,350 joules per square metre per second (or 1.35 kilowatts of power per square metre); it is sometimes alternatively calculated as 1.9 calories per square centimetre per minute.

solar cycle [astronomy; time] The 11.1-year cycle of activity on the Sun's surface, during which sunspots appear and fade, the frequency of solar flares and solar prominences waxes and wanes, and at the end of which the polarity of the Sun's magnetic field reverses. The frequency of an aurora, as sighted on Earth, is similarly linked to the solar cycle.

solar day [astronomy; time] The time it takes for a place on Earth directly facing the Sun to make a complete revolution (because of the Earth's rotation) and return to the same place facing the Sun: 23 hours 56 minutes.

solar parallax [astronomy] The average value of the angle subtended at the Sun by a line the length of the Earth's radius, equal to 8.79405 seconds of arc.

solar rotation [astronomy] At its equator, the Sun rotates once every 24.65 days, whereas nearer its poles the period is approximately 34 days. The time it takes a point on the Sun's equator to return to the same place – the synodic period as viewed from Earth – is 27 days.

Solar System [astronomy] The Solar System comprises one star – the Sun – and a number of much smaller bodies (planets and asteroids, comets) held by gravitational attraction in orbit around it. The planets all orbit the Sun in approximately the same plane (the ecliptic). Light from the Sun travels to the outermost planet (generally Pluto, but occasionally Neptune) in less than six hours. Farther out still, at a distance of about one light year, there is thought to be a huge 'cloud' of cometary material and dust, called the Oort Cloud, individual parts of which at times assume eccentric orbits.

For statistical information on the Sun, *see* SUN.

The planets are often categorized in two groups. Those nearer the Sun (Mercury, Venus, Earth, and Mars) are the *terrestrial planets* – small but dense, with a solid outer surface and formed mostly of silicates and heavier elements. The *Jovian planets* (Jupiter, Saturn, Uranus, and Neptune), on the other hand, are much larger but less dense, surrounded by an atmosphere composed mainly of hydrogen and helium, retaining many of the characteristics of the gaseous envelope from which they formed. The planet Pluto resembles the terrestrial planets in its small size but the Jovian planets in its low density.

For statistical information on the planets, *see* PLANETS OF THE SOLAR SYSTEM. For statistical information on the moons orbiting those planets, *see* SATELLITES OF THE SOLAR SYSTEM.

The Solar System (probably) originated as a swirling cloud of gas and dust which, for one reason or another, began to contract under its own weight. Influenced by gravity, the contracting cloud gradually took the form of a flattish rotating disc, its mass, temperature, and density greater at the centre and less at the outer edge. Some 4,600 million years ago, the Sun finally condensed out at the centre of the disc, then by far the hottest and densest part of the whole. Once the thermonuclear reactions began at the Sun's core, its luminosity dimmed and the material that had formed the surrounding disc cooled as its gaseous envelope expanded. Parts of this envelope solidified in discrete masses – of heavier elements near the Sun, of frozen gases farther away. Through accretion and multiple collisions, the discrete masses gradually built up into bodies of planetary dimensions, the process halting once the bodies reached a diameter of about 1,000 kilometres (600 miles). The planetary bodies at that point took on independent growth, largely due to the effects of gravitational interaction, resulting eventually – after some 100 million years – in the planets we know today.

solar year [astronomy] The time it takes for two successive passages of the Sun through the First Point of Aries – or for the Earth to make one complete revolution around the Sun, beginning and ending at the vernal equinox – equal to 365.24219 solar days (or 365 days 5 hours 48 minutes 45.51 seconds). It is known alternatively as the *tropical year* or an *astronomical year*.

soldo [comparative values] Former unit of currency in Italy, in the form of a copper coin.

$$20 \text{ soldi} = 1 \text{ lira}$$

The name of the unit derives from Latin *solidus* 'the big (solid coin)', and the term is thus cognate with the French SOL and SOU.

sol-fa, solfeggio *see* SCALE

solid angle [maths] The three-dimensional angle subtended at the centre of a sphere by a given area on its surface: *see* STERADIAN.

Similarly, but without reference to a sphere, the angle at a point where three or more planes intersect.

solid of revolution [maths] A three-dimensional (that is, solid) figure generated by rotating a plane figure about an axis. For example: a circle rotated about its diameter generates a sphere and a rotating ellipse forms an ellipsoid. The areas and volumes of solids of revolution can be found by using integral calculus.

solidus [comparative values] Unit of currency in ancient Rome, in the form of a gold coin introduced by the Emperor Constantine (ruled AD 313-37). Not part of the standard Roman currency system, its value varied according to the value of the gold in the coins, but in general it was worth around 25 DENARII.

It was described in Latin as *solidus nummus* 'the big-and-solid coin', a name that later in part was applied to the Italian SOLDO and the French SOL and SOU. After Constantine moved his capital from Rome to Byzantium (now Istanbul) in AD 330, the solidus took on the alternative name (now spelled) *bezant*.

solidus, oblique, virgule [literary] Three names for the symbol / , used, for example, to separate alternatives (1.6 kilometre/1 mile), to denote fractions (3/4, 4/5), and formerly to denote shillings and pence (2 shillings and sixpence = 2/6). *See also* DATE.

solmization *see* SCALE

solo [music] Piece of music for one voice or one instrument only (with or without accompaniment).

The term is the medieval (and modern) Italian for 'alone', cognate with English *sole*.

solstice [astronomy; time] Either of the two times every year when the Sun reaches its most north-westerly (mid-winter) or most south-westerly (mid-summer) position away from the equator, corresponding in the first case to the 'shortest day' of the year (in the Northern Hemisphere, 21 December), and in the second case to the 'longest day' (in the Northern Hemisphere, 21 June).

Also, either of a pair of points on the ecliptic that are midway between the equinoxes: *see* EQUINOX.

The related adjective is *solstitial*.

The first element of the word *solstice* is the Latin for 'the Sun'; the other element derives from a Latin version of the Greek *stasis* 'a standing still'.

solubility [chemistry] The quantity of a substance (the solute) that will dissolve in a liquid (the solvent) at a particular temperature. It may be expressed in any of the units of concentration (*see* CONCENTRATION).

solubility product [chemistry] The product of the activities (concentrations) of the ions in a solution that is saturated with an electrolyte (a substance that dissociates into ions in solution).

solute *see* SOLUBILITY

solution [maths] The answer to any arithmetical problem.

Also, more technically, the ascription of a (whole) number to one or more of the algebraic symbols in one or more linked algebraic equations. For example: in the equation $x^2 + 3y = 25$, a solution for x is 4 if the solution for y is 3.

solution [chemistry] A liquid consisting of a solid (the solute) dissolved in a liquid (the solvent). Solutions may also be formed by dissolving a liquid in another liquid; solid solutions also exist. *See also* SOLUBILITY.

solvent *see* SOLUBILITY; SOLUTION; CONCENTRATION

somatotype [biology; medicine] Any of the three body shapes discerned as characteristic of human morphology: *see* ECTOMORPH; ENDOMORPH; MESOMORPH.

Also, in comparative anthropology, any unusual but constant human body shape characteristic of a group or community.

some, -some [quantitatives] A number (of), a few – but not all.

By derivation, <u>some</u> represent an as<u>sembl</u>y (of elements that are) all the <u>same</u> or at least <u>sim</u>ilar. It is in the sense of 'assembly' that the word is used as a suffix after numerals – for example, twosome, threesome, foursome.

sonar [physics; navigation] The use of reflected ultrasonic waves for echolocation, especially in detecting objects or measuring depth under water.

The term derives from the first letters of <u>so</u>und <u>na</u>vigation <u>r</u>anging. *See also* RADAR.

sonata form [music] Standard form of a movement (often the first movement) of a Classical symphony or concerto, comprising the *exposition* (in which the themes or 'subjects' are introduced, and may be repeated), the *development* (in which variations on the themes may be introduced, with subordinate or new 'subjects'), and a *recapitulation* (in which the original themes are reintroduced, in the original scoring or in a more complex scoring), the whole then (usually) being rounded off with a *coda* (a final section which may or may not contain elements of the original subjects).

sonde [meteorology] A device carried in a balloon, rocket, or satellite for sending back data (telemetry).

The word is the French for 'lead', the plumbline with which the depth of water may be gauged or *sounded*. *See also* RADIO SONDE; ROCKET SONDE.

sone [physics] Unit of subjective loudness, equal to a tone of frequency 1 kilohertz at a level 40 decibels above the listener's threshold of hearing.

$$1 \text{ sone} = 40 \text{ phons}$$

The term derives as an adaptation of the Latin *sonus* 'sound'. *See also* PHON.

sonic boom *see* SPEED OF SOUND; MACH NUMBER

sonnet *see* VERSE FORMS

sonograph [physics] A device that produces a visual representation of sound (a sonogram or sound spectrograph); also, the representation of sound that the device produces. If the sound is speech, the representation may (for instance) be in the form of phonetic symbols. Probably the most complex form of sonograph is the three-dimensional 'picture' of sound in which the coordinates are frequency, intensity, and time.

soprano range [music] The highest range of the human voice, natural to many females and to young boys before their voices 'break' (although the boys may instead be described as 'trebles'), generally considered to span from middle C to the note F above the next higher C: a range of about an OCTAVE and a half.

A mezzo-soprano has a vocal range between those of a soprano and a contralto.

Various families of instruments – recorders and saxophones, for instance – include a soprano member with a range above that of the alto member; saxophones and trumpets even have a range above the standard soprano known as the *sopranino*.

For the etymology of *soprano*, *see* SOVEREIGN.

soroban [maths] The Japanese form of abacus. Meaning both 'abacus' and 'account' in Japanese, the word nonetheless probably derives ultimately from Chinese *suan p'an*, literally 'counting-board', but actually a bamboo abacus.

sorting [comparative values] Unit of value in bartering between European maritime traders and coastal West African dealers in the 1600s and 1700s. A sorting commonly comprised an established quantity of cloth or oil, but the barter value of the sorting varied according to contemporary supply and demand.

sostenuto [music] Musical instruction: extend the duration of a note, or progressively slow the tempo so as to hold each note in a passage by an increasing proportion. Italian: 'sustained'.

Sothic year, Sothic cycle [time] The Sothic year was the ancient Greek term for the ancient Egyptian year of 365 days, as based on careful measurement of the exact return of the Dog Star, Sirius (Greek *Sothis*), to its position in the night sky at the end of that period. By this measurement, the ancient Egyptians were well aware that the actual year was near enough 365¼ days, but they reckoned that the loss of one day every four years made little or no difference during an individual lifetime (at a time when the average lifespan was less than forty years), and in any case they knew that in 1,460 years – the Sothic cycle (or Sothic period: 4 x 365 years) – the year would have revolved completely back to its original calendar position. *See also* YEAR.

sou [comparative values] Any of quite a number of former units of currency in France, initially in the form of gold coins, but thereafter decreasing in value as silver, then copper, then bronze coins. Finally, after corresponding to the name for a 5-centime piece (one-twentieth of a franc), the word is now used of virtually anything of little or no value. How very different from its inception as the Latin *solidus* 'the big (solid coin)'. *See also* SOL; SOLIDUS.

sound [physics] The sensation that results when waves of alternate compression and rarefaction of air strike the eardrum; regular mechanical vibrations in a gas, liquid, or elastic solid (sound waves will not travel in a vacuum). The ears of a young adult can detect sounds in the approximate frequency range 20 hertz to 20 kilohertz. *See also* INFRASOUND; ULTRASOUND.

sound, speed of *see* SPEED OF SOUND

sound intensity [physics] For a travelling sound wave, its flux of power per unit area (at right-angles to the direction of travel).

sound intensity level [physics] The intensity of a sound, measured in decibels, above a standard level of 10^{12} watts per square metre (equivalent to a sound pressure of 20 micropascals, or 20×10^{-5} newton per square metre, in air).

sound level meter [physics] An apparatus for measuring total sound intensity (in decibels) above a standard level, usually consisting of a microphone, amplifier, and a meter (for example, a millivoltmeter) calibrated directly in decibels.

sound pressure [physics] The mean of the pressure variations (for example, in air or water) accompanying a sound wave, expressed in decibels.

sound pressure level [physics] The pressure of a sound, measured in decibels, above a standard level of 20 micropascals (20×10^{-5} newton per square metre).

sound spectrograph *see* SONOGRAPH

sousaphone *see* BRASS INSTRUMENTS' RANGE

south *see* POINTS OF THE COMPASS

southing [astronomy; geography: navigation; surveying] In astronomy, a (coordinate's) distance southwards from the celestial equator; a negative declination.

In navigation, the distance travelled southwards, usually expressed as a difference in LATITUDE.

In surveying, a southerly displacement from the observer or the origin (reference point); a south latitude.

south magnetic pole *see* MAGNETIC POLE

sovereign [comparative values] Former unit of currency in England, in the form of a gold coin of various values at different times.

During the reign of Henry VII (acceded 1485) the coin was worth 22 shillings and sixpence (one and one-eighth pounds, 270 pence). Within thirty years, however, the coin's value had slipped to 11 shillings and then down to 10 shillings (half of one pound, 120 pence). But issue of the coin ceased only on the execution of Charles I (January 1649), when the whole notion of a 'sovereign' had become politically incorrect. It was then not until around 1817 that a coin named a sovereign was minted once more, this time worth one pound (twenty shillings, 240 pence), and was legal tender thereafter for at least a century, although increasingly available only by special request from the Bank of England.

The name of the coin derives through medieval French from a late Latin adjective *superanus*, a frequentive form of *super* 'above', thus 'high', thus 'ruler', in reference to the depiction of the ruler's head on the actual coin. (The same late Latin word via medieval Italian gave rise to the English word *soprano*, referring to the 'high' voice register.)

Soxhlet scale [physics] An obsolete unit of specific gravity for use with a hydrometer for measuring milk (in lactometry). Gravities were expressed in degrees Soxhlet.

sp, sp., spp. *see* SPECIES

space-time [physics] Three-dimensional space with the added dimension of time, important in relativistic physics.

space velocity [astronomy] For a star moving in space (in three dimensions), its rate and direction of movement. Its movement to or from an observer can be determined by spectroscopy *(see* RED SHIFT*)*, and its lateral movement is revealed by its proper motion (its angular movement in one year on the celestial sphere). These can be used to calculate the vector that is its space velocity.

spades *see* SUITS OF CARDS

span [linear measure] From ancient times, a measure of the average distance between the tip of a man's thumb and the tip of his little finger, with the hand outstretched.

In the Bible,

1 Hebrew span (*zeret*)	=	12 digits, 3 palms
	=	22.5 centimetres, 225 millimetres
	=	8.858 inches
2 spans	=	1 ordinary CUBIT (*ammah*)
8 spans, 4 cubits	=	1 'fathom'

In ancient Greece,

1 span	=	12 digits, 3 palms
	=	23.1 centimetres, 231 millimetres
	=	9.095 inches
2 spans	=	1 cubit
8 spans, 4 cubits	=	1 'fathom'

Informally, in the English-speaking world,

1 span	=	9 inches, one-quarter of a yard
		(1 foot = 1.3333 span)
	=	22.86 centimetres, 228.6 millimetres
4 spans	=	36 inches, 1 yard
		(1 metre = 4.3745 spans)
8 spans	=	1 fathom

As the farthest distance that can be measured by one hand, it is strangely enough related only incidentally to the measure known as the HAND, but corresponds precisely to the unit used in the measurement of the width of flags, the BREADTH.

If the tip of thumb and forefinger can be said to link the two ends of the measurement, it is easy to see how the verb *span* means 'to bridge', 'to unite', and thus derive also the noun 'a span' of (two) oxen, united in their yoke. Moreover, a device for drawing things together and making them tight fast (in the manner of a nut and bolt) is thus a *spanner*.

span loading [weight] A measure of how much an aircraft can safely carry, equal to the gross weight of the plane divided by its wingspan.

spark spectrum [physics] A SPECTRUM produced when an electric spark is passed between metal electrodes. It is characteristic of – and may be used to identify – the metal: *see* SPECTROMETRY.

spat [astronomy] An obsolete unit of distance proposed in 1944 for use in astronomy, equal to 10^{12} metres.

$$
\begin{aligned}
1 \text{ spat} \quad &= \quad 1 \text{ UK thousand million/1 US billion kilometres} \\
&= \quad 621,371,200 \text{ miles} \\
&= \quad 6.68449 \text{ astronomical units} \\
&\quad\quad (1 \text{ light year} = 9,465.4147 \text{ spat} \\
&\quad\quad\, 1 \text{ parsec} = 30,857.253 \text{ spat})
\end{aligned}
$$

The name of the unit may or may not derive from an abbreviation for 'space-time (unit)'.

special theory of relativity *see* RELATIVITY: THE MASS-ENERGY EQUATION

specie [comparative values] Currency in coins (especially coins of comparatively high individual value), as opposed to gold or other precious metal (bullion), or paper money (bills and bonds).

The term derives from a medieval (Latin) description of coinage *in specie* 'in kind' – that is, in the form of coinage or coins.

species [biology: taxonomy] In the taxonomic classification of life-forms, the lowest usual major rank in the biological classification (taxonomy) of organisms, into which a genus is divided. Only one type of interbreeding organism is classified as a species, although a species may be subdivided into subspecies.

The word *species* is the Latin for 'type', 'kind', 'sort', although the word originally referred to appearance (as in English a*spect*, pro*spect*, *spect*acular, etc.).

For full list of taxonomic categories *see* CLASS, CLASS.

specific activity [physics: radiology] Rate at which nuclear transformations occur in the nucleus of a radioisotope, expressed in becquerels, thus providing a measure of the amount of isotope.

specific capacitance [physics] Another name for RELATIVE PERMITTIVITY.

specific charge [physics] For an elementary particle, the ratio of its charge (in coulombs) to its mass (in kilograms): *see*, for example, ELECTRON CHARGE/MASS RATIO.

specific dielectric strength [physics] For an insulator, its dielectric strength expressed in volts per millimetre: *see* DIELECTRIC STRENGTH.

specific fuel consumption [engineering] The amount of fuel used by an engine or motor for each unit of usable energy produced, usually expressed in kilograms per megajoule or pounds per brake horsepower-hour.

specific gravity [physics] For any substance, the mass of a given volume divided by the mass of an equal volume of water (at its maximum density – that is, at 4°C) for solids and liquids, and of hydrogen or air for gases. Its alternative (and preferred) name is *relative density*.

specific heat, specific heat capacity [physics] The amount of heat needed to raise the temperature of unit mass of a substance by 1 degree, expressed in joules per kilogram per kelvin (SI units), or expressed as a ratio in relation to the amount of heat needed to raise the same mass of another substance by 1 degree. Formerly, in CGS units, the number of calories of heat required to raise the temperature of 1 gram of a substance by 1 degree Celsius. *See also* MECHANICAL EQUIVALENT OF HEAT.

For a gas, there are two specific heat capacities. There is the specific heat capacity at constant pressure (when a gas is heated and expands against a constant pressure) and the specific heat capacity at constant volume (when a gas is heated while its volume is kept constant).

specific humidity [physics; meteorology] For moist air, the mass of water vapour in a sample divided by the total mass of the sample.

specific impulse [physics] The thrust of a rocket (in kilograms or pounds force) divided by the rate of fuel consumption (in mass of propellant consumed in unit time), expressed in seconds. It is also equal to the exhaust velocity divided by the acceleration of free fall (acceleration due to gravity): *see* EXHAUST VELOCITY.

specific latent heat [physics] The heat needed to change unit mass of a substance from a solid to a liquid (at its melting point) or from a liquid to a gas or vapour (at its boiling point). The first is the specific latent heat of fusion, and the second is the specific latent heat of vaporization. *See also* LATENT HEAT.

specific permeability [physics] Another name for RELATIVE PERMEABILITY.

specific resistance [physics] Another (incorrect) name for RESISTIVITY.

specific surface [physics] For a solid substance, its surface area per unit mass or per unit volume.

specific volume [physics] The volume occupied by unit mass of a substance – that is, the reciprocal of its DENSITY.

specimen, medical *see* URINALYSIS

spectral series [physics] A group of lines in the atomic (line) spectrum of an element, representing electron 'jumps' from their initial energy levels to a common final energy level. The various series were named after the physicists who first identified and studied them (such as, for the hydrogen atom, the Balmer series, Bracket series, Lyman series, and Paschen series). The frequencies of the lines can be formulated in terms of their wave number (reciprocal wavelength) and RYDBERG'S CONSTANT.

spectral types of stars [astronomy] Method of classifying stars in terms of their temperatures as revealed by their spectra, ranging from O (hottest and blue-white) to S (coolest, with band spectra). The full list, corresponding to the *Main Sequence* of the HERTZSPRUNG-RUSSELL DIAGRAM, is O, B, A, F, G, K, M, R, N, S; each class is further divided into subclasses numbered from 0 to 9, and the luminosity may also be classified from I to VII.

Our Sun is accordingly categorized as a star of type G2V – that is, its absorption lines show that singly ionized calcium is prominent, with weaker hydrogen bands and the presence of neutral metals (G), that its effective temperature is around 5,800 K (2), and that it is a dwarfish member of the Main Sequence (V).

spectrograph [physics] Type of spectroscope that records spectra on photographic film. *See also* MASS SPECTROGRAPH.

spectroheliograph [astronomy] Type of SPECTROSCOPE used for photographing the spectrum of the Sun in light of a single wavelength.

spectrometer [physics] Type of SPECTROSCOPE used for measuring wavelengths or energies in a beam of radiation (that has been split into a spectrum).

spectrometry [astronomy; physics; chemistry] The production, recording, and measurement of spectra, using various types of spectroscopic instruments *(see above and below)*. The technique has many applications, for example in astronomy (studying stars), physics (studying electromagnetic radiation), and chemistry (identifying elements).

spectrophotometer [physics] Type of SPECTROSCOPE used for measuring the intensity of light in each colour (wavelength) of a spectrum, from the infra-red region through visible light to ultraviolet, or for comparing the intensities of light in two spectra.

spectroscope [physics] The basic apparatus of SPECTROMETRY: *see* the various types of instruments *described above*.

spectrum [physics] Bands of coloured light produced by splitting visible light into various wavelengths (by means of a prism or diffraction grating in a spectroscope); a rainbow is a visible spectrum.

By extension the term also includes the range of other types of electromagnetic radiation, ionic species, or even sound according to frequency, mass, or energy.

The word *spectrum* is the Latin for 'appearance'.

See also ELECTROMAGNETIC SPECTRUM *and* the various spectroscopic instruments *described above*.

speech sound [literary: phonetics] Another name for the now outdated concept of the minimum speech unit otherwise known as a PHONE.

speed [physics; speed] The distance travelled by a moving object in unit time; the rate at which it moves. It is measured in such units as metres per second, kilometres per hour (km/h), or miles per hour (mph). It is a scalar quantity, unlike VELOCITY (which is a vector quantity).

speed limit, speed restriction [speed] In most countries there are statutory speed limits for motor vehicles that locally take account of the number of pedestrians customarily in proximity to the highway, the number of lanes for motor vehicles, the state or condition of the highway surface, and other factors generally to do with overall safety. But, as with all measurements of speed, there are two major world systems of units: kilometres per hour (km/h) and miles per hour (mph), and drivers familiar with one set of units are mostly unfamiliar with the other set.

mph limits	km/h equivalent	km/h limits	mph equivalent
10	16.1	20	12.4
15	24.1	30	18.6
20	32.2	40	24.9
30	48.3	50	31.1
40	64.4	55	34.2
50	80.4	60	37.3
60	96.5	70	43.5
70	112.6	80	49.7
80	128.7	100	62.1
90	144.8	110	68.4
		120	74.6

speed of light [speed; physics; astronomy] Finite velocity at which light travels through a vacuum (such as space), linked also to theories of relativity. The closer to the speed of light an object travels, the more its mass increases, and, from the point of view of a stationary observer, the more its time sense is elongated.

The speed of light is:186,283 miles per second; 670,619,880 miles per hour; 299,792.46 kilometres per second; 1,079,252,956 kilometres per hour; 7.2 ASTRONOMICAL UNITS per hour; 63,271.47 astronomical units (1 LIGHT YEAR) per year.

speed of rotation [speed; physics] For a rotating object, the number of rotations in unit time, expressed in such units as revolutions per minute or radians per second. It is measured using a TACHOMETER or rev(olution) counter.

speed of sound [speed; physics] The speed of sound (Mach 1) in dry air at sea-level is: 1,229 kilometres per hour; 341.389 metres per second; 763.67 miles per hour; 1,120 feet 0.6 inches per second.

The speed of sound is different in moist air, or at altitude. In water, for example, the speed of sound is considerably faster: 5,076 kilometres per hour; 1,410 metres per second; 3,154 miles per hour; 4,626 feet per second.

Approaching the speed of sound (at about 335 metres per second/1,100 feet per second in dry air at sea-level), an aircraft or a projectile encounters a sudden increase in air resistance (the sound barrier) and as it exceeds the speed of sound creates a shock wave, heard by stationary observers below as a 'sonic boom'. *See also* INFRASOUND; MACH NUMBER; ULTRASOUND.

speedometer [engineering; speed] An instrument for measuring the road speed of a vehicle (in kilometres per hour, miles per hour, or both). It usually consists of a TACHOMETER, connected directly or indirectly to the final drive from the engine, calibrated in speeds.

speed-skating competitive distances [sport] There are two forms of track at which speed skating takes place: an international speed-skating track and a short-track speed-skating rink. The events that relate to each type of track are very different.

Competitors on an international speed-skating track are obliged to enter at least two, and often four, races over different distances. Aggregate times are then calculated for individual placings. The events are:

for men:	for women:
500 metres	500 metres
1,500 metres	1,000 metres
5,000 metres	1,500 metres
10,000 metres	3,000 metres

On the short track, there are three different types of race: as individuals, with up to four (shorter races) or six (longer races) competitors on the track at a time; in teams of two or four members as a relay race; or in a pursuit race (much like a cycle pursuit race) with the two competitors starting simultaneously on opposite sides of the track and competing against each other and the clock. In individual and relay race events, the distances are:

> 500 metres
> 1,000 metres
> 1,500 metres
> 3,000 metres
> 5,000 metres

In pursuit races, the maximum distance is 10 laps of the track.

Speed skating is thought to have begun as a winter contest in Holland, on frozen canals.

Spenserian sonnet, Spenserian stanza *see* VERSE FORMS

sperm count [medicine] Microscopic analysis of a specimen of a male's seminal fluid (semen) to determine how many viable sperms it contains per millilitre, in turn to establish whether it is possible for the male to father offspring (a low sperm count makes the chances of pregnancy by normal methods less likely, although in-vitro fertilization using the sperm remains quite feasible).

Normal human seminal fluid contains between 20 and 200 million sperm per millilitre. A low sperm count (oligospermia) corresponds to less than 18 million sperm per millilitre. The condition in which there is a complete absence of viable (motile) sperm in the seminal fluid is known as azoospermia.

SPF *see* SUN PROTECTION FACTOR

sp. gr. *see* SPECIFIC GRAVITY

sphere [maths] A solid figure generated by rotating a circle about a diameter – a ball or globe. Its surface area is 4π times the square of its radius ($4\pi r^2$); its volume is $\frac{4}{3}\pi$ times the cube of the radius ($\frac{4}{3}\pi r^3$).

The word derives in English through Latin from Greek *sphaira* 'ball', which may well be akin to Greek *speira* 'rotation', 'revolution', and thus to English *spiral*.

spherical angle [maths] In geometry, on the surface of a sphere, the angle made by two intersecting lines which are circumferences.

spherical candle power [physics] For a sphere of unit radius with a point source of light at its centre, the illumination on its surface.

spherical coordinates [maths] Three-dimensional polar coordinates, in which the position of a point in space is given in terms of two angles (from fixed lines at right-angles to each other) and its distance from a fixed point (the origin). *See also* CYLINDRICAL COORDINATES; POLAR COORDINATES.

spherical triangle [maths] A triangle formed on the surface of a sphere by three lines. Just as various measurements in right-angled triangles in two dimensions are calculable by means of trigonometry, certain measurements of right-angled spherical triangles are likewise calculable through spherical trigonometry.

spheroid [maths] A solid figure generated by rotating an ellipse about its minor axis (giving an oblate spheroid) or its major axis (giving a prolate spheroid).

spherometer [physics] An instrument for measuring the curvature of a surface, such as the surface of a lens or curved mirror.

sp. ht *see* SPECIFIC HEAT

sphygmograph [medicine] Another name for the *pulsimeter*: *see* PULSE.

The first element of the term is the ancient Greek *sphygmos* 'pulse'.

sphygmomanometer [medicine] The usual instrument for measuring a person's blood pressure, consisting of an inflatable 'collar' that is fitted around the upper arm and inflated so as to tighten it there. The physician or medical assistant then uses a stethoscope in the bend of the elbow to listen to the pulse in the brachial artery, noting the value on the pressure dial when the pulse begins again as the collar is gradually released.

For normal blood pressure values, *see* BLOOD PRESSURE.

spin [physics] The quantized angular momentum of a spinning subatomic particle, characterized by a quantum number (equal to a whole number multiple of half a

DIRAC UNIT). Bosons (for example: photons) have even spin quantum numbers; fermions (for example: electrons, neutrons, protons) have odd ones. *See also* QUANTUM NUMBERS.

spindle [textiles] The textiles industry measures its wares in yards (or less commonly metres) of yarn in SKEINS (or, in the United States, LEAS), seven of which together form a HANK, 20 of which in turn form a BUNDLE. However, there is a measure intermediate between the hank and the bundle, and that is the spindle:

1 spindle   =   18 hanks

The term evidently derives from the standard size of spool or cylinder around which the textiles were wound.

spindle [physics] An informal name for a HYDROMETER.

spinode [maths] A point where two branches of a curve meet and the tangents to the branches coincide; it is alternatively known as the *cusp*. *See also* CRUNODE.

spiral [maths] A curve that resembles a coiled rope, mathematically described as the locus of a point that winds around a fixed point from which it continuously moves away. There are various types of spirals, best characterized by their polar equations – generalizations of their polar coordinates (r, θ). Some of the equations are:

Archimedes' spiral: $r = a\theta$ (a is a constant)
equiangular, or logarithmic, spiral: $\log r = a\theta$
hyperbolic spiral: $r\theta = a^2$
parabolic, or Fermi's, spiral: $r^2 = a\theta$

Spirals occur in nature, particularly in molluscs that enlarge their shells as they grow (for example: nautilus, periwinkle, whelk, and snail). Most are equiangular spirals, in which the tangent to a point on the curve makes a constant angle with the line joining that point to the centre.

spirit level [physics] An instrument containing a coloured liquid and an air bubble in a glass or plastic tube at the centre, used to ascertain if a surface is precisely horizontal. The air bubble is central in the tube if the surface is horizontal; if the surface slopes, the air bubble floats to one end.

spirometer [medical] An instrument for measuring the volume of air breathed in and out of the lungs.

split [volumetric measure] Technical term in dispensing drinks in a bar: a measure of lemonade or other form of carbonated (soda) mixer, equal to about one ordinary cupful (or mixer bottle), half a standard soda bottle.

1 split   =   6 to 8 US fluid ounces (6.3 to 8.5 UK fluid
                          ounces)
            =   178 to 237 millilitres, 17.8 to 23.7 centilitres
            =   three-eighths to one-half US pint,
                          0.313 to 0.417 UK pint

It is as half a standard soda bottle that the measure gets its name: a bartender can split a whole bottle between two customers, who then get a split each. The name can, however, be used for the the same measure in relation to non-carbonated drinks, especially cocktails.

spondee, spondaic foot *see* FOOT [literary]

spoonful *see* DESSERTSPOON; TABLESPOON(FUL); TEASPOON(FUL)

spot test [chemistry] An investigation of a substance by means of a chemical analysis of it after it has been mixed with a reagent (on a filter paper or a glass dish), to identify the presence of constituent components.

spp. *see* SPECIES

spread [comparative values] In North America, a general term for the going rate of net profit or loss in the buying and selling of one type of goods.

Also (in North America), a term for the range of betting odds quoted on the likelihood of winning of either team in a team game (such as American football) when the teams involved are known to be of different standards.

spread [literary; square measure] In publishing, a term for two facing pages in a book, magazine, or newspaper (originally called 'a two-page spread').

spring balance [weight] A weighing device consisting of a spring slung beneath a vertical linear scale or a dial and a movable pointer; a weight placed on the lower end of the spring moves the pointer to the correct value shown on the scale or dial.

Another name for the spring balance is the *Newton meter*.

spring tide [geography; astronomy] An extra-high tide that coincides with a new Moon or full Moon, when the Moon and the Sun are in line with the Earth and their gravitational attractions combine. It is often preceded and followed by an extra-low tide. *See also* NEAP TIDE.

sprint races [sport] The track (running) and hurdles races between 60 metres and 400 metres, or between 60 yards and 440 yards: *see* ATHLETICS TRACK EVENTS RACE DISTANCES.

The term is also used of short-distance horse and greyhound races, car and cycle races, rowing, and swimming races.

spur royal [comparative values] Former unit of currency in Britain, in the form of a gold coin. Issued during the reign of James I of England (VI of Scotland, ruled Scotland 1567-1625 initially by regent having acceded to the Scottish throne at the age of one, ruled England and Scotland 1603-25), it was worth 15 shillings (three-quarters of one pound, 180 pence), and was so called because it was a RYAL or REAL (minted by *royal* command at the *royal* mint) that had what looked like the rowel of a spur depicted on it (although it might have been meant as a spiky star).

squad [military] In the army (and less formally in the navy and airforce), a small team of men with a leader (a corporal or a sergeant) that is a basic unit for any particular drill, duty, or inspection. In the United States, a squad ordinarily comprises twelve men, including the leader, and four squads make up a PLATOON. In Britain, the men who make up any particular squad (for example, a squad of military police) may be known laconically as 'squaddies'.

The term *squad* derives from the Latin verb *ex-quadrare* 'to separate a quarter', 'to make a fourth' (hence four squads to a platoon), and is thus cognate with French *cadre* and English *square*.

squadron [military] A basic tactical unit of an airforce, a navy, or a motorized or cavalry army regiment.

An airforce squadron consists of between eight and twelve aircraft, in two or three 'flights', detailed for particular operations (such as attack or reconnaissance).

A naval squadron constitutes a section of a fleet that is allocated special duties (such as defending a convoy); it is usually made up of two or more divisions of ships of the same class and type.

An army squadron of motorized or cavalry personnel is the equivalent of a BATALLION of foot-soldiers, most often consisting of between 100 and 240 men under the command of a major or lieutenant colonel.

The term derives from the diminutive of the medieval Italian form of the Latin word which in English became 'squad' (*see above*).

squadron leader [military rank] Middle rank in the airforces of Britain, Australia, Canada, and some other countries.

A squadron leader ranks above a flight lieutenant but below a wing commander, and on a parallel with a major in the army and a lieutenant commander in the navy. The equivalent rank in the US Airforce is major.

For the etymology of *squadron, see* SQUADRON *above*.

square [maths] In arithmetic, a number that equals another number multiplied by itself; a number raised to the power 2. For example: 7 x 7 (or 7 squared, written 7^2) equals 49, so 49 is the square of 7.

In geometry, a four-sided plane figure (quadrilateral) with four equal sides and four right-angles. Its diagonals are equal, and intersect at right-angles. Its area equals the length of one side multiplied by itself – that is, the length of one side squared – and the result is in square units. So, for example, a square in which each side is 4 blogs long occupies an area of (4 x 4 =) 16 square blogs.

The term derives through medieval French ultimately from the Latin verb *ex-quadrare* 'to separate a quarter' in the sense 'to form out of four (sides)', and is thus cognate with French *cadre* and English *squad*.

square [square measure] Area unit in the measurement of finished timber.

1 square	=	100 square feet, 11.1111 square yards
	=	9.2903 square metres

But, rather than corresponding to finished timber 10 feet x 10 feet in dimensions, the unit is usually applied to lengths of wood (planks) that together have a total surface area of 100 square feet.

square centimetre, square centimeter *see* CENTIMETRE, SQUARE

square chain *see* CHAIN, SQUARE

square degree [maths] Unit of solid angle equal to $\pi/180$ squared – $(\pi/180)^2$ – expressed in steradians.

square division [military] Large military troop unit basic to infantry warfare from the 1800s until the 1930s, comprising three BRIGADES: two of infantry soldiers and one of field artillery support.

square foot *see* FOOT, SQUARE

square furlong *see* FURLONG

square grade [maths] Unit of solid angle equal to $\pi/200$ squared – $(\pi/200)^2$.

square inch *see* INCH, SQUARE

square kilometre *see* KILOMETRE, SQUARE

square law *see* INVERSE SQUARE LAW

square metre, square meter *see* METRE, SQUARE

square mile *see* MILE, SQUARE

square millimetre *see* MILLIMETRE, SQUARE

square number *see* SQUARE [maths]

square rod *see* ROD, SQUARE

square root [maths] Of a given number, another number that multiplied by itself equals the given number; it is indicated by the symbol $\sqrt{}$ before the given number, or by the power (index) ½. For example:
$$\sqrt{49} = 7 \qquad \text{or } 49^{½} = 7$$

See also ROOT.

square yard *see* YARD, SQUARE

squash rackets measurements and units [sport] Squash is played with racket and ball in an enclosed four-walled court marked with various lines representing where the ball may or may not bounce. The object of the game is to score points by making it impossible for an opponent to return the ball. A number of points (9 in Europe and in international matches, 15 in North America) makes up a game; five games make a match.

The dimensions of a singles court:

    overall length: 32 feet (9.75 metres)

    overall width: international – 21 feet (6.4 metres) North America – 18 feet
        6 inches (5.64 meters)

    short line (front of service box) distance from front wall: 18 feet (5.49 metres)

    front wall:

        height of out-of-court line (to serve below):

            international – 15 feet (4.57 metres)

            North America – 16 feet (4.88 meters)

        height of service line (to serve above):

            international – 6 feet (1.83 metre)

            North America – 6 feet 6 inches (1.98 meter)

        height of telltale board (to play above):

            international – 19 inches (48 centimetres)

            North America – 17 inches (43 centimeters)

    back wall:

        height:

            international – 7 feet (2.13 metres)

            North America – 12 feet (3.66 meters)

    side walls:

        international – diagonal line from front wall out-of-court line (15 feet/
            4.57 metres) to top of back wall (7 feet (2.13 metres)

        North America – stepped line at 16 feet (4.88 meters) to above short line, and
            from short line at 12 feet (3.66 meters) to back wall

    a doubles court is proportionately larger in all dimensions

    Points scoring:

        international – only the server scores points; failure to do so loses the serve

North America – server and receiver can both score

Dimensions of equipment:

the racket:

overall length: 35½ inches (90 centimetres)

handle length: 27 inches (68.5 centimetres)

head length: 8½ inches (21.5 centimetres)

head width: 7¼ inches (18.4 centimetres)

the ball:

diameter: 39.5-41.5 millimetres (1.555-1.634 inches)

weight: 23.3-24.6 grams (0.822-0.868 ounce)

The name of the game derives, perhaps slightly sardonically, from the force with which the very light ball is hit – enough, indeed, to *squash* it – etymologically from late Latin *ex-quassare* 'to beat out (of shape)'.

squid, SQUID [physics] An acronym for superconducting quantum interference device – an apparatus for measuring very small electric currents, voltages and magnetic fields. One type consists of a thin layer of niobium surrounding a small rod of quartz.

stack [cubic measure] Apart from the general sense of the word – meaning a pile or assembly of similar elements (notably of roulette or poker chips, and of aircraft awaiting their turn to land) – a stack is also an obsolete English measure of fuel for burning, especially of cut wood or coal.

    1 stack = 4 cubic yards, 108 cubic feet

           = 3.058 cubic metres (3,058 litres by volume)

In this sense, the word dates from around the mid-1600s and probably referred originally to a cartload.

stade [time; geology] In describing a period of glaciation, a division of time less than a glacial stage but which has some feature or event to distinguish it.

The word is the French for 'stage', but ultimately derives through Latin from the ancient Greek word represented in English by *stadium* (*see below*).

stade, stadion, stadium [linear measure] In ancient Greece and Rome, a 'fixed' length or distance corresponding to a set number of 'feet'. The term is now used also of a similar measure as applied in ancient Israel.

In ancient Greece,

    1 stadion = 600 'feet', 100 'cables' (*orguiai*), 6 plethra

             = 608 feet, 202.6666 yards

                (600 feet, 200 yards = 0.9868 stadion)

             = 185.318 metres

                (200 metres = 1.0792 stadion)

    4 stadia = 1 'ride'

The athletics track and horse race course at Olympia was exactly 1 stadion in length, and it is from that usage that English now derives the word *stadium* as a term for such a venue.

In ancient Rome,

    1 stadium = 625 'feet', 125 (double-)paces

             = 608.333 feet, 202.7777 yards

                (600 feet, 200 yards = 0.9863 stadium)

             = 185.420 metres

                (200 metres = 1.0786 stadium)

    1.6 stadium = 1 mille passus ('mile')

Until the 1960s it was customary to use the French variant *stade* for both the ancient Greek and the ancient Roman measures; the practice has since become increasingly rare, but the form remains in most dictionaries.

Ironically, the higher measures of the ancient Israelite distance measurement system have traditionally been given Greek names.

In ancient Israel,

    1 stadion = 360 cubits, 60 reeds, 6 plethra

             = 161.856 metres

                (150 metres = 0.9267 stadion)

             = 531 feet (exactly), 177 yards

(200 yards = 1.1299 stadion)

It was as a 'fixed' distance that the stadion got its name, cognate with the English adjectives *staid* and *steady*, it has *stood* ever since.

stadia, stadia rod [linear measure] A type of levelling staff, used in surveying, that has prominent graduations that can be seen (using a telescope) over long distances, as is usually required in the technique called *stadia tacheometry*.

stadiometer [maths] An instrument for measuring the lengths of curves, consisting of a toothed wheel (which is run around the curve) connected to a calibrated dial.

stadion, stadium *see* STADE, STADION, STADIUM

staff, stave *see* MUSICAL NOTATION

staff sergeant [military rank] A non-commissioned army officer equivalent to or ranking next above a sergeant, and with duties biased towards executive and administrative responsibilities rather than drill, discipline, or combat operations.

stage [geology] In rock strata made up of beds of related rocks, a set of two or more strata representative of a period of time less than a geological epoch, but with distinctive and common features. Two or more stages in turn make up a *series* – which does represent a geological epoch or major subdivision thereof – and a number of series make up a *system*.

Also, in relation to the ice ages, a period of time in which climatic conditions were relatively stable (either as a period of glaciation or of intermission), the effects of which are distinct although each stage might contain a number of *stades* representing peaks or troughs sustained for various much shorter periods of time.

The term derives through medieval French from Latin *staticum* 'a level/location at which conditions remain static (for a time)', akin thus to English *station* and the verb 'to *stand*'.

stake [square measure] In the (Mormon) Church of Jesus Christ of Latter-Day Saints, a large administrative area made up of smaller territories known as wards, and for which an official called a president is responsible. (It is the equivalent, perhaps, of an episcopal see in other denominations.)

stake [sporting term] The money or premium put at risk in gambling.

By extension, a share in a potentially winning gamble or profitable deal.

stalagmometry [physics] The measurement of SURFACE TENSION, usually by weighing a drop that falls from a hole of known size.

The first element of the word is ancient Greek *stalagmos* 'dripping'.

stalemate *see* CHESS PIECES AND MOVES

stalling speed [aeronautics] For a fixed-wing aircraft, the speed at which the flow of air over the wing breaks down and there is no longer any lift (upward force on the wing). It occurs in level flight as the lowest speed an aircraft can fly without losing altitude, and at higher speeds when the maximum safe angle of attack (pulling up the nose) is exceeded.

stamp duty, stamp tax [comparative values] A means of imposing a government tax by requiring the purchase of special stamps when buying certain goods (such as real estate) or when rendering official certain documents (such as title deeds).

standard [comparative values] A valuable metal (such as gold) or combination of metals (such as gold and silver) used as the fixed standard of value against which the value of a national monetary system rises or falls.

Also the fixed ratio of metal and alloy used in a country's coinage.

standard [cubic measure; volumetric measure] Unit of volume in the measurement of timber/lumber.

$$
\begin{aligned}
1 \text{ standard} \ &= \ 1{,}980 \text{ board feet} \\
& \qquad \text{(equivalent to 330 planks 6 feet long, 1 foot} \\
& \qquad \text{wide and 1 inch thick)} \\
&= \ 165 \text{ cubic feet, } 6.1111 \text{ cubic yards by volume} \\
&= \ 4.339 \text{ cubic metres}
\end{aligned}
$$

Why this measure should be a 'standard' unit is not known; some reference sources suggest that, in former times, there were other measures of timber/lumber also known as standard.

standard atmosphere [physics] Unit of pressure, equal to that exerted by a column of mercury 760 millimetres tall at 0°C.

$$
\begin{aligned}
\text{1 standard atmosphere} \quad &= \quad 760 \text{ torr} \\
&= \quad 101{,}325 \text{ newtons per square metre} \\
&= \quad 101{,}325 \text{ pascals} \\
&= \quad 1{,}013{,}250 \text{ dynes per square centimetre} \\
&= \quad 1{,}013.25 \text{ millibars} \\
&= \quad 14.72 \text{ pounds per square inch}
\end{aligned}
$$

It is the pressure referred to in *standard temperature and pressure* (s.t.p.*).

Standard Book Number (SBN) *see* INTERNATIONAL STANDARD BOOK NUMBER

standard candle [physics] Former unit of luminous intensity, which was superseded by the international candle (itself superseded by the CANDELA).

standard currency unit [comparative values] The contemporaneous value of a currency unit in gold. A *standard dollar* is thus the value of 1 dollar in gold; a *standard pound* is thus the value of 1 pound in gold.

standard deviation [maths] In statistics, for a set of values (data) the square root of the mean of the squared deviations from the mean value. That is, it is the root mean square deviation from the mean value (equal to the square root of the variance).
  An example may help to clarify:
    Set of values: 5.0, 8.6, 6.4, 4.4, 7.0, 6.4 (sum 37.8)
    Mean value: 37.8/6 = 6.3
    Deviations from mean: −1.3, +2.3, +0.1, −1.9, +0.7, +0.1
    Squares of deviations: 1.69, 5.29, 0.01, 3.61, 0.49, 0.01
      (sum 11.1)
    Mean of the squares: 11.1/6 = 1.85
    Square root of this mean (that is, standard deviation): 1.36
  The standard deviation is alternatively known as the *standard error*.

standard dollar *see* STANDARD CURRENCY UNIT

standard electrode [chemistry] An electrode (an element in contact with a solution of its ions) used as a standard for measuring electrode potentials. The usual standard is the hydrogen electrode, which is assigned a potential of zero.

standard electrode potential [chemistry] An electrode potential measured or specified with reference to a STANDARD ELECTRODE.

standard error *see* STANDARD DEVIATION

standard form *see* SCIENTIFIC NOTATION

standard gravity [physics] Another name for the ACCELERATION OF FREE FALL (acceleration due to gravity).

standard mortality ratio (SMR) [medicine] In a group of organisms, the observed number of deaths divided by the expected number of deaths.

standard mean chord [aeronautics] The average of the chord (width) of an aircraft's wing, equal to the total wing area divided by the span.

standard pressure *see* STANDARD ATMOSPHERE

standard score [medine] In a group test of educational achievement, psychological development, or attitudinal disposition, a score attained by an individual that is different from the average, expressed in units of standard deviation (*see above*) in relation to the overall distribution.

standard solution [chemistry] A solution of precisely known concentration, used in volumetric analysis. *See also* NORMALITY.

standard temperature and pressure (s.t.p.) [physics] A temperature of 0°C (273.16 K) and a pressure of 101,325 newtons per square metre or 760 millimetres of mercury: *see* STANDARD ATMOSPHERE.

standard time [time] For a particular time zone, the internationally agreed time in the region, usually given as a whole number of hours ahead of or behind Greenwich mean time (GMT).

standard volume [physics] The volume occupied by a kilogram molecular weight (1,000 moles) of a gas at 0°C and 1 standard atmosphere pressure, equal to 22.414 cubic metres (kilolitres).

standing wave [physics] A wave for which the amplitude at any point divided by the amplitude at any other point does not vary with time. It is alternatively known as a *stationary wave*.

Stanford Binet test [medicine] Revised version of the BINET-SIMON TEST for determining the mental age of a subject, with specific adaptation to conditions and dispositions in the United States, as produced at Stanford University, California.

Stanton number [engineering] A dimensional number used in calculations involving the convection of fluids, equal to the reciprocal of the PRANDTL NUMBER. It was named after the British physicist Thomas Stanton (1865-1931).

stanza [literary] In poetry, a number of lines of set form and length (also called a *verse*), used as a pattern for all or most of a poem. In Classical poetry, specifically in very long (and intendedly epic) poems, a number of stanzas make up a *canto*, and a number of cantos make up a *volume* or *book*.

The word is the medieval Italian for 'stance', meaning that the end of each stanza represents a place to pause or stand still briefly.

stapp [medicine; aeronautics] In the testing of a pilot's or astronaut's physical endurance by means of simulated space flights with or without a centrifuge, a unit of force equal to the acceleration of free fall (gravity: 1 g) acting on the body for 1 second.

The unit is named after John P. Stapp (1911-), the US Airforce medical officer who pioneered biological research into rocket sled acceleration and deceleration during the 1950s and 1960s.

star magnitude *see* MAGNITUDE; STARS, THE BRIGHTEST IN THE SKY

stars, the brightest in the sky [astronomy] The ten brightest stars (excluding the Sun) are:

usual name	technical name	apparent magnitude	spectral type	distance (light years)
Sirius	Alpha Canis Majoris	−1.4	A1V	8.6
Canopus	Alpha Carinae	−0.7	F0Ib	190
'Rigilkent'	Alpha Centauri	−0.3	G2V	4.3
Arcturus	Alpha Boötis	0	K2III	36
Vega	Alpha Lyrae	0	A0V	26.5
Capella	Alpha Aurigae	+0.1	G8	45
Rigel	Beta Orionis	+0.2	B8Ia	660
Procyon	Alpha Canis Minoris	+0.4	F5IV	11.4
Achernar	Alpha Eridani	+0.5	B5IV	130
Hadar/Agena	Beta Centauri	+0.6	B1II	390

See also MAGNITUDE.

stars, the nearest [astronomy] The ten stars nearest to us (excluding the Sun) are:

star name	constellation	distance (light years)	apparent magnitude	spectral type
Proxima Centauri	Centaurus	4.26	11.0	M5
Alpha Centauri A	Centaurus	4.35	−0.0	G2V
Alpha Centauri B	Centaurus	4.35	1.3	K2V
Barnard's Star	Ophiuchus	6.0	9.5	M5V
Wolf 359	Leo	7.7	13.5	M8
Lalande 21185	Ursa Major	8.2	7.5	M2V
Luyten 726-8 A	Cetus	8.4	12.5	M5
star name	constellation	distance (light years)	apparent magnitude	spectral type
Luyten 726-8 B or UV Ceti	Cetus	8.4	13.0	M6
Sirius A	Canis Major	8.6	−1.4	A1V
Sirius B	Canis Major	8.6	8.6	WD*

* WD = white dwarf

star types *see* HERTZSPRUNG-RUSSELL DIAGRAM; SPECTRAL TYPES OF STARS

stat- [physics: prefix] A prefix on units in the (obsolete) electrostatic system of electrical CGS units. Examples include the statampere (current), statcoulomb (charge), statfarad (capacitance), statohm (resistance), and statvolt (electromotive force). Often they have inconvenient values for practical use – for example: 1 statampere = 0.333 x 10^{-9} ampere and 1 statvolt = 300 volts. The whole confused area of such units was rationalized with the introduction of SI units. *See also* ABSOLUTE UNITS.

-stat [physics: suffix] A suffix indicating a device that regulates or controls: thus, a *thermostat* controls heat; a *rheostat* regulates flow (of electric current).

stater [comparative values] Unit of currency in ancient Greece, differing in metallic composition – and therefore in value – between various city states. The first stater was probably that of the Lydians under the legendary (but historical) King Croesus. The earliest stater at Athens was made of silver; a later Athenian stater was made of gold and was the fiscal equivalent of 20 drachmae.

But the name of the coin betrays its origin: *stater* was at first simply the Greek for 'a weight' that could be *stood* on a balance – of no particular size but any of a set of such weights used in market trade.

static pressure [physics] The pressure at right-angles to an object moving through a fluid.

stationary orbit *see* GEOSTATIONARY ORBIT

stationary point [maths] For a curve in coordinate geometry, a point on it at which the tangent to the curve is horizontal (parallel to the x-axis). At that point, the derivative (obtained by differentiation) of the equation representing the curve is equal to zero. Maxima and minima are stationary points (*see* MAXIMUM; MINIMUM).

stationary wave *see* STANDING WAVE

statistical mechanics [physics] The application of statistics to the study of subatomic particles and other microscopic members of a macroscopic system. Wave mechanics recognizes the dual particle/wave nature of such components (*see* SCHRÖDINGER EQUATION; WAVE EQUATION). The inclusion of quantum theory enlarges the subject to quantum mechanics.

statistics [maths] The area of mathematics that collects, classifies, and analyses large quantities of numerical and quantitative data, dealing with such things as distributions, averages, and probabilities.

The word derives in English via a German word apparently intended to mean 'the study of statuses' (cf. *Technik* for 'technology').

statute acre, mile, ton [square measure; linear measure; weight] Another term for the ordinary (standard) ACRE, MILE, and TON, also known as the imperial or English acre, mile, and ton, but referring to the fact that the measures were defined and fixed by law (statute).

Staudinger value [chemistry] A number that relates to – but does not equal – the relative molecular mass (molecular weight) of a polymer, used in classifying types of plastics and fibres. It was named after the German chemist Hermann Staudinger (1881-1965).

stave *see* MUSICAL NOTATION

steady-state theory [physics; astronomy] Theory that suggests that the quantity of matter that makes up all the atoms in the Universe is pretty well constant, and that, as atoms are lost or consumed by black holes, other atoms are being created or generated elsewhere in precisely proportional amounts. It is the proposition that atoms are somehow being generated that militates against the other major current theory of how the Universe came into being: the Big Bang theory, which posits that all the matter the Universe has ever held exploded in one unimaginably violent cataclysm some 15,000-20,000 million years ago, and has been expanding ever since (inevitably losing atoms all the time).

steam point [physics; engineering] The boiling point of pure water, equal to 100°C (212°F) at normal atmospheric pressure.

steel band *see* PERCUSSION INSTRUMENTS' RANGE

steelyard [weight] Type of bar scale for weighing items, consisting of a near-horizontal bar suspended from a chain close to one end. The longer end of the bar has a calibrated scale and a weight that is movable along it; from the shorter end of the bar is suspended the object to be weighed. By balancing the movable weight along the bar against the weight of the subject item, the latter can be read off from the calibrated scale.

steerageway [speed] The lowest speed of a ship that makes it responsive to the helm (wheel, rudder); at a speed less than the steerageway, the ship is not steerable, which is one of the reasons that larger liners and tankers require tugs to manoeuvre them into harbour. In general, the larger the ship, the faster is the steerageway, although it depends also on the means of propulsion and the size and nature of the steering mechanism.

Stefan-Boltzmann law [physics] For a black body (that is, a perfect) radiator, the total energy radiated per unit area per unit time is proportional to the fourth power of its absolute temperature. The proportionality constant – the *Stefan-Boltzmann constant* – equals 5.6696 x 10^{-8} watt per square metre per fourth power of temperature. It was named after the Austrian physicists Joseph Stefan (1835-93) and Ludwig Boltzmann (1844-1906).

stein [volumetric measure] A beer mug in German-speaking countries. The classic stein has a capacity of around 75 centilitres (1.32 UK pint, 1.585 US pint), which is a considerably greater quantity than that of most mugs elsewhere in the world. But there are also smaller steins, the normal smallest of which is about one-third the classic size – about half a pint – and the largest stein is the 1-litre size (1.76 UK pints, 2.1134 US pints).

The word is an abbreviation of *Steinkrug*, literally 'stone-jug', for the original steins were made of stoneware, a form of ceramic that looks like stone.

stella [comparative values] Former unit of currency in the United States, in the form of a gold coin worth 4 dollars. Minted only in 1879 and 1880, it was so called because it had a star depicted on the back.

step, whole step [music] Technical term in the United States for what in Britain and most of Europe is called a *tone*, corresponding on the piano to the interval between two white notes separated by a black note, or between two black notes separated by a white one. It is thus twice the interval of a half-step or semitone (which corresponds to the interval on the piano between a black and an adjacent white note).

The SCALE of any KEY is made up of a series of steps and half-steps.

steradian [maths] Unit of solid angle in the SI system of units, equal to the angle subtended at the centre of a sphere by an area on its surface equal to the square of its radius. The whole surface subtends an angle of 4π steradians at the centre.

The term is made up of the ancient Greek prefix *ster(eo-)* 'solid' and the English word *radian*.

stere [cubic measure] A historic unit of volume (originating in revolutionary France as a *stère*) equal to 1 cubic metre, still sometimes used for stacked timber.

$$
\begin{aligned}
1 \text{ stere} \quad &= \quad 1 \text{ metre x 1 metre x 1 metre} \\
&= \quad 1 \text{ kilolitre by volume} \\
&= \quad 35.31467 \text{ cubic feet, } 1.307951 \text{ cubic yard} \\
&\qquad (1 \text{ cubic yard} = 0.764555 \text{ stere}) \\
&= \quad 27.495 \text{ UK bushels, } 28.378 \text{ US bushels} \\
&\qquad (30 \text{ UK bushels} = 1.091 \text{ stere} \\
&\qquad 30 \text{ US bushels} = 1.057 \text{ stere}) \\
&= \quad 219.97 \text{ UK dry gallons, } 227.027 \text{ US dry gallons}
\end{aligned}
$$

The term derives from the ancient Greek adjective *stereos* 'solid'.

stereocomparator [linear measure; astronomy] An instrument by which two photographs of part of the night sky taken at different times (with some fair interval between) are superimposed on each other to detect any movement of stars or other celestial objects.

stereoisomer [chemistry] One of two or more isomers that have the same molecular formula but a different arrangement of their atoms in space.

stereoplotter [geography] In making maps from aerial photographs, a device that records the coordinates of the terrain beneath in three dimensions.

sterling [comparative values; metals] Standard of the British pound unit of currency. The term was first applied to a silver penny issued by the Normans in England but known by the English as a *steorling* because it had a little star ('star-ling') depicted on it. Later the term was applied to the pure form of silver used in other coins.

Today, sterling silver is 92.5 per cent pure silver. By English law, however, the fineness of silver in coins has only to be 0.500 – but there is no silver at all in current British coinage anyway.

The *sterling area* is a collective name in fiscal contexts for those countries that use British currency either as their money, or that use it as a medium of trade (as do certain Commonwealth countries).

stethometer [medicine] A medical instrument for measuring the (exterior) expansion of the chest as a person breathes in.

The first element of the term is ancient Greek *stethos* 'chest'.

stethoscope [medicine] An instrument that amplifies the sound of a person's breathing or heartbeat via a pair of earpieces, so assisting a physician's diagnosis in such cases as lung infection or pulse irregularity. It can also be used to detect the heartbeat of a foetus.

The first element of the term is ancient Greek *stethos* 'chest'.

sthène [physics] Unit of force in the obsolete metre-tonne-second system (formerly known as a *funal*), equal to the force needed to accelerate a mass of 1 tonne by 1 metre per second per second.

$$
\begin{aligned}
1 \text{ sthène} \quad &= \quad 1{,}000 \text{ newtons} \\
&= \quad 101.97 \text{ kilograms force} \\
&= \quad 224.9 \text{ pounds force} \\
&= \quad 6.984 \text{ poundals}
\end{aligned}
$$

The term derives from ancient Greek *sthenos* 'strength'.

stich [literary] A line of verse or poetry. Occasionally, however, the word is used instead (inaccurately) to mean a couplet or even a verse.

The term is the ancient Greek for 'file' (in the sense that people file in and file out in a line), and thus 'line'.

sticheron [literary] A stanza or verse of a hymn, particularly in the Greek Orthodox Church; it is known alternatively as a *troparion*.

The term is an ancient Greek word meaning 'a thing of lines'.

stiffness [physics] For a system that is vibrating mechanically (such as a spring), the restoring force needed per unit displacement; the reciprocal of *compliance*.

stilb [physics] Unit of luminance (formerly called brightness) in the centimetre-gram-second (CGS) system, equal to the emitted luminous intensity per unit area in a given direction.

$$
\begin{aligned}
1 \text{ stilb} \quad &= \quad 1 \text{ candela per square centimetre} \\
&= \quad 10{,}000 \text{ candela per square metre}
\end{aligned}
$$

The name of the unit derives from the ancient Greek verb *stilbein* 'to shine', 'to glisten'. *See also* APOSTILB.

Stirling's formula [maths] A mathematical formula for arriving at the approximate values of higher factorials (a factorial number corresponds to the product of all the integers from the given number down to 1); it involves the use of the transcendental numbers π and e.

It was named after the Scottish mathematician James Stirling (1692-1770), although originally formulated by the Frenchman Abraham de Moivre (1667-1754).

stiver [comparative values] Medieval silver coin in Holland and Dutch Belgium (the Low Countries), the name of which was for centuries afterwards slang in the Netherlands (in the form *stuiver*) for a 5-cent piece, until that value was reduced to practically nothing. It is with the meaning of '(something – especially a coin – worth) practically nothing' that the term can be used in English.

stocks and shares [comparative values] Certificates representing equal shares in the capital invested in a public company or corporation, and entitling their owners to annual interest and other premium offers as part-owners of the company or corporation. A number of shares in a single company or corporation which are owned by one person is called his or her 'stock'; such stocks are highly tradable, and it is upon such marketable wares that the stock markets of the world exist.

That a share is indeed a share is self-evident, but a stock is so called because, in the very early days, the 'receipt' paid to an investor and representing an exact quantification of how much he or she had invested was the portion of a tally-stick ('stock') equivalent to his or her input.

stokes, stoke [physics] Unit of kinematic viscosity of a fluid in the centimetre-gram-second (CGS) system, equal to the coefficient of dynamic viscosity divided by the density. Known in Britain and Europe as a stokes, it is more often called a stoke in North America, and was named after the British physicist George Stokes (1819-1903).

$$
1 \text{ stokes} \quad = \quad 10^{-4} \text{ square metre per second (SI units)}
$$

Stokes' law [physics] For a small spherical object falling through a fluid, its terminal velocity equals twice the product of the acceleration of free fall (acceleration due to gravity), the square of the radius of the sphere, and the difference in densities of the sphere and the fluid, divided by 9 times the coefficient of viscosity. The law may be stated mathematically thus: $v = 2gr^2(d_1 - d_2)/9\eta$. It was named after the British physicist George Stokes (1819-1903).

stone [weight] Measure of weight in Britain and some other countries (although not ordinarily in North America), used particularly in relation to one's own bodyweight (which is referred to then, if not an exact number of stone, as '*n* stone *m* pounds'). Even in Britain, however, this mode of weight measurement is now being superseded by units of the metric system (kilograms).

1 stone	=	14 pounds
	=	6.35026 kilograms
		(10 kilograms = 1.5747 stone)
2 stone	=	1 quarter (i.e. of a hundredweight)
8 stone	=	1 (long) hundredweight or quintal, 112 pounds
160 stone	=	1 (long) ton, 2,240 pounds

In former times the stone was used more widely, and varied in weight according to the commodity that was being weighed. In these circumstances the value of 1 stone could be anything from 8 pounds to 24 pounds.

The use of the term derives from the weight of an actual stone or rock used as a weight on the scales that weighed out the commodity for which that particular stone was applicable. *See also* WEIGHT, HUMAN BODY (STANDARDS AND NORMS).

Stone Age [time] The earliest form of human culture, in which the mode of subsistence was largely hunting and gathering, although, by the end of the Stone Age, the rudiments of agriculture were being learned and there was some domestication of animals. But what particularly distinguishes such a culture is that the tools and implements were of wood or stone: the use of metals came later in first the Bronze Age and then the Iron Age. The Stone Age itself is customarily divided into the Palaeolithic, Mesolithic, and Neolithic periods.

stop [photography] An informal way of describing the aperture of a camera lens and the *f*-number that indicates its setting: *see* F-NUMBER. To *stop down* is to reduce the aperture.

stopwatch [time] A watch on which the two hands describe minutes and seconds (rather than hours and minutes) in great detail, and that can be stopped at a precise moment (on some watches, to the nearest hundredth of a second), used to time races or events. It can then be reset at zero (the 12 o'clock position) by a spring to be ready for the next timing. Modern stopwatches give (electronic) digital displays of times. The first stopwatches appeared in the 1730s.

storm force [meteorology] An informal term for a wind of speed 91-104 kilometres per hour/64-72 miles per hour, 11 on the BEAUFORT SCALE OF WINDSPEED. It is known technically as 'whole gale'.

stotinka [comparative values] Unit of currency in Bulgaria: plural *stotinki*.

100 stotinki = 1 lev

The term is a form of diminutive of Bulgarian *stotna* 'one-hundredth'. *See also* COINS AND CURRENCIES OF THE WORLD.

s.t.p., STP *see* STANDARD TEMPERATURE AND PRESSURE

strabismometer [medicine] An ophthalmic instrument used to measure the difference in visual direction of the eyes of a patient who suffers from strabismus (a squint).

The first element of the term is ancient Greek *strabismos* 'squint'.

straight, straight flush *see* POKER SCORES

straight-line depreciation [comparative values] In insurance and in tax assessment, the perception of electronic or mechanical equipment (especially office equipment or motor vehicles acquired new) as an asset of which the value depreciates at a fixed rate per year over the number of years of expected or actual use. Logically, after the initial and most severe degree of depreciation that results from first taking possession of the equipment (after which it becomes 'second-hand' to sell), depreciation in value might be expected to be minimal for the first couple of years, and then gradually to increase as the signs of age manifest themselves and as new models progressively outdate the equipment. Straight-line depreciation makes the mathematics easy for everyone, aided by valuation charts available to insurance agents and tax assessors.

strain [physics] The deformation of an object under stress, equal to the ratio of the amount of deformation to its original (undeformed) dimension. For an elastic material, the various strains are represented by the elastic constants (moduli of elasticity): *see* BULK MODULUS; RIGIDITY MODULUS; YOUNG'S MODULUS.

strain gauge [physics; engineering] Any of a number of instruments, most of them electrical in operation, that monitor and measure strain and the effects of pressure (especially in metals).

Straits dollar [comparative values] Unit of currency in the Straits Settlements, the former British Crown Colony occupying parts of the mainland of the Malay Peninsula and various islands off its coast, until the colony was incorporated into the Malayan Federation in 1957. The unit took the form of a silver coin.

$$1 \text{ Straits dollar } = 100 \text{ cents}$$

strangeness [physics] In nuclear physics, a property ascribed to certain elementary particles in which the peculiarly slow rate of radioactive decay seemed at the time to require some explanation. Such particles – heavy mesons and hyperons of the more unstable elements – were then collectively known as *strange particles*. Leptons and gauge bosons have a strangeness of zero.

strata, geological *see* GEOLOGICAL STRATA

stratificational grammar [literary] Analysis of a language not by the formal separation of words into parts of speech according to their semantic meaning and their position within a sentence, but into *strata*, each interrelated stratum comprising such units as PHONEMES, MORPHEMES, SEMEMES, and LEXEMES.

stratopause, stratosphere *see* ATMOSPHERE, COMPOSITION OF

stratum, geological *see* GEOLOGICAL STRATA

streak [metals and minerals] When a mineral is rubbed or scratched by a harder surface, a line of coloured powder is scraped off to form a 'streak' that may or may not be of a different colour from the rest of the outside of the base. Such a streak may be a valuable identifying characteristic of a mineral. The streak colour of (yellow) pyrite, for example, is black.

stremma [square measure] Obsolete unit of area in (modern) Greece, now in name assimilated in any case to a fairly unusable metric measure – the decare – which is nonetheless quite close to one-quarter of an acre.

$$
\begin{aligned}
1 \text{ stremma } &= 10 \text{ ARES, } 0.1 \text{ hectare} \\
&= (31.623 \times 31.623 \text{ metres}) \text{ } 1,000 \text{ square metres} \\
&= 1,196 \text{ square yards, } 10,764 \text{ square feet} \\
&= 0.2471 \text{ acre, } 2.471 \text{ square chains} \\
&\quad (1 \text{ acre} = 4.047 \text{ stremma})
\end{aligned}
$$

stress [physics] For a material under strain, the force per unit area that tends to change its dimensions – usually either shear stress or tensile (or compressive) stress – measured in such units as pascals, newtons per square metre, bar, or pounds-force per square inch (or their multiples): *see* STRAIN.

strike [minerals; geology] The *strike* is the linear (horizontal, compass) direction of a vein of ore within rock layers, or of the layers themselves; the *dip* is the angular direction upwards or downwards.

strike [chemistry] In former centuries, a term for the proportion of malt in an ale or beer, and thus for its 'strength' of taste (if not of alcohol). (The word *strike* is etymologically akin to *strong* and *strength*.)

strike, strike zone, strike-out [sporting term] In baseball, a strike is a miss by the batter when trying to hit a ball that is a valid pitch within the strike zone – the cubic space within reach of a full swing of the bat at a height between the batter's armpits and kneecaps, as judged by the referee behind the catcher – or a partial hit that sends the ball out of fair territory under the rules of the game. After two strikes, a batter may legally hit a ball that is not within the strike zone. Three strikes, and a batter is out for the inning: he is/has 'struck out'.

In ten-pin bowling, a strike is the knocking down of all ten pins with the first ball bowled.

stringed instruments' range [music] The virtue – or perhaps from a player's point of view, the major defect – of a stringed instrument is that it technically has no limit to the highest note attainable, either by stopping the string as high up the neck as possible, or through using harmonics. The range can therefore be defined only at the lower end, by the note produced by the lowest-pitched open string.

Even then, whereas many instruments are manufactured to play only at a certain well-defined pitch (such as brass or woodwind instruments), stringed instruments can be deliberately tuned well below the normal pitch, if desired for any specific reason, simply by loosening all the strings in proportion.

In the table below, the note to which the lowest-pitched string is ordinarily tuned, and the general range over which music for the instrument is or was ordinarily scored, is given.

violin	about 3 octaves, from G below middle C
viola	about 4 octaves, from C an octave below middle C
cello, violoncello	about 4 octaves, from C two octaves below middle C
rebec	about 3 octaves, middle and upper register
viol da gamba	about 2 octaves, middle register
double bass	about 3 octaves, from E an octave below the bass clef stave
zither, psaltery	usually 4 octaves, from C an octave below middle C
autoharp	usually 2 octaves plus basses, from middle C
kantele	usually 3 octaves, tuned to register required
koto	usually 2 octaves, middle register
sitar	usually 3 octaves, tuned to register required
cembalom, dulcimer	usually 3 octaves, middle and upper register
lyre	about 1 octave, upper register
harp	5½ octaves, from C two octaves below middle C
Welsh/Irish harp	about 3 octaves, middle and upper register
banjo	about 3 octaves, from C an octave below middle C
ukulele	about 2 octaves, from middle C
guitar	about 3 octaves, from E below middle C
cittern	about 2 octaves
theorbo	about 3 octaves, around middle register
lute	about 2 octaves
mandolin	usually 2 octaves, from middle C
balalaika	usually 2 octaves, from middle C
bandoura	usually 2 octaves

See also KEYBOARD INSTRUMENTS' RANGE.

stringendo [music] Musical instruction: speed up a little to add the flavour of a climax. Italian: 'applying pressure'.

strob [physics; speed] Unit of velocity relating to objects that move in a circular path or orbit.

$$1 \text{ strob } = 1 \text{ radian per second}$$

(A radian is a length of arc – a distance around the circular path or orbit – identical to that of the radius of the circle described.)

The name of the unit derives from ancient Greek *strobos* 'rotating'.

stroboscope [physics] An instrument that can be used to measure speed of rotation of a rotating object. It consists of a rapidly flashing lamp in which the rate of flashing can be adjusted until the rotating object appears to be stationary (when its speed is a whole number multiple or submultiple of the flashing speed).

The first element of the term is the Greek *strobos* 'rotating'.

strong interaction [physics] A rapid interaction between subatomic particles (baryons and mesons) that binds together protons and neutrons in an atomic nucleus.

strontium unit [medicine] Unit of radioactive concentration for the radioisotope strontium-90 in calcium (in food or, after absorption, in bones).

1 strontium unit $= 10^{-12}$ curie per gram

The element strontium was named after the village of Strontian, Highland Region, Scotland, where the mineral was first discovered in 1790.

strophe [literary] In ancient Greek choral and lyric poetry, a set of lines in a strict metrical pattern that is then 'answered' by another set of lines in the same pattern (the *antistrophe*). During dramatic performances of such verse, the ancient Greek chorus would declaim the strophe while moving from right to left on stage, and would declaim the antistrophe while moving back from left to right.

By extension, the term also describes any one stanza in more modern English lyric poetry where there are more than two other stanzas of similar metrical pattern.

The term is the ancient Greek for 'turning', referring to the chorus's movement in one direction or the other.

stroud [comparative values] Unit of value for bartering with North American Indians in the late seventeenth century (1680s-90s): it comprised a fairly large, coarse blanket, of a type that may or may not have been made especially for the purpose. The material of which it was made may or may not have been named after a similar material formerly woven in Stroud, Gloucestershire, England.

Strouhal number [physics] For an aeolian tone – a note caused by a draught of air vibrating a stretched wire or string – the product of the wire's thickness and its frequency of vibration divided by the speed of the airflow. It was named after V. Strouhal (1850-1922).

structural formula [chemistry] A description of a chemical compound that provides a shorthand description of the arrangement of the atoms in its molecules (as well as its composition). For example, ethoxyethane (ether) has the structural formula $C_2H_5.O.C_2H_2$, whereas that of the isomeric *n*-butanol (butyl alcohol) is C_4H_9OH . *See also* EMPIRICAL FORMULA; MOLECULAR FORMULA.

structural isomerism [chemistry] A type of isomerism in which two chemical compounds have the same molecular formula but different structural formulae. The examples in the article on STRUCTURAL FORMULA (*above*) are structural isomers which, like other such pairs, have entirely different physical and chemical properties.

Stuart [time] Describing the period in Scottish and English history between AD 1371 and 1714, when members of the Stuart family were on the throne of one or both countries. The monarchs involved were:

Robert II of Scotland	ruled	1371-90	(19 years)
Robert III		1390-1406	(16 years)
James I		1406-37	(31 years)
James II		1437-60	(23 years)
James III		1460-88	(28 years)
James IV		1488-1513	(25 years)
James V		1513-42	(29 years)
Mary, Queen of Scots		1542-67	(25 years)
James VI of Scotland		1567-1625	(58 years)
= James I of England		1603-25	(22 years)
Charles I of Scotland and England		1625-49	(24 years)
[the republic or Commonwealth]			
Charles II		1660-85	(25 years)
James II		1685-88	(34 months)
Mary II (with William of Orange)		1689-1702	(13 years)
Anne		1702-14	(12 years)

The first Stuart king was Robert II, who was son of Walter the Steward of Scotland (technically holding the country as the agent of the English crown – and hence the surname, also sometimes spelled Stewart).

The last Stuart monarch, Anne, underwent no fewer than seventeen childbirths, but virtually all those babies who were not stillborn lived only for hours at a time and, in due course, she was succeeded by her second cousin George, the Elector of Hanover, so ending the Stuart dynasty.

Student's distribution, Student's t-distribution *see* T-DISTRIBUTION

stunde [linear measure] Obsolete unit of distance in Switzerland.

$$1 \text{ stunde} = 4{,}800 \text{ metres, } 4.8 \text{ kilometres}$$
$$(5 \text{ kilometres} = 1.0417 \text{ stunde})$$
$$= 2.9826 \text{ miles (2 miles 1,729.4 yards)}$$
$$(3 \text{ miles} = 1.0058 \text{ stunde})$$

In most Germanic languages (for example, Norwegian and Swedish), the word is an imprecise measure of time, not distance, meaning 'period' or 'a while'. German, however, has made it much more precise: in German it is the normal word for 'hour'.

Stuttgart pitch *see* INTERNATIONAL PITCH

suan pan *see* SOROBAN

sub- [quantatives: prefix] Prefix denoting 'under', and thus 'less than', 'smaller than'. For examples *see entries beginning* SUB- *below*.

The prefix derives directly from Latin; the Greek equivalent is *hyp-*.

subatomic particle [physics] A particle that is smaller than an atom, often forming part of an atom, such as an electron, a neutron, or a proton. *See also* ELEMENTARY PARTICLE.

subaudio frequency [physics] Describing a sound frequency that is too low to be conveniently reproduced by a sound system. *See also* SUPERAUDIO FREQUENCY.

subcalibre, subcaliber [linear measure] Describing a projectile, such as a shell or bullet, that has a diameter less than the bore of the weapon from which it is fired. It is fired either having been fitted with a metal disk of the correct bore, or from within a tube of its own bore inserted down the bore of the weapon. The reason for using subcalibre ammunition when practising is generally to reduce the cost.

Subclass, subclass *see* CLASS, CLASS

subcritical mass [physics] A mass of fissile material that is less than the CRITICAL MASS (and cannot therefore take part in a self-sustaining nuclear chain reaction).

Subdivision, subdivision *see* CLASS, CLASS

subfactorial [maths] The number of ways of arranging n items so that none of them is in its original position. It is symbolized as n¡ . *See also* FACTORIAL.

Subfamily, subfamily *see* CLASS, CLASS

subform *see* CLASS, CLASS

Subgenus, subgenus *see* CLASS, CLASS

subharmonic [physics] The frequency of a wave, such as a sound wave, that is a (simple) fraction of the frequency of the fundamental wave. *See also* HARMONIC.

Subkingdom, subkingdom *see* CLASS, CLASS

sublieutenant [military rank] In the British Royal Navy (and in the navies of some other countries), a junior commission corresponding to a rank above that of midshipman but below a (full) lieutenant. A sublieutenant is the equivalent of a lieutenant in the British army and a flying officer in the Royal Air Force, a first lieutenant in the US Army and Airforce and a lieutenant (junior grade) in the US Navy.

sublunar point [astronomy] A point at which a line from the Moon to the centre of the Earth cuts the Earth's surface – that is, a point at which the Moon is directly overhead.

sub-millimetric waves [physics] Radio waves (microwaves) whose wavelength is less than 1 millimetre (corresponding to a frequency in excess of 300 gigahertz).

submultiple [quantitatives; maths] A number or quantity that will divide exactly into another number or quantity, also known as a *factor*.

subnormal [maths] For a curve in coordinate geometry, the normal to the curve at any point crosses the y-axis somewhere. The subnormal is the projection, on to the y-axis, of the part of the normal between the curve and the y-axis. *See also* SUBTANGENT.

Suborder, suborder *see* CLASS, CLASS

Subphylum, subphylum *see* CLASS, CLASS

Subsection, subsection *see* CLASS, CLASS

Subseries, subseries *see* CLASS, CLASS

subset [maths] In set theory, if the elements in one set are members also of another set, the first set is a subset of the second one. The empty set is a subset of all sets, and every set is a subset of the universal set. Thus

$$\{a, b, c\} \text{ is a subset of } \{a, b, c, d, e, f\}$$

If $\{a, b, c\} = A$ and $\{a, b, c, d, e, f\} = B$, the relationship is written $A \subset B$.

subsolar point [astronomy] A point at which a line from the Sun to the centre of the Earth cuts the Earths surface – that is, a point at which the Sun is directly overhead.

subsonic [physics; aeronautics] Describing a speed that is less than the speed of sound in a given medium, or an aircraft speed that is less than Mach 1: *see* MACH NUMBER; SPEED OF SOUND.

subspecies [biology] In the taxonomic classification of life-forms, a rank that is below that of SPECIES. It describes a group within a species that has characteristics that are not shared with other members of the species (such as geographical location), and includes races, varieties, and breeds in the animal world, and varieties and forms in the plant world.

For a full list of taxonomic categories, *see* CLASS, CLASS.

substantive [literary] In grammar, another term for a NOUN or a PRONOUN.

As an adjective, describing a verb in relation to which the subject is in apposition to the 'object' – a verb, that is, corresponding to the verb 'to be' or an equivalent.

substellar point [astronomy] The point at which a line from a star to the centre of the Earth cuts the Earth's surface – that is, a point vertically below the star. Its latitude is the same as the star's declination.

subtangent [maths] For a curve in coordinate geometry, the tangent to the curve at any point crosses the *x*-axis somewhere. The subtangent is the projection, on to the *x*-axis, of the part of the tangent between the curve and the *x*-axis. *See also* SUBNORMAL.

subtend [maths] How a line, arc, or part of a surface defines an angle. For example: an arc of a circle subtends an angle at the centre of the circle; an area on a sphere subtends a solid angle at the centre of the sphere. *See also* RADIAN; STERADIAN.

subtotal [maths] Either a cumulative total up to a certain point in a substantial calculation, or an independent total representative of a specific part of a calculation which, when added to the independent totals of other parts, will make up the overall total.

subtraction [maths] Mathematical operation that produces the difference between two quantities, usually denoted by the symbol – . The *subtrahend* is subtracted from the *minuend* to give the difference. The operation is the inverse of ADDITION but, unlike addition, is not commutative – the order in which the two quantities are subtracted is critical (a – b is not the same as b – a; for example: 8 – 5 is not the same as 5 – 8).

subtrahend [maths] The number that is subtracted from the *minuend* in a subtraction sum: *see* SUBTRACTION.

Subtribe, subtribe *see* CLASS, CLASS

subunit [quantitatives] A SUBMULTIPLE: a unit of a unit.

subvariety *see* CLASS, CLASS

subzero [quantitatives] Below zero on any scale, particularly of the Celsius scale of temperature.

sucre [comparative values] Unit of currency in Ecuador, in the form of a silver coin and a banknote.

                    1 sucre = 100 centavos

The unit is named after the South American general and popular liberator Antonio José de Sucre (1795-1830). *See also* COINS AND CURRENCIES OF THE WORLD.

suits of cards [sporting term] In cards, the four suits are:

| Hearts (red) | Spades (black) |
| Diamonds (red) | Clubs (black) |

Hearts were first named as a suit during the AD 1520s, although the stylized design (intended to represent a human heart) had by then been in existence for some sixty years. The shape of the spade represents the type of short sword or dagger (poniard) used up until the mid-1700s; the name of the suit comes through Italian ultimately from Greek *spathe* 'sword' and thus has nothing at all to do with the agricultural implement. The diamond had been a heraldic device for centuries although, in that context, generally known as a lozenge. Nonetheless, it was made a suit of cards only during the 1590s, taking on the red colour of the hearts. The clubs derive their name in English as a translation of the Spanish *basto* 'club', 'battle-mace', but the design of the figure is in fact taken directly from the French form on playing cards, the *trèfle* or trefoil.

It is often suggested that the form of playing cards – including the idea of four suits – derives from the *minor arcana* of the tarot cards. The minor arcana consists of fifty-six cards in four suits, each suit numbered ace to 10 and with four court cards: king, queen, knight, and page. Two of the tarot suits correspond with the playing-card suits: clubs (also called staves, staffs, or batons) and swords ('spades', in French the cognate *épées*); the other two tarot suits are cups and coins.

In view of the presence of court cards, the word *suit* is etymologically entirely appropriate (suitable), for it derives from Middle English *sywte* 'courtiers', 'followers at court' (ultimately from Latin *sequita* 'followers'), and thus also '*suite*'. It is in reference to the fine apparel and liveries of those at court that one's own best clothes may today correspond to a *suit*.

suk [volumetric measure] Unit of dry capacity in Korea, used mainly in the measurement of fruit and vegetable market produce.

1 suk	=	about 6.25 cubic feet
	=	4.868 UK bushels, 5.024 US bushels
	=	176,900 cubic centimetres, 0.1769 cubic metre
	=	176.9 litres by volume
	=	38.944 UK dry gallons, 40.192 US dry gallons

sultchek [volumetric measure] Unit of liquid and dry capacity in Turkey, equal to 1 litre but apprehended as a cubic measure based on a *parmak*, a decimetre.

1 sultchek	=	1 parmak x 1 parmak x 1 parmak
	=	10 centimetres x 10 centimetres x 10 centimetres
	=	1,000 cubic centimetres
	=	1 litre
	=	1.76 UK pint, 2.1134 US liquid pints
	=	1.76 UK dry pint, 1.816 US dry pint

sum [maths] The result of addition, or to carry out an addition (*see* ADDITION).

For some reason we talk about adding *up* and totalling *up*, although the total that we add up to is usually at the bottom of the calculation; this idea of ascent, however, is part of the derivation of the word *sum*, from Latin *summa* '(thing) at the top', 'the highest'.

summa cum laude [comparative values] Highest grade in the marking of university papers, theses, and dissertations, as conferred by the authorities of many universities in continental Europe and some in North America and elsewhere.

The ordinary grade, the equivalent of a pass-mark, is given *cum laude* 'with praise'; the next grade is given *magna cum laude* 'with great praise'; and the top grade is given *summa cum laude* 'with the highest praise'.

summand [maths] Technical term for one of two or more numbers or quantities to be added together.

summation sign [maths] The Greek capital sigma, Σ, used as a symbol to denote the sum (total) of a series of quantities.

Sumner unit [biology] Unit of enzyme activity (referred to urease, the enzyme that catalyses the breakdown of urea), equal to the amount of enzyme that causes the liberation of 1 milligram of ammonia-nitrogen in 5 minutes (at 20°C).

$$1 \text{ Sumner unit} = 14.28 \text{ international units}$$

It was named after the American biochemist James Sumner (1887-1955).

Sun [astronomy] Our local star. All the energy on the Earth derives from energy given out at one time or another by the Sun. This energy, which causes the Sun to shine, is a result of nuclear fusion reactions taking place at the Sun's core, by which hydrogen is converted to helium. The nuclear reactions are represented at the Sun's surface – the photosphere – by sunspots and the brighter areas known as faculae; in the thin layer of rarefied gas above – the chromosphere – they cause solar flares and prominences; and in the Sun's upper atmosphere – the corona – they produce the stream of atomic particles known as the solar wind.

Solar statistics

mean diameter	(kilometres)	1,392,000
	(miles	865,000
mass	(tonnes)	2×10^{27}
	(long tons)	1.968×10^{27}

	(short tons)	2.205×10^{27}
	(Earth = 1)	332,946
volume (Earth = 1)		1,303,600
mean density (water = 1)		1.4
mean distance from Earth	(astronomical units)	1.0
	(million kilometres)	149.6
	(million miles)	89.8
surface gravity (Earth g = 1)		27.9
surface temperature	(°C)	6,000
	(°F)	10,800
core temperature	(°C)	14-15,000,000
	(°F)	25-27,000,000
mean rotation period	(sidereal)	25.38 days
	(synodic)	27.28 days
mean apparent magnitude	−26.8	
absolute magnitude	+4.83	
spectral type	G2V	
distance from centre of Galaxy (light years)		32,000
Galactic orbital velocity	(kilometres per second)	2,150
	(miles per second)	1,310
duration of Galactic orbit (million years)		approx. 200

For information about the eleven-year cycle of thermal activity on the Sun's surface, *see* SOLAR CYCLE.

For information about the origins and composition of the Solar System (including the Sun), *see* SOLAR SYSTEM.

For information about the planets of the Solar System, *see* PLANETS OF THE SOLAR SYSTEM.

See also GREAT YEAR.

Sunday letter [literary; time] Together with the GOLDEN NUMBER, the Sunday letter is a means of finding the date of Easter (which is a movable feast timed according to the full Moon) in any year.

Until the year 2099 (inclusive), the Sunday letter may be determined by the following method:

divide the year by 4 (and ignore any remainder)

add this quarter on to the year

add 6 (for a year between 2100 and 2199 add 5)

divide by 7

note any remainder against the following table

remainder	Sunday letter
no remainder	A
1	G
2	F
3	E
4	D
5	C
6	B

For example: the Sunday letter for the year 2000 is $(2,506 \div 7 = 358$, no remainder) A . *See also* EASTER DAY, DATE OF.

sundial [time] A clock by which the shadow of a rod or pointer (a gnomon), as cast by the Sun, falls on a dial calibrated with the hours of the day. The main disadvantages of this timepiece are that it is unusable except during sunlit periods, and that as the day lengthens and shortens according to the season of the year, so do the hours as measured.

The earliest sundials known date from Egypt in the eighth century BC.

Sung dynasty [time] In China, the time between AD 960 and 1279. Now regarded as one of China's golden ages, it was a period notable for its artistic works on ceramics and in painting, and for the philosophical dominance of Confucianism.

sun protection factor (SPF) [physics; medicine] Unit of classification by the US Food and Drug Administration of anti-sunburn preparations, corresponding to the length of time for which a preparation continues to protect the skin once applied. Factors are numbered from 1 to 12 (or more): a factor of 1 allocated to a preparation (barrier, sun-block, sunscreen, etc.) represents protection for 10 minutes: an SPF of 7 therefore guarantees skin protection for a minimum of 70 minutes if the preparation is applied correctly.

sunspot cycle Another name for the SOLAR CYCLE.

sunspot number [astronomy] Number describing sunspot activity, equal to 10 times the number of disturbed regions on the Sun's surface plus the total number of sunspots, multiplied by a constant (which is a characteristic of the measuring instrument used). *See also* SOLAR CYCLE.

Sun-synchronous orbit [geology] An orbit for an Earth satellite that makes it pass over the same point on the Earth at the same time every day.

super- [quantitatives: prefix] Prefix denoting 'over', and thus 'more than', 'larger than'. For examples *see* entries beginning SUPER- *below*.

The prefix derives directly from Latin; the Greek equivalent is *hyper-*.

super-8 [photography] System of (amateur) cine photography that uses 8-millimetre film with an enlarged frame (image area), which extends between the widely spaced sprocket holes, unlike regular-8. It gives better image quality on projection. *See also* EIGHT MILLIMETRE.

super-16 [photography] System of cine photography that uses 16-millimetre film with an enlarged frame (image area) so that it can be enlarged to make prints on 35-millimetre film for commercial projection with little loss of quality. *See also* SIXTEEN MILLIMETRE.

superaudio frequency [physics] Describing a sound frequency that is too high to be conveniently reproduced by a sound system. *See also* SUBAUDIO FREQUENCY.

Superclass, superclass *see* CLASS, CLASS

superconductivity [physics] The property of a metal to conduct an electric current with no resistance at all (therefore requiring no voltage to transmit it farther). A number of metals – such as lead and tin – act in this way at temperatures not far removed from absolute zero but, in recent years, the hunt has been on, with increasing success, to find a metal or combination of metals that has this property at something near ordinary room temperatures.

supercontinent [geology; time] In prehistoric times, the two enormous landmasses now called Laurasia and Gondwanaland that together comprised virtually all the present-day continents of the world. Also, the single even greater landmass, Pangaea or Pangea, that existed before Laurasia and Gondwanaland split apart. Laurasia eventually divided to become Asia (without India), Europe, Greenland, and North America; Gondwanaland separated into Africa (and Madagascar), Antarctica, Australasia, India, and South America.

supercooled [physics] Describing a liquid that is still in liquid form despite being (or having been) cooled to well below its normal freezing point.

supercritical mass [physics] A mass of fissile material that is more than the CRITICAL MASS (and can therefore take part in a self-sustaining nuclear reaction).

superelevation *see* CANT

Superfamily, superfamily *see* CLASS, CLASS

superfecta [sporting term] An extension of the EXACTA or perfecta system of betting on horses (or greyhounds) in North America, by which the better must nominate the first, second, third, and fourth animals in the correct order to win the bet.

superfluidity [physics] The property of a fluid (especially liquid helium) to flow without friction through even the minutest orifice because of the virtually total lack of viscosity at temperatures not far removed from absolute zero.

supergiant, supergiant stars *see* HERTZSPRUNG-RUSSELL DIAGRAM

superheated [physics] Describing a liquid that is still in liquid form despite being (or having been) heated to well above its normal boiling point (for example, by heating under pressure).

superhigh-frequency [physics: radio] Describing radio frequencies of between 3,000 and 30,000 megahertz, corresponding to wavelengths of between 10 centimetres and one-thousandth of a millimetre.

Superorder, superorder *see* CLASS, CLASS

supersaturated [chemistry] Describing a solution that contains more dissolved substance (solute) than a saturated solution at the same temperature. It is unstable and readily crystallizes to reduce its concentration to that of a saturated solution.

supersonic [physics] Describing a speed that is faster than the SPEED OF SOUND (in a particular medium). *See also* MACH NUMBER; ULTRASONIC.

supplemental chords [maths] Lines that join any point on a circle, ellipse, or hyperbola to the ends of a diameter. The supplemental chords of a circle are at right-angles to each other (they form the angle in a semicircle).

supplementary angle [maths] One of a pair of angles that add to 180 degrees. For example, the angles on a straight line on each side of a line that meets it are supplementary, as are the opposite angles of a cyclic quadrilateral (any quadrilateral inscribed within a circle).

supply frequency [physics] The frequency of the mains electricity supply – and therefore also called the mains frequency. The two most common supply frequencies are 60 hertz (as in the United States and Canada) and 50 hertz (as in Britain and most of continental Europe).

surd [maths; literary] In number theory, a number that involves an irrational root, and is therefore itself an irrational number – for example: $\sqrt{2}$, $2 + \sqrt{3}$: *see* NUMBER; ROOT.

In phonetics, a consonant that does not require any voicing (any vocalization by the vocal cords) – for example: f, k, p, and s (as opposed to v, g, b, and z, which are the voiced equivalents).

In both cases, the word derives from Latin *surdus* 'unhearing', thus 'unheard' (and in the first case above, 'unheard of' or 'ab-surd').

surface [maths] In geometry, a technical term for an area that has the dimensions of length and breadth but no depth (or thickness).

surface conductivity [physics] The reciprocal of SURFACE RESISTIVITY.

surface energy [physics] For any surface, its free potential energy, equal to the product of its surface tension and area: *see* SURFACE TENSION.

surface measure [timber] A way of measuring timber in terms of the area of one face (or many faces), irrespective of thickness. *See also* SQUARE.

surface resistivity [physics] For a unit square on the surface of a conductor, the resistivity between opposite sides of the square; the reciprocal of *surface conductivity*.

surface tension [physics] The force per unit length that acts on the surface of a liquid at right-angles to a line drawn on it, measured in units of force per unit length, such as dynes per centimetre (CGS units) or newtons per metre (SI units). It is the property of a liquid, caused by attractive forces between the molecules in the surface, that makes it appear to have a thin skin on its surface (which will support light objects such as aquatic insects). It is also responsible for phenomena such as capillarity and the spherical shape of (free) soap bubbles.

surface wind [meteorology] The wind at a (standard) height of 10 metres (33 feet) above the ground. It is the subject of meteorological measurements at this height to avoid any ground effects.

surveyors' measurements [linear measure; square measure] Surveyors have by tradition used their own system of measurements of both length and area although, in major measurements, they correspond also to imperial or English standard units.

In linear measure:

1 link	=	7.92 inches
	=	20.117 centimetres
100 links	=	1 chain
	=	66 feet, 22 yards
	=	20.117 metres
10 chains	=	1 furlong
	=	220 yards (one-eighth of a mile)
	=	201.168 metres
8 furlongs	=	1 mile
	=	1.609344 kilometre, 1,609.344 metres

In square measure:

1 square link	=	62.73 square inches

	=	404.69 square centimetres
625 square links	=	1 square rod or 1 square pole
	=	30.25 square yards
	=	25.292 square metres
16 square rods	=	1 square chain, 1,000 square links
	=	484 square yards
	=	404.69 square metres
10 square chains	=	1 acre
	=	4,840 square yards
	=	4,046.9 square metres, 0.40469 hectare
10 acres	=	1 square furlong
		(16 acres = 1 quarter section)
64 square furlongs	=	1 square mile or 1 section
	=	2.5899 square kilometres
36 sections	=	1 township
	=	36 square miles
	=	93.2364 square kilometres

See also PERCH.

susceptance [physics] For an AC circuit, the negative ratio of the reactance to the sum of the squares of the resistance and reactance; the imaginary part of the ADMITTANCE.

susceptibility [physics] The capacity of a substance to be magnetized. It corresponds to the ratio between the intensity of magnetization and the magnetizing force (applied field). It is also the amount by which the substance's relative permeability differs from unity. Paramagnetic substances have positive susceptibilities; diamagnetic materials have negative ones.

svedberg, svedberg unit [physics; time] For particles forming a sediment under the influence of a gravitational field (as in a centrifuge), the sedimentation coefficient, equal to the speed of the boundary between the particle-containing solution and the solvent, divided by the distance of that boundary from the axis of rotation times the square of the angular velocity (in radians per second). The term is also used as the name of a time unit used in measuring sedimentation rates, equal to 10^{-13} second. In both cases, the unit was named after the Norwegian biologist T. Svedberg (1884-1971).

swarm of bees, of locusts, of insects [collectives] This collective is of obscure derivation: most commentators confess that they simply have to guess at an etymology. But there are three main possibilities. A swarm may be a swarm because it makes a loud buzzing or shushing noise (akin therefore to Sanskrit *svarati* 'it sounds', Latin *susurrus* 'whispering'), because the insects in it are nagging and persistent (akin thus to Latin *severus* 'importunate', 'getting on one's nerves', Dutch *zwaar* 'nagging', 'heavy', English *sore* and *severe*), or because the cloud of insects blackens the sky (akin thus to Old English *sweart* 'black', English *swarthy*, Latin *sordidus* 'filthy').

sweep [aeronautics] For a fixed-wing aircraft, the angle between the span line or the rear edge of the wing and the fuselage. In high-speed aircraft an acute sweep angle – sweepback – is more usual than an obtuse one (forward sweep).

sweepstake, sweepstakes [sporting term] A type of gambling in which the prize money, together with the fees for organizing the event, are found from the money contributed by those who wish to gamble – although the gamblers themselves have no say whatever in the ticket number, horse, greyhound, or other subject of the gamble that they are allocated (if they are allocated one at all: in some sweepstakes, the premium is first of all risked on getting or not getting a share in the actual gamble that is the primary event).

Sweepstakes are extremely popular in Ireland and continental Europe, and in South America, in which countries national sweepstakes and those organized for the benefit of charitable causes generally give out many prizes in each competition. The original point of a sweepstake – known to Shakespeare as a swoopstake, incidentally – however, was that one person scooped the lot – there was only one prize, and the winner made a clean *sweep* (or swoop) of all the *stakes*.

swell [music] An increase in volume (crescendo) followed by a corresponding decrease in volume (diminuendo), within the symbols < and > , sometimes on a single sustained note. Also, the device on an organ (generally operated by one foot) that produces such an increase and decrease in volume.

swimming disciplines and competitive distances [sport] Swimming competitions are for individuals (competing against each other and the clock) and for teams. There are four principal strokes: the breaststroke; the backstroke; the butterfly; the freestyle (virtually always the crawl).

The usual length of the pools in which competitions take place is 33.333 metres (109 feet 4.323 inches) or, more often, 50 metres (164 feet ½ inch). Usual lane width per competitor is 2.5 metres (8 feet 2 inches). In breaststroke, butterfly, and freestyle races, the height of the box or podium from which competitors take off in their initial dive is 75 centimetres (2 feet 6 inches) above the water. The rope that is dropped into the water across the pool to signify a false start is usually 15 metres (49 feet 2½ inches) from the starting line.

The most common competition events are: 100 metres breaststroke; 200 metres breaststroke; 100 metres backstroke; 200 metres backstroke; 100 metres butterfly; 200 metres butterfly; 100 metres freestyle; 200 metres freestyle; 400 metres freestyle; 800 metres freestyle (usually only for women); 1,500 metres freestyle (usually only for men); 400 metres individual medley (in turn over 100 metres: backstroke, breaststroke, butterfly, and crawl); 4 x 100 metres freestyle relay, for women; 4 x 200 metres freestyle relay, for men; 4 x 100 metres medley relay (each team member a different stroke, in the medley order).

Other events that involve a swimming pool (and associated equipment) are diving, synchronized swimming (for pairs, quartets, and larger groups), and water polo.

sycee [comparative values] A lump of pure silver, onced used as a medium of value for trading in China. Sycees were of different sizes corresponding to different values, and authenticated as such by being officially stamped by an assayer or a banker.

The term is the English spelling of the Cantonese *sai-si* 'fine silk', referring to the fact that pure silver when melted can be drawn out into long thin 'threads' like silk.

syli [comparative values] Between 1973 and 1986, a unit of currency in Guinea.

1 syli = 100 caury

Adopted as a unit of currency some fifteen years after the country's independence from France, the syli was an attempt to go back to a truly native system of values, involving the use of (coins named after) shells (cowries) for trading and bartering.

syllable [literary: phonetics] A word or part of a word that is a single sound, whether long or short, stressed or unstressed.

A syllable that ends in a vowel is said to be an open syllable; a syllable that ends in a consonant (or a consonantal sound, such as the second syllable in *subtle*) is called a closed syllable.

The term derives from ancient Greek elements meaning 'taken together'.

symbol, chemical *see* PERIODIC TABLE OF ELEMENTS

symbol, mathematical *see* ADDITION; DIVISION; MULTIPLICATION; SUBTRACTION

symmetrical [maths] In geometry, describing a shape that is consistent and/or balanced (especially one that is identical in two or four aspects).

The term derives from ancient Greek elements meaning 'measuring together'.

synchronometer [physics] For an AC electricity supply, an apparatus that counts the number of cycles in a given time. It is a digital frequency meter if the time is made unity (that is, if it counts cycles per second and it measures the frequency in hertz).

synchronous orbit [astronomy] For a moon or satellite orbiting a planet, an orbit that has the same period (that takes the same time for one complete orbit) as the period of rotation of the planet. It therefore appears to remain stationary over one place on the planet's surface. For the Earth, it is also called a *geostationary orbit*.

synodic month [astronomy] The time that elapses between two successive passages of the Moon through conjunction (or opposition); the total time occupied by the Moon's phases. It equals 29.53059 days.

synodic period [astronomy] The time that elapses between two similar positions of

a moon or planet, such as between one conjunction (or opposition) and the next.
system *see* GEOLOGICAL STRATA
systole *see* BLOOD PRESSURE

T

t *see* TEASPOON(FUL); TESLA; TON; TONNE, METRIC TON
tablespoon(ful) [volumetric measure] Unit of volume used in cookery and in
mixing drinks (and some medicines), in North America the equivalent of 3
teaspoon(ful)s but in Europe the equivalent of 4-4½ teaspoon(ful)s.

1 tablespoon(ful)	=	around 14.5 millilitres
	=	about half a fluid ounce
	=	about one-tenth of a UK gill
	=	about one-eighth of a US gill
	=	about 4 fluid drams

The tablespoon is the largest spoon in normal household cutlery.
table tennis measurements and units [sport] Table tennis is a game for two
players playing singles, or four players playing doubles, and requires speed of
action and of thought.

The dimensions of the table:
overall length: usually 9 feet (2.74 metres)
width: usually 5 feet (1.52 metres)
net height: 6 inches (15.24 centimetres)
net overhang outside table: 6 inches (15.24 centimetres)
Points scoring:
opponent fails to return ball: 1 point
opponent serves into the net or off table: 1 point
the service changes every five points
first player or team to score 21 wins a game; a score of 20-20 usually requires
playing on until one player or team wins two clear points
matches usually comprise (the best of) three games
Dimensions of equipment:
the bat or racket:
no weight or size restrictions
striking surface usually rubber
the ball:
diameter: 1½ inches (38.2 millimetres)

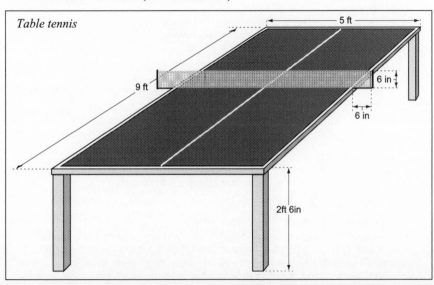

Table tennis

5 ft

9 ft

6 in

6 in

2ft 6in

weight: 2.4-2.53 grams (0.085-0.09 ounce)

Other than having a net at the centre of the 'court', table tennis has very little in common with (lawn) tennis.

Tacan, TACAN [geography; electronics] Electronic device in an aircraft that displays the aircraft's distance and direction from a fixed point by detecting a constant pulse transmitted from a station at that point.

The term is derived from the first letters of the words T̲actical A̲ir N̲avigation.

tacheometry, tachymetry [surveying] A method of surveying using a tacheometer (or tachymeter), which is a theodolite equipped with a pair of crosshairs that enables distances to a levelling staff to be measured.

Ta Ch'ing [time] Another name for the MANCHU DYNASTY in China.

tachograph [engineering] An instrument for measuring and graphically recording the journey times and speeds of a road vehicle, such as a truck or bus.

tachometry [physics; engineering] The measurement of speeds of rotation, using a tachometer – better known (in road vehicles) as a rev(olution) counter. Speeds are usually expressed in revolutions per second or per minute.

The first element of the term is the ancient Greek *tachos* 'speed'.

tachycardia [medicine] Abnormally fast heartbeat: *see* PULSE.

tachymeter, tachymetry *see* TACHEOMETRY, TACHYMETRY

tachyon [physics] A hypothetical subatomic particle that can move faster than the speed of light.

tactometer [medicine] A device that gauges the sensitivity of a person's sense of touch.

tael, tahil [weight; comparative values] Ancient unit of weight in China, corresponding also once to a silver coin of that same weight. The unit was then assimilated into the system of weights used throughout the Far East during the colonial times of the mid-1800s especially for taxation and the imposition of customs duties.

In China,

1 tael	=	10 chi'en (in silver, 10 mace)
	=	1.3333 ounce avdp.
		(1 ounce avdp. = 0.750 tael)
	=	37.798 grams
		(100 grams = 2.6456 taels)
16 taels	=	1 CATTY or chin
1,600 taels, 100 chin	=	1 tan

These values were consistently followed by the unitary systems of Hong Kong and Korea, except that the Chinese *chin* was in Hong Kong called the *kan*, and in Korea called the *kon*. It was much the same in Malaysia.

In Malaysia,

1 tahil	=	10 chee, 100 hoon
	=	1.333 ounce avdp.
	=	37.798 grams
16 tahils	=	1 kati or gin
1,600 tahils, 100 gins	=	1 picul

The Japanese, however, assimilated the weight to one of their own system while nonetheless adopting the name.

In Japan,

1 tael	=	10 momme, 100 fun
	=	1.323 ounce avdp.
		(1 ounce avdp. = 0.7559 Japanese tael)
	=	37.5 grams
		(100 grams = 2.6666 taels)
16 taels	=	1 kin
100 taels	=	1 kwan
1,600 taels, 16 kwan	=	1 picul

The name of the unit is not, however, Chinese but seems to derive ultimately from Sanskrit *tula* 'a suspension-balance', and thus 'a weight (to be used on that balance)'. Presumably Indian traders visited China long before Europeans arrived there. On the other hand, it seems to have been taken up as a Chinese measure by the Spanish and Portuguese merchants who introduced the unit to areas such as Malaysia and Korea. *See also* TALENT.

tagmeme [literary] Technical term in stratificational grammar intended to represent the smallest meaningful unit of grammatical form. It is on the foundation of such basic units that the stratified nature of linguistic expression may be analysed (in the study known as *tagmemics*).

Coined as a term by the US linguist Leonard Bloomfield (1887-1949), the tagmeme remains a little-known concept.

tahil *see* TAEL, TAHIL

taka [comparative values] Unit of currency in Bangladesh.

1 taka   =   100 paisa

The name of the unit probably derives from the Sanskrit *tanka*, originally a weight, and then the value of a specific weight of precious metal. *See also* COINS AND CURRENCIES OF THE WORLD.

tala [comparative values] Unit of currency in Western Samoa.

1 tala   =   100 sene

The name of the unit is an attempt at the English word *dollar* (made up, here as elsewhere, by 100 *cents*). *See also* COINS AND CURRENCIES OF THE WORLD.

talari [comparative values] Unit of currency in Ethiopia, in the form of a silver coin modelled in shape (and in name) on the Levant dollar: *see* LEVANT DOLLAR.

talbot [physics] Unit of luminous energy in the metre-kilogram-second (MKS) system, equal to the energy that each second is associated with a flux of 1 lumen. A luminous output of 1 lumen per watt from 1 joule of radiant energy has a luminous energy of 1 talbot. The unit was named after the pioneer British photographer William Fox Talbot (1800-77).

talent [weight; comparative values] Unit of weight – especially in relation to precious metals and coins – in the ancient Hebrew and ancient Greek unitary systems, but ultimately deriving as a measure from an ancient Babylonian unit.

In ancient Israel,

1 talent (*kikkar*)   =   60 MINAS, 3,600 shekels (3,000 of metal)
                     =   about 30.24 kilograms, 66.666 pounds avdp.
                         (25.19 kilograms, 55.555 pounds of metal)

In ancient Greece,

1 talent   =   60 minas, 6,000 drachma
           =   about 25.80 kilograms, 56.880 pounds avdp.

By the time of the early Roman Empire, the value of the Greek mina had slipped to the stage that there were 80 to the talent.

The name of the unit is Greek, from a truly ancient verbal stem meaning 'to weigh down on', and thus 'to bear', that is represented in the name of the giant Atlas (who bore the world on his shoulders), the first syllables of the English *tolerate* and the word *toll*, and even the seemingly Chinese unit of weight, the TAEL.

taler *see* THALER, TALER

tambala [comparative values] Unit of currency in Malawi.

100 tambala   =   1 kwacha

The term means 'cockerel' in the local language, and *kwacha* means 'dawn', so that there are in effect 100 cockerels to 1 dawn – which is a pleasantly rural thought.

tan [weight] Ancient unit of weight in China, but assimilated into the system of weights used throughout the Far East during the colonial times of the mid-1800s especially for taxation and the imposition of customs duties.

1 tan   =   100 CATTY or chin, 1,600 taels
        =   133.3333 pounds avdp.
            (140 pounds, 10 stone = 1.05 tan)
        =   60.4775 kilograms
            (60 kilograms = 0.9921 tan)

Tang dynasty, T'ang dynasty [time] Period in China between AD 618 and 907. A time of literary, especially poetic, flowering, during which Buddhism reached the country and took on its localized forms, it was also an era in which the rulers expanded the national territory towards central Asia.

tanga [comparative values] The name of several coins of southern and central Asia in former centuries, of various values. The word is sometimes still used locally in Iran and India as a slang expression for a standard coin of either copper or silver.

The term seems to derive (in English through Portuguese) from Sanskrit *tanka*, originally a weight, and then the value of a specific weight of precious metal.

tangent [maths] In geometry, a line that touches (Latin *tangens* 'touching') a curve, whose slope (gradient) is equal to the slope of the curve at the point of contact; also a plane that touches a curved surface.

Also, for an angle in a right-angled triangle, the length of the side opposite the angle divided by the length of the side adjacent to it.

tanh [maths] *see* HYPERBOLIC FUNCTIONS

tank [weight] Former small unit of weight in the Bombay region of India.

$$
\begin{aligned}
1 \text{ tank} &= 4.4129 \text{ grams, } 68.1 \text{ grains} \\
&= 0.1556 \text{ ounce avdp.} \\
72 \text{ tanks} &= 1 \text{ Bombay seer} \\
&= 317.73 \text{ grams, } 11.2 \text{ ounces avdp.}
\end{aligned}
$$

The Bombay seer was thus rather different from the governmentally defined seer (*see* SEER) used elsewhere in India.

The name of the unit derives from the Sanskrit *tanka*, originally a weight, and then the value of a specific weight of precious metal.

tanka *see* VERSE FORMS

tare [weight] The weight of a vehicle, vessel, or container. Subtracting it from the *gross weight* results in the *net weight* of a load, cargo, or product.

The term derives through medieval Italian from Arabic *tarha* 'something (to be) deducted'.

tariff [comparative values] A list of prices, taxes, or customs duties, or any one price, tax, or duty listed.

The word derives (through Italian) from Arabic *ta'rif* 'public notice'.

tasimetry [physics] The measurement of pressure, sometimes by also measuring temperature or electrical conductivity.

The first element of the term is ancient Greek *tasis* 'tension'.

tauon [physics] An unusually massive subatomic particle belonging to the group called leptons, and alternatively known as a *tau particle*.

tautomer [chemistry] One of a pair of ISOMERS that is in equilibrium with its other form – that is, tautomers coexist (in the phenomenon known as tautomerism).

taxeme [literary] A device in language that is the minimum unit by which the sense of a word or statement may be changed in a grammatical or semantic way – such as a verbal ending denoting tense or number, word order to denote statement or question, the case-ending of a noun to denote plural or possessive, and so forth.

tax gallon [volumetric measure] By regulation in the United States, a (US) gallon comprising a capacity of 231 cubic inches (3.786 litres, 6.66 UK pints) exactly half of which by volume is ethyl alcohol.

taxon [biology] In biological classification (taxonomy), a group of organisms that belongs to a given taxonomic rank (such as Family, genus, or species).

Taylor series, Taylor's series [maths] For an infinitely differentiable mathematical function, a method of representing it as a power series.

It is named after the English mathematician Brook Taylor (1685-1731), who devised the theorem on which it is based.

tbs, tbs. *see* TABLESPOON(FUL)

t-distribution [maths] In statistics, a form of distribution that uses the STANDARD DEVIATION of a small sample as the basis for an estimated standard deviation of an entire population.

The first element of the term may or may not derive from the word *trial* or *test*. In any case, the t-distribution is at least as well known as Student's distribution (or Student's t-distribution), after the pen-name used by the British statistician William Sealy Gosset (1876-1947).

tea caddy *see* CATTY

tea chest [cubic measure] The ordinary size of a tea chest as used initially for the transportation of tea (and later for the removal of household or office goods and light furnishings) is 24 inches x 24 inches x 30 inches – a total capacity of 10 cubic feet (0.37 cubic yard, 0.283 cubic metre).

teacup(ful) [volumetric measure] A problem with teacups as a measure – as opposed to the standard 'cup' in cooking and in making some cocktails – is that different countries have different standard sizes according to the popular local and historical view of tea. The teacup in Britain is thus commonly regarded as the ordinary form of cup, whereas in Scandinavia and in parts of North America the teacup is a smaller form of rather more delicate container, much like that which the British might call a 'coffee-cup', about half the size of the British teacup.

In Britain, then,

1 teacup(ful)	=	about 2.5 decilitres, 25 centilitres
	=	about 0.44 UK pint, 0.53 US pint
	=	8.80 UK fluid ounces, 8.48 US fluid ounces

In Scandinavia and parts of North America,

1 teacup(ful)	=	about 12.5 centilitres, 125 millilitres
	=	about 0.22 UK pint, 0.265 US pint
	=	4.40 UK fluid ounces, 4.24 US fluid ounces

tear factor [paper] A measure of the ability of paper to resist tearing, equal to its lateral tearing strength divided by its thickness expressed in grams per square metre (gsm).

teaspoon(ful) [volumetric measure] Unit of volumetric measure in cookery, in making some cocktails, and in dispensing doses of some medicines, based on the standard size of a teaspoon – which is not the same in North America as it is in Europe.

In North America,

1 teaspoon(ful)	=	one-third of a tablespoon(ful)
	=	1.3333 US fluid dram
	=	4.833 milliliters

In Europe, and some other areas of the world,

1 teaspoon(ful)	=	about one-quarter of a tablespoon(ful)
	=	1.0208 UK fluid dram
	=	3.625 millilitres

In dispensing liquid medicines, the teaspoon has largely been replaced by a spoon with a capacity of 5 millilitres.

The teaspoon is ordinarily the smallest spoon in household cutlery, except for those made especially for serving mustard or other condiments.

teens [quantitatives] Ages thirteen to ninteen inclusive, all of which have the suffix -teen representing a form of the word *ten*.

telemetry [telecommunications; surveying] The sending of measured data (suitably coded) over a distance by telephone line or radio (*see also* SONDE; RADIO SONDE).

In surveying, the measurement of distances using a TELLUROMETER.

telephone traffic *see* ERLANG; TRAFFIC FLOW

tellurometer [surveying] An electronic instrument for measuring distances by timing the echoes of high-frequency radio signals (like radar). It is used in surveying for distances of up to 65 kilometres (about 40 miles).

temperament in music *see* TEMPERED SCALE

temperate zone *see* ZONE

temperature [physics; medicine] A measure of the quantity of heat in an object, expressed in degrees on one of the established temperature scales. In practice, it is a comparison between the hotness (or coldness) of two systems.

In medicine, temperature is in many families a loose synonym for fever. Normal human body temperature is 37°C (98.6°F). The average body temperatures of most other mammals are higher: horse 38°C (100.5°F), cat 38.5°C (101.5°F), cattle 38.5°C (101.5°F), dog 39°C (102°F), pig 39°C (102°F), and sheep 39.5°C (103°F).

temperature coefficient [physics] For any property that changes with temperature, its fractional increase for each degree temperature rise. For example: the resistance of a conductor at a given temperature (say T) divided by its resistance at a lower temperature (say t) is equal to 1 minus a constant times the temperature difference $(T - t)$. In this relationship, the constant is the temperature coefficient of resistance.

temperature gradient [physics; engineering] The rate of temperature change (difference) between two points or locations.

temperature-humidity index [physics; meteorology] A means of expressing temperature and humidity as a single unit, equal to the sum of the temperature (in degrees) and the relative humidity (as a percentage) divided by 2. Known also as the *discomfort index*, it is published and broadcast on a daily basis in the United States during the summer by the US Weather Service. European meteorological offices tend still to keep forecasts of temperature and humidity separate. (In Britain, for example, humidity rarely features in meteorological broadcasts at all.)

temperature scales *see* ABSOLUTE SCALE; CELSIUS SCALE; FAHRENHEIT SCALE; FIXED POINT; FUNDAMENTAL INTERVAL; KELVIN SCALE; RANKINE SCALE; RÉAUMUR SCALE.

tempered scale [music] The musical scale on the modern piano and other keyboard instruments (and by extension on the guitar and other fretted stringed instruments) in which all semitones (half-steps) are in the frequency ratio $1:^{12}\sqrt{2}$ (a ratio of 1 to the twelfth root of 2 – that is, $1:1.0594631$); 12 identical semitones (half-steps) between 13 successive keys on the keyboard add to one complete octave.

This is a sort of happy compromise. There remain notes that are unobtainable on such keyboards, in the form of quarter-tones – notes that are audibly neither one semitone nor the next, but something in between. Moreover, the semitones that are available in a tempered scale are not precisely all those that accord with the harmonic frequencies of an instrument such as a woodwind or brass instrument that relies solely on harmonics to obtain notes other than fundamental tones.

tempo [music] A term that is often taken to mean simply the speed at which a piece of music is performed, but which properly includes also an indication of the rhythm involved. A waltz, for example, is a movement that is not particularly fast but is characterized by a marked rhythm of three beats to the bar: its tempo can thus be expressed as 'slow 3/4'. Classic rock music is generally faster, and has four beats to the bar: it thus has a tempo of medium-fast 4/4. South American music often has a tempo of medium 8/8.

As for speed by itself, Classical music often includes an indication at the beginning of each movement of the intended average duration of a crotchet (quarter-note) as the number of crotchets to the minute (in the form $\quarternote = 90$), as can be regulated by a METRONOME.

temporary hardness (of water) *see* HARDNESS OF WATER

ten [quantitatives] 10; in Roman numerals, X .
$$^{10}\!/_{10} = 1$$
$$10\% = 0.1$$
A group of ten is a decad or decade.

A POLYGON with ten sides (and ten angles) is a decagon.

A solid figure with ten plane faces is a decahedron.

The number system based on 10 and powers of 10 is called the *decimal system*.

The Ten Commandments are known also as the Decalogue.

As is evident from the above, the prefix meaning 'ten-' in English corresponds to 'dec-' or 'deca-'.

The name for the number derives from common Indo-European, attested over the entire range of languages from Sanskrit to Welsh, although the form in English represents an unusually elided variant. It is possible – even likely – that by etymology the number's name derives from an expression meaning 'two hands', referring to the ten fingers, although it must be remembered that very early systems of counting with the fingers tended to work in fours, not fives, as the thumbs counted off the other fingers of each hand (*see* EIGHT; NINE). The suffix -teen, as in thirteen to nineteen inclusive, is another variant [meaning '(added to) ten']; and the suffix -ty, as in twenty, thirty, forty, fifty, sixty, seventy, eighty, and ninety, is yet another variant [meaning '(times) ten'].

teng [volumetric measure; weight] The *teng* is a rice basket in Burma (Myanmar), its capacity defined officially as 2,218.2 cubic inches or 1.2837 cubic foot (1 UK bushel, 1.032 US bushel, 36.352 litres by volume). But the actual weight of rice that it contains depends on local standards, which differ around the country. In general, a teng includes between 48 and 51 pounds (21.77 and 23.13 kilograms) of rice.

tennis measurements and units [sport] Lawn tennis – which may be played on tarmac or any other hard surface as well as on grass – is an energetic court game for two (singles) or four (doubles) players. The object of the game is to get the ball to bounce in the opponent's half of the court in such a way that the opponent is unable to return it, and so score points.

The dimensions of the court:
>   overall length: 78 feet (23.77 metres)
>   overall width: 36 feet (10.97 metres)
>   distance of service line from baseline: 18 feet (5.49 metres)
>   distance of service line from net: 21 feet (6.4 metres)
>   width of doubles sidelines: 4 feet 6 inches (1.37 metre)
>   area of service court: 21 feet x 13 feet 6 inches (6.4 x 4.11 metres)
>   height of net at centre: 3 feet (91 centimetres)
>   net overhang outside court: 3 feet (91 centimetres)

Points scoring:
>   winning 4 points usually completes a game
>   6 games make a set
>   3 or 5 sets make a match

the four points of a game have individual names: '15', '30', '40' and 'game'; a score of zero within a game is called 'LOVE'; a score of 40-40 is called 'DEUCE' and requires a player to win two clear points after it to win the game, the first point of

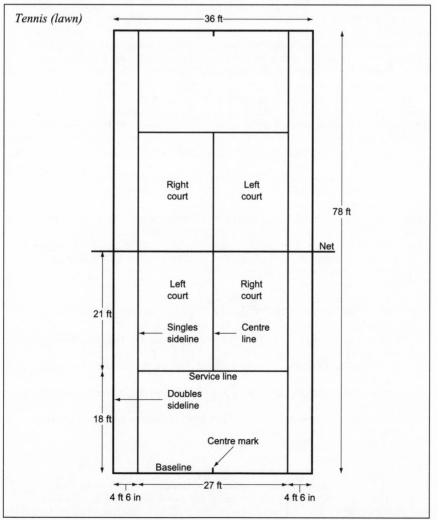

Tennis (lawn)

which is called the 'ADVANTAGE' the score of 5-5 games in a set may require a player to win two clear games after it to win the set; in some competitions, alternatively, a TIE-BREAK(ER) system is operated in these circumstances except in the final set

Dimensions of equipment:

the racket/racquet:

size restrictions in some countries, not in others

usual weight: 13½-14 ounces (382.7-396.9 grams)

the ball:

diameter: 2½-2⅝ inches (6.35-6.67 centimetres)

weight: 2-2¹⁄₁₆ ounces (56.7-58.5 grams)

The name of the game was first applied to what is now called real or court tennis, and comes to English from French. It is either a corrupt version of the French command *tenez!* 'hold (this)!', as apparently once uttered by the server on serving, or a version of the Arabic name of the medieval city of Tinnis in Lower Egypt, where a form of cloth was manufactured that may once have been used to make the balls for the game. (Note that the word *racket* derives from the Arabic for 'the palm of the hand'.)

tenor range [music] The tenor range is generally considered to extend from about D below middle C to about G above middle C – about an OCTAVE and a half. The lower half of this range is the province also of a baritone, and overlapping farther below that still is the bass register.

A *counter-tenor* is a man who sings in the ALTO RANGE (the male equivalent of a contralto).

Various families of instruments – recorders and saxophones, for example – contain a tenor member with a range between the alto and bass members of the same family. In a peal of bells, however, the tenor is the bell with the lowest pitch.

The term derives from the fact that female voices were rarely heard to make formal music in medieval times, and the melody line was thus sung by the highest male voice: the tenor is the re*tainer* and sus*tainer* of the *tune* which he sings with *tone*.

tenpence, tenpence piece [comparative values] Coin worth 10 pennies first issued in Great Britain in 1971 on the introduction of decimal currency, replacing the former 2-shilling piece (or florin) which nonetheless remained valid at the new value until phased out finally in 1993.

Surprisingly, this was not the first tenpence piece issued in the British Isles: the Bank of Ireland issued a tenpenny coin (24 to the pound) in the early 1800s.

tenpenny nail *see* NAILS

tenpin bowling units and scoring [sport] Tenpin bowling is played either by two individuals or by teams of two or more (up to five). The object is to roll a heavy ball down a flat, narrow lane, to knock down as many of the ten skittles or pins at the far end as possible at a time. One player's turn is known as a 'frame', and ordinarily comprises two 'throws'.

Points scoring:

for every pin down per two throws: 1 point

a strike (10 pins down on first throw): 10 points plus the score of the next two throws

a spare (10 pins down after two throws): 10 points plus the score of the next (one) throw

a strike in the final frame allows a player an extra two throws to complete the bonus

a spare in the final frame allows a player an extra (one) throw to complete the bonus

the player or team with the greater number of points at the end of the game wins

Dimensions of equipment:

the pins:

height: 1 foot 3 inches (38.1 centimetres)

weight: 3 pounds 2 ounces to 3 pounds 10 ounces (1.417-1.644 kilogram)

distribution: 1-2-3-4 in a equilateral triangle of sides each 3 feet (91 centimetres) long

the bowl:
    full-size diameter: 8½ inches (21.6 centimetres)
    weight: up to 16 pounds (7.258 kilograms)
  Tenpin bowling is a North American variant of an originally British game that used only nine pins (or ninepins). The mechanized sport became extremely popular in Europe in the 1960s, but gradually lost much of its following thereafter.

tensimeter [chemistry] An instrument for measuring the TRANSITION POINT (of a crystalline substance) by showing the temperature at which the two forms have the same vapour pressure.

tensiometer [physics] An instrument for measuring the surface tension of a liquid, usually consisting of a sensitive balance that records the force needed to lift a wire ring from the surface of the liquid under test.

tensiometry [physics; textiles] The measurement of the tensile strength of a wire or fibre, using a TENSOMETER.
  Also, the measurement of the surface tension of a liquid, using a TENSIOMETER.

tensometer [physics] An apparatus for measuring the tensile strength of a material (by gradually applying an increasing force to stretch a sample of the material until it breaks).

tensor [maths] A set of components (related to a system of coordinates) that represent a point in space and which transform in a linear way between various coordinate systems. It is therefore the generalization of a VECTOR.

tenth [quantitatives] ¹⁄₁₀ (one-tenth), one of ten equal constituent parts; the element in a series between the ninth and the eleventh, or the last in a series of ten.
  The prefix denoting 'one-tenth' in English is 'deci-'. *See also* TEN; TITHE.

tenth [music] The interval of a tenth is equivalent to an OCTAVE and a third, is a harmony, and can thus be either major or minor (*see* KEY, KEYNOTE; SCALE).

tera- [quantitatives: prefix] Prefix in the metric system which, when it precedes a unit, multiplies the unit by 1 UK billion/1 US trillion times – that is, by 1 million million (10^{12}) times.
    Example: 1 teracycle = 1,000,000,000,000 cycles (per second)
  The prefix derives from the ancient Greek *teras* 'monster'.

terabit [electronics; telecommunications] Unit of information in computing and electronic telecommunications, equal to 1 million million bits.
                1 terabit = 1 UK billion/1 US trillion bits (10^{12} bits)

terawatt [physics] Unit of power, equal to 1 million million watts.
                1 terawatt = 1 UK billion/1 US trillion watts (10^{12} watts)

tercentenary, tercentennial [quantitatives; time] Of 300 years, describing a 300th anniversary.

term day [time] A less common expression for any of the four QUARTER DAYS.

terminal velocity [physics] The constant final velocity acquired by an object falling through a fluid (gas or liquid), at which the gravitational force downwards equals the frictional forces upwards.
  The terminal velocity of a person falling through air from an aircraft at sufficient height has been reckoned to be about 190 kilometres per hour/120 miles per hour (less than the combined speed of many car crashes). For comparison, a cat falling from the same height would have a (maximum) terminal velocity of 'only' about 65 kilometres per hour/40 miles per hour because of its much greater air resistance in relation to its weight.

ternary [quantitatives] Third in rank or in order; with three constituent parts, threefold.
  The word derives as a derivative of Latin *terni* 'three apiece'.

ternary [chemistry] Describing a compound that consists of three different elements, or a system (such as an alloy) with three components. For example: sodium hydroxide (NaOH) is a ternary compound, and nickel (German) silver is a ternary alloy (containing copper, zinc, and nickel).

ternary system [maths] A number system with a base (radix) of 3; its digits are 0, 1 and 2. Thus, for comparison,

decimal notation	binary system	ternary system
0	0	0
1	1	1

decimal notation	binary system	ternary system
2	10	2
3	11	10
4	100	11
5	101	12
6	110	100
7	111	101
8	1000	102
9	1001	110
10	1010	111

For convenience it may be useful to remember that in place of decimal's units, tens, hundreds, thousands (and so forth), the ternary system has units, threes, nines, twenty-sevens (and so forth). For example: the number 2101 in ternary equals 64 in decimal notation, because ternary $2101 = (2 \times 27) + (1 \times 9) + (0 \times 3) + (1 \times 1) =$ decimal 64.

terpolymer [chemistry] A POLYMER formed from three different MONOMERS.

terrella [geography; physics] Just as a globe may have represented upon it the geographical formations of lands and oceans of the world, together with lines of latitude and longitude to make measurement easy, a terrella is a representation of the Earth's surface in terms of its magnetic formations and phenomena, with meridians and parallels to make measurement easy.

The term is a modern version of a Latin diminutive of *terra* 'Earth'.

terrestrial poles [geography] The two points (opposite each other) at which the Earth's axis cuts its surface. Lines of longitude run between the terrestrial poles. *See also* MAGNETIC NORTH; TRUE NORTH.

tertian [time; medicine] Recurring every third day, inclusively – that is, with one day's interval between recurrences, or every forty-eight hours. Fevers symptomatic of some tropical diseases (notably specific forms of malaria) follow this pattern.

The term derives from the medieval Latin expression *tertiana febris* 'fever (every) third (day)'.

tertiary [quantitatives] Third in rank or in order; following the second.

tertiary colour [physics] A colour obtained by mixing two secondary colours. For example: the secondary colours orange and green mix (subtractive process) to give olive or khaki. *See also* PRIMARY COLOUR; SECONDARY COLOUR.

Tertiary period [time] The geological period immediately before the present (QUATERNARY) period, corresponding to between roughly 65 million years ago and 3 million years ago. Within the Tertiary period were the Palaeocene, Eocene, and Oligocene epochs – sometimes together known as the Eogene – and the Miocene and Pliocene epochs – sometimes together known as the Neocene. The cold end of the Pliocene and the beginning of the icy Pleistocene epoch of the Quaternary period are difficult to distinguish, but in any case the Tertiary and Quaternary periods together are all part of the CENOZOIC ERA.

The period is called the Tertiary because it follows the Mesozoic era, which is known also as the Secondary era.

See chart of the Cenozoic era *under* PALAEOCENE EPOCH.

tervalent [chemistry] An alternative term for TRIVALENT.

terza rima *see* VERSE FORMS

tesla [physics] Unit of magnetic flux density in SI units, equal to 1 weber per square metre, and to the magnetic induction which, on a current of 1 ampere, produces a force of 1 newton. It was named after the Croatian-born American physicist Nikola Tesla (1856-1943).

teston, tester, testoon [comparative values] In the France and England of the late 1400s to 1600s, any of various silver coins of various values, so called because each had the monarch's head (medieval Latin *testa* 'head', original meaning 'pot', 'cauldron', 'crucible'; present-day French *tête*) depicted on one side.

The most famous in England was the silver shilling issued by Henry VIII, but this became so debased a coin that, within a century, the word (in any of the forms above) had become a slang term for 'sixpence' – half its original value.

tetartohedral [minerals] Describing a crystal that has only one-quarter of the faces in its system that the highest degree of symmetry demands.

The word combines ancient Greek elements together meaning 'fourth-sided'.

tetra-, tetr- [quantitatives: prefix] Prefix which, when preceding a unit, implies a fourfold unity.

Examples: tetragonal – having four angles

tetradactyl – having four fingers

tetrasyllable – a word with four syllables

The prefix corresponds to the combining form of the ancient Greek *tessares* 'four', an unusual variant of the Indo-European form better evident in Latin *quadr-* and Welsh *pedwar*.

tetrad [quantitatives] Group or set of four elements. *See also* QUARTET.

tetradrachm [comparative values] Unit of currency in ancient Greece, in the form of a silver coin.

1 tetradrachm   =   4 drachme, 24 oboloi

25 tetradrachms   =   1 mina

See also DRACHME.

tetragon [maths] A closed plane figure (POLYGON) with four sides (and four angles) – also known as a *quadrilateral*. A regular tetragon is a square. Other special cases include the rectangle (oblong), parallelogram, diamond, rhombus, and trapezium.

The corresponding adjective is *tetragonal*.

tetrahedron [maths] A solid figure (POLYHEDRON) with four triangular plane faces. A regular tetrahedron, which resembles a triangular-based pyramid, has four equilateral triangles as its faces.

tetralogy [quantitatives] A set of four (books, plays, operas, medical conditions, etc.) all linked in some form of unity.

tetralogy of Fallot *see* FALLOT'S TETRALOGY

tetramer [chemistry] A POLYMER formed of four similar MONOMERS.

tetrameter [literary] In verse, a line of poetry consisting of four metrical feet: *see* FOOT.

tetrapod [biology] An organism with four feet (or legs). The word is the Greek-based equivalent of the Latin-based *quadruped*.

tetravalent [chemistry] Having a valence (valency) of 4; an alternative description is *quadrivalent*.

tex [physics; textiles] Unit of linear density (mass per unit length) of textile fibres or yarn, equal to the mass of a 1,000-metre length of the fibre, expressed in grams.

The other commonly used yarn count/number, the *denier*, equals the mass in grams of 9,000 metres of fibre. So 1 tex is equivalent to 9 denier. *See also* DREX.

thaler, taler [comparative values] Former unit of currency in Germany, in the form of a rather large silver coin, based in name and in value on the *Joachimsthaler* issued in Sankt Joachimsthal (north of Prague, Czech Republic) from 1519, which was generally known as the thaler or DOLLAR.

The name is also sometimes applied in the present day to the LEVANT DOLLAR.

thaw point [physics] The temperature at which a solid just begins to melt.

thebe [comparative values] Unit of currency in Botswana.

100 thebe   =   1 pula

See also COINS AND CURRENCIES OF THE WORLD.

theodolite [surveying] An instrument for making accurate measurements of angles and, by extension, measuring distances.

The name of the instrument was apparently coined by the Englishman who was probably its inventor, one Leonard (or Thomas) Digges, in 1571, who spelled it *theodelitus* but told nobody why he called it that.

theorem [maths] A statement of a mathematical or geometrical relationship that has been proved or that can readily be proved.

The word derives through Latin from ancient Greek *theorema* 'something seen (to be true)'. A theorem is thus very different from a *theory*, which by derivation is 'something that still needs looking at'.

therapeutic index [medicine] In the dispensing of powerful drugs (especially to treat cancer), the dose that is lethal to a patient divided by the dose that is most effective.

therapeutic ratio [medicine] In the use of radiation to treat tumours, the radiation dose that is lethal to a tumour divided by the tissue tolerance.

therblig [engineering] In time-and-motion studies intended to increase efficiency in the use of personnel and equipment, a unit of physical activity (equal to a 'motion in time' as recorded by time-and-motion supervisors).

The name of the unit was devised as the virtual reverse of his own surname by the pioneering American time-and-motion expert Frank B. Gilbreth (1868-1924).

therm [physics] Unit of quantity of domestic gas, related to its energy content (which is released on burning) and equal to the amount of heat needed to raise the temperature of 1,000 pounds (453.5924 kilograms) of water through 100°F (55.55°C).

$$
\begin{aligned}
1 \text{ therm} &= 100,000 \text{ British thermal units} \\
&= 25,200 \text{ kilocalories} \\
&= 105.5 \text{ megajoules (SI units)}
\end{aligned}
$$

The term is based on the ancient Greek *thermē* 'heat' (cognate with English *warm*).

thermal capacity [physics] Thermal property of an object equal to the amount of heat needed to raise its temperature by 1 degree, expressed in units such as calories per degree Celsius (CGS units) or joules per kelvin (SI units). *See also* SPECIFIC HEAT CAPACITY.

thermal conductivity [physics] A measure of a material's ability to conduct heat, equal to the negative value of the rate at which heat is transmitted across it divided by the product of its area and the temperature gradient (the difference in temperature between each side of the material), expressed in watts per metre per kelvin (SI units). *See also* THERMAL DIFFUSIVITY.

thermal cross-section [physics] In nuclear physics, a measure of the probability that a given nucleus will interact with a thermal (that is, slow) neutron, a necessary precursor to a fission and a chain reaction.

thermal diffusivity [physics] For a material involved in the transfer of heat, the ratio of its thermal conductivity to the product of its density and specific heat capacity, expressed in square metres per second.

thermal efficiency [physics; engineering] For a fuel-burning engine or a steam engine, the work done by the engine divided by the mechanical equivalent of the heat supplied in the fuel or steam: *see* MECHANICAL EQUIVALENT OF HEAT.

thermal leakage factor [physics] For a nuclear reactor, the number of thermal neutrons that are lost through leakage divided by the number absorbed in the core. It is a measure of the efficiency of the reactor design and operation. *See also* THERMAL UTILIZATION FACTOR.

thermal neutron [physics] In nuclear physics, a neutron with the comparatively low energy of about 0.025 electron volt, which can be captured by a nucleus of a fissile material to initiate or maintain a chain reaction. Such a neutron can alternatively be called a *slow neutron*.

thermal resistance [physics] By analogy with electrical resistance (the resistance of a conductor to the flow of electricity), thermal resistance is the resistance of a material to the flow of heat, equal to the resistance that needs a temperature difference of 1°C to make heat travel through it at the rate of 1 watt, expressed in thermal ohms.

There is also a relationship analogous to that expressed by Ohm's law: thermal resistance equals temperature difference divided by rate of heat transfer.

thermal units [physics] The properties of a material with respect to the flow of heat can be made analogous to the properties of a conductor with respect to the flow of electricity. Thus: temperature difference is comparable with potential difference (voltage); quantity of heat flowing is comparable with current; resistance to heat flow is comparable with electrical resistance.

Thermal capacitance and inductance are similarly analogous to their electrical counterparts. The units of resistance, capacitance, and inductance are the same as the electrical ones prefixed by the word *thermal* (*see* THERMAL RESISTANCE *above* for a definition of one of these units).

Units of heat energy, such as the British thermal unit, calorie, and joule, may also be termed thermal units.

thermal utilization factor [physics] For a nuclear reactor, the fraction of absorbed thermal neutrons that are actually absorbed by the fuel (and therefore theoretically able to contribute to sustaining the chain reaction). *See also* THERMAL LEAKAGE FACTOR.

thermie [physics] Unit of heat in the obsolete metre-tonne-second system, equal to the heat needed to raise the temperature of 1 tonne of water by 1 degree Celsius.

$$1 \text{ thermie } = 4.185 \times 10^6 \text{ joules (SI units)}$$
$$= 3.967 \text{ British thermal units}$$

The term is based on the ancient Greek *thermē* 'heat' (cognate with English *warm*).

thermion [physics] An ion (with positive or negative charge) emitted from a material at incandescence.

thermobarograph [meteorology] A combination of THERMOGRAPH and BAROGRAPH, used for making continuous records of atmospheric temperature and pressure.

thermocline [physics: oceanography] A stratum within a large body of water that represents a comparatively fine boundary between volumes of water of markedly different temperatures.

thermodynamics, laws of [physics] Thermodynamics is the study of heat and its relationship to mechanical energy and other forms of energy. There are four laws of thermodynamics.

The first law states that energy can be neither created nor destroyed (also known as the law of conservation of energy).

The second law states that heat will not, of itself, travel from an object to a hotter object. This may also be expressed as: during a spontaneous process, the entropy of a closed system increases (*see* ENTROPY).

The third law states that the entropy of a perfect solid (crystal) is zero at the absolute zero of temperature.

And the zeroth law states that if two objects are in thermal equilibrium with a third, then all three objects are in thermal equilibrium with each other.

thermodynamic scale of temperature *see* ABSOLUTE SCALE; KELVIN SCALE

thermodynamic temperature [physics] Another name for ABSOLUTE TEMPERATURE.

thermograph [meteorology] A thermometer, used in meteorology, that makes a continuous recording of temperature.

thermometer [physics] An instrument for measuring temperature. There are many types, the most familiar being the liquid-in-glass thermometer (a graduated capillary tube containing mercury or dyed alcohol which moves along the tube from a bulb reservoir). Almost any physical property that varies with temperature can be employed to indicate changes in temperature. The movement of a coil of dissimilar metals (bimetallic strip), for example, is commonly used in a thermograph (*see above*). The change of a metal's electrical resistance with temperature is used in a platinum resistance thermometer (for high temperatures), and the generation of an electromotive force (voltage) at a junction between two dissimilar metals is used in a thermocouple-type thermometer. *See also* PYROMETER.

thermometer scales *see* TEMPERATURE SCALES

thermopause, thermosphere *see* ATMOSPHERE, COMPOSITION OF

thermostat [physics] A device for maintaining a system (such as a heated room or oven) at or about a particular chosen temperature. The most common types of thermostat are used with electric heating, when they sense the temperature and interrupt (break) or restore (make) the current flow when the temperature rises higher or lower than that required.

THI *see* TEMPERATURE-HUMIDITY INDEX

thick space [printing] In typesetting, the normal standard space between words, equal to a third of the type size. For example: in 12-point type, a thick space is 4 points wide (*see* POINT). *See also* THIN SPACE.

thimbleful [volumetric measure] An informal measure used commonly in Victorian times for dispensing very small amounts of brandy or strong wine (especially to ladies or to the elderly). It seems that – just as with teaspoon(ful)s – thimbles are and were apparently larger in the United States than in Europe anyway.

1 modern European
thimble(ful)   =   about 0.5 UK fluid dram

$$= \quad \text{half a UK teaspoon(ful)}$$
$$= \quad 1.8 \text{ millilitre}$$

But in American dictionaries, a thimbleful is often defined as a synonym for 'dram', in which case

$$1 \text{ US thimbleful} \quad = \quad 1 \text{ US fluid dram}$$
$$= \quad \text{three-quarters of a US teaspoon(ful)}$$
$$= \quad 3.7 \text{ milliliters}$$

and the American thimbleful is thus more than twice the European thimbleful.

thin space [printing] In typesetting, a spacing between words equal to a fifth of the type size. For example: in 10-point type, a thin space is 2 points wide (*see* POINT). *See also* THICK SPACE.

third [quantitatives] ⅓ (one-third), one of three equal constituent parts; the element in a series between the second and the fourth, or the last in a series of three.

$$⅓ \quad = \quad 33.3333\% \quad = \quad 0.3333$$
$$⅔ \quad = \quad 66.6666\% \quad = \quad 0.6666$$

The word is a form of 'three-th' (compare 'four-th' and 'six-th') that took on its final spelling as late as in Middle English (AD 1200s to 1400s).

third [music] An interval either of two tones or steps (for example, C to E on a keyboard) or of a tone (step) and a half (for example, A to C on a keyboard). The notes as played constitute a harmony: the larger interval is the major third (contributing to a major chord), the smaller interval a minor third (contributing to a minor chord). It is the converse of – and has much in common with – a SIXTH. *See also* PICARDY THIRD.

third-degree burn *see* BURNS

third person singular/plural *see* PERSON

Third World [comparative values] Describing a country or a group of countries that is neither a major force in international politics or finance (a First World nation) nor one of a group of advanced and powerful but less prosperous states (a Second World nation, such as any country of the former Eastern bloc). By common understanding, many of the inhabitants of a Third World country live at or below a level of extreme poverty.

thirteen [quantitatives] 13; a baker's dozen.

Why the number should have attracted to itself an unhealthy aura of superstition is not known. So common is the superstition, however, that there is even a technical term for a person who is abnormally fearful of its connotations: a *triakaidekaphobe* (or sometimes *triskaidekaphobe*).

The name of the number derives through Germanic sources and is literally '3, 10'.

thirteen [comparative values] Unit of currency in Ireland in the 1810s-30s, in the form of a silver coin.

$$1 \text{ thirteen} \quad = \quad 13 \text{ copper pennies}$$

The coin was known also as 'the Irish shilling'.

thirty [quantitatives] 30.

$$30\% \quad = \quad 0.30 \quad = \quad ³⁄₁₀$$
$$30 \text{ seconds} \quad = \quad \text{half a minute}$$
$$30 \text{ minutes} \quad = \quad \text{half an hour}$$
$$30 \text{ centimetres} \quad = \quad 11.811 \text{ inches (close on 1 foot)}$$
$$30 \text{ miles per hour} \quad = \quad 48.3 \text{ kilometres per hour}$$
$$30°\text{C} \quad = \quad 86°\text{F}$$

By tradition in newspaper and magazine journalism, the figure 30 is written at the end of an article, especially a news story, to denote that the copy is complete.

The name of the number derives from the Germanic variants of the common Indo-European term, combining elements that mean '3 (times) 10'.

thirty-five millimetre, 35-mm [photography] The standard width of sprocketed film used for professional motion picture cameras, and projectors, and a popular size for still cameras (professional and amateur).

thirty-second note *see* DEMI-SEMIQUAVER

thirty-twomo, 32mo [paper sizes] A paper size that results from folding a designated sheet five times (to form thirty-two pages).

thou [quantitatives; physics; engineering] An informal abbreviation in speech and writing for '1,000' or 'one-thousandth', especially in the former case referring to 1,000 currency units (such as dollars or pounds).

As an abbreviation for 'one-thousandth', however, the term is used mainly as a colloquial expression in engineering for one-thousandth of an inch (also called a *mil*).

$$1 \text{ thou} = 0.001 \text{ inch}$$
$$= 0.0254 \text{ millimetre}$$
$$(1 \text{ millimetre} = 39.37008 \text{ thou})$$

thousand [quantitatives] 1,000; ten hundred, 10 x 10 x 10, 10^3. One-thousand years is a millennium.

As is evident from the above, the prefix that most commonly denotes 'one-thousand times' a unit is 'kilo-', which derives through French as an adaptation of the ancient Greek *chilioi*: *see* KILO-. In some compounds, however, the prefix MILLI- may be used instead (although it more often refers to 'thousandths').

To multiply by 1,000 in the decimal system, take the decimal point three places to the right; to divide by 1,000 in the decimal system, take the decimal point three places to the left. Examples:

$$12.345678 \text{ x } 1,000 = 12,345.678$$
$$123,456.78 \div 1,000 = 123.45678$$

The square root of 1,000 is 31.622776 (to six decimal places).

One-thousand is one-thousand – except, formerly, in the building trade, where the word *thousand* was a unit number of roofing slates equal to 1,200 slates! The measure was in due course reduced to 1,000 slates, but had then to be described as 'thousand actual', so that the customer did not expect any more. (The original extra quantity was presumably to insure against breakages during transportation, and thus to guarantee a full 1,000 slates on arrival.)

The term derives through Germanic sources. Etymologists agree that the second syllable -sand is a variant of what in Latin is *cent-* 'hundred', but there is some debate about whether the first syllable corresponds to 'ten' or whether in fact it represents an otherwise unattested Germanic variant of a comparatively rare Indo-European word element meaning 'greater', 'stronger'. The plural is also *thousand* if preceded by a number or an expression standing in for a less than general number, *thousands* if not (and followed by the word of).

See also THOUSANDTH.

thousandth [quantitatives] $\frac{1}{1,000}$, 10^{-3}: one of 1,000 equal constituent parts; the element in a series between the 999th and the 1,001st, or the last in a series of 1,000.

$$\frac{1}{1,000} = 0.001 = 0.1\%$$

The prefix that most commonly denotes 'one-thousandth' of a unit is 'milli-', although this can on occasions alternatively mean 'one-thousand times' (*see* MILLI-).

thousandth of an inch [linear measure; engineering] Small unit of length used mainly in engineering contexts and sometimes known as a *thou* or *mil*: *see* THOU.

three [quantitatives] 3.

$\frac{3}{10} = 0.3 = 30\%$		$3\% = 0.03$
3 feet = 1 yard		3 tierces = 1 pipe

A group of three is a trio, triad, triplet, trilogy, trinity, or threesome.

A closed plane figure (POLYGON) with three sides (and three angles totalling 180°) is a triangle.

Words meaning 'three times' or 'threefold' are thrice, treble, triple, triplex, triplicate, tripartite, and ternary.

The ordinal of *three* is third, tertiary, or ternary.

The principal prefix meaning 'three-' in English is 'tri-', adopted from compound forms in both ancient Greek and Latin.

The name for the number derives from common Indo-European, attested over the entire range of languages from Sanskrit to Welsh. It is likely, however, that the basic meaning was not so much numerical as to distinguish between one, two, and more-than-two – so that where one was the individual, two corresponded to 'that', 'there', 'thou' (as opposed to this, here, I), three was effectively the 'other(s)'.

three-day eventing measurements and units [sport] The three-day event is effectively three separate equestrian competitions held on three successive days, in which competitors – as individuals and/or in teams – amass a cumulative total of points. The competitions are: dressage (obedience and style); endurance (cross-country); show jumping.

 Horses competing generally have to be at least 14.2 hands (144.27 centimetres/ 4 feet 8.8 inches) tall, must be six years old or older, and must be registered with a national equestrian federation.

 Points scoring:

  the three competitions are scored according to a ratio of importance that can be stated mathematically as dressage 3 to endurance 12 to show jumping 1

  dressage scoring:

   each judge first allocates an overall mark, then gives 0 to 6 marks for each compulsory movement, then deducts penalty points for error or time faults; the marks of all the judges for the rider are then averaged, and the score subtracted from the maximum obtainable (so converting it to a 'penalty score'); the penalty score is then multiplied by 3 (or by the ratio of importance of the event in respect of the other events)

  endurance scoring:

   mostly penalties for taking too long to complete the course or for errors committed at individual fences and hazards (a refusal costs 20 faults, a fall costs 60 faults), but there are also bonus points awarded for completing certain sections of the course in measurably less than the time restriction

  show jumping scoring:

   penalties for knocking down a fence or for putting a foot in the water at the water jump (10 faults), for disobedience (first time 10 points, second time 20 points), for a fall (30 points), and for every second or fraction thereof in excess of the time restriction set (0.25 faults each)

three-dimensional [maths; physics] Of a solid object that can be described in terms of the three dimensions of height, length, and width. Mathematically, any point on a solid (that is, the location of any point in space) can be described in terms of three coordinates – such as, for example, the three Cartesian coordinates x, y, and z related to a set of three axes at right-angles to each other. *See also* FOURTH DIMENSION.

three-four time (3/4 time) [music] In music, a style that stresses the first of every three beats (and that therefore has three beats to the bar, each of a duration of a crotchet or quarter-note. A classic bar in 3/4 time thus comprises

♩ ♩ ♩

although of course it may be broken up into different units:

♩ ♪♪♪   ♪♪♪ ♩

The best-known style of 3/4 music is the waltz, although its derivative six-eight time (six beats in the bar, each of the duration of a quaver or eighth-note) is probably better attested in popular and folk music around the world.

three-mile limit [geography] The distance out to sea that a country's legal, fiscal, and security obligations remain in force, under international law. It is only when a ship has reached that distance offshore, for example, that duty-free goods may be sold, or the captain may lawfully act as celebrant in a wedding ceremony.

 The three-mile limit has little to do with territorial waters, and does not apply in any case to air travel (with the result that duty-free goods may be purchased at the airport even before boarding a flight) or international travel by train via undersea tunnel (as between England and France).

three of a kind *see* POKER SCORES

threepenny bit, threepenny piece [comparative values] Former unit of currency in Great Britain, from 1589 to the early 1940s in the form of a small silver coin, and from the 1940s to 1971 in the form of a rather heavy, yellowish, hexagonal, cupronickel coin. In both forms, the unit was of course worth 3 pence (one-quarter of a shilling, one-eightieth of a pound). *See also* TICKEY.

three-quarters [quantitatives] ¾, less commonly known as three-fourths.

 ¾ = 0.75 = 75%

 Three-quarters of an hour is 45 minutes

 Three-quarters of a mile is 1,320 yards (1,207 metres)

Three-quarters of a foot is 9 inches (22.86 centimetres)
Three-quarters of a pound is 12 ounces (340.2 grams)
Three-quarters of a stone is 10 pounds 8 ounces
Three-quarters of a (long) ton is 1,680 pounds
Three-quarters of a short ton is 1,500 pounds
Three-quarters of an acre is (60.25 x 60.25 yards) 3,630 square yards,
3 roods (0.3035 hectare)

three score, three-score years and ten [quantitatives] Three score is three
times twenty, sixty: *see* SCORE.

Three-score years and ten – that is, seventy – is often quoted as a sort of folk
idiom expressing the average human lifespan. But it has been only since the 1960s,
even in Europe and North America, that life expectation has actually reached that
sort of average.

threshold dose [physics: radiology] The smallest possible dose of radiation that
will cause a required effect. *See also* X-RAYS.

threshold frequency [physics] The minimum sound frequency that is audible
under specified conditions (for instance, of sound pressure).

threshold of hearing [physics] The minimum sound pressure that can just be heard
by an average adult human being, at a particular frequency; for example at 1,000
hertz (1 kilohertz) it is 2×10^{-5} pascal.

throwing events *see* ATHLETICS FIELD EVENTS AREA, HEIGHT, AND DISTANCE MEASURE-
MENTS

throw weight [physics] The thrust (power) of a ballistic missile or rocket, measured
in terms of the explosive force of the maximum weight of warheads it can deliver
accurately (thus usually measured in megatons).

thrust [physics] For a rocket motor or vehicle, the reaction force produced by the
high-speed exhaust gases; the force that moves the rocket. It is equal to the mass of
the rocket times its acceleration, and should be expressed in newtons (although
rocket thrusts are often stated in kilograms or even pounds).

Similarly, for the propeller of a ship or aircraft, the reaction force exerted
backwards by the propeller that pushes the ship or aircraft forwards.

thrust-to-weight ratio [physics] The thrust of a rocket motor or vehicle divided by
its weight. If both are stated in kilograms or pounds (*see* THRUST), this ratio must be
greater than 1 for the rocket to leave the ground.

tical [comparative values] Former principal unit of currency in Thailand. Until 1909,
1 tical = 4 salung, 8 fuang, 16 songpy, 32 peinung, 64 att
but this system was at that time simplified so that
1 tical = 100 satang
After only three years, however, the system was again modified, this time to the
system that still pertains today (which does not include the tical): *see* COINS AND
CURRENCIES OF THE WORLD.

The name of the unit is in fact Malay, and originally referred to a unit of weight
equal to 15 grams (just over half an ounce avdp.), a local variant of the ancient
Sanskrit *tanka* that was originally the weight of a coin in precious metal.

tickey [comparative values] The silver threepenny piece (coin) of South Africa during
the time (until 1961) the national currency was modelled on that of Great Britain.

But, despite the British connection, the name of the coin in fact derived from
earlier Dutch antecedents – a coin representing one-tenth of a guilder (gulden), thus
10 cents, known as a *tientje* [in Afrikaans *tientjie* (*tien* '10', -*tjie* diminutive suffix),
pronounced 'teenki'].

tidal volume *see* LUNG CAPACITY

tides *see* NEAP TIDE; SPRING TIDE

tie [sporting term] In games, a contest that remains undecided at the end because the
score of each competitor or team is identical. Also, less commonly, another term for
an eliminator game within a specific competition or series.

In both senses the term derives from the expression 'a tied game', meaning in the
first case a game in which the competitors have been metaphorically tied together at
(or to) the same point, and in the second case a game in which the competitors are
bound to meet each other. *See also* DRAW.

tie-break, tie-breaker [sporting term] In tennis, mainly in British professional competitions, a method of completing any but the final set after the score has reached 6 games all. Taking two serves in a row each, the players play for the best over nine points – the first player to reach 5 points wins the set, the score of which is then given as 7-6.

In any other game, competition, or quiz, the accepted method of determining a winner from participants of equal attainment. *See also* SUDDEN DEATH PLAYOFF.

tierce [volumetric measure] Unit of wet and dry capacity formerly used mostly in marine freight transportation. It represented a measure half-way between a BARREL and a HOGSHEAD – and thus varied, as the barrel does still, according to the nature of the contents.

But as a standard unit,

1 tierce	=	42 old wine (or modern US) gallons
	=	34.97 imperial (UK or English) gallons
		(35 UK gallons = 1.009 tierce)
	=	158.98 litres
		(150 litres = 0.944 tierce)
3 tierces	=	1 PIPE, 2 hogsheads, 4 barrels

It is as one-third of a pipe that the unit gets its name (via French, ultimately from Latin *tertius* 'third').

tiltmeter [geology] Another name for a clinometer: *see* CLINOMETRY.

timber/lumber measurement *see* BOARD FOOT; CORD; CORD FOOT; STERE

time [time; physics; astronomy] The fundamental unit of time in all current systems of measurement is the second (*see* SECOND). In astronomy, there are various forms of time (*see* MEAN SOLAR TIME; SIDEREAL TIME), as there are in everyday life (*see* GREENWICH MEAN TIME; STANDARD TIME). *See also* DAY; MINUTE; MONTH; HOUR; YEAR.

time dilation [physics] A consequence of the special theory of relativity, which states that elapsed time appears to be longer for a system that is moving faster than that of the observer. If the observer is stationary, the time on a clock moving at a velocity v is equal to 'stationary' time multiplied by the square root of 1 minus the square of the ratio between v and the speed of light. Stated mathematically,

$$t_m = t_0[1 - (v^2/c^2)]^{\frac{1}{2}}$$

The time difference becomes significant only at speeds approaching that of light. *See also* RELATIVISTIC MASS.

time-faults *see* FAULT

time immemorial [time] Not just an expression meaning '(from) before living memory', or '(from) prehistoric times', but actually a definition at law of the period before the legal code came into force (at a specific date).

In England, for example, time immemorial is technically all of prehistory and history up to AD 1189 – when King Richard I acceded to the throne and English law is regarded to have been established.

time signature [music] Indication at the beginning of a piece of music of the TEMPO of the piece – that is, not just the speed (which may be suggested in the form of a metronome value for a crotchet or quarter-note) but how many beats there are in the bar, and of what duration.

The time signature takes the form of one number immediately on top of another number within the lines of a stave. The upper number represents the number of beats in the bar; the lower represents the fraction of a SEMIBREVE that is the *time value* of the notes corresponding to the beats. If the two numbers are (for example) 6 and 8 – which outside musical scores may be written '6/8' and pronounced 'six-eight' – then there are six beats in the bar, each bar corresponding to six eighth-notes or quavers. *See also* FOUR-FOUR TIME (4/4 TIME); THREE-FOUR TIME (3/4 TIME).

time trial [sporting term] A race in which competitors leave the starting-line serially – not all at once – and are timed individually; the competitor with the shortest time on finishing, wins the race.

time value *see* TIME SIGNATURE

tissue-typing [medicine] The classification of specific body tissues to achieve the greatest degree of conformity for the surgical transplantation of body tissues from one person to another. The tissues of donor and recipient must match as closely as possible to minimize the risk of tissue rejection by the recipient's immune system. One major element in tissue-typing is BLOOD GROUPING.

titer *see* TITRE, TITER

tithe [quantitatives; comparative values] A tenth portion; one-tenth.

In former centuries in Europe, the Church expected to be presented each year with one-tenth of all farm produce (grains and animals) or of any professional financial profits of those who did not work on the land. Each parish had its own tithe barn in which the cereals contributed could be collected and stored, and the funds so raised went in large measure to the upkeep of both clergy and local churches.

The term derives from the Anglo-Saxon *teoth*, a denasalized version of the modern word *tenth*.

Titius-Bode law [astronomy] *see* BODE'S LAW

titles *see* NOBILITY

titrant [chemistry] A solution of accurately known concentration, added from a burette to perform titrations in volumetric analysis: *see* STANDARD SOLUTION.

titration [chemistry] The chief technique in VOLUMETRIC ANALYSIS, in which a solution is added from a burette (a vertical graduated tube with a tap at the bottom) to a flask containing a known volume of another solution. When the reaction between the two is complete – at the end point, usually denoted by a change to an indicator – the concentration of one of the solutions can be calculated from a knowledge of the concentration of the other solution and the two reacting volumes.

titre, titer [chemistry; weight] The weight of a quantity of a chemical element in a pure form that reacts with, is equivalent to, or is contained in, a unit volume of a reagent solution, most commonly expressed in milligrams per millilitre.

titrimeter [chemistry] An apparatus for performing titrations automatically by monitoring changes in electrode potentials. *See also* POLAROGRAPHY.

TME [physics] Abbreviation of *Technische Mass Einheit* (Engineering Mass Unit), a mass that is given an acceleration of 1 metre per second per second by a force of 1 kilogram weight. It was formerly used in Germany and Switzerland.

to [volumetric measure] Unit of both liquid and dry capacity in Japan.

1 to	=	10 SHÔ, 100 GO, 1,000 shaku
	=	18.0391 litres
		(20 litres = 1.1087 to)
	=	3.9687 UK gallons, 4.7575 US (liquid) gallons
		[4 UK gallons = 1.008 to
		5 US (liquid) gallons = 1.0510 to]
	=	18,039.1 cubic centimetres, 1,100.8 cubic inches
	=	3.9687 UK dry gallons, 4.0954 US dry gallons
		(4 US dry gallons = 0.9767 to)
	=	1.008 UK peck, 2.0477 US peck
		(2 UK pecks = 1.0315 to
		2 US pecks = 0.9767 to)
10 to	=	1 KOKU

tod [weight] Obsolete unit of weight in the measurement of wool in England.

1 tod	=	28 pounds, 2 stone, 1 quarter
	=	12.7005 kilograms
4 tod	=	1 (long) hundredweight

But the word originated as a term merely for 'bundle' and was thus not normally regarded as an exact measurement. In certain English dialects, moreover, it was more common to hear the word used as a verb (rather than as a noun) in relation to the number of sheep shorn: 'to tod threes', for example, meant to produce a bundle of the standard weight of wool (a tod) from every three sheep in the flock.

tog [physics] Unit of thermal insulation, on a scale of numbers from 1 to about 15, as used for duvets, etc. The tog value of a garment is technically defined as ten times the temperature difference between the inner and outer surfaces (in °C) when the flow of heat from one surface to the other is equal to 1 watt per square metre. A better idea of the system, however, is provided by an example: a duvet of 10.5 tog is described as 'warm', 12 tog as 'extra warm', 13.5 as 'superwarm', and 15 tog as 'superwarm plus'.

$$1 \text{ tog} = 0.645 \text{ CLO}$$

The term is evidently a back-formation from *togs* 'clothing', especially 'uniform', 'gear', which is itself an abbreviated form of the old London thieves' and coachmen's cant word *tog(e)mans* – in which the second element is equivalent to the abstract noun ending -ment, and the first element may or may not be akin to *duck* 'linen canvas' (German *Tuch*, Dutch *doek* 'cloth', ultimately akin to Latin *tegere* 'to cover', 'to clothe', hence the English expression 'to be *decked* out').

togrog *see* TUGRIK, TOGROG

toise [linear measure] Obsolete unit of length in France, introduced there in AD 790 by Charlemagne and used until superseded by the metre in the late 1700s. It approximates closely in notional value to the FATHOM, but only reasonably in actual value.

$$
\begin{aligned}
1 \text{ toise} &= 6 \text{ PIEDS-DE-ROI, 72 pouces} \\
&\quad (1 \text{ fathom} = 6 \text{ feet, 72 inches}) \\
&= 1.949 \text{ metres} \\
&\quad (2 \text{ metres} = 1.0262 \text{ toise}) \\
&= 6 \text{ feet 4.74 inches} \\
&\quad (1 \text{ fathom} = 0.9382 \text{ toise})
\end{aligned}
$$

The term derives from the late Latin expression *tesa* (*brachia*) 'outstretched (arms)' – and is very similar in root meaning therefore to *fathom*.

tola [weight] Unit of weight in India, equal (by law) to the standard weight of 1 rupee.

$$
\begin{aligned}
1 \text{ tola} &= 180 \text{ grains} \\
&= 11.664 \text{ grams, } 0.4113 \text{ ounce avdp.} \\
&\quad (10 \text{ grams} = 0.8573 \text{ tola} \\
&\quad 1 \text{ ounce avdp.} = 2.4313 \text{ tola}) \\
80 \text{ tolas} &= 1 \text{ (government) SEER}
\end{aligned}
$$

The name of the unit derives from Sanskrit *tula* 'a suspension-balance', and thus 'a weight (to be used on that balance)'. It is therefore cognate with the Chinese-seeming TAEL, and akin to the ancient Greek TALENT.

tolerance dose [radiology] The maximum safe dose of radiation to a particular tissue when neighbouring tissues are also irradiated. *See also* X-RAYS.

toman [quantitatives; comparative values] A Persian (Farsi) word used in Iran and in Turkey to mean '10,000', a modern equivalent of the ancient Greek *myrias*.

The word is especially used of an army of soldiers numbering about that quantity, but it was at one stage also the name of an Iranian coin worth 10,000 dinars (or 10 krans).

It is likely that the old Arabian volumetric measure used specifically in weighing out rice, the *tomand* (equal to 84.899 kilograms/187.17 pounds avdp.), which does not apparently relate to other measures used locally, also derives in some way from the Persian unit.

tomography *see* SCANNING SYSTEMS

ton [weight; volumetric measure; quantitatives] Standard large unit of weight in the imperial (English) measurement system, coincidentally approximating fairly closely to the metric unit now called a tonne or metric ton. The standard ton in use in North America, however, is a variant of the imperial measure that is considerably easier to work with mathematically, but which does not correspond anything like as closely to the metric unit.

In Europe and some other areas of the world,

$$
\begin{aligned}
1 \text{ (long) ton} &= 20 \text{ hundredweight or quintals} \\
&\quad \text{(or, in some countries, centners or sentners)} \\
&= 2,240 \text{ pounds, 160 stone} \\
&= 1,016.0416 \text{ kilograms, } 1.0160416 \text{ tonne} \\
&\quad (1 \text{ tonne} = 0.9842116 \text{ ton}) \\
&= 1.120 \text{ short ton}
\end{aligned}
$$

In North America,

$$
\begin{aligned}
1 \text{ (short) ton} &= 20 \text{ short hundredweight or centals} \\
&= 2,000 \text{ pounds} \\
&= 907.18474 \text{ kilograms, } 0.90718474 \text{ tonne} \\
&\quad (1 \text{ tonne} = 1.1023113 \text{ short ton}) \\
&= 0.8928571 \text{ (long) ton}
\end{aligned}
$$

For comparison,

1 tonne or metric ton	=	10 metric quintals, 1,000 kilograms
	=	2,204.634 pounds
		(2,204 pounds 10.144 ounces)
	=	0.9842116 (long) ton
	=	1.1023113 short ton

A short ton is sometimes alternatively known as a *net ton*.

But a ton has by tradition always additionally been a measure of volume. This has particular relevance to shipping and the stowage of freight on board seagoing vessels: *see* DISPLACEMENT TON; FREIGHT TON; REGISTER TON.

In former centuries, the volumetric ton – like the barrel (to which it is historically related, *see below*) – varied in its dimensions, and therefore its overall weight, according to the nature of the goods involved. One ton of timber by weight could be considered exactly the same as a freight ton as defined above: 40 cubic feet of timber.

In stone, however,

1 ton	=	16 cubic feet
	=	0.4528 cubic metre
	=	12.46 UK bushels, 12.86 US bushels

Whereas in wheat,

1 ton	=	24.88 cubic feet
	=	0.704766 cubic metre, 704.766 (dry) litres
	=	19.38 UK bushels, 20 US bushels

Much more surprisingly, the ton is also a unit of refrigeration, especially in the United States, equal to the heat consumed in freezing 1 short ton (2,000 pounds) of ice in twenty-four hours, calculated as 288,000 British thermal units per day, or 200 Btu per minute. The metric equivalent – the refrigerant tonne – is calculated as 1 kilocalorie per second (equal to 342,000 Btu per day, or 1.1875 times the United States value).

Finally, the word *ton* is also used colloquially for the number 100, especially a score of 100 (for example, in cricket or darts) or a speed of 100 miles per hour.

That the word *ton* is at least as much a measure of volume as it is of weight is really only to be expected, in that the term derives as a variant of the TUN, a large unit of volume equal to 2 PIPES or 8 barrels.

tønde [volumetric measure] In Denmark a former unit of dry and liquid goods, although the dry and liquid measures were different.

In liquid measure,

1 tønde	=	144 Danish pots
	=	131.4 litres
		(130 litres = 0.9893 tønde)
	=	28.908 UK gallons, 34.713 US gallons
		(30 UK gallons = 1.0378 tønde
		35 US gallons = 1.0083 tønde)

In dry measure,

1 tønde	=	144 Danish pots, 8 skaeppe, 4 fjerdings
	=	139.1 (dry) litres
	=	0.1391 cubic metre
		(1 cubic metre = 7.189 tønde)
	=	3.825 UK bushels, 3.947 US bushels
		(4 UK bushels = 1.0458 tønde
		4 US bushels = 1.0134 tønde)

The name of the unit is evidently akin to the English TUN and TON.

See also KORNTONDE.

tone [music] Technical term in Britain and other English-speaking areas of the world for what in North America is called a *step* or a *whole step*, corresponding on a piano to the interval between two white notes separated by a black note, or between two black notes separated by a white one. It is thus twice the interval of a semitone or half-step (which corresponds to the interval on a piano between a black note and an adjacent white note).

The SCALE of any KEY is made up of a series of tones and semitones.

tonelada [weight] Former unit of weight in Spain and Portugal and their colonies in South America and elsewhere – but the Spanish unit was not the same as the Portuguese.

1 Spanish tonelada	=	2,000 Castilian libras, 20 quintals
	=	2,028 pounds avdp. [0.905 (long) ton]
		(1 ton = 1.1045 Spanish tonelada)
	=	919.886 kilograms
		(1 tonne = 1.0871 Spanish tonelada)
1 Portuguese tonelada	=	1,728 arratels or libras, 13.5 quintals
	=	1,748.601 pounds avdp. [0.78 (long) ton]
		(1 ton = 1.2811 Portuguese tonelada)
	=	793.15 kilograms
		(1 tonne = 1.2608 Portuguese tonelada)

The fact that there were 1,728 arratels to the Portuguese tonelada suggests that the unit was only half-assimilated from some other system of weight measurement (cf. 1,760 yards to the mile). But 1,728 is 12 x 12 x 12, and perhaps it is only the subunits that have changed value: *see* QUINTAL.

The name of the unit derives initially as a diminutive (*tonnel*) of the French variant of what in English is TUN and TON, to which an adjectival (participular) suffix has been added.

toneme [literary] Unit of stratificational grammar in relation to languages that rely on intonation (the rising or falling of the pitch of speech in individual words) to distinguish meaning. Most Chinese languages and dialects employ four tonemes: an even pitch (also called 'high'), a rising pitch, a gently falling pitch, and a distinct drop in pitch.

tonic, tonic sol-fa *see* KEY, KEYNOTE; SCALE

ton-mile [comparative values] In estimating the cost and efficiency of transporting freight, a unit relating the cost of moving 1 ton (in Europe, usually 2,240 pounds/ 1,016.05 kilograms; in North America, usually 2,000 pounds/907.19 kilograms) of freight (sometimes also including the weight of the vehicle) a distance of 1 mile (1.609344 kilometres). That unit cost can then be used to calculate the cost of moving any weight any distance.

tonne, metric ton [weight] Metric unit of mass equal to 1,000 kilograms and the fundamental unit in the obsolete metre-tonne-second system of units. Known in Europe (and elsewhere) as the tonne (or occasionally as the tonneau), it is more commonly called the metric ton in North America.

1 tonne or metric ton	=	10 metric quintals, 1,000 kilograms
	=	2,204.634 pounds (2,204 lb 10.144 oz)
	=	0.9842116 (long) ton
		(1 ton = 1.016042 tonne)
	=	1.1023113 short ton
		(1 short ton = 0.907185 tonne)

The French name of the unit is at least as ancient as the English form: *see* TUN.

tonometer [physics; medicine] A name for several different types of instrument:
An instrument for measuring vapour pressure.
A simple apparatus for finding the frequency of a note (tone), consisting of a series of tuning forks.
An instrument for measuring the fluid pressure within the eyeball (used in the diagnosis and treatment of glaucoma).
The first element of the word is ancient Greek *tonos* 'tension'.

topology [maths] The mathematical study of shapes that do not change (certain geometrical properties) with distortions such as those that occur during mapping and other transformations.

toroid, torus [maths] A ring-doughnut-shaped solid, alternatively known as an *anchor ring*. If the radius of the solid part is r and the radius of the ring (to the centre of the solid part) is R, its volume is $2\pi^2 r^2 R$, and its surface area is $4\pi^2 rR$.

The word *toroid* is technically the adjectival form of *torus*, both of which derive etymologically on account of the roundness of the shape, not because of the presence of the hole: Latin *torus* 'lump', 'swelling'.

torque [physics] When a force acts tangentially at a distance from an axis of rotation, the turning moment is the torque, equal to the magnitude of the force times the distance.

The term derives through French from the Latin *torques* 'a metal necklet with a twisted design on it', itself derived from the verb *torquere* 'to twist'. *See also* MOMENT OF A FORCE.

torr [physics] Unit of atmospheric pressure equal to 1 millimetre of mercury.

$$1 \text{ torr} = \frac{1}{760} \ (0.001316) \text{ standard atmosphere}$$
$$= 1.33322 \text{ millibar}$$
$$= 1{,}333 \text{ dynes per square centimetre (CGS units)}$$
$$= 133.3 \text{ newtons per square metre (SI units)}$$

The unit was named after the Italian physicist Evangelista Torricelli (1608-47; *see below*).

Torricelli's law, Torricelli's theorem [physics] In hydrodynamics, the velocity v at which a liquid flows through a (small) opening at the lower end of a container, when the opening is a distance h below the surface of the liquid, is equal to the velocity at which an object falls the distance h in air, under the influence of gravity. It is a special case of Bernoulli's equation (*see* BERNOULLI'S LAW), and the velocity can be calculated from the equation

$$v = (2gh)^{\frac{1}{2}} \quad (\text{or: } v = \sqrt{2gh})$$

where g is the acceleration of free fall (the acceleration due to gravity).

The principle was first stated by the Italian physicist Evangelista Torricelli (1608-47), one-time assistant to Galileo and later Galileo's successor as Professor of Mathematics at Florence Academy.

torrid zone *see* TROPICS; ZONE

torsion balance [physics] A very sensitive balance that measures forces (electric, gravimetric, or magnetic) by the twisting action they exert on a fibre or wire suspended vertically.

The word *torsion* is another derivative (through French) from the Latin verb *torquere* 'to twist'.

torus *see* TOROID

tossing the caber *see* CABER-TOSSING

tot [volumetric measure] A small measure of liquid – especially spirits – as sold in a bar. Although the word does not strictly apply to a specific quantity, in England a standard tot of whisk(e)y or brandy is generally equal to one-sixth of a UK gill.

$$1 \text{ English tot} = \frac{1}{6} \text{ UK gill, } \frac{1}{24} \text{ UK pint}$$
$$= 2.3677 \text{ centilitres, } 23.677 \text{ millilitres}$$
$$= 6.667 \text{ UK fluid drams, } 6.405 \text{ US fluid drams}$$
$$= \text{just over } \frac{1}{20} \text{ US pint}$$

As a non-specific measure, however, a tot is frequently held to be a synonym for a *dram*.

The word *tot* derives as a dialectal English term akin to other Germanic bywords meaning '(something) really small' – hence the tot who is an infant – and akin therefore also to *tit* as in <u>tit</u>lark and <u>tit</u>mouse, <u>tit</u>bit, and the old slang nickname <u>Tit</u>ch (perhaps additionally cf. the informal American *tad*).

total electron binding energy [physics] The energy needed to remove all the electrons from an atom (and separate them from each other).

total heat [physics] The amount of heat needed to raise a unit mass of a substance at or near its freezing point to a temperature at which it is capable of vaporizing under constant pressure. Defined also as the amount of heat energy in a substance, and expressed in joules, it is effectively the same as ENTHALPY or heat content.

total impulse [physics] For a rocket or rocket motor, the mean thrust multiplied by the firing time, expressed in newton seconds.

total internal reflection [physics] When a light beam passes from one transparent medium to another, complete reflection of the incident beam at the interface between the two media. In these circumstances the angle of refraction is 90 degrees and the angle of incidence is the CRITICAL ANGLE.

totalizator, tote [maths] A calculating machine that can take account of many subtotals to derive complex statistical information.

The most famous form of totalizator is that used at horse and dog races to calculate the odds (and pay out the winners) according to mathematically precise calculations relating to the actual bets placed. The odds as propounded by a totalizator may thus be quite different from those advertised by a bookmaker.

tou [volumetric measure] A unit of liquid capacity in China apparently assimilated from the metric system, and equal to the decalitre.

$$1 \text{ tou} = 10 \text{ litres}$$
$$= 2.200 \text{ UK gallons, } 2.64179 \text{ US gallons}$$

The use of this unit is not compatible with the use of the units of the earlier Chinese system of volumetric measure.

The name of the unit may or may not be etymologically related to that of the Japanese volumetric unit the TO.

touchdown *see* AMERICAN FOOTBALL MEASUREMENTS, UNITS, AND POSITIONS

tour cycling *see* CYCLING DISCIPLINES AND DISTANCES

tovar [weight] Unit of weight in Bulgaria approximating closely to one-eighth of a ton or tonne (metric ton).

$$1 \text{ tovar} = 100 \text{ (Bulgarian) oka}$$
$$= 128.20 \text{ kilograms}$$
$$(100 \text{ kilograms} = 0.7800 \text{ tovar})$$
$$= 282.6327 \text{ pounds (282 pounds 10.12 ounces)}$$
$$(280 \text{ pounds, 20 stone} = 0.9907 \text{ tovar})$$

The name of the unit seems to relate to the Bulgarian (and common Slavonic) word for 'goods', 'merchandise' [akin to English *ware(s)*], possibly as the sort of quantity to be stocked in a warehouse. In this connection it may be worth pointing out that the subunit of the tovar, the oka, is in addition a unit of volume: *see* OKA, OKE.

township *see* SURVEYORS' MEASUREMENTS

track and field events *see* ATHLETICS FIELD EVENTS AREA, HEIGHT, AND DISTANCE MEASUREMENTS; ATHLETICS TRACK EVENTS RACE DISTANCES; CYCLING DISCIPLINES AND DISTANCES; HIGHLAND GAMES EVENTS; HURDLES EVENTS

tractive effort, tractive force [engineering] For a railway locomotive, the maximum pull that it can exert at the drawbar, equal to the weight on its coupled wheels multiplied by the coefficient of friction between the wheels and the rails.

trade balance *see* BALANCE OF PAYMENTS, BALANCE OF TRADE

trade dollar [comparative values] A special silver dollar minted and issued in the United States for use in transactions with Far Eastern markets.

trade winds [geography; meteorology] The trade winds are the constant breezes that blow towards the Equator – from around 30° north of the Equator north-east to south-west, and from around 30° south of the Equator south-east to north-west.

The term has nothing to do with trading or traders, but derives from the dialectal English expression of around 1600 'to blow trade', describing a wind that is of constant direction and speed (*trade* in this sense being cognate with *tread*).

traffic flow [engineering; telecommunications] For a road system or a single road, the (average) maximum number of vehicles on it per unit distance per unit time, often calculated or measured to check that a road is capable of accommodating an expected increase in traffic. The usual monitoring method is a physical count made with one or two rubber cables containing a numerator placed at a strategic point across the road. (One cable can make a simple count; two cables in conjunction can also discern traffic direction.)

For a telephone exchange, the number of calls it is handling at a given moment. It can be measured by a *traffic meter*, which totals all calls passing through the exchange.

A *traffic pattern* is the arrangement of aircraft in the air around an airport, either on the way to land or having just taken off, as shown on the radar screen in the control tower.

trajectory [physics] The path followed by a missile, rocket, or spacecraft from its launch point to its destination. For a missile (such as a shell) fired from one point to another on land, the trajectory is a close approximation to a parabola.

transactinides, transactinide series [chemistry] Elements beyond lawrencium (the last of the actinides) in the PERIODIC TABLE – that is, elements with atomic numbers greater than 103; for example: rutherfordium (104) and hahnium (105).

transcendental number [maths] An irrational number that cannot be derived by solving a polynomial equation that has rational coefficients – that is, which is algebraic: for example, π, e. The square root of 2 ($\sqrt{2}$), for example, while irrational, is not transcendental because it can be found by solving the algebraic equation $x^2 = 2$.

transference number *see* TRANSPORT NUMBER

transformation [maths] In algebra, a change in an equation or expression brought about by substituting one set of variables by another set.

In coordinate geometry, a change in the direction or position of the axes without any change in the angle(s) between them.

transition point [chemistry; physics] For a substance that can exist in two crystalline forms, the temperature at which it changes from one form to the other (at which the two forms coexist).

transition temperature [chemistry; physics] The temperature at which a substance changes from one phase (*see* PHASE) or allotropic form into another.

Also, the (normally very low) temperature at which a metal becomes a superconductor.

transmission coefficient *see* TRANSMITTANCE

transmission loss [physics; telecommunications] For the transmission of sound through materials, 10 times the logarithm of the ratio of the power of the transmitted sound to the power of the incident sound (the power ratio).

transmission ratio [physics] For light striking a transparent medium, the ratio of the transmitted luminous flux to the incident luminous flux. Also known as the *transparency*, it is the reciprocal of opacity.

transmissometer [physics; meteorology] An electronic device that measures the amount of obstruction in the atmosphere between a light transmitter and a light receiver, both at fixed points, and so gives an indication of general visibility in a locality (especially useful at airport control towers).

transmittance [physics] For an object that reflects energy, the ratio of the energy transmitted by it to the energy incident on it. The transmittance is known alternatively as the *transmission coefficient*.

transmutation [chemistry; physics] The process by which a radioactive chemical element becomes another element through the spontaneous decay (emission) of subatomic particles and/or radiation. It generally happens continuously, and the time it takes for half of the primary element to decay – its HALF-LIFE – is used as a characteristic for identification and comparison (*see also* DECAY CONSTANT).

It was by 'transmutation' that the old alchemists hoped to change base metals into gold or silver – but their primary (and inevitably unsuccessful) method of transmutation was chemical or physical reaction.

transonic [physics; aeronautics] Describing aircraft speeds in the approximate range Mach 0.8 to Mach 1.4 (980-1,720 kilometres per hour/610-1,070 miles per hour), in which subsonic and supersonic airflows play a part.

transparency *see* TRANSMISSION RATIO

transport number [chemistry] For an electrolyte (a solution of a salt or an acid which is dissociated into ions), the fraction of the total current flowing that is carried by a given ion. The transport number is alternatively known as the *transference number*.

transuranic elements [chemistry] Elements beyond uranium in the PERIODIC TABLE – that is, elements with atomic numbers greater than 92.

transversal [maths] In geometry, a line that cuts two or more other lines.

transverse wave [physics] A wave motion whose oscillations are at right-angles to the direction of travel – for example, electromagnetic waves or waves on a vibrating string.

trapezium, trapezoid [maths] An imprecise term, because of different American and British usages. In the United States, a trapezium is a quadrilateral that has no sides parallel (called a trapezoid in Britain). In the United states, a trapezoid is a quadrilateral with one of its pair of opposite sides parallel (called a trapezium in Britain). The area of the parallel-sided figure (UK trapezium, US trapezoid) equals half the product of the sum of the lengths of the parallel sides and the distance between them.

The term (in both variants) derives ultimately from the ancient Greek for a small irregularly shaped table [*trapezion*, itself named as an object that had four feet, *(te)tra-peza*] that was in common use.

treble [quantitatives] Three times, threefold, triple (with the last two of which the word is cognate).

In betting, therefore, a treble is an accumulator or parlay involving three races.

treble [music] The SOPRANO RANGE, and anyone (particularly a boy) who sings in it (as scored in the *treble clef*). Soprano members of the families of musical instruments that comprise soprano, alto, tenor, and bass members, may also be called trebles. And in a peal of bells, the treble is the bell with the highest pitch.

The use of the word in this sense may or may not derive from apprehending *treble* as somehow 'higher' than *double* when *double* is already apprehended as 'higher' than *single*.

trecento *see* -CENTO

trental [quantitatives] In the Roman Catholic Church, a service or series of services all together adding to thirty (French *trente*, Latin *triginta*) Requiem masses celebrated on a single person's behalf.

tret [comparative values; weight] In buying and selling by weight in Norman (eleventh- to twelfth-century) England or France, an extra discount of $\frac{1}{26}$ (0.0385) of the value of goods after the value of the weight of the goods' container (the tare) had also been discounted.

It would seem that the container was believed not merely to have its own weight (that must be discounted in the sale by weight), but that with its contents it somehow effected an extra force (Old French *tret*, from Latin *tractus* 'pull') on the scales, albeit a slight one, which had thus also to be discounted.

trey [quantitatives] Old-fashioned term for the three on a dice, a domino, or a card, or a total of three on more than one dice.

tri- [quantitatives: prefix] A prefix denoting a threefold unity.

Examples: trimaran – a boat with three hulls

tripod, trivet – an apparatus with three legs (feet)

The prefix derives from ancient Greek and Latin equivalents.

triad [music] A chord of three notes, usually all within an octave and comprising a harmony.

A triad is regarded as a basic unit of harmony because all harmonic chords consist of three notes which may or may not be accompanied also by other notes that are octave reinforcements of any of the three notes or additional harmonics. A basic chord in the KEY of C major, for example, comprises the notes C, E, and G in any order or inversion; a basic chord of C minor comprises the notes C, E♭, and G in any order or inversion. Any of those notes may be given its octave, and the addition of B♭, for example, would turn the chord into a SEVENTH.

triakaidekaphobe, triskaidekaphobe *see* THIRTEEN

triangle [maths] A three-sided plane figure; the area of a triangle equals half the length of its base times its perpendicular height (altitude). Various types of triangle have specific names: scalene – with no sides equal; isosceles – with two sides equal; equilateral – with three sides equal; right-angled – with one angle of 90 degrees.

The word derives from Latin elements meaning 'three-angled'.

triangle of error [geography: navigation; surveying] In surveying or navigation, a position can be located by plotting back to it three bearings from visible objects of known position (TRIANGULATION). Almost always, the three back-bearings form a small triangle, the triangle of error, inside which the position is somewhere located. Because of its shape, it was formerly known in the Royal Navy as a cocked hat (in reference to the tricorn hat once worn by officers).

triangle of forces [physics] Three forces in equilibrium can be drawn as a triangle in which the length and direction of each side represents a force vector. If only two of the forces are known, the triangle can be drawn (constructed) to determine the other one. *See also* PARALLELOGRAM OF FORCES.

triangulation [geography: navigation; surveying] In surveying or navigation, the plotting of one's position using three bearings from visible, audible, or radio-emitted marks of known location. *See also* TRIANGLE OF ERROR.

Triassic period [geology; time] The first geological PERIOD during the mesozoic or Secondary era, after the permian period of the palaeozoic era but before the jurassic period, corresponding roughly to between 248 million years ago and 213 million years ago. The actual geological boundary between the Permian and the Triassic is often difficult to distinguish because the great natural disaster that rendered so many marine creatures – particularly corals, foraminifera, and brachiopods – extinct at the end of the Palaeozoic era did not affect the deposition of rock strata as such. But the final stage of the Triassic (the Rhaetian or Rhaetic) is marked by marine sedimentation deposited on previous continental strata, indicating yet another comparatively sudden and sizable rise in sea-levels. During the drier stages, in which considerable volcanic activity took place, reptiles flourished – the period saw the emergence of the first true dinosaurs. As the waters rose, molluscs and echinoderms became more profuse.

The stages shown in the chart are meant as examples only: they represent

Triassic period

Era	Period			Stages
S e c o n d a r y o r M e s o z o i c	Cretaceous	Upper		Senonian Turonian Cenomanian
		Lower		Albian Aptian Barremian Neocomian
	Jurassic	Upper (Malm)	O o l i t h i c	Portlandian Kimmeridgian Oxfordian
		Middle (Dogger)		Callovian Bathonian Bajocian Aalenian
		Lower (Lias)		Toarcian Pliensbachian Sinemurian Hettangian
	Triassic			Rhaetian Keuper Muschelkalk

predominantly north-west European terms for which there are sometimes corresponding east European and North American equivalents.

The name of the period was derived by the German geologist Friedrich von Alberti in 1834 from a triple-layered stratum of rocks (*Trias*) of the period visible at a specific site in Germany.

triathlon [sport] An exhausting race combining three events that are undertaken one after the other without rest. The first element is a swim of (usually) 3,800 metres (2 miles 704 yards); it is followed at once by a cycle ride of (usually) 180 kilometres (112 miles) generally over a relatively flat course on roads; and the final stint is a run over the usual marathon distance of 26 miles 385 yards (42.195 kilometres). Triathlons for men and for women are normally held separately, the women frequently starting their first leg (the swim) as the men begin their second (the cycling). Occasionally, however, men and women compete together, although the women are then given a handicap start of around thirteen minutes.

triatomic [chemistry] Describing a compound that has three identical atoms in its molecules: for example, ozone (O_3).

Tribe, tribe *see* CLASS, CLASS

tribometer [engineering] A device for measuring the force of friction in surfaces that slide across one another.

The first element of the term is ancient Greek *trib-* 'rub'.

tricentenary, tricentennial [quantitatives; time] Of 300 years, describing a 300th anniversary.

trick [sporting term] In card games, such as whist and contract bridge, a set of cards representing one round of a hand (one card from each player), won by the player or side who had the card of the highest value. In many games, the number of tricks won is crucial to the game.

The word *trick* in this sense would seem to derive from the trick in expressions such as 'that'll do the trick', meaning to have the desired effect, whereas the idea of trickery (and the cognate *treachery*), although early, is an extension on this basic meaning.

triennial [time] Once every three years; lasting for three years.

trifecta [sporting term] An extension of the EXACTA or perfecta system of betting on horses (or greyhounds) in North America, by which the better must nominate the first, second, and third animals in the correct order to win the bet. *See also* SUPERFECTA.

trigon, trine [astronomy; astrology] In astronomy, the aspect of two planets as seen from Earth to be 120 degrees apart in the night sky.

In astrology, one-third part of the (twelve signs of the) zodiac, or three individual signs of the zodiac 120 degrees apart from each other (also known as a *triplicity*), regarded as a particularly benign or favourable combination.

The word *trigon* derives from ancient Greek *tri-gonos* 'three-angled'; *trine* comes from Latin *trinus* 'threefold'.

trigonometrical ratio, trigonometrical function [maths] For an angle in a right-angled triangle, there is a series of relationships given by the ratio of the lengths of two of the sides of the triangle. If the longest side is called the hypotenuse, the (non-hypotenuse) side next to the angle is called the adjacent side, and the third side is called the opposite side, the three most important trigonometrical ratios are:

$$\text{sine (sin)} = \text{opposite/hypotenuse}$$
$$\text{cosine (cos)} = \text{adjacent/hypotenuse}$$
$$\text{tangent (tan)} = \text{opposite/adjacent}$$

Other ratios include *secant* (= 1/cosine), *cosecant* (= 1/sine), and *cotangent* (= 1/tangent). *See also* HYPERBOLIC FUNCTION.

trihedron [maths] An 'open' three-dimensional figure formed by three planes that meet at a single point (like three sides of a four-sided pyramid).

trillion [quantitatives] In the UK a trillion is 1 million million million (10^{18}) – and, because of such quantification, is a relatively rare concept; in North America and parts of continental Europe, notably France, a trillion is 1 million million (10^{12}) and therefore much more common.

trilogy [quantitatives; literary] A group or set of three, especially three books or plays in which most of the same characters appear or over which the exposition of a subject is fully worked and completed.

trim [linear measure] The degree of horizontal balance in a ship or in an aircraft in motion, especially any difference in vertical height between the prow or nose and the stern or tail. It may also be expressed as an angle.

trimer [chemistry] A POLYMER whose molecules are made by the combination of three MONOMERS. *See also* DIMER.

trimester [time] A period of three months, especially as a division of a school or university year or of pregnancy.

The word derives through French from Latin *tri-mestris* 'of three months'. *See also* SEMESTER.

trimeter [literary] In verse, a line of poetry consisting of three metric feet: *see* FOOT.

trim size [paper sizes; literary] The size of the pages of a book after they have been properly trimmed by a finishing machine, before the cover (outer binding) is put on.

trine *see* TRIGON, TRINE

trinity, Trinity [quantitatives] A group of three apprehended as a unit or unity, especially a group of three forms of divine nature regarded as manifestations of Deity (thus in the Christian Church, God as Father, Son, and Holy Spirit/Ghost, and in Hinduism, Brahma, Vishnu, and Shiva as together embodying the essence of the divine, also called the *Trimurti* 'three forms').

The term derives from Latin *trinitas* 'threefold-ness'.

trinomial [maths] An algebraic expression (polynomial) with only three terms; for example: $4y^2 - 6y + 5$ is a trinomial.

trio [quantitatives; music] A group of three, especially three musicians.

Also, in Classical music, the central part of a symphonic movement (such as a minuet or scherzo) in which the first and third parts are identical (although the third part may be an abbreviated version of the first part) and scored for full orchestra, whereas the central trio is for a much smaller section of the orchestra (sometimes indeed, as originally, for only three players).

triolet *see* VERSE FORMS

trip [quantitatives; collectives] A word that by derivation in this sense is probably the same as TROOP, meaning initially a community, and thereafter a small flock of countryside animals (goats, hares, etc.) or birds (especially wildfowl), and later still even a small group of seals.

trip, tripmeter *see* ODOMETER

tripartite [quantitatives] Comprising three parts, triple, threefold.

triphthong [literary: phonetics] Technically, in phonetics, a triple vowel sound which when, written phonetically, combines three vowels but which may or may not be represented by three vowels in the conventionally written word.

the iao in miaow [miau] is a triphthong

the ia in mediate [miidieit] is a triphthong

the io in studio [studiou] is a triphthong

The term derives from ancient Greek *tri-phthongos* 'three-syllabled'.

triple [quantitatives] Three times (in size, quantity, or number), threefold, comprising three elements.

triple bond [chemistry] A covalent bond between two atoms that involves the sharing of three pairs of electrons (three electrons from each atom): *see* COVALENCE.

triple jump *see* ATHLETICS FIELD EVENTS AREA, HEIGHT, AND DISTANCE MEASUREMENTS

triple point [physics] For a substance that can exist in three different phases (*see* PHASE), the temperature and pressure at which they can all coexist. For example: ice, liquid water, and water vapour coexist at water's triple point (at a temperature of 273.16 K and pressure of 610 newtons per square metre).

triplet [music; literary] In music scoring, a group of three equal notes to be played or sung over the normal duration of two.

In poetry, a verse or stanza of three lines (which are usually of the same length, and may or may not rhyme).

triple time [music] Another expression for THREE-FOUR TIME (or occasionally for SIX-EIGHT TIME).

triple X *see* X (MALT LIQUOR)

triplex [quantitatives] Threefold, triple, comprising three elements – especially of the number of floors in a house.

The word is Latin for 'threefold'.

triplicate [quantitatives] Multiplied by three, threefold, corresponding to three identical elements.

The word derives from the past participle of the Latin verb formed from *triplex* (*see above*).

triskaidekaphobe *see* THIRTEEN

trivalent [chemistry] Describing an atom with a valence (valency) of three. An alternative description is *tervalent*.

trochee, trochaic foot *see* FOOT [literary]

trochoid [maths] A curve traced by a point on the radius of a circle (or its extension) as the circle rolls along a straight line. A point on the circumference of a circle traces a CYCLOID; a point inside the circle traces a curtate trochoid (also called a contracted cycloid); and a point on an extended radius traces a prolate trochoid (or extended cycloid).

The term derives from ancient Greek elements meaning 'wheel-shaped'.

troland [physics] Unit of retinal illumination produced by an illuminated surface (with a luminance of 1 candela per square metre) through a pupil aperture of 1 square millimetre. It was named after the US physicist and psychologist Leonard T. Troland (1889-1932).

trombone *see* BRASS INSTRUMENTS' RANGE

troop [quantitatives; collectives; military] The various uses of this word all stem from its original meaning of 'settlement', 'community' (cognate with modern German *Dorf*, Dutch *dorp*, English place-names *-thorp(e)*, Welsh *tref*) and thus 'throng', 'herd', 'pack', and 'band' (as in French *troupe*, borrowed in a special sense in English).

As a military term, a troop can be said to be the equivalent in some units – notably in armoured cavalry – to a company or battery in other units, mostly under the command of a captain, and to be a body of men numbering between eighty and 200.

In the Scouts (or Guides), a troop is two, three, or four patrols, depending on the locality, thus numbering altogether between sixteen and thirty-two members. *See also* TRIP.

troparion [literary] A stanza or verse of a hymn, particularly in the Greek Orthodox Church; it is known alternatively as a *sticheron*.

The term derives as a diminutive of the ancient Greek *tropos* 'a turning'.

tropical month [astronomy] The time it takes for the Moon to make one complete revolution on its axis (with reference to the equinox), equal to 27.32158 days. *See also* SYNODIC MONTH.

tropical year *see* SOLAR YEAR

tropics [geography] The zone of the Earth between the Tropic of Cancer (23 degrees 45 minutes north) and the Tropic of Capricorn (23 degrees 45 minutes south), in the centre of which is the Equator. It is known also as the torrid zone, outside which (both north and south) are the temperate zones.

tropopause, troposphere *see* ATMOSPHERE, COMPOSITION OF

trot *see* GAIT

Trouton's rule [chemistry] For a non-associated liquid, the molar latent heat of vaporization divided by the boiling point is a constant. At atmospheric pressure it equals approximately 88 joules per mole per kelvin.

troy weight system [weight] Early standard system of weights, partly incorporated by the APOTHECARIES' WEIGHT SYSTEM, but superseded in Britain (and therefore in English-speaking countries) by the actually even earlier AVOIRDUPOIS WEIGHT SYSTEM, and now used only in the measurement of the weights of precious stones and metals.

1 grain	=	0.0020833 troy ounce
	=	0.06479891 gram, 0.002286 ounce avdp.
24 grains	=	1 pennyweight
	=	1.5552 gram, 0.054857 ounce avdp.
20 pennyweight	=	1 troy ounce (480 grains)
	=	1 ounce apothecaries' (= 24 scruples)
	=	31.103475 grams
		(30 grams = 0.9645 troy ounce)
	=	1.09709 ounce avdp.
		(1 ounce avdp. = 0.9115 troy ounce, 437.5 grains)
12 troy ounces	=	1 troy pound (1 pound apothecaries')
	=	373.2417 grams, 0.3732417 kilogram
		(400 grams = 1.0717 troy pound)
	=	13.1657 ounces avdp., 0.82286 pound avdp.
		(1 pound avdp. = 1.2153 troy pound)

The pennyweight is still customarily used only in North America; in Britain and elsewhere it is more usual to divide the troy ounce into decimal fractions (tenths; 0.1 troy ounce = 2 pennyweight). In any case, the grams and kilograms of the METRIC SYSTEM are now taking over even from the avoirdupois system.

The name of the troy system is thought to derive from the city of Troyes, some 140 kilometres/90 miles south-east of Paris, France where, in medieval times, great trade fairs were held specifically for dealers in precious stones and metals, for whom the city authorities provided scales and weights that had been verified and authenticated by the local guilds.

true altitude [astronomy] The altitude of a celestial object obtained by applying various corrections to its apparent altitude (angular height above the horizon).

true azimuth [surveying] Azimuth measured with respect to true north (*see* AZIMUTH; TRUE NORTH).

true bearing [surveying] A bearing measured clockwise round from true north (*see* TRUE NORTH).

true horizon [astronomy; surveying] On the celestial sphere, the great circle that is parallel to the (Earth's) horizon and passes through the centre of the Earth.

true north [geology] The direction to the geographical North Pole (as opposed to magnetic north, the direction in which a compass needle points).

trug [cubic measure] Obsolete North American local volumetric unit of dry capacity used particularly in the measurement of wheat.

1 trug	=	two-thirds of a (US) bushel
	=	2.6666 US pecks, 5.3333 US dry gallons
	=	23.4918 liters by volume
	=	0.6459 UK bushel, 2.5836 UK pecks
3 trugs	=	2 (US) bushels

trumpet *see* BRASS INSTRUMENTS' RANGE

truth table [maths; electronics] In mathematics and computer engineering, a list of every single possible combination of TRUE and FALSE values that can be attributed to a proposition: *see* TRUTH VALUE.

truth value [maths: logic] In Boolean algebra, TRUE and FALSE, often given by the letters T and F or the binary digits 1 and 0 (*see* BINARY SYSTEM).

try *see* RUGBY LEAGUE/UNION MEASUREMENTS, UNITS, AND POSITIONS

tsp, tsp. *see* TEASPOON(FUL)

t'sun [linear measure] Linear unit in China adopted by European colonial authorities during the 1700s to late 1800s when trading with China. The unit varied at different Chinese ports. In general, nonetheless, and for Customs purposes,

1 t'sun	=	1.41 inch, 3.5814 centimetres
		(2 inches = 1.4184 t'sun
		10 centimetres = 2.7922 t'sun)
10 t'sun	=	1 CH'IH
50 t'sun, 5 ch'ih	=	1 PU
100 t'sun, 10 ch'ih, 2 pu	=	1 chang

tu [linear measure] Linear unit in China adopted by European colonial authorities during the 1700s to late 1800s when trading with China. The unit varied at different Chinese ports because the unit it was based on (the CH'IH) also varied. In general, however,

1 tu	=	250 LI, 45,000 chang, 90,000 pu, 450,000 ch'ih
	=	100.142 miles (100 miles 250 yards)
	=	161.163 kilometres

tub [volumetric measure] An old term used by liquor smugglers running brandy, rum, or whisk(e)y into the coastal United States, and referring more to the keg or cask that the drink was in than to the volume of drink inside. For all that, the keg or cask was of generally a consistent size.

1 tub	=	4 (US) gallons
	=	3.3307 UK gallons
	=	15.1412 litres

The word derives from Germanic sources but is probably akin also to the (Latin-based) English word *tube*.

Tudor times [time] The period in British history between AD 1485 and 1603, when members of the Tudor (*Tudwr*, a Welsh version of the name Theodore) family were on the throne. The monarchs involved were:

Henry VII	ruled	1485-1509	(24 years)
Henry VIII		1509-47	(38 years)
Edward VI	ruled	1547-53	(6 years)
Mary I		1553-58	(5 years)
Elizabeth I		1558-1603	(45 years)

tug-of-war, tug o' war *see* HIGHLAND GAMES EVENTS

tugrik, togrog [comparative values] Unit of currency in Mongolia.

1 tugrik or togrog = 100 mung or mongo

See also COINS AND CURRENCIES OF THE WORLD.

tun [volumetric measure] Old unit of liquid volume, corresponding to the capacity of a giant barrel or cask used especially in the transportation of beers or wines.

$$1 \text{ tun} = 2 \text{ pipes, 4 hogsheads, 6 tierces, 8 barrels}$$
$$= 252 \text{ old wine (or modern US) gallons}$$
$$= 209.82 \text{ imperial (UK or English) gallons}$$
$$(200 \text{ UK gallons} = 0.9532 \text{ tun})$$
$$= 953.88 \text{ litres, } 0.95388 \text{ kilolitre}$$
$$(1 \text{ kilolitre} = 1.0483 \text{ tun})$$

The word is ancient, probably borrowed even in late Latin (*tunna*) from Celtic maritime sources, but found particularly in northern and western European variants that have to do both with volume (as the Danish *tønde*) and with weight (as the English *ton* and Spanish/Portuguese *tonelada*).

tun [time] A year in the calendar of the Mayas of Mexico and Central America, comprising 360 days.

tuning-fork *see* PITCH

turbidimeter [physics] An apparatus for measuring the surface area of a powder by measuring the amount of light scattered by its suspension (for example, in water). *See also* NEPHELOMETRY.

turbidity in liquids, turbidimetric analysis *see* NEPHELOMETRY

turning point (of a curve) [maths] In coordinate geometry, a maximum or minimum, at which the slope of a tangent to the curve is zero. At that point, the DERIVATIVE of the function representing the curve is also zero.

turnover [comparative values] The total income of a business per unit time, not taking into account any expenditures or overhead costs necessary to achieve that income.

turns ratio [physics] For the windings on a transformer, the number of turns on the primary divided by the number of turns on the secondary. It is less than 1 for a step-up transformer, and more than 1 for a step-down transformer. The voltages in the primary and secondary windings are in the same ratio; the currents in the two windings are in the inverse of the turns ratio.

twelfth [quantitatives; music] $\frac{1}{12}$ (one-twelfth); one of twelve equal or identical constituents; the element in a series between the eleventh and thirteenth, or the last in a series of twelve.

$\frac{1}{12}$			=	0.08333	=	8.333%
$\frac{2}{12}$	=	$\frac{1}{6}$	=	0.16666	=	16.666%
$\frac{3}{12}$	=	$\frac{1}{4}$	=	0.25	=	25%
$\frac{4}{12}$	=	$\frac{1}{3}$	=	0.33333	=	33.333%
$\frac{5}{12}$			=	0.41666	=	41.666%
$\frac{6}{12}$	=	$\frac{1}{2}$	=	0.5	=	50%
$\frac{7}{12}$			=	0.58333	=	58.333%
$\frac{8}{12}$	=	$\frac{2}{3}$	=	0.66666	=	66.666%
$\frac{9}{12}$	=	$\frac{3}{4}$	=	0.75	=	75%
$\frac{10}{12}$	=	$\frac{5}{6}$	=	0.83333	=	83.333%
$\frac{11}{12}$			=	0.91666	=	91.666%

In music, the interval of a twelfth corresponds to an octave and a perfect fifth: for example, from middle C to the G above the next C (and the sound of those two notes together). It is not in itself a harmony, but may contribute to a harmonic chord.

twelve [quantitatives] 12; a dozen.

$$12\% \;=\; 0.12$$

    12 inches make 1 foot

    12 months make 1 year

A plane figure (POLYGON) with twelve sides (and twelve angles) is a duodecagon. A solid figure (POLYHEDRON) with twelve plane faces is a duodecahedron.

Something that has 12 parts, notably a valve (electron tube) that has twelve pins, may be described as duodecal.

The numbering system that has base (radix) of 12 is the DUODECIMAL SYSTEM.

As is evident from the above, the prefix meaning 'twelvefold' in English is 'duodec-' or 'duodecim-', based on the Latin *duodecem* 'twelve' (literally '2, 10').

The name of the number in English derives solely through Germanic elements (unlike the names of all the numbers from 1 to 10 inclusive, which are common Indo-European in origin). In Old English it was *twelf*, an abbreviated form of a compound corresponding in present-day words to 'two leave' – that is, the number is '(ten plus) two leave (over)' . . . just as (more obviously) eleven is 'one leave'. *See also* DOZEN.

twelve-eight time (12/8 time) *see* TIME SIGNATURES

twelvemo, 12mo *see* DUODECIMO

twelve-tone (system of) music [music] Method of writing music devised by Arnold Schoenberg in 1924. The principal intention was that the music should not be linked to older tone values – above all, there should be no such concept as a home key. Instead of the familiar notes of the key chord and harmonies related to it, all twelve notes (the semitones or half-steps) of the chromatic scale were given equal status, and were to be used one after the other in a consistently recurrent sequence (a 'tone-row', or 'note-row') of the composer's own devising; only the timing was to change as the piece progressed, although pauses could be introduced.

twentieth [quantitatives] $\frac{1}{20}$ (one-twentieth); one of 20 equal or identical constituents; the element in a series between the nineteenth and the twenty-first, of the last in a series of twenty.

$$\tfrac{1}{20} \;=\; 0.05 \;=\; 5\%$$

twenty [quantitatives] 20; a score.

$$20\% \;=\; 0.20 \;=\; \tfrac{2}{10}, \tfrac{1}{5}$$
$$20 \text{ (long) hundredweight} \;=\; 1 \text{ (long) ton}$$
$$20°C \;=\; 68°F$$

Until 1971 in Britain, 20 shillings = 1 pound.

Something that has twenty constituent parts may be described as *vicenary*. Something that occurs every twenty years, or lasts for twenty years, may be described as *vicennial*. Something that is based on the number 20 or comes in groups of twenties may be described as *vigesimal*.

The name of the number derives from the Germanic variants of the common Indo-European term, combining elements that mean '2 (times) 10'. *See also* SCORE.

twenty-one, vingt-et-un, blackjack *see* PONTOON SCORES

twenty-twenty vision, 20/20 vision [biology; medicine] Visual capacity of the normal (perfect) human eye, corresponding to the ability of one eye to distinguish a symbol or character $\frac{1}{3}$ inch (8 millimetres) in diameter from a distance of 20 feet (6.096 metres).

If the vision of the eye is less acute than this, it is the second number that differs proportionately. An eye that fails to distinguish symbols at that distance until they are five times greater than the size of the normal, for example, is said to have $\frac{20}{100}$ vision.

two [quantitatives] 2.

$$\tfrac{2}{10} \;=\; 0.2 \;=\; 20\% \qquad\qquad 2\% \;=\; 0.02 \;=\; \tfrac{1}{50}$$
$$2 \text{ dry gallons} \;=\; 1 \text{ peck}$$
$$2 \text{ barrels} \;=\; 1 \text{ hogshead}$$

A group of two is a pair, a couple, a brace, a duo, a duet, a double(t), or a twosome. In cards, a two may be called a deuce.

Words meaning 'twofold' are double, duple, duplex, dual, twice, and binary.

The ordinal of 'two' is second or binary.

The principal prefixes meaning 'two-' in English are 'bi-', 'di-', and 'duo-', adopted from compound forms in both Greek and Latin. The prefix meaning 'second' is 'deutero-'.

The name for the number derives from common Indo-European, attested over the entire range of languages from Sanskrit to Welsh. It is virtually certain, however, that the basic meaning was not so much numerical as to distinguish between 'one' (the singular, here, I) and 'something other than one' (that, there, thou, and so beginning with the consonantal sound corresponding to differentiation, in English represented most by th-). *See also* THREE.

two-dimensional [maths] Of a figure in one plane only, having dimensions of height and length but no depth (although depth may be suggested through the use of perspective). *See also* THREE-DIMENSIONAL.

twopence [comparative values] In Britain in former centuries, a unit of currency equal in value to 2 (old) pence (or half a groat) when issued as legal tender in the form of a silver coin (in 1450). Since 1662, however, the twopence has been minted as a silver coin for special distribution by the sovereign to deserving elderly folk as part of the Maundy Money every year on Maundy Thursday (the day before Good Friday) in an amount corresponding to the sovereign's personal age. There was a copper coin of the value minted in the 1790s, which remained current for a couple of decades, but not until 1971 and the introduction of decimal currency (when 2 new pence became $\frac{1}{50}$ of a pound as opposed to its previous value of $\frac{1}{120}$) was a modern coin of the value reintroduced.

tyndallimetry [physics] For a material suspended in a liquid, a method of finding its concentration by measuring the amount of light scattered by it (the Tyndall effect). It was named after the British physicist John Tyndall (1820-93). *See also* TURBIDIMETER; TURBIDITY IN LIQUIDS, TURBIDIMETRIC ANALYSIS.

type sizes [literary: printing] In Britain, the United States, and many parts of the English-speaking world, type sizes are usually specified in points (*see* POINT); the equivalent unit in continental Europe is the Didot point (*see* DIDOT POINT SYSTEM). In both regions, large typefaces are increasingly being specified in millimetres, referring either to the cap (capital) height or the x-height.

typp [physics] Unit for the thickness of yarn, equal to the mass in pounds of 1,000 yards of the yarn.

The term derives from the first letters of 'thousand yards per pound'. *See also* DENIER; DREX; TEX.

U

udometer *see* RAIN GAUGE

UHF *see* ULTRA-HIGH FREQUENCY (UHF)

ullage [volumetric measure] The amount by which a container of alcoholic drink, of grain, of flour, or of any other perishable goods falls short of being completely full. In former times, when such a shortfall occurred through leakage, spillage, or absorption, the purveyor of the goods would have to make up the difference to the purchaser, and it was this operation that actually gave rise to the term, in the alcohol trade. The purveyor had to refill the cask up to the 'eye' (French *oeuil*) or bunghole into which the tap was placed: the quantity inserted would then be the *oeuillage* or ullage.

ult. [time] Abbreviation – meaning 'of last month' – of the word *ultimate*. Until perhaps the 1970s, a common method of referring to dates of the previous month in formal correspondence, the expression is now rapidly becoming obsolete and is generally regarded as stylistically unacceptable.

The term derives from the Latin superlative *ultimus* of *ulter, ultra* 'beyond', thus 'outermost', 'most distant'.

ultra- [quantitatives: prefix] Prefix meaning 'beyond', 'surpassing', 'outside'.
Examples: ultraviolet – beyond the violet (end of the spectrum)
ultrasonic – outside the frequency of audible sound
The term derives directly from the Latin.

ultradian [time] Recurring in a cyclic pattern completed in well over twenty-four hours. The word is used especially in relation to body rhythms such as regular digestive processes and other neurologically organized functions of a periodic nature.

The term is as crassly derived as the related CIRCADIAN, in this case supposedly coming from Latin elements *ultra* 'outside' and *dies* 'day'.

ultra-high frequency (UHF) [physics; telecommunications] A range of radio frequencies between 300 million (3×10^8) and 3,000 million (3×10^9) hertz (corresponding to wavelengths of 10 centimetres to 1 metre).

ultramicroscope [physics: optics; biology] A high-power microscope used for viewing objects that are too small to be seen with an ordinary optical microscope (for example, smoke particles). *See also* ELECTRON MICROSCOPE.

ultra-short wavelengths [physics; telecommunications] Imprecise term for radio waves of less than about 10 metres wavelength (corresponding to frequencies of 3×10^7 hertz and less).

ultrasonic [physics] Describing sound, above the upper limit of human hearing, of frequencies over 2×10^9 hertz. Such sounds are used in medical scanning systems and in sonar. The description 'ultrasonic' is sometimes replaced by others (especially in aeronautics) such as 'supersonic' and 'hypersonic' (the latter particularly of mach number in excess of 5), although these terms are generally applied to speeds.

ultrasound scan *see* SCANNING SYSTEMS

ultraviolet (UV) [physics] Describing electromagnetic radiation beyond the violet end of the visible spectrum, of wavelengths between 4 and 400 nanometres (4×10^{-9} to 4×10^{-7} metre). It thus occupies the part of the electromagnetic spectrum between visible light and X-rays.

uncertainty principle *see* HEISENBERG UNCERTAINTY PRINCIPLE

uncia [quantitatives; linear measure; weight] Measure in ancient Rome, originally a unit of linear distance but within a relatively short time, as a fraction of a larger unit, applied also as a quantitative, and used as a unit of weight.

1 uncia	=	one-twelfth of a (Roman) FOOT (*pes*)
	=	0.97 (modern) inch
		(1 inch = 1.0309 uncia)
	=	2.464 centimetres
		(5 centimetres = 2.0292 unciae)
3 unciae	=	1 palm
12 unciae	=	1 (Roman) foot, 16 DIGITS
18 unciae	=	1 (Roman) CUBIT

Akin in Latin to the word *unus* 'one', the uncia in English is found in the ordinary form *inch*, still one-twelfth of a foot.

As a quantitative, meaning simply 'one-twelfth', the unit is found in English in such expressions as QUINCUNX (Latin *quinqu-* 'five', *unciae* 'twelfths').

And as a unit of weight, the ancient Roman measure was also one-twelfth of a larger measure:

1 uncia	=	one-twelfth of a (Roman) pound (LIBRA)
	=	3 duellae, 4 siculi, 6 solidi, 48 oboli
	=	0.9627 OUNCE avdp.
		(1 ounce avdp. = 1.0388 uncia)
	=	27.2875 grams
		(25 grams = 0.9162 uncia)
12 unciae	=	1 libra

The uncia in this sense is found in English in the ordinary form *ounce* although, on the modern scales, there are 16 to the pound.

undecagon [maths] A POLYGON that has eleven sides (and eleven angles).

The term derives from Greek/Latin elements meaning 'eleven [literally, one(-and-)ten] angles'.

ungula [maths] In geometry, a cone, a cylinder, a cube, or any other solid figure that has had its top part sliced off at an angle oblique to the horizontal base.

The term derives from Latin *ungula* 'little hoof' because a cone cut in this way looks rather like one.

uni- [quantitatives: prefix] Prefix meaning 'one-', 'single-'.

It derives directly from Latin.

union [maths] In set theory, an operation that combines two sets to form another that contains all elements from each combining set. For example: the union of {A, B, C, D, E,} and {C, D, E, F, G} is {A, B, C, D, E, F, G}.

unit [quantitatives] In number theory, a single whole number less than 10; also, in whole numbers above 10, the figure on the farthest right of the number (in the 'units column', as opposed to the tens, hundreds, or thousands, etc.).

In weights and measures, a unit is 1 of whatever weight or measure is involved.

The term derives from Latin *unitus* '(made) one', 'united', 'a unity': actually a past participle of the verb *unire* 'to unite'.

unit [physics] An alternative name for KILOWATT-HOUR, used for measuring electricity consumption.

unit cell [chemistry; crystallography] The smallest set of atoms, ions, or molecules, regular repetition of which (in three dimensions) produces the lattice of a crystal.

unit charge *see* UNIT ELECTRIC CHARGE

unit cost [comparative values] The cost of production of one saleable item, calculated from the overall cost of production of the run or batch of items divided by the total number of items produced.

unit electric charge [physics] In the SI system of units, 1 COULOMB. In MKS units, an electric charge that experiences a force of repulsion of 1 newton when it is 1 metre away from a similar charge in a vacuum. In CGS units, it is defined as a charge that experiences a force of 1 dyne when 1 centimetre from a similar charge.

unit magnetic pole [physics] A (theoretical) magnetic pole that experiences a force of repulsion of 1 newton when it is 1 metre away from a similar pole in a vacuum. In CGS units, it is defined as a pole that experiences a force of 1 dyne when 1 centimetre from a similar pole.

unit of blood/plasma/saline [medicine] Sachet of tissue for intravenous drip-feeding into a medical patient, usually in a quantity sufficient to take about six hours' drip at a slow but constant rate, or for immediate large-scale replacement of body fluids in an emergency.

unit pole *see* UNIT MAGNETIC POLE

units, unitary systems *see* CENTIMETRE-GRAM-SECOND (CGS) SYSTEM; METRE-GRAM-SECOND (MKS) SYSTEM; SI SYSTEM

univalent, univalence [chemistry] Having a valence (valency) of 1. An alternative (and more common) description is *monovalent/monovalence*.

Universal Decimal Classification (UDC) system *see* DEWEY DECIMAL SYSTEM

universal law of gravitation *see* NEWTON'S LAW OF GRAVITATION

unkindness of ravens [collectives] Seeing the peaceable ravens as watchbirds at the Tower of London, it is difficult to believe that this collective was initially coined not because the birds are deliberately ill-disposed to others (the modern meaning of the word), but because, from medieval times up to the mid-nineteenth century, they were regarded as unnaturally evil, sinister portents of death and plague – not a kind of creature to be viewed without a shudder ('unkind' = 'not proper to/consistent with its kind', 'unnatural'). There are elements of this superstition even in *The Raven*, written by Edgar Allen Poe in 1845.

unknown quantity [mathematics] The number or amount in a sum, equation, or mathematical problem for which a value has to be found. It is generally given the symbol x; where there are more than one unknown quantities, the next ones are generally given the symbols y and z.

unnil- [quantitatives: prefix] Prefix allocating a number to chemical elements isolated so recently or under such disputed circumstances that they have yet to be given an official name. The intended meaning is 'one hundred and . . . ' in the form *un* '1', *nil* '0' plus another word indicating a number, corresponding to a term representing each element's atomic weight.

Elements of this order to date are:

unnilquadium	atomic weight	104	(isolated 1964/1969)
(also now known as rutherfordium or kurchatovium)			
unnilpentium		105	(isolated 1967/1970)
(also now known as hahnium)			
unnilhexium		106	(isolated 1974)

unnilseptium	107	(isolated 1976)
unniloctium	108	(isolated 1984)
unnilennium	109	(isolated 1982)

It is to be regretted that this practical method of terminology, temporary as it is, conforms to no formal pattern of etymological derivation from the Greek or the Latin numeral names. (There is also the question of what the 110th element can be called . . .)

U particle [physics] A subatomic particle that is a lepton (other leptons include electrons) of a type that has twice the mass of a proton.

upthrust [physics] The upward force that acts on an object when it floats or is immersed in a fluid (equal to its apparent loss in weight). *See also* ARCHIMEDES' PRINCIPLE.

uranium-lead dating [geology] A method of estimating the age of rocks and minerals (in years) by measuring the extent to which naturally occurring uranium isotopes in them have decayed to lead. Uranium-235 changes to lead-207, and uranium-238 changes to lead-206. *See also* CARBON DATING.

Uranus: planetary statistics *see* PLANETS OF THE SOLAR SYSTEM

urinalysis [medicine] Analysis of urine as an aid to diagnosis or to monitor treatment. Typical substances looked for include sugar (glucose), proteins (albumen), acetone, and blood cells. Urine may also be analysed for alcohol and its breakdown products to measure alcohol consumption: *see* ALCOHOL CONTENT.

urna [volumetric measure] In ancient Rome, a unit measure of liquid capacity.

1 urna	=	4 CONGII, 24 SEXTARII
	=	12.75 litres
	=	2.805 UK gallons, 3.368 US gallons
2 urnae	=	1 AMPHORA

The term is evidently the origin of the English word *urn*.

UV *see* ULTRAVIOLET (UV)

V

v, V (Roman numeral) [quantitatives] As a numeral in ancient Rome, the symbol V corresponded to 5 – yet, in this sense, it did not derive from the twenty-second letter of the Latin alphabet, the V that was in turn derived in a roundabout way from the Greek letter *upsilon* (which, to the Greeks as a numerical symbol, signified 400 or 400,000). Instead, it is thought that it was devised in very early times as a representation of a hand, the thumb and little finger outstretched.

Certainly, the Roman numeral X, representing 10, was a diagrammatic representation of two Vs together, and similarly not an alphabetical character.

vac [physics] Obsolete unit of pressure equal to 1,000 dynes per square centimetre, and therefore identical to the MILLIBAR (= 100 newtons per square metre), the unit in contemporary use.

valence, valency [chemistry] The combining power of an atom or ion (expressed as a positive number). For an atom of a given element, it is equal to the number of hydrogen atoms (or their equivalent) that the atom can combine with. For an ion, it is equal to its charge. Some atoms (for example, those of the transition elements) have more than one possible valence. The various valences have their own names (many with alternatives). In the following list, the preferred term is given first:

valence	preferred	alternative
1	monovalent	univalent
2	divalent	bivalent
3	trivalent	tervalent
4	tetravalent	quadrivalent
5	pentavalent	quinquevalent
6	hexavalent	
7	heptavalent	
8	octavalent	

See also OXIDATION NUMBER.

Van Allen (radiation) belts [astronomy; space science] Regions of intense radiation encircling the Earth at 3,000 to 20,000 kilometres from the surface. They result from high-energy particles that have been trapped by the Earth's magnetic field, and were named after the American physicist James Van Allen (1914-).

van der Waals' equation of state [physics] Equation relating to a gas (in terms of its pressure, volume, and temperature) that takes into account the volume of the gas molecules and the forces of attraction between them (which ideal gas equations do not). It therefore better represents the behaviour of a real gas. The equation is:
$$(P + a/V^2)(V - b) = RT$$
where P = pressure, T = absolute temperature, R is the gas constant, and a and b are also constants for the gas concerned.

It was named after the Dutch physicist Johannes van der Waals (1837-1923).

van der Waals' force [physics] A weak force due to the attraction between the atoms (or non-polar molecules) of a gas, incorporated in the VAN DER WAALS' EQUATION OF STATE.

van 't Hoff's law [physics] The OSMOTIC PRESSURE of a solution equals the pressure that would be exerted by the dissolved substance (solute) if it were a gas occupying the same volume as the solution, at the same temperature. It was named after the Dutch chemist Jacobus van 't Hoff (1852-1911).

vapour density [physics; chemistry] The mass of a volume of gas divided by the mass of an equal volume of a reference gas (usually hydrogen) at the same temperature and pressure. Its significance in chemistry is that it is equal to half the relative molecular mass (molecular weight), and so provides a method of determining this for a gas.

vapour pressure [physics] The pressure at which a liquid is in equilibrium with its vapour, also known as the saturation vapour pressure.

var, VAr [physics] Unit of reactive electrical power.
$$1 \text{ var} = 1 \text{ VOLT-AMPERE}$$
The unit derives its name from the initials of 'volt-ampere, reactive'.

vara [linear measure] The vara is a linear measure mostly in Spain, Chile, Peru, Argentina, Mexico, and other South American, Spanish-speaking countries, also in ex-Spanish-speaking Texas and California (United States), and in Portugal. In many of those countries it corresponds to the same unit length, but there are also different versions. Basically, however, it corresponds to a very ancient form of the YARD, made up of three feet, the FOOT in this case being the old Roman foot (Latin *pes*, *ped-*), taken in medieval Spain to be equivalent to 10.97 inches in today's imperial measure, and from that time called the PIE, comprising 12 PULGADA ('inches').

1 vara in Spain, Peru, Chile, and elsewhere
$$= 3 \text{ pie, 4 (Spanish) palmos, 36 pulgada}$$
$$= 32.91 \text{ imperial inches, 2 ft 8.91 in}$$
$$= 83.59 \text{ centimetres, 0.8359 metre}$$
(1 imperial yard = 1.0939 Spanish vara
1 metre = 1.1963 Spanish vara)

1 vara in Mexico
$$= 32.99 \text{ imperial inches, 2 ft 8.99 in}$$
$$= 83.80 \text{ centimetres, 0.8380 metre}$$
(1 imperial yard = 1.0912 Mexican vara
1 metre = 1.1933 Mexican vara)

1 vara in California
$$= 33.00 \text{ imperial inches, 2 ft 9 in}$$
$$= 83.82 \text{ centimetres, 0.8382 metre}$$
(1 imperial yard = 1.0909 Californian vara
1 metre = 1.1930 Californian vara)

1 vara in Texas
$$= 33.33 \text{ imperial inches, 2 ft 9.33 in}$$
$$= 84.67 \text{ centimetres, 0.8467 metre}$$
(1 imperial yard = 1.0801 Texan vara
1 metre = 1.1811 Texan vara)

1 vara in Argentina

$\quad$ = $\quad$ 34.09 imperial inches, 2 ft 10.09 in
$\quad$ = $\quad$ 86.60 centimetres, 0.8660 metre
$\qquad$ (1 imperial yard = 1.0560 Argentine vara
$\qquad$ 1 metre = 1.1547 Argentine vara)

1 vara in Portugal

$\quad$ = $\quad$ 5 (Portuguese) palmos
$\quad$ = $\quad$ 43.31 imperial inches, 3 ft 7.31 in
$\quad$ = $\quad$ 110 centimetres, 1.10 metre
$\qquad$ (1 imperial yard = 0.8312 Portuguese vara
$\qquad$ 1 metre = 0.9091 Portuguese vara)

Surprisingly, the Italians and the French seem to have no equivalent to this otherwise Latinate yard, although there are well-established equivalents for the foot. *See also* PIE (Italian); PIED.

variable [maths] In algebra, a symbol (such as x, y, p, q) that can have any number of values.

variance [maths] In statistics, the arithmetic mean (average) of the squares of the deviations of each number in a set of numbers from the mean of the whole set. The square root of the variance is the standard deviation.

variation [physics] In geography and navigation, the angular difference between geographic north and magnetic north at a specified place – also known as the magnetic variation.

variety [physics: taxonomy] In the taxonomic classification of life-forms, the highly distinctive category of plant life between a species and a form. *See* full list of taxonomic categories *under* CLASS, CLASS.

vector [maths; physics] A quantity that has both magnitude and direction (for example, velocity). A quantity with magnitude only is called a scalar (for example, speed).

$\quad$ The term *vector* is the Latin for 'transporter', 'carrier', but by derivation also 'one who follows a course with method'.

vector meson [physics] Any of a class of subatomic particles that have a mass of more than 1.2 UK thousand-million/1.2 US billion electron volts (1.2×10^6 eV). They include omega mesons, phi mesons, and rho mesons.

vedro [volumetric measure] In Russia and Bulgaria, different units of volume evidently originally the same.

$\quad$ 1 Russian vedro $\quad$ = $\quad$ 100 CHARKI, 16 BOUTYLKI, 10 SCHTOFFS
$\qquad\qquad\qquad\qquad$ = $\quad$ 12.3 litres
$\qquad\qquad\qquad\qquad$ = $\quad$ 2.7057 UK gallons, 3.2497 US gallons
$\qquad\qquad\qquad\qquad\qquad$ (1 litre = 0.0813 Russian vedro
$\qquad\qquad\qquad\qquad\qquad$ 1 UK gallon = 0.3696 Russian vedro
$\qquad\qquad\qquad\qquad\qquad$ 1 US gallon = 0.3077 Russian vedro)
$\quad$ 1 Bulgarian vedro $\quad$ = $\quad$ 10 litres
$\qquad\qquad\qquad\qquad$ = $\quad$ 2.1997 UK gallons, 2.6420 US gallons
$\qquad\qquad\qquad\qquad\qquad$ (1 litre = 0.100 Bulgarian vedro
$\qquad\qquad\qquad\qquad\qquad$ 1 UK gallon = 0.4546 Bulgarian vedro
$\qquad\qquad\qquad\qquad\qquad$ 1 US gallon = 0.3785 Bulgarian vedro)
$\quad$ 2 Bulgarian vedros $\quad$ = $\quad$ 1 Bulgarian krina

The Russian measure is obviously the older, whereas the Bulgarians have standardized to the metric system.

velar consonants [literary: phonetics] Consonants that are pronounced by raising the back of the tongue towards or against the soft palate (Latin *velum*). In English they correspond to:

$\quad$ c [k], $\qquad$ g, $\qquad$ k, $\qquad$ ng, $\qquad$ qu, $\qquad$ w, $\qquad$ x [ks]

velocity [physics] The rate of change of position in a given direction, expressed as the distance travelled in unit time (for example, as mph, metres per second, and so on). If distance is not specified, the rate of change of position equals the speed (velocity is a vector, speed is a scalar).

$\quad$ The term derives from the Latin abstract noun *velocitas*, a derivative of *velox* 'rapid', 'quick-moving'.

velocity ratio *see* DISTANCE RATIO, VELOCITY RATIO

Venn diagram [maths] In set theory, a graphical way of representing sets as circles (within a rectangle representing the universal set). Subsets are shown as smaller circles within those representing sets. If two circles (sets) overlap, the elements in the overlapping region belong to both sets. It was named after the British mathematician John Venn (1834-1923).

Venturi (tube) [physics] A device for measuring fluid flow, consisting of a cylindrical tube with a constriction in the middle. The rate of flow of a fluid flowing through the tube increases, and its pressure decreases, in the constriction. The pressure difference between that at the ends of the tube and in the constriction gives a measure of the rate of flow. It was named after the Italian physicist Giovanni Venturi (1746-1822).

The effect of this device was put into very practical use during the 1980s, when Venturi ducts (channels acting as Venturi tubes, using air as the fluid) were incorporated on the undersides of high-performance motor cars and racing cars to increase downward pressure ('downforce') and so improve traction when cornering at speed.

Venus: planetary statistics *see* PLANETS OF THE SOLAR SYSTEM

verb [literary] The verb, in a sentence, is the word that implies action (even if the action is only 'being' or 'experiencing'). For a sentence to be a sentence at all, it requires the presence of a subject and a verb; an object of the verb is common but not essential – a verb that requires an object is called a *transitive verb*; one that does not require an object is an *intransitive verb*. The difference between a clause and a phrase is similarly the presence of a verb in a clause.

An active verb (a verb in the active voice) corresponds to action undertaken by the subject of the sentence:

I *hit* him   He *was speaking* to the press   She *rode* for an hour

A passive verb (a verb in the passive voice) implies that the subject undergoes the action of the sentence:

I *was struck* by him   They *were murdered*   She *has been elected*

According to some authorities, there is another class of verbs known as *factitive verbs* – verbs that use the active voice and which take both a direct and indirect object, the latter in apposition – as, for example,

We *thought* him mad   The queen *dubbed* him knight

In some foreign languages there are verbs that are passive in form but active in meaning: these are known as *deponent verbs*.

From the examples above it is evident that the verb also gives some idea of the timing of the action – whether it was in the past, is happening in the present, or will happen in the future. To do this, the verb is expressed in any of a number of tenses. The tenses not only define when the action occurs but whether it was, is, or will be continuous, or was, is, or will be finite; there are several continuous tenses that are said to give a verb its *durative aspect*.

Moreover, to express these tenses, verbs require secondary verbal elements – generally parts of the verbs 'to be' and 'to have', although there are others – known as *auxiliary verbs*. An auxiliary verb that assists in the expression of mood – indicative, subjunctive, imperative (jussive), or (according to some) optative – is known as a modal auxiliary. Classic examples of these are the modal auxiliaries 'may', 'might' and 'let' in the subjunctive mood.

verchok [linear measure] In Russia, an old linear measure not normally now used.

1 verchok	=	4.445 centimetres, 1.75 inches
		(1 centimetre = 0.225 verchok
		1 inch = 0.5714 verchok
		1 foot = 6.857 verchoki)
16 verchoki	=	1 ARSHIN
	=	71.12 centimetres
	=	28 inches (exactly), 2 ft 4 in
48 verchoki	=	3 arshin = 1 SADZHEN
	=	2.1336 metres
	=	84 inches (exactly), 7 feet (exactly)
24,000 verchoki	=	500 sadzhen' = 1 VERST or versta

vernal equinox *see* EQUINOX

vernier scale (gauge) [linear measure] A device for making more accurate measurements than can be achieved with a simple graduated scale. The vernier is a small scale graduated in intervals that are 9/10 the size of those on the main scale, along which it can slide. It allows the main scale to be read to a tenth of a division. A circular vernier scale is employed on a MICROMETER GAUGE. It was named after the French mathematician Pierre Vernier (1580-1637).

vers *see* VERSINE, VERS

verse [literary unit in poetry or scriptures] In poetry (also known in the abstract as *verse*), a number of lines of set form and length, used as a pattern for all or most of a poem (also called a *stanza*). In Classical poetry, specifically in very long (and intendedly epic) poems, a number of verses make up a *canto*, and a number of cantos make up a *volume* or *book*.

The major difference between poetry/verse and prose is the former's regular form, whether rhyming is included (rhyming verse), whether the style is of assonance within lines (as in Anglo-Saxon and some medieval Latin verse), or whether there is absence of both (blank verse). Modern verse is considerably freer of literary restraints, relying more for its effect on the use of individual ideas, associations, and shades of nuance than on rhythm and form.

From medieval times – especially as the Judaeo-Christian Bible began to be translated into languages other than Greek or Latin – religious scriptures have also been divided into verses. In most texts, a number of verses make up a *chapter*, and a number of chapters make up a *book*.

The word *verse* derives from the past participle of the Latin verb *vertere* 'to turn', that is *versus* 'turned' – originally of earth turned over to create an agricultural furrow, thus a 'line', thus a 'verse'. It is thus cognate with the Russian linear measure, the VERST (*see below*).

versed sine [engineering] For an arch, the distance from the centre of the span (in line with the springings of the arch) to the centre of the soffit (the underside of the apex of the arch). It is alternatively known as the *rise*.

verse forms [literary] The best-known formal verse forms are (in alphabetical order):

ballad: quatrains: 1st and 3rd lines 8 syllables (4 feet), 2nd and 4th lines 6 syllables (3 feet); rhyming *a b a b* or more often *a b c b*

ballade: three stanzas of 7 or 8 (or occasionally 10) lines, all with an identical last line, plus a 2- or 4-line *envoi* (moral or footnote); only three or four rhymes throughout

ballade royal: see rhyme royal *below*

blank verse: unrhyming verse, especially pentameters (lines of 5 metric feet or 10 syllables)

Classic iambic pentameters: lines of 5 iambic feet (pentameters; *see* FOOT), usually rhyming couplets (also known as HEROIC VERSE or *elegiac couplets*) and usually not divided into stanzas; the standard form in the age of Classicism

clerihew: humorous doggerel generally consisting of, and rhyming in, two couplets, starting with a very short line, and each line thereafter at least one syllable longer; mostly on the subject of some person of note

elegiac couplets: see Classic iambic pentameters *above*; but the term occasionally applied instead to hexametric couplets

elegiac verse: two stanzas of 4 iambic pentameters; rhyming *a b a b*

epic poetry: no set form, more a style of expression; long poem about heroes or national events, couched in grandiose and majestic language

free verse: poetry that does not conform to any set verse form but is nonetheless written within a set pattern albeit with variations of cadence and rhythm; usually no rhyming

haikai: long (Japanese) poem in stanzas in which alternate lines are 14 and 17 syllables long; often humorous

haiku: Japanese form of free verse in a set pattern that comprises 17 syllables in three phrases (5 syllables, 7 syllables, 5 syllables), and includes a word or expression implying one of the four seasons

heroic verse: see HEROIC VERSE *and* Classic iambic pentameters *above*

Italian or Petrarchan sonnet: a sonnet in two parts: the first 8 lines are in two quatrains with only two rhymes between them, the last 6 lines rhyming in alternate lines or in triplets; rhyming scheme thus *a b b a a b b a c d c d c d* or *a b b a a b b a c d e c d e*; form much used by Dante and by Petrarch

limerick: humorous single stanza of 5 lines in a dactylic rhythm, the 1st, 2nd, and 5th of 3 metric feet, the 3rd and 4th of only 2 metric feet; rhyming scheme *a a b b a*, and the last word of the last line may be identical to the last word of the first line

ode: elaborate and often lengthy poem of no set form, usually addressed to some person of note (or beauty) in lyrical terms

ottava rima: a stanza of 8 lines (the expression is Italian for 'rhyming octave') with only 3 rhymes between them; rhyming *a b a b a b c c*; in the original Italian, each line comprised 11 syllables: in English versions lines were/are more often of 10 syllables

rhyme royal: a sort of ballade (in fact, alternatively called ballade royal): (one, two or) three stanzas of 9 lines; rhyming *a b a b b c b c*

rondeau: a poem of 13 lines divided into three unequal stanzas, the last comprising 6 lines; the first words (but not necessarily the whole first line) of the poem are repeated before and after the last stanza as a sort of pointed refrain

rondel: a highly stylized verse form consisting of 13 (or sometimes 14) lines with only two rhymes between them; the 1st, 7th, and 13th lines are identical; if there is a 14th line, it is identical with the 2nd and 8th

rondelet: a short poem of 5 or 7 lines with only 2 rhymes between them; as in the rondeau (*see above*), the first word or phrase is generally inserted between lines once or twice, as appropriate, as a sort of pointed refrain

sestina: a poem consisting of 6 stanzas of 6 lines followed by a final shorter verse of 3 lines; the final words of each stanza are the same, but in a different order each time, and the shorter verse at the end includes the same words yet again

sonnet: a classic verse form comprising 14 lines, generally iambic pentameters, in which a stylized rhyming scheme is apparent; the Shakespearean or Elizabethan sonnet, for example, may be described as three quatrains and a couplet: it has a rhyming scheme *a b a b c d c d e f e f g g*; the Spenserian sonnet in its way is also three quatrains and a couplet, although the quatrains are additionally linked in rhyme: it has a rhyming scheme *a b a b b c b c c d c d e e*

Spenserian stanza: a verse of 9 lines, all iambic pentameters except the last, which is an iambic alexandrine (of 6 metric feet), with only 3 rhymes between them; rhyming *a b a b b c b c c*

tanka: Japanese free verse form of 5 lines comprising a total of 31 syllables in lines of 5, 7, 5, 7, and 7 syllables; the term is Japanese for 'short verse'

terza rima: a verse form comprising three-line iambic stanzas of lines of 10 or 11 syllables (5 iambs), each stanza after the 1st linked to the previous one by rhyme: rhyming *a b a b c b c d c . . .*

triolet: a highly stylized verse of 8 lines with only two rhymes between them, and in which the 1st, 4th, and 7th lines are identical, and the 2nd and 8th lines are identical; rhyming thus *A B a A a b A B*

villanelle: a highly stylized (and potentially strongly emotive) verse form of 19 lines in 5 triplets and 1 quatrain with only two rhymes between the lot, and in which the 1st, 6th, 12th, and 18th lines are identical, and the 3rd, 9th, 15th, and 19th are identical

virelay: complex French verse form in stanzas of long and short lines, the long lines rhyming and the short lines rhyming separately, but the rhyme of the short lines in one stanza recurring as the rhyme of the long lines of the following stanza

verse metre *see* METRE IN VERSE

versine, vers [maths] A spherical trigonometrical ratio, equal for a given angle to 1 minus its cosine. It is known also as *versed sine.*

verst [linear measure] An old Russian measure of distance not normally now used.

1 verst or versta = 500 SADZHEN' = 1,500 ARSHIN

= 1.067 kilometre, 1,067 metres
= 1,166.67 yards, 3,500 feet (exactly)
(1 kilometre = 0.9372 verst
1 mile = 1.5086 verst)

The Russian word originally meant 'line', 'row', but its meaning then extended in two directions – both to emcompass the linear measure and to mean 'age (in years of a person's life)'. In the latter sense it is akin to the English verbal forms of the verb 'to be', *were* and *wert* (cf. German *werden* Dutch *worden* 'to become'), and also to *worth*.

vertex [maths; physics; astronomy] In geometry, a point at which two sides of a polygon (plane figure) or three faces of a polyhedron (solid figure) meet. The pointed tip of a pyramid or cone is also called a vertex.

In physics, the point on a lens at which the optical axis meets the surface.

In astronomy, the highest point reached by a heavenly body.

The term derives from the Latin *vertex* 'whirlpool', thus 'the axis of the heavens', 'the celestial pole', thus 'straight upwards (vertically) to the top'.

vertical [maths] The direction at right-angles (perpendicular) to the horizontal (the plane of the horizon). Vertical angles are measured in the vertical plane (for example, as by a theodolite in surveying).

The term derives from the adjectival form of the Latin *vertex* in its extended meaning (*see above*).

vertimeter [physics: aeronautics] An instrument that measures and displays an aircraft's rate of climb or descent in flight.

very high frequency (VHF) [physics; telecommunications] Describing radio frequencies between 30 and 300 megahertz (3×10^7 to 3×10^9 hertz), corresponding to very short wavelengths in the range 10 metres to 10 centimetres. VHF is used for high-quality radio broadcasts and communications between locations that are in sight of each other (line-of-sight communications).

very low frequency (VLF) [physics; telecommunications] Describing radio frequencies between 10 and 30 kilohertz (10^4 to 3×10^4 hertz), corresponding to very long wavelengths in the range 10,000 to 30,000 metres.

VHF Abbreviation of VERY HIGH FREQUENCY.

vice-admiral [military rank] Senior commissioned rank in the navy.

In the Royal (British) Navy, a vice-admiral ranks between a rear admiral and an admiral, and is the equivalent of a lieutenant general in the British army and an air marshal in the Royal Air Force.

In the US Navy, a vice-admiral ranks between a rear admiral (on the upper half of the list) and an admiral, and is the equivalent of a lieutenant general in the US Army and Airforce.

vicenary [quantitatives] Consisting of, or having, twenty (parts, members, elements, factors, etc.)

The term derives from the Latin *viceni* 'twenty times'.

vicennial [time] Occurring once every twenty years; lasting for twenty years at a time.

Vicker's test (for hardness) [engineering] A method of measuring the hardness of a metal by using a standard load to impress a diamond pyramid into it. The size of the indentation gives a measure of hardness.

Victorian age/era [time] Period from 1837 to 1901, during which Queen Victoria occupied the British throne.

viertel [volumetric measure] A measure of volume in Switzerland and Denmark but, as a unit, very different in those countries.

1 Danish viertel = 8 Danish POTS
= 7.73 litres
= 1.7006 UK gallon, 2.0421 US gallons
(10 litres = 1.2937 viertel
1 UK gallon = 0.5880 viertel
1 US gallon = 0.4897 viertel)
17 Danish viertels = 1 TØNDE
1 Swiss viertel = 40 schoppen, 10 immi, Swiss maassen, or Swiss pots

$$= \quad 15 \text{ litres}$$
$$= \quad 3.300 \text{ UK gallons, } 3.9627 \text{ US gallons}$$
$$(10 \text{ litres} = 0.6667 \text{ viertel}$$
$$1 \text{ UK gallon} = 0.3030 \text{ viertel}$$
$$1 \text{ US gallon} = 0.2523 \text{ viertel})$$

It is strange that the unit, meaning 'one-quarter' in both countries, seems to have no corresponding unit four times the size.

vigesimal [quantitatives] Twentieth; one-twentieth; in groups of twenty.

villanelle *see* VERSE FORMS

vinometer [physics] An instrument that measures and displays the alcoholic strength or purity of a wine.

vingt-et-un *see* PONTOON SCORES

viola, viol family *see* STRINGED INSTRUMENTS' RANGE

virelay *see* VERSE FORMS

virgate [square measure] Old measure of land in rural England, now no longer in use as a unit. It was one-quarter of a HIDE, a measure reckoned as the area of land that could support one free family and its dependants over the agricultural year, using one plough. The actual area thus varied somewhat.

$$1 \text{ virgate} \quad = \quad \text{between 25 and 30 ACRES}$$
$$(= \text{between 10 and 12 HECTARES})$$
$$4 \text{ virgates} \quad = \quad 1 \text{ hide}$$
$$(= \text{between 100 and 120 acres})$$

virgule *see* SOLIDUS, OBLIQUE, VIRGULE

viscometer [physics; engineering] An instrument for measuring VISCOSITY.

viscosity [physics] The property of a fluid that makes it resist flow – known alternatively as internal friction. It results from there being different rates of flow at different points in the fluid (liquid or gas). The viscosity is expressed as the coefficient of viscosity, measured in poise, equal to $\frac{1}{10}$ newton per square metre-second (in SI units), or 1 dyne per square centimetre-second (in CGS units). There are also various empirical viscosity indexes or scales, particularly for common fluids such as oil and sugar solutions.

VLF *see* VERY LOW FREQUENCY

voiceprinting [physics] The recording (usually on a computer) of an audio spectrum that displays the subtle variations in the sound patterns unique to an individual person's voice. The record can then be used to identify that person (by a comparison of his or her spoken voice with the recorded audiogram).

Elements of this system are used in computer voice-recognition programs for security or control operations.

vole [sporting term] In card games in which tricks are taken, the taking of all available tricks in a hand is called a *slam* (English via Scandinavian Germanic) or *vole* (English via French).

volleyball measurements, units and positions [sport] The six players on a volleyball team all move round one position clockwise per serve when their team has service. Any part of the body above the waist may be used to send the ball over the net, the object of the game being to bounce the ball in the opponents' half of the court or to oblige an opponent to touch the ball before the ball goes directly out of court. Up to three players in one team may touch the ball (without allowing it to touch the ground) before it must go back across the net.

The dimensions of the court:

overall length: 18 metres (59 feet)

overall width: 9 metres (29 feet 6 inches)

net height: (top) 2.43 metres (7 feet 11 inches)

net depth: 1 metre (3 feet 3 inches)

Timing:

the game comprises (the best of) 5 sets

2-minute intervals between sets, but a 5-minute interval before a fifth set

each team may request two 30-second time-outs per set

Points scoring:

only the serving team scores points; failure to win a point loses serve the first team to reach 15 points wins a set

a score of 14-14 in a set may require a team to win two further clear points to win the set

Dimensions of the ball:

circumference: 65-67 centimetres (25.6-26.4 inches)

weight: 260-280 grams (9.17-9.88 ounces)

volt [physics] Unit of potential difference or electromotive force (EMF), equal to the potential difference between two points (in an electric field) when 1 joule of work is done in making 1 coulomb of electricity go from one of the points to the other (in other words, a power of 1 watt is dissipated by a constant current of 1 ampere). It was named after the Italian physicist Alessandro Volta (1745-1827). *See also* ABVOLT.

voltage [physics] The value of an electromotive force (EMF) or potential difference expressed in volts. It is commonly measured with a voltmeter.

voltameter [physics] An electrolytic cell used to measure (DC) electric current in terms of the amount of gas liberated at, or metal deposited on, the cathode in a given time. It makes use of Faraday's law of electrolysis: the mass of an element liberated during electrolysis is proportional to the amount of electricity used.

volt-ampere [physics] Unit that describes the apparent power in an AC electric circuit, equal to the product of the root-mean-square (rms) voltage and the current. A reactive volt-ampere – abbreviated as 'VAr' or 'var' for 'volt-ampere, reactive' – is a measure of the 'wattless' power in such a circuit, equal to the product of the reactive voltage and the current, or the reactive current and the voltage. *See also* POWER; WATT; KILOVOLT-AMPERE.

volt-metre [physics] Unit of electric flux equal to the rate of flow of electric field across a given surface (normal to the surface).

volume [maths; volumetric measure] The amount of space occupied by an object, or the capacity of a container. Volumes of solid objects are usually calculated by multiplying together three lengths, and are generally reckoned in units of length cubed (such as cubic feet, cubic centimetres). Capacities generally use different units, such as pints or litres for liquids.

The following formulae may be used to calculate the volumes of some common geometrical figures (l = length, b = breadth, h = height or altitude; r = radius]:

cone	$\frac{1}{3}\pi r^2 h$
cube	l^3
cuboid	lbh
cylinder	$\pi r^2 h$
pyramid	$\frac{1}{3}lbh$
sphere	$\frac{4}{3}\pi r^3$

The term derives in a roundabout way from Latin *volumen* '(something) rolled up', 'a scroll' – thus 'a large book', 'a book containing much information' (which is thus *voluminous*); the medieval French adopted the notion of amplitude and overall capacity to give the word its present meaning.

volume (of sound) [physics] An imprecise term for the general loudness of a sound, one measurement of which is the integrated value of the peak amplitudes over a short time period.

volumeter [physics] An instrument for measuring VOLUME.

volumetric analysis [chemistry] A method of chemical analysis that uses accurate measurements of the volumes of solutions that react together. Its principal technique is *titration*, in which a solution of known concentration is added (using a burette – an accurately graduated measuring tube fitted with a tap) to a flask containing a known volume of another solution of unknown concentration (measured using a pipette – a calibrated glass tube, usually with a swelling at its centre). The reaction is complete at the end-point of the titration, which is usually detected using an appropriate indicator. From the reacting volumes and the known concentration, the unknown concentration can be calculated.

vowels [literary] In alphabetical terms, the vowels in English are

a   e   i (and y)   o   u

In terms of phonetics, these vowels have short and long forms, the latter of which ordinarily involve combinations of two short forms (DIPHTHONGS).

voyage recorder [navigation] Multiple device on board every commercial ocean-going vessel that comprises 350 or more sensors which record virtually every aspect of the mechanical operation and human navigation of a ship's passage. The presence of the device has been a requirement of maritime law since 1989. Like an aircraft's *flight recorder*, its purpose is to provide clues to what actually happened in the event of a collision, particularly if there is a subsequent capsize or sinking, and especially if the ship's crew are unable themselves to give evidence. The equipment is therefore made as outwardly strong, waterproof, and pressure resistant as possible.

V-particle [physics] A subatomic particle – a type of HYPERON – produced in nuclear disintegrations, so called because of the V-shaped track it leaves in a cloud- or bubble-chamber.

vulgar fraction [maths] A fraction in which the numerator (number above the line) and the denominator (number below the line) are whole numbers (integers); for example: $3/4$, $11/15$, $73/100$. In the United States, this type of fraction is also known as a *common fraction*.

Common or even vulgar, it may yet be improper too: an *improper fraction* is a vulgar fraction expressing a quantity greater than 1; for example: $4/3$, $15/11$, $100/73$.

W

W *see* WATT

waka *see* TANKA

walking race distances and measurements [sport] The usual distances for walking races are 20 kilometres (12.099 miles) for men and, much less commonly, 10 kilometres (6.093 miles) for women. A competitor who is 'lifting' (who has both feet off the ground at any point) is warned up to two times, and may thereafter be eliminated/disqualified if it continues.

walk of snipe [collectives] This collective at first seems strange: these rather rare wading birds have short legs and, if seen at all, tend to run or take to the air in zigzag flight. But, during the 1600s, the word 'walk' could mean both 'procession' and 'bird-run', either of which would be fairly appropriate to snipe on the ground.

A collection of snipe in flight is described as a WISP.

Wallace's line [biology] An imaginary line between southern Asia and Australasia that separates the habitats of the placental mammals of Asia from the indigenous monotremes and marsupials (egg-laying and pouched mammals) of Australasia. It was named after the British naturalist Alfred Wallace (1823-1913).

waltz time *see* THREE-FOUR TIME (3/4 TIME)

wapentake [square measure] In certain counties of England, the former division of a county known in other counties as a HUNDRED, and originally reckoned to contain about 100 families and to require its own court. But, whereas the hundred was an administrative division for the purposes of taxation and social discipline, the more Scandinavian-based wapentake apparently required its inhabitants to take weapons (*taecan waepen*) along when attending the annual assembly (*waepentaec*) for the area, possibly to show defensive capacity, but more probably to swear oaths of fealty upon while touching the blades. The word was later applied solely to the local lawcourt, and then to the court's chief officer, the bailiff (whose mace or staff of office in present-day courts may represent the last vestige of the weapons once paraded in this way).

watch [time] On board ship the 24-hour day is divided into seven watches, five of them 4 hours long and the remaining two 2 hours long. The latter two 'dog watches' are from 16.00 (4 p.m.) to 18.00 (6 p.m.) and 18.00 (6 p.m.) to 20.00 (8 p.m.), with the purpose that those who regularly alternate duty on watch automatically get a daily change of hours and at a time when an evening meal is available.

On British naval ships, the BELLS that sound the time every half-hour – culminating in eight bells rung in paired chimes at 4 o'clock, 8 o'clock, and 12 o'clock, a.m. and p.m. – follow the watch system, thus beginning again at one bell (rather

than five) at 18.30 (6.30 p.m.). This practice derives from the time of the naval
mutiny at the Nore (1797) when the signal for the mutiny was the five bells at the
beginning of the second dogwatch. Five bells at that time of day has since then
never been sounded on a British naval ship.

By derivation, however, a watch was a period of duty during the night, when a
person had to stay awake (Old English *waeccan*, both 'watch' and '(stay) awake').
The ancient Hebrews had three watches during the night; the ancient Greeks
observed four, or, in some cities five, watches; and the Romans had four.

water-clock *see* CLEPSYDRA

water equivalent [physics] For any object, the mass of water that needs the same
amount of heat (as the object does) to increase its temperature by 1 degree, equal to
its thermal capacity divided by 4.186 (the specific heat capacity of water in
kilojoules per kilogram per kelvin).

water-inch [speed/flow] A unit of the flow of water through a circular orifice of
1 inch (2.54 millimetres) diameter per unit time.

waterline [linear measure] On the side of a ship, any of several lines denoting the
average level to which the sea surface comes up under a specific condition: when
the ship is fully laden, for example, or half-laden, or empty. *See also* PLIMSOLL LINE.

water polo measurements, units, and positions [sport] Water polo is a game
for two teams of seven players, whose object is to throw the ball into the goal at the
opponents' end of the swimming pool.

The dimensions of the pool:
usual overall length 33.333 metres (109 feet 4.323 inches)
distance between goal lines: 30 metres (98 feet 5.1 inches)
the goal:
width: 3 metres (9 feet 10 inches)
height: 90 centimetres (2 feet 11½ inches) above water surface if water
depth more than 1.5 metre (4 feet 11 inches), 2.4 metres (7 feet 10 inches)
above floor surface if water depth less than 1.5 metre (4 feet 11 inches)
Timing: 20 minutes actual playing time, in four 5-minute periods separated by 2-
minute intervals
in the event of a tie in a league competition, two 3-minute periods of extra time
may be played, separated by a 1-minute interval; if the tie persists, further
extra time periods may be played
Points scoring:
the team that has scored the greater number of goals when the game finishes
wins
Dimensions of the ball:
circumference: 68-71 centimetres (26.77-27.95 inches)
weight: 400-450 grams (14.1-15.9 ounces)
Despite its name, water polo has never had anything to do with the (field) game
of polo, from which it is in any case vastly different.

water skiing disciplines and measurements [sport] The three major disciplines in water skiing competitions are:
the slalom
water-ski jumping
tricks
It is common practice for competitors to enter at least two of the three events; in
some competitions, competing in all three is compulsory.

Slalom
dimensions of the course:
usual overall (perpendicular) length: about 260 metres (284 yards)
perpendicular distance between object buoys (as travelled by towing boat):
between gate and 1st buoy – 27 metres (29 yards)
between 1st and 2nd buoy, 2nd and 3rd buoy, 3rd and 4th buoy, 4th and
5th buoy, 5th and 6th buoy – 41 metres (44.8 yards)
between 6th buoy and exit gate – 27 metres (29 yards)
distance of buoys from centre of course: 11.5 metres (12½ yards)
the first rounds are taken at successively increased speeds

the towing line is shortened by a specified length after each further round
Jumping
ramp length: 20 feet (6.1 metres)
ramp width: 12 feet (3.66 metres), plus apron
ramp maximum height:
for men – 6 feet (1.83 metres) on a 1-foot (30-centimetre) base
for women and veterans – 5 feet (1.52 metres) on a 1-foot (30-centimetre)
base
three jumps per competitor
longest distance jumped wins
Tricks
two 20-second sequences of tricks while skiing down the course at a speed
requested by the competitor
each trick scored (by judges) according to a tariff of difficulty
highest total score wins

watt [physics; engineering] Unit of power (rate of doing work), equal to an energy
expenditure of 1 JOULE per second.

$$745.7 \text{ watts} = 1 \text{ horsepower}$$
$$1,000 \text{ watts} = 1 \text{ KILOWATT}$$

It was named after the British engineer James Watt (1736-1819).

watt-hour [physics; engineering] Measure of the consumption of electric power: *see*
KILOWATT-HOUR; UNIT.

wattmeter [physics] An instrument for measuring the power in an electric circuit (in
watts).

wave [physics] A regular disturbance in a material or in space. For example: electromagnetic waves (such as light, radio waves, and X-rays) have electric and magnetic
waves at right-angles to each other. A sound wave in air causes alternate regions of
high and low pressure to travel along (although the air itself does not move). Waves
on water also move along, while the water is only displaced up and down as the
wave passes. The three chief parameters of a wave are its WAVELENGTH (the distance
from wave crest to wave crest – or trough to trough – usually in metres or
nanometres), AMPLITUDE (the maximum displacement from the mean position), and
FREQUENCY (the number of waves per unit time, usually in hertz or its multiples).

waveband [physics; telecommunications] The range of radio wavelengths used by a
particular group of transmissions. For example: AM radio broadcasts are generally
allocated to the short waveband (10 to 200 metres), medium waveband (200 to
1,000 metres), or long waveband (above 1,000 metres).

wave equation [physics] For harmonic waves passing through a given medium, a
differential equation that describes the waves. It thus provides a way of expressing
wave mechanics in mathematical form. *See also* SCHRÖDINGER (WAVE) EQUATION.

wave function [physics] An equation that describes the variations in amplitude (in
time and space) for a wave system. *See also* SCHRÖDINGER (WAVE) EQUATION.

wavelength [physics] The distance between successive crests (or troughs) of a wave
– that is, between two successive points at which the wave has the same phase. It
equals the ratio of wave velocity to frequency. Thus, for electromagnetic waves,
wavelength equals the velocity of light divided by the frequency; for moving
subatomic particles considered as a wave motion, wavelength equals Planck's
constant divided by the particle's momentum (*see* DE BROGLIE WAVELENGTH). The
wavelengths of radio waves are generally expressed in centimetres or metres; those
of light and X-rays in nanometres (metres x 10^{-9}).

wave mechanics [physics] In modern quantum theory, the way in which interactions between atoms or subatomic particles are treated as wave systems: *see* WAVE
EQUATION; SCHRÖDINGER (WAVE) EQUATION.

wave meter [physics] An instrument for measuring (radio) wavelength, generally by
determining its frequency by comparison with a standard frequency (usually from
an oscillator).

wave number [physics] For an electromagnetic wave, the reciprocal of wavelength
– and thus alternatively known as the reciprocal wavelength. It equals the number
of waves in unit distance, and is therefore related to FREQUENCY.

weak interaction [physics] The interaction between neutrinos (or their antiparticles), mediated by intermediate vector bosons, which is involved in radioactive decay of an atomic nucleus with the emission of beta particles (electrons). *See also* STRONG INTERACTION.

weather map [meteorology; geography] Meteorological maps are intended in particular to show the 'contours' of atmospheric pressure – from high-pressure areas to low-pressure areas, thus indicating the speed and strength of winds (the closer the contours, the more forceful the wind) – and the presence and location of cold fronts (black triangles on a line) or warm fronts (black semicircles on a line), thus indicating the onset of wet, dry, or unsettled conditions.

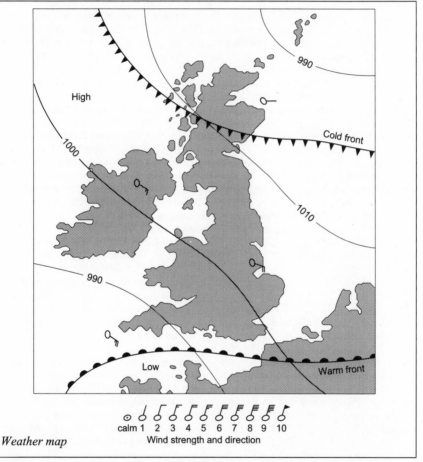

Weather map

weber [physics] Unit of magnetic flux, defined as the flux through a surface over which the total normal component of the magnetic induction is 1 tesla per square metre. A magnetic flux change of 1 weber per second in a circuit induces an EMF of 1 volt in it.

$$
\begin{aligned}
1 \text{ weber} &= 1 \text{ volt-second} \\
&= 1 \text{ joule per ampere} \\
&= 10^8 \text{ maxwells (CGS units)}
\end{aligned}
$$

The weber is also an obsolete unit of magnetic pole strength (1 weber produces a magnetic field of 1 gauss at a distance of 1 centimetre in air): *see* UNIT MAGNETIC POLE.

Wechsler-Bellevue scale, Wechsler test [medicine: psychology] A series of tests to determine intelligence and mental age, suitable for adults and for children. It was named after its inventor, David Wechsler (b.1896), a psychiatrist on the staff at the Bellevue Psychiatric Hospital in New York City.

weight [physics] Property of an object caused by the force of gravity acting on its mass. For a stationary object, *weight* equals mass multiplied by the acceleration of free fall (acceleration due to gravity, 9.80665 metres per second per second, or 32.1740 feet per second per second, at the Earth's surface). Weight is thus itself a force, measured in such units as newtons (SI units), dynes (CGS units), poundals, or pounds-force. It decreases with distance from the Earth's centre, because the force of gravity also decreases with distance from the Earth's centre.

In everyday usage, weight is synonymous with mass (and expressed in such units as kilograms, pounds, and tons), although the difference between weight and mass is critically significant in physics and astronomy (objects weigh less on the Moon because its force of gravity is less than that of the Earth, although of course their mass does not change). The term is also sometimes confused with density, which is the mass of an object divided by its volume (mass per unit volume). A dense object is not necessarily heavy or massive – it depends on how much of it there is (hence the schoolboy's conundrum: 'Which is heavier, a ton of feathers or a ton of coal?').

weight [sporting term] In bowls (and in games such as snooker and pool), the term weight perversely means 'length' – the distance the bowl (or ball) goes. A bowl 'of good weight' goes the required distance; a 'light' bowl falls short, a 'heavy' bowl goes too far.

weight, human body (standards and norms) [medicine; biology] As the average human body height has gradually increased in the English-speaking world since 1900 (*see* HEIGHT, HUMAN BODY), so has the average weight. Weight is a much more complex measurement, however: it depends on many more factors than simply genetic, and may fluctuate quite considerably with a person's physical health and mental stability. Moreover, there is quite a large difference between the average weight of a man and the average weight of a woman (even of the same body height).

The tables below set out the average weights of men and women in relation to their heights. Individual readers may find that the figures listed bear no relation to their own vital statistics.

MEN						WOMEN				
height		*average weight*				*height*		*average weight*		
in	cm	lb	st./lb		kg	in	cm	lb	st./lb		kg
61	155	124	8	12	56	56	142	102	7	4	46
62	158	127	9	1	58	57	145	104	7	6	47
63	160	130	9	4	59	58	147	107	7	9	49
64	163	133	9	7	60	59	150	110	7	12	50
65	165	137	9	11	62	60	152	113	8	1	51
66	168	141	10	1	64	61	155	116	8	4	53
67	170	145	10	5	66	62	158	120	8	8	54
68	173	149	10	9	68	63	160	123	8	11	56
69	175	153	10	13	69	64	163	128	9	2	58
70	178	158	11	4	72	65	165	132	9	6	60
71	180	162	11	8	74	66	168	136	9	10	62
72	183	167	11	13	76	67	170	140	10	0	64
73	185	171	12	3	78	68	173	144	10	4	65
74	188	176	12	8	80	69	175	148	10	8	67
75	191	181	12	13	82	70	178	152	10	12	69

weight density [weight, volume] The weight of a substance per unit volume.

weightlifting bodyweights and disciplines [sport] There are two principal types of weightlifting competition:

  the snatch
  the clean-and-jerk

The snatch involves lifting the bar directly from the ground above the competitor's head while squatting, straightening the legs until upright, and maintaining the bar there for at least two seconds. The clean-and-jerk involves lifting the bar from the ground up to shoulder height while squatting, straightening the legs until upright, then raising the bar over the competitor's head and maintaining it there for at least two seconds.

Weightlifters are categorized according to body weight in any of nine classes:

		weight up to		
class	kilograms	pounds	stone	pounds
flyweight	52	114½	8	02½
bantamweight	56	123½	8	11½
featherweight	60	132¼	9	06¼
lightweight	67.5	148¾	10	08¾
middleweight	75	165¼	11	11¼
light heavyweight	82.5	181¾	12	13¾
middle heavyweight	90	198½	14	02½
heavyweight	110	242½	17	04½
super heavyweight	more	more	more	

A belt, if worn, must not be more than 10 centimetres (4 inches) wide. The weights are loaded on to the barbell with the largest on the inside and the smallest on the outside, leaving a minimum of 1.3 metre (4 feet 3 inches) of bar free for gripping. The weight of the bar itself, plus the collars used to attach the weights on to it, is precisely 25 kilograms (55 pounds 1.85 ounce).

welterweight *see* BOXING WEIGHTS

Wentworth-Udden scale *see* PARTICLE SIZE

wergild [comparative values] In the medieval Germanic-speaking countries, the blood money paid by a murderer or accidental killer (or his family) to the victim's family in full and complete satisfaction for the death, so that no further punishment or obligation would be imposed or sought.

Known as *wergild* ('man-money'), the sum payable was calculated according to the rank and status of both killer and victim. Moreover, other crimes against the person that did not result in death but that could start a vendetta were also open to being settled on payment of wergild.

The medieval Irish equivalent was the ERIC.

west *see* POINTS OF THE COMPASS

Weston (standard) cell [physics] A source of a standard EMF (voltage) consisting of a primary electrolytic cell (anode: mercury or mercury/platinum; cathode: cadmium/mercury; electrolyte: saturated cadmium sulphate solution). Its EMF, always reproducible, is 1.018636 volt at 20°C. It is used for calibrating VOLTMETERS and other voltage-measuring apparatus.

wet and dry bulb hygrometer [physics; meteorology] An instrument for measuring the relative humidity of air, consisting of a pair of thermometers – one with its bulb enclosed by a wick that dips into a container of water. As water evaporates from the wick it cools the wet bulb, and the fall in temperature (compared with the other thermometer) can be correlated with humidity.

Wheatstone bridge [physics] An apparatus for measuring electrical resistance. It consists of a diamond-shaped circuit with three resistors of known values in three of the sides and the unknown resistor in the fourth. A voltage source (such as a cell, or battery) is connected across two opposite points of the diamond, and a current-detecting instrument (galvanometer) connected across the other two points. The values of one or two of the known resistors are varied until no current flows through the galvanometer. The bridge is then 'balanced' and the value of the unknown resistor can be calculated in terms of the other three. It was named after the British physicist (and inventor of the concertina) Charles Wheatstone (1802-75).

white dwarf, white giant *see* HERTZSPRUNG-RUSSELL DIAGRAM

white heat [physics; engineering] Loose term for the temperature of a hot metal or ceramic, in excess of 1,000°C. *See also* RED HEAT.

white-wine glass *see* DRINKING-GLASS MEASURES

whole note *see* SEMIBREVE

whole number [maths] A number (positive or negative) with no fractional part; an integer: *see* NUMBER.

whole step *see* TONE

wicket *see* CRICKET MEASUREMENTS, UNITS, AND POSITIONS

wide [sporting term] In cricket, a ball bowled that is so inaccurately aimed at the wicket that the batsman not only makes no move to hit it but might be unable to hit it if he did try. The wide counts as one run if the wicketkeeper fields it, but the run is not counted in the batsman's own personal score but is included in the list of Extras in the total score. If the wicketkeeper fails to field the ball and the ball reaches the boundary line, the score instead of being one wide is four wides. If the wicketkeeper half-fields the ball and the batsmen run one while it is being collected, the score is one wide and one BYE (two extras altogether).

wide receiver *see* AMERICAN FOOTBALL MEASUREMENTS, UNITS, AND POSITIONS

Wiedmann-Franz law [physics] For a metallic conductor, the ratio of its electrical and thermal conductivities is proportional to its absolute temperature. This law is alternatively known as the *Lorentz relation*.

wild card [sporting term] In cards, a card that can have its own value and the value of any other card, or of other specified cards.

The team that qualifies for the quarter-finals or semi-finals of a competition by virtue of winning a sudden-death play-off between teams which are next highest down the list of the season's competing sides is said to have got a wild card through to these preliminary finals.

wind chill factor [meteorology] A factor applied to temperatures that makes allowance for cold winds and humidity, and thus better represents the 'feel' and effect of low temperatures on people. The most significant contributing parameter is the windspeed.

windspeed, wind speed [meteorology] The speeds of winds, measured by an ANEMOMETER, are generally expressed in metres per second, miles per hour, or knots, and are classified accordingly on the BEAUFORT SCALE OF WINDSPEED.

wine gallon [volumetric measure] The wine gallon is the British fifteenth- to seventeenth-century measure of wine that was taken by the rebellious United States as their standard GALLON.

$$1 \text{ wine gallon} = 1 \text{ US gallon} = 0.8326 \text{ UK gallons}$$
$$= 3.7853 \text{ litres}$$
$$= 231 \text{ cu.in, } 3{,}785.3 \text{ cc}$$
$$(1 \text{ UK gallon} = 1.201 \text{ wine gallon}$$
$$1 \text{ litre} = 0.2642 \text{ wine gallon})$$

wine glass *see* DRINKING-GLASS MEASURES

wing commander [military rank] In the Royal Air Force and various other English-speaking air forces – but not the United States Airforce – a commissioned officer of senior rank between squadron leader and group captain. The equivalent in the United States Airforce is a LIEUTENANT COLONEL (who, as in the army, ranks between a major and a colonel). *See also* MILITARY RANKS.

wing load, wing loading [weight] The weight per unit square measure on wings and other supporting surfaces in a fully laden aircraft. It is calculated as the gross weight of the fully laden aircraft divided by the total area of the supporting surfaces (in square feet or square metres).

wire gauges [engineering] Early methods of calibrating wire used various gauges, such as the Birmingham wire gauges (based on those used by various manufacturers in that UK city) and the Standard Wire Gauge (SWG), also employed in Britain. Most were based on the number of times the wire had to be forced through a die to reach a particular diameter, and so the higher the gauge number, the thinner the wire.

In the Standard Wire Gauge (SWG):

$$0 = 0.324 \text{ inch (8.229 millimetres) diameter}$$
$$36 = 0.0076 \text{ inch (0.193 millimetre) diameter}$$

In the American Wire Gauge (AWG):

$$0 = 0.325 \text{ inch (8.255 millimetres) diameter}$$
$$36 = 0.0050 \text{ inch (0.127 millimetre) diameter}$$

and in that system, each number stands for wire with a cross-sectional area 20 per cent smaller than the number below it.

France and much of continental Europe has (or had) the Paris Gauge, a metric system with numbers from 1 to 30; in this case the larger the number, the thicker the wire.

In the Paris Gauge:

$$1 = 0.6 \text{ millimetre (0.0236 inch) diameter}$$
$$30 = 10 \text{ millimetres (0.3937 inch) diameter}$$

Nearly all of these gauges have been superseded by the straightforward metric system of specifying wire diameters in millimetres.

wisp of snipe [collectives] This collective applies only to snipe in flight (for snipe on the ground, *see* WALK OF SNIPE). The characteristic flight pattern of these wading birds is a zigzag course of complex spirals – and it is to that that the collective applies for, by etymology, *wisp* in this sense derives from exactly the same source as the word *whisk* (defined in most authoritative dictionaries primarily as 'a light but rapid sweeping movement . . . ' or similar).

witch of Agnesi [maths] A peculiar name for a fairly ordinary mathematical curve that has a complicated definition – the curve resembles the shape (in section) of a large sheet of flexible material laid over a horizontal cylinder resting on a plane. That is, it is symmetrical about the (vertical) y-axis and asymptotic to the (horizontal) x-axis. Its equation (in Cartesian coordinates) is $x^2y = 4a^2(2a - y)$, and it has the alternative name *versiera*, from its Italian name *versiera di Agnesi*: the versine (versed sine) of (the Italian mathematician Maria Gaetana) Agnesi. By coincidence, *versiera* is also the Italian for 'witch', hence the strange English name for the curve. *See also* VERSINE, VERS.

Wolf number *see* SUNSPOT NUMBER

won [comparative values] Unit of currency in North Korea and South Korea.

In North Korea:
$$1 \text{ won} = 100 \text{ jun}$$

In South Korea:
$$1 \text{ won} = 100 \text{ chon (formerly hwan)}$$

See also COINS AND CURRENCIES OF THE WORLD.

wood screws [engineering] A wood screw is described by its material (for example, steel, brass, cadmium-plated steel), its style of head (for example, countersink, cheese-head, round-head), its type of driver (for example, slotted, Phillips, posidrive), its length (in millimetres or inches), and its maximum diameter. Traditionally, wood screw diameters are specified by a number system which, like wire gauges, varies from country to country (but in which, unlike many wire gauges, the higher the number the thicker the screw).

For the purposes of example and comparison, the British sizes 0, 10, and 20 stand for diameters of 0.060 inch (1.524 millimetre), 0.192 inch (4.877 millimetres), and 0.332 inch (8.433 millimetres) respectively. The corresponding American sizes 0, 10, and 20 represent diameters of 0.058 inch (1.473 millimetre), 0.189 inch (4.801 millimetres), and 0.321 inch (8.153 millimetres) respectively.

Increasingly, such number systems are being replaced by diameters specified in millimetres.

woodwind instruments' range [music] Many woodwind instruments formerly made primarily of wood are now, in fact, mostly or completely made of metal (or even plastic) but are nonetheless still classified as woodwind instruments.

	-3C	-2C	-1C	middle C	+1C	+2C	+3C
soprano (descant) recorder				C —(2½ octaves)—			
piccolo					C —(3 octaves)		
(ordinary/concert) flute			C —(3 octaves)——				
G alto flute				G —(3 octaves)——			
F alto flute				F —(3 octaves)——			
oboe				B♭—(2½ octaves)—			
cor anglais (English horn)				E—(2½ octaves)-			
baritone oboe			B♭ —(2½ octaves)—				
heckelphone (bass oboe)			B♭ —(2½ octaves)—				
E♭ clarinet				G —(3¼ octaves)—			
B♭ clarinet			D —(3¼ octaves)—				
A clarinet				C# —(3¼ octaves)—			
bass clarinet			D —(3¼ octaves)—				
bassoon		B♭ —(3½ octaves)——					

	-3C	-2C	-1C	middle C	+1C	+2C	+3C
double (contra-)bassoon	B♭ —(3½ octaves)——						
E♭ sopranino saxophone					D♭ —(2½ octaves)—		
B♭ soprano saxophone					G♭ —(2½ octaves)—		
E♭ alto saxophone				D♭ —(2½ octaves)—			
C melody saxophone				C —(2½ octaves)—			
B♭ tenor saxophone			G♭ —(2½ octaves)—				
E♭ baritone saxophone			D♭ —(2½ octaves)—				

word [literary] A unit of language in both talking and writing.

By extension, in computing a word is a unit of meaning stored or transmitted in the form of a set of symbols or characters.

word association test [medicine: psychology] A test in which a subject is required to respond to a word supplied by a psychologist with another word that, in the subject's mind, is closely associated. In this way, a whole sequence of words may provoke a range of corresponding responses from the subject, and may give an indication both of the subject's level of verbal skills and of the subject's mental outlook.

work [physics] The energy transferred when a force causes an object to move, equal to the product of the force and the distance. It is measured in joules (SI units).

$$\begin{aligned} 1 \text{ joule} \quad &= \quad 1 \text{ newton-metre} \\ &= \quad 1 \text{ watt per second} \\ &= \quad 10^7 \text{ ergs (CGS units)} \end{aligned}$$

It is thus a manifestation of energy: *see* ENERGY.

work function [physics] The minimum amount of energy needed completely to remove an electron from an atom (to form an ion). The work function is alternatively known as the electron affinity.

working capital [comparative values] In company finance, the working capital is calculated as the positive total resulting from the subtraction of current liabilities from current assets. Alternatively, the working capital is defined as the assets of a business that are immediately available (liquid) as opposed to assets that may not immediately be realized in the form of cash (fixed assets, such as the company premises).

W particle [physics] An alternative name for an INTERMEDIATE VECTOR BOSON.

wrestling [sport] There are two internationally recognized codes for wrestling competitions: freestyle, and Gr(a)eco-Roman. The major difference between codes is that in Gr(a)eco-Roman wrestling, the use of the legs is prohibited.

The dimensions of the mat:

12 metres (39 feet 3 inches) square, with a circular contest area diameter
9 metres (29 feet 6 inches) marked on it

Timing: a bout lasts three minutes

Points scoring:

a fall wins outright
holding an opponent down: 1 point
a referee's caution to the opponent: 1 point
placing an opponent 'in danger' (of a fall): 2 points
a momentary/accidental fall: 2 points
keeping an opponent 'in danger' for 5 seconds: 3 points
series of momentary falls: 3 points

As in boxing and weightlifting, wrestlers are categorized by their body weight in any of 10 classes:

class	kilograms	*weight up to* pounds	stone	pounds
light flyweight	48	105¾	7	07¾
flyweight	52	114½	8	02½
bantamweight	57	125½	8	13½
featherweight	62	136½	9	10½
lightweight	68	149	10	09
welterweight	74	163	11	09

class	*weight up to*			
	kilograms	*pounds*	*stone*	*pounds*
middleweight	82	180¾	12	12¾
light heavyweight	90	198½	14	02½
heavyweight	100	220½	15	10½
super heavyweight	more	more	more	

X

X (Roman numeral) [quantitatives] As a numeral in ancient Rome, the symbol X corresponded to 10 – yet, in this sense, it did not derive from the twenty-third letter of the Latin alphabet, the X that was, in turn, derived from the Greek letter *chi* (which, to the Greeks as a numerical symbol, more commonly signified 600 or 600,000 although it could in text also stand as an abbreviation of *chilioi* '1,000'). Instead it was a way of representing two Vs – that is, twice 5. *See also* ROMAN NUMERALS; V, v.

X (malt liquor) [quantitatives] In defining comparative strengths and quality of malt liquors, manufacturers use XX (or 'double X') to refer to medium-quality liquor, and XXX ('triple X') to refer to the strongest quality.

A similar system was once used to denote the comparative quality of tin plate.

x (unknown quantity) [maths] As a term or symbol for an unknown in algebra, x is thought to derive as a medieval Latin abbreviation of the word *xei*, itself an adaptation of the contemporary Arabic *shei* 'thing', 'item', 'object', 'whatever'. It was from the Arabs, or at least through Arabic, that much of mathematics was opened to the medieval European scientific world.

It was probably through the choice of x as the initial unknown, that the next two letters in the alphabet, y and z – almost as little used as the letter x in many languages – also took on the role of unknown quantities. *See also* AXIS; VARIABLE.

x-axis [maths] In two-dimensional coordinate geometry, the horizontal axis (perpendicular to the vertical y-axis).

x-height [literary: printing] In typography and calligraphy, the x-height is the standard height of characters (letters and symbols) that have neither ascenders nor descenders. The x is chosen as the standard because it sits on the base line (unlike characters with rounded bases such as a, c, e, o, u, which very slightly infringe it) and also has a standardly level top (unlike i, m, n, r).

Xing dynasty *see* CHING DYNASTY, CH'ING DYNASTY, XING DYNASTY

X particle [physics] An alternative name for a MESON.

X-ray diffractometer [physics; crystallography] An instrument that records the patterns created by X-ray beams diffracted by the atoms in a crystal lattice; formerly called an X-ray spectrometer.

X-ray fluorescence spectrometer [chemistry] In chemical analysis, an instrument that records the wavelengths of secondary radiations emitted by a substance bombarded with high-energy X-rays or gamma-rays. The secondary radiations are characteristic of different elements and can therefore be used to identify them.

X-ray photon [physics] A quantum of X-ray energy: *see* PHOTON.

X-rays [physics; medicine] X-rays are electromagnetic radiation in the very short wavelength range 10 to 10^{-3} nanometre (rays of even shorter wavelength are gamma rays). X-rays are produced by bombarding a heavy metal target, in a vacuum, with high-energy electrons. The rays are characteristic of the metal and can be used to identify it – by its X-ray spectrum.

The rays are able to penetrate matter to varying degrees, and this property is used in medicine, both in diagnosis (radiography) and treatment (radiology). As an ionizing radiation, X-rays can be damaging to health (by inducing cancers or causing adverse genetic effects) and so X-ray doses are carefully monitored for both patients and radiographers, often using a DOSEMETER (dosimeter).

There are various measures of quantity of radiation in terms of its possible effects on people, including:

Absorbed dose, the energy absorbed from a radioactive isotope implanted in a patient (measured in GRAYS = joules per kilogram).

Collective dose equivalent, the product of the average effective dose equivalent and the number of people exposed to a source of radiation (measured in SIEVERTS).

Dose equivalent, the product of the absorbed dose and a factor that takes into account the varying (harmful) effects on tissues, depending on the type of radiation (measured in sieverts).

Effective dose equivalent, the sum of the products of the dose equivalents to various tissues and a risk factor appropriate to each tissue (measured in sieverts).

Genetically significant dose, the dose which, given to everyone before they become parents, would cause the same genetic harm as the actual doses received by each person (measured in sieverts).

In physics, use is made of the very short wavelengths of X-rays to determine the structure of crystals (X-ray crystallography). The atoms or ions in a crystal lattice are so close together – of the order of a few nanometres – that they can function as a diffraction grating for X-rays, and information about crystal structure is obtained by studying such diffraction patterns recorded photographically.

X-ray spectrometer *see* X-RAY DIFFRACTOMETER

xu [comparative values] Unit of currency in Vietnam.

$$100 \text{ xu or sau} = 10 \text{ hao} = 1 \text{ dong}$$

See also COINS AND CURRENCIES OF THE WORLD.

X unit *see* SIEGBAHN UNIT

xylophone *see* PERCUSSION INSTRUMENTS' RANGE

Y

y (unknown quantity) [maths] As a symbol for an unknown quantity, the *y* probably took on the role (with the *z*) solely because it was the next letter in the alphabet to the *x*, which was the primary symbol for an unknown quantity. *See also* AXIS; VARIABLE.

yachting classes and dimensions [sport] International yacht racing authorities recognize a large number of different classes, and definitions change from time to time in any case. But the customary classes for inshore yachting (as in the Olympic Games) are:

Tornado, a plywood or fibreglass catamaran crewed by two;
overall length 20 feet (6.096 metres);
beam 10 feet (3.048 metres);
minimum weight 295 pounds (133.8 kilograms)

Finn, a fibreglass dinghy with a centreboard crewed by one;
overall length 14 feet 9 inches (4.5 metres);
beam 4 feet 11½ inches (1.51 metres);
minimum weight 319 pounds (145 kilograms)

470, a fibreglass dinghy with a centreboard crewed by two;
overall length 15 feet 4¾ inches (4.7 metres);
beam 7 feet 7¾ inches (2.33 metres);
weight 260 pounds (118 kilograms)

Laser, a fibreglass dinghy with a centreboard crewed by one;
overall length 13 feet 10½ inches (4.23 metres);
beam 4 feet 6 inches (1.37 metres);
minimum weight 130 pounds (59 kilograms)

Flying Dutchman, a plywood or fibreglass dinghy with a centreboard crewed by two;
overall length 19 feet 10 inches (6.04 metres);
beam 5 feet 10½ inches (1.79 metres);
minimum weight 384 pounds (174 kilograms)

Tempest, a fibreglass keel yacht crewed by two;
overall length 21 feet 11¾ inches (6.7 metres);
beam 6 feet 3½ inches (1.92 metres);
weight 1,299 pounds (589 kilograms)

Soling, a fibreglass keel yacht crewed by three;
overall length 26 feet 9 inches (8.16 metres);
beam 6 feet 3 inches (1.91 metres);
weight 2,200 pounds (998 kilograms)

The customary classes for ocean or offshore racing, involving larger boats and a more numerous crew, are according to overall length.

Class V	21 feet to 22 feet 11 inches (6.4-7 metres)
Class IV	23 feet to 25 feet 5 inches (7.01-7.75 metres)
Class III	25 feet 6 inches to 28 feet 11 inches (7.77-8.8 metres)
Class II	29 feet to 32 feet 11 inches (8.84-10.03 metres)
Class I	33 feet to 70 feet (10.05-21.34 metres)

yarborough [sporting term] A hand in whist or in bridge containing no card of value higher than a nine. By tradition, the term derives from an Earl of Yarborough who habitually wagered against its occurrence, giving odds of 1,000 to 1.

yard [linear measure] An imperial measure of extremely common usage, although now in Europe largely overtaken by the slightly larger METRE of the eponymous metric system.

1 yard	=	3 feet, 36 inches
	=	91.44 centimetres, 0.9144 metre
		(1 metre = 1.093613 yard)
2 yards	=	1 fathom (1.8288 metre)
5.5 yards	=	1 rod, pole, or US perch, 16½ feet
22 yards	=	1 chain, the length between cricket wickets
220 yards	=	1 furlong, one-eighth of a mile (201.17 metres)
1,760 yards	=	1 (statute) mile, 8 furlongs, 5,280 feet
	=	1.609344 kilometre
		(1 kilometre = 1,093.613 yards)

An old measure of the Germanic peoples, the yard as a length derives etymologically independently from the yard as an enclosure or court before or behind the house. The measure comes from Old English *gierd* or *gyrd* 'rod', 'stick', 'rule', whereas the enclosure comes as much from Germanic as from Celtic and Latinate sources: it is the same word as *court*, *gard(en)*, *horti(culture)*, and even *earth*. Ultimately, however, the word as both measure and enclosure stems from an ancient root corresponding to the present-day English word *growth*. *See also* VARA.

yard, cubic [cubic measure] An imperial measure now in Europe largely overtaken by the CUBIC METRE of the metric system.

1 cubic yard	=	3 ft x 3 ft x 3 ft = 27 cubic feet
	=	36 in. x 36 in. x 36 in. = 46,656 cubic inches
	=	0.764555 cubic metre, 764,555 cc
		(1 cubic metre = 1.307951 cubic yard
		= 35.31467 cubic feet)
	=	764.555 LITRES by volume

yard, square [square measure] An imperial measure now in Europe largely overtaken by the SQUARE METRE of the metric system.

1 square yard	=	3 ft x 3 ft = 9 square feet
	=	36 in. x 36 in. = 1,296 square inches
	=	0.83613 square metre, 8,361.3 cm^2
		(1 square metre = 1.19599 square yard
		= 10.76391 square feet)
30.25 square yards	=	1 square rod or UK perch
1,210 square yards	=	1 rood , 40 square rods or UK perches
	=	1,011.717 square metres
4,840 square yards	=	1 ACRE, 4 roods, 10 square chains, 160 square rods or UK perches
	=	4,047 square metres, 0.4047 HECTARE
		(1 are = 119.5964 square yards
		1 hectare = 11,959.64 square yards)
3,097,600 square yards	=	1 SQUARE MILE
	=	258.999 hectares, 2.58999 SQUARE KILOMETRES

yarn measurement [textiles] The yarn industry measures its wares in yards (or less commonly metres) of yarn in SKEINS, seven of which together form a HANK, twenty of which in turn together make up a BUNDLE. There is a further but less useful measurement in that 18 hanks make up a SPINDLE (nine-tenths of a bundle). The actual number of yards (or metres) in the skeins differs according to the composition of the yarn.

In the United States a skein is alternatively called a LEA.

See also AUNE; BOLT; ELL; NAIL.

yaw [aeronautics; shipping] The angular rotation of a craft (aircraft, rocket, ship) about a vertical axis. *See also* PITCH; ROLL.

y-axis [maths] In two-dimensional coordinate geometry, the vertical axis (perpendicular to the horizontal *x*-axis).

year [time; astronomy] The ordinary calendar credits the year with 365 days, plus an extra day (29 February) every fourth year to compensate for accumulated time over. This fourth year, the LEAP YEAR, is reckoned to occur in every year that is divisible by the number 4, *except* full centuries *unless* the century figure itself is divisible by 4. Technically, however (*and see below*), the solar year has a duration of precisely 365 days 5 hours 48 minutes and 45.51 seconds – which means that, when we add an extra twenty-four hours on at the end of the four years, we are actually 'compensating' for 23 hours 15 minutes and 2.04 seconds.

How is it we do not actually gain about forty-five minutes every four years (one whole day every 128 years)? The answer is that we actually would, but the loss of a leap day in the full-century years that are not divisible by 4 compensates almost exactly for the time surfeit. This means that the present calendar year is only 0.0003 mean solar days (about 25 seconds) inaccurate over any four years.

This incredibly accurate means of calculating years was first promulgated by decree of Julius Caesar in 45 BC (at which time January was also decreed to be the first month, rather than March as hitherto). It was perfected by papal bull issued by Pope Gregory XIII in 1582, whose mathematicians had come up with the idea of losing the leap day in century years non-divisible by 4. In that year, too, the ten extra days that had accrued over the centuries since Julius Caesar's time were eliminated – they had caused chaos among the ecclesiastical authorities who had to decide on the dates for the Church's movable feasts.

The Hindu calendar is very similar to the Gregorian, but divides the year into twelve months of either thirty or thirty-one days – there is no twenty-eight-day month. (Its emergent year is AD 78.)

The lunar year comprising twelve months of twenty-nine-and-a-half days consists of a total of 354 days; comprising thirteen months of twenty-nine-and-a-half days, it consists of a total of 383½ days. Either of these used as the basis for a calendar would, within a short number of lunar 'years', move out of phase with the observable seasons on the planet, and so are completely useless for consistent time measurement unless something like a lunar month is fitted in (intercalated) at regular intervals in the same way as an extra day is fitted in to make the leap (solar) year.

And the lunar year with such an intercalated month is, in fact, the basis of the Jewish calendar. A thirteenth month – II Adar, or Veadar – lasting thirty days is intercalated in the third, sixth, eighth, eleventh, fourteenth, seventeenth, and ninteenth year of a nineteen-year cycle. Furthermore, the number of days in a Jewish year that does not have the intercalated month varies between 353, 354, and 355 days (the last of which is known as a perfect, full, complete, or abundant year) depending on which day of the week the New Year occurs. The day must be manipulated, if necessary, to avoid the infelicitous timing of other feasts and fasts dependent on it.

The era in use with the Jewish calendar is based on a date for the Creation of (what non-Jews might reckon as) 3761 BC. The Gregorian year AD 2000 will thus contain parts of the Jewish years 5760 and 5761 AM (*anno mundi*). The Jewish New Year (as adopted at the time of the Babylonian Exile) – Rosh Hashana, the first day of the month Tishri – generally begins in the Gregorian month of September.

The lunar calendar formerly used in China was of sixty years' duration, each year of twelve months of twenty-nine or thirty days, with an intercalary month added after each half-cycle (thirty years). Originally, the emergent year was 2637 BC, but the Chinese adopted the Gregorian calendar in 1912 (the first year of the Chinese Republic), and in China years are now counted from then.

On the other hand, the Islamic calendar relies almost totally on a year of twelve lunar months alternately of twenty-nine days and thirty days, comprising altogether 354 days but with an occasional extra day to bring it up to 355. There is no intercalated period, the months retrogress through the planetary seasons year by year, and the cycle restarts every 32½ years. By decree of the second caliph Omar I, the first day of the first year of the Islamic calendar was (in the Julian calendar) 16 July 622.

In astronomy there are various technical definitions of a year.

An *anomalistic year* is the time that elapses between two consecutive passages of the Sun through perigee, equal to 365.25964 mean solar days.

An *eclipse year* is the time that elapses between two successive passages of the Sun through the same node on the orbit of the Moon (*see* NODE), equal to 365.62003 days.

A *sidereal year* is the time that elapses between two successive passages of the Sun through the same point on its orbit (relative to the fixed stars), equal to 365.25636 days.

A *solar*, or *tropical*, *year* is the time that elapses between two successive passages of the Sun through the First Point of Aries, equal to 365.242194 mean solar days.

See also HOUR; SOTHIC YEAR, SOTHIC CYCLE.

yen [comparative values] Unit of currency in Japan, its name originally adopted from the Chinese *yüan* 'round object'.

$$1 \text{ yen} = 100 \text{ sen}$$

See also COINS AND CURRENCIES OF THE WORLD.

yield point [physics] When a material is being gradually stretched, the stress at which sudden extensive stretching (plastic deformation) takes place for a constant or even reducing load. *See also* ELASTIC LIMIT.

Young's modulus [physics] Measure of elasticity equal to the longitudinal (that is, tensile) stress in a material being stretched, divided by the longitudinal strain; usually expressed in meganewtons per square metre. It was named after the British physicist Thomas Young (1773-1829). *See also* STRAIN; STRESS.

yüan [comparative values] Unit of currency in the People's Republic of China. The word actually means 'round object', 'circular thing'.

$$1 \text{ yüan} = 10 \text{ chiao}$$

The same name is also given to the monetary unit otherwise known as a Taiwan dollar. *See also* COINS AND CURRENCIES OF THE WORLD.

yuga [time] In Hindu cosmology, an age: any of the four ages of the duration of the world, each comprising 4.32 million years.

Z

zaire [comparative values] Unit of currency in Zaïre.

$$1 \text{ zaire} = 100 \text{ makuta (singular: likuta)}$$

See also COINS AND CURRENCIES OF THE WORLD.

zak [volumetric measure] In the Netherlands an ancient measure – presumably of dry goods, in that the word literally means 'sack' or even 'pocket' – now assimilated to a measure in the metric system.

$$
\begin{aligned}
1 \text{ zak} &= 100 \text{ litres, } 22.00 \text{ UK gallons,} \\
&\quad\quad 22.703 \text{ US (dry) gallons} \\
&= 2.75 \text{ UK bushels, } 2.84 \text{ US bushels} \\
&= 6{,}102 \text{ cu. in., } 3.53 \text{ cu.ft}
\end{aligned}
$$

Zener cards [medicine] A set of twenty-five cards comprising five sets of five very distinct designs, used in testing a subject's powers of extra-sensory perception, in terms either of precognition or of telepathy.

The cards were devised by the US psychologist Karl E. Zener during the early 1900s.

zenith [physics: astronomy] Point on the celestial sphere – the half-orb that is the sky – precisely above the observer; it is directly opposite the nadir. Alternatively, the zenith of an astronomical object is the highest altitude above the horizon that it reaches in relation to an observer. The distance in degrees of arc between that zenith and the zenith of the observer is known as the *zenith distance*.

zenithal projection *see* AZIMUTHAL PROJECTION

zenith distance [astronomy] For a celestial object, its angular distance (measured southwards) from the zenith along the great circle that passes perpendicularly through the object; the complement of its ALTITUDE.

zero [quantitatives] For the numerical contexts of zero, *see* NOUGHT, NAUGHT.

The word *zero* means both 'nothing', 'nil', 'void', and 'nought' (or 'naught'), the figure 0 that represents successive powers of the number 10. Possibly the most significant mathematical figure of all, therefore, it nonetheless came into mathematics at a comparatively late stage, through the medieval Arabs. It was the Italians who took on the Arab knowledge and adapted the Arab word *tsifr* 'empty' as their word for the symbol: medieval Italian *zefiro*. The French at different times took this into their language as *zéro* and *chiffre*, which we now have in English as both *zero* and *cipher*.

zero, absolute *see* ABSOLUTE ZERO

zero cloud height [physics: meteorology] In meteorology and air navigation, *zero* refers to a (cloud, flying) ceiling of not more than 50 feet, 15.24 metres. If the horizontal visibility is also zero (*see* ZERO VISIBILITY *below*), the conditions are together known as 'zero-zero'.

zero-point energy [physics] The energy possessed by a system at absolute zero, a finite amount (half an energy quantum per particle plus its kinetic energy) demanded by quantum theory.

zero-point entropy [physics] The ENTROPY possessed by a system at absolute zero which, as required by the third law of thermodynamics, equals zero.

zero temperature [physics] Absolute zero – nominally the coldest temperature possible – is reckoned as:

–273.16 degrees Celsius	–459.67 degrees Fahrenheit
0 kelvin	0 degrees Rankine

In the major temperature scales, Celsius (°C), Fahrenheit (°F), kelvin (K) and Rankine,

		°C	°F	K	°Rank.
0°C	=		32	273.16	491.67
0°F	=	–17.78		255.38	459.67
0 kelvin	=	–273.16	–459.67		0
0° Rankine	=	–273.16	–459.67	0	

zero-valent [chemistry] Having a valence (valency) of zero – and therefore incapable of forming chemical bonds with other atoms. An alternative description is nonvalent.

zero visibility [physics: meteorology] In meteorology and air navigation, zero refers to a horizontal visibility of not more than 165 feet (55 yards), 50.29 metres. If the cloud ceiling or maximum flying height is also zero (*see* ZERO CLOUD HEIGHT *above*), the conditions are together known as 'zero-zero'.

In road traffic control, the horizontal visibility in *zero* conditions means a maximum visible distance of more like 15 feet (5 yards), 4.57 metres, or less.

Zip code [quantitatives] A five-figure number that plots exactly where in the United States any specific address is located, at least to the nearest mail-delivery area. The coding is so called because it formed part of the 1963 Zone Improvement Plan.

zither *see* STRINGED INSTRUMENTS' RANGE

zloty [comparative values] Unit of currency in Poland, named after a golden coin (Polish *zloto* 'gold', *zloty* 'golden').

1 zloty = 100 groszy

See also COINS AND CURRENCIES OF THE WORLD.

zodiac, zodiacal constellations *see* HOROSCOPE

zoll [linear measure] An old Swiss measure that was formerly just under 1 inch, 2.54 centimetres, and is generally translated as 'inch' in modern dictionaries. When

actually used today, however, the measure is ordinarily rounded up to 3 centimetres, 1.18 inch.

$$12 \text{ zolls} = 1 \text{ fusz ('foot')}$$
$$= 36 \text{ centimetres, } 14.16 \text{ inches}$$

zolotnik [weight] One of the smallest Russian units of weight.

$$1 \text{ zolotnik} = 65.8306 \text{ grains}$$
$$= 21.33 \text{ carats}$$
$$= 4.266 \text{ grams}$$
$$= 0.1505 \text{ ounce}$$

(1 gram = 0.2344 zolotnik
1 ounce = 6.6445 zolotniks)

zone [geography; maths] In geography, the five zones represent the greater divisions
of the planet's surface: the torrid zone between the tropics, the north and south
temeprate zones outside the torrid zone, and the north and south frigid zones around
the poles.

In mathematics, a zone represents part of the surface of a sphere contained
between two parallel planes.

zwitterion [chemistry] An ion that carries both positive and negative charges,
commonly derived from an organic compound that has both acidic and basic groups
in its molecules. For example, the amino acid glycine (aminoacetic acid),
$H_2N.CH_2.COOH$, yields the zwitterion $^+H_3N.CH_2.COO^-$.

The term derives from the German *Zwitter* 'hermaphrodite' (akin to German *zwei*
'two' – that is, 'both').

zymometry [chemistry] Measuring the progress of fermentation (zymolysis), usually
in brewing, using a *zymometer*.

The first element of the term is ancient Greek *zume* 'leaven', '(cause of) fermentation', as also found in English *enzyme*, a substance that causes digestive fermentation in the stomach and duodenum.

Index to Selected Themes

Note that a single headword may appear in two or more categories.